CHICAGO PUBLIC LIBRARY
BUSINESS / SCIENCE / TECHNOLOGY
400 S. STATE ST. 60605

REF
QD
480
.S36
1998

HWLCTC

Chicago Public Library

R0128852286

Molecular modeling on the PC.

MOLECULAR MODELING
ON THE PC

MOLECULAR MODELING ON THE PC

Matthew F. Schlecht
DuPont Agricultural Products
Stine Haskell Research Center
Newark, Delaware

WILEY-VCH

New York · Chichester · Weinheim · Brisbane · Singapore · Toronto

This book is printed on acid-free paper. ∞

Copyright © 1998 by John Wiley & Sons, Inc. All rights reserved.

Published simultaneously in Canada.

No part of this publication may be reproduced, stored in a retrieval system or transmitted in any form or by any means, electronic, mechanical, photocopying, recording, scanning or otherwise, except as permitted under Sections 107 or 108 of the 1976 United States Copyright Act, without either the prior written permission of the Publisher, or authorization through payment of the appropriate per-copy free to the Copyright Clearance Center, 222 Rosewood Drive, Danvers, MA 01923, (978) 750-8400, fax (978) 750-4744. Requests to the Publisher for permission should be addressed to the Permissions Department, John Wiley & Sons, Inc., 605 Third Avenue, New York, NY 10158-0012, (212) 850-6011, fax (212) 850-6008, E-mail: PERMREQ@WILEY.COM.

Library of Congress Cataloging-in-Publication Data:

Schlecht, Matthew F.
   Molecular modeling on the PC / by Matthew F. Schlecht.
     p. cm.
   Includes bibliographical references and index.
   ISBN 0-471-18567-1 (alk. paper)
    1. Molecular structure--Data processing. 2. Molecules--Models-
-Computer simulation. 3. Microcomputers. I. Title.
QD480.S36 1998
541.2'2'0113416--dc21   97-18387

Printed in the United States of America.

10 9 8 7 6 5 4 3 2 1

Dedicated to My Wife, Meryl, and My Son, Isaac

CONTENTS

Preface	xiii
1. Introduction to Molecular Modeling	**1**
1.1. Historical Overview of Molecular Modeling	3
1.2. Molecular Mechanics—The Force Field Model	10
1.3. Molecular Modeling Strategy and Limitations	12
1.4. Molecular Modeling Resources	16
1.4.1. Printed Literature	16
1.4.2. Electronic Literature	16
1.4.3. Structure Databases	17
1.4.4. Structures and Data from the World Wide Web	17
1.5. Molecular Modeling in Industry	18
1.6. Molecular Modeling in Education	19
2. PC Molecular Modeling Hardware and Software	**21**
2.1. PC Usage Conventions	21
2.2. Hardware Operations	22
2.3. Overview of PCMODEL	22
2.4. PCMODEL Features	23
2.4.1. The Opening Window	23
2.4.2. The Menu Bar Commands	25
2.4.3. The Tool Bar Commands	32
2.4.3.1. The Periodic Table Menu	34
2.4.3.2. The Metals Menu	35
3. Input and Output	**36**
3.1. Structure Input from the Graphics Interface	36
3.1.1. General Information for Drawing Structures	36
3.1.2. Drawing Tools and Commands in PCMODEL	37
3.2. Structure Input from Files	49
3.2.1. The Rings Template Menu	51
3.2.2. The Amino Acids Template Menu	51
3.2.3. The Sugars Template Menu	53

3.2.4. The Nucleosides Template Menu	59
3.2.5. User-Defined Templates	59
3.2.6. Substructures	63
3.3. Structure Input from Physical Data	75
3.3.1. X-ray Diffraction Data	75
3.3.2. Neutron Diffraction Data	77
3.3.3. Electron Diffraction Data	78
3.3.4. Microwave Spectroscopy	79
3.4. Mathematical Representations of Structure	79
3.4.1. One-Dimensional Structures	80
3.4.2. Two-Dimensional Structures	80
3.4.3. Three-Dimensional Structures	82
3.4.4. Automated Generation of Three-Dimensional Structures	85
3.5. Data Output	89
3.6. Graphic Output and Display Tools	96
Graphics Capture Utilities	96
The Display of Atoms	98
The Display of the Molecular Skeketon	101
The Display of Whole-Molecule Properties	108
3.6.1. Rotations and Perspective	116
3.6.2. Queries	117
3.6.3. Structural Comparisons of Molecules	119
4. Input File Formats	**128**
4.1. PCM File Format	129
4.2. MMX File Format	133
4.3. MDL MOL File Format	142
4.4. X-Ray and Neutron Diffraction File Formats	148
4.5. Other File Formats and Interconversion	152
4.5.1. The Mopac inp Format	152
4.5.2. The Alchemy MOL Format	157
4.5.3. The SYBYL MOL/SYB Format	160
4.5.4. The MacroModel MAC Format	161
4.5.5. The Chem-3D C3D Format	165
4.5.6. Interconversion between File Formats	169
4.6. Modification of the Input File	174
4.7. Using Added Constants	185
5. The Molecular Mechanics Force Field	**200**
5.1. Components of the Molecular Mechanics Force Field	201
5.1.1. Valence Terms	202
5.1.1.1. Bond Stretching Potential	202

5.1.1.2. Angle Bending Potential	205
5.1.1.3. Torsional Interaction Potential	206
5.1.1.4. Out-of-Plane Angle Bending Potential	208
5.1.1.5. Improper Torsional Interaction Potential	210
5.1.2. Valence Cross Terms	210
5.1.2.1. Stretch–Bend Cross-Term Potential	212
5.1.2.2. Other Cross-Term Potentials	212
5.1.3. Nonbonded Interaction Potentials	216
5.1.3.1. van der Waals Nonbonded Interaction Potential	216
5.1.3.2. Urey–Bradley Geminal Nonbonded Interaction Potential	219
5.1.3.3. Electrostatic Interaction Potentials	220
5.2. The MMX Force Field	222
5.3. Atom Types in the MMX Force Field	238
5.4. Other Force Fields	241
5.4.1. The Allinger MM Force Fields	242
5.4.1.1. The MMI Force Field	242
5.4.1.2. The MM2 Force Field	245
5.4.1.3. The MM3 Force Field	250
5.4.1.4. The MM4 Force Field	254
5.4.2. The Consistent Force Field (CFF)	262
5.4.2.1. The Urey–Bradley Consistent Force Field (UBCFF)	263
5.4.2.2. The Valence Consistent Force Field (CVFF)	268
5.4.2.3. The Lyngby CFF	272
5.4.2.4. The Quantum Mechanical Force Field (QMFF)	276
5.4.2.5. CFF93	281
5.4.3. The AMBER Force Field	282
5.4.4. The CHARMM Force Field	288
5.4.5. The Tripos Force Field	296
5.4.6. The MMFF94 Force Field	299
5.4.7. The OPLS Force Field	303
5.4.8. Miscellaneous Force Fields	306
5.4.8.1. MacroModel	306
5.4.8.2. The ECEPP and UNICEPP Force Fields	307
5.4.8.3. The EAS Force Field	308
5.4.8.4. The DREIDING Force Field	309
5.4.8.5. The UFF Formulation	310
5.4.8.6. The EFF Formulation	312
5.4.8.7. The PIMM Force Field	313
5.4.8.8. The YETI Force Field	314
5.5. Parameters—Their Sources and Utility	315
5.6. Electrostatics and Solvation	316

5.6.1. Electrostatics	317
5.6.2. Solvation	329
5.6.2.1. Explicit Solvent Model	329
5.6.2.2. Solvent Continuum Model	330
5.7. Energy Calculations	334
5.7.1. The PCMODEL Output Window	335
5.7.2. Single-Point Energy	336
5.7.3. Energy Minimization	339
5.7.3.1. Steepest Descent Algorithm	342
5.7.3.2. Conjugate Gradient Algorithm	342
5.7.3.3. Newton–Raphson Algorithm	343
5.7.3.4. Block-Diagonal Newton–Raphson Algorithm	344
5.7.4. Minimum-Energy Structures	344
5.8. Conformation Searching and Global Minimization	345
5.8.1. Strategies and Methods	346
5.8.1.1. Systematic Search	347
5.8.1.2. Stochastic Search	347
5.8.1.3. Molecular Dynamics	348
5.8.1.4. Genetic Algorithms	349
5.8.1.5. Distance Geometry	349
5.8.1.6. Directed Tweak	349
5.8.1.7. Other Methods	350
5.8.2. The GMMX Program	350
5.8.2.1. The Input File with GMMXINP	351
5.8.2.2. The Conformation Search with GMMX	367
5.9. Molecular Dynamics	376
6. Applications	**385**
6.1. Modeling with Geometric Constraints	385
6.2. Surfaces and Volumes	421
6.2.1. Surface Analysis and Surface Comparisons	441
6.2.2. Numerical vs. Analytical Calculations of Molecular Surfaces and Volumes	445
6.3. Dihedral Driver Calculations	455
6.3.1. Rigid Rotor	457
6.3.2. Minimized Step Dihedral Driver	459
6.4. NMR Correlations	464
6.5. π-Calculations	488
6.6. Hydrogen Bonding	508
6.7. Metal Atom Bonding and Coordination	526
6.7.1. Covalent Bonding	529
6.7.2. Coordination Bonding	531

6.7.3.	Metal–Carbon Bonds	539
6.7.4.	Metal–Oxygen Bonds	540
6.7.5.	Metal–Hydrogen Bonds	540
6.7.6.	Metal–Nitrogen Bonds	541
6.7.7.	Metal–Metal Bonds	541
6.7.8.	Metal–Sulfur Bonds	541
6.7.9.	Metal–Phosphorus Bonds	542
6.7.10.	Metal–Halogen Bonds	542
6.7.11.	MMX Algorithms for Metallic Compounds	542
6.8.	Docking of Substructures	571
6.9.	Reactive Intermediates and Transition States	582
6.9.1.	Reactive Intermediates	583
6.9.2.	Transition State Atoms	584

Appendices   599

A. General Information   599
   A.1. PC Filenames   599
   A.2. Unit Abbreviations, Unit Conversions, and Sample Values   600
   A.3. Conformer Population Analysis   603
   A.4. Derivations   604
      A.4.1. Atomic Charges for Chloroethane Calculated from Bond Moments   604
      A.4.2. Distance between Ring Centroids from Distance Constants   604
         A.4.2.1. Derivation for the Face-to-Face Benzene Dimer   604
         A.4.2.2. Derivation for the T-shaped Perpendicular Benzene Dimer   607
B. Glossary of PCMODEL Commands   609
C. Atom Types in PCMODEL   622
D. Glossary of Computer and Molecular Modeling Terms   626

References   704

Index   739

PREFACE

I think that nature has ... some sort of arithmetical geometrical coordinate system, because nature has all kinds of models. What we experience of nature is in models, and all of nature's models are so beautiful.
—R. Buckminster Fuller, from *Profiles: In the Outlaw Area*
C. Tompkins, *New Yorker*, Jan. 8 1996

The drive inherent in the human race to use mathematical models in their understanding of Nature, so eloquently captured above by Fuller, likely predates recorded civilization. Poised upon the brink of a new millennium, this drive survives undiminished, though many of its objects are now more obscure. Our modern conception of molecular structure had its origins in the eighteenth and nineteenth centuries. At the end of the twentieth century, nearly everyone who has completed their secondary education or attended college has at least a qualitative concept of a molecule; until recently, though, a quantitative or even semi-quantitative appreciation of molecular structure has remained obscure to all but small fraction of practicing scientists.

The growing familiarity of the general population with PCs, due to the popularity of word-processing programs and computer games, set the stage for the democratization of computer-aided molecular modeling seen over the last ten years. Relatively inexpensive personal computers operating with an intuitive, graphically oriented interface are now widely available, and one among myriad opportunities which this situation presents is the power for any educated person to carry out molecular structure calculations.

Three goals drove the design of this book. It is firstly a primer, facilitating an entry to molecular modeling techniques on desktop PCs for those with little or no prior experience in the field. Molecular modeling concepts are explained, placed in a context, and exemplified with computer modeling experiments. A diskette which accompanies this book contains the structure files for each such experiment. The second goal is to provide the novice with a bridge to practical applications. Thus, the material has been paced from the introductory to an intermediate level. Knowledgable readers may wish to skim the early treatment in a section and then skip ahead. The third goal is to provide lasting value as a reference work, in order to aid the journeyman modeler in further study and practice. To this end, advanced material is surveyed and documented with citations to the print and electronic literature, current to mid-1997. Given the growing importance of the World Wide Web as a source of information, an HTML-format document entitled "Molecular Modeling Resources on the World Wide Web" is also provided on the accompanying diskette. This resource contains hyperlink pointers to structure files, tutorials, information on software and modeling algorithms, commerical software suppliers, and to the World Wide Web citations in the text. An extensive glossary of computer and chemistry terms relevant to modeling is provided in Appendix D.

The scope of this book should be made clear at the outset. The number of molecular modeling techniques, and of new hybrids of the existing techniques, is ever increasing. However, they still fall more or less into three categories: *ab initio* methods, semiempirical methods, and empirical force field methods. This book deals only with empirical force field methods, and all of the exercises are force field experiments. To cover all molecular modeling techniques adequately and intelligently would exceed both the experience of the author and the size limitations for this book.

In the past, many novices began modeling at the force field level with user-friendly, graphically oriented programs. Then, if their needs or interest dictated, they could graduate to higher levels of modeling theory. Ten years ago, these higher levels of theory were simply not accessible to the novice using a graphically oriented desktop computer, but the situation has changed dramatically since. Nowadays, many introductory computer modeling courses, or chemistry courses which incorporate modeling methods, will include or focus exclusively on semiempirical methods or *ab initio* methods. This work cannot serve to introduce the reader to those techniques, and the reader will need to look to other sources, many of which are cited herein.

In *Molecular Modeling on the PC*, the pattern used is to introduce an area, followed by one or more exercises which exemplify the important concepts. The book is structured to provide the amateur modeler with a bridge to the intermediate level, by providing additional context to the area under study. A moderate familiarity with PCs is assumed; many features and simple operations are described, but this treatment does not substitute for basic working knowledge of PCs. Some basic knowledge of physical organic chemistry is also assumed, equivalent to an advanced undergraduate or a first-year graduate course. To aid in comprehension, many chemistry, computer and molecular modeling technical terms and jargon appear in **boldface** and are defined in the Glossary (Appendix D), although reading these definitions doesn't substitute for a basic knowledge of chemistry. Where force field atom types are referred to by number in this book, the number is preceeded by an @; it is hoped that this convention will aid clarity and not be annoying.

The molecular modeling exercises employ the popular desktop program PCMODEL® by Serena Software. In the author's experience, PCMODEL provides the most functionality, and the most friendly front end and drawing utility among the affordable commercial packages available today. This book supplements, but does not replace, direct experience in computer-aided molecular modeling, and it is strongly recommended that the reader use PCMODEL in order to work through the examples.

Writing a book such as this with the aim of using more than one modeling program would render the treatment tedious and unpleasantly complicated, and there are limitations attending the choice of any single program for this purpose. The choice of PCMODEL was dictated by several factors: the author's familiarity with the use of this program, its low cost relative to comparable programs, its broad applicability to many structural types, the intuitiveness or friendliness of the graphics-oriented program interface, and the supportive response of the program's developers to requests for help. Not every reader will choose to remain with PCMODEL, but there are sufficient similarities of operation across such programs that this approach seemed workable.

The emphasis here is on employing computer-aided molecular modeling as a tool to develop a more accurate mental picture of molecules and their properties. Computational methods constitute some of the most important tools available to the chemist today, just as crucial to modern research and teaching in chemistry as NMR spectroscopy or X-ray diffraction crystallography. An additional advantage of computer chemistry is that molecules can be studied which do not yet exist, in far less time than it would take to prepare them. Computational techniques are not perfect, and certainly they cannot substitute completely for chemical intuition. There are always uncertainties, and the best results are obtained when it is possible to make some comparisons with experiment measurements, to gauge the accuracy in the context of the problem at hand. Nevertheless, there are uncertainties in any advanced analytical method. This book places molecular modeling in a technical and experiental context which should allow the average scientist, acting alone or with fellow novices, to approach many problems of interest.

There are three main categories of problems which can be addressed with PC molecular modeling: simple model building for geometry and property predictions; global energy minimum of flexible molecules, conformational equilibria, and docking; and comparison of molecules, reactive intermediates, and transition states. These are the kinds of problems presented in the exercises contained herein, and are described briefly below:

Simple Model Building: Generating a reasonable first-pass three-dimensional structure for a molecule which approximates its shape and relative geometries. This approach works best on relatively rigid molecules; molecules with significant conformational freedom will require further analysis. The results can be used to predict physical or chemical properties, or to correlate the structure or conformation with known properties. The result can be visualized through various depictions of the molecule (ball and stick, CPK, etc.), as a molecular surface, or in terms of relative geometries.

Global Energy Minimum of Flexible Molecules, Conformational Analysis, and Docking: Discovering the lowest energy single conformation, or group of conformations, for a flexible molecule, with a quantitative energy ranking. These results can also be used to predict or correlate properties, with the added dimension of considering a molecule not as a single structure but as a weighted average of several structures. Docking experiments examine the optimization of intermolecular association, driven by specific interactions such as hydrogen bonding, dipole or charge attractions, metal atom coordination, or van der Waals attraction. This can be used to study solvation, the structure of aggregates, or specific binding interactions.

Comparison of Molecules, Reactive Intermediates, and Transition States: Either between conformations of the same molecule or between parts of distinct molecules or molecular species. Comparisons of conformations can determine identity or the extent of differences. Comparisons of portions of different molecules can aid in mapping structure–property relationships. Comparisons of the relative stabilities of intermediate anions, cations, or radicals correlate with product ratios. Comparison of transition states aids in the analysis of reaction pathways and the prediction of product ratios.

In most cases, these modeling techniques can be used to answer questions more quickly than a lab experiment, and thus can be used to direct the strategies of scientific inquiry. The challenge arises of whether calculations such as these qualify as experiments, or even whether they are Science! Our assumption going in is that *calculations are scientific experiments,* and as such are believable to the extent that proper techniques are used, the proper controls are exercised, and the limitations are recognized. Just as in a wet lab procedure, the better the experimental design, the better the quality of the result. Computer-aided molecular modeling doesn't exist for its own sake, but to contribute to scientific endeavor, and to enable the scientist to work smarter.

Matthew F. Schlecht

ACKNOWLEDGMENTS

As this book has been developing in one way or another for ten years, there are many to whom thanks are due.

First and foremost, I owe a great deal to the generosity, patience, and interest shown toward me and this project by the developers of the PCMODEL program: Kevin Gilbert, Joe Gajewski, Mark Midland, and Kosta Steliou. I hope that I've presented their program in the favorable light it deserves. As I progressed through sections of this book, contacts with experts in the pertinent fields have been very helpful in ensuring accuracy in what is presented herein. Notable among these are Dr. Wolf-Dieter Ihlenfeldt of Computer Chemistry Center, University of Erlangen–Nuernberg; Mike Burnett and Carroll Johnson of Oak Ridge National Laboratory; George Lie of MDL; Kevin J. Koboldt of Tripos; Ulrich W. Suter of the ETH; and surely others whom I've forgotten to name. However, any remaining flaws or inconsistancies are the sole responsibility of the author.

I must also thank those who had an important influence on me in my younger days. My parents, who instilled in me the confidence to achieve what I set out to do. Carol Clark, whose talent was bringing science alive in a grade school classroom. Larry Haskins, who probably doesn't realize the pivotal role he played in shaping my scientific identity. Barry Trost and Gilbert Stork, my most important scientific mentors, who sculpted me into a synthesis chemist and provided me with my own set of mental molecular models. Clark Still and Wayne Guida, who fostered my first experiences in computer-aided molecular modeling. Last of this group are the half dozen students who participated in my special topics course in molecular modeling at Polytechnic University in the Fall of 1987. They attended the birth of the concept which led to this book, and helped me nurse it through its first few months of life.

I must thank the members of Chemical Discovery, DuPont Agricultural Products, and especially the members of the Computational Chemistry Group, for providing a healthy and stimulating atmosphere in which to put my fledgling talents in the modeling area to good use.

This book would never have seen the light of day had it not been for the extraordinary patience of my editor, Dr. Edmund Immergut, and his staff, notably Mr. Jon Glover. Ed and Jon were supportive when possible, demanding when necessary, but never seemed to waver in their belief in the successful completion of this project. For that, I am grateful.

I have saved the most heartfelt thanks for last, and these go to my wife,

Meryl, and my son, Isaac. In spite of the void created in our family life by my preoccupation with completing the manuscript for this book, they accepted my devotion to it and wished with me for a speedy finish. As it was not finished swiftly, this proved to be an ordeal for them, as well as for me. I can remember their excitement about the book which would result at times when I felt totally inundated by it, and I remember their optimism about the project at times when I was absolutely drained from the effort. That they loved me all the way through it took an enormous effort that I hope not to ask of them again soon.

MATTHEW F. SCHLECHT

MOLECULAR MODELING
ON THE PC

CHAPTER 1

Introduction to Molecular Modeling

Molecular modeling in the broadest sense is the use of graphical, mathematical or physical representations to help understand and predict the properties of molecules. These models serve best as aids in the development of our own internal mental models for and conceptualization of molecules. However, they will always be imperfect and incomplete, and whoever takes them too seriously risks the sin of revering a graven image.

The major intent of this book is to provide an entry to the area of force field molecular modeling for people who know little or nothing about it. It is also intended as a reference work, pointing to key citations, providing definitions, and containing relevant data to enable further study and practice in the field. To provide perspective, the limitations of the method are brought up at the appropriate points. In this regard, it should form a bridge, enabling the initiate to become a novice, and the novice to become a journeyman.

There are many approaches to this subject, and a gradation of levels of theory in the understanding of molecules through computer modeling. This book is not a survey of all of these methods; that type of treatment has been realized in other works.[1] Although many references are made here to the use of Internet or World Wide Web resources for computer molecular modeling, the present work is also not an authoritative source on that subject, and the interested reader is directed elsewhere.[2-6] The subject of this book is force field molecular modeling, and the focus is on learning the basics, doing computer experiments to reinforce that learning, and gaining information on how to make further progress. For better or for worse, from this point forward, when molecular modeling is mentioned in the book with no further qualification, it is intended to mean the application of empirical force fields to calculating molecular structure.

This chapter serves to introduce the book, and place the present treatment in context within a number of comparable and related works. It begins with a historical overview of molecular modeling. The history of molecular modeling and the force field model, and of their associated computer methodologies, has been told in many contexts and in various levels of detail.[7-26] The treatment here focuses on those developments which, in the author's opinion, shaped the course of the field most significantly. Some ellipses are intentional, in order not to duplicate the contents of other referenced works, and the reader interested in further study may obtain more information from the listed citations. The

2 INTRODUCTION TO MOLECULAR MODELING

author regrets that this brief historical review, like any piece of this type, may well overlook some important contributions, or may attach by implication an inappropriate precedence or importance to one area or research group.

Chapter 1 next examines the force field model briefly, and then proceeds to outline strategy and limitations in this modeling method. This is followed by a description of what further resources are available for molecular modeling, in both print and electronic media, and more specifically the availability of molecular structural data in electronic form, both in local databases and from the Internet. The chapter closes with sections on the use of molecular modeling in education and in industry.

Chapter 2 deals with the computer hardware and software which will be used in the experimental portions of the book. Again, for better or for worse, the author has selected a single commercial modeling package to use in this work, PCMODEL® by Serena Software. This chapter covers the basics of computer use and serves as an introduction to the look and feel of PCMODEL.

Chapter 3 covers input and output for the PCMODEL program. Topics covered relating to input are: how to create structures with the graphic user interface, how to use and build upon templates, and how to adapt empirically derived structural input for use in PCMODEL. Topics covered relating to output are: an overview of mathematical representations of structure, data output from the program, and graphical output and the display of structures.

Chapter 4 is concerned with input file formats. It covers four major types in detail: the PCM, MMX, MDL MOL, and crystal diffraction (X-ray and neutron) formats. The issue of conversion between file formats is addressed. Finally, modification of the input file and the use of added constants or parameters are explained.

Chapter 5 deals with the molecular mechanics force field in detail. First, the components of force fields are discussed in general. This is followed by a series of sections explaining the makeup of selected individual force field formulations, starting with the PCMODEL MMX force field, proceeding through the Allinger MM-type force fields, the Lifson CFF force field and its offspring, the Kollman AMBER force field, the Karplus CHARMM force field, the Tripos force field, the Halgren MMFF94 force field, and the Jorgensen OPLS force field. A number of other force fields are then examined in less detail.

Some general aspects of force field modeling are examined next: parameters, electrostatics and solvation, energy calculations, conformational searching, and molecular dynamics.

Chapter 6 has nine sections, each dealing with a specific application of force field modeling: geometric constraints; surfaces and volumes; dihedral driver calculations; NMR correlations; π-calculations; hydrogen bonding; metal atom bonding and coordination; docking of substructures; comparisons of reactive intermediates; and transition state atoms and transition state modeling.

Appendix A contains general information on PC filenames, unit conversions, a brief description of a Boltzmann population analysis (with examples), and some derivations for geometrical relationships used in other parts of the book.

Appendix B contains an alphabetical index of all of the commands and tools in PCMODEL for Windows. For each, a brief statement characterizes its function, and its location (menu bar or tool bar) is given. Then, the command or tool is either described in more detail, or a reference is given to the main text where this description is given.

Appendix C contains a list of the atom types used in the MMX force field, with some of their characteristics.

Appendix D contains a glossary of technical terms and jargon from chemistry, computers, and molecular modeling.

There are two additional resources in the electronic medium which accompany the book and are contained on diskette: a structure input file library containing structures and other associated computer files for use with the exercises given in the book; and an HTML-format document which contains pointers to molecular modeling resources on the World Wide Web.

1.1. HISTORICAL OVERVIEW OF MOLECULAR MODELING

By copying, the ancient models should be perpetuated.
—Hsieh Ho, fl. 500, from *Spirit of the Brush*

Drawn representations of chemical structures are among the earliest forms of molecular model, having arisen in the mid-nineteenth century in the course of the development of the structural theory of organic chemistry. From the concept of the molecular formula and a knowledge of functional groups, or "radicals," molecules were depicted as "rational formulas." These followed a variety of conventions in grouping the "radicals" together with brackets or parentheses, but essentially failed to provide a sense of molecular structure as we know it today.[27] Between 1858 and 1861, the three chemists Archibald Scott Couper, Friedrich August Kekulé von Stradonitz, and Aleksandr Mikhailovich Butlerov independently introduced the general rules of valence for organic chemistry, and the first written structures involving chains of carbons with lines drawn as "bonds" to substituent atoms and groups.[28,29] The term "chemical structure" was first used at this time.

Also in 1861, Johann Josef Loschmidt produced an amazingly prescient collection of 368 molecular structures as graphic displays with atomic domains. These included the first correct structure of benzene, showing a uniform cyclic arrangement of the ring carbon atoms. Also described were 120 other aromatic compounds, cyclopropane, and representations of the vinyl and allyl moieties. Loschmidt's structures depicted relative atomic size, and varied bond distance with bond multiplicity. Although we can appreciate its value from our modern vantage point, Loschmidt's contribution was published obscurely and never properly reached the attention of the chemical community of his day. In his seminal 1865 paper on the structure of benzene, Kekulé mentions Loschmidt only in passing, and decades would pass before structural theory would catch up.[30]

Starting in 1860, Louis Pasteur embellished the theory of molecular structure with the third spatial dimension as a result of his work on optically active compounds. The first recorded use of a physical molecular model in organic chemistry was by August Wilhelm Hofmann in 1865. In a lecture entitled "On the Combining Power of Atoms" before the Royal Society of Great Britain, he used the metaphor of croquet balls joined by sticks to describe methane, chloroform, and other compounds of carbon.[31]

The color scheme which he established is still widely used today, with occasional variations: white for hydrogen, black for carbon, red for oxygen, blue for nitrogen, and green for chlorine. The Institute of Physics has attempted to promulgate a uniform color code for molecular models which, in addition to the Hofmann usage mentioned above, stipulates yellowish-green for fluorine, light green for chlorine, midgreen for bromine, dark green for iodine, yellow for sulfur, purple for phosphorus, black or gray for silicon, and brown, silver or gold for metals.[32]

Two-dimensional drawn versions of ball-and-stick models were employed by Crum-Brown in 1865 and by Sir Edward Frankland and B. F. Duppa in 1867, although they continued to equivocate on the significance of such structural representations. Further elaborations on three-dimensional molecular structure came with the suggestions in 1869 by E. Paterno, and in 1874 by Jacobus Henricus van't Hoff and Joseph Achille LeBel that carbon compounds have tetrahedral geometry. Paterno had published structural drawings depicting rotational isomers of 1,2-dibromoethane, but did not anticipate interconversion. C. A. Bischoff suggested energy barriers to rotation about carbon–carbon bonds in 1890, and in 1898 van't Hoff elaborated upon this with the idea of there being a "favored configuration" about a torsional angle, which we would understand today as the most stable conformation.[27]

The concept of the **force field** originated in the first half of the twentieth century from vibrational spectroscopy, which considered the forces acting between every pair of atoms in the molecule, or in a lattice in the case of ionic crystals. A formulation which later had a significant effect on molecular modeling was that of Urey and Bradley,[33] in which they wrote quadratic **Hooke's law** potential equations to describe some of the harmonic vibrations in simple molecules, but found the **Morse potential** to give the best fit to empirical data for bond stretching. This early history of force fields is covered well in Burkert and Allinger.[34]

The force field concept did not have a great influence outside the physical chemical community until the year 1946, when the first suggestions for the use of a new method for modeling molecules in a more quantitative way arose. It was based upon a combination of steric interactions and a Newtonian mechanical model of bond stretching, angle bending, and torsional vibrational modes. Three separate groups proposed versions of what would be later known as the **molecular mechanics** method, or the empirical force field method, for calculating molecular structures. Hill's force field had a Lennard–Jones 6–12 potential to handle the repulsive and attractive nonbonding terms, and quadratic terms for bond stretch and angle bend.[35] Dostrovsky, Hughes, and Ingold focused on the repulsive and attractive nonbonding terms, in a study on substitution and elimination reactions.[36] Westheimer and Mayer analyzed the conformations of hindered biphenyls, formulating the potential energy as a function of the displacements of individual atoms from their equilibrium positions (to handle bond stretch and angle bend), along with a nonbonding interaction potential which included repulsive and attractive terms.[37,38] From our modern vantage point, the Hill force field is the one that most closely resembles the current formulations. However, the force field model became more widely associated with the work of Westheimer, and this approach was for a while referred to as the "Westheimer method."

Despite the inspired work in the nineteenth century on the three-dimensional nature of molecules, this concept did not yet widely influence the thinking of the chemical community. That changed abruptly in 1950 with the publication of Barton's short note on how the conformations of steroids affect their chemistry,[39] which laid the foundation of **conformational analysis**. From then onward, an appreciation of the three-dimensional aspect became crucial to understanding structure, stability, conformation, and reactivity. In the following year, Pitzer described the torsional potential as a cosine function with structure-dependent periodicity, and presented an analysis of the structural features in simple hydrocarbons which should govern the energetics of this interaction.[40]

Occasionally, appreciation of a scientific discovery will expand beyond the scientific community to be shared by the entire world. Such was the case with Watson and Crick when in 1953,[41] building upon the work of at least a dozen other contemporary scientists, they presented their structure of DNA. This structure arguably constitutes the most famous physical molecular model of all time.

In the same year, a group of scientists from Los Alamos published their studies of "Equation of State Calculations by Fast Computing Machines."[42] This work, carried out on the advanced MANIAC computer, laid the groundwork for computer-based **Monte Carlo** methods, established the **Metropolis algorithm** (from the first-named author) for **simulated annealing**, and was the ancestor of **molecular dynamics** calculations.

In 1959, Corey and Bailar reported[43] the mathematical derivation of models for isomeric cobalt *bis*- and *tris*-diamine complexes from bond length and angle span input, using a vector analysis method[44] and admittedly ignoring solvation effects. Assuming that the potential energy differences between conformers were due primarily to nonbonded interactions between the ligands, they calculated the relative steric energies of the isomers by applying the Mason–Kreevoy[45] nonbonding H—H repulsion potential function to both the H—H and C—H nonbonding interactions present in their models.

Applications of nonbonding potentials to organic structure modeling began to appear from Kitaigorodsky in 1960.[46-48] In his model, hard sphere nonbonding potentials were used to calculate interactions between any proximate atom pairs not directly bonded to each other. These interactions were superimposed upon initial structures having ideal bond lengths and angle spans, with results which were of good quality for that period.

The first published use of a computer for empirical force field calculations of molecular structure was in 1961 by Hendrickson,[49] who examined the conformations of medium-sized rings. His force field contained a quadratic potential for angle bending, a threefold cosine potential for torsional strain, and a modified Buckingham potential for nonbonding steric interactions. He reasoned that changes in the bond length would be negligible, so these were frozen at the equilibrium values. Hendrickson applied his technique widely to the problem of conformational analysis.[50-55]

In 1963, Schachtschneider and Snyder reported the derivation of a set of transferable parameters for a valence force field for hydrocarbons,[56] from a vibrational analysis of the *n*-alkanes. Transferability is crucial to the development of a general force field with a limited number of parameters.

Wiberg employed computer force field calculations to address conformational analysis in 1965.[57] He developed a **steepest descent algorithm** as a scheme for geometry optimization. Wiberg also published algorithms for transforming Cartesian coordinates to internal coordinates, to calculate bond lengths, bond angles, and torsional angles from these coordinates. In addition, he provided a geometric algorithm to predict the location of the hydrogens on a methylene carbon.

In a paper in 1968 in which he reported force field calculations on some cyclophanes, Boyd presented the modified **Newton–Raphson method** for strain energy minimization,[58] which reached convergence more rapidly than the Wiberg algorithm.

The force field treatment of metal-containing compounds was evolving from the pioneering work of the previous decade, when in 1969 the first in a series of publications from the Snow group appeared.[59-61] They described the results of careful geometry optimization in cobalt pentaamine chelates using Boyd's method[58] and constants culled from the literature.

Over the latter half of the twentieth century, force field calculations of molecular structure have developed largely in pace with the development of fast computing machines to carry out the extensive calculations required. Early growth was slow, though, due to unfamiliarity of the average chemist with the programming and use of mainframe computers, and also due simply to lack of access. Faced with laborious trigonometric calculations, many researchers would still employ carefully constructed scale models to study molecular structure.[62] U.S. government scientists did have access to computers, and made seminal contributions in this area, such as the Monte Carlo work mentioned above, and the archetypal molecular drawing program, **ORTEP** (Oak Ridge Thermal Ellipsoid Plotter), developed at Oak Ridge

National Laboratory in 1965.[63,64] Also with the government's support, the first networked computer (the ancestor of the Internet) was demonstrated in October 1969, by establishing a linkage between a computer at UCLA in Los Angeles and another at SRI in Menlo Park.[65] This latter event went virtually unnoticed at the time, but subsequent advances in this area would revolutionize the study of molecular structure, and nearly every other human endeavor which depends upon the transfer of information.[5]

By the early 1970s, the major force field formulations were being reported: the work leading up to the **ECEPP** and later the **UNICEPP** force field by Scheraga et al.,[66–70] the Consistent Force Field (**CFF**) by Lifson and Warshel,[71] the **MMI** force field by Allinger et al.,[72] the **EAS** force field by Schleyer et al.,[7,73] Boyd's force field,[58,74] and the **MUB** force field by Bartell.[75–79] This stage of the history of molecular modeling is well documented elsewhere,[7–8,10] and some details will be discussed in the case of specific force fields in Chapter 5. These force field models were based on different assumptions, and though they used many of the same or similar potential functions, those components were weighted and parametrized differently. For example, Allinger's MMI and Engler, Andose, and Schleyer's EAS force field would both replicate the *gauche*-butane interaction well, but for different reasons: MMI has hard hydrogens which gave rise to greater repulsions in the *gauche* form, whereas the EAS field with softer hydrogens and harder carbons was still able to reproduce the desired interactions.[7]

In 1971, Lee and Richards described the **molecular surface** in the context of protein structure, and provided an algorithm to derive it.[80] This involved constructing spheroids about each atom scaled by the van der Waals radius, and considering as the surface those portions of these spheroids which were not contained within any other such spheroids. This analysis proceeded through the molecule in a series of planar sections, upon each of which the corresponding projection of the surface was "plotted." In those days before the advent of modern computer graphics, the surface projection for each planar section was imprinted on a transparency sheet, and these were stacked atop one another to visualize the molecular surface. Expanding upon this work two years later, Shrake and Rupley came up with the first dot surface formulation applied to protein structure, finding that 92 points would adequately define an atomic spheroid for the purposes of determining a surface.[81]

In 1972 Wiberg and Boyd presented an algorithm for exploring conformational interconversions based on systematic modifications torsional angle.[82] This technique subsequently enjoyed broad usage as the **dihedral driver** method.

The year 1974 saw the first report of computer-aided molecular modeling of oligosaccharides from the Lemieux group.[83–85] They began with crystal diffraction structures for the individual saccharide units, employed the Kitaigorodsky hard sphere nonbonding potentials, and developed torsional potentials for the φ**-angles** and ψ**-angles** which were parametrized to account for the *exo*-anomeric effect. This hard sphere, *exo*-anomeric (**HSEA**) formulation was far simpler than the other force fields of the time. However, it was computationally rapid and gave quite satisfactory correlations with the conformational equilibria derived from NMR studies, and continues to be popular with carbohydrate chemists.

Around this same time, force field formulations for synthetic macromolecules began to appear, such as was used in the conformational analysis of polypropylene by Suter and Flory.[86]

The early 1980s saw the beginning of the personal computer industry, the consequent increase in the accessibility of computers in general, and the use of the **graphic user interface** (GUI). As force field modeling was married to graphic display on desktop units, the era of personal molecular modeling for the average chemist had begun. A convenient demarcation is the year 1983, when the distribution of the MODEL (later MACROMODEL) program by the Still group at Columbia began.[22] This program could be configured to run on a mainframe with a variety of desktop platforms as the user interface. At about the same time, the development of high-quality real-time color graphics and Connolly's molecular surface program[87] contributed to the rapid evolution of this technique.[88] The graphic presentation of color-coded molecular surfaces which displayed at the same time shape, charge, and hydrophobicity provided the modeler with a much higher information density.

In 1986, M. Saunders refocused attention on the central force field model with his STRFIT molecular modeling program.[16] In this method, bond angles and torsional angles are not considered; only forces between pairs of individual atoms are calculated. The resulting structures are quite close to those obtained from MM2. In the following year, Saunders presented the random kick algorithm for conformation searching, which was especially helpful in searching over ring conformations.[89]

The history of the World Wide Web begins in 1993, when the first graphical browsers became available. Nongraphical Internet applications, such as electronic mail and file transfer protocol (FTP), had been developing since 1969, but the vast potential of the graphical user interface can encompass and integrate these functions, and provide the additional enhancements of graphic images.

Within the past ten years, so much has occurred that it is difficult to sort out which advances will truly stand out over time. One is the advent of the so-called **class II force fields**,[90,92,93] which contain anharmonic potentials, and utilize explicit off-diagonal terms from the force constant matrix. A related development is the use of structural data obtained from high-end *ab initio* calculations to parametrize these new force fields (see for example Sections 5.4.1.4, 5.4.2.4, and 5.4.6). The quantum mechanically derived potential surface is then scaled to reproduce experimental data.[90-92,94]

The prediction has been made that the **class III force fields** will be able to model the influence of chemical effects, electronegativity, and hyperconjugation on molecular structure and properties.[93]

Another ongoing trend is the development and implementation of virtual

1.1. HISTORICAL OVERVIEW OF MOLECULAR MODELING 9

reality for molecular modeling.[6,95,96] Basically, this will entail enhancing the visual third dimension, further developing and enhancing audio sensation, and introducing tactile sensation. It is unlikely that this type of modeling will be implemented on the desktop platform any time soon.

The best (in the author's opinion) demonstration of virtual reality molecular modeling has been achieved through a group collaboration.[97] In this Virtual Biomolecular Environment (**VIBE**) system, a massively parallel computational engine runs an interactive molecular dynamics simulation under the CHARMM force field, and this is interfaced with a 3-m cubic "theater," called the **CAVE** (Cave Automatic Virtual Environment). The modeler enters this theater and interacts with the three-dimensional simulation by means of a wand-link remote control device. The modeler's head and hand movements are tracked with a six-degree-of-freedom electromagnetic sensor to maintain the correct perspective in the virtual environment. The wand may be used to dock a ligand into a binding site, and bumps and other interactions are experienced as sounds.

Before leaving the subject of the history of modeling, some additional comments on graphical (written) and physical models may be made. In drawn molecular structures today, a relatively uniform international standard exists with the common two-dimensional line and polygon orthography of organic chemistry, despite numerous embellishments and variations. The stylistic features of orientation and the depiction of stereochemistry can still vary a great deal between subfields within organic chemistry (for example carbohydrates and terpenoids), as well as from group to group and from scientist to scientist. A number of conventions have been developed to enhance the information conveyed by drawings, for example the Newman and sawhorse projections to indicate dihedral geometry,[98] the Haworth depiction of carbohydrates, and the myriad of techniques for depicting peptide chains. Each of these emphasizes certain features of the molecule, while distorting or ignoring others. For example, a conformational analysis which depends upon Newman projections can succeed only at a very crude, qualitative level and is highly dependent upon the graphic skill and perspective of the artist.

Connolly has reviewed the physical molecular models from the mid-twentieth century onward.[25] A comprehensive treatment of the subject of physical molecular models for both organic and inorganic substances is found in the monograph by Walton,[32] and a more compact compilation of information on the commercially available molecular model sets is provided by Gordon and Ford.[99] The better physical molecular models are somewhat more successful, in that they normally give at least one type of information accurately. Bond lengths, bond angles, internuclear distances, and torsional angles (but not torsional barriers) can be modeled semiquantitatively with skeletal stereomodels, such as the Dreiding, Darling, or Fieser type, or any of the many varieties of ball-and-stick or ball-and-spoke models (e.g. Cenco–Petersen, HGS–Maruzen). The Dreiding stereomodels have perhaps found the widest usage. Molecular models need not be commercial or costly to prepare, and it is possible to produce useful models from inexpensive materials, such as colored wire.[100] Molecular surfaces, some torsional barriers, and crude docking and molecular recognition phenomena can be modeled with CPK or other kinds of space-filling models. These models all have the advantage that they can provide simultaneous visual and tactile information about the subject molecule.

As the field of molecular modeling continues to evolve, ways are being developed to couple the higher accuracy of the computer-based methods with the intuitive feel of physical models. In one recent report,[101] **stereolithography**[102] is used to translate computer-designed structures into physical molecular models. The data from molecular orbital calculations or from neutron diffraction studies was used to guide the fabrication of plastic models by laser curing of an acrylate resin blend. The stereolithography apparatus works by building the model iteratively through a series of "slices," from the bottom up. The ideally spherical atoms are approximated by polyhedrons, and the number of facets can be varied to differentiate atom types while remaining low enough to give manageable data file sizes. The scales used were 1 Å = 0.19–0.89 cm, depending upon the overall size of the molecule. Molecules, transition state complexes, and "molecular impressions" or shape-complementary surfaces were prepared. The fabrication process, including a postcuring, takes from 10 to 27 h depending on the complexity of the model. The authors point out that the availability of such custom-made models can serve a general need, and can be especially helpful to the visually impaired. Another recent patent claims the preparation by stereolithography of a physical model of an inhibitor plus part of the receptor, guided by the use of modeling software.[103]

1.2. MOLECULAR MECHANICS—THE FORCE FIELD MODEL

> We are perhaps not far removed from the time when we shall be able to submit the bulk of chemical phenomena to calculation.
> —Joseph Louis Gay-Lussac, *Mémoires de la Société d'Arcueil* 2 207 (1808)

> The underlying physical laws necessary for the mathematical theory of a large part of physics and the whole of chemistry are thus completely known, and the difficulty is only that the application of these laws leads to equations much too complicated to be soluble.
> —P. A. M. Dirac, *Proc. Roy. Soc. (London)* 123 714 (1929)

These two oft-quoted statements by Gay-Lussac and Dirac form a bridge from the confidence of the nineteenth century scientist to the cynicism of his twentieth century counterpart. In fact, they are both correct: we can submit the phenomena to calculation, but must accept that we will never obtain an exact answer. The first principle in the art and craft of computer-aided molecular modeling is knowing the limitations implicit in the assumptions which form the basis of the model being employed.

In this section we describe the molecular mechanics or empirical force field model in general terms. The components of the force field model, along with the details of PCMODEL's MMX force field and many other force fields, are discussed at greater length in Chapter 5.

The force field model and the molecular mechanics method have been the subject of many review articles, book chapters, and books,[1,10-11,19-20,104-121] and several tutorials on the World Wide Web can be found through the pointers in HTML document "Molecular Modeling Resources on the World Wide Web," found on the accompanying diskette.

Some molecules have little or no flexibility, in which case the method will produce a single **conformation**. In many examples, however, modeling must allow for the fact that a single molecule can assume two or more different spatial arrangements or conformations. These conformations are interconverted by rotations and by stretching bonds without breaking them, and will possess a higher or lower relative energy. Nearly all molecules will have more than one minimum-energy conformation. The simplest interpretation is that the conformation lowest in energy (the global minimum) will provide the best model of the molecule's behavior. A more realistic perspective is that the behavior of the molecule is better modeled by considering a weighted average of the conformers based on the Boltzmann distribution.

The molecular mechanics method uses a force field, which is described by a set of equations which relate the motions and energies of portions of the molecule to simple models from classical physics. To limit the number of considerations necessary to construct accurate models, the **Born–Oppenheimer approximation** dictates that the motions of the electrons and of the nuclei can be considered separately. In the case of molecular mechanics, this means that the nuclei may be moved and it is assumed that the electrons will adopt an optimum distribution around them.

The parameters (constants, coefficients, and variables) in the equations can be adjusted to give the best fit for a particular basis set of molecules, and the assumption is made that these results will apply to any conceivable molecule related to the basis set. In practice, a number of different force fields have been developed, some focusing on a particular structural area of chemistry, and others designed for broader applicability.

The force field is used to govern how the parts of a molecule relate to each other, how each atom and collection of atoms is affected by its atomic environment, and how these forces contribute to the structure of the molecule under a certain set of constraints. Usually the calculations are based on a gas phase model, and so the question of how well they relate to the solution state or solid phase is relevant.

It is important to remember that the force field is simply a method to produce a model. That model should be scrutinized and compared with experimental results, but the components of the force field should not. This is because the force field itself is not intended to be a model of molecular forces—it is an approximation which has been fine tuned to produce molecules which do mimic their natural counterparts. The equations for the component potentials and their associated parameters have been "tweaked" in order to reproduce experimentally determined molecular properties. There are a number of different ways to tweak the equations, and the developer of a force field is primarily concerned with the closest approximation of real molecules, not with consistency of the component potentials across other existing force fields.

For example, one force field may use a hard hydrogen and a soft carbon to model the behavior of hydrocarbons, while another force field uses a softer hydrogen and a harder carbon. In the balance, both systems produce good results, but the building blocks of these two force fields are not mutually interchangeable.

As will be seen in Section 3.4.4, computer molecular models can also be constructed with a rule-based algorithm. This is an expert system application, which has access to a large library of geometric structure parameters and applies these to the flat structure of an input molecule to derive a three-dimensional structure. The resulting structure is not necessarily the (or even a) minimum-energy conformation, but it is a conformation that makes sense in view of the rules stored for molecules of the type modeled. This method has the advantage that it is much faster than molecular mechanics, since no minimization takes place.

The interactions included in the force field model can be considered in two groups: bonded and nonbonded. Which of the components in each group are used in a force field, and what weight they are given, depends upon the makeup of that force field. Among the bonded interactions are bond stretching, angle bending, and torsional strain. Among the nonbonded (or noncovalent) interactions are van der Waals (with both repulsive and attractive components), electrostatic, hydrogen bonding, metal atom coordination, polarization, and charge transfer.[122,123] Each type of interaction will be represented by a potential function or equation, for which there will be appropriate constants and parameters depending upon which atoms or groups are involved in the interactions. In some cases, the interactions will be coupled, and the force field will need to include cross terms in order to allow for this. An example is the stretching of the two bonds which make up a bond angle; in this case, a stretch–bend cross term will be needed to produce an accurate model of the molecule. Chapter 5 introduces the components found in most force fields, and further details may be found there.

1.3. MOLECULAR MODELING STRATEGY AND LIMITATIONS

Rules and models destroy genius and art.
—William Hazlitt, "On Taste"

1.3. MOLECULAR MODELING STRATEGY AND LIMITATIONS 13

There will be almost as many strategies for molecular modeling as there are modelers. However, there are some general approaches and areas of application which may be mentioned. There are a great many published reports on, or including, the application of molecular modeling, and the quality varies from excellent to mediocre or worse. One must have an appreciation of the field in order to judge the results of modeling experiments, and more to the point, the authors must describe clearly what they have done and what methodology they have used to do it. Just as standards for the descriptions of wet chemistry had to be established and enforced by journal editors and referees, the same must now be done with experimental descriptions of modeling work.[124] The best preparation for a critical approach, though, is gaining some skills in the art.

There are several good reviews of the application of molecular modeling to organic synthesis.[7,22,88,106,110,117,125–128] Many applications are mentioned briefly below, and a number of these will be taken up in greater detail in Chapter 6. All of the applications mentioned can be carried out using the PCMODEL® modeling program by Serena Software. Many other of the commercially available modeling packages should be able to do the same, more or less.

Perhaps the most obvious use of the method is that one can generate, fairly rapidly, an accurate minimum-energy molecular model. The model can be built up by the user, or imported from some other source. Unlike a drawing on paper, which is symbolic or representational (or **iconic**[127]), this model is "alive" (or **enactive**[127]). We can obtain from it information on bond lengths and angles, internuclear distances, torsional angles, and the geometric relationships of nonbonded atoms as well. The structure can be manipulated, rotated, or viewed in stereo; a molecular surface can be prepared for a better intuitive grasp of shape and molecular recognition. The electron distributions in conjugated systems can be modeled by carrying out π-calculations iteratively with the energy minimization of the σ-framework. Predictions of ^{1}H-NMR coupling constants for certain three-bond (H–C–C–H) systems can be made. Dipole moments may be calculated within the limitations of the electrostatic model. Strained compounds which are not accessible by standard physical models, such as bridgehead olefins or hexaphenylethane, can be modeled.

Atomic positions, internuclear distances, or torsional angles can be fixed (locked) in order to determine the effect of such constraints on the conformational energy of the molecule. Two molecules can be compared in whole or in part to see how closely related they are in structural and conformational terms. Two molecules can be docked or complexed together to explore quantitatively the aspects of molecular recognition due to van der Waals repulsion/attraction, electrostatic repulsion/attraction, hydrogen bonding interactions and metal atom coordination bonding.

Transition states and reactive intermediates can be modeled in a limited way, to measure the steric effects on stability. Rotational barriers in compounds such as hindered biphenyls can be modeled, and an estimate of the energy barrier to rotation can be derived. There are methods to do conformational searching, in order to discover families of related conformations, find the global energy minimum, and determine populations based on the Boltzmann partitioning relationship.

Lipkowitz and Peterson distinguish between *a priori* and *a posteriori* applications of molecular modeling.[22] Much of the published modeling work is of the *a posteriori* type, in that a result is obtained and then rationalized or confirmed with an after-the-fact modeling experiment. The *a priori* applications potentially can have a far greater contribution to make for the practicing synthesis chemist. Since modeling a compound is far faster and cheaper than preparing it, doing this sort of pretesting seems to make sense as long as one is confident of the result. In nearly all of the cases of an *a posteriori* modeling-based rationalization of the failure of an experiment to yield the desired result, it is true that the same modeling would have given the same result *a priori*, and might possibly have served to recommend an alternative strategy.

Three basic types of *a priori* uses of modeling have been described.[22] In computer-aided molecular design (**CAMD**), molecular models are prepared to test a property or feature by computation before committing to the preparation of a physical sample. The stereochemical outcome of a synthetic reaction can be predicted by using molecular modeling to study the transition states or reaction pathways, or to compare the properties of the products. Synthesis planning with the aid of molecular modeling should enable one to choose between different preparative pathways when one intermediate can be shown to be far less stable or inaccessible.

Every computational method has its limitations, and molecular mechanics modeling is no exception. A few important limitations are: lack of appropriate parameters; caveats about transition states; difficulty in modeling kinetically controlled reactions; size limitations; the global minimum problem; unnatural geometries; and inaccurate heats of formation. Other limitations will be mentioned in other parts of the book.

The lack of appropriate parameters can be frustrating, but except for some fairly esoteric functional groups, molecular fragments, or elements, most modern force fields will have resident parameters or can make a best guess with generic parameters. A thorough knowledge of the parameter set of the force field one employs is very desirable. However, even with little familiarity with the software, a glance at the text log file from a modeling experiment will reveal any limitations or problems with the system being studied. Many modeling programs, including PCMODEL, will display messages on screen to alert the user, such as 'Gnrl Param' when generalized constants have been used. If the software has a serious problem modeling the system at hand, often a related system may be substituted for which parameters do exist. For example, there are no MMX force field parameters for a protonated or alkylated epoxide (i.e. oxiranium ring), but for most purposes the corresponding isoelectronic aziridine serves as a good model, and this is parametrized in MMX.

There are significant limitations to the molecular modeling of transition

1.3. MOLECULAR MODELING STRATEGY AND LIMITATIONS 15

states with the molecular mechanics method, because this technique only assesses steric and limited electrostatic interactions. The process in which bonds are being made or broken cannot be modeled well with just a few parameters. A reaction coordinate calculated from molecular mechanics, even with a correction for bond order, will be accurate only by accident. If discrete charges are being created or neutralized along the coordinate, the problems multiply. The best case for transition state modeling with this method is in comparing two stereoisomeric transition states. Here, the user must make the assumption that the bond rehybridization, electronic, and solvation effects roughly cancel out, and that the relative steric energies will be determining.

Unlike in thermodynamically controlled reactions, in which the relative stability of the products is of primary importance, in kinetically controlled reactions it is the relative stability of the transition states and the activated complexes approaching them which will decide the course of the reaction. Thus, the same caveats expressed above must be observed, and the user must be able to assume that relative steric energy will be controlling.

Depending upon the particular modeling package, there is usually a size limitation (number of atoms), which can become important in many applications. For the Windows version of PCMODEL, the limit is 500 atoms with 100 π-atoms. This works well for most small-molecule work, but will fail with many applications in biochemistry, polymer science, surface science, etc. Time limitations for running the minimization routines also become apparent as the upper limits on the number of atoms are reached.

A limitation that is shared with all molecular mechanics programs is that the energy optimization routines always seek local minima related to the input geometry. This is inherent in the method, and can be ameliorated only with a judicious analysis of the output geometries and resort to conformation sampling techniques as mentioned in Section 5.8. Except in the simplest systems, one doesn't ever know that the global minimum has been found; there is simply a level of confidence that it has.

A related problem is minimization to unnatural geometries (e.g. square planar carbon), or acceptance of unnatural input structures (pentavalent carbons). Both input and output geometries should be checked for such inconsistencies. Such unnatural structures are usually accompanied by unusually high energies after minimization, which can be a tipoff, especially if closely related structures are minimizing to much lower energies.

The heats of formation (ΔH_{form}) are still problematic in some systems. This problem is one of the main reasons that Allinger et al. have continued to develop the newer force field formulations (MM3, MM4; see Section 5.4.1). The ΔH_{form} values are fairly good for saturated hydrocarbons, but when a π-calculation is involved they are good only to within 2 kcal/mol. Some bond types are outside the limits of the parametrization of the ΔH_{form} calculation, and a warning to this effect will be printed in the output file. As was the case with transition state energies above, the user is on much safer ground when comparing the ΔH_{form} values of two stereoisomers with the same connectivity.

1.4. MOLECULAR MODELING RESOURCES

This book has been designed as a resource for the beginning modeler, but realistically it cannot serve that need indefinitely. There are many sources of data and background information available which are pertinent to molecular modeling, and this section serves as an introduction to them.

1.4.1. Printed Literature

The printed literature on molecular modeling and computer-aided molecular design consists of reports in the primary journals, review articles,[7,19,22,88,106,110,125,126,129] book chapters,[8,19,117,128,130] and monographs.[1,10,26,131] The computational chemistry literature as a whole has been reviewed.[132] As can be seen from scanning the list of references at the end of this book, publications which report modeling results appear in many of the standard scientific research journals. There are now a number of journals focused on or devoted to molecular modeling: *Computers in Chemistry, Journal of Chemical Information and Computer Science, Journal of Computational Chemistry, Journal of Computational Physics, Journal of Computer-Aided Molecular Design, Journal of Molecular Graphics, Journal of Molecular Structure, Molecular Simulations,* and *Tetrahedron Computer Methodology*. Other journals frequently feature modeling-related articles—for example, the *Journal of Chemical Education* runs a feature entitled "Computer Series" which often addresses modeling issues.

There are two review series which cover topics in computer molecular modeling: *Reviews in Computational Chemistry* and *Advances in Molecular Modeling*.

1.4.2. Electronic Literature

Electronic literature did not exist in its current form ten years ago. It is a product of the Internet, and more specifically of the World Wide Web, which enables the broad availability of text and graphics, and even live structures, as will be seen below. There are several articles,[2-4,6,133-135] and now a book,[5] which discuss chemistry sites and the availability of chemical information on the Internet and the World Wide Web. The hypertext document which accompanies this book, "Molecular Modeling Resources on the World Wide Web," provides links to many sites devoted to molecular modeling and graphics.

There are pointers to software reviews, catalogs, and companies; these sites very often give information about the capabilities of modeling programs and how to obtain them. There are pointers to home pages for the programs themselves, such as Allinger's MM2/MM3, Kollman's AMBER, and Karplus's CHARMM force fields and programs. These sites contain a wealth of information about modeling, and often permit the downloading of parameters, manuals, etc. Most academic research groups have World Wide Web sites, which can provide information about their activities, sometimes Web-format research papers, data, bibliographies, etc.

There are links to a number of sites which contain course notes and tutorials on the subject of molecular modeling. One example is the tutorial "Evaluation of the Potential Energy Function for n-Butane," a thorough force field analysis of that molecule, maintained by R. L. DeKock, T. M. Gray, and J. G. Pruis of Calvin College. Another is the pair of tutorials, "The Representation of Molecular Models" and "Rendering Techniques," containing a wealth of information and examples of modeling graphics, maintained by F. Savary of the University of Geneva. There are also many general resources and reference materials, such as WebElements, maintained by M. Winter of the University of Sheffield.

Nearly all of the print journals mentioned above maintain Web sites, and the links may be found in "Molecular Modeling Resources on the World Wide Web." These sites provide access to keyword-searchable tables of contents, supplementary material, and in some cases full text retrieval, and there are at least two strictly electronic journals: *The Journal of Molecular Modeling*, which contains reports of original research, and *Network Science*, which features review articles that often involve aspects of molecular modeling or chemical information retrieval.

1.4.3. Structure Databases

There are a number of commercial structure databases, and commercial offerings of public structure databases available, which contain 3D structures in various accessible formats. The Cambridge Crystallographic Structural Database[136] and the Fachinformationszentrum (FIZ) are repositories for crystallographic structures. They operate commercially, offering access to 3D structures and data for a fee.

MDL offers access to a number of 3D-format structure databases as part of its product line: the Cambridge Crystallographic Structural Database (CSD), National Cancer Institute (NCI), Available Chemicals Directory (ACD), and many others. The major suppliers of modeling software (referenced in the hypertext document which accompanies this book, "Molecular Modeling Resources on the World Wide Web") provide libraries of 3D-format structures to enhance the usefulness of their modeling systems.

1.4.4. Structures and Data from the World Wide Web

A variety of World Wide Web sites offer access to data and structures of interest to molecular modelers (see the hypertext document which accompanies this book, "Molecular Modeling Resources on the World Wide Web," under "Structures"). These may be accessed with any standard Web browser. Helper programs, or *plug-ins*, have been developed to work with Web browsers in order to visualize molecules in this environment. These molecule viewer applications *enliven* the molecules, permitting rotation, enabling different viewing formats such as ball and stick, space-filling, ribbon backbone for peptides, etc., and mono- or stereoscopic presentation. Essentially, they provide access to 3D structures posted on the World Wide Web, with graphics performance comparable to what is found in molecular modeling programs. The number of these programs is increasing rapidly, but specific mention can be made of two: **RASMol** and **Chime**™.

RASMol is a public domain molecular visualization program developed by R. A. Sayle, which can operate either as a browser plug-in or standalone program. It is intended for the visualization of proteins, nucleic acids, and small molecules, and is aimed at display, teaching, and generation of publication quality images. It consists of a main window for display and a child window for script commands. The input and output formats handled by RASMol are discussed in Section 4.5.

Chime™ is a commercial molecular visualization program developed by MDL, which operates as a browser plug-in. It incorporates the RASMol 3D rendering code, and has many of the same capabilities. The input formats handled by Chime™ are discussed in Section 4.5.

The Cambridge Crystallographic Structural Database is accessible in this medium, but the payment of a fee is required to receive structures. The Protein Data Bank,[137] administered by Brookhaven National Laboratory, offers structure files of biomacromolecules at no cost. A number of other sites offer searching capability and access to the PDB structures with various protocols.

There are two groups who have "collected" structures from across the World Wide Web, categorized and indexed them, and now offer access at no charge: the WWW Chemical Structures Database, maintained by the Computational Chemistry Center at the Universität Erlangen,[138] and ChemFinder, maintained by CambridgeSoft. The Erlangen group also offers access to two useful utilities: one which creates GIF or PNG format graphic images from 2D plots of chemical structures,[139] and another which uses their proprietary CORINA algorithm to convert 2D chemical structure formats to 3D formats.[140]

1.5. MOLECULAR MODELING IN INDUSTRY

Analyses of the use of molecular modeling in the workplace have appeared,[141] which provide some idea of how this technique is being employed in the industrial sphere. It is estimated that there are between 300 and 450 commercial groups worldwide which actively employ some form of computational chemistry in their research programs. This estimate may be low, since it focuses on the pharmaceutical and biotechnology areas, and companies involved in agrochemicals, specialty chemicals, polymers, and materials are using these techniques increasingly.

Definitions also vary, so that besides traditional molecular modeling (structure prediction/confirmation, conformational analysis, reaction product prediction, etc.), computational chemistry is applied to structural biochemistry (i.e. prediction of biomacromolecular structure, and visualizing the results of physical measurements such as X-ray diffraction, NMR, CD, etc.) and chemical information (which can include 3D structural databases, assessment of structural diversity, management of the results of combinatorial research, etc.).

In the industrial research setting, the use of computer modeling techniques ranges over a broad spectrum.[88,127,142] In a medium-to-large company, one will generally find a computer modeling or computational chemistry group, a subset of the general scientific staff which is highly conversant in the techniques, a larger group of people who have some idea of how to use computer modeling or how to interpret modeling results, and the rest of the staff, whose feelings may range from slight interest to apathy to antipathy.

The ratios of chemical research staff to computational staff in the industrial setting typically range from 3:1 to 10:1.

Probably the greatest use of molecular modeling in industry is in computer-aided molecular design. This topic has been addressed in a number of articles,[126,143–173] book chapters,[174–179] and monographs,[180,181] and there is a journal devoted to the topic (see above).

1.6. MOLECULAR MODELING IN EDUCATION

In recent years, offerings of courses in molecular modeling have flourished as the educational system responds to the need for their graduates to have this skill. This trend has virtually paralleled the growth in the use of computational methods in general, and molecular modeling in particular, in the technical workplace. These may be standalone university courses, may constitute a portion or a focus within a standard chemistry course (general chemistry, organic chemistry, inorganic chemistry, etc.), or may be short courses or seminars open to a wide variety of interested parties.

Six objectives of a college course in molecular modeling have been proposed:[182] (1) computer graphics visualization and design of molecules from structural data using the facility of a computer; (2) use of computerized structural databanks to identify molecular systems with common points; (3) use of empirical force fields to determine molecular properties as well as atomic distances; (4) ability to correlate molecular properties with a given electronic structure; (5) ability to gain information on dynamic molecular movements and their energies; and (6) application of the design of molecules by computer to molecular recognition in organic, bioorganic, and medicinal chemistry as well as material sciences.

A large number of reports have appeared describing laboratory experiments which explicitly use molecular modeling,[100,183–210] or describe approaches to teaching the subject in the classroom or laboratory,[106,107,182,211–230] mostly in the *Journal of Chemical Education*. This journal also runs a feature entitled "Computer Series" which often addresses modeling issues.

Many university educators now use the Internet in their teaching, and this is true for course notes and tutorials on the subject of molecular modeling. Links to a number of these sites may be found in the hypertext document which accompanies this book, "Molecular Modeling Resources on the World Wide Web."

CHAPTER 2
PC Molecular Modeling Hardware and Software

2.1. PC USAGE CONVENTIONS

The focus of this book is on the usage of molecular modeling programs on desktop computers. A basic knowledge of desktop computers in general is assumed. Moreover, at least an introductory level working knowledge of the Windows™ environment on IBM and related hardware platforms, and of the graphic user interface (GUI), is required to be able to understand and work through the examples. No attempt is made to teach these skills beyond what will be gleaned from the description of the basic level exercises. Some definitions of terms from basic computer usage are provided in the Glossary (Appendix D), but those with limited experience on such desktop platforms may want to keep available as a reference one of the many standard primers on Windows. Standard nomenclature will be used as much as possible, with references to the mouse cursor, the menu bar, "clicking" on a command button, "pulling down" or "bringing up" a menu or dialog box, etc. There are many figures in the book which are screen captures from the actual operation of the PCMODEL program, and these should be helpful to the less experienced.

The names of menu commands and tools available in PCMODEL are given in bold type, as in **Minimize**, and may be nested as in **File\Open**. These commands are described in the course of the exercises, and are documented fully in Appendix B in alphabetical order. The titles of windows, boxes, or options within a box are set off by single quotation marks, as in 'Output'. Filenames for data files are capitalized and set within braces, as in {FILENAME.OUT}. Terms which may be unfamiliar to the reader also appear in bold, and definitions for these will be found in the Glossary (Appendix D).

For carrying out many of the operations and experiments described in this book, a number of commercial programs would suffice. For the sake of simplicity and manageability, the descriptions in this book generally will deal with a single commercial package, PCMODEL® by Serena Software. This treatment should provide the least ambiguity, and one should be able to adapt the instructions given here to the use of a favorite alternative package without too much difficulty. It is beyond the scope of this book to document every molecular modeling program; several reviews have been published which accomplish this.[219,231-239] Some are described in Chapter 5, on force fields, and the names of many appear with brief descriptions in the Glossary in Appendix D. In the hypertext document which accompanies this book, "Molecular Modeling Resources on the World Wide Web," pointers are provided to many World Wide Web sites which give information on these programs, and this approach will yield the most timely information.

In many of the following chapters, input files mentioned in the exercises are available on the diskette which accompanies this book. This diskette has directories corresponding to the sections of chapters where the exercises are found, and the reader will likely wish to access these files in order to get the maximum benefit from working through the exercises. When these files are referred to in the text, the names are contained within braces for example {MOLECULE}. The user may wish to prepare edited versions of these files, and in this case the file first should be copied onto another diskette or to the hard drive. It would be prudent to keep an unaltered backup copy of the entire diskette.

2.2. HARDWARE OPERATIONS

The specific hardware requirements for running the different versions of PCMODEL are discussed at length in the documentation which accompanies the programs, and no attempt is made to reproduce the official guide on this subject. Briefly, the successful desktop platform will have at least 4 MB of volatile RAM and 2.5-4 MB on the hard drive to enable the use of PCMODEL. The standard peripherals for the graphic user interface are required: a two-button mouse or equivalent, such as a trackball, and the keyboard for text entry.

2.3. OVERVIEW OF PCMODEL™

The discussions here focus on the MicroSoft Windows™ version 5.13 (March 1995) of the PCMODEL program. As of mid-1996, version 6.0 has been available, but it was not used extensively in preparing the book.

PCMODEL can handle molecules of up to 500 atoms, with up to 100 atoms involved in π-calculations. Parameters are resident in the program for 17 main group elements, including some charged species and different hybridizations, transition state atoms, and 40 metallic element atoms (see Section 5.3 and Appendix C). The treatment of standard valence interactions includes bond stretching, angle bending, and torsional strain, along with a stretch-bend cross term. The treatment of nonbonding interactions includes van der Waals repulsions, electrostatic effects, specific hydrogen bonding attractions, and metal-ligand coordination effects. Up to 16 named substructures can be created and manipulated independently.

Structures can be drawn using the **graphic user interface** (GUI), or imported from standard templates or input files in ten formats, and new input files may written out for any structure in seven formats (see Chapter 4). Standard templates are available from pop-up menus for ring systems, amino acids, sugars, and nucleotides. These templates can be modified, and new templates can be defined and stored by the user. Input structures can be subjected to geometry optimization by structure-perturbation–energy-minimization protocols. Conformational space can be sampled with angle-driver utilities, or with the DOS-environment GMMX program. Dynamics simulations can be carried out.

Displays of the structures can be prepared in various formats, such as line drawings, ball-and-stick or CPK models. Molecular surfaces (van der Waals or water solvation) can be calculated and displayed. Peptide chains can be represented in the "backbone ribbon" format. Geometry and charge characteristics of structures can be queried and displayed (see Section 3.6).

While earlier PCMODEL versions needed graphics and printer drivers tailored to the hardware configuration, currently all of the graphics and printing is handled by Windows. In summary, PCMODEL is a graphics-oriented molecular modeling and presentation package, and has the capability of generating structure input files via a graphic drawing interface, reading in structural files and parameters in various formats, performing energy minimization on structures or ensembles of structures, and displaying structures in various visual formats for analysis or comparison.

In addition to PCMODEL, users will need two other types of programs to carry out the exercises described in this book, and also for the most efficient routine practice of modeling.

A file-handling program will be needed for viewing, copying, and renaming files and for effecting file transfers. The standard MicroSoft Windows™ FILE MANAGER (or an equivalent such as XTREE™ for Windows) serves this function. XTREE has an additional advantage in making available file compression utilities such as the ZIP format, which help to save space in storage media.

A text processing program will be needed to carry out editing functions on the PCMODEL input files. The author has had success with MicroSoft Word™ for Windows. Almost any word-processing program will suffice, as long as the file remains in ASCII text format. If used in their document mode, most word processors will insert binary formatting flags into the modeling input file. In this case, PCMODEL's input algorithms will be derailed as they **parse** the binary data, rendering the file unreadable. In some file formats such as MMX (see Chapter 4), the meaning of the data is highly position-specific, and so when modifying such files it is best to use a nonproportional screen font for greater ease in maintaining column specificity.

2.4. PCMODEL FEATURES

2.4.1. The Opening Window

Launching PCMODEL for Windows brings up the PCMODEL opening window, shown in Fig. 2-1. Consistent with the Windows environment, the title bar

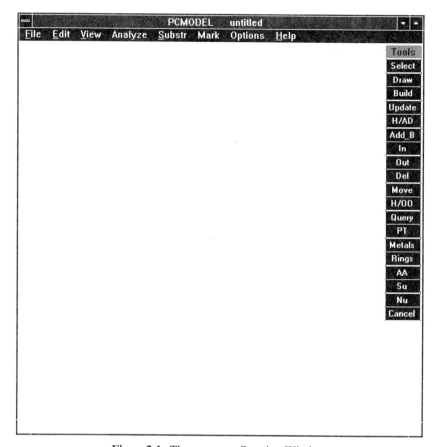

Figure 2-1. The PCMODEL Opening Window.

gives the name of the application, PCMODEL, followed by the name of the open file. In this case it is untitled because no named file is open.

Just below this is the menu bar, with eight command menu headers: **File**, **Edit**, **View**, **Analyze**, **Substr**, **Mark**, **Options**, and **Help**. When these headers are activated with a mouse cursor click, a pull-down menu appears listing the available commands (which are described in detail in Appendix B). Below the menu bar, and occupying most of the area of the window, is what we have termed the work space, where structure drawing and manipulation takes place.

Near the right-hand side of the work space is found a vertical tool bar listing nineteen choices: **Select**, **Draw**, **Build**, **Update**, **H/AD**, **Add_B**, **In**, **Out**, **Del**, **Move**, **H/OO**, **Query**, **PT**, **Metals**, **Rings**, **AA**, **Su**, **Nu**, and **Cancel**. This tool bar can be moved around the workspace by using the mouse cursor to drag the 'Tools' title bar to the desired location. Most of these tools are command buttons, the functions of which are described in detail in Appendix B. Six of these tools (**PT**, **Metals**, **Rings**, **AA**, **Su**, and **Nu**) will bring up addi-

2.4. PCMODEL FEATURES   25

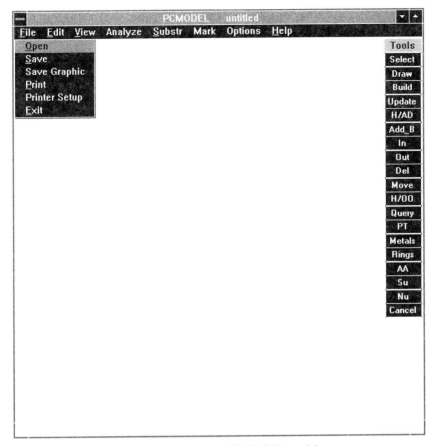

Figure 2-2. The PCMODEL File Pull-Down Menu.

tional option windows. The first two of these will be discussed below, and the last four are covered in Section 3.2 in the discussion of templates.

2.4.2. The Menu Bar Commands

The menu bar contains eight commands, which are really command headings. According to Windows convention, commands which appear with one of their letters underlined can be activated by entering from the keyboard at the same time ⟨Alt⟩ plus that letter. Clicking on the command headings with the mouse cursor brings up a pull-down menu of commands, each of which can be activated by clicking on them with the mouse cursor. The commands on these pull-down menus are described briefly below—they are covered in greater detail in Appendix B.

Clicking on the **File** command gives a pull-down menu containing six commands, shown in Fig. 2-2: **Open**, which is used to read input files; **Save**, which is used to write input files; **Save Graphic**, which is used to write a graphic image of the window to a file; **Print**, which prints the structure(s) and other material from the current screen to the default printing device; **Printer Setup**, which allows the user to make changes in the printer settings; and **Exit**, which closes PCMODEL.

Clicking on the **Edit** command gives a pull-down menu containing 11 commands, shown in Fig. 2-3: **Draw**, which reactivates the vertical tool bar; **Erase**, which clears the work area of all structures and substructures; **Structure Name**, which allows the user to enter or edit a name for the structure(s) currently active in the work space; **Rotate_Bond**, which rotates an indicated bond by a specific amount; **Epimer**, which allows epimerization of a stereocenter; **Enantiomer**, which draws the mirror image of the structure in the work space; **Copy_BW_To_Clipboard**, which reduces the color image of the workspace to black-and-white, and copies it to the Windows clipboard; **Copy_CO_To_Clipboard**, which copies the color image of the workspace to

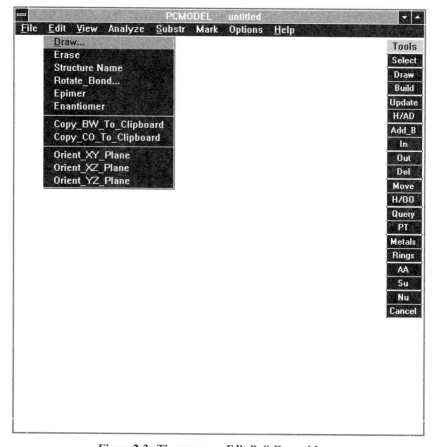

Figure 2-3. The PCMODEL **Edit** Pull-Down Menu.

the Windows clipboard; **Orient_XY_Plane**, which redraws the structure with any three selected atoms in the XY plane; **Orient_XZ_Plane**, which redraws the structure with any three selected atoms in the XZ plane; and **Orient_ZY_Plane**, which redraws the structure with any three selected atoms in the ZY plane.

Clicking on the **View** command gives a pull-down menu containing twelve commands, shown in Fig. 2-4: **Control_Panel**, which brings up a window with scroll bar controls affecting size, translation, and rotation of structures; **Labels**, which brings up a window to choose between different atom-labeling formats; **Mono/Stereo**, which toggles the display between monoscopic and stereoscopic graphic representations; **Stick Figure**, which renders the displayed molecule(s) with lines for bonds, corners for atoms, and labels for all atoms save carbon; **Ball_Stick**, which renders the displayed molecule(s) with lines for bonds and spheres for atoms; **Ortep**, which renders the displayed molecule(s) with lines for bonds and spheres for atoms; **Tubular Bonds**, which renders the displayed

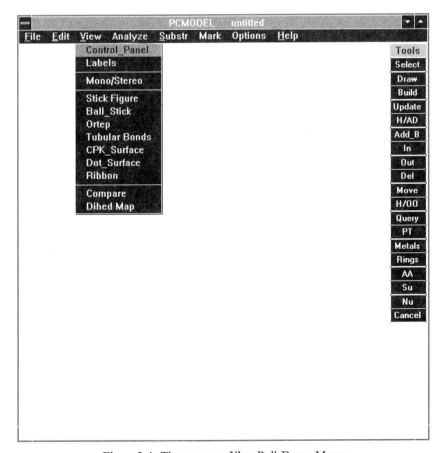

Figure 2-4. The PCMODEL **View** Pull-Down Menu.

molecule(s) with thickened cylindrical lines for bonds and corners or ends for atoms; **CPK_Surface**, which renders a solid surface depiction of the displayed molecule(s); **Dot_Surface**, which renders a dotted surface depiction of the displayed molecule(s); **Ribbon**, which renders a backbone for a displayed peptide or polynucleotide as a ribbon; **Compare**, which allows the user to compare some or all atomic positions of two or more molecules; and **Dihed Map**, which graphically represents the results of a dihedral driver calculation.

Clicking on the **Analyze** command gives a pull-down menu containing ten commands, shown in Fig. 2-5: **Minimize**, which initiates the search process for the energy minimization of the displayed molecule(s); **Single Point E**, which calculates the energy of all the displayed molecule(s) with no minimization search; **Mopac**, which links up with the semiempirical molecular orbital program of the same name to carry out a calculation on the displayed molecule; **Dynam**, which initiates a molecular dynamics calculation; **Dock**, which initiates a simulated annealing process to dock one structure with another;

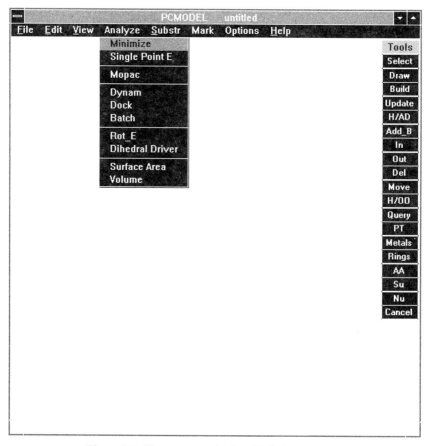

Figure 2-5. The PCMODEL **Analyze** Pull-Down Menu.

2.4. PCMODEL FEATURES

Batch, which allows the user to submit an appended-format file for of many structures sequentially; **Rot_E**, which carries out a rigid rotation with energy calculations for a selected bond in a molecule in steps and over a range selected by the user; **Dihedral Driver**, which carries out energy minimization calculations for rotation of one or two selected bonds in a molecule, in steps and over a range selected by the user; **Surface Area**, which calculates the surface area of the displayed molecule(s); and **Volume**, which calculates the molecular volume of the displayed molecule(s).

Clicking on the **Substr** (Substrate) command gives a pull-down menu containing seven commands, shown in Fig. 2-6: **Read**, which initiates a "read input file" sequence for a substructure; **Create**, which defines one of several structures, or a selected portion of a single structure, as a substructure; **Move**, which moves a selected substructure; **Connect**, which connects one substructure to another; **Erase**, which erases the designated substructure; **Hide**, which

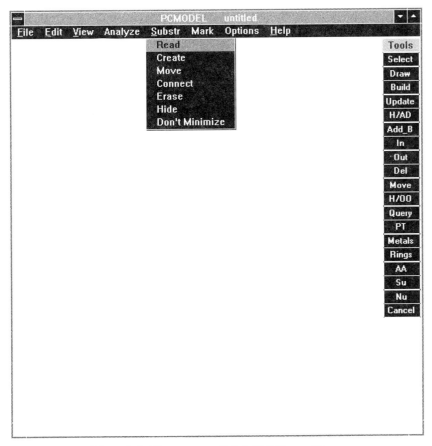

Figure 2-6. The PCMODEL **Substr** Pull-Down Menu.

hides a selected substructure from view; and **Don't Minimize**, which exempts a selected substructure from the energy minimization routine.

Clicking on the **Mark** command gives a pull-down menu containing eight commands, shown in Fig. 2-7: **H-Bonds**, which flags the appropriate hydrogens for inclusion of the attractive hydrogen-bonding potential; **Piatoms**, which flags the appropriate heavy atoms for inclusion in the π-calculation; **Metal_Coord**, which flags the appropriate metal atom(s) and donor ligand atoms for inclusion of the attractive metal-coordination potential; **TS_Bond Orders**, which allows the user to select a value for the bond order in a bond between transition state atoms; **Fix_Atoms**, which allows the user to hold designated atoms at a fixed position; **Fix_Distance**, which allows the user to maintain a fixed distance between designated atoms; **Fix_Torsions**, which allows the user to maintain a fixed value for a designated torsional angle; and **Reset**, which resets all or selected marked atoms.

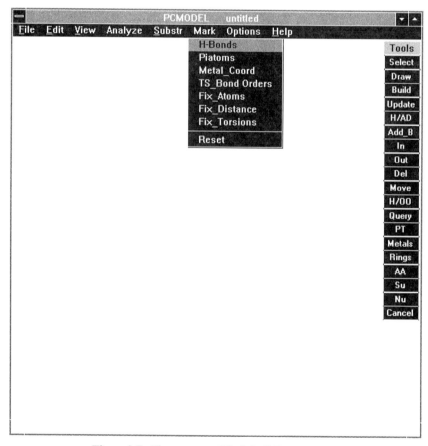

Figure 2-7. The PCMODEL **Mark** Pull-Down Menu.

2.4. PCMODEL FEATURES 31

Clicking on the **Options** command gives a pull-down menu containing eight commands (see Fig. 2-8): **Printout**, which allows the user to choose the format for the output file generated during minimization; **Dielc**, which allows the user to choose a nonstandard value for the dielectric constant used in the calculation of electrostatic interactions; **DP_DP**, which toggles between the dipole–dipole and charge–charge models for electrostatic interactions; **MMX_PI_Calc**, which allows the user to choose parameters for the π-calculation; **Standard Constants**, which returns to default constants; **Added Constants**, which allows the user to input nonstandard constants; **Stereo**, which switches the sense of rotation in stereo display; and **Pluto**, which prepares a high-quality CPK rendering of the displayed molecule.

Clicking on the **Help** command gives a pull-down menu containing one command, shown in Fig. 2-9: **About PCModel**, which gives the version number and date for the user's copy of PCMODEL.

Figure 2-8. The PCMODEL **Options** Pull-Down Menu.

32 PC MOLECULAR MODELING HARDWARE AND SOFTWARE

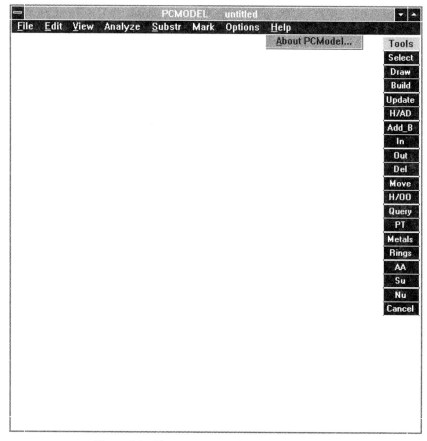

Figure 2-9. The PCMODEL **Help** Pull-Down Menu.

2.4.3. The Tool Bar Commands

The tool bar contains nineteen tools, and is depicted in Fig. 2-10. When PCMODEL is opened, the tool bar appears as a column near the middle of the full window, which can be relocated to a more convenient place on the screen by clicking and holding the mouse cursor on the 'Tools' heading and moving it as desired. The tools are activated by clicking on them with the mouse cursor. According to Windows convention, commands which appear with one of their letters underlined can be activated by entering from the keyboard at the same time ⟨Alt⟩ plus that letter. Each of the tools is described briefly below; more detail can be found in Appendix B. Six of the tools pop up new menus when activated; the first two of these are discussed in the following sections, and the last four are covered in Chapter 3:

Figure 2-10. The PCMODEL Tools Bar Menu.

Select allows the user to select atoms within the structure to flag them for other operations;

Draw sets up the mouse cursor for drawing new atoms in the work space;

Build converts attached hydrogens to sp^3-hybridized carbons;

Update repaints the screen incorporating any recent changes;

H/AD toggles between 'H-ADD' (add hydrogens and lone pairs) and 'H-DEL' (delete hydrogens and lone pairs);

Add_B increases the multiplicity of an indicated bond by one;

In moves an atom along the Z-axis into the screen in 0.25-Å increments;

Out moves an atom along the Z-axis out of the screen in 0.25-Å increments;

Del deletes undesired atoms or decreases bond multiplicity by one;

Move moves atoms in the workspace, within the plane of the screen;

H/OO toggles on and off the depiction of hydrogen atoms without deleting their positions;

Query allows the user to obtain information on the relative geometry of the structure(s) in the workspace; e.g. position, distance, or angles;
PT pops up the 'periodic table' menu (see Section 2.4.3.1);
Metals pops up the 'metals' menu (see Section 2.4.3.2);
Rings pops up the 'rings' menu (see Section 3.2.1);
AA pops up the 'amino acids' menu (see Section 3.2.2);
Su pops up the 'sugars' menu (see Section 3.2.3);
Nu pops up the 'nucleosides' menu (see Section 3.2.4);
Cancel removes the tool bar from the work space.

2.4.3.1. The Periodic Table Menu   Clicking on the **PT** on the tool bar pops up the menu shown in Fig. 2-11, which contains buttons bearing the symbols for 38 general or specific atom types, and a **Cancel** button. The 'Periodic Table' menu will remain active in the work space until it is dismissed by clicking on the **Cancel** button.

The symbols for the atom types correspond to those described in Section 5.3 and in Appendix C. Clicking on one of these buttons *loads* the mouse cursor with the corresponding atom type, which can be deposited in a novel location in the work space, or be used to convert an existing atom to the new atom type. The mouse cursor remains loaded with this atom type until another such button on the 'Periodic Table' menu is activated, or until the **Cancel** button is activated.

2.4.3.2. The Metals Menu   Clicking on the **Metals** on the tool bar pops up the menu shown in Fig. 2-12, which contains buttons bearing the symbols for 40 specific metal atom types, and an **Other** and a **Cancel** button. The 'Metals' menu will remain active in the work space until it is dismissed by clicking on the **Cancel** button.

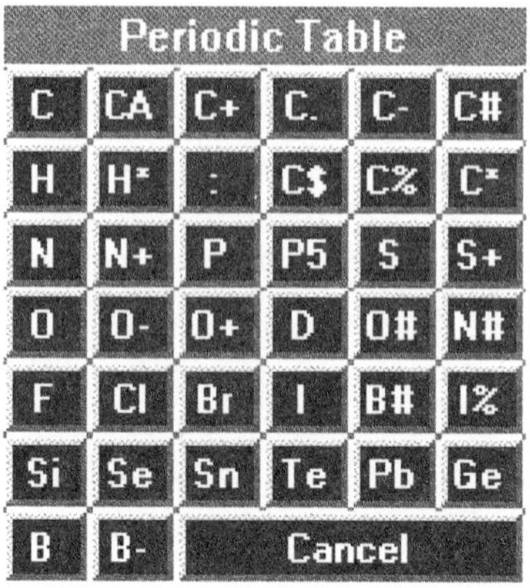

Figure 2-11. The PCMODEL 'Periodic Table' menu Window.

2.4. PCMODEL FEATURES   35

The arrangement of the metal atom type buttons embodies a subset of the periodic table, including the first five alkali metals and alkaline earths, along with the transition series. The MMX treatment of metal atoms is discussed at length in Section 6.7. The **Other** button is intended to enable user-defined metal atoms, and will be implemented in a future version. Clicking on one of these buttons loads the mouse cursor with the corresponding metal atom type, which can be placed in an unoccupied location in the work space, or be used to convert an existing atom to the new metal atom type. The mouse cursor remains loaded with this metal atom type until another such button on the 'Metals' menu is activated, or until the **Cancel** button is activated.

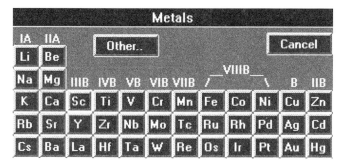

Figure 2-12. The PCMODEL 'Metals' Menu Window.

CHAPTER 3

Input and Output

3.1. STRUCTURE INPUT FROM THE GRAPHICS INTERFACE

... Molecular modeling must be primarily a kind of communication language between [hu]man and computer...[113]

The most direct way to prepare a molecular input file for molecular modeling is through the graphic user interface (GUI), i.e. the combination of graphics monitor, mouse and keyboard as described in Section 2.3. In this way, one may examine any structure that can be drawn, without needing to find and import a template or a model structure. In this section, first some general features of using the GUI are covered. Then, each of the PCMODEL tools and commands used in structure drawing will be explained. Finally, two drawing exercises are given: for 1,1-difluoroethene and adamantane.
A disadvantage is that the input geometry of a drawing created by the user will often be of poor quality, even when care is taken to make bond lengths and angles close to ideality. This is because a two-dimensional (2D) drawing in good style only rarely will project into an accurate three-dimensional (3D) structure, and conversely, a good 3D structure when directly projected into 2D most often gives an unsatisfying rendering. There are a number of tools and commands available in PCMODEL to tailor such a computer-drawn structure with depth corrections, as will be discussed below. There are also commercial programs which use rule-based algorithms to clean up a poorly drawn 2D structure, or to carry out a 2D → 3D translation, such as **CONCORD**™ discussed in Section 3.4.

3.1.1. General Information for Drawing Structures

PCMODEL uses a Cartesian frame of reference, where the X-axis is horizontal in the plane of the screen, the Y-axis is vertical in the plane of the screen, and the Z-axis is perpendicular to the plane of the screen. When the mouse cursor is positioned over an atom, a small box appears around that atom; when positioned over a bond, × crosshairs appear over the midpoint of the bond. The default atom produced by a **Draw** operation is carbon. A carbon can be converted to any MMX atom type by loading the mouse cursor from the appropriate menu and then clicking on the atom of interest to effect the desired change. The main group atoms, including reactive intermediate and transition state atoms are available from the 'Periodic Table' menu, and metal atoms are available from the 'Metals' menu, as discussed in Section 2.4.3. In most cases, an operation described as affecting a structure on screen will have an analogous effect upon all structures present, if there is more than one.

3.1.2 Drawing Tools and Commands in PCMODEL

There are 13 tools and commands in PCMODEL used in the preparation of input structures with the graphical user interface. Four are found under **Edit** on the command bar: **Enantiomer, Epimer, Erase,** and **Rotate_bond**. One is found under **Substr** on the command bar: **Connect**. Eight are found on the tool bar: **Add_B, Build, Del, Draw, H/AD, In, Move,** and **Out**. The use of each of these, in alphabetical order, is described below.

The **Add_B**, or add bond, tool increases the multiplicity of a bond by one, up to a total of three. When this tool is activated, the mouse cursor may be clicked at the midpoint of the target bond. This bond is then overprinted with a symbol for the change, a 'D' for a double and a 'T' for a triple bond. Clicking on the **Update** button redraws the structure with bond having the new multiplicity. Once activated, any number of **Add_B** operations can be carried out sequentially. Clicking on **Update** or any other command will end the add bond operation. After a series of **Add_B** operations have been made, it is recommended to process a structure through two cycles of the **H/AD** tool, deleting and re-adding the hydrogens and lone pairs. This is to ensure proper valences prior to energy minimization.

When the **Build** tool is activated, the mouse cursor is loaded with the ability to "grow" methyl groups onto existing structures in the work space. Clicking on a hydrogen atom in the current structure converts it into the carbon of a new methyl group. The carbon and its attached hydrogens are added according to a geometry-based algorithm which is related to the generation of 3D structures and is described in Section 3.4. This drawing tool can be used to elaborate a carbon skeleton rapidly. It can be used many times sequentially, and is only canceled when another tool or command is activated, such as **Draw** or **Update**. **Build** will only convert a hydrogen which is attached to a carbon, so that an alcohol or amine X—H will not be affected by this tool.

Substr\Connect joins two substructures by removing a selected hydrogen on each substructure and joining the two open valences. This command is found under **Substr** on the menu bar. Before activating this command, the user should have both desired substructures on screen. The **Select** tool is first employed to mark a hydrogen atom on each of the substructures. Then, clicking on **Connect** causes the two substructures to form a bond at the points where the hydrogens were bonded. Some reorientation of the structures normally takes place, and the atom file numbers of the new combination structure are appropriately reordered.

The **Del**, or delete, tool erases undesired atoms or bonds during a drawing operation. After this tool is activated, clicking on the undesired atom or bond causes an 'X' to be written over that atom or bond. If the indicated position doesn't map to an atom or bond midpoint, a '?' will be printed at that location on the screen. It is only necessary to activate the **Del** command once if a sequential series of deletions are to be carried out. An **Update** will repaint the screen taking account of all of the deletions. After a series of deletions have been made, it is recommended to process a structure through two cycles of the **H/AD** tool, deleting and re-adding the hydrogens and lone pairs. This is to ensure proper valences prior to energy minimization.

There are two locations where a **Draw** button is found: the **Edit\Draw** command brings up the tool bar window if it has been closed, whereas the **Draw** tool found on the tool bar loads the mouse cursor for a drawing sequence, either to build new structures or to edit existing ones. Once the **Draw** tool is activated, clicking the mouse cursor in the work space draws a default carbon atom, the next click draws a second atom bonded to the first, and so on, so that any number of atoms can be drawn in sequence. This series may be ended by clicking on the **Draw** button again, and the user can then start another **Draw** sequence. The atoms are numbered in the input file in the same order that they are drawn. All atoms drawn in this way will be added in the plane of the screen, and when new bonds are drawn to an existing atom, the new atoms are drawn in the same plane as the existing atom. The coordinates of the first few heavy atoms are subject to rescaling of the XYZ coordinates so that bond distances average 1.53 Å. Drawing over an existing bond has the same effect as the **Add_B** tool.

The **Edit\Enantiomer** command generates the mirror image of the structure in the work space. When this command is activated, a dialog option box pops up, prompting the user for the desired axis of reflection, as shown in Fig. 3-1.

The default is reflection about the X-axis, corresponding to a reflection across the YZ-plane. In practice, this is effected by reversing the algebraic sign on each of the X-coordinates. Selecting the other axes of reflection will give an analogous result. The user should read in one or two molecules from the

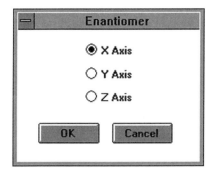

Figure 3-1. The 'Enantiomer' Dialog Option Box.

'Amino Acids' or 'Sugars' template menus and try this command. The reversal of coordinates will be easy to see, and the enantiomeric switch can be detected visually by employing a stereopair of images.

The **Edit\Epimer** command carries out an epimerization of two substituents bonded to a common stereocenter atom. Prior to invoking this command, three atoms of a molecule in the work space must be marked in the proper order with the **Select** tool: first the atom of the stereocenter, and then the two attached atoms corresponding to the substituents which will be epimerized. Then when the **Epimer** command is activated, the structure will be redrawn with the stereochemistry of these two atoms and their attachments reversed. In practice, this is effected by setting the three selected atoms in a plane, then reflecting the coordinates of the two attached atoms (and all of their attached groups) about an orthogonal plane which bisects the bond angle of the three selected atoms.

The use of **Epimer** is straightforward unless the stereocenter and at least one of the attached atoms is part of a ring. Some distortion will occur if the stereocenter and one or both of the attached atoms are all part of the same ring, but a reasonable conformation may result after energy minimization. If the stereocenter is contained in two rings (e.g. fused or bridged, or spiro bicyclic) and at least one attached atom is part of a ring, a severely distorted structure may result from an application of **Epimer**, in some cases unremediable through energy minimization. In these cases, a bond in the affected ring may be broken with the **Del** tool and reattached after the **Epimer** operation has been carried out.

There are two **Erase** commands for removing structures in the work space. One is **Edit\Erase** for global erasure, and the other is **Substr\Erase** for selective erasures of substructures (which will be addressed in Section 3.2.3). Activating the **Edit\Erase** command in the menu brings up a confirmation prompt, as shown in Fig. 3-2. Confirming this will erase all currently active structures from the work space.

Note that erased structures are not recoverable once the erasure has taken place, unless they have been written to a file, or run through an energy minimization and thus saved as the default {PCMOD.BAK} file.

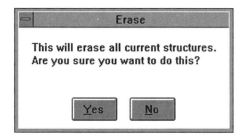

Figure 3-2. The 'Erase' Confirmation Prompt.

The **H/AD** tool toggles between (1) adding hydrogens and lone pairs to a structure which has none, and (2) deleting hydrogens and lone pairs from a structure which already has them. When this tool is activated, the screen will be repainted in the alternative format. When applied to a heavy-atom-only structure, the **H/AD** algorithm begins with atom file number 1, adds the appropriate number of hydrogens and/or lone pairs according to a standard geometry recipe, and then proceeds through all the heavy atoms in order of increasing atom file number. To see which heavy atoms are assigned lone pairs by PCMODEL, see Appendix C. Since the geometry of the added hydrogens and lone pairs is a function of the cross vectors of the existing bonds between heavy atoms, distorted starting geometries will lead to eccentric placements. It is nearly always better to carry out even a partial energy minimization on the heavy atom skeleton before using **H/AD**. Since the **H/AD** algorithm is related to the generation of 3D structures, it is described in more detail in Section 3.4.

This tool saves time during minimization, since the heavy atom skeleton can be minimized more rapidly, then embellished with hydrogens and lone pairs for a final minimization. Using **H/AD** can be a shortcut during the building or modification of a structure, since the added hydrogens can be converted to new atoms from the 'Periodic Table' template menu, which will furnish the desired structure with a reasonable 3D geometry. This can also be accomplished by converting the introduced hydrogen atoms into carbons with the **Build** tool.

Note: Hydrogens added to a heteroatom and deuterium (D) have different atom types, and they will not be affected by the **H/AD** tool. The **Del** tool must be used to remove these atoms. For an energy-minimized structure, multiple uses of **H/AD** to carry out a sequential deletions and additions will not return the hydrogens and lone pairs to their original, minimum energy positions, since the hydrogens and lone pairs are added according to the rule-based algorithm described in Section 3.4. Hydrogens may be hidden for viewing or comparison purposes, but not deleted, by using the **H/OO** tool described in Section 3.6.

The **In** tool is one of two which can be used to add 3D aspect to a flat structure drawn in the work space in the default XY plane (the other is **Out**). When this tool is activated, the mouse cursor is clicked over the atom to be moved. This decrements the atom's Z-axis coordinate (perpendicular to the plane of the screen) by ≈ 0.275 Å, and a small '$-$' appears just above the atom. Additional clicks on the mouse button repeat this effect on the position of the atom, which will be cumulative. An **U**pdate, or any operation which repaints the screen, will return the structure to a normal appearance with the new coordinates. When the **In** tool is used on a heavy atom, its attached hydrogens will be repositioned concurrently.

There are two locations where a **Move** button is found: the **Substr\Move** command enables the movement of a substructure to another position in the work space, whereas the **Move** tool found on the tool bar allows the user to move the atoms in a structure in the work space within the XY plane of the screen. The **Substr\Move** command will be discussed in Section 3.2.3.

When the **Move** tool is activated, the mouse cursor is used to click on the atom to be moved, which will be marked with a small dot. Then the cursor is relocated to the new position for the atom and the mouse button is clicked again, where another marking dot will be drawn. If desired, a number of consecutive **Move** operations may be carried out. A final **Update** repaints the screen with the structure reflecting the changes made.

The **Out** tool is one of two which can be used to add 3D aspect to a flat structure drawn in the work space in the default XY plane (the other is **In**). When this tool is activated, the mouse cursor is clicked over the atom to be moved. This increments the atom's Z-axis coordinate (perpendicular to the plane of the screen) by ≈ 0.275 Å, and a small '+' appears just above the atom. Additional clicks on the mouse button repeat this effect on the position of the atom, which will be cumulative. An **Update**, or any operation which repaints the screen, will return the structure to a normal appearance with the new coordinates. When the **Out** tool is used on a heavy atom, its attached hydrogens will be repositioned concurrently.

The **Edit\Rotate_Bond** command is used to carry out a measured rotation on the chosen bond. This works best when the molecule has been through an energy minimization sequence. Before starting this operation, the user must mark the four contiguous atoms which define the torsional angle to be modified with the **Select** tool. When this is done and **Rotate_Bond** is activated, a dialog scroll bar box entitled 'Rotate_bond,' pops up as shown in Fig. 3-3.

In the first line is the title 'Atoms', with the atom file numbers for the four atoms which define the torsional angle. In the second line is the title 'Dihedral', and the current dihedral angle value in degrees is reported in the box. Below this is a scroll bar which governs the angle rotation, and each click gives a rotation of 2°. The **Rotate_Bond** command works by rotating the first

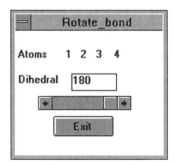

Figure 3-3. The 'Rotate_bond' Dialog Scroll Bar Box.

42 INPUT AND OUTPUT

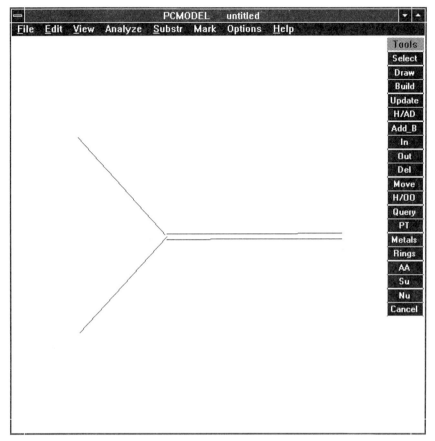

Figure 3-4. Drawing the Input Structure for 1,1-Difluoroethene, Part 1.

two atoms selected (and all their attachments) with respect to the third and fourth atoms, which are kept in their original orientation. At the bottom is an 'Exit' button, which can be clicked on when all rotations have been finished. The screen is then repainted with the edited structure.

With an acyclic dihedral (i.e. second and third atoms not contained in a ring), any value can be selected for the new angle, and the structure will be redrawn to reflect the new dihedral angle. With an endocyclic dihedral (i.e. second and third atoms contained in a ring), a large angle might distort the ring so much that the structure will not recover upon minimization. In order to carry out a large-magnitude rotation of a cyclic torsional angle, one should proceed stepwise as follows: introduce a small-magnitude (i.e. 5–10°) change in the angle, freeze it with **Mark\Fix_Torsions**, and then run the structure through **Analyze\Minimize**. Next, release the torsional constraint in **Mark\Reset**, alter the angle again, freeze it again, and carry out another minimization.

3.1. STRUCTURE INPUT FROM THE GRAPHICS INTERFACE   43

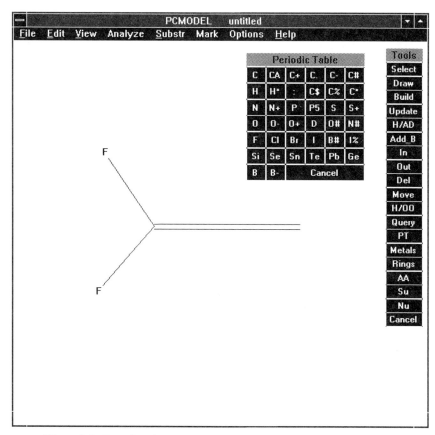

Figure 3-5. Drawing the Input Structure for 1,1-Difluoroethene, Part 2.

This can be repeated until the desired value for the torsional angle is achieved. This is the equivalent of a manual dihedral driver operation (see Section 6.3).

Drawing Exercise 1: 1,1-Difluoroethene. As the first drawing exercise, we will use the graphic interface to prepare the input file for 1,1-difluoroethene. In the PCMODEL window, the **Draw** tool is activated with the mouse cursor, which is then used to draw in the heavy atoms for the target molecule. The first click, just to the left of center, establishes C1, the second one just to the right of center creates C2. Clicking back on C1 makes this a carbon–carbon double bond, indicated at first as a 'D' superimposed on the bond. The first fluorine atom is drawn in by clicking above and to the left of C1. The **Draw** default atom is carbon, and we will convert this to fluorine in a later step. Next, the **Draw** tool button is activated again to interrupt the continuously drawing mode, and then a bond to the second fluorine atom is drawn in by clicking again on C1, and then below and to the left. Activating the **Update** tool

44 INPUT AND OUTPUT

redraws the structure with a true double bond; the screen should resemble the depiction in Fig. 3-4.

The two substituents may be changed to fluorine atoms by clicking on the **PT** tool (see Section 2.4.3.1), which pops up the 'Periodic Table' menu, then clicking on the **F** button in this menu to load the mouse cursor with fluorine, and then clicking on each of the two substituent atoms of the structure in the work space. At this point the screen should resemble the depiction in Fig. 3-5.

To finish up, the 'Periodic Table' menu is dismissed by clicking on its 'Cancel' button, and then the **H/AD** tool is used to add hydrogens to C2. The name of the molecule can be entered by activating the **Edit\Structure Name** command, under on the menu bar. This pops up the 'Edit Structure Name' dialog box, where the user may enter the name '1,1-Difluoroethene' from the keyboard. This dialog box is closed by clicking on its OK button, and the final screen should resemble the depiction in Fig. 3-6.

Finally, we write this structure out to an input file, to save it for future use. This is done by activating the **File\Save** command. In the 'Save File' dialog

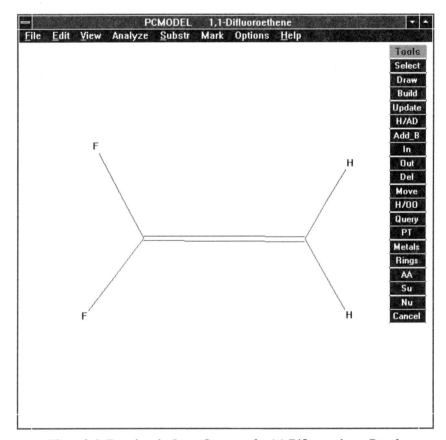

Figure 3-6. Drawing the Input Structure for 1,1-Difluoroethene, Part 3.

3.1. STRUCTURE INPUT FROM THE GRAPHICS INTERFACE   45

box which pops up (see Fig. 3-25), we use the default PCMODEL format and enter the filename {F2ETHENE.PCM} from the keyboard. This structure will be used for an exercise in Section 3.5, in which it will be submitted to energy minimization, and the output data file will be examined.

Drawing Exercise 2: Adamantane. This exercise requires the use of some display commands discussed later in this chapter, and it may be helpful to read through Section 3.6 if the steps become confusing. The first step is to activate the **Draw** tool, and use the mouse cursor to sketch a complete hexagon, starting near the top of the work space and proceeding in a clockwise sense. The result should resemble Fig. 3-7.

The next step is to modify this flat hexagon to more closely approach the standard chair form of cyclohexane. Activate the **In** tool, then click once each at the top carbon (C1), at the bottom right-hand carbon (C3), and at the bottom left-hand carbon (C5). Next, activate the **Out** tool and click once each

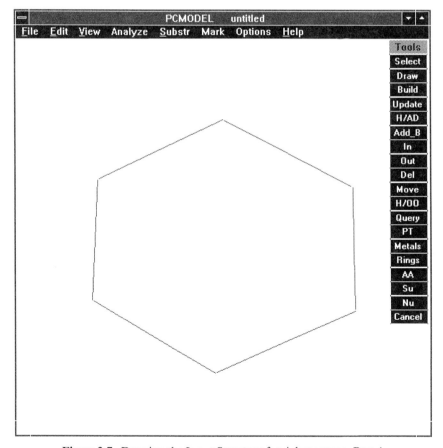

Figure 3-7. Drawing the Input Structure for Adamantane, Part 1.

at the upper right-hand carbon (C2), at the bottom carbon (C4), and the upper left-hand carbon (C6). These actions will move C1, C3, and C5 to a distance of ≈ 0.275 Å below the XY plane, and C2, C4, and C6 to a distance of ≈ 0.275 Å above the XY plane.

To get the structure in a better perspective, activate the **View\Control Panel** command (see Section 3.6). In the 'Dials' pop-up dialog box, move to the 'Rotate' scroll bars. Click on the right-hand arrow of the 'X' scroll bar until the numerical value is ≈ 260, and then click on the right-hand arrow of the 'Y' scroll bar until the numerical value is ≈ 210. This will give a perspective approximately along the average plane of the ring atoms, and it can be seen that our input geometry is slightly distorted. Activate the **H/AD** tool to add hydrogens, giving a structure such as is depicted in Fig. 3-8.

The distortions in the ring structure are mirrored in the geometry of some of the newly attached hydrogens. At this stage a vast improvement in the geometry will be obtained by energy minimization, so activate the

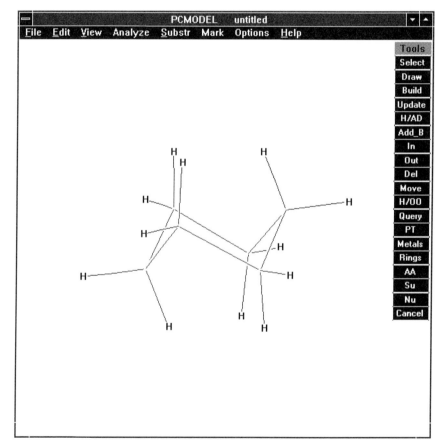

Figure 3-8. Drawing the Input Structure for Adamantane, Part 2.

3.1. STRUCTURE INPUT FROM THE GRAPHICS INTERFACE 47

Analyze\Minimize command. This procedure will be discussed in greater detail in Section 5.5. The structure which results is much improved, and ready for further elaboration. Activate the **Build** tool, and then click in turn on each of the axial hydrogens above the ring. A methyl group will sprout at each position, and the result should look like Fig. 3-9.

At the new methyl group on the right, click on one of the outward-pointing hydrogens to sprout in a fourth methyl group. This will become C10, the final one needed to complete the adamantane structure. To approach the best geometry, we will use the **Edit\Rotate_bond** command to rotate this last methyl group in toward the center of the ring. First, designate the desired dihedral angle by activating the **Select** tool, and then use the mouse cursor to mark in sequence this C10 methyl group, its attached methylene carbon, the methine ring carbon next in line, and the tertiary hydrogen at that same position. The marked atoms will be covered by a small dot. Next, activate the

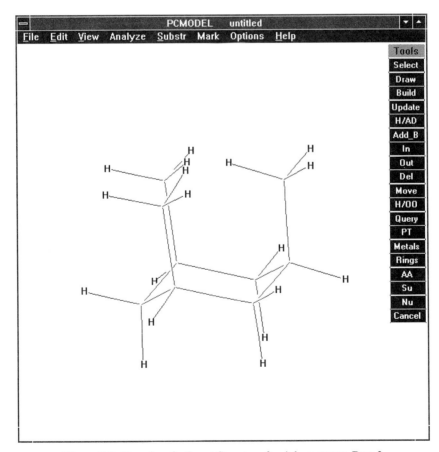

Figure 3-9. Drawing the Input Structure for Adamantane, Part 3.

Edit\Rotate_bond command, which pops up the 'Rotate_bond' dialog box as in Fig. 3-10.

This shows the atom file numbers of the marked atoms: 10, 7, 2, and 13; it also gives a starting value of $\approx 60°$ for this torsional angle. The desired geometry will have this angle at $\approx 180°$, so either click on the arrow buttons until this is so, or grab the scroll cursor with the mouse and slide it to one side. There will be severe overlap of attached hydrogens, but this is remedied by removing them with the **H/AD** tool. To complete the skeleton, first double click on the **Select** tool to remove the marks, then activate the **Draw** tool and link C10 with the remaining methyl groups C8 and C9. The fastest way to do this will be to click on them in the order C8, C10, C9 or C9, C10, C8. The result will is shown in Fig. 3-11.

The exercise can be finished off by adding hydrogens with the **H/AD** tool, and passing the structure through energy minimization by activating the **Analyze\Minimize** command. The structure may be named by invoking the

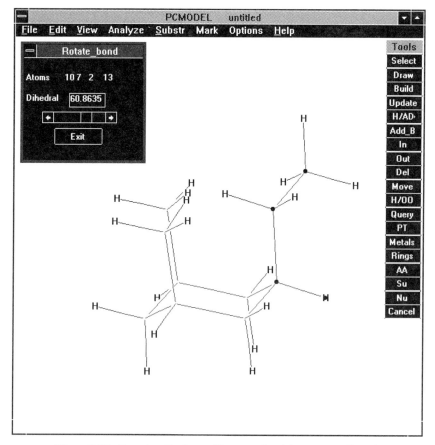

Figure 3-10. Drawing the Input Structure for Adamantane, Part 4.

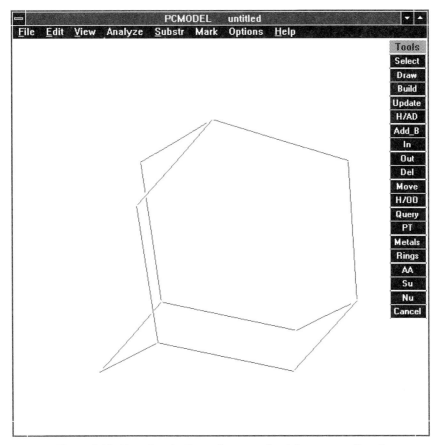

Figure 3-11. Drawing the Input Structure for Adamantane, Part 5.

Edit\Structure Name command and entering 'Adamantane' and then using the File\Save command to write an input file such as {ADAMANTN.PCM}. The final result is depicted in Fig. 3-12.

3.2. STRUCTURE INPUT FROM FILES

In this section we will discuss importing structures into PCMODEL from input files and templates. After some general comments, each of the four PCMODEL template menus is described: 'Rings Menu', 'Amino Acids Menu', 'Sugars Menu', and 'Nucleosides Menu'. User-created input files are considered next, followed by descriptions of the PCMODEL substructure (**Substr**) commands.

Often, it is much more expedient to read a structure into PCMODEL from an input file or template than to build it up from scratch. A number of templates are available in PCMODEL, as will be discussed below. Input files prepared by

50   INPUT AND OUTPUT

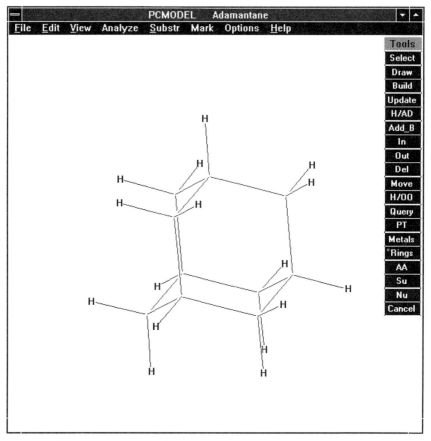

Figure 3-12. Drawing the Input Structure for Adamantane, Part 6.

the user previously can be used, as well as input files obtained from colleagues or the literature or adapted from literature data, or obtained from commercial sources such as 2D or 3D structure databases. As was discussed in Section 1.4, many 3D structures are available from electronic databases on the local platform, or via the Internet. Such imported structures can form a foundation which may be elaborated to more complex structures; they may be used as is, or modified as need be. Chapter 4 describes the details of the input file formats handled by PCMODEL.

As was seen in the adamantane drawing exercise above, it is a good strategy to proceed in a stepwise fashion. Carrying out a few building and editing operations on the structure and optimizing the geometry at each stage is a good way to reach a minimum-energy structure efficiently. For larger molecules, it is best to construct the core heavy atom skeleton first and bring it to minimum energy. Afterward, the additional rings and functional groups can be added, other necessary substitutions can be made, and the completed structure can be run through final minimizations.

Prior to energy minimization, especially on larger structures or those for which the input geometry is questionable, it is also prudent to employ the **Control_Panel** command to rotate the structure and look over the starting geometry before running an energy minimization sequence. When significant deviations are discovered in bond lengths or angles, or severe van der Waals repulsions are detected, one may use the structure editing tools discussed in the previous section to manipulate the initial geometry. A geometry check should also be done if unusual energy values are obtained from minimization.

3.2.1 The Rings Template Menu

The 'Rings' menu is activated from the Tool Bar by clicking on the **Rings** button, and the menu box which comes up is depicted in Fig. 3-13. Clicking on the appropriate button brings up the corresponding structure in the work space. These structures are contained in an appended format file entitled {RINGS.SST}, which comes as part of the commercial PCMODEL package.

There are 24 ring systems available in this menu: **C5** is cyclopentane, **C6** is chair cyclohexane, **C6 boat** is boat cyclohexane, **C7** is cycloheptane, **C8** is cyclooctane, **C9** is cyclopentane, **Ph** is benzene, **Naph** is naphthalene, **Anth** is anthracene, **Phen** is phenanthrene, **Corann** is corannulene, **C60** is buckminsterfullerene, **B210** is bicyclo[2.1.0]pentane, **B310** is bicyclo[3.1.0]hexane, **B220** is bicyclo[2.2.0]hexane, **B221** is bicyclo[2.2.1]heptane, **B222** is bicyclo[2.2.2]octane, **B320** is bicyclo[3.2.0]heptane, **B420** is bicyclo[4.2.0]octane, **c-Decalin** is *cis*-decalin, **t-Decalin** is *trans*-decalin, **steroid** is the standard steroid nucleus, **steroid-c** is the A–B *cis*-fused steroid nucleus, and **indole** is indole.

As will be discussed in Section 3.2.5, another 42 ring templates are available in the file {MORINGS.SST}, which may be found on the input file library diskette under the CHAP3\SECT2 subdirectory.

3.2.2. The Amino Acids Template Menu

The 'Amino Acids' menu is activated from the tool bar by clicking on the **AA** button, and the menu box which comes up is depicted in Fig. 3-14. Clicking on

Rings					
C5	C6	C6 boat	C7	C8	C9
Ph	Naph	Anth	Phen	Corann	C60
B210	B310	B220	B221	B222	B320
B420	c-Decalin	t-Decalin	Steroid	Steroid-c	Indole
			Cancel		

Figure 3-13. The 'Rings' Template Menu.

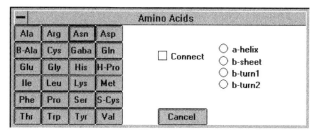

Figure 3-14. The 'Amino Acids' Template Menu.

the appropriate button brings up the corresponding structure in the work space. These structures are contained in an appended format file entitled {AA.SST}, which comes as part of the commercial PCMODEL package.

There are 24 amino acid residues available by clicking on the appropriate box in this menu: **Ala** is alanine, **Arg** is arginine, **Asn** is asparagine, **Asp** is aspartic acid, **B-Ala** is β-alanine, **Cys** is cysteine, **Gaba** is γ-aminobutyric acid, **Gln** is glutamine, **Glu** is glutamic acid, **Gly** is glycine, **His** is histidine, **H-Pro** is hydroxyproline, **Ile** is isoleucine, **Leu** is leucine, **Lys** is lysine, **Met** is methionine, **Phe** is phenylalanine, **Pro** is proline, **Ser** is serine, **S-Cys** is cystine, **Thr** is threonine, **Trp** is tryptophan, **Tyr** is tyrosine, and **Val** is valine.

As will be discussed in Section 3.2.5, another two amino acid templates and two pentapeptide templates are available from the file {BIOMOL.SST}, which may be found on the input file library diskette in the CHAP3\SECT2 subdirectory.

Several options are also provided in the 'Amino Acids' menu. The 'Connect' choice allows amino acid residues to be concatenated starting at the N-terminal end. The 'a-helix', 'b-sheet', 'b-turn1', and 'b-turn2' options control the secondary structure of the peptide produced, giving respectively an α-helix, β-sheet, or a type 1 or type 2 β-turn structure. In order to generate these structural types, PCMODEL uses a geometric recipe in terms of the peptide backbone torsional angles for joining the amino acids, which is summarized in Fig. 3-15, using the IUPAC descriptors ϕ, ψ, and ω.[240]

The 'Connect' utility is not limited to use within a single session; it will operate at any time when an appropriate peptide structure is present in the work space. The success of this operation depends on the peptide structure not being modified too much. For example, if the N-terminal end is acetylated, the 'Connect' algorithm will fail to recognize the structure as an incipient peptide chain.

Template Drawing Exercise 1: The Octapeptide $H_2N–(Gly–Ala)_4–CO_2H$. This exercise describes the preparation of an input structure for an octapeptide in the α-helix conformation, and exemplifies several graphics features. The first step is to bring up the 'Amino Acids' template menu by clicking on the **AA** tool. Clicking in the 'Connect' choice box also selects the 'a-helix' option as the

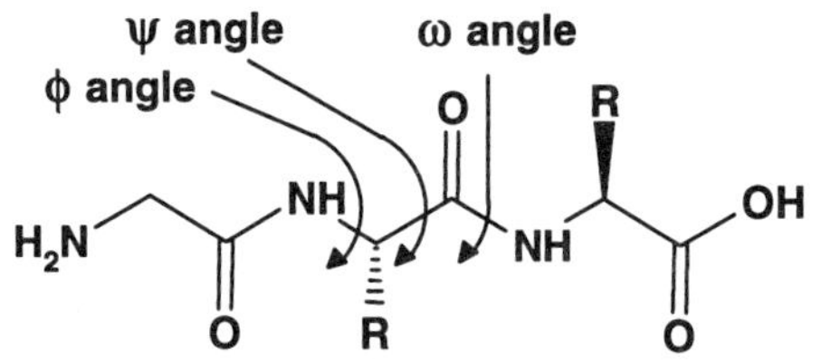

ψ angle ω angle
φ angle

H$_2$N—CH$_2$—C(=O)—NH—CH(R)—C(=O)—NH—CH(R)—C(=O)—OH

Structure	φ-angle	ψ-angle	ω-angle
α-helix	-53°	-53°	180°
β-sheet	180°	180°	180°
type 1 β-turn	-58.5°	180°	180°
type 2 β-turn	-20°	-90°	180°

Figure 3-15. Definitions of Peptide Backbone Angles and Conformational Types.

default mode. Next, the mouse cursor is used to click first on the **Gly** button, then on the **Ala** button, and then once again on **Gly**. The tripeptide H$_2$N–Gly–Ala–Gly–CO$_2$H is now in the work space, as shown in Fig. 3-16.

Now, the remaining residues may be added, by successive clicks on **Ala**, **Gly**, **Ala**, **Gly**, and **Ala**. The 'Amino Acids' template menu should be canceled, and the **View\Control Panel** command used to give the octapeptide a better perspective for viewing. Fig. 3-17 shows the structure with the axis of the α-helix roughly parallel to the Y-axis.

Some of the confusing detail can be cut out by hiding the carbon-bound hydrogens with the **H/OO** tool, and the result is given in Fig. 3-18. In the discussion of the graphical display formats later in the chapter, this octapeptide structure will be used to illustrate the **View\Ribbon** graphical display format. The molecular input file may be found on the input file library diskette in the CHAP3\SECT2 subdirectory, under the name {OCTPEPTD.PCM}.

3.2.3. The Sugars Template Menu

The 'Sugars' Menu is activated from the tool bar by clicking on the **Su** button, and the menu box which comes up is depicted in Fig. 3-19. These structures are contained in an appended format file entitled {SU.SST}, which comes as part of the commercial PCMODEL package.

There are 25 sugar residues available by clicking on the appropriate button in this menu: **Gly** is D-glyceraldehyde, **Ery** is D-erythrose, **Thr** is D-threose, **Tar** is D-tartaric acid, **Ara** is β-D-arabinofuranose, **Lyx** is β-D-lyxofuranose, **Rib** is β-D-ribofuranose, **Xyl** is α-D-xylofuranose, **Alo** is α-D-allopyranose, **Alt** is α-D-altropyranose, **Glc** is α-D-glucopyranose, **Man** is α-D-mannopyranose, **Gul** is α-D-gulopyranose, **Ido** is α-D-Idopyranose, **Gal** is α-D-galactopyranose, **Tal** is α-D-talopyranose, **Fru** is α-D-fructopyranose, **Psi** is α-D-psicopyranose, **Sor** is α-D-sorbopyranose, **Tag** is α-D-tagetopyranose, **FruF** is α-D-fructofuranose,

54 INPUT AND OUTPUT

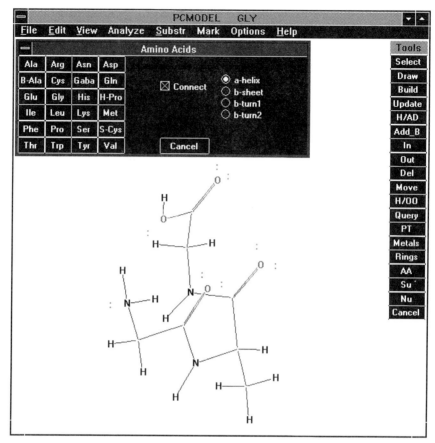

Figure 3-16. Preparing the Octapeptide $H_2N—(Gly—Ala)_4—CO_2H$, Tripeptide.

PsiF is α-D-psicofuranose, **SorF** is α-D-sorbofuranose, **TagF** is α-D-tagetofuranose, and **GlcF** is β-D-glucofuranose. Of these, the subset of **Alo, Alt, Glc, Man, Gul, Ido, Gal**, and **Tal** is available in either the α- or the β-anomer by selecting the corresponding radio button in the 'Sugars' menu window.

As will be discussed in Section 3.2.5, another four sugar templates are available in the file {BIOMOL.SST}, which may be found on the input file library diskette in the CHAP3\SECT2 subdirectory.

A 'Connect' option is also provided in the 'Sugars' menu, which allows the members of a subset of the sugar templates (**Alo, Alt, Glc, Man, Gul, Ido, Gal, Tal**, and **Fru**) to be concatenated. The first sugar to be read in will be glycosylated (joined at its anomeric carbon) with the C4 hydroxyl of the second sugar to be read in, and so forth. The stereochemistry of the glycosyl linkage will be the same as that of the first sugar, and the next sugar being read in will have the anomeric carbon stereochemistry dictated by the currently selected menu radio button.

3.2. STRUCTURE INPUT FROM FILES   55

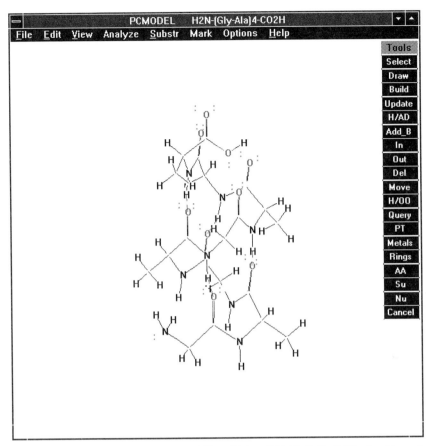

Figure 3-17. The Octapeptide H_2N—$(Gly—Ala)_4$—CO_2H, Default Display Format.

A system for characterizing the glycosidic conformation in terms of angles ϕ and ψ has been described,[241] which is somewhat analogous to that in current use for peptides. Here, the angle ϕ refers to the torsional angle defined by the ring oxygen of the glycosyl donor, the anomeric carbon of the glycosyl donor, the glycosidic oxygen, and the attached carbon of the glycosyl acceptor. The angle ψ refers to the torsional angle defined by the anomeric carbon of the glycosyl donor, the glycosidic oxygen, the attached carbon of the glycosyl acceptor, and the carbon atom one lower in numbering than the attached carbon. This is depicted for the example of lactose in Fig. 3-20.

Template Drawing Exercise 2: The Disaccharide Lactose. This exercise describes the preparation of an input structure for the disaccharide lactose (4-O-β-D-galactopyranosyl-β-D-glucopyranose); it exemplifies the saccharide concatenation option and the **Rotate_bond** command. The first step is to bring up the 'Sugars' template menu by clicking on the **Su** tool. The 'beta' radio button should be clicked, followed by the **Gal** button to read in β-D-galactopyranose

56 INPUT AND OUTPUT

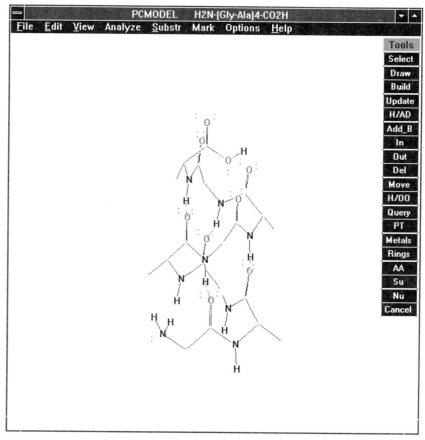

Figure 3-18. The Octapeptide H_2N—(Gly—Ala)$_4$—CO_2H, **H/OO** Display Format.

as the glycosyl donor. Use **View\Control Panel** to manipulate the structure for better viewing, if desired. To aid in sorting out a confusing structure, we will label the anomeric carbon with deuterium. Activate the **PT** tool, pick up the **D** atom, and then click on the anomeric hydrogen. The result will resemble Fig. 3-21.

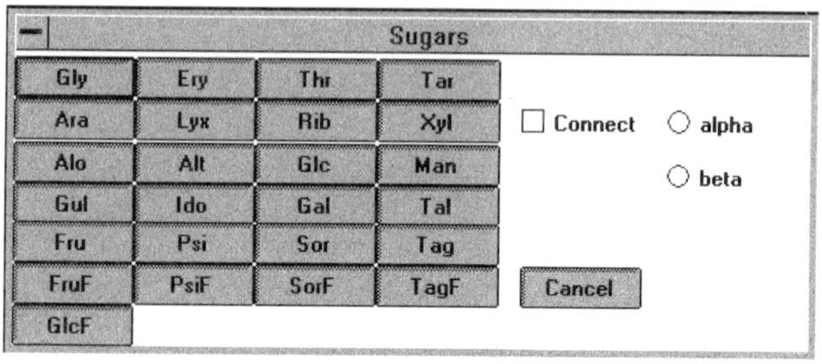

Figure 3-19. The 'Sugars' Template Menu.

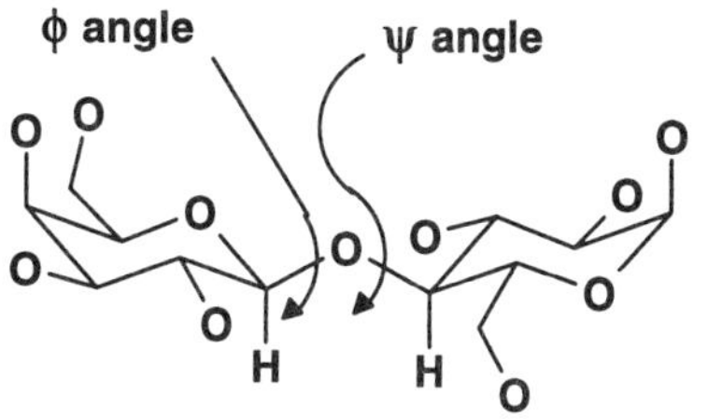

3.2. STRUCTURE INPUT FROM FILES   57

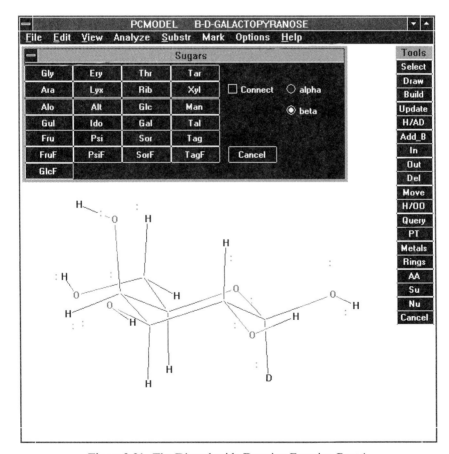

Figure 3-20. Definitions of the ϕ-Angle and the ψ-Angle for the Lactose Structure. The hydroxyl hydrogens have been omitted for clarity.

Next, click on the Connect choice box, and on the alpha radio button, followed by a click on the **Glc** button to select α-D-glucopyranose as the glycosyl acceptor. This will link the C4 hydroxyl of the glucose residue to the anomeric carbon of the galactose molecule. The orientation of the new disaccharide does not enable us easily to determine which part is which. Careful inspection will

Figure 3-21. The Disaccharide Drawing Exercise, Part 1.

reveal the location of the deuterated anomeric carbon of galactose, but some adjustment of the ϕ and ψ angles of the structure will improve the display.

Use the **Select** tool to mark the glucose C4, glycosidic oxygen, galactose C1, and D atom, in that order. Then, activate **Edit\Rotate_bond**, and use the scroll bar to adjust this ϕ-angle to 60°. Next, double-click on **Select** to remove the first set of marks, and then mark glucose C—H hydrogen, glucose C4, glycosidic oxygen, and galactose C1, in that order. Then, adjust this ψ-angle to −60° using **Rotate_bond**. Now, finish up by using **View\Control Panel** to rotate the structure until both rings are side by side and the substituents are all readily discernible. During some of the rotations, it may be helpful to employ the **H/OO** tool to hide the hydrogens and view the skeletons more clearly. Finally, replace the deuterium with a hydrogen from the 'Periodic Table' menu. This final result should resemble Fig. 3-22. This lactose structure is available on the input file library diskette in the CHAP3\SECT2 subdirectory under the name {LACTOSE.PCM}.

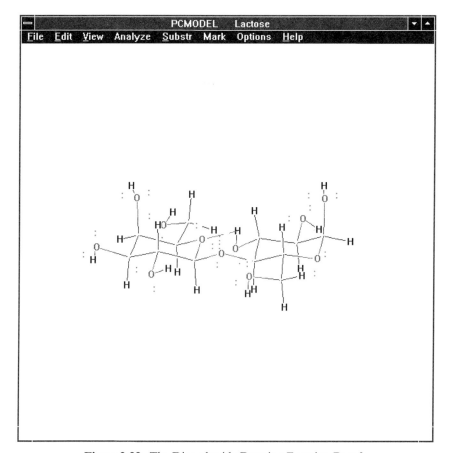

Figure 3-22. The Disaccharide Drawing Exercise, Part 2.

3.2.4. The Nucleosides Template Menu

The 'Nucleosides' template menu is activated by clicking on the **Nu** button on the tool bar, and the menu box which comes up is depicted in Fig. 3-23.

There are five nucleoside residues available by clicking on the appropriate box in this menu: **A** is for adenosine-5'-monophosphate, **T** is for thymidine-5'-monophosphate, **G** is for guanosine-5'-monophosphate, **C** is for cytidine-5'-monophosphate, and **U** is for uridine 5-monophosphate. Either the DNA or RNA versions of these nucleosides can be produced by selecting the appropriate radio buttons.

As will be discussed in Section 3.2.5, another three nucleoside templates are available in the file {BIOMOL.SST}.

3.2.5. User-Defined Templates

In this section we will discuss the use of input files as templates. The 'Open File' and 'Save File' operations are described, and the use of appended format input files is covered. Four new input files of template substructures are introduced.

The template menus just described provide a shortcut to importing structures into PCMODEL. Any input file may be used as a template and modified to reach a desired final structure. Since the template or input file will usually have been preminimized, using templates in preparing structure input files will save time.

Most input files will be read in using the **File\Open** command. Reading in an input file involves several steps. PCMODEL must first recognize the format; this can be flagged in the input file itself, as in the case of the PCM and MMX file formats, but it should be selected as an option in the 'Open File' menu. Currently, the program reads input files in any of twelve formats, and writes input files in any of seven formats.

When the **File\Open** command is activated, a pop-up menu dialog box entitled 'Open File' comes up, as shown in Fig. 3-24. This menu has dialog boxes for the 'Current Path' and 'Filename', and a directory browser to find

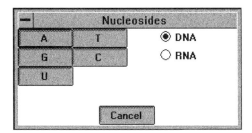

Figure 3-23. The 'Nucleosides' Template Menu.

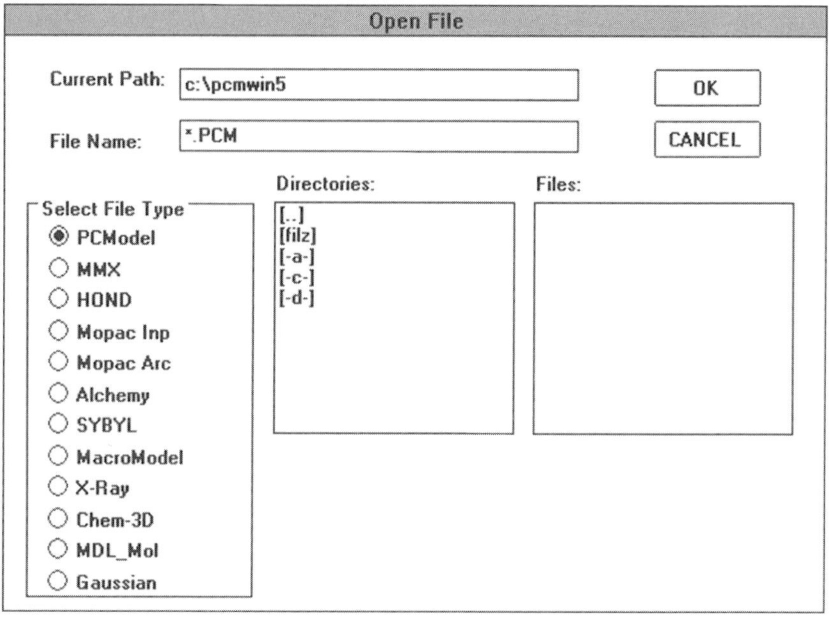

Figure 3-24. The **File\Open** 'Open File' Dialog Box.

the file of interest. It also presents an array of the readable formats in an area of the menu entitled 'Select File Type', where the user selects the desired format by clicking on the appropriate radio button. This places a filter on the files in the directory so that only those files bearing the proper filename extension are presented for selection. In some cases, the extension in the 'Filename' file filter may need to be changed before PCMODEL will recognize the desired file and show it in the directory. An example is the appended PCM format substructure files with an {*.SST} extension. However, files with a different or no filename extension may be read if the user is able to enter the filename from the keyboard. There follows a brief description of each file format; more information can be found in Chapter 4.

PCModel is the free format developed by Serena Software, and is the default file format used in PCMODEL. These files generally have the form {*.PCM}, although appended format substructure files will have the {*.SST} extension (see Section 4.1).

MMX is an enhanced version of the Allinger MM2 format developed by Serena Software, and was formerly the PCMODEL standard (see Section 4.2).

Hond is an *ab initio* Hückel frontier molecular orbital calculation package, developed by M. DuPuis, *et al.* of IBM. It is available from the **QCPE**.

Mopac inp is the Z-matrix format used as input for the semiempirical molecular orbital calculation program MOPAC, developed by J. J. P. Stewart.

Mopac arc is the Mopac archive file, which contains the results generated by the semiempirical molecular orbital calculation program MOPAC, developed by J. J. P. Stewart.

Alchemy is the file format used by the desktop molecular modeling package of the same name by Tripos Associates.

SYBYL is the file format used by the workstation molecular modeling package of the same name by Tripos Associates.

MacroModel is the file format for the workstation and mainframe platform molecular modeling program of the same name developed by the research group of W. C. Still (Columbia University).

X-Ray is the file format containing title, cell parameters and Cartesian coordinates derived from an X-ray diffraction crystal structure (see Section 4.4).

Chem-3D is the file format for the molecular modeling program of the same name developed by Cambridge Scientific Computing, Inc.

MDL_Mol is the file format for structures used in MDL Information Systems products, such as MACCS, REACCS, and ISIS. These files may be either two- or three-dimensional (see Section 4.3).

Gaussian is the file format for the *ab initio* quantum mechanics calculational package developed by J. A. Pople *et al.* and available from the QCPE and commercially in various forms.

When the format and filename have been selected, an algorithm called a **parser** then reads through the file and assembles the contained values and parameters into a representation of the described file on screen. From this point onward, the file may be written to another format if desired. Once an energy minimization sequence has been run, the modified structure will be stored temporarily in the PCM format in a file named {PCMOD.BAK}. This file is rewritten each time an energy minimization sequence is run.

If a user makes constant use of variations on or analogs of a core structure, it is wise to make and preminimize this core structure, and save it with a descriptive filename (e.g. {FENOL} or {FNOL} for phenol). If there is a single structure with a number of conformations, one could carry out a conformational search (see Section 5.6) and store the family of conformations in a single file in the appended format, possibly in the order of increasing energy. Such a strategy is particularly helpful in building an oligomeric structure.

Once an interesting structure has been prepared and run through energy minimization, it is prudent to save the corresponding input file for future reference and use. PCMODEL routinely generates a backup input file {PCMOD.BAK} in the PCModel format after an energy minimization. This may be renamed and used in lieu of explicitly saving a file, but if not renamed it will be overwritten (and thus lost) after each new minimization.

Writing an input file is accomplished with the **File\Save** command. Any changes to be made, such as **Edit\Structure Name**, or any of the parameters available under **Options**, must be completed prior to saving the file. Activating this command brings up the 'Save File' dialog box, shown in Fig. 3-25.

In the lower left portion is found a series of radio buttons for selecting from among the available file formats (see Chapter 4). In PCMODEL v. 5.13 the acceptable output file formats are 'PCModel,' 'MMX,' 'Mopac,' 'Alchemy,' 'SYBYL,' 'MacroModel,' and 'Chem-3D.' In PCMODEL v. 6.0, two additional formats are available for output: 'MM2,' 'MM3,' and 'PDB.' Also found are standard Windows prompts for accessing directories and files. After clicking on the radio button indicating the desired format, the name of the file is entered at the 'File Name:' prompt. If the filename entered already exists, another dialog box entitled 'File Exists' appears (see Fig. 3-26), giving the option of appending the new structure to the existing file (**Append** button) or overwriting the existing file (**Replace** button).

In previous versions of PCMODEL, template files containing a large number of structures accompanied the program. The most popular of these have been incorporated into the template menus mentioned in earlier sections. The additional structures have now been sorted and assembled into four appended PCM format files which can be read in as substructures from a 'Structure List' menu (see Section 3.2.6). The contents of these lists are shown in Tables as follows: {MORINGS.SST}, 42 more ring structures, listed in Table 3-1; {BIOMOL.SST}, 24 more biological molecule and natural product structures,

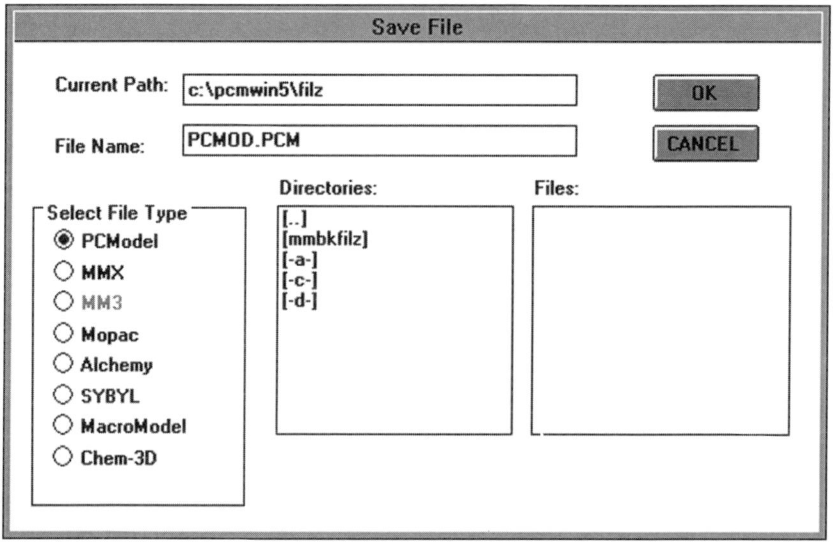

Figure 3-25. The **File\Save** 'Save File' Dialog Box.

Figure 3-26. The 'File Exists' Dialog Box.

listed in Table 3-2; {TRNZSTAT.SST}, 14 transition state structures (discussed at greater length in Section 6.9), listed in Table 3-3; and {MISC.SST}, 8 miscellaneous structures, listed in Table 3-4. These files are all found on the input file library diskette in the CHAP3\SECT2 subdirectory.

3.2.6. Substructures

A general definition of substructure is a separate fragment or a portion of an intact structure. In the PCMODEL context, substructure specifically means a designated collection of atoms, which may or may not be connected and which are defined as a separate group, differentiated from the remaining molecule in

TABLE 3-1. Structure Input Files in {MORINGS.SST}

Adamantane	1,5-Cycloheptadiene
Benzocycloheptane	Cycloheptatriene
Benzocycloheptane, twist boat	Cycloheptane
Benzocyclohexane	1,3-Cyclohexadiene
Benzocyclopentane	1,4-Cyclohexadiene
Benzofuran	1,3-Cyclooctadiene
Bicyclo[1.1.1]pentane	1,5-Cyclooctadiene
Bicyclo[2.1.1]hexane	Cyclooctatetraene
Bicyclo[3.1.1]heptane	Cyclooctatriene
Bicyclo[3.2.1]octane	Cyclooctene
Bicyclo[3.2.2]nonane	Cyclopentadiene
Bicyclo[3.3.0]octane	Cyclopentadienyl
Bicyclo[3.3.1]nonane	Cyclopentene
Bicyclo[3.3.2]decane	Cyclopropane
Bicyclo[3.3.3]undecane	Cyclopropene
Cubane	Cyclotetradecane
Cyclobutane	Cyclotridecane
Cyclobutene	Cycloundecane
Cyclodecane	Dodecahedrane
Cyclododecane	cis-Hydrindane
1,3-Cycloheptadiene	trans-Hydrindane

TABLE 3-2. Structure Input Files in {BIOMOL.SST}

β-D-Arabinopyranose	Inosine-5'-monophosphate
Arachidonic acid	Lanthionine
(+)-Biotin	β-D-Lyxopyranose
Cephalosporin C	Morphine
Cholesterol	Penicillin-N
Cortisone	Progesterone
Cyclic adenosine-5'-monophosphate	β-D-Ribopyranose
Cyclic guanosine-5'-monophosphate	5-α-Steroid
Leu-enkephalin (Tyr–Gly–Glu–Phe–Leu)	5-β-Steroid
Met-enkephalin (Tyr–Gly–Gly–Phe–Met)	Testosterone
Estradiol	Thienamycin
Gliotoxin	β-D-Xylopyranose

the work space. The substructure definition is used primarily to permit separate handling of different structures in the work space, but may also be used to define a portion of the molecule which will be excluded from the energy minimization process.

Substructure identity may be established either by reading a molecule in separately as a substructure with the **Substr\Read** command, or by using the **Select** tool to mark the included atoms (in a separate structure or a portion of a structure) and labeling them with the **Substr\Create** command, as described below. The substructure membership is recorded in the input file. Substructure membership can be removed in the 'Reset' dialog box under **Mark\Reset**. Other commands may accommodate or require a substructure, such as the

TABLE 3-3. Structure Input Files in {TRNZSTAT.SST}

Diels–Alder cycloaddition TS	1,3-Dipolar nitrone cycloaddition TS
N#C# Diels–Alder cycloaddition TS	Hydroboration TS
C#N# Diels–Alder cycloaddition TS	Claisen rearrangement TS
O#C# Diels–Alder cycloaddition TS	Ene reaction TS
C#O# Diels–Alder cycloaddition TS	SN2 C-nucleophile TS
Cope rearrangement TS	SN2 O-nucleophile TS
1,3-Dipolar nitrile oxide cycloaddition TS	1,5-Hydrogen shift TS

TABLE 3-4. Structure Input Files in {MISC.SST}

18-Crown-6, K complex	Phosphoric acid
Ammonia	Triphenylphosphine (Ar atoms)
Tetradeuteromethane	Triphenylphosphine (Pi atoms)
Trideuterophosphine	Water

Analyze\Dock command which docks one substructure into another. These substructures will retain all previously associated flags, such as **H-Bonds**, **Piatoms**, or **Metal_Coord**. Substructures may be displayed in different colors, either as stick figures ('Color by Substructure' in **View\Labels**) or CPK representations ('Color by: Substructure' in **View\CPK_Surface**). Up to 32 substructures can be defined.

LIMITATIONS: When defining the substructure as a portion of a single structure, the boundary point of the definition is limited to acyclic bonds. Cyclic structures currently cannot be dissected into substructures. If this is attempted, there is no warning but the results will be odd. If the **Compare** command is activated while substructures are defined, only the first substructure will be used in the **Compare** operation, and the other substructures will vanish from the work space. In order to use **Compare** on only a portion of a structure, that portion should be indicated using the **Select** tool, and the 'Selected' option of **Compare** should be invoked.

The **Substr** (substructure) command on the menu bar governs a pull-down menu of seven options for operating on substructures: **Read, Create, Move, Connect, Erase, Hide**, and **Don't Minimize** (see Fig. 2-6).

The **Substr\Read** command is used to bring in named structure files as substructures. When it is activated, a standard 'Open File' dialog box comes up (see Fig. 3-24), which prompts the user for the file type and filename of the new substructure to be read in. These substructures will retain all previously associated flags, such as **H-Bonds**, **Piatoms**, or **Metal_Coord**.

The **Substr\Create** command is used to define all or a portion of a structure in the work space as a substructure. Either a single structure or a portion of a structure can be defined as a substructure. Before activating the **Create** command, the **Select** tool must be used to mark either one or two atoms in a structure in the work space. Marking any one atom in a separate structure will allow that entity to be defined as a substructure. To define a portion of a structure as a substructure, the two atoms joined by the bond which will separate the second substructure from the first substructure must be marked. In this case, the order is important—the first atom marked (and all of its attachments) will be in the second substructure, while the second atom marked (and all of its attachments) will remain in the original (first) substructure. Once this is done, the **Create** command can be activated. If there are no errors, a dialog box entitled 'SubStructure Name' will come up with the prompt 'Enter Substructure Name:' allowing the user to enter the name for the substructure, as shown in Fig. 3-27.

The substructure membership will be annulled when the pair of structures is written to an input file, but may be reestablished with another application of **Substr\Create**.

There are two locations where a **Move** button is found: the **Substr\Move** commands enables the movement of a substructure to another position in the

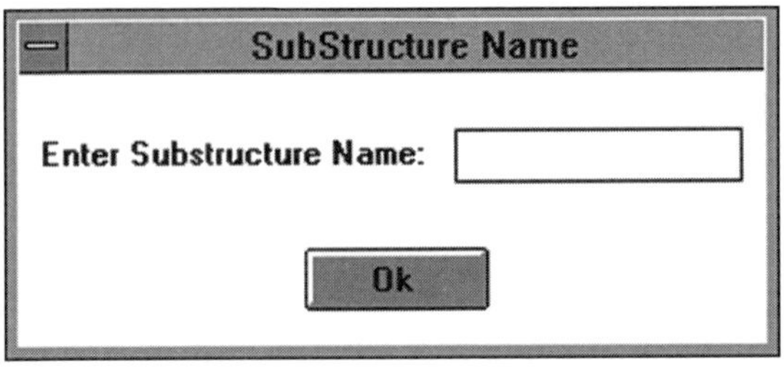

66 INPUT AND OUTPUT

Figure 3-27. The 'SubStructure Name' Dialog Box.

work space, while the **Move** tool found on the tool bar allows the user to move the atoms in a structure in the work space within the XY plane of the screen. The use of the **Move** tool was discussed in Section 3.1. One common reason for using the **Substr\Move** command is to optimize the viewing perspective separately on the different substructures in the work space. Other uses are to preorient the substructures for using **Dock**, or for carrying out a manual docking exercise, or to prepare for a **Substr\Connect** operation.

When two or more defined substructures in the work space and the **Substr\Move** command is activated, a dialog box entitled Structure List appears, allowing the user to select 'All' or individual substructures to be moved by substructure name, as shown in Fig. 3-28.

When the selection has been made, another dialog box entitled 'Dials' appears, as shown in Fig. 3-29, which contains seven scroll bars and three buttons to effect translations and rotations.

If the representation in the work space is too crowded or confused to carry out commands such as **Query**, it may be altered within the **Control_Panel**. The 'Scale' function may be used to expand or shrink the molecule on screen, or the 'Translate' and/or 'Rotate' functions may be used to change their perspective and allow easier viewing and access to the portion of the molecule under scrutiny.

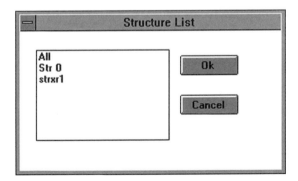

Figure 3-28. The 'Structure List' Dialog Option Box.

3.2. STRUCTURE INPUT FROM FILES

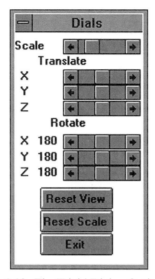

Figure 3-29. The 'Dials' Dialog Option Box.

The topmost scroll bar is labeled 'Scale'. Clicking on the right-hand arrow button will increase the size of the representation, and clicking on the left-hand arrow button will decrease its size. Clicking on **Reset Scale** near the bottom returns the representation to its original size.

The next group of three scroll bars appear under the label 'Translate' and are individually identified with the Cartesian axes 'X', 'Y', and 'Z'. These are used to effect the translation of the structure in the work space, and each click moves the structure by 0.25 Å in the indicated direction. Clicking on the right-hand 'X' arrow button moves the structure to the right, clicking on the right-hand 'Y' arrow button moves the structure upwards, and clicking on the right-hand 'Z' arrow button moves the structure out of the plane of the screen. Clicking on the left-hand arrow button in each case gives a translation in the opposite sense.

The last group of three scroll bars appear under the label 'Rotate' and are individually identified with the Cartesian axis labels 'X', 'Y', and 'Z'. Each of these axes has a number between the axis label and the scroll bar—this number begins at '180', and will report the current angular displacement. These are used to effect the rotation of the structure in the work space, and each click alters the rotation by 5°. Clicking on the right-hand 'X' arrow button rotates the structure in the counterclockwise when viewed from the origin along the X-axis in the positive direction. Clicking on the right-hand 'Y' arrow button rotates the structure in the counterclockwise when viewed from the origin along the Y-axis in the positive direction. Clicking on the right-hand 'Z' arrow button moves the structure clockwise in the plane of the screen. Clicking on the left-hand arrow button in each case gives a rotation in the opposite sense. Clicking on **Reset View** near the bottom returns the original representation, i.e., it cancels all rotations carried out during the current **Substr\Move** session. Once the 'Dials' menu is canceled, the existing settings become the new defaults.

A powerful menu tool for joining molecular fragments is the **Substr\Connect** command, which links one substructure to another. When two substructures are active in the work area, the **Select** tool can be used to mark a hydrogen atom from each structure. When the **Substr\Connect** command is activated, the hydrogens will be removed and the two structures will be joined at those points with a reasonable but not energy-minimized orientation. The first substructure will retain its original perspective, and the second substructure will be reoriented to accommodate this operation.

If desired, one may remove the structure–substructure differentiation in a composite structure using the **Mark\Reset** command as described above, or if the new, larger structure is defined as a single substructure by marking an atom with **Select** and using the **Create** command used to define the entire new molecule as its own, single substructure. Otherwise, individual substructure portions will retain their substructure identity during a PCMODEL session.

There are two **Erase** commands for removing structures in the work space. One, **Edit\Erase**, is for global erasure and was addressed in Section 3.1, and the other, **Substr\Erase**, is for selective erasures of substructures. When this command is activated with two or more defined substructures in the work space, the 'Structure List' dialog box appears (see Fig. 3-28), allowing the user to select between substructures by structure title or name. Selecting the 'All' option erases all current structures. If there is only one active structure, it will be erased by **Substr\Erase** without any confirmation prompt. Note that, unless they have been written to a file, or run through an energy minimization and thus saved as the default {PCMOD.BAK} file, erased structures are not recoverable once the erasure has taken place.

The **Substr\Hide** command is a toggle to render visible or invisible the graphic representation of a selected substructure in the work space. The hidden substructure has not been deleted, and can be recalled with another application of the **Hide** command. Activating this command brings up the 'Structure List' dialog box (Fig. 3-28), allowing the user to select between substructures by structure title, and the status of each substructure is also shown. The structure title is followed after a few spaces by a 'v' to indicate that it is currently visible, or an 'i' to indicate that it is currently invisible or hidden. This command is most often used to provide a less cluttered view.

The **Substr\Don't Minimize** command allows the user to exempt the designated substructure from the energy minimization process. It operates as a toggle, and activating it a second time and selecting the same substructure cancels the exemption. Activating this command brings up the 'Structure List' dialog box (Fig. 3-28), allowing the user to select between substructures by structure title, and the status of each substructure is also shown. The structure title is followed after a few spaces by an 'm' to indicate that it will be included

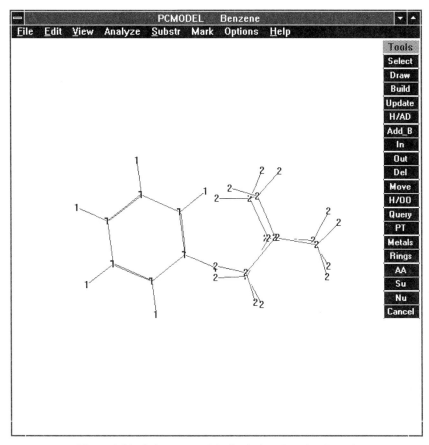

Figure 3-30. Substructure Manipulation Exercise, Part 1.

in the minimization, or an 'x' to indicate that it is currently excluded from minimization.

This command can be used to freeze the components of the substructure while obtaining a force field minimization of the other substructures. One common use of this feature is when the substructure geometry has been optimized in another way, such as from a different type of calculation (semiempirical or *ab initio*), or from physical measurements such as diffraction crystallography. In this way, the geometry of that substructure could be maintained while the remainder of the molecule would undergo MMX geometry optimization.

Substructure Manipulation Exercise 1: 1-Phenylbicyclo[2.2.2]octane. This exercise will demonstrate the substructure-handling capabilities of PCMODEL, and familiarize the user with these tools.

70 INPUT AND OUTPUT

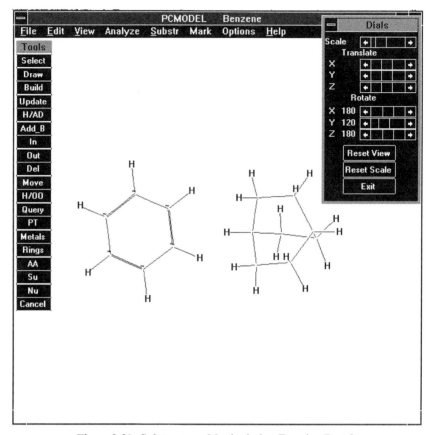

Figure 3-31. Substructure Manipulation Exercise, Part 2.

From the PCMODEL window, activate the **Rings** tool to bring up the 'Rings' template menu. Click first on **Ph** to read in a benzene and then on **B222** to read in bicyclo[2.2.2]octane. Depending on the graphics resolution, there may be some overlap of these structures, but in any case we will first verify the substructure membership. Activate the **View\Labels** command, and select either 'Substructure Numbers' or 'Color by Substructure'. Figure 3-30 shows the result of labeling by substructure numbers, where the benzene ring has every atom marked with a '1' and the bicyclic structure has each atom marked with a '2'.

Activate **Substr\Move**, and select the 'benzene' structure from the 'Structure List' pop-up window. On the 'Dials' dialog box which pops up next, click several times on the left-hand arrow of the 'Translate' 'X' scroll bar. The benzene substructure will be seen to move leftward in small jumps. Exit the 'Dials' dialog box, reactivate **Substr\Move**, and then select the 'bicyclo[2.2.2]octane' structure, and now click on the left-hand 'Rotate' 'Y' scroll bar until the angle is reported to be '120'. The result is shown in Fig.

3-31, in which the 'Dials' box and the tool bar have been repositioned for clarity.

Next, use the **Select** tool to mark the bridgehead hydrogen of the bicyclic substructure which is nearest the benzene, and the lower right-hand hydrogen on the benzene. Then, activate the **Substr\Connect** command. The connected structure will be redrawn as it appears in Fig. 3-32.

For clarity, use **View\Control Panel** to rotate the new combined structure on the Z-axis to be more horizontally oriented. The user may wish to experiment with some of the **View\Labels** options, to see how the substructures are differentiated by numbers or by coloring. Electing the 'Atom Numbers' label option shows that renumbering has occurred so that the heavy atoms have the lowest numbers. **Substr\Hide** should also be tried, to demonstrate how each substructure (or both) can be hidden from view.

Substructure Selective Minimization Exercise 2: 1-Phenylbicyclo[2.2.2]octane.
This more advanced example covers restricted minimization and structural

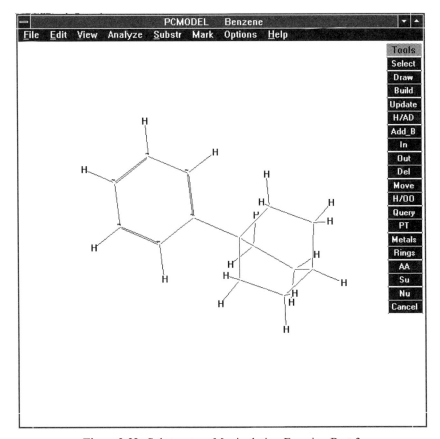

Figure 3-32. Substructure Manipulation Exercise, Part 3.

72 INPUT AND OUTPUT

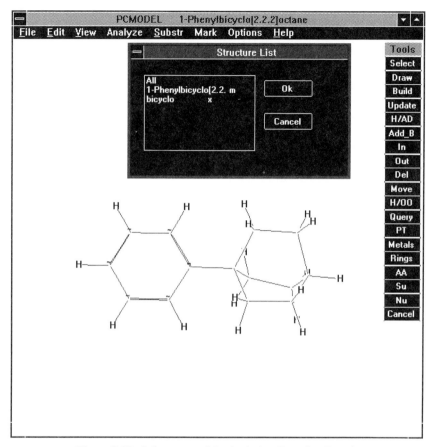

Figure 3-33. Substructure Manipulation Exercise, Part 4.

comparisons. These techniques are discussed in Sections 3.6.3 and 5.5, which may be referred to in case the steps are unclear.

First, the structure should be run through **Analyze\Minimize** several times to arrive at a low-energy conformation. The resulting structure may be named as '1-Phenylbicyclo[2.2.2]octane', and should be written to an input file with a name such as {1PB222O.PCM}, which is available on the input file library diskette in the CHAP3\SECT2 subdirectory. At this point, the substructure membership will be annulled.

From the PCMODEL window, read in {1PB222O.PCM}. It can be seen with the **View\Label** substructure viewing options that the substructure membership has been annulled. To designate the bicyclic part of the structure as a substructure, use the **Select** tool to mark in order first the bicyclic bridgehead atom attached to the phenyl ring, and secondly the attached phenyl ring atom. Activate the **Substr\Create** command, and to the prompt in the 'SubStructure Name' dialog box answer 'bicyclo' or some other identifying name. Now that the substructure has been designated, double-click on the **Select** tool to unmark the atoms, and activate the **Substr\Don't Minimize** command. In the resulting 'Structure List' dialog box, both substructures will have an 'm', indicating that currently they are both set to minimize. Highlight the line for 'bicyclo' and click on the **OK** button. The flag has now been set, and to check, activate **Substr\Don't Minimize** again. Figure 3-33 shows what the screen should look like at this point.

Now click on the **PT** tool, pick the **Si** from the 'Periodic Table' menu, and place it on the upper bridge atom closest to the phenyl ring, i.e. atom 14. Run the resulting structure through **Analyze\Minimize** several times to arrive at a low-energy conformation, to produce a screen like Fig. 3-34.

Before carrying out the comparison, the substructure membership must be annulled (see above) with **Mark\Reset**. Now activate the **View\Compare** command, which brings up the 'Compare' dialog box (see Fig. 3-71); select the

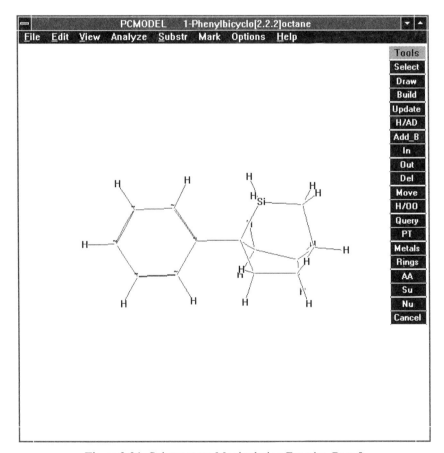

Figure 3-34. Substructure Manipulation Exercise, Part 5.

74 INPUT AND OUTPUT

'Heavy Atoms' option, then click on the 'Next Struct' button and read in the file written above, {1PB222O.PCM}. With the two structures on screen, click on the 'Calculate' button to carry out the least squares fit between the sets of atoms. The result will be given in the 'Output' window, and the RMS deviation should be on the order of 0.089–0.099 Å, a very small deviation. Since we exempted the bicyclic portion from minimization, the presence of the silicon ring atom made no difference in this geometry optimization process.

The control experiment for this will be to carry out energy minimization without constraints, and see what difference this makes. Run this structure through **Analyze\Minimize** several times to arrive at a low-energy conformation, and now Fig. 3-35 indicates that some distortion of the bicyclic portion has occurred.

Run the **View\Compare** operation as before, to find the RMS deviation is now on the order of 0.19–0.23 Å, with the largest deviations in atoms 14 and

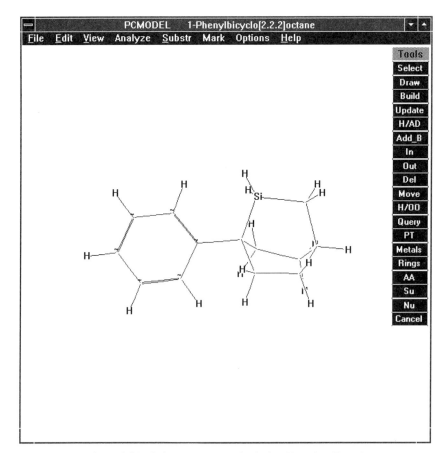

Figure 3-35. Substructure Manipulation Exercise, Part 6.

13, which correspond to the silicon and the carbon with which it makes a two-atom bridge.

3.3. STRUCTURE INPUT FROM PHYSICAL DATA

Physical data from diffraction and spectroscopic methods can provide a wealth of structural information of use to the molecular modeler.[242,243] Many force field parameters were derived using physical data as a starting point, and these data also provide a source of input geometry for partial or whole structures. In this section we will discuss four of the most common sources of physical data for modeling: X-ray diffraction, neutron diffraction, electron diffraction, and microwave spectroscopy.

3.3.1. X-Ray Diffraction Data

X-ray diffraction in the present context is the technique of elucidating structural information by interpreting the interference patterns produced when electromagnetic radiation from the soft X-ray portion of the spectrum is diffracted or scattered when passing through a substance. Among several diffraction techniques, the one of interest here is the determination of molecular structure by diffraction with a single crystal of the pure substance, with X-ray wavelengths on the order of 0.1–1 nm (1–10 Å), which not coincidentally is in the same range as common molecular bond lengths. Diffraction occurs because the X-rays are scattered by the electrons of, or by the surface of the electron clouds of, the atoms which make up a molecule. While this feature makes X-ray diffraction a powerful technique for determining structure, the interpretation may be difficult when the **deformation density** is not centered on the conventional depiction of the bond, or in the case of locating hydrogen (or deuterium) atoms exactly. In the former case, usually with strained or small-ring compounds, **bent bonds**[244] are invoked to explain the data. In the latter case, because the only electrons around the hydrogen are the ones involved in bonding, this atom doesn't visualize well in X-ray diffraction. Such a distortion of the electron cloud is not significant except for hydrogen, but this limitation must be kept in mind when comparing X-ray results with those from other techniques, such as neutron diffraction, in which scattering takes place off the nuclei. The effects of this distortion upon force field modeling may be compensated by employing a reduced hydrogen correction (see Sections 3.5 and 5.1). In this case, the hydrogen in a C—H bond is moved closer to the carbon to assume a new, virtual position which is used to calculate van der Waals and electrostatic interactions.

X-ray diffraction analysis of a crystalline molecular substance provides X-ray intensity data, which can be interpreted in terms of a map of the electron density of the molecule(s) in the unit cell. The electron density map for a molecule may be translated into a set of Cartesian coordinates for the constituent atoms, also called crystal coordinates. Crystal coordinates from X-ray diffraction may be obtained from the literature, or by searching a computerized X-ray structure database. The premier example of such a database is the Cambridge Crystallographic Data Centre,[136,245,246] which is searchable in a variety of ways, including text strings and whole or partial chemical structures. The Protein Data Bank is a good source of data on large biomolecules.[137] The interested reader is referred to other sources for further information on X-ray diffraction or X-ray crystallography.[247]

X-Ray Format Input File Exercise: Importing the Imidazole Structure. For importing X-ray structures into PCMODEL, the input file may need to be created or modified by the user. This process is explained at greater length in Section 4.4. By way of demonstration, we will import a set of crystal coordinates for the molecule imidazole, and then make some measurements on the bonds. This structure will be used later on in Section 6.5 on π-calculations. These data come from the X-ray diffraction structure of imidazole as the hydrogen-bonded solid crystal at room temperature,[248] and the corresponding input file is seen in Fig. 3-36. This file may be found on the input file library diskette in the CHAP3\SECT3 subdirectory as {IMDAZL01.XRA}.

From the PCMODEL opening window, activate the **File\Open** command, and from the dialog box activate the radio button for the 'X-Ray' file format. Select the appropriate disk drive and directory from the file list displayed under 'Files:'; then select {IMDAZL01.XRA}. The structure will then come on screen and should look like Fig. 3-37.

The **View\Control_Panel** command may be used to rotate the structure to give the desired perspective, and **View\Labels** may be used to label each atom with its number in the input file. The **Query** tool may be used to show a slight difference between the two C—N bond lengths (1–2 > 2–3), and also between the two corresponding N—H bonds (1–6 < 3–8). The resulting screen should look like Fig. 3-38.

```
Imidazole X-Ray Structure/Omel'chenko
7.746 5.447 9.750 90.000000 117.5 90.00000
 .2197  .1700  .0880 N
 .1600  .2767  .1812 C
 .2095  .1525  .3102 N
 .3080 -.0435  .2998 C
 .3141 -.0361  .1611 C
 .225   .188  -.015 H
 .086   .433   .145 H
 .197   .169   .412 H
 .352  -.147   .386 H
 .375  -.145   .100 H
```

Figure 3-36. The XRA Format Input File for Imidazole.[248]

3.3 STRUCTURE INPUT FROM PHYSICAL DATA

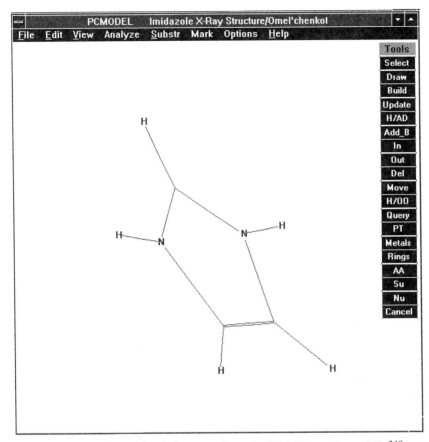

Figure 3-37. The Imidazole Structure from the XRA Format Input File.[248]

This would lead us to use **Add_B** to place a double bond between C2 and N3, and it would also be prudent either to remove H8 or to adjust the atom type of N3 to the @41 'N+' before carrying out further modeling.

3.3.2. Neutron Diffraction Data

Neutron diffraction in the present context is the technique of elucidating structural information by interpreting the interference patterns produced when high-energy thermal neutrons (kinetic energy ≈ 0.025 eV) are diffracted or scattered when passing through a substance. As with X-ray diffraction, the technique of interest here is the determination of molecular structure by diffraction with a single crystal of the pure substance. Thermal neutrons have energies equivalent to wavelengths around 0.1 nm (1 Å), which is in the same range as common molecular bond-lengths. In the case of neutron diffraction, and unlike X-ray diffraction, the neutrons are scattered by the nuclei of the

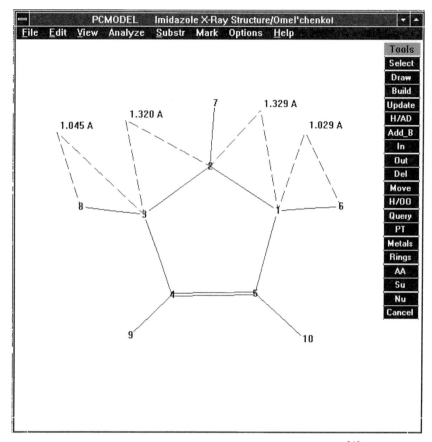

Figure 3-38. Bond Lengths for the Imidazole Structure.[248]

atoms which make up a molecule. This difference makes neutron diffraction far more precise in determining the positions of light atoms such as hydrogen and deuterium.

In practical terms, the diffraction data and the parameters and the derived coordinate sets are very similar to those obtained in X-ray diffraction, and PCMODEL can read the associated input files in the same way. Neutron diffraction data is found in many of the same places that X-ray data is. The interested reader is referred to other sources for further information on neutron diffraction.[249]

3.3.3. Electron Diffraction Data

Electron diffraction in the present context is the technique of elucidating structural information by interpreting the interference patterns produced when high-energy electrons (kinetic energy ~5–500 keV) are diffracted or scattered when striking a substance. Unlike the case with X-ray and neutron diffraction, electrons penetrate into matter only a short distance. Because of energy loss in such a collision, the wavelength of the incident electron changes, and this loss of coherence prevents the production of interference patterns obtained with the other diffraction techniques. The most successful application of electron diffraction is in the gas phase, where the lower density permits the analysis of diffraction by single atoms or molecules, as opposed to a crystalline lattice. From such measurements can be obtained the equilibrium distances between atoms and the average amplitude of vibration associated with that distance, which translates into bond lengths and angles. A comprehensive, but now quite dated, collection of molecular data from electron diffraction is available.[250] The interested reader is referred to other sources for further information on electron diffraction.[251]

3.3.4. Microwave Spectroscopy

Microwave spectroscopy is quite different from the other techniques mentioned here, in that the energies of the radiation employed are much lower. In terms of the kind of data obtained, though there are significant similarities to electron diffraction. The microwave spectra are obtained in the gas phase, and the analysis of the vibrational and rotational spectra provides measurements of bond lengths and bond angles. The absorption of microwave energy requires an electric dipole moment, and this is a drawback to the study of many symmetrical, nonpolar organic molecules, such as hydrocarbons, which have extremely small electric dipoles. This problem can be overcome by using deuterium-substituted substrates, to break the symmetry, or by using Raman rotational spectroscopy. More recently, pulsed microwave Fourier transform spectroscopy has proven capable of detecting spectra for molecules with very small electric dipole moments.[252]

3.4. MATHEMATICAL REPRESENTATIONS OF STRUCTURES

There are three basic format types for structure representations, based upon dimensionality: one-, two-, and three-dimensional. There are many utilities for interconverting them, and they find significant usage as the media for searching structural databases. They also serve as input for the calculation of physical properties, molecular similarity indices, and many other useful measures.

Many molecular modeling programs and even chemical drawing programs contain an "on the fly" translation utility, in which the display of structure graphics is driven by the electronic mathematical representation in a structure file. Such structure images may be embedded in documents as *live structures*, or enactive,[127] chemically significant objects, insofar as they may be extracted from the document and used. On the other hand, if the structure graphics have been reduced to a bitmapped (raster) or vector format image (e.g. GIF, JPEG, EPS), it is far more difficult to retrieve the chemical significance. There is substantial interest in developing an effective program to translate from graphics formats to chemical structure formats especially from scanned images. The commercial program Kekule® does this with some success. There is also a major collaboration currently underway entitled CLiDE, for Chemical Literature Data Extraction.[253,254] This purpose of this project is to have a program which accepts a raster image and converts it to one of several two-dimensional chemical structure formats.

3.4.1. One-Dimensional Structures

The best-known example of a one-dimensional file format is SMILES code (Simplified Molecular Input Line Entry System), an ASCII text-only line notation which codes for the connectivity of chemical structures.[255-258] It is readily converted into a two-dimensional structure by a set of rules, and may be used as input for the calculation of physical properties such as lipophilicity measures or molar refractivity.

In SMILES code, atoms are indicated by their elemental symbols in an uppercase (or an initial uppercase) letter, and hydrogens are generally not specified. Aromatic atoms are indicated in lowercase letters. Single bonds are generally not specified, while a double bond is indicated by '=' and a triple bond by '#'. Branches are specified by enclosure within parentheses. Cyclic structures are depicted by breaking open the ring to a linear string, and marking the two formerly bonded atoms by appending a number to their atom designators, using the same number for both atoms.

The SMILES code string for the calix[4]arene molecule depicted in Fig. 3-40 may be found under CHAP3\SECT4 on the input file library diskette as {CALIXREN.SMI}, and is:

Oc1c2cccc1Cc1cccc(c1O)Cc1c(c(ccc1)Cc1cccc(c1O)C2)O

Another one-dimensional structure format in current use is ROSDAL (Representation Of Structure Description Arranged Linearly), developed for the Beilstein Institute for use in handling the structures in the Beilstein and Gmelin databases.[259] Like SMILES, it codes for the chemical structure, and atom charge, isotope and coordinate values are interpreted if present.

3.4.2. Two-Dimensional Structures

Several good reviews of the mathematical representations of two-dimensional structures are available.[113,257] In the present context, we will define a two-dimensional structure as a representation of a molecule which codes for the connectivity: which atoms are bonded to which other atoms by bonds of what

```
        -ISIS-  02239608292D

     32 36  0     0  0  0  0  0  0  0  1 V2000
        5.5167   -6.2916    0.0000 C   0  0  0  0  0  0  0  0  0  0  0  0
        5.5167   -6.9833    0.0000 C   0  0  0  0  0  0  0  0  0  0  0  0
        6.1162   -7.3333    0.0000 C   0  0  0  0  0  0  0  0  0  0  0  0
        6.7199   -6.9833    0.0000 C   0  0  0  0  0  0  0  0  0  0  0  0
        6.7199   -6.2916    0.0000 C   0  0  0  0  0  0  0  0  0  0  0  0
        6.1162   -5.9416    0.0000 C   0  0  0  0  0  0  0  0  0  0  0  0
        3.8910   -4.1416    0.0000 C   0  0  0  0  0  0  0  0  0  0  0  0
        3.5405   -4.7379    0.0000 C   0  0  0  0  0  0  0  0  0  0  0  0
        3.8800   -5.3434    0.0000 C   0  0  0  0  0  0  0  0  0  0  0  0
        4.5778   -5.3476    0.0000 C   0  0  0  0  0  0  0  0  0  0  0  0
        4.9283   -4.7513    0.0000 C   0  0  0  0  0  0  0  0  0  0  0  0
        4.5852   -4.1437    0.0000 C   0  0  0  0  0  0  0  0  0  0  0  0
        4.9195   -5.9490    0.0000 C   0  0  0  0  0  0  0  0  0  0  0  0
        6.7375   -3.2041    0.0000 C   0  0  0  0  0  0  0  0  0  0  0  0
        6.7375   -2.5125    0.0000 C   0  0  0  0  0  0  0  0  0  0  0  0
        6.1380   -2.1625    0.0000 C   0  0  0  0  0  0  0  0  0  0  0  0
        5.5343   -2.5125    0.0000 C   0  0  0  0  0  0  0  0  0  0  0  0
        5.5343   -3.2041    0.0000 C   0  0  0  0  0  0  0  0  0  0  0  0
        6.1380   -3.5541    0.0000 C   0  0  0  0  0  0  0  0  0  0  0  0
        8.3632   -5.3542    0.0000 C   0  0  0  0  0  0  0  0  0  0  0  0
        8.7136   -4.7579    0.0000 C   0  0  0  0  0  0  0  0  0  0  0  0
        8.3741   -4.1523    0.0000 C   0  0  0  0  0  0  0  0  0  0  0  0
        7.6764   -4.1482    0.0000 C   0  0  0  0  0  0  0  0  0  0  0  0
        7.3259   -4.7445    0.0000 C   0  0  0  0  0  0  0  0  0  0  0  0
        7.6689   -5.3521    0.0000 C   0  0  0  0  0  0  0  0  0  0  0  0
        7.3346   -3.5468    0.0000 O   0  0  0  0  0  0  0  0  0  0  0  0
        7.3194   -5.9489    0.0000 O   0  0  0  0  0  0  0  0  0  0  0  0
        4.9350   -3.5494    0.0000 O   0  0  0  0  0  0  0  0  0  0  0  0
        6.1151   -5.2500    0.0000 O   0  0  0  0  0  0  0  0  0  0  0  0
        5.6199   -4.7573    0.0000 C   0  0  0  0  0  0  0  0  0  0  0  0
        6.1390   -4.2458    0.0000 O   0  0  0  0  0  0  0  0  0  0  0  0
        6.6342   -4.7385    0.0000 C   0  0  0  0  0  0  0  0  0  0  0  0
      1  6  2  0  0  0  0
     38  1  2  1  0  0  0  0
     39  2  3  1  0  0  0  0
     40  3  4  2  0  0  0  0
     41  4  5  1  0  0  0  0
     42  5  6  2  0  0  0  0
     43  1  5  1  0  0  0  0
     44  7  8  1  0  0  0  0
     45  7 12  2  0  0  0  0
     46  8  9  2  0  0  0  0
     47  8 10  1  0  0  0  0
     48  9 11  1  0  0  0  0
     49 10 11  2  0  0  0  0
     50 11 12  1  0  0  0  0
     51 12 13  1  0  0  0  0
     52  5 27  1  0  0  0  0
     53 14 19  2  0  0  0  0
     54 14 15  1  0  0  0  0
     55 15 16  2  0  0  0  0
     56 16 17  1  0  0  0  0
     57 17 18  2  0  0  0  0
     58 18 19  1  0  0  0  0
     59 14 26  1  0  0  0  0
     60 20 25  2  0  0  0  0
     61 20 21  1  0  0  0  0
     62 21 22  2  0  0  0  0
     63 22 23  1  0  0  0  0
     64 23 24  2  0  0  0  0
     65 24 25  1  0  0  0  0
     66 23 26  1  0  0  0  0
     67 25 27  1  0  0  0  0
     68 18 28  1  0  0  0  0
     69  6 29  1  0  0  0  0
     70 11 30  1  0  0  0  0
     71 19 31  1  0  0  0  0
     72 24 32  1  0  0  0  0
     73       M  END
```

Figure 3-39. MDL MOL Format Structure File for Calix[4]arene, {CALIXINP.MOL} Folded Over into Two Columns.

81

multiplicity. A two-dimensional structure may have stereochemical cues, but does not give a quantitative representation in three-dimensional space. Handwritten structural formulas, some of the history of which was discussed in Section 1.1, qualify as the earliest mathematical representation of two-dimensional structure. As was noted, these include a wide variety of conventions and individual styles, and so do not always code for a distinct molecule unambiguously. Before the advent of personal computers in the early 1980s, some standardization in the appearance of these structures was possible by using dry transfers.

The advent of graphics-oriented personal computers, and in particular the structure drawing program ChemDraw™ for the Apple Macintosh® environment, revolutionized chemistry presentation graphics. Now, there was a standard toolkit from which two-dimensional structures could be assembled, and stereochemical cues such as stereo wedges and hatched lines could also be incorporated. Many other structure-drawing programs have been developed in the intervening time. This trend toward computer structures was reinforced by the development at that time of the molecule database programs MACCS® (for molecular structures) and REACCS® (for reaction schemes) by Molecular Design Ltd. (now MDL). These enabled users to draw structures and store them in an indexed fashion in databases with accompanying data, so that they could be searched and sorted.

In terms of mathematical representations of structure which may be stored, retrieved and manipulated by computer, there are two basic types: the structure input file format, and the coded data string. Structure-drawing programs store structures in a file format which represents the two-dimensional structure.

As an example, a structure file for the molecule calix[4]arene was prepared in ISISDraw® in the MDL MOL file format. The structure file is reproduced in Fig. 3-39, and may be found under CHAP3\SECT4 on the input file library diskette as {CALIXINP.MOL}. In Fig. 3-40 is shown the structure as it is read into PCMODEL from this input file.

This file format is discussed in detail in Section 4.3, but several items are noteworthy here. In lines 5–36, we see that the first three columns of data consist of five-digit rational numbers, which represent the Cartesian X, Y, and Z coordinates for each atom in the structure. Note in particular that the third column has only the value '0.0000'. Only the X- and Y-coordinates are used to describe this two-dimensional structure. In lines 37–72 are found the connectivity data, the descriptions of which atoms are bonded to which other atoms and with what bond multiplicity.

3.4.3. Three-Dimensional Structures

A good review of the mathematical representations of three-dimensional structures is given by Marsili.[113] In the present context, we will define a three-dimensional structure broadly as a representation of a molecule which codes

3.1. STRUCTURE INPUT FROM THE GRAPHICS INTERFACE   83

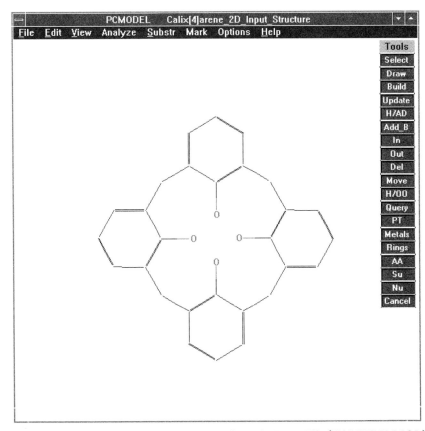

Figure 3-40. The Two-Dimensional Structure from the Input File {CALIXINP.MOL}.

for the relative positions of the atoms of that molecule in some coordinate system (Cartesian, rotational, etc.), or some coding or index which uniquely describes a three-dimensional molecule. Most molecule input files will be in a three-dimensional format (see Chapter 4). A number of novel formats for the latter type of description have been reported.[260-262] The various types of molecular surface may also be considered as a three-dimensional structure (see Section 6.2).

Three-dimensional structures are useful as starting points for molecular modeling, and in studying properties and interactions, but the major use of them currently is in 3D-QSAR, in creating a biophore description from a set of 3D-format active structures, and then using the biophore to screen large numbers of 3D-format structures to discover novel active compounds.

As an example, the three-dimensional structure file in the MDL MOL file format for the molecule calix[4]arene, {CALIXCR1.MOL}, is reproduced in Fig. 3-41. This structure was prepared by the operation of **CORINA**[140,262] on the 2D file {CALIXINP.MOL}. Both of these structure files may be found

```
  Calix[4]arene_Corina_Conversion_from_2D
  wicorina   02249603003D
  Corina  1.6 0000  28.08.1995
 32 36  0     0  0  0  0  0  0  0  1 V2000
   -0.7882    2.1834    1.3283 C   0  0  0  0  0  0  0  0  0  0  0  0
    0.2101    3.0759    1.6697 C   0  0  0  0  0  0  0  0  0  0  0  0
    1.5251    2.6533    1.7355 C   0  0  0  0  0  0  0  0  0  0  0  0
    1.8433    1.3349    1.4649 C   0  0  0  0  0  0  0  0  0  0  0  0
    0.8479    0.4384    1.1335 C   0  0  0  0  0  0  0  0  0  0  0  0
   -0.4705    0.8613    1.0541 C   0  0  0  0  0  0  0  0  0  0  0  0
   -3.2173    3.5740   -2.7931 C   0  0  0  0  0  0  0  0  0  0  0  0
   -2.8333    4.5794   -1.9248 C   0  0  0  0  0  0  0  0  0  0  0  0
   -2.5100    4.2771   -0.6154 C   0  0  0  0  0  0  0  0  0  0  0  0
   -2.5660    2.9707   -0.1710 C   0  0  0  0  0  0  0  0  0  0  0  0
   -2.9436    1.9593   -1.0420 C   0  0  0  0  0  0  0  0  0  0  0  0
   -3.2756    2.2664   -2.3537 C   0  0  0  0  0  0  0  0  0  0  0  0
   -2.2229    2.6364    1.2581 C   0  0  0  0  0  0  0  0  0  0  0  0
   -0.6901   -0.8404   -4.3713 C   0  0  0  0  0  0  0  0  0  0  0  0
   -0.3052   -0.1634   -5.5131 C   0  0  0  0  0  0  0  0  0  0  0  0
   -1.0111    0.9496   -5.9306 C   0  0  0  0  0  0  0  0  0  0  0  0
   -2.1075    1.3850   -5.2079 C   0  0  0  0  0  0  0  0  0  0  0  0
   -2.4980    0.7094   -4.0701 C   0  0  0  0  0  0  0  0  0  0  0  0
   -1.7844   -0.4004   -3.6420 C   0  0  0  0  0  0  0  0  0  0  0  0
    2.9301   -0.8847   -0.9325 C   0  0  0  0  0  0  0  0  0  0  0  0
    3.3255   -1.0068   -2.2524 C   0  0  0  0  0  0  0  0  0  0  0  0
    2.4094   -1.3796   -3.2175 C   0  0  0  0  0  0  0  0  0  0  0  0
    1.0966   -1.6285   -2.8679 C   0  0  0  0  0  0  0  0  0  0  0  0
    0.6953   -1.4965   -1.5467 C   0  0  0  0  0  0  0  0  0  0  0  0
    1.6189   -1.1288   -0.5779 C   0  0  0  0  0  0  0  0  0  0  0  0
    0.0955   -2.0463   -3.9143 C   0  0  0  0  0  0  0  0  0  0  0  0
    1.1860   -1.0053    0.8596 O   0  0  0  0  0  0  0  0  0  0  0  0
   -3.7054    1.1688    0.2915 O   0  0  0  0  0  0  0  0  0  0  0  0
   -1.4483   -0.0170   -0.7086 O   0  0  0  0  0  0  0  0  0  0  0  0
   -2.9877    0.6710   -0.6119 O   0  0  0  0  0  0  0  0  0  0  0  0
   -2.1564   -1.0546   -2.5104 O   0  0  0  0  0  0  0  0  0  0  0  0
   -0.5990   -1.7260   -1.2017 O   0  0  0  0  0  0  0  0  0  0  0  0
  1  6  2  0  0  0  0
```

```
 38  1  2  1  0  0  0  0
 39  2  3  2  0  0  0  0
 40  3  4  1  0  0  0  0
 41  4  5  2  0  0  0  0
 42  5  6  1  0  0  0  0
 43  6 13  1  0  0  0  0
 44  7  8  2  0  0  0  0
 45  7 12  1  0  0  0  0
 46  8  9  1  0  0  0  0
 47  9 10  2  0  0  0  0
 48 10 11  1  0  0  0  0
 49 11 12  2  0  0  0  0
 50 10 13  1  0  0  0  0
 51 12 28  1  0  0  0  0
 52  5 27  1  0  0  0  0
 53 13 19  1  0  0  0  0
 54 14 15  2  0  0  0  0
 55 14 19  1  0  0  0  0
 56 15 16  1  0  0  0  0
 57 16 17  2  0  0  0  0
 58 17 18  1  0  0  0  0
 59 18 19  2  0  0  0  0
 60 14 26  1  0  0  0  0
 61 20 21  2  0  0  0  0
 62 20 25  1  0  0  0  0
 63 21 22  1  0  0  0  0
 64 22 23  2  0  0  0  0
 65 23 24  1  0  0  0  0
 66 24 25  2  0  0  0  0
 67 23 26  1  0  0  0  0
 68 18 29  1  0  0  0  0
 69  6 30  1  0  0  0  0
 70 11 30  1  0  0  0  0
 71 19 31  1  0  0  0  0
 72 24 32  1  0  0  0  0
 73  M  END
```

Figure 3-41. MDL MOL 3D Fromat Structure File for Calix[4]arene, {CALIXCR1.MOL}, Converted by CORINA from the 2D Format Structure File, {CALIXINP.MOL}, Folded Over into Two Columns.

in CHAP3\TSECT4 on the input file library diskette. The two structures, the 2D {CALIXINP.MOL} and the 3D {CALIXCR1.MOL} are displayed in Fig. 3-42, in the following subsection.
This file format resembles closely that found in Fig. 3-39. In lines 5–36, we see that the first three columns of data consist of five-digit rational numbers, which represent the Cartesian X, Y, and Z coordinates for each atom in the structure. Unlike the case of the 2D structure file depicted in Fig. 3-39, each atom from the file shown in Fig. 3-41 has nonzero values for all three Cartesian coordinates. Lines 37–72 again contain the connectivity data.

3.4.4. Automated Generation of Three-Dimensional Structures

The automated generation of 3D structures from 1D or 2D inputs (or equivalent) is an area of high current interest.[177,263–266] In the present context we will discuss this technique as a way of preparing 3D structural input geometry, in order to arrive more quickly at a stage where the "interesting" part of modeling can take place. In a related area, these techniques are used to produce 3D structures which are consistent with NMR spectral determinations in solution.

One important area where advances in 3D structure prediction are having an influence is that of biological macromolecules, and in particular the ability to develop structural information on proteins[267] and nucleic acid polymers such as DNA.[268,269] Another is the generation and use of 3D structures in programs which model reactivity and predict the outcome of organic reactions, such as CAMEO.[265]

The major push in this area, though, comes from the use of these techniques to convert entire databases of 2D structures into reliable 3D structures, to be used as the feedstocks for 3D-QSAR analytic molecule discovery efforts. In other words, when a computational structure–activity or structure–property relationship with 3D characteristics has been devised, attempts are made to exploit the predictive capability of such relationships. One way to do so is to use the relationship to screen through a large and diverse set of molecules, in order to arrive a much smaller subset which should optimize the desired activity or property. More and more, 3D formats for many of the existing molecule databases are being developed. Still, the problem is not an easy one, and no one method is completely reliable. Given the scope of this task, empirical force field geometry optimization methods themselves require input geometries, and they are simply not quick enough to do the job. The preferred tradeoff is more structures of lower quality. A recent review[263] has discussed the problems inherent in this kind of approach, such as how to choose the coordinate system, whether to represent flexible molecules by more than one conformation, how to find and express the members of a family of conformations for the flexible molecules, and what to do about very large cyclic molecules. The most successful solutions fall into two categories: the rule-based methods and the data-based methods.

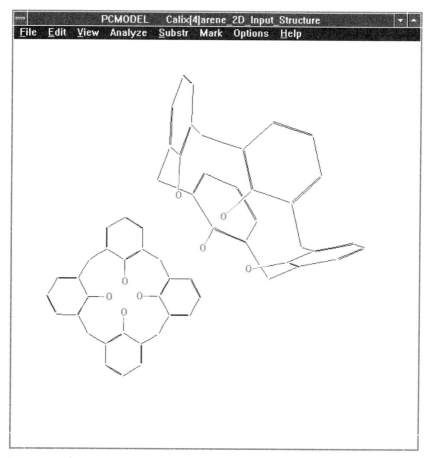

Figure 3-42. {CALIXINP.MOL} Input File for Calix[4]arene (lower left), and the {CALIXCR1.MOL} Output File Generated by CORINA (upper right).

The rule-based methods involve applying rules governing what the bond length should be between an atom of type X and an atom of type Y, what the angle should be between any three atoms, and what is the best torsional angle to represent a particular four-atom string. The algorithm iterates through the molecule in this way, until it has converted all of it. There may be additional checks and balances to avoid the bad interactions which result from this method of assembly. One of the earliest examples of this type of method was the algorithm published by Wiberg[57] to add hydrogens to a secondary carbon in a heavy atom skeleton. More recent examples of this approach are WIZARD,[270] COBRA,[264] CONCORD and CORINA.[263,266] These programs operate by similar methods. They parse the structure input in SMILES or other common file formats, and intelligent algorithms recognize chain and ring patterns from a library base. The fragments from the library are assembled, and

3.1. STRUCTURE INPUT FROM THE GRAPHICS INTERFACE 87

then a limited force field optimization (**COSMIC** force field[271] in the case of COBRA, proprietary force fields in the cases of CONCORD and CORINA) is employed to fine-tune the resulting 3D structures.[140,263]

The data-based methods build models by comparing the 2D *molecule graph* of the structure in question against the corresponding 2D graphs of molecules in a large database of 3D structures, such as the Cambridge database. They utilize close analogies between the 2D structures or substructures to recommend all or a portion of a 3D structure for the molecule of interest. Examples of this approach are AIMB and the Chem-X Builder.

As an example of the automated generation of three-dimensional structures, the results of applying the CORINA program to the pair of two-dimensional input files {CALIXINP.MOL} and {CALIXREN.SMI} for the calix[4]arene molecule are presented here. The computational chemistry group at Universität Erlangen provides a limited number of free accessions to this algorithm via the Internet to introduce the method to the worldwide chemical community.[140] In the present cases, each of the abovementioned two-dimensional input files was submitted, and the corresponding three-dimensional output files were received in the MDL MOL format. Figure 3-42 shows both the {CALIXINP.MOL} input file and the {CALIXCR1.MOL} output file in the work space, and Fig. 3-43 shows the {CALIXCR2.MOL} output file derived from {CALIXREN.SMI}. As will be seen in Section 3.6.3, these molecules are close in geometry, but not exactly superimposable. This is possibly the result of {CALIXCR1.MOL} being generated with heavy atoms only and {CALIXCR2.MOL} being generated with a full complement of hydrogens.

PCMODEL does not include a utility for automatic conversion from 2D to an approximate 3D structure, beyond the computational energy optimization approach. It does, however, implement two structure-building tools which operate on geometry-based algorithms which will be described here. The **Build** tool is an example of such an algorithm, and was mentioned in Section 3.1 as a way to elaborate a carbon skeleton rapidly. It converts a hydrogen atom on a structure to a methyl group according to a geometric rule described below. The other example is the **H/AD** tool toggle, also mentioned in Section 3.1, with which hydrogens and lone pairs are added to (or deleted from) a heavy atom skeleton. Both require a preexisting 3D substrate in order to work properly.

The **Build** tool works most reproducibly by replacing hydrogens which have been added with the **H/AD** tool. Operating on the hydrogens in a minimized structure may give longer C—C bonds, but an energy minimization will correct any discrepancies. In general, the **Build** algorithm replaces a C—H bond of 1.06–1.12 Å with a C—C bond of 1.54 Å. The new attached hydrogens will have C—H distances of 1.108–1.110 Å. For conversion of a hydrogen attached to an sp^3 carbon, the new torsional angles will be such that the hydrogens on the new carbon will be staggered with respect to the attached groups on the original carbon. For conversion of a hydrogen attached to an

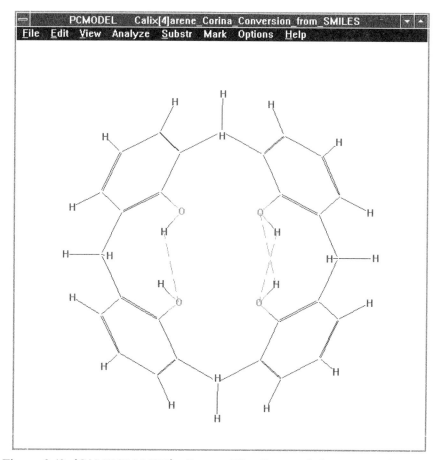

Figure 3-43. {CALIXCR2.MOL} Output File Generated by CORINA from the {CALIXREN.SMI} Input File.

sp^2 carbon, the torsional angles will be such that one of the attached hydrogens will be antiperiplanar to the other group attached to the sp^2 carbon, and the angle between the remaining two new hydrogens will be bisected by the plane containing the double bond and the new carbon atom.

The **H/AD** tool is the other example of an algorithm that generates 3D structural features. In this case the 'add' part of the toggle adds hydrogens and/or lone pairs to an extant 3D heavy atom structure according to a geometry-based recipe. It begins with atom file number '1', and then proceeds through all the heavy atoms in order of increasing atom file number.

Since the geometry of the added hydrogens and lone pairs is a function of the cross vectors of the existing bonds between heavy atoms, distorted starting geometries will lead to eccentric hydrogen placements. It is always better to carry out an energy minimization on the heavy atom skeleton before using **H/AD**.

With a preoptimized geometry in an unstrained heavy atom skeleton, the following bond length parameters are used for adding hydrogens and lone pairs (LPs). The (sp^3)C—H methyl and methylene bonds, the N—H and N—LP distances in primary and secondary amines, and the O—H and O—LP distances for alcohols, are all 1.11 Å. The (sp^3)C—H methine and the (sp^2)C—H and (sp)C—H bonds are 1.10 Å. The O—LP distance for ethers and the N—LP distance in tertiary amines are 0.6 Å.

In an unstrained, preoptimized structure, the following bond angle parameters are used for adding hydrogens and lone pairs. The (sp)C—(sp)C—H angle is 180°. The LP—O—LP angle is 131°. The (sp^3)C—(sp^2)C—H angle is 125°. The (sp^2)C—(sp^2)C—H angle in mono- and disubstituted alkenes is 113.6°. The H—(sp^2)C—H angle is 119°. The LP—N—C angle in tertiary amines is 111.2°. The H—N—C and LP—N—C angles in secondary amines are 111°. The (sp^3)C—(sp^3)C—H angle for methylenes is 110.6°. The (sp^3)C—(sp^3)C—H angle for methyls, the H—N—C and LP—N—C angles in primary amines, and the H—O—C and LP—O—C angles in alcohols are 110°. The (sp^3)C—(sp^3)C—H angle for methines is 109.5°. The H—(sp^3)C—H angle for methyls, the H—N—H and LP—N—H angles in primary amines, and the H—O—LP and LP—O—LP angles in ethers are 108.9°. The H—(sp^3)C—H angle for methylenes and the LP—N—H angle in secondary amines are 105°. The LP—O—C angle in ethers is 104.3°.

3.5 DATA OUTPUT

There are two types of output from an energy minimization experiment carried out with a molecular modeling program. One is the molecule description file, which can serve as a new input file for further work on that molecule or ensemble. In PCMODEL, this file is written by the program as the default file {PCMOD.BAK}. This file will be overwritten with each subsequent energy minimization; in order to capture it, the **File\Save** command should be used to write it to an input file. The other output file documents the input file used, describes the course of the energy minimization process, and provides the energy parameters resulting from the calculations. In PCMODEL, this file is written by the program as the default file {PCMOD.OUT}. If it exists at the beginning of a new PCMODEL session, this file will be overwritten with the first energy minimization, but within a session the output results from subsequent energy minimizations are appended to this file. For the purposes of this discussion, we will use the term "output file" to indicate the latter.

In PCMODEL, the user may select from two output file formats which differ in the amount of detail provided. The selection is made with the **Options\Printout** command, which when activated brings up a dialog box entitled 'Print Level', shown in Fig. 3-44. The default is for the 'Minimal Printout', but if the more

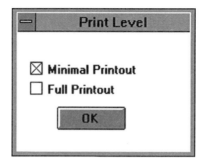

Figure 3-44. The 'Print Level' Dialog Option Box.

extensive 'Full Printout' is desired, then the mouse cursor is used to click on the corresponding box. If the print level is changed to 'Full Printout', that mode remains in effect until changed back, or until the current session is ended.

The content of the PCMODEL output file in these two formats is shown and contrasted below for a simple case, the molecule 1,1-difluoroethene, the input file {F2ETHENE.PCM} for which was prepared in Section 3.1. These examples of output are shown in Figs. 3-45 and 3-46, and are discussed below.

The 'Minimal Printout' version is found in Fig. 3-45. The first six lines of the file detail the input of constants from two files, {MMXCONST.PAR} and {GENCONST.PAR}. This report is only generated for the first energy minimization of the session. Next come two lines reporting the use of generalized parameters for two of the F—C—C—H torsional angles: one cisoid and one transoid angle.

The next six lines give some details for the six atoms in the input file: the input file atom number ('Atom #'), the atom type number ('mmtype'), and the charge, which is calculated from the resident parameters for the bond dipoles as described in Section 5.2. Then come two lines reporting the use of generalized parameters for the two F—C—C angles.

Following this comes a five-line report on the initial energy parameters, prior to minimization. The 'MMX Energy' is the total energy for the system, which can be seen to be quite large for the input structure used. This is broken out into constituents: 'Str' is the bond stretch term, 'Bnd' is the angle bending term, 'S-B' is the stretch–bend cross term, 'Tor' is the torsional strain term, 'Vdw' is the van der Waals repulsion term, and 'Q-Q' is the electrostatic term. 'Dip Mom' is the calculated dipole moment for the input structure.

The next lines in the output file have been added after the energy minimization process is complete. The maximum energy difference between the last iterations is given, followed by the newly calculated value of the total energy (both in kilocalories per mole), and the 'Accumulated movement' or maximum Cartesian displacement of atoms between the last iterations (in angstroms per atom).

```
The number of constants read in C:\PCMWIN5\mmxconst.par was:
 Bonds: 398    Angles 290   Torsions 577   Eheat 107   Eheatpi 25
 VDW: 75   3 Mem Angles 9   4 Mem Angles: 31  4 Tors 0   OOP 47   EVHF 55
The number of constants read in C:\PCMWIN5\genconst.par was:
 Bonds: 398    Angles 352   Torsions 659   Eheat 107   Eheatpi 25
 VDW: 75   3 Mem Angles 30  4 Mem Angles: 51  4 Tors 0   OOP 47   EVHF 55

Start of : 1,1-Difluoroethene
Generalized constants used for angle 3 1 2 5 of types 11 2 2 5
Generalized constants used for angle 3 1 2 6 of types 11 2 2 5
Atom #: 1 mmtype: 2 charge:  0.460
Atom #: 2 mmtype: 2 charge: -0.076
Atom #: 3 mmtype: 11 charge: -0.230
Atom #: 4 mmtype: 11 charge: -0.230
Atom #: 5 mmtype: 5 charge:  0.038
Atom #: 6 mmtype: 5 charge:  0.038
Generalized constants for angle 2 1 3 of types 2 2 11 are used
Generalized constants for angle 2 1 4 of types 2 2 11 are used
MMX Energy = -2.161
Str = 0.004       Bnd = 0.398
S-B = -0.012      Tor = 0.000
Vdw = 0.123       Q-Q = -2.674
Dip Mom = 1.869
Energy is minimized within 0.0006 kcal/mol.
MM2 energy is -2.16072  Accumulated movement is 0.0003064738

Minimization Done
Heat of Formation, Strain Energies and Entropies at 300 k (units are kcal or eu.)
Bond Enthalpy (be) and Entropy:
#   Bond or Structure    Each      Total     Tot S contrib.
1   C=C SP2 SP2          22.840    22.840      0.000
2   C-H OLEFINIC         -3.205    -6.410     27.600
2   F-CSP2              -44.000   -88.000     37.200

                       BE= -71.570    S= 64.800

3 & 4 Ring corrections to entropy are included w/o symmetry corrections.
    for each 5-ring add 26 eu.; for each 6 &7-ring add 16 eu.
    for each 8-ring add 14 eu.; for higher rings add 12 eu. each.
    there are no symmetry corrections to the entropy.

Heat of Formation Calculation
Partition Function Contribution (PFC):
Conformational Population Increment (POP):   0.000
Torsional Contribution (TOR)             :   0.000
Translation/Rotation Term (T/R)          :   2.400
                          PFC =          :   2.400

Heat of Formation (hf0) = energy + be + pfc :    -71.331

Strain Energy (energy+environment corrs.)=  :     -2.161
MMX Energy = -2.161
Str = 0.005       Bnd = 0.398
S-B = -0.013      Tor = 0.000
Vdw = 0.124       Q-Q = -2.674
Dip Mom = 1.868
```

Figure 3-45. The 'Minimal Printout' Format Output File for 1,1-Difluoroethene, {F2ETHENE.PCM}.

After a blank line, this is followed by the statement 'Minimization Done', and then the calculations of heat of formation, strain energy and entropy. The first section here displays the constituent fragments and their contribution to the bond enthalpy and entropy. In the next four lines comes a statement of ring and symmetry corrections to the entropy. Then, come the heat-of-formation calculation and summary statement, and the output file is concluded with the final summary of the strain energy and its components, and the calculated dipole moment.

```
The number of constants read in C:\PCMWIN5\mmxconst.par was:
Bonds: 398    Angles 290   Torsions 577   Eheat 107   Eheatpi 25
VDW: 75   3 Mem Angles 9   4 Mem Angles: 31  4 Tors 0   OOP 47   EVHF 55
The number of constants read in C:\PCMWIN5\genconst.par was:
Bonds: 398    Angles 352   Torsions 659   Eheat 107   Eheatpi 25
VDW: 75   3 Mem Angles 30  4 Mem Angles: 51  4 Tors 0   OOP 47   EVHF 55

Start of : 1,1-Difluoroethene
Generalized constants used for angle 3 1 2 5 of types 11 2 2 5
Generalized constants used for angle 3 1 2 6 of types 11 2 2 5
Atom #: 1 mmtype: 2 charge:  0.460
Atom #: 2 mmtype: 2 charge: -0.076
Atom #: 3 mmtype: 11 charge: -0.230
Atom #: 4 mmtype: 11 charge: -0.230
Atom #: 5 mmtype: 5 charge:  0.038
Atom #: 6 mmtype: 5 charge:  0.038
Generalized constants for angle 2 1 3 of types 2 2 11 are used
Generalized constants for angle 2 1 4 of types 2 2 11 are used
MMX Energy = -2.161
Str = 0.004       Bnd = 0.398
S-B = -0.012      Tor = 0.000
Vdw = 0.123       Q-Q = -2.674
Dip Mom = 1.869
Energy is minimized within 0.0006 kcal/mol.
MM2 energy is -2.16072  Accumulated movement is 0.0003064738

Minimization Done

    Bond and Non-Bonded Energies

       at1        at2         r          r0         bk         eb
   C ( 1) - C ( 2)         1.336      1.337      9.600      0.000
   C ( 1) - F ( 3)         1.348      1.350      5.750      0.002
   C ( 1) - F ( 4)         1.348      1.350      5.750      0.002
   C ( 2) - H ( 5)         1.101      1.101      4.600      0.000
   C ( 2) - H ( 6)         1.101      1.101      4.600      0.000
 Q F ( 3) - H ( 5)        -0.230      0.038      3.273     -0.594
 V F ( 3) - H ( 5)         3.273      3.360      3.150      0.061     -0.068
 Q F ( 3) - H ( 6)        -0.230      0.038      2.616     -0.743
 V F ( 3) - H ( 6)         2.616      2.657      3.150      0.061      0.130
 Q F ( 4) - H ( 5)        -0.230      0.038      2.616     -0.743
 V F ( 4) - H ( 5)         2.616      2.657      3.150      0.061      0.130
 Q F ( 4) - H ( 6)        -0.230      0.038      3.273     -0.594
 V F ( 4) - H ( 6)         3.273      3.360      3.150      0.061     -0.068

       ANGLE DATA    number of angles =    6

              at1        at2        at3      Theta       Theta0
   Angle  C ( 2) - C ( 1) - F ( 3)         123.882     120.000
   Str-Bnd                                   0.130      -0.004
   in-pln    2 -     1 -     3             123.882     120.000    0.450   0.149
   out-pln   1 -     4 -     1               0.000       0.600    0.000
   Angle  C ( 2) - C ( 1) - F ( 4)         123.882     120.000
   Str-Bnd                                   0.130      -0.004
   in-pln    2 -     1 -     4             123.882     120.000    0.450   0.149
   out-pln   1 -     3 -     1               0.000       0.600    0.000
   Angle  F ( 3) - C ( 1) - F ( 4)         112.237     109.100
   Str-Bnd                                   0.130      -0.005
   in-pln    3 -     1 -     4             112.237     109.100    0.450   0.097
   out-pln   1 -     2 -     1               0.000       0.600    0.000
```

Figure 3-46. The 'Full Printout' Format Output File {F2ETHENE.PCM} for 1,1-Difluoroethene.

The 'Full Printout' version of the output file for this minimization is found in Fig. 3-46. This version contains everything found in the 'Minimum Printout', with the addition of the specific details of the deviation from ideal geometry. In this example, these details are tabulated in turn for two-atom

```
Angle   C (   1) - C (   2) - H (   5)    120.795    120.500
Str-Bnd                                      0.080      0.000
in-pln   1 -      2 -      5              120.795    120.500    0.360    0.001
out-pln  2 -      6 -      2                 0.000      0.600    0.000
Angle   C (   1) - C (   2) - H (   6)    120.795    120.500
Str-Bnd                                      0.080      0.000
in-pln   1 -      2 -      6              120.795    120.500    0.360    0.001
out-pln  2 -      5 -      2                 0.000      0.600    0.000
Angle   H (   5) - C (   2) - H (   6)    118.411    119.000
in-pln   5 -      2 -      6              118.411    119.000    0.320    0.002
out-pln  2 -      1 -      2                 0.000      0.600    0.000

Torsional Energy Contribution from    4 angles

      at1         at2        at3        at4    dihed       V1        V2        V3       Energy
F (   3)-C (   1)-C (   2)-H (   5)-180.000    0.000   15.000     0.000    0.000
F (   3)-C (   1)-C (   2)-H (   6)   0.000    0.000   15.000     0.000    0.000
F (   4)-C (   1)-C (   2)-H (   5)   0.000    0.000   15.000     0.000    0.000
F (   4)-C (   1)-C (   2)-H (   6) 180.000    0.000   15.000     0.000    0.000
Heat of Formation, Strain Energies and Entropies at 300 k (units are kcal or eu.)
Bond Enthalpy (be) and Entropy:
#   Bond or Structure      Each        Total     Tot S contrib.
1  C=C SP2 SP2             22.840     22.840       0.000
2  C-H OLEFINIC            -3.205     -6.410      27.600
2  F-CSP2                 -44.000    -88.000      37.200

                 BE=  -71.570    S=   64.800

3 & 4 Ring corrections to entropy are included w/o symmetry corrections.
   for each 5-ring add 26 eu.; for each 6 &7-ring add 16 eu.
   for each 8-ring add 14 eu.; for higher rings add 12 eu. each.
   there are no symmetry corrections to the entropy.

Heat of Formation Calculation
Partition Function Contribution (PFC):
Conformational Population Increment (POP):    0.000
Torsional Contribution (TOR)             :    0.000
Translation/Rotation Term (T/R)          :    2.400
                                   PFC = :    2.400

Heat of Formation (hf0) = energy + be + pfc  :    -71.331

Strain Energy (energy+environment corrs.)=   :     -2.161
MMX Energy = -2.161
Str = 0.005       Bnd = 0.398
S-B = -0.013      Tor = 0.000
Vdw = 0.124       Q-Q = -2.674
Dip Mom = 1.868
```

Figure 3-46. (*Continued*)

interactions (bonded and nonbonded), three-atom angle interactions (stretch–bend cross terms, in-plane bending, and out-of-plane bending), and four-atom angle interactions (torsional strain).

For the two-atom interactions, the first column indicates the type of interaction described with a letter symbol. No letter means a bonded interaction (stretch or compression), 'Q' indicates a charge-charge non-bonded interaction, and 'V' indicates a van der Waals nonbonded interaction. The second and third columns give the identifiers for the atoms involved: 'at1' (atom 1) and 'at2' (atom 2), and contain the atom label (and atom file number in parentheses) for the each of the two atoms involved in the interaction. The meaning of the values in the remaining columns is quite different, depending upon the type of interaction, and each of these three cases will be considered in turn.

1. *Bonded Interactions.* The next two columns, entitled 'r' and 'r0', give respectively the actual and the ideal internuclear distance for the pair in question. After this comes a column entitled 'bk', which gives the bond stretching constant used for bonded interactions (i.e. 'ks' in the MMX force field). The last column, entitled 'eb', gives the enthalpy contributions, in kilocalories per mole, for the corresponding interactions. We see that in this example, the enthalpy contributions from bond perturbations is quite small, and sums to 0.005 kcal/mol in the final report.

2. *Electrostatic Interactions.* For the electrostatic interactions, the columns entitled 'r' and 'r0' give the charges on atom 1 and atom 2, respectively. The column entitled 'bk' gives the reduced (or effective) internuclear distance for the pair in question. These reduced distances will differ from the value measured directly from the structure when one of the atoms is a hydrogen bonded to carbon. The reduced hydrogen position was discussed in Section 3.3.1 as a correction factor employed to compensate for the distortion of the electron cloud about hydrogen toward the C—H internuclear region. The MMX force field employs a reduction factor of 8.5%, which is applied to both electrostatic and van der Waals interactions with hydrocarbon C—H. The column entitled 'eb' gives the electrostatic potential contribution for the interaction.

3. *van der Waals Interactions.* For the van der Waals nonbonded interaction, the next two columns entitled 'r' and 'r0' give the reduced (or effective) internuclear distance and measured internuclear distance, respectively, for the pair in question. These distances will differ when one of the atoms is a hydrogen bonded to carbon. For the reasons mentioned above and discussed in Section 3.3.1, the force field model gives better results when the component of the van der Waals surface due to such a hydrogen is 8.5% closer to the carbon atom. The column entitled 'bk' gives the sum of the van der Waals radii for the two atoms. The column entitled 'eb' gives the value for 'eps', a measure of the "hardness" of the van der Waals contact derived from the square root of the product of the hardness coefficients or ε of the individual atoms (see Section 5.1.3.1). The far right column gives the van der Waals potential contribution for the interaction.

For the next section on three-atom interactions, a header line reads, 'ANGLE DATA number of angles = ' followed by the number of angles analyzed, which in this case is 6. There are five labeled columns, 'at1', 'at2', 'at3', 'Theta', and 'Theta0'. Each angle is described over four rows, which will be considered in turn. The first row begins with the title 'Angle' and is followed by the atom label (and atom file number in parentheses) for the each of the three atoms involved in the interaction, under the appropriate identifier. In the first such entry in Fig. 3-46, this is given as 'C (2) - C (1) - F (3)'. The

3.5 DATA OUTPUT 95

fourth column ('Theta') contains the value for the angle in the energy-minimized molecule, in this case 123.882, and the fifth column ('Theta0') contains the equilibrium value for that angle, in this case 120.000.

The next row down is entitled 'Str-Bnd' for the stretch–bend cross term, and two values are given here. The first, roughly under the 'Theta' title, is the stretch–bend force constant for the angle, 0.130 in this case, and the second value, roughly under the 'Theta0' title, gives the stretch–bend potential contribution for the interaction, in this case −0.004.

The next row down is entitled 'in-pln' for the in-plane component of the angle bending. The atom file numbers for the angle are repeated (as '2 - 1 - 3'), as are the value for the angle in the energy-minimized molecule ('Theta', 123.882), and the equilibrium value for that angle ('Theta0', 120.000). The next value is the force constant for bending the current angle, 0.450 in this case, and the last value given is the contribution from in-plane angle bending potential, 0.149.

The last row in each group is entitled 'out-pln' for the out-of-plane component of the angle bending. This will be a characteristic of the central atom of the angle, and so, of the numbers found in the next three columns, the first and third are the atom file number for the central atom of the angle, while the second is the atom file number for a reference atom. Here this is given as '1 - 4 - 1'. This reference atom ('4') defines (with the two angle-terminal atoms, '2' and '3') the plane out of which the central atom of the angle may (or may not) bend. The fourth value over (under 'Theta') gives the actual out-of-plane displacement, 0.000 in this case. The fifth gives the out-of-plane bending parameter, which is a property of two-atom types, and here is 0.600. The last value in this line gives contribution from the out-of-plane angle bending potential, 0.000 here.

For the next section on four-atom interactions, a header line reads, 'Torsional Energy Contribution from' # 'angles', where # is the number of angles analyzed, which in this case is 4. There are nine columns of data displayed. The first four are the identifiers for the atoms making up the torsional angle of interest. For the first of these four, these data are given as, 'F (3) - C (1) - C (2) - H (5)'. The fifth column is entitled 'dihed', and gives the value for the dihedral angle in the energy-minimized molecule, which here is −180.000. The next three columns are entitled 'V1', 'V2', and 'V3', and give respectively the force constants for onefold, twofold, and threefold torsional barrier components of the torsional potential. Here, these values are, 0.000, 15.000, and 0.000. The last column, entitled 'Energy', gives the contribution from torsional angle strain potential, 0.000 here.

The remainder of the output file is identical to the summary statements and data seen in the short form output.

While the output files for more complex structure minimizations will contain other elements, notably in the case of π-calculations (see Section 6.5) and hydrogen bonding (see Section 6.6), the foregoing provides a general key to interpreting these files.

3.6. GRAPHIC OUTPUT AND DISPLAY TOOLS

The principal information about molecular structure can be perceived and understood more easily when presented in the chemist's most familiar language: by a picture of its structure by means of molecular graphics techniques, rather than by a list of numbers. The human brain is the best pattern recognizer, and a picture is better than a thousand words.

—Mario Marsili[113]

There are a number of tools which may be used to manipulate the graphic output and display the structures obtained in PCMODEL. These tools may be considered in four main groups in terms of their primary objectives—graphics capture and printing utilities (**Save Graphic**, **Print**, **Copy_BW_To_Clipboard**, and **Copy_CO_To_Clipboard**), the display of atoms (five of the options under **Labels**), the display of the molecular skeleton (**Stick Figure**, **Ball_Stick**, **Ortep**, **Tubular Bonds**, and **H/OO**), and the display of whole-molecule properties (two of the options under **Labels**, **Mono/Stereo**, **CPK_Surface**, **Dot_Surface**, and **Ribbon**). The graphics capture commands are found under the **Edit** command on the menu bar and are described below. The rest are found under the **View** command on the menu bar, and the features and use of each of these will be discussed and demonstrated in turn using the example of the molecule tryptophan. The user should follow along with these descriptions by activating the **AA** tool and importing the template structure for tryptophan, listed as **Trp**. Then, the procedure given in each case may be followed to produce the desired display.

Graphics Capture Utilities **File\Save Graphic** writes an image of the PCMODEL window to a graphics format file. Activating this command pops up an 'Open File' dialog box shown in Fig. 3-47.

This dialog box resembles the one used to read in input files (see Fig. 3-24), differing in the 'Select File Type' file format area. In this case, the options are 'Windows Metafile', 'HPGL' (Hewlett–Packard graphics language), and 'Postscript'. The user who wishes to save the graphics display to such files will understand the relative advantages of each format.

File\Print sends an image of the PCMODEL window to a suitable printer. Activating this command pops up a dialog box entitled 'Print', shown in Fig. 3-48.

This dialog box displays the identifier for the 'Current Printer:' in a box at the upper left, and provides two groups of options. The first group is entitled 'Print:' and allows the user to select which of the active components of the PCMODEL window to print. The choices include the 'Structure Window', the 'Energy Window', and the 'Compare Window'. The second group is entitled 'Print As:' and allows the user to choose whether to print in 'Black and White' or in 'Color/Gray Tones'. Gray tones will retain some of the distinguishing characteristics of colors, depending upon the graphics configuration and the

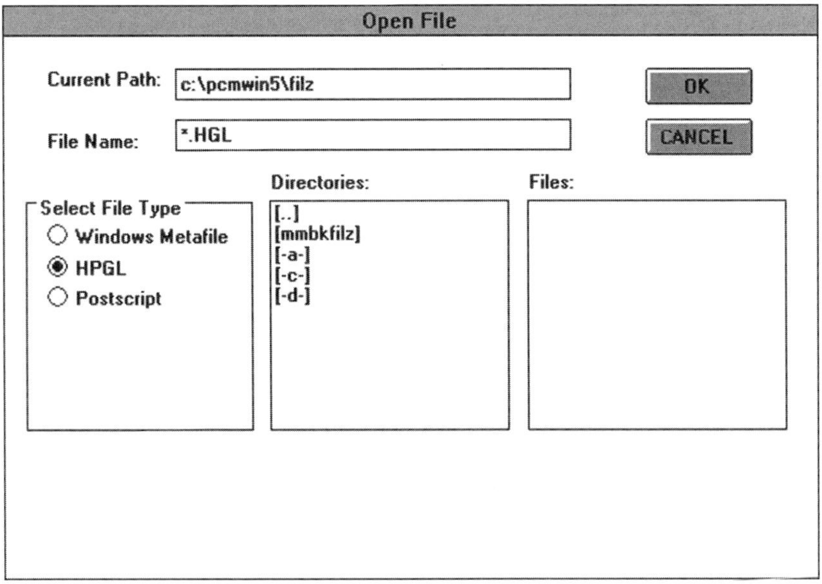

Figure 3-47. The 'Save Graphic' Dialog Box Menu.

printer drivers. Clicking on the **OK** button will then initiate the print job. If the user wishes to change printers, the **Edit\Printer Setup** command (see Appendix B) will allow them to do so without exiting PCMODEL.
   Edit\Copy_BW_To_Clipboard, i.e. copy black-and-white image to the clipboard, copies a rendering of the work space in black and white to the Windows clipboard. Activating this command causes the PCMODEL display momentarily to shift to a black-and-white viewing mode, with the background

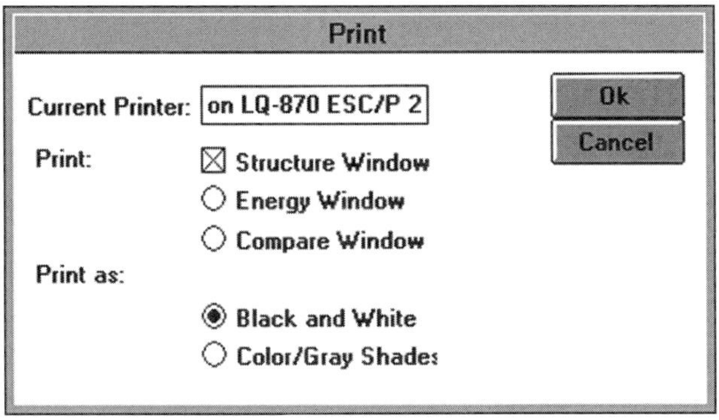

Figure 3-48. The 'Print' Dialog Box Menu.

inverted from the normal black to white, and then copies this graphic onto the Windows Clipboard. There are no gray tones to differentiate the colors of different atoms. The image is saved in a bitmap-type format which is memory intensive: ≈ 375–450 KB for a screen capture.

Edit\Copy_CO_To_Clipboard, i.e. copy color image to the clipboard, copies the work space in color to the Windows clipboard. Activating this command causes the PCMODEL display to repaint, and then copies this graphic onto the Windows clipboard. The image is saved in a bitmap-type format which is memory intensive: ≈ 375–453 KB for a screen capture.

The Display of Atoms The **View\Labels** command activates a pop-up menu entitled 'Labels' (shown in Fig. 3-49) which provides the user with seven different formats for identifying the atoms or substructures present in the work area. Of these, five are specific to characterizing the constituent atoms. The molecule tryptophan will be used as an example to demonstrate these 'Labels' formats.

The first format is 'Hydrogens and Lone Pairs'. This is the default setting, which gives the molecular skeleton as a stick figure with each atom and the closest half of all of its bonds coded by the characteristic color for that element (see Section 5.3). It shows atom labels for all atoms save carbon: 'H' for a hydrogen, ':' for a lone pair of electrons, and the corresponding element symbols for the noncarbon heavy atoms. Figure 3-50 shows tryptophan with the 'Hydrogens and Lone Pairs' display option. The tildes '∼' on the atoms of the indole ring indicate that they are π-atoms (see Section 6.5).

The second format is 'Bonds Only'. This gives the stick figure seen in the previous example, but with no atom labels. In this format the display shows only the colored bond segments of the skeleton of the molecule, like a Dreiding Stereomodel®. Besides its providing a less cluttered picture, operations in

Figure 3-49. The 'Labels' Options Box.

3.6. GRAPHIC OUTPUT AND DISPLAY TOOLS   99

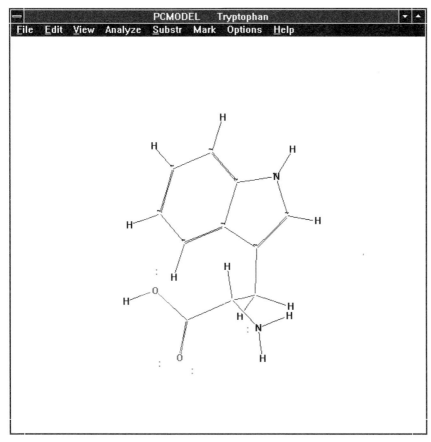

Figure 3-50. Tryptophan with the 'Hydrogens and Lone Pairs' Display Format.

which screens are rapidly repainted (such as rotations) are faster in the 'Bonds Only' display. Figure 3-51 shows tryptophan with the 'Bonds Only' display option.

The third format is 'Atom Numbers'. This gives the standard stick figure for the molecular skeleton, and for each atom label is given the atom file number indicating its sequential position in the corresponding input file (see Chapter 4 on input file formats). For heavy atoms this is generally the order in which they are drawn; for hydrogen atoms and lone pairs added through **H/AD**, the file numbers are assigned based on the program's priority for adding these components to the structure. This information can be helpful in several ways: reconciling the graphic with the atom connectivity tables in the input file, responding to queries or error messages which mention atoms by their atom file number, or preparing the input for the GMMX conformational search routine (see Section 5.8.2). Figure 3-52 shows tryptophan with the 'Atom Numbers' display option.

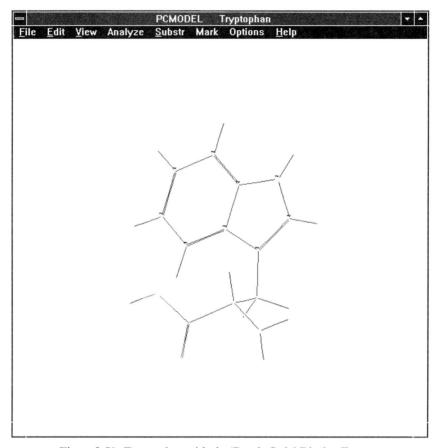

Figure 3-51. Tryptophan with the 'Bonds Only' Display Format.

The fourth format is 'Atom Types'. This gives the standard stick figure for the molecular skeleton, and for each atom label is given the atom type number, which is a function of the element, the hybridization, etc., within the MMX force field (see Appendix C and Section 5.3). This information is helpful when the user wishes to check or change the atom types in a molecule. Figure 3-53 shows tryptophan with the 'Atom Types' display option.

The fifth format is 'MM3 Atom Types'. This gives the standard stick figure for the molecular skeleton, and for each atom label is given the atom type number, which is a function of the element, the hybridization, etc., within the MM3 force field (see Section 4.5). This option is not fully implemented at this writing—a future version of PCMODEL will have an interface to MM3 force field calculations, and the developers plan to complete the implementation of this file conversion at that time. Currently, an input file converted to 'MM3 Atom Types' cannot be backconverted to MMX faithfully.

3.6. GRAPHIC OUTPUT AND DISPLAY TOOLS 101

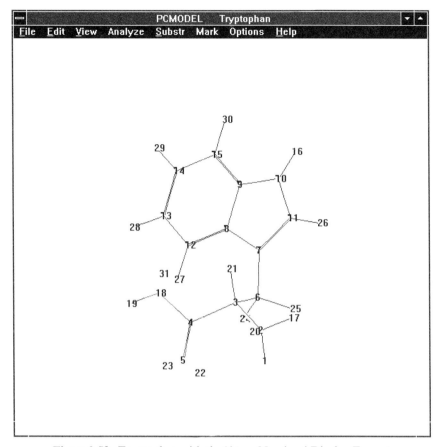

Figure 3-52. Tryptophan with the 'Atom Numbers' Display Format.

The Display of the Molecular Skeleton These display options give different ways to depict the skeleton of the molecule, depending upon what aspects the user wishes to convey. Each of the five choices will be discussed, and exemplified with the molecule tryptophan for comparison. The user should follow along with these descriptions by activating the **AA** tool and importing the template structure for tryptophan, listed as **Trp**. Then, the procedure given in each case may be followed to produce the desired display.

The **View\Mono/Stereo** command is a toggle between monoscopic and stereoscopic representations of the structure present in the work area. It may be used in conjunction with any other display option. The monoscopic representation is the default. The stereoscopic involves two depictions on screen which differ slightly from each other in perspective. This stereoscopic pair may be viewed with stereoscopic glasses, or by properly rotating and focusing the unaided eyes. For most individuals, rotating the eyes inward (cross-eyed) is

102 INPUT AND OUTPUT

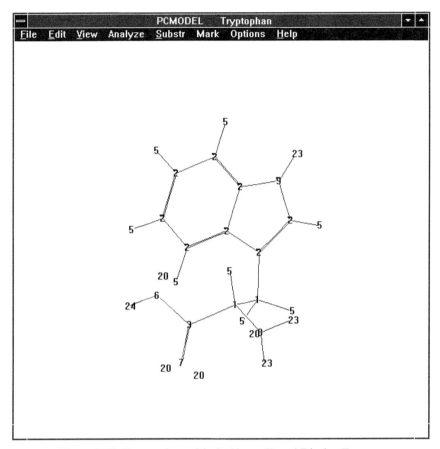

Figure 3-53. Tryptophan with the 'Atom Types' Display Format.

easier than rotating them outward (wall-eyed). Using stereo glasses is easier still for all but the most proficient, and is equivalent to rotating the eyes outward.

The **Options\Stereo** command may be used to invert the perspective of the stereoscopic pair. Activating this command pops up a dialog option box entitled 'Stereo Options' which allows the user to toggle between the default 'Rotate Out' and 'Rotate In' (see Fig. 3-54). The latter selection will give the proper perspective when the stereo pair is viewed with crossed eyes.

With the monoscopic depiction of the molecule tryptophan in Fig. 3-50 for comparison, the two stereoscopic perspective options are shown below. The first gives the default stereoscopic depiction with the perspective rotated out (Fig. 3-55). If this pair is viewed with crossed eyes, the indole ring appears to be in the foreground, with the carboxylic acid and amine group in the background.

3.6. GRAPHIC OUTPUT AND DISPLAY TOOLS   103

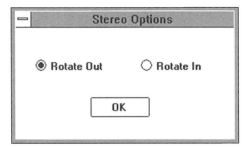

Figure 3-54. The 'Stereo Options' Dialog Option Box.

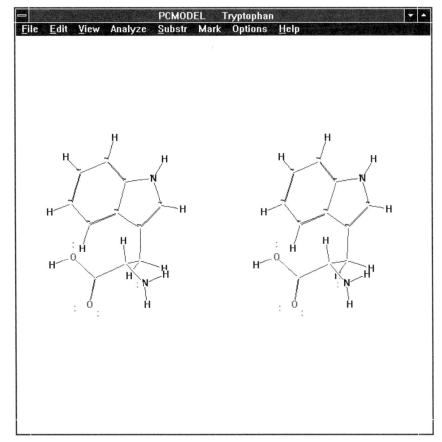

Figure 3-55. The Stereoscopic Pair of Images for the Molecule Tryptophan, with the Perspective Rotated Out.

The second gives the stereoscopic depiction with the perspective rotated in (Fig. 3-56). If this pair is viewed with crossed eyes, the indole ring now appears to be in the background, and the carboxylic acid and amine group in the foreground.

The **View\Stick Figure** command gives the default representation of the structure present in the work area, which was seen above for the example of the molecule tryptophan in Fig. 3-50. This command may be used to return to the default if another presentation option is currently active.

The **View\Ball_Stick** (i.e. ball-and-stick) command gives a **PLUTO**-style display option which renders the molecule as shaded balls for the atoms and thickened sticks for the bonds. Activating this command changes the representation of the structure(s) in the work space to one with colored atomic spheroids linked by white cylindrical bonds. This viewing option imitates the ball-and-stick physical models of the same name. The atomic spheroids are

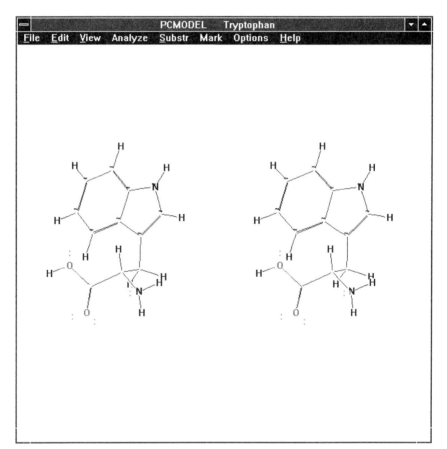

Figure 3-56. The Stereoscopic Pair of Images for the Molecule Tryptophan, with the Perspective Rotated In.

3.6. GRAPHIC OUTPUT AND DISPLAY TOOLS 105

given light and dark shading in the PLUTO style to enhance the perception of depth and perspective. Multiple bonds are not explicitly indicated. This representation of the molecular skeleton may be kept when a dot surface is drawn (see **Dot_Surface**), to aid in relating the surface to the skeleton. Figure 3-57 shows tryptophan in the **Ball_Stick** display format.

The features of this display are highly tunable with the **Options\Pluto** command. The parameters contained in the 'Pluto Options' dialog box (shown in Fig. 3-58) may be adjusted before or after the 'ball-and-stick' display has been drawn.

Nine options are presented. The first three are contained in a group labeled 'Shade Ang:' for shade angle. The box labeled 'Ang1:' contains a default value of 75.0000, and this field governs the rotation of the "illuminated" area about the Y-axis in a fashion that resembles the phases of the moon. At a value of 0 or 360, the entire sphere is illuminated area, while at values near 180 only the outline of the circle remains. The box labeled 'Ang2:' contains a default value

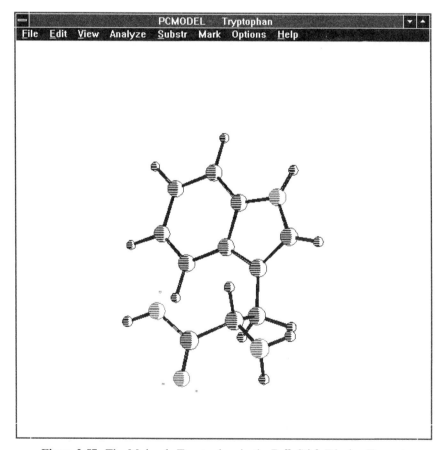

Figure 3-57. The Molecule Tryptophan in the **Ball_Stick** Display Format.

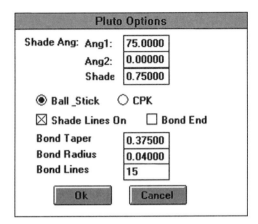

Figure 3-58. 'Pluto Options' Dialog Box.

of 0.00000, and this field governs the rotation of the "illuminated" area about the Z-axis. The box labeled 'Shade' contains a default value of 0.75000, and this field governs the spacing between the colored lines which make up the illuminated portion. In this case, a smaller value gives a denser coloring.

The next portion of the 'Pluto Options' dialog box offers the option of either 'Ball_Stick' (default) or 'CPK'. In the latter case, the spheroids drawn will occupy approximately 0.6 of the van der Waals radii of the atoms. The bonds will no longer be visible. Figure 3-59 shows tryptophan in the 'CPK' display format, with the other Pluto options set at their defaults.

Two choices are offered next: 'Shaded Lines On' (the default) and 'Bonds End'. If 'Shaded Lines On' is deselected, the illuminated hemisphere will disappear, and only the outline of the circle will remain. The 'Bonds End' option changes how the ends of bonds are drawn, but this is not apparent in most common desktop graphics configurations.

The last three options appear at the bottom of the 'Pluto Options' dialog box. 'Bond Taper' has a default value of 0.37500 and governs the degree of depth perspective cueing in the bonds by tapering them narrower as they recede in the distance. Higher values increase the degree of taper. 'Bond Radius' has a default value of 0.04000, and governs the width of the bond. A larger value will give a thicker bond. 'Bond Lines' has a default value of 15, and dictates the number of lines drawn to make up the bond. A larger value will give a denser representation of the bond (more lines).

The **View\Ortep** command gives a rendering of the molecule as a type of ball-and-stick presentation in which a one-eighth section (octant) is removed and the hole shaded to aid the eye in developing a three-dimensional perspective.

The name is an acronym for Oak Ridge Thermal Ellipsoid Plot Program.[64] Figure D-2 in Appendix D shows how the **ORTEP** presentation is assembled. The ORTEP display format finds its most frequent use in the presentation of a

3.6. GRAPHIC OUTPUT AND DISPLAY TOOLS   **107**

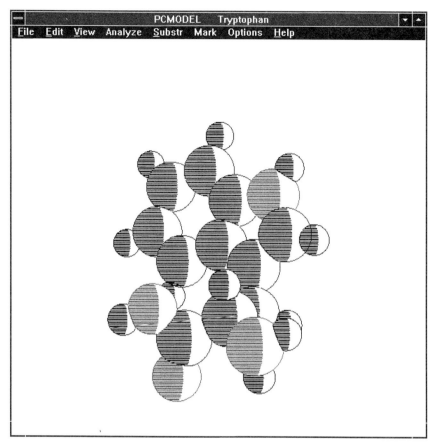

Figure 3-59. Tryptophan in the **Ball_Stick** Display Format in the PLUTO 'CPK' Format.

molecule derived from its diffraction crystal structure data. In that case the "balls" will be ellipsoids which encompass the area swept out by the constituent atoms in the course of their averaged thermal motions. Figure 3-60 shows tryptophan in the ORTEP display format.

The **View\Tubular Bonds** rendering is a type of stick figure presentation of a molecule in which the thickness of the bonds has been increased. This aids the eye in focusing on the bonds and the skeleton instead of primarily on the atoms. Figure 3-61 shows tryptophan in the **Tubular Bonds** display format.

The **H/OO** tool toggles on and off the display of nonacidic hydrogens. When the **H/OO** command is activated, PCMODEL redraws the structure in the work space in the alternative format. Unlike **H/AD**, which actually deletes or adds the hydrogens, **H/OO** simply hides the hydrogen atoms while preserving their scaling and positions. Lone pairs and functional hydrogens, i.e. hydrogens attached to heteroatoms such as nitrogen and oxygen, are not affected by

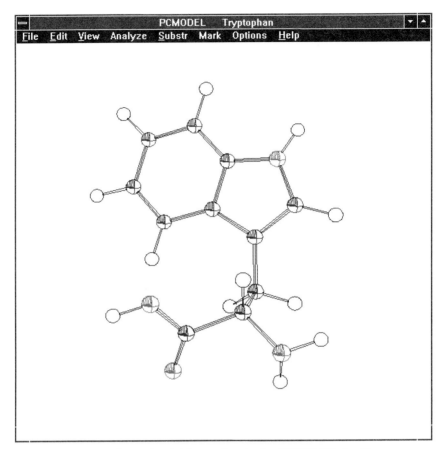

Figure 3-60. The Molecule Tryptophan in the ORTEP Display Format.

this command. Figure 3-62 shows tryptophan with the skeletal hydrogens toggled off using **H/OO**.

The Display of Whole-Molecule Properties These representations give different ways to consider the molecule as a whole, emphasizing such things as substructure membership, three-dimensional aspects, surfaces, and the backbone of biopolymers. Each of the six choices will be discussed, and exemplified for comparison with the molecule tryptophan, a combination of tryptophan and benzene, or an oligopeptide as appropriate. The user should follow along with these descriptions by activating the **AA** tool to import the template structure for tryptophan, listed as **Trp**. For the displays involving substructures, the benzene molecule may be imported by activating the **Rings** tool and selecting **Ph**. For the **Ribbon** display option, the alternating octapeptide $H_2N-(Gly-Ala)_4-CO_2H$ will used as an example. This structure was prepared in a right-handed α-helix with the 'Amino Acids' menu (see Section 3.2.2), and is available on the input file library diskette in the CHAP3\SECT2 sub-

3.6. GRAPHIC OUTPUT AND DISPLAY TOOLS 109

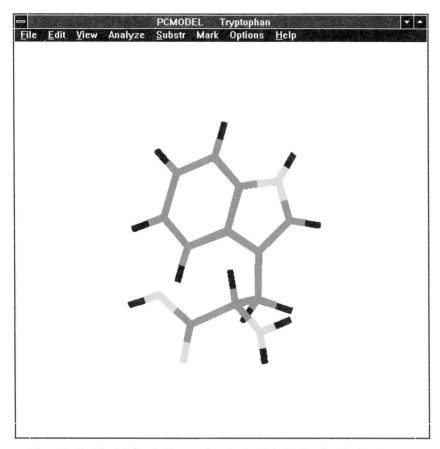

Figure 3-61. The Molecule Tryptophan in the **Tubular Bonds** Display Format.

directory as {OCTPEPTD.PCM}. The procedure given in each case may then be followed to produce the desired display.

As seen above, the **View\Labels** command activates a pop-up menu entitled 'Labels' which provides the user with seven different formats for identifying the atoms or substructures present in the work area (see Fig. 3-49). Of these, the last two are specific to characterizing whole molecule properties, and specifically to substructures (see Section 3.2.3). The molecules tryptophan and benzene will be used as an example to demonstrate these 'Labels' formats. Figure 3-63 depicts this pair of molecules in the default representation.

The sixth format under **View\Labels** is 'Substructure Numbers'. This display format only makes sense if there are at least two active structures on screen. The display is the standard stick figure for the molecular skeleton, and the each atom label in a particular structure is given the substructure number. The numbering system used gives a value of 1 to the first structure, and each additional structure takes a number which increases as powers of two: 2, 4, 8,

110 INPUT AND OUTPUT

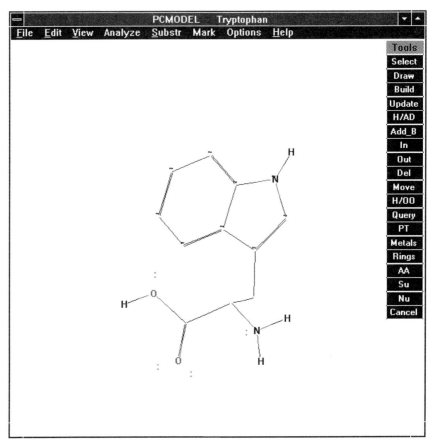

Figure 3-62. Tryptophan with Nonacidic Hydrogens Toggled Off (H/OO Tool).

16 ... The numbers are retained even if a structure with an intermediate number is deleted. The molecules tryptophan and benzene in this representation are shown in Fig. 3-64.

The seventh format under View\Labels is 'Color by Substructure'. This display format only makes sense if there is are least two active structures on screen. The display is the standard 'Stick Figure' with 'Bonds Only' for the molecular skeleton, and each structure is color-coded to indicate its substructure membership. The coloring system will depend upon the graphics configuration used, but in the 16-color mode has the first structure in dark blue, and additional structures in green, cyan, red, yellow, violet, gray, dark gray, light blue, light green, light cyan, light red, light violet, light yellow, white. After these fifteen variations, the color scheme repeats for additional structures. The assigned colors are retained even if a structure with an intermediate color is deleted. The molecules tryptophan and benzene in this representation are shown in Fig. 3-65.

3.6. GRAPHIC OUTPUT AND DISPLAY TOOLS   111

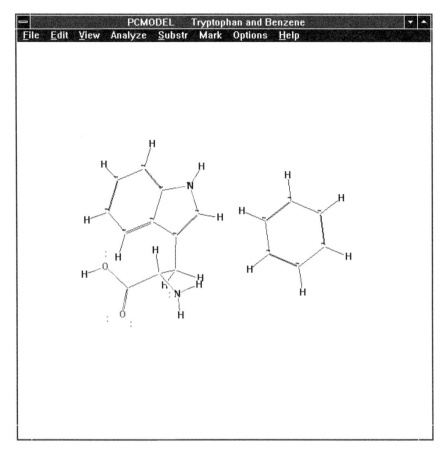

Figure 3-63. The Molecules Tryptophan and Benzene in the Default Display Format.

The next two display options are used to represent molecular surfaces, to gain an appreciation for how the molecule "appears" as a whole, and thereby how it will interact physically with other species in its environment. Molecular surfaces and volumes are discussed in more detail in Section 6.2.

The **View\CPK_Surface** command gives a rendering of a solid molecular surface as the sum of the overlapping spheres corresponding to the constituent atoms. The graphic representation produced by **CPK_Surface** approximates the physical CPK models (see Section 1.1). Activating this command brings up a dialog check and option box entitled 'CPK Models', which is shown in Fig. 6-37. Figure 3-66 shows tryptophan in the **CPK_Surface** display format using the default parameters.

The surface may be removed by reactivating the **CPK_Surface** command and then clicking on the 'Cancel' button in the dialog box, or by selecting another option within the **View** menu on the Menu Bar.

112 INPUT AND OUTPUT

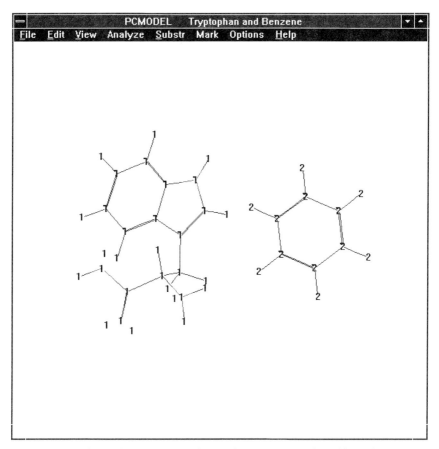

Figure 3-64. The Molecules Tryptophan and Benzene, Numbered by Substructure.

The **View\Dot_Surface** command draws a diaphanous surface around the structure in the work area to depict the molecular skeleton and a selected surface simultaneously. Activating this command will bring up a dialog options box entitled 'Dot Surface Display', which is shown in Fig. 6-40. Figure 3-67 shows tryptophan in the **Dot_Surface** display format using the default parameters.

The **View\Ribbon** command draws a spine ribbon consisting of four parallel lines representing the backbone of amino acid residues in a polypeptide or of nucleic acid residues in a polynucleotide. The complete polypeptide or polynucleotide must be on screen prior to invoking this command, which brings up a dialog options box entitled 'Ribbon Information' as shown in Fig. 3-68, to select the parameters for the display.

There are three types of options to choose from. In 'View:' the user chooses whether to view both the 'Ribbon and Structure' or the 'Ribbon Only'. In 'Type:' the user indicates whether an 'Acyclic Peptide', a 'Cyclic Peptide', or a

3.6. GRAPHIC OUTPUT AND DISPLAY TOOLS   113

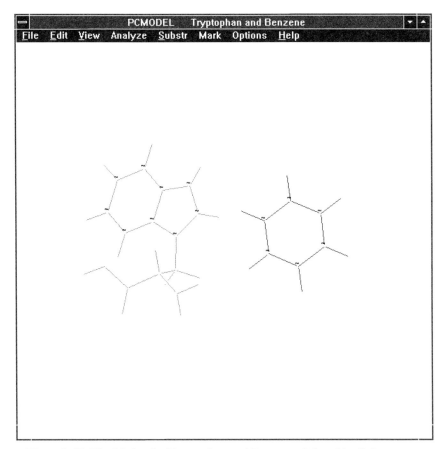

Figure 3-65. The Molecules Tryptophan and Benzene, Colored by Substructure.

'DNA/RNA' fragment is being viewed. In 'Orient:' a choice is presented of whether to 'Swap CO vectors', i.e. exchange the carbonyl vectors, but the implementation of this option is incomplete at this writing. The choices are verified by clicking on 'OK,' which brings up the desired display, a ribbon consisting of four lines colored red, cyan, green, and blue. The red line roughly follows the peptide carbonyl oxygens, and the blue line roughly follows the peptide nitrogens.

The alternating octapeptide $H_2N-(Gly-Ala)_4-CO_2H$ will used as an example. This structure was prepared in a right-handed α-helix with the 'Amino Acids' menu (see Section 3.2.2). It is available on the input file library diskette in the CHAP2\SECT2 subdirectory as {OCTPEPT.PCM}, and is presented in the PCMODEL default display (stick figure with atom labels) in Fig. 3-17. Note first of all the confusion of detail with this display. To help in orientation, the N-terminal end (Gly1) is at the lower left of the chain, and the C-terminal end (Ala8) is at the top. There are hydrogen bonds between the

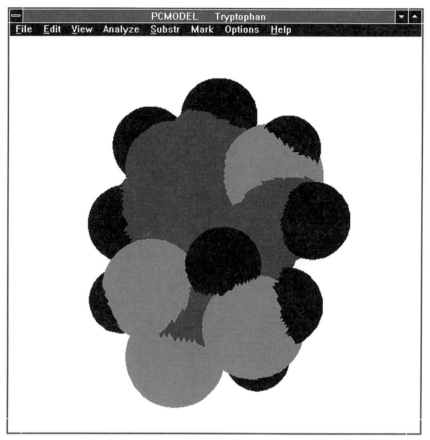

Figure 3-66. The Molecule Tryptophan in the **CPK_Surface** Display Format.

Gly1 C=O and the Gly5 NH, the Ala2 C=O and the Ala6 NH, the Gly3 C=O and the Gly7 NH, and the Ala2 C=O and the Ala6 NH.

Several steps will simplify this display prior to creating the ribbon display. First, the **H/OO** tool is used to hide all nonacidic hydrogens; this view was shown in Fig. 3-18. Then the **Tubular Bonds** display format (see above) is invoked. Finally, the **Ribbon** display is activated, using the default options. The result is shown in Fig. 3-69, which gives a much better sense of the α-helix than does Fig. 3-17 or 3-18. In the gray-tone representation here, the top line is the red one (along the carbonyl oxygens), and the bottom is blue (along the peptide nitrogens).

LIMITATIONS: An acyclic peptide must have a basic nitrogen or an ammonium at the N-terminus. If the N-terminus is acetylated, for example, it will not be found by the N-terminus search routine and no ribbon will be

3.6. GRAPHIC OUTPUT AND DISPLAY TOOLS   115

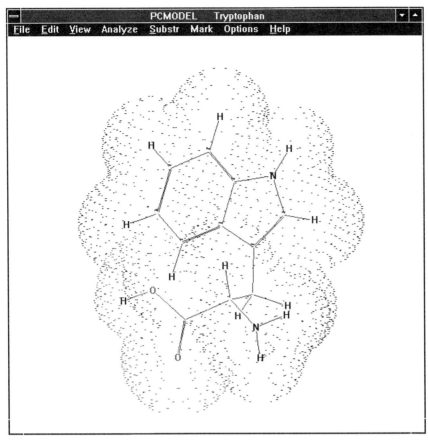

Figure 3-67. The Tryptophan Molecule in the **Dot_Surface** Display Format.

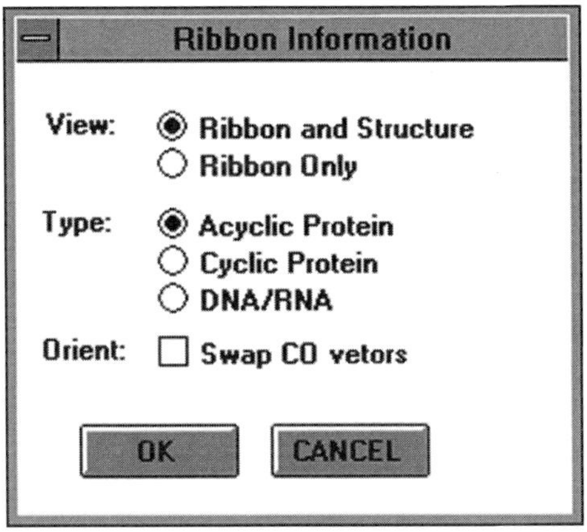

Figure 3-68. The 'Ribbon Information' Dialog Box.

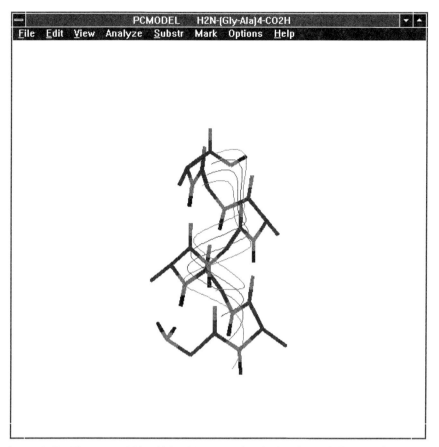

Figure 3-69. The H_2N–$(Gly$–$Ala)_4$–CO_2H Molecule in the **Tubular Bonds** Display Format Overlaid with the **Ribbon** Display Format.

displayed. As soon as the substituent is removed from the N-terminus, the ribbon will be displayed.

3.6.1. Rotations and Perspective

The position or perspective of the representation in the work space may be adjusted with the **View\Control_Panel** command, which can change the size of, translate, or rotate the structure(s). **Control_Panel** normally operates on all on screen substructures simultaneously. The operation can be limited to individual substructures by using the **Select** tool. If all of the heavy atoms in the substructure of interest are so marked, then the scroll bars will affect only that substructure. Substructures may be operated upon independently with the **Substr\Move** command, as was discussed in Section 3.2.6. Activating the **Control_Panel** command brings up the 'Dials' dialog box, which was shown in Fig. 3-29.

If the representation in the work space is too crowded or confused to carry out commands such as **Query**, it may be altered within the **Control_Panel**. The 'Scale' function may be used to expand or shrink the molecule(s) on screen, or the 'Translate' and/or 'Rotate' functions may be used to change their perspective, and allow easier viewing and access to the portion of the molecule under scrutiny. Clicking on 'Reset View' near the bottom returns the original representation, i.e., it cancels all rotations carried out during the current **Control_Panel** session.

The user should draw or read in some simple molecules and experiment with the **Control_Panel** utilities until they get a good feel for how the various scroll bars aid in the manipulation of structures. Proficiency in carrying out translations and rotations greatly speeds up modeling work.

There are three other commands used to position the structure, in this case to orient it along one of the three Cartesian planes. The commands are found under the **Edit** command on the menu bar, and are **Orient_XY_Plane**, **Orient_XZ_Plane**, and **Orient_YZ_Plane**. All of these commands work in a similar fashion, and so they will be demonstrated with the example of **Edit\Orient_XY_Plane**. This command enables the user to reorient a structure by relocating three marked atoms in the XY plane (the plane of the screen). The definitions of these axes are that the X-axis is horizontal in the plane of the screen, the Y-axis is vertical in the plane of the screen, and the Z-axis is orthogonal to the screen.

Prior to invoking this command one must use the **Select** tool to mark three atoms in the structure. If no structure is present in the work space, or if fewer than three atoms have been selected, a reminder windows pops up to prompt the user in the right way to proceed. The order in which the atoms are marked is significant, since the first atom will be placed at the Cartesian origin, the second will be placed along the positive X-axis, and the third will occupy the positive quadrant (upper right-hand) of the XY-plane. If other substructures are present, they will also suffer a reorientation. The resulting display will often be off center; carrying out an **Update**, or any other operation which repaints the screen, will reposition the center of mass at the origin. The **View-\Mono/Stereo** display may be used either before or after the reorientation, and in this case two images remain side by side, each having the first marked atom at its own stereoscopic Cartesian origin.

As an example, we will read in the tryptophan molecule from the 'Amino Acids' template menu, resulting in a screen that looks like Fig. 3-50. Activate the **Select** tool, and then use the mouse cursor to mark in turn the indole N1, C2, and C3. Next, activate the **Edit\Orient_XY_Plane** command. The molecules will be repositioned to look like Fig. 3-70.

3.6.2. Queries

Quantitative information on the geometry of a structure may be extracted from the graphical display with the **Query** tool. This tool operates without

118   INPUT AND OUTPUT

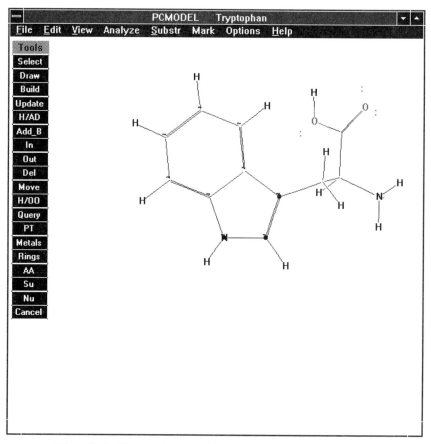

Figure 3-70. Tryptophan after the **Orient_XY_Plane** Operation.

premarking of the atoms concerned, and can be used on any structure, whether imported, drawn, or produced from any of the **Analyze** functions. After **Query** is activated, the mouse cursor is used to select between one and four atoms of interest, and then clicking on a blank space produces a report line for the appropriate quantities. The number of atoms selected determines what quantity is being measured. After clicking on the atoms they appear marked with a small dot. When the blank space is then clicked on, a dotted line (for one atom) or dotted caliper lines (for two to four atoms) appear which connect the indicated atoms to the space selected for the report message. At the same time, the appropriate status message prints on screen.

If one atom is selected, the report line gives the Cartesian coordinates in angstroms and charge in atomic charge units for that atom.

If two atoms are selected, the report line gives the bond length or through-space distance between those two atoms in angstroms.

If three atoms are selected, the report line gives the bond angle between those three atoms in degrees.

If four atoms are selected, the report line gives the dihedral angle (for four contiguous atoms) or the "flap" angle (for four noncontiguous atoms) in degrees. Vicinal hydrogens are a special case. If the four atoms correspond to an H—C—C—H chain (within constraints described in Section 6.4), then the report will also contain a prediction of the ^{1}H-NMR spin–spin coupling constants J_{HH} in hertz.

Query may be reused sequentially until another command or tool is activated. Any number of these reports may be displayed together clearly on a screen as long as care is taken not to overlap them. After this, each dotted line or calipers and its report line will persist through other operations until the user double-clicks on **Update**, which will clear the screen of all **Query** reports. Thus, a report line feature may be queried prior to an energy minimization sequence, and the effect of the minimization on that measure can be followed throughout the course of the minimization sequence. If the structure in question is crowded, the clarity of the representation may be improved by using the **View\Control Panel** command. If rotations and translations are made while **Query** reports are on the screen, the representation may get confused.

Query Display Exercise: Lactose. This exercise illustrates the reporting capabilities of the **Query** tool with the lactose molecule as the object of study. This molecule was prepared in Section 3.2.3.

From the PCMODEL window, read in the file {LACTOSE.PCM} (CHAP3\SECT2 subdirectory on the input file library diskette). If the perspective in not clear, carry out the necessary rotations with **View\Control Panel**. Activate the **Query** tool, click on one to four atoms, then click in an empty space to display the report line. Some sample reports are seen in the screen depicted in Fig. 3-71.

In this example, the anomeric oxygen of the galactose unit was selected, and the report line gives the Cartesian coordinates and the charge: 'xyz: -0.75 0.16 -0.73 Q:0.03.' The two atoms of an O—H group were selected, and the report line gives the bond length as '0.984 A'. The three atoms of the galactose C1, linking oxygen, and glucose C4 were selected, and the report line gives the bond angle as '108.0 DEG'. The four atoms of the galactose C4 hydrogen, C4, C5, and C5 hydrogen were selected, and the report line gives the coupling constant and torsional angle as '0.47 Hz (-63.0 DEG)'.

3.6.3. Structural Comparisons of Molecules

Structural comparisons can be made between molecules in several ways. The methods described in this section will concentrate on the molecular skeleton. The use of molecular surfaces for structural comparisons is described in Section 6.2.

120 INPUT AND OUTPUT

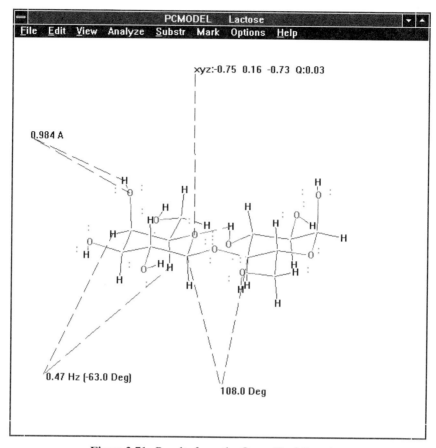

Figure 3-71. Results from the **Query** Tool Exercise.

The use of the **Query** tool was just demonstrated, and this method may be used to compare discrete geometric measures such as interatomic distances, bond lengths, bond angles or torsional angles. These could be carried out on each structure in turn and afterwards the results could be analyzed.

Overall comparisons between different molecules or between different conformations of the same molecule can be carried out with the **View\Compare** command. The first or "standard" structure for the comparison must be present in the workspace before beginning the sequence, and must not be partitioned into substructures (see Section 3.2.6). When the **View\Compare** command is activated, this first structure is rendered in dark blue in the 'Bonds Only' representation (see **Labels**), and a menu entitled 'Compare' pops up, as shown in Fig. 3-72.

The user clicks in one of the three radio buttons near the top of the box to indicate which atoms should be used in the comparison: 'All Atoms', 'Heavy Atoms' (i.e. nonhydrogen), or 'Selected' atoms. Clicking on the 'Next Struct'

3.6. GRAPHIC OUTPUT AND DISPLAY TOOLS 121

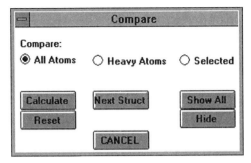

Figure 3-72. The 'Compare' Dialog Box.

button brings down a standard **File\Open** dialog box to read in the next structure for comparison. This structure will also be in the 'Bonds Only' representation, and will be colored green. This next and subsequent substructures will be colored by the standard scheme described in Section 3.2.6.

If either the 'All Atoms' or the 'Heavy Atoms' option is chosen, the program carries out atom-to-atom mapping based on the atom file numbers. Note that the structures are treated as substructures and the atom numbering will be consecutive. Thus, if the main structure has a total of n atoms, then the atom numbers for the first comparison structure will be $n + 1$, $n + 2$, etc. This means that the atom numbers listed in the results window will not be the same for equivalent atoms in the two structures, but the second set will be higher than the first set by an additive factor of n.

If the 'Selected' atoms option has been chosen, the user must use the **Select** tool to indicate a particular atom-to-atom mapping. After activating this tool, the mouse cursor is used sequentially to mark the atoms in the order in which they are to be compared. This may be done in two different ways: marking first an atom in one structure and then the corresponding atom in the second structure; or marking all of the desired atoms in the first structure, and then all of the corresponding atoms in the second structure, in the same order. For the proper comparison to be carried out, it is only necessary that the order of atoms in the lists for the two structures be correlated.

A least squares correlation of atomic positions is carried out by clicking on the 'Calculate' button. When this is finished the numerical results come up in a new window entitled 'Compare', and the two structures are redrawn superimposed upon one another with red and white hatched lines between the compared atoms, most visible when the interatomic distances are largest. The display of these red and white hatched lines can be canceled by clicking on the 'Reset' button.

Three columns are presented in the 'Compare' window, entitled 'At1' (i.e. atom numbers in structure 1), 'At2' (i.e. atom numbers in structure 2), and 'Dist' (least squares minimized interatomic distance in angstroms). Beneath these columns are given two statistical reports: 'Ave dev' (average distance between corresponding atoms) and 'RMS dev' (root mean square of these distances) in angstroms. At this point, other viewing commands may be employed, such as **Control_Panel**, **Query**, etc., to analyze the pair of structures further.

The user has the option of having all comparison structures additively superimposed (the default), or canceling one comparison structure before adding the next. This latter is accomplished by clicking on the 'Hide' button. As indicated by the name, the first comparison structure is still available, just hidden. When finished with the first comparison structure, another may be brought in by clicking again on the 'Next Struct' button and reading in the appropriate file. At any later point, all of the cumulated comparison structures can be superimposed upon the main structure by clicking on the 'Show All' button.

One exercise involving the **View\Compare** command was carried out in Section 3.2.6. Two more will be described below: the first involves the comparison of two different conformations of the same molecule, and the second involves *mapping* a smaller molecule onto a larger one.

Structure Comparison Exercise 1: Conformations of 1,2-Dichloroethane. This exercise demonstrates how the 'All Atoms' option can be used to compare different conformations of the same molecule, in this case the *anti* and *gauche* conformers of 1,2-dichloroethane. In the PCMODEL window, activate the **Draw** tool, and use it to draw a zigzag of four atoms in a line as a lopsided N-shape. Bring up the 'Periodic Table' menu and click on the chlorine **Cl**; then use this to convert the two end atoms to chlorines. Add hydrogens with the **H/AD** tool, and run the resulting molecule through several cycles of energy minimization. **View\Control_Panel** can be used to do some rotations and improve the perspective. This will be the *anti* conformer, as shown in Fig. 3-73. Name this molecule as 'anti_1,2-dichloroethane', and write an input file as {DICLETHN.PCM}.

Now mark the four heavy atoms in order with the **Select** tool, and activate the **Edit\Rotate_Bond** command. From the 'Rotate_Bond' dialog box, use the scroll bar to adjust the dihedral angle from $\approx 180°$ to $\approx 60°$. Cancel out of the dialog box, and run this new conformation through several cycles of energy minimization. Name this structure 'gauche_1,2-dichloroethane', and write it to the same input file created above, {DICLETHN.PCM}. In answer to the prompt that the 'File Exists', click on the 'Append' button. This will now store both conformations in the same input file.

Erase the current structure, and read in the file {DICLETHN.PCM}. From the 'Structure List' dialog window, select the *anti* conformer. Now, activate the **View\Compare** command, and ensure that the option is set to 'All Atoms'. Click on the 'Next Struct' button, and read in {DICLETHN.PCM} again, this time selecting the *gauche* conformer from the list. At this point, the stick drawing of the *anti* conformer will be in dark blue and that of the *gauche* conformer will be green. Click on the 'Calculate' button to carry out the least

3.6. GRAPHIC OUTPUT AND DISPLAY TOOLS   123

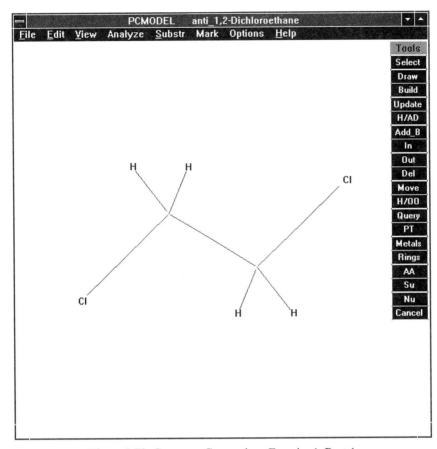

Figure 3-73. Structure Comparison Exercise 1, Part 1.

squares atom fit. It quickly becomes clear that these conformers are not superimposable. The work space shows the disparities between the atomic positions (Fig. 3-74), while the numerical results in the 'Compare' window (Fig. 3-75) emphasize this fact with an RMS deviation of 1.719 Å.

Structure Comparison Exercise 2: Mapping Thiophene onto Benzene. This exercise demonstrates how the 'Selected' atoms option can be used to compare one molecule with a portion of another, in this case thiophene onto benzene. In the PCMODEL window, bring up the 'Rings' menu, select the **Ph** button to import the benzene molecule, and cancel the 'Rings' menu. Run benzene through energy minimization; then save it as {BENZENE.PCM}. Bring up the 'Rings' menu again, and this time import cyclopentane with **C5**. Remove the hydrogens, bring up the 'Periodic Table' menu and click on the sulfur **S**, and then use this to convert one of the ring atoms to sulfur. Use the **Add_B**

124 INPUT AND OUTPUT

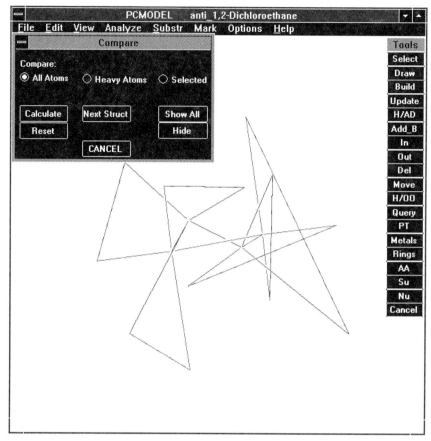

Figure 3-74. Structure Comparison Exercise 1, Part 2.

tool to make the two double bonds of thiophene; then add back the hydrogens and lone pairs with **H/AD**. Use the **Mark\Piatoms** command to flag 'All' ring carbon atoms for π-calculations. Run this through energy minimization a few times; then name the molecule and write it to a file, {THIOFENE.PCM}. The result should look like Fig. 3-76.

Now, activate the **View\Compare** command, choose the 'Selected' option, and then click on 'Next Struct' and read in the previously prepared {BENZENE.PCM}. Activate the **Select** tool and use it to map the carbons of the thiophene onto four contiguous atoms of the benzene ring. When this is finished, click on the 'Calculate' button. These molecules map closely onto each other, as can be seen in Fig. 3-77, and the RMS deviation is also quite small, as seen in the 'Compare' data output window in Fig. 3-78. This tells us that thiophene is a reasonably good isostere for benzene.

3.6. GRAPHIC OUTPUT AND DISPLAY TOOLS 125

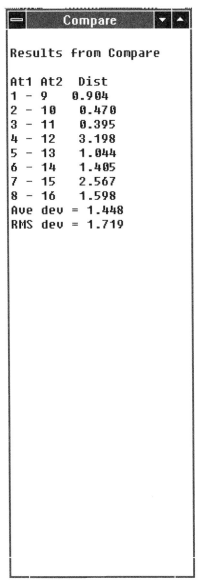

Figure 3-75. Structure Comparison Exercise 1, 'Compare' Window.

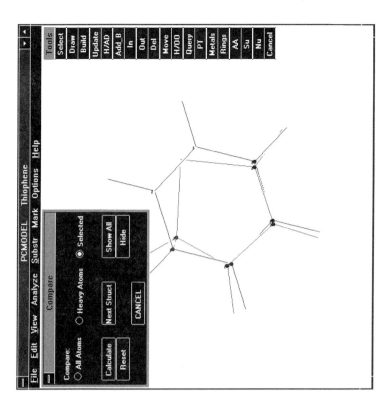

Figure 3-77. Structure Comparison Exercise 2, Part 2.

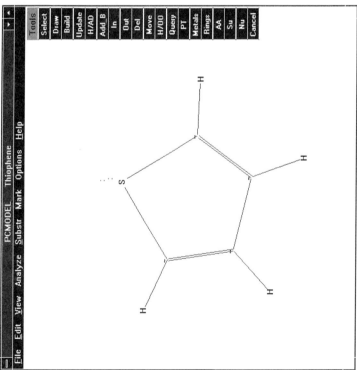

Figure 3-76. Structure Comparison Exercise 2, Part 1.

3.6. GRAPHIC OUTPUT AND DISPLAY TOOLS

Figure 3-78. Structure Comparison Exercise 2, 'Compare' Window.

CHAPTER 4
FILE FORMATS

In this chapter, we will discuss the file formats which can be used for molecular structure files to be imported into or exported from PCMODEL. The four file formats deemed most important are treated in the greatest detail: the PCMODEL formats PCM, MMX, the MDL MOL format, and a simplified X-ray format. Following this, briefer descriptions are given of the MOPAC format, the Tripos Alchemy MOL format, and the new Alchemy 2000 AL2 format, the Tripos SYBYL MOL/SYB format, the MACROMODEL MAC format, the CambridgeSoft CHEM-3D C3D format, and the Brookhaven PDB format.

Having so many different molecule file formats can be a serious hindrance to the free flow of information. Guidelines for a general molecule file format have been proposed,[124] but these are seldom actually followed. Following the discussion of formats, the subject of interconversion between formats is addressed. Provided here are brief descriptions of the ExoGraphics CONSYSTANT/CHEMELEON, BABEL, and MOLMOL conversion programs, and the World Wide Web viewers RASMOL and CHIME, along with tables of the file formats handled by each.

Following this comes a section on modifying input files, with two exercises. The chapter concludes with a discussion of using added constants and the form which MMX parameter files take, along with an illustrative exercise.

Molecular structure input files generally consist of four parts, although these features are not presented in the same way nor in the same order in all formats. In many cases, information of different types may be commingled. It is helpful to make these distinctions between the four types of information conveyed, to gain more insight into input files and how the different formats are interrelated.

> *Header/Footer Information.* This includes all or some of "start read" and "stop read" cues, an indication of the format which the file follows, the name of the program of origin, its version number, the users name or initials, the title of the structural fragment, molecule or ensemble, comments, the date, etc.
>
> *Coordinates.* This section gives the relative positions of the constituent atoms in three-dimensional space. This generally consists of X-, Y-, and Z-axis Cartesian coordinates, most often in angstroms, although rotational coordinate systems are also possible.

Connection Tables. This section, often in tabular form, contains information which establishes the connectivity of the structure, i.e. which atoms are connected or attached to which other atoms.

Property Values and Flags. This section gives numerical values or flags for some properties of the structure or of its constituent atoms or fragments. Examples of numerical values are the charges on atoms, their covalent or van der Waals radius, the multiplicity of bonds, and any added or modified parameters. Examples of flags are asymmetry or chirality, hydrogen bond donor or acceptor character, coordination bond donor or acceptor character, atoms involved in extended conjugation, substructure membership, etc.

PCMODEL for Windows currently reads input files in 12 formats, and writes input files in seven (v. 5.13, or ten in v. 6.0) formats (see Section 3.2, and **Open** and **Save** in Appendix B). We will examine four of the most important formats in detail with the example of the molecule L-4-hydroxyproline: the PCM format, the MMX format, the MDL MOL format and the X-ray format. Following this, the remaining formats will be covered briefly, along with a discussion of file format interconversion. Then, modifications of the input file will be covered, and finally the use of added constants will be discussed.

4.1. PCM FILE FORMAT

The PCM structure input files have a free format devised by the Serena Software developers specifically for PCMODEL. It does not have the requirement for Fortran-style position-specific data fields which the MMX format has, but rather the file-reading **parser** recognizes data by the sequence in which it appears, or by data prefixes which identify the data field. These data labels and the data values are separated from each other by one or more spaces. A number of the data fields in the MMX format have become vestigial (see the following section). These are not retained in the PCM format file, and nearly all of the defaults are assumed in the absence of the associated flag or data field. This format is also simpler and easier to modify than the others. To be recognized as an PCM-format input file when the PCMODEL **File\Open** command is invoked, the default filename extension is {*.PCM}. A PCM format input file in the appended format with an {*.SST} extension will contain a number of named substructures, and when the file is opened the desired structure may be selected from a 'Structure List' menu (see Section 3.2.6).

The following guide addresses the input file for L-4-hydroxyproline, {HPROLINE.PCM}, which can be found in the input file library in the CHAP4 directory. The structure of this molecule with standard atom labels is seen in Fig. 4-1, and the same structure with the atom file numbers indicated is shown in Fig. 4-2, to aid in identifying the atom file numbers with particular

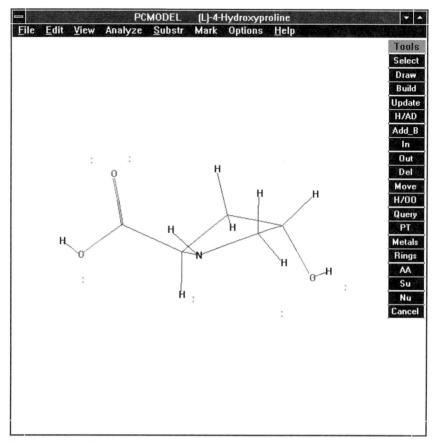

Figure 4-1. L-4-Hydroxyproline in PCMODEL PCM Format, from {HPROLINE.PCM}.

atoms in the structure. The PCM-format input file for this molecule is seen in Fig. 4-3.

The entire input file is contained within braces { }, with the left-hand brace starting on the first line, and the right-hand brace as the sole character on the final line of the file. These are signals, or **tokens**, prompting the parser to start and stop reading the input file parameters. We will proceed through the file line by line, to explain how the necessary information is presented. Only those data fields utilized by the current example will be stressed—other aspects of the PCM input file format will be discussed in other sections as the occasion arises. A more complete discussion of this file format may be found in the documentation which accompanies the PCMODEL program.[303]

Line 1 contains a left-hand brace followed immediately by the designation PCM. This cues the PCMODEL parser that the file conforms to the PCM format. After an empty space comes the title of the file, up to a maximum of 58 characters.

4.1 PCM FILE FORMAT   131

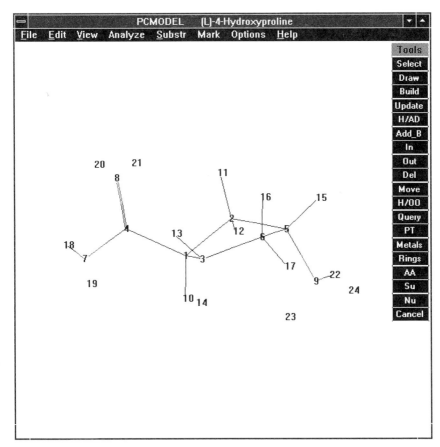

Figure 4-2. L-4-Hydroxyproline in the PCMODEL PCM Format from {HPROLINE.PCM}, with Atom File Number Labels.

Line 2 begins with 'NA', followed by a number which gives the number of atoms in the subject molecule. In the example, there are 24 atoms (9 heavy atoms, 9 hydrogens, and 6 lone pairs).

Line 3 begins with 'FL', which denotes flags for some of the options. 'EINT4' refers to the model used for electrostatic interactions, and the possibilities are given in Table 4-1. In the example, the default option 4 in used, for the charge–charge model with charges calculated from dipoles and added π-atom charges. 'UV1' refers to the protocol for π-calculations, and the possibilities are given in Table 4-2. No π-calculations will be carried out in the example, but the default option 1 in invoked, for the singlet RHF.

In most of the remaining lines 4–27, each begins with 'AT' to signify that details will be given about each atom in the input file. After the 'AT', and a

```
 1  {PCM (L)-4-Hydroxyproline
 2  NA 24
 3  FL EINT4 UV1
 4  AT  1  1   -0.512   -0.218    0.432 B 2 1 3 1 4 1 10 1 S 0
 5  AT  2  1    0.866    0.634    0.579 B 1 1 5 1 11 1 12 1 S 0
 6  AT  3  8   -0.727   -0.634   -1.050 B 1 1 6 1 13 1 14 1 S 0
 7  AT  4  3   -1.749    0.653    0.938 B 8 2 1 1 7 1 S 0
 8  AT  5  1    1.770    0.021   -0.627 B 2 1 9 1 6 1 15 1 S 0
 9  AT  6  1    0.638   -0.349   -1.735 B 3 1 5 1 16 1 17 1 S 0
10  AT  7  6   -2.519    0.210    2.000 B 4 1 18 1 19 1 S 0
11  AT  8  7   -2.032    1.770    0.359 B 4 2 20 1 21 1 S 0
12  AT  9  6    2.525   -1.209   -0.179 B 5 1 22 1 23 1 24 1 S 0
13  AT 10  5   -0.460   -1.147    1.019 B 1 1 S 0
14  AT 11  5    0.704    1.728    0.476 B 2 1 S 0
15  AT 12  5    1.341    0.507    1.575 B 2 1 S 0
16  AT 13 23   -1.554   -0.066   -1.523 B 3 1 S 0 H
17  AT 14 20   -1.021   -1.701   -1.141 B 3 1 S 0
18  AT 15  5    2.596    0.656   -0.978 B 5 1 S 0
19  AT 16  5    0.512    0.471   -2.472 B 6 1 S 0
20  AT 17  5    0.936   -1.233   -2.337 B 6 1 S 0
21  AT 18 24   -2.866    0.632    2.249 B 7 1 S 0 H
22  AT 19 20   -2.449   -0.371    2.133 B 7 1 S 0
23  AT 20 20   -2.421    2.196    0.591 B 8 1 S 0
24  AT 21 20   -1.725    2.007   -0.112 B 8 1 S 0
25  AT 22 21    3.290   -0.931    0.576 B 9 1 S 0 H
26  AT 23 20    1.824   -1.942    0.273 B 9 1 S 0
27  AT 24 20    3.030   -1.682   -1.046 B 9 1 S 0
28  }
```

Figure 4-3. The PCM Format Input File for L-4-Hydroxyproline, {HPROLINE.PCM}.

TABLE 4-1. Flags for Electrostatic Models in the PCM Format Input File EINT Variable

EINT	Electrostatic Model Used
0	Dipole–dipole model, no interactions for geminal dipoles
1	Dipole–dipole model, all interactions
2	Dipole–dipole interactions suppressed
3	Charge–charge model, no interactions for charges on geminal atoms; charges must be read in
4	Charge–charge model, charges calculated from dipoles plus π-atom charges

TABLE 4-2. Flags for π-Calculation Protocols in the PCM Format Input File UV Variable

UV	Electrostatic Model Used
0	Prompt user for π-calculation parameters
1	Singlet RHF π-calculation
2	Doublet UHF π-calculation (for monoradicals)

space, comes the atom file number, having values between 1 and 24 in this example and corresponding to the atoms as labeled in Fig. 4-2. Another space, and then comes the atom type designator (for descriptions of the MMX atom types see Section 5.3 and Appendix C). Next, after three or four spaces, come the Cartesian X-, Y-, and Z-coordinates of the atom, with three or four spaces in between each value. The Cartesian coordinates in the PCM and MMX input file formats are in angstroms. After another space comes the descriptor 'B' which alerts the parser that bond connection information will be read in. Next, pairs of numbers appear, separated by spaces, which give first the atom file number of an attached atom (1–24 in this example), followed by the multiplicity of the bond (1–3). After this, and separated by another space, comes an 'S' descriptor alerting the parser that substructure membership information will be read in. A space comes after the 'S', followed by an integer indicating the substructure group. Since there are no substructures in the example, these indicators all read 'S 0'. Finally, some of the atoms have an 'H' descriptor at the end of the line. This designates that atom as a hydrogen-bonding hydrogen; in this example the atom file numbers 13, 18, and 22 are so designated.

There are several other descriptors which are not found in the example, but may appear in the atom definition lines of other input files. A 'C' signals that a value for the charge on that atom will be read next. An 'M' signals that a flag will be read describing the electronic state of a metal atom (see Section 6.7). A 'P' signals that the atom is to be included in π-calculations.

Three specific lines will be interpreted as examples:

Line 6   describes the atom with file number 3, which is an @8 sp^3 nitrogen. Its Cartesian coordinates are $X = -0.727$, $Y = -0.634$, $Z = -1.050$. It is bonded to atom 1 (single bond); atom 6 (single bond); atom 13 (single bond); and atom 14 (single bond). It is not a member of any substructure.

Line 11   describes the atom with file number 8, which is an @7 sp^2 oxygen. Its Cartesian coordinates are $X = -2.032$, $Y = 1.770$, $Z = 0.359$. It is bonded to atom 4 (double bond); atom 20 (single bond); and atom 21 (single bond). It is not a member of any substructure.

Line 16   describes the atom with file number 13, which is an @23 amine hydrogen. Its Cartesian coordinates are $X = -1.554$, $Y = -0.066$, $Z = -1.523$. It is bonded to atom 3 (single bond). It is not a member of any substructure, but it will act as a hydrogen bond donor.

Line 28   the final line of the file, contains only the right-hand brace.

4.2. MMX FILE FORMAT

The MMX structure input file is a modification of the MM2 file format devised by Allinger and coworkers.[272] The MM2 file format was developed when card-fed mainframe computers were the only platforms available. Many descriptions of the file format for MM2 and related programs such as MMX still refer to individual "cards", which are represented by lines of code in the input files used by DOS- and Windows-based systems. The following description refers to lines or the equivalents of lines, since groups of "sub lines" may be nested into the file for some types of input.

MMX is the name of the force field used by PCMODEL, and it is also the name of the DOS-based nongraphic molecular mechanics program available through Serena Software. Most users will find the graphics-based PCMODEL program far easier to use, but occasions may arise when it is desirable to use the MMX file format. The MMX format is one of the options in the **File\Save** command in PCMODEL, and a more thorough explanation of the format can be found in the documentation which accompanies PCMODEL.[303] To be recognized as an MMX format input file when the PCMODEL **File\Open** command is invoked, the default filename extension is {*.MMX}.

The following guide addresses the input file for L-4-hydroxyproline, {HPROLINE.MMX}, which can be found in the input file library in the CHAP4 directory. To aid in identifying the atom file numbers with particular atoms in the structure, refer to Fig. 4-1 and 4-2 in the previous section, where this structure was shown with standard atom labels and with atom file number labels, respectively. The MMX-format input file itself is seen in Fig. 4-4. MMX input files follow Fortran format, and so part of the meaning of the data in the file is conveyed by its position in the line of code; thus it is critical for data to appear in the appropriate column(s) in order for it to be interpreted correctly. The data fields, or locations where a particular variable is found, will be described, and the name of the MMX variable will be given in capital letters. In some cases, the data field is one column unit or space; these fields are most often "flags" or toggles, that is, they signal an on–off condition. In the fields which are two or more columns in width, numerical data is always right-justified. In the explanation given below, the "address" of the data is referred to in terms of row (line) and column, and with this key the reader may follow at the intended point of the input file in the figure.

{HPROLINE.MMX} has 22 lines. There are additional lines and fields for other options in the MMX format which are not shown in this example, to treat features such as metal atom coordination, added constants, dihedral driver parameters, symmetry sets, fixed atoms, etc. These are explained in detail in the documentation to PCMODEL[303] and in the QCPE 395 documentation for the MM2 program.[272]

Line 1: Columns 1–60 are the ID field. The first three spaces are filled with the specification 'MMX', which identifies the format of the file so that PCMODEL can read its data properly. This is followed by an empty space, and then comes the title of the file, in this case 'L-4-Hydroxyproline'. The title occupies columns 5–60, and thus can be a maximum of 56 characters.

Columns 61–65 contain the N field, which gives the total number of atoms including lone pairs. In this case it is 24 (9 heavy atoms, 9 hydrogens, and 6 lone pairs).

```
1   5    10   15   20   25   30   35   40   45   50   55   60   65   70   75   80
....:....|....:....|....:....|....:....|....:....|....:....|....:....|....:....|
MMX (L)-4-Hydroxyproline                                       24 0  0 0 00  10.
FFFFFFFFFFFFFFFFFFFFFFFFFFFFFFFFFFFFFFFFFFFFFFFFFFFFFFFFFFFF 0          101
    4 0     1.5    0    1   15    0    0    0    0    4    0    1         1    0
    1    2    5    9    0    0    0    0    0    0    0    0    0    0    0    0
    1    3    6    5    0    0    0    0    0    0    0    0    0    0    0    0
    1    4    8    0    0    0    0    0    0    0    0    0    0    0    0    0
    4    7    0    0    0    0    0    0    0    0    0    0    0    0    0    0
    1   10    2   11    2   12    3   13    3   14    5   15    6   16    6   17
    7   18    7   19    8   20    8   21    9   22    9   23    9   24
  -0.55572  -0.26377   0.38604    1    0   0.81084   0.40863   0.59982    1    0
  -0.62567  -0.64549  -1.01736    8    0  -1.71314   0.63452   0.77224    3  801
   1.65394  -0.15799  -0.54875    1    0   0.60482  -0.23222  -1.66277    1    0
  -1.88680   0.67391   2.09909    6    0  -2.41491   1.24929   0.00372    7    0
   2.10860  -1.45713  -0.22764    6    0  -0.61936  -1.20208   0.99003    5    0
   0.72055   1.51432   0.47888    5    0   1.24967   0.19902   1.60152    5    0
  -1.42547  -0.21973  -1.47834   23    0  -0.68701  -1.24124  -1.06120   20    0
   2.53772   0.47044  -0.80752    5    0   0.44968   0.78185  -2.10215    5    0
   0.88301  -0.93429  -2.48102    5    0  -2.61163   1.21548   2.45242   24    0
  -1.52557   0.36830   2.47223   20    0  -2.86204   1.59069   0.21290   20    0
  -2.31722   1.21394  -0.58782   20    0   2.79548  -1.37043   0.41218   21    0
   1.68720  -1.72746   0.10215   20    0   2.42374  -1.66080  -0.69650   20    0
FFFFFFFFFFFFTFFFFTFFFTFFFFFFFFFFFFFFFFFFFFFFFFFFFFFFFFFFFFFFFFFFFFFFFFFFFFFFFFFF
```

Figure 4-4. The MMX-Format Input File for L-4-Hydroxyproline, {HPROLINE.MMX}.

Column 67 is the IPRINT field, which controls the amount of printout that will be generated at different stages of the minimization process, as described in Table 4-3 (see also Section 3.5). In the example the value is 0, which is the default. PCMODEL will not write any different value for this variable, which may be altered by modifying the input file (see Section 4.6).

Column 70 is the NRSTR field, which is a flag for restricted atoms (see Section 6.1). A 0 means no restricted atoms, and a 1 means restricted-atom data will be read in later. The default value of 0 is given here.

Column 72 is the INIT field, which is a flag to indicate whether energy minimization will be carried out (value 0), or whether the initial energy only will be calculated (value 1). The value 0 is given here.

Column 74 is the IDOCK field, which is a flag for docking interactions (see Section 6.8). The value of 0 in the example means no docking interactions.

Column 75 is the NCONST field, which is a flag for added constants (see Section 4.7). A value of 0 means that no constants will be read in later; a value of 1 means that constants will be read in later. This option may be invoked in PCMODEL with the **Options\Added Constants** command, which prompts the user for the name of the file containing the added constants. If flagged, the PCMODEL program will look the constants file {CONST.STO}. If this is not found, the user will be prompted for a new filename for the constants file. The value 0 is given in the example.

Columns 76–80 are the TMAX field, which indicates the maximum number of minutes that the job should be allowed to run. This flag is an obsolete vestige from the MM2 format; its value is ignored in PCMODEL, but is set at '10' to forestall the interference of a bug.

TABLE 4-3. Levels of Printout Signaled in the MMX Format Input File IPRINT Variable

IPRINT	Print out Level		
	Initial Stage	Minimize (Part 1)	Final (Part 3)
0	Simple	Minimum	Full
1	Simple	Minimum	Simple
2	Full	Full	Full
3	Full	Simple	Simple
4		Minimum	Minimum
5		No print	No print

For the stages of the minimization process, "Initial" refers to the initial calculation of energy, prior to any energy minimization; "Minimize" refers to the calculations which take place during the energy minimization; "Final" refers to the final results after energy minimization is finished. "Simple" omits van der Waals interactions which are less than 0.1 kcal/mol. "Miminum" gives only the summary results, not all of the bond stretch/compress, bond angle bend, and torsional angle strain values. "Full" gives all of these values. Option 5 ("No print") is for operations, such as the dihedral driver, which generate their own printout.

Line 2: Columns 1–60 constitute the LOGARY field, and indicate which atoms are to be included in π-calculations. A 'T' in a column means that the atom having the same number as the column is a π-atom, and an 'F' means that it is not. If there are more than 60 atoms in the molecule, additional lines will be added after line 2, with more T's and F's to document all the atoms of the molecule. All atoms are given 'F' in the example.

Column 62 is the NPLANE field, which is a flag for the planarity of the π-system. A value of 0 means the π-system should be maintained planar, and a value of 1 means that the π-system may become nonplanar. Our example has a value of 1.

Column 66 is the JPRINT field, which controls the amount of printout that will be generated for the **VESCF** π-calculation process, as described in Table 4-4 (see also Section 6.5). This field is left blank in our example.

Columns 69–70 are the LIMIT field, which gives the value that defines the limit for self-consistency in the π-calculation. For a value x given in this field, self-consistency will be achieved when successive calculations differ by 10^{-x} eV (electron volts) or less. The default is 5, and this field is left blank in our example.

Columns 71–72 are the ITER field, which gives the value for the maximum number of iterations of the **VESCF** π-calculation which will be tried in one pass to achieve self-consistency. The default is 8, but this flag is ignored in PCMODEL with open shell. This field is left blank in our example.

Column 73 is the IHBD field, which contains a flag for hydrogen bonding (see Section 6.6). A value of 0 means no array will be read in to indicate which hydrogen atoms will participate in hydrogen bonding, and a value of 1 instructs PCMODEL to read in such an array, immediately following the Cartesian coordinates. This array will have the same form as the LOGARY field (see line 22 below). This field has a value of 1 in our example.

Column 74 is the ICOV field, which contains a flag for metal atoms coordinated to ligands (see Section 6.7). A value of 0 means that no array will be read

TABLE 4-4. **Levels of Printout Signaled in the MMX Format Input File JPRINT Variable**

	Printout Level	
JPRINT	Initial VESCF Calculation	During Energy Minimization
0	π-Energy only	π-Energy only
1	π-Energy and final matrices	π-Energy only
2	π-Energy and matrices for each SCF iteration	π-Energy and final matrices
3	π-Energy and matrices for each SCF iteration	π-Energy and matrices for each SCF iteration

For the stages of the minimization process, "Initial VESCF Calculation" refers to the initial calculation of π-energy, prior to any energy minimization; "During Energy Minimization" refers to reporting on the π-energy during the standard force field energy minimization.

in to indicate which metal atoms will participate in coordination bonding, and a value of 1 instructs PCMODEL to read additional pertinent data fields following the hydrogen bonding array (if any) or after the Cartesian coordinates. These data fields, for each metal atom involved in coordination, will be contained in a line which gives the giving the metal atom label and radius, followed by a logical array (in the LOGARY-type TF format) of those atoms which are coordinated to that metal. This field has a value of 0 in our example.

Column 75 is the IUV field, the value for which determines the type of π-calculation to be carried out (see Section 6.5). A value of 0 directs the program to prompt the user for input on the options desired; a value of 1 results in a singlet **RHF** π-calculation; a value of 2 gives a doublet **UHF** π-calculation, e.g. for a monoradical. These options may be set in the PCMODEL dialog box menu 'MMX Pi Calc Options', accessed with the **Options\MMX_Pi_Calc** command. This field has a value of 1 in our example.

Column 76 is the IHUCK field, which is a flag that determines whether the first π-calculation carried out prior to force field energy minimization is the full **VESCF** treatment (for a value of 0), or a simple Hückel treatment (for a value of 1; see Section 6.5). This option may be set in the PCMODEL pop-up menu 'MMX Pi Calc Options', accessed with the **Options\MMX_Pi_Calc** command. This field does not have an assigned value in our example.

Column 78 is the NSETAT field, the value for which gives the number of sets of equivalent π-atoms (see Section 6.5). If NSETAT is nonzero, additional data must be read later. If IUV is 1 or 2, this field is ignored. This field does not have an assigned value in our example.

Column 80 is the NSETBD field, the value for which gives the number of sets of equivalent π-bonds (see Section 6.5). If NSETBD is nonzero, additional data must be read later. If IUV is 1 or 2, this field is ignored. This field does not have an assigned value in our example.

Line 3: Columns 1–5 constitute the NCON field, the value for which gives the number of connected atom lists which will be read in, with a maximum of 30. This field has a value of 4 in our example.

Column 7 is the IDYN field, the value for which is a flag for running molecular dynamics. A value of 0 means no dynamics, and a value of 1 means that molecular dynamics will be run interactively (see Section 5.7). This field has a value of 0 in our example.

Columns 11–15 contain the DIELE field, the value for which gives the dielectric constant to be used in the force field calculations (see Section 5.4). If the value is a positive real number, that value is taken as the dielectric constant. If a value of 0 is found, the default dielectric constant of 1.5 is used. If the value is any negative number, a charge–charge electrostatic model is assumed, and a distance-dependent dielectric constant will be used in the Coulomb potential function. This field has a value of 1.5 in our example.

Columns 16–20 contain the JSTART field, the value for which gives the atom list number which begins the substructure for docking. The value for the IDOCK field (*vide supra*) must be nonzero if JSTART is nonzero. This field has a value of 0 in our example.

Column 25 is the NBUT field, the value for which is a flag for the presence of small rings. A value of 0 means no small rings present, and a value of 1 invokes special parameters necessary to treat three- and four-membered rings. PCMODEL invokes this flag by default. This field has the value 1 in our example.

Columns 26–30 are the NATTCH field, the value for which gives the number of attached atoms in the attached atom lists found later on. This field has the value 15 in our example.

Columns 31–35 contain the NSYMM field, the value for which gives the number of symmetry matrices to be read in. PCMODEL writes a value of 0 in this field by default, and this field has a value of 0 in our example.

Columns 36–40 constitute the NX field, the value for which gives the number of coordinate calculation or replacement lines to be read in. This field is vestigial from MM2, where it was used to generate more complex structures from simpler ones in the days before the graphic user interface. This field has a value of 0 in our example.

Column 45 is the NROT field, the value for which is a flag for whether the final structure should be rotated before the final Cartesian coordinates are written. A value of 0 means the coordinates of the output structure will not be reoriented, and a value of 1 means that the output structure will be reoriented so that its center of mass is at the origin and the principal axes of inertia will correspond to the Cartesian axes. In this case, the components of the dipole moment along the principal axes will also be reported. This field has a value of 0 in our example.

Columns 46–50 contain the LABEL field, the value for which is a flag to indicate that different names and atomic weights will be supplied later for some atom types. A value of 0 means no changes, and a value that is a positive integer specifies the number of atom types for which the resident names and weights will be adjusted. PCMODEL writes a value of 0 for this field by default, and this is so in our example.

Column 55 is the NDC field, the value for which is a flag for how the electrostatic potential will be calculated (see Section 5.4). A value of 0 means that dipole–dipole interaction will be used and the energy will be calculated for all interactions save those between dipoles which share a common carbon atom (the DD option); a value of 1 means that all of the dipole–dipole interactions will be included in the calculation; a value of 2 means that dipole–dipole interaction calculation will be suppressed; a value of 3 means that charge–charge interaction energy will be used and the energy will be calculated for all interactions save those between charges on atoms bound to a common carbon atom; and a value of 4 means that all of the charge–charge interactions will be included, with additions for π-atom charges (the QQ option). This flag can be set in PCMODEL in the 'Electrostatic of Dipole-D' dialog box menu available through the **Options\DP_DP** command, and will write a 0 for the DD electrostatic model and a 4 for the electrostatic QQ model. This field has a value of 4 in our example.

Column 60 is the NCALC field, the value for which is a flag which determines whether crystal conversion and/or reorientations to be performed on input coordinates. A value of 0 means that these options are not used; a value of 1 means that the input coordinates are reduced crystal coordinates and that they may be converted to Cartesian coordinates by employing the values for unit cell parameters given later; a value of 2 means that the input coordinates should be reoriented according to parameters given later; a value of 3 is a combination of options 1 and 2; a value of 4 means the initial coordinates are determined after the molecular rotation. If any of these options are invoked, NX must be 0. PCMODEL writes a value of 0 for this field as a default. This field has a value of 0 in our example.

Column 65 is the HFORM field, the value for which is a flag which governs the calculation of the heat of formation. A value of 0 means that no heat of formation is calculated; a value of 1 means that the heat of formation is calculated; a value of -1 means that the heat of formation will be calculated and an experimental heat of formation will be read in later; a value of 2 means the heat of formation is calculated using values of a partition function which will be read later; a value of -2 means that the heat of formation will be calculated, and both the experimental heat of formation and the partition function values will be read later. PCMODEL writes a value of 1 for this field as a default, and this is so in our example.

Column 75 is the MVDW field, the value for which is a flag for how the energy of van der Waals interactions are calculated. A value of 0 directs the program to use approximate van der Waals energies during minimization, which speeds up minimization by about $\frac{1}{3}$; a value of 1 directs that the van der Waals energies are calculated exactly. If there are fewer than 30 atoms, option 1 is automatically invoked. This field has a value of 1 in our example.

Column 80 is the NDRIVE field, the value for which is a flag for the dihedral driver calculation (see Section 6.3). A value of 0 means no dihedral driver calculations are carried out; a value of 1 invokes the endocyclic dihedral driver; a value of -1 invokes the side-chain dihedral driver. This field has a value of 1 in our example.

Lines 4–7 are the four connected atom lists implied in the value for the NCON field. Each of these lines is considered part of an array for the ICONN field. For ICONN(i, j), i is the number of the line within this set (1–4 in this case) and j is the number of entries on that line (minimum value 2, maximum value 16). Each line gives a chain of contiguous heavy atoms, and the lists taken together provide the logical skeleton of the molecule. The atoms are identified by their atom file numbers, where each number occupies a group of five columns (this dictates a maximum of $80/5 = 16$ numbers per line). For any given structure, there are a variety of possible atom connection list sequences. Zeros are entered at the unused positions to avoid incorrect input:

Line 4 has ICONN(1, 4), and in this example gives the sequence of atom file numbers 1, 2, 5, 9, which corresponds to the sequence C2, C3, C4, C4 hydroxyl oxygen in the L-4-Hydroxyproline structure (see Fig. 4-2).

Line 5 has ICONN(2, 4), and in this example gives the sequence of atom file numbers 1, 3, 6, 5, which corresponds to the sequence C2, C1, C5, C4 in the structure.

Line 6 has ICONN(3, 3), and in this example gives the sequence of atom file numbers 1, 4, 8, which corresponds to the sequence C2, carbonyl carbon, carbonyl oxo oxygen in the structure.

Line 7 has ICONN(4, 2), and in this example gives the sequence of atom file numbers 4, 7, which corresponds to the sequence carbonyl carbon, carbonyl hydroxyl oxygen in the structure.

Lines 8–9 are the two attached-atom lists, containing a total of 15 pairs of atom file numbers implied in the value of 15 for the NATTCH field. Each of these lines contains pairs with one member from the JATTCH(i) set and the other member from the KATTCH(i) set, where i will vary from a value of 1 up to the value of NATTCH. There can be a maximum of eight pairs per line, and additional lines are used until all attached atoms are accounted for.

Line 8 contains pairs from JATTCH(1), KATTCH(1) (in this case 1, 10) to JATTCH(8), KATTCH(8) (in this case 6, 17). In each case, JATTCH(i) is a heavy atom, and KATTCH(i) is a hydrogen or lone pair.

Line 9 contains pairs from JATTCH(9), KATTCH(9) (in this case 7, 18) to JATTCH(15), KATTCH(15) (in this case 9, 24).

Lines 10–21 are the Cartesian coordinate tables. Each line contains descriptions for two atoms in terms of the following variables: for the atom having atom file number i, X(i) gives the X-axis coordinate (ten columns), Y(i) gives the Y-axis coordinate (ten columns), Z(i) gives the Z-axis coordinate (ten columns), ITYPE(i) gives the MMX atom type (five columns), and MATOM(i) gives information on multiple bonds or bonds to metals (five columns). For descriptions of the MMX atom types see Section 5.3 and Appendix C. Since each line will contain descriptions of two atoms, each of these fields will appear twice in a line, leading to the following column assignments: X in columns 1–10, 41–50; Y in columns 11–20, 51–60; Z in columns 21–30, 61–70; ITYPE in columns 31–35, 71–75; MATOM in columns 36–40, 76–80.

The MATOM field is subdivided a bit further. Columns 36–38 or 76–78 give the atom file number of the atom to which the multiple bond is directed. Column 39 or 79 is a flag for the type of metal bonding where a value of 0 indicates coordinative saturation, 1 indicates coordinative unsaturation, 2 indicates a metal with more than 18 electrons, and 3 indicates a square planar geometry for the complex. Column 40 or 80 holds a value equal the multiplicity of the bond described, minus one. Most atoms, which are not a metal and are not part of a multiple bond, will have 00000 or 0 in the MATOM field.

A few examples will illustrate:

Line 10 contains descriptions of the atoms with file numbers 1 and 2. Atom 1 has an X-coordinate of -0.55572, a Y-coordinate of -0.26377, and a Z-coordinate of 0.38604; it is an @1 sp^3 carbon atom with no multiple bonds. Atom 2 has an X-coordinate of 0.81084, a Y-coordinate of 0.40863, and a Z-coordinate of 0.59982; it is an @1 sp^3 carbon atom with no multiple bonds.

Line 11 contains descriptions of the atoms with file numbers 3 and 4. Atom 3 has an X-coordinate of -0.62567, a Y-coordinate of -0.64549, and a Z-coordinate of -1.01736; it is an @8 sp^3 nitrogen atom with no multiple bonds. Atom 4 has an X-coordinate of -1.71314, a Y-coordinate of 0.63452, and a Z-coordinate of 0.77224; it is an @3 sp^2 carbon atom with a multiple bond (double) to atom number 8.

Line 22 contains the logical (T–F) flags for hydrogen bonding, where columns map to atom file numbers. In this example, atom file numbers 13, 18, and 22 are flagged. These signify that the hydrogens attached to the sp^3 nitrogen (file number 3), the carboxylic acid oxygen (file number 7), and the sp^3 oxygen (file number 9), are acidic and will act as hydrogen bond donors.

4.3 MDL MOL FILE FORMAT

Another very useful input file format is the **MDL_Mol**, which was developed by MDL Information Systems Inc. to represent molecular structures in their MACCS® (structure) and REACCS® (reaction scheme) database programs. The files have filenames of the form {*.MOL} and contain Cartesian coordinates for each atom, and bond connection information. To be recognized as an MOL format input file when the PCMODEL **File\Open** command is invoked, the default filename extension is {*.MOL}.

The *.MOL format accommodates both 2D and 3D structures; the user should either ensure that the imported file is in the 3D format, or use PCMODEL editing tools to improve the starting geometry if a 2D structure is imported. Files in the *.MOL format can be created using the **Draw** utilities in the MACCS and REACCS programs, or within the newer ISISDraw®, or can be found by searching various public and proprietary structure databases with MACCS, REACCS, or the newer ISISBase®. It is this latter resource which makes the importability of *.MOL files so useful. MDL refers to these input files as Ctab's, or **connection tables**, although more generally the use of the term "connection table" is more restrictive (see Appendix D).

A characteristic of the **MDL_Mol** format is that the structures are most often represented by heavy atoms only, that is, no hydrogens or lone pairs. However, PCMODEL can add these easily with the **H/AD** tool.

The following guide uses the appropriate input file for L-4-hydroxyproline, {HPROLINE.MOL}, which can be found in the input file library in the CHAP4 directory. The structure of this molecule with standard atom labels is

4.3. MDL MOL FILE FORMAT

seen in Fig. 4-5, and the same structure with the atom file numbers indicated is seen in Fig. 4-6, to aid in identifying the atom file numbers with particular atoms in the structure. The MOL-format input file for this molecule is seen in Fig. 4-7.

The MOL-format input files, like the MMX input files, follow Fortran format, and so part of the meaning of the data in the file is conveyed by its position in the line; thus it is critical for data to appear in the appropriate column(s) in order for it to be interpreted correctly. The data fields, or locations where a particular variable is found in this format, will be described in terms of the example. MOL files may extend to a maximum of 80 columns. For more detailed information, the reader is directed to the publication detailing this format,[273] or the MDL documentation booklet "Ctfile Formats."[274,275]

The first three lines are known as the *header block*. The data in lines 1 and 3 is in free format, while line 2 has some formatting. These lines are not required

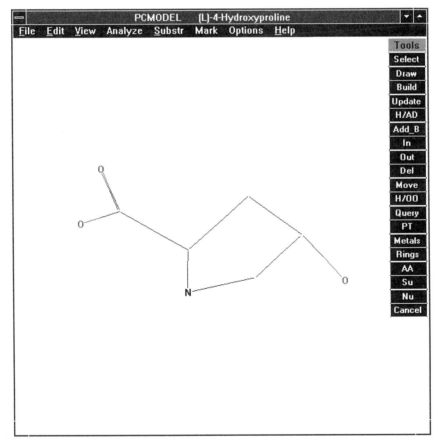

Figure 4-5. The MDL MOL-Format Input File, {HPROLINE.MOL}, with Atom Labels.

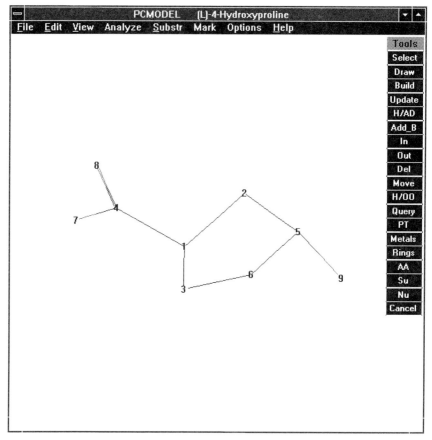

Figure 4-6. The MDL MOL-Format Input File with Atom File Number Labels, {HPROLINE.MOL}.

to hold any data, but in that case the blank lines must still appear in order to fulfill the format requirements.

Line 1: the name of the molecule, 'L-4-Hydroxyproline' in the example.

Line 2: this line may contain the user's first and last initials, the name of the program used to generate the file, the date, scaling and dimensional codes, the energy value, and the internal registry number. In this example, only the name of the program, '-ISIS-' is found.

Line 3: a line for comments. In this example, it is left blank.

Line 4 is the *counts line.* This line gives numerical parameters and flags which help characterize the subject molecule. The numerical values are all right-justified within their fields:

4.3. MDL MOL FILE FORMAT

```
      1    5   10   15   20   25   30   35   40   45   50   55   60   65   70
      ....:....|....:....|....:....|....:....|....:....|....:....|....:....|
 1   (L)-4-Hydroxyproline
 2   -ISIS-
 3
 4    9  9  0  0  1  0  0  0  0  0  1 V2000
 5       7.1356   -4.7124   -0.0449 C   0  0  3  0  0  0  0  0  0  0  0  0
 6       8.2951   -3.6992   -0.0736 C   0  0  0  0  0  0  0  0  0  0  0  0
 7       7.1307   -5.5104   -1.2794 N   0  0  0  0  0  0  0  0  0  0  0  0
 8       5.8198   -3.9873    0.1068 C   0  0  0  0  0  0  0  0  0  0  0  0
 9       9.3327   -4.4301   -0.9458 C   0  0  3  0  0  0  0  0  0  0  0  0
10       8.4304   -5.2281   -1.9054 C   0  0  0  0  0  0  0  0  0  0  0  0
11       5.0596   -4.2205    1.1479 O   0  0  0  0  0  0  0  0  0  0  0  0
12       5.4646   -3.1843   -0.7403 O   0  0  0  0  0  0  0  0  0  0  0  0
13      10.1553   -5.2974   -0.1611 O   0  0  0  0  0  0  0  0  0  0  0  0
14    1  2  1  0  0  0  0
15    1  3  1  0  0  0  0
16    1  4  1  0  0  0  0
17    2  5  1  0  0  0  0
18    3  6  1  0  0  0  0
19    4  7  1  0  0  0  0
20    4  8  2  0  0  0  0
21    5  9  1  0  0  0  0
22    5  6  1  0  0  0  0
23   M  END
```

Figure 4-7. The MDL MOL Format Input File for L-4-Hydroxyproline, {HPROLINE.MOL}.

Columns 1–3 give the number of atoms, with a maximum of 255. In this example the value is 9.

Columns 4–6 give the number of bonds, with a maximum of 255. In this example the value is 9.

Columns 7–9 give the number of atom lists, with a maximum of 30. In this example the value is 0.

Columns 10–12 are now obsolete. In this example the value is 0.

Columns 13–15 give the chiral flag, where a value of 0 means not chiral, and a value of 1 means chiral. In this example the value is 1.

The next five triples, columns 16–18, 19–21, 22–24, 25–27, and 28–30, contain descriptors which are used only in MDL's Chemist's Personal Software Series programs.[274] In this example values of 0 appear for these fields.

Columns 31–33 gives the number of lines of additional properties, including the 'M END' line. Since this latter is the only additional properties line in this example, the value is given as 1.

Columns 34–39 gives the current Ctab version number. In this example the value is given as 'V2000' for version 2.000.

Next comes the group of lines known as the *atom block*. There is one line for each atom in the molecule, and they are in the order of the atom file number. For the example shown in Fig. 4-7, this corresponds to lines 5–13. We will first discuss the format, and then examine several specific lines as examples. For each such line, the format is as follows:

Columns 1–10, 11–20, and 21–30 give the X-, Y-, and Z-axis Cartesian coordinates for the atom, respectively. These are given to four decimal places, in angstroms.

Columns 32–34 give the atom label for the atom, left-justified.

Columns 35–36 give the mass difference for the atom, in integers, if it is a nonstandard isotope. In newer versions, this information is conveyed in an additional property line at the end of the file.[274]

Columns 37–39 give a flag for the charge on an atom, or indication that it is a radical. In newer versions, this information is conveyed in an additional property line at the end of the file.

Columns 40–42 give a flag for the atom stereo parity. The values and their meaning are given in Table 4-5.

The next two triples, columns 43–45 and 46–48, contain descriptors which are used only in MOL files used as queries for database searches.[274] In this example, values of 0 appears for these fields.

Columns 49–51 give a flag for marked valence. In this example, values of 0 appear for this field.

The next three triples, columns 52–54, 55–57, and 58–60, contain descriptors which are used only in MDL's Chemist's Personal Software Series programs.[274] In this example values of 0 appear for these fields.

The last three triples, columns 61–63, 64–66, and 67–69, contain descriptors which are used only in MDL's reaction files.[274] In this example values of 0 appear for these fields.

There follow four specific lines from the atom block as examples:

Line 5 describes atom 1 in the input file, a carbon atom with the coordinates $X = 7.1356$, $Y = -4.7124$, and $Z = -0.0449$. This atom is flagged as a stereo atom of undetermined parity (3 in column 42);

TABLE 4-5. Values for the Stereo Parity Flag in the MDL MOL Format Input File

Value	Significance
0	Not a stereo atom
1	Odd parity
2	Even parity
3	Either odd or even parity, or unmarked stereo atom

This flag defines which stereochemical descriptor will be used in a 2D rendering of the molecule: the solid wedge for "up" or the hatched line for "down." A stereo atom is essentially the same as a chiral or asymmetric atom. For purposes of determining the stereo parity of a tetrahedral carbon atom, the four attached atoms are assigned priority 1–4 in parallel with increasing atom file number. If one substituent is a hydrogen, either implicit or explicit, it automatically becomes priority 4. The molecule is then viewed with the bond to the priority-4 atom projecting away from the viewer into the plane determined by the other three atoms. The sense of the rotation of the remaining three atoms, $1 \rightarrow 2 \rightarrow 3$, is then assessed. A clockwise rotation gives an odd parity, and the value for this flag is 1; a counterclockwise rotation gives an even parity, and the value for this flag is 2. A stereo parity value of 3 means a stereochemical descriptor is not defined for this stereo atom, commonly the case with a three-dimensionally displayed molecule.[274]

4.3. MDL MOL FILE FORMAT

Line 7 describes atom 3 in the input file, a nitrogen atom with the coordinates $X = 7.1307$, $Y = -5.5104$, and $Z = -1.2794$;

Line 8 describes atom 4 in the input file, a carbon atom with the coordinates $X = 5.8198$, $Y = -3.9873$, and $Z = 0.1068$;

Line 12 describes atom 8 in the input file, an oxygen atom with the coordinates $X = 5.4646$, $Y = -3.1843$, and $Z = -0.7403$.

Next comes the group of lines known as the *bond block*, which corresponds to what is more generally referred to as a connection table. This information gives the connectivity in the molecule, with a flag for bond multiplicity and some other flags for use in structural queries and reaction representations. The number of these lines is equal to the number of bonds in the molecule. For the example shown in Fig. 4-7, there are nine bonds corresponding to lines 14–22. Each bond is described once.

We will first discuss the format, and then examine the bond information pertaining the four atoms mentioned above as examples. For each such line, the format is as follows:

Columns 1–3 give the atom file number of the first atom in the bond described.

Columns 4–6 give the atom file number of the second atom in the bond described.

Columns 7–9 give a flag for the bond type. For the MOL files of interest here, this will give the bond multiplicity, with a value of 1 meaning single bond, 2 meaning double bond, or 3 meaning triple bond. This field can hold other values (4–8), which are only used for queries for database searching.

Columns 10–12 give a flag for the bond stereo, and determines which direction a stereochemical descriptor will point in a 2D representation. This field has a value of 0 for the current example.

Columns 13–15 code for a field which is now obsolete. This field has a value of 0 for the current example.

Columns 16–18 give a flag for bond topology, used only in queries for database searching. This field has a value of 0 for the current example.

Columns 19–21 give a flag for reacting center status, used only in queries for reaction database searching. This field has a value of 0 for the current example.

There follow four specific lines from the bond block as examples:

Line 14 connects atom 1 to atom 2 with a single bond;
Line 17 connects atom 2 to atom 5 with a single bond;
Line 20 connects atom 4 to atom 8 with a double bond;
Line 22 connects atom 5 to atom 6 with a single bond.

Line 23 contains the stop token, 'M END'.

```
 1 | Blank Form for X-ray Coordinates Input
 2 | 00.000000  00.000000  00.000000  00.000000  00.00000  00.00000
 3 |   .000000    .000000    .000000 A
 4 |   .000000    .000000    .000000 A
 5 |   .000000    .000000    .000000 A
 6 |   .000000    .000000    .000000 A
 7 |   .000000    .000000    .000000 A
 8 |   .000000    .000000    .000000 A
 9 |   .000000    .000000    .000000 A
10 |   .000000    .000000    .000000 A
11 |   .000000    .000000    .000000 A
12 |   .000000    .000000    .000000 A
13 |   .000000    .000000    .000000 A
14 |   .000000    .000000    .000000 A
15 |   .000000    .000000    .000000 A
16 |   .000000    .000000    .000000 A
17 |   .000000    .000000    .000000 A
18 |   .000000    .000000    .000000 A
19 |   .000000    .000000    .000000 A
20 |   .000000    .000000    .000000 A
21 |   .000000    .000000    .000000 A
22 |   .000000    .000000    .000000 A
```

Figure 4-8. The Dummy Input File in the XRA Format, {BLANK.XRA}.

4.4. X-RAY AND NEUTRON DIFFRACTION FILE FORMATS

This section describes the PCMODEL format for the structure input files from X-ray and neutron diffraction crystallography known collectively as XRA-Format files. In Section 3.3, we learned about structure input from these physical data. The present format was devised by the Serena Software developers, and is a modification of the standard X-ray diffraction formats such as that used by the Cambridge Crystallographic Data Centre.[136,245,246] It is a *free format*, that is, the meaning of the data is not dependent upon its occupying

```
Title
a b c α β γ
  x(1)    y(1)    z(1)  A(1)
  x(2)    y(2)    z(2)  A(2)
  x(3)    y(3)    z(3)  A(3)
  x(4)    y(4)    z(4)  A(4)
  x(5)    y(5)    z(5)  A(5)
  x(6)    y(6)    z(6)  A(6)
  x(7)    y(7)    z(7)  A(7)
  x(8)    y(8)    z(8)  A(8)
  x(9)    y(9)    z(9)  A(9)
  x(10)   y(10)   z(10) A(10)
(etc.)
```

Figure 4-9. The PCMODEL XRA Input File Format.

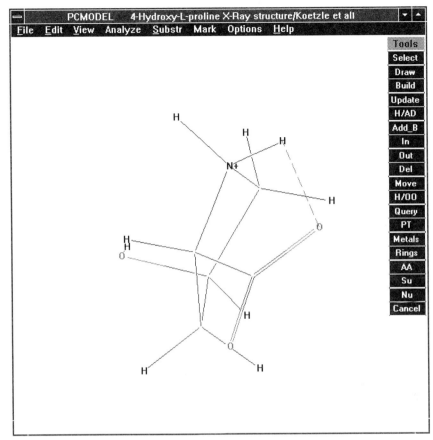

Figure 4-10. Zwitterionic L-4-Hydroxyproline from the XRA Format Input File, {HPROLINE.XRA}, with Atom Labels (from Neutron Diffraction Structure.)[276]

certain columns within the line. The order of the data is critical, though, and the values must be separated by at least one space. All numbers must have a decimal point, since they are read as real numbers. The input file should contain only standard ASCII characters (i.e. those found on the keyboard), no higher or control characters. When using a text processor to prepare an XRA-format file from a paper copy, or to modify an existing file in electronic form, it is convenient to employ a nonproportional font with the text processor, so that the relationships within the columns of data can be seen readily.

This format is an important one, because in excess of 10^5 small-molecule 3D structures are available from the Cambridge database. A dummy file in the XRA format is available in the input file library in the CHAP4 directory as {BLANK.XRA}, to serve as a guide to preparing an appropriate input file from external data. This input file is displayed in Fig. 4-8, and Fig. 4-9 gives a description of contents of the file.

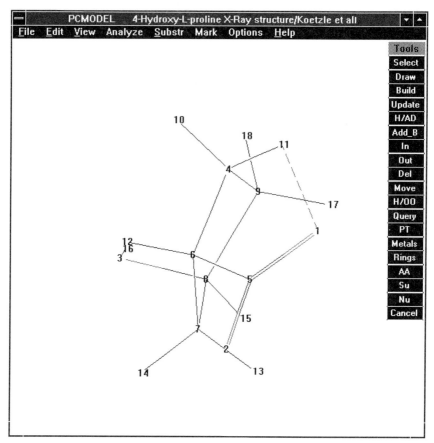

Figure 4-11. Zwitterionic L-4-Hydroxyproline from the XRA Format Input File, {HPROLINE.XRA}, with Atom File Number Labels (from Neutron Diffraction Structure.)[276]

The first line contains the title of the file. The second line contains the cell parameters, where **a**, **b**, and **c** are the X-, Y-, and Z-dimensions of the unit cell, in angstroms; α, β, and γ are the angles of the principal vertices of the unit cell. The third and succeeding lines contain Cartesian coordinates and an atom label for each atom in the file, where the values $x(n)$, $y(n)$, and $z(n)$ are the X-, Y-, and Z-axis Cartesian coordinates of the nth atom; $A(n)$ is the atom label of the nth atom.

If the coordinates are available only as hard copy (e.d. from a journal article), the user must transcribe the input file into electronic form with a text processor, for example by overwriting the {BLANK.XRA} format. If the data are already in electronic form, editing with a text processor will be necessary to convert the file to the proper XRA format. To be recognized as an X-ray

4.4. X-RAY AND NEUTRON DIFFRACTION FILE FORMATS

format input file when the PCMODEL **File\Open** command is invoked, the filename must bear the extension {*.XRA}.

The following guide uses the input file {HPROLINE.XRA}, which can be found in the input file library in the CHAP4 directory. This input file was prepared from literature data for the neutron diffraction crystal structure of L-4-hydroxyproline.[276] The structure of this molecule with standard atom labels is seen in Fig. 4-10, and the same structure with the atom file numbers indicated is seen in Fig. 4-11, to aid in identifying the atom file numbers with particular atoms in the structure. The XRA-format input file for this molecule is seen in Fig. 4-12.

Line 1 contains the title of the file; up to 80 alphanumeric characters are accepted. In this example, the title is 'L-4-Hydroxyproline.'

Line 2 contains the cell parameters, the unit cell dimensions **a**, **b**, and **c** in angstroms, and the angles of the principal vertices of the unit cell, α, β, and γ. In this example, **a** = 4.995, **b** = 8.307, **c** = 14.193, and $\alpha = \beta = \gamma = 90.00000°$. The inclusion of these parameters is critical to obtaining proper bond lengths and angles for the imported structure. If values are not provided for the angles of the principal vertices in a publication, they are assumed to be orthogonal (i.e. 90°). These values should still be placed in the PCMODEL X-ray format input file to avoid corrupting the importation.

The remaining lines contain the Cartesian coordinates X, Y, and Z, and the atomic symbol A for each atom in the molecule (1, 2, 3, etc.). This corresponds to lines 3–20 in the example. The Cartesian coordinates are given as fractions

```
 1 | (L)-4-Hydroxyproline X-Ray structure/Koetzle et al
 2 | 4.995  8.307 14.193 90.000000 90.00000 90.00000
 3 | .9913  .5227 .8090 O
 4 | .7234  .3163 .8371 O
 5 | .4045  .4734 .5022 O
 6 | .7255  .6301 .6594 N
 7 | .7927  .4372 .7918 C
 8 | .6227  .4803 .7057 C
 9 | .6396  .3504 .6299 C
10 | .6617  .4384 .5363 C
11 | .8146  .5908 .5622 C
12 | .5713  .7153 .6598 H
13 | .8838  .6730 .6979 H
14 | .4187  .5029 .7280 H
15 | .8112  .2770 .6409 H
16 | .4679  .2735 .6311 H
17 | .7734  .3687 .4841 H
18 | .4202  .4907 .4348 H
19 | 1.0276 .5681 .5647 H
20 | .7744  .6887 .5160 H
```

Figure 4-12. The XRA Format Input File {HPROLINE.XRA} for Zwitterionic L-4-Hydroxyproline, (from the Neutron Diffraction Crystal Structure.)[276]

of the corresponding edge of the unit cell. To illustrate, the details of a few lines will be presented.

Line 3 corresponds to atom 1 in the input file, an oxygen atom with $X = 0.9913$, $Y = 0.5227$, and $Z = 0.8090$;

Line 10 corresponds to atom 8 in the input file, a carbon atom with $X = 0.6617$, $Y = 0.4384$, and $Z = 0.5363$;

Line 17 corresponds to atom 15 in the input file, an oxygen atom with $X = 0.7734$, $Y = 0.3687$, and $Z = 0.4841$.

It can be seen from inspection of Fig. 4-12 that there is no explicit information in the file on connectivity or multiplicity of bonds for the imported structure. In the XRA format, the sole criterion for drawing single, double, or triple bonds is interatomic distance. This is a crude estimate, and the bond multiplicity may often be in error; the bonds can be edited with the drawing and editing commands (**Draw**, **Add_B**, and **Del**). If a crystal structure is obtained which gives data for the heavy atoms only, the **H/AD** tool may be used to add hydrogens and lone pairs.

4.5. OTHER FILE FORMATS AND INTERCONVERSION

In PCMODEL, files can be imported in twelve formats. Four of these formats have been considered in detail with examples above: PCM (Section 4.1), MMX (Section 4.2), MDL_Mol (Section 4.3), and X-ray (Section 4.4). The remaining eight input file formats are: MOPAC, ALCHEMY, SYBYL, MACROMODEL, CHEM-3D, and GAUSSIAN. PCMODEL can write structure files in seven of these formats: PCM, MMX, MOPAC, ALCHEMY, SYBYL, MACROMODEL, and CHEM-3D.

At this writing, support of the MM3 file format[277–279] is planned for a future release of PCMODEL. The PDB structure format used by the Brookhaven Protein Data Bank[137,280,281] is described here even though it is not readable by PCMODEL, because the availability of so many structures in this format over the World Wide Web makes it an important one to be aware of (see Section 1.4.4).

4.5.1 The Mopac inp Format

Mopac inp is the Z-matrix format used as input for the semiempirical molecular orbital calculation program MOPAC. *Mopac arc* is the MOPAC archive file, which contains the results generated by the semiempirical molecular orbital calculation program MOPAC. When PCMODEL reads or writes files in the Mopac-inp-format, the default filename extension is MOP. When PCMODEL reads Mopac arc format, it expects a filename of the type FOR* where * is a three-digit number. In practice, the MOPAC program will accept almost any filename extension, but expects the file {FOR005} as input, and generates {FOR005.INP} as the output structure file as well as several archival files.

An example of a Mopac-inp-format input file is shown in Fig. 4-13; Fig. 4-14 shows the structure as it looks when read into PCMODEL with the regular atom labels, and Fig. 4-15 shows the structure with the atom file numbers to aid in understanding the construction of the input file. This file was created by PCMODEL from the file {HPROLINE.PCM} discussed in Section 4.1, and may be found in the CHAP4 directory of the structure input file library diskette as {HPROLINE.MOP}. It will be seen that the structure has been reoriented with respect to the depiction in Fig. 4-1, that some reassignment of the atom file numbers has taken place, and that the lone pairs have been removed since these pseudoatoms are unnecessary in molecular orbital calculations. The reorientation occurs because the Mopac input file begins with atom 1 at the origin, the bond between atom 1 and atom 2 is aligned in the positive direction on the X-axis, and the bond angle between atoms 1, 2, and 3 is placed within the upper right-hand quandrant of the XY plane. The depiction in Figs. 4-14 and 4-15 has been tilted slightly for clarity.

This Mopac-inp-format input file contains 22 lines. The data is arranged to appear position-specific, but PCMODEL will interpret a MOPAC file in free format as long as at least one space separates each field. The first three lines in the input file seen in Fig. 4-13 constitute the header information for this file. The first line consists of keywords which control the calculation. The next two lines may contain comments or titles for the input file will be carried out. When the MOPAC structure file is created within PCMODEL, the first three lines are set during the write file procedure. A 'MOPAC Data' dialog box pops up enabling the user to enter this data, as seen in Fig. 4-16.

1	PM3									
2	(L)-4-Hydroxyproline									
3										
4	C	0.000000	0	0.000000	0	0.000000	0	0	0	0
5	C	1.626775	1	0.000000	0	0.000000	0	1	0	0
6	N	2.607511	1	34.040726	1	0.000000	0	2	1	0
7	C	1.553601	1	69.480988	1	-166.614243	1	3	2	1
8	C	1.626649	1	110.660286	1	17.090065	1	4	3	2
9	C	1.595257	1	110.012383	1	-121.333679	1	1	2	3
10	O	1.384555	1	119.989456	1	-119.301949	1	6	1	2
11	O	1.511168	1	111.375610	1	91.995407	1	5	4	3
12	H	1.100143	1	110.732666	1	117.893112	1	1	2	3
13	H	1.110716	1	109.386017	1	-101.898949	1	2	3	1
14	H	1.110752	1	139.982468	1	45.737972	1	2	3	1
15	H	1.109181	1	118.222313	1	88.832352	1	3	2	1
16	O	1.289581	1	120.044945	1	60.661205	1	6	1	2
17	H	1.099410	1	115.897072	1	-153.414307	1	5	4	3
18	H	1.109705	1	109.185089	1	-105.752029	1	4	3	2
19	H	1.110254	1	109.197914	1	139.966782	1	4	3	2
20	H	0.600412	1	114.495934	1	163.626205	1	7	6	1
21	H	1.110195	1	110.002563	1	-179.972031	1	8	5	4
22										

Figure 4-13. The MOPAC Format Input File for L-4-Hydroxyproline, {HPROLINE.MOP}.

154 FILE FORMATS

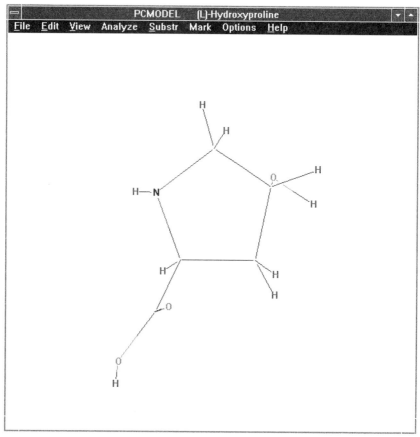

Figure 4-14. MOPAC MOP Structure Input File for L-4-Hydroxyproline, {HPROLINE.MOP}.

In this case, the keyword 'PM3' was entered into the data entry box beside the label 'Keywords', so that this newer Hamiltonian will be employed.[282,283] Below this there are two text entry boxes beside the labels 'Title1' and 'Title2', for entering the title or comment, which in this case was 'L-4-Hydroxyproline'. After this come three check boxes for setting the optimization flags in the input file. The default has the 'Optimize' check boxes selected for 'Bond Lengths', 'Bond Angles', and 'Dihedrals'.

Starting on line 4 and continuing through line 21 we find the **Z-matrix**, which constitutes the coordinate information for this file. In this section, there are ten columns of data, of which the first identifies the atom by a element symbol. Columns 2–10 contain numbers. The second, fourth, and sixth columns are floating point numbers having six decimal places, while the third, fifth, and seventh columns hold single-digit integers, and the remaining eighth

4.5. OTHER FILE FORMATS AND INTERCONVERSION 155

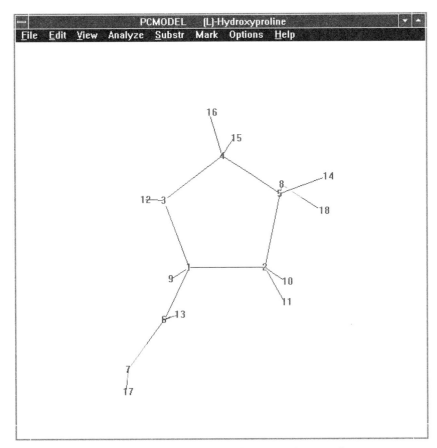

Figure 4-15. MOPAC MOP Structure Input File for L-4-Hydroxyproline, {HPROLINE.MOP}, with Atom File Number Labels.

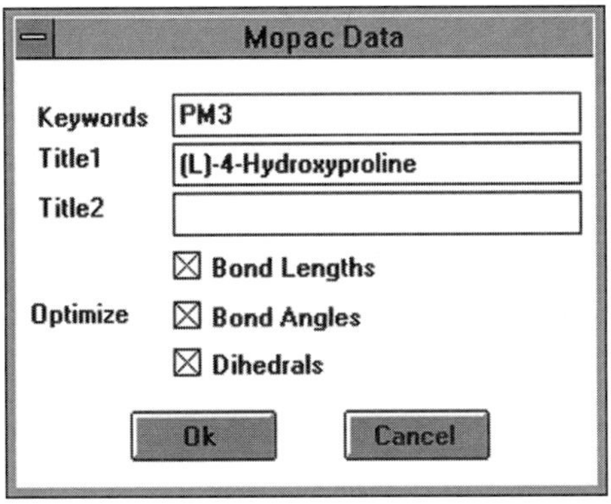

Figure 4-16. The 'Mopac Data' Dialog Box.

through tenth columns will hold one- or two-digit integers. We examine lines 4–7 in detail:

```
C    0.000000 0    0.000000 0    0.000000 0 0 0 0
```

In line 4 above, atom 1, a carbon, occupies the origin in the internal coordinates system, and the numbers following it are all zeros.

```
C    1.626775 1    0.000000 0    0.000000 0 1 0 0
```

In line 5, atom 2, also a carbon, is followed by the value 1.626775, which gives the distance between it and atom 1 in angstroms. The next integer, a 1, indicates that this bond length should be optimized (a 0 would indicate that this bond length should *not* be optimized). The value of 1 in the eighth column identifies by number the other atom in the bond described.

```
N    2.607511 1   34.040726 1    0.000000 0 2 1 0
```

In line 6, atom 3, a nitrogen, is followed by the value 2.607511, which gives the distance between it and atom 2 (eighth column) in angstroms. The '1' in the third column indicates that this bond length should be optimized. The fourth column gives a value of 34.040726, which describes the "bond angle" for atoms 3–2–1, and the 1 in the fifth column indicates that this bond angle should be optimized. Note that this doesn't constitute a real bond angle. In MOPAC structure files, the order in the file implies connectivity. The present result is an artifact from deriving the MOPAC file from a PCMODEL file, where connectivity information is provided in a connection table. If difficulties were to be encountered in an eventual MOPAC calculation, a more rational reassignment of the atom file numbers would be a first suggestion for improvement.

```
C    1.553601 1   69.480988 1 -166.614243 1 3 2 1
```

Line 7 describes atom 4, a carbon. The value 1.553601 gives the bond length between atoms 4 and 3, the value 69.480988 gives the bond angle for atoms 4–3–2, and the value −166.614243 gives the torsional angle for atoms 4–3–2–1. The '1' in the third, fifth, and seventh columns indicates that all geometric quantities are to be optimized.

This pattern is continued until all atoms have been registered into the internal coordinate system. When the last atom has been so specified, there must be a blank line (line 22 in this example) in order to terminate the geometry definition. In special cases, additional lines may be added after the geometry specification, such as for symmetry data, to define a reaction path, or a second complete data set may be added when defining a saddle point. More information is available from the MOPAC manual.[284]

The PCMODEL documentation indicates that the routine which reads MOPAC files skips the first three lines of the file and then reads the Z-matrix. As long

```
1   │ 18 ATOMS,    18 BONDS,      0 CHARGES,  (L)-4-Hydroxyproline
2   │  1 C3   -0.5119  -0.2181   0.4320      0.0000
3   │  2 C3    0.8661   0.6339   0.5790      0.0000
4   │  3 N3   -0.7269  -0.6341  -1.0500      0.0000
5   │  4 C2   -1.7489   0.6529   0.9380      0.0000
6   │  5 C3    1.7701   0.0209  -0.6270      0.0000
7   │  6 C3    0.6381  -0.3491  -1.7350      0.0000
8   │  7 O3   -2.5189   0.2099   2.0000      0.0000
9   │  8 O2   -2.0319   1.7699   0.3590      0.0000
10  │  9 O3    2.5251  -1.2091  -0.1790      0.0000
11  │ 10 H    -0.4593  -1.1459   1.0205      0.0000
12  │ 11 H     0.7036   1.7272   0.4759      0.0000
13  │ 12 H     1.3412   0.5066   1.5742      0.0000
14  │ 13 H    -1.5539  -0.0661  -1.5230      0.0000
15  │ 14 H     2.5957   0.6570  -0.9789      0.0000
16  │ 15 H     0.5131   0.4713  -2.4724      0.0000
17  │ 16 H     0.9371  -1.2329  -2.3365      0.0000
18  │ 17 H    -2.8659   0.6319   2.2490      0.0000
19  │ 18 H     3.2901  -0.9311   0.5760      0.0000
20  │  1   1    2  SINGLE
21  │  2   1    3  SINGLE
22  │  3   1    4  SINGLE
23  │  4   1   10  SINGLE
24  │  5   2    5  SINGLE
25  │  6   2   11  SINGLE
26  │  7   2   12  SINGLE
27  │  8   3    6  SINGLE
28  │  9   3   13  SINGLE
29  │ 10   4    8  DOUBLE
30  │ 11   4    7  SINGLE
31  │ 12   5    9  SINGLE
32  │ 13   5    6  SINGLE
33  │ 14   5   14  SINGLE
34  │ 15   6   15  SINGLE
35  │ 16   6   16  SINGLE
36  │ 17   7   17  SINGLE
37  │ 18   9   18  SINGLE
```

Figure 4-17. The Alchemy-III MOL File Format for L-4-Hydroxylproline, {HPRLALCH.MOL}.

as a space separates all but the three reference atoms for distance, angle, and dihedral, the variables will be read. The routine will also read charges as a floating point variable one space beyond the last reference atom but not beyond column 80. These charges may be included in the MOPAC archive file. Some editing of these files may be necessary to conform to the requirements of the PCMODEL file-reading routine. If charges are read in, the 'Charge' options in 'CPK_Surface' and 'Dot_Surface' may be used to prepare an electrostatic surface representation. In a Mopac-inp- or Mopac-arc-format file, the criterion for drawing single, double, or triple bonds is distance. Since this is a crude estimate, the bond multiplicity may often be in error. The bonds can be edited with the drawing tools described in Section 3.1.

4.5.2. The Alchemy MOL Format

The Alchemy MOL file format is used by the desktop molecular modeling package ALCHEMY-III distributed by Tripos Associates. When PCMODEL reads or writes files in the Alchemy file format, the default filename extension is MOL. Unless care is taken, this can be confused with the MDL MOL format, and also with the SYBYL MOL format. In this work, the Alchemy format file is given a different filename as a way to keep the files differentiated.

PCMODEL was used to prepare the input file {HPRLALCH.MOL} for L-4-hydroxyproline in the Alchemy MOL file format from the file {HPROLINE.PCM}. This input file is shown in Fig. 4-17, and the structure is displayed with atom file number labels in Fig. 4-18.

Comparing Figs. 4-18 and 4-1, we see that some renumbering has also taken place here. This arises because the Tripos force field does not employ lone pair pseudoatoms, so these have been stripped from the structure. The data fields in this format appear to be position-specific.

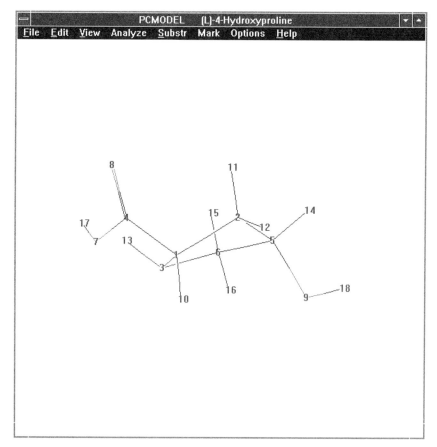

Figure 4-18. L-4-Hydroxyproline in the Alchemy-III MOL File Format {HPRLALCH.MOL}, with Atom File Number Labels.

4.5. OTHER FILE FORMATS AND INTERCONVERSION   159

Line 1 of this file gives the header information, in this case the atom count as '18 ATOMS', the bond count as '18 BONDS', the charge count as '0 CHARGES', and the name of the molecule, '(L)-4-Hydroxyproline'.
Lines 2–19, having six columns of data, provide the coordinate information. The first column gives the atom file number; the second, the atom type; the third, fourth, and fifth, the Cartesian coordinates for that atom; and the sixth, the charge. For example, in line 2 we have

```
1   C3    -0.5119   -0.2181    0.4320    0.0000
```

The atom file number is 1, the atom type is 'C3', an sp^3 carbon; -0.5119, -0.2181, and 0.4320 are the X-, Y-, and Z-coordinates, and 0.0000 is the charge. Among the other atom types employed in this file are 'N3' for sp^3 nitrogen, 'C2' for sp^2 carbon, and 'O2' for sp^2 oxygen.

Lines 20–37 give the connection table, or bond list. Each line here contains four columns of data: the bond file number; atom 1 of the bond; atom 2 of the bond; and, the bond multiplicity. For example, in line 20 we have:

```
1   1    2   SINGLE
```

Here, the bond file number is 1, the atom file number for atom 1 is 1, the atom file number for atom 2 is 2, and the multiplicity is 'SINGLE'.

Tripos has released a newer version of this desktop modeling program, ALCHEMY 2000. The native file format for this version has changed substantially, and it will be instructive to compare it with ALCHEMY-III. In Fig. 4-19 is found the input file {HPROLINE.AL2} for L-4-hydroxyproline in the

```
 1  [CHEMFILE] V100 1
 2  ;(L)-4-HYDROXYPROLINE
 3  [MOLECULA] 1 18 SMALL 0 (L)-4-HYDROXYPROLINE
 4  [A] 1  1  C C3  0.07062 -0.65880 -0.45631  0.55412  4 2 s 3 s 4 s 10 s
 5  [A] 2  2  C C3 -0.01908  0.65330  0.38497  0.61710  4 1 s 5 s 11 s 12 s
 6  [A] 3  3  N N3 -0.33259 -0.81071 -0.85796 -0.87459  3 1 s 6 s 13 s
 7  [A] 4  4  C C2  0.25905 -1.85385  0.34900  1.02434  3 1 s 8 d 7 s
 8  [A] 5  5  C C3  0.03168  1.51695 -0.26290 -0.50127  4 2 s 9 s 6 s 14 s
 9  [A] 6  6  C C3 -0.00806  0.46016 -0.58055 -1.60170  4 3 s 5 s 15 s 16 s
10  [A] 7  7  O O3 -0.38170 -2.50184 -0.14194  2.11362  2 4 s 17 s
11  [A] 8  8  O O2 -0.26661 -2.20580  1.37544  0.45474  1 4 d
12  [A] 9  9  O O3 -0.43771  2.10180 -1.49140 -0.01639  2 5 s 18 s
13  [A] 10 10 H H   0.07928 -0.52996 -1.37850  1.15177  1 1 s
14  [A] 11 11 H H   0.04003  0.42403  1.42991  0.33417  1 2 s
15  [A] 12 12 H H   0.04003  1.13853  0.35100  1.60968  1 2 s
16  [A] 13 13 H H   0.16474 -1.64005 -0.32805 -1.33931  1 3 s
17  [A] 14 14 H H   0.08287  2.31782  0.40005 -0.86638  1 5 s
18  [A] 15 15 H H   0.05825  0.30809  0.31228 -2.23503  1 6 s
19  [A] 16 16 H H   0.05825  0.77112 -1.44085 -2.22104  1 6 s
20  [A] 17 17 H H   0.28808 -3.26161  0.37167  2.40238  1 7 s
21  [A] 18 18 H H   0.27285  2.71563 -1.37376  0.70617  1 9 s
22  [ENDM] 1
23  [ENDFILE]
```

Figure 4-19. The Alchemy-2000 AL2 Format Input File for L-4-Hydroxyproline, {HPROLINE.AL2}.

ALCHEMY 2000 AL2 file format, which was produced by importing the PCMODEL file {HPROLINE.PCM} into ALCHEMY 2000 and writing the structure to an input file in the native format. The data in this file are in free format.

The first three lines of this file form the header section. Line 1 identifies the contents as chemical ('[CHEMFILE]'), gives the version number of the program used ('V100'), and indicates that there is only one molecule in this structure file ('1').

Line 2 is optional, and is flagged with an initial semicolon as a comment line. Here this line contains the title of the file, '(L)-4-HYDROXYPROLINE'.

In line 3, '[MOLECULA] 1' is a start token indicating to the parser the beginning of data for *molecular* structure 1. The '18' gives the number of atoms, 'SMALL' characterizes the structure as a small molecule (as opposed to a protein or polymer), '0' gives the number of residues in the context of a protein-type molecule, and the title of the structure is included in this line, '(L)-4-HYDROXYPROLINE'.

Lines 4–21 include the coordinate and connection table information for each atom. There are between 12 and 18 columns of data, depending upon the valence of the atom being described. For the example of line 4,

[A]   1   1   C   C3   0.07062   -0.65880   -0.45631
                      0.55412   4   2 s   3 s   4 s   10 s

In column 1 is found '[A]', indicating that this is an atom record; in column 2 the '1' is the atom file number, in column 3 the '1' is the atom number (which may be different than the atom file number if atom renumbering has been carried out; the atom numbers are used as atom identifiers during root mean square fitting in structure comparisons); in column 4 the 'C' is an atom label; in column 5 the 'C3' indicates the atom type is sp^3 carbon; in column 6 the number 0.07062 is a calculated charge; in columns 7–9 the numbers −0.65880, −0.45631, and 0.55412 are the *X*-, *Y*-, and *Z*-coordinates; in column 10 the '4' gives the number of attached atoms. The remaining columns 11–18 are grouped pairwise; the first of each pair gives the atom file number of an attached atom, and the following letter indicates the kind of bond. Thus, in this case our atom 1 has a single bond to atom 2 ('2 s'), a single bond to atom 3 ('3 s'), a single bond to atom 4 ('4 s'), and a single bond to atom 10 ('10 s'). Other descriptors in use for bonds are 'd' for double, 't' for triple, and 'a' for aromatic.

The '[ENDM] 1' in line 22 is a stop token indicating to the parser the end of data for structure 1, and the '[ENDFILE]' in line 23 is a stop token indicating to the parser the end of all data in this file.

4.5.3. The SYBYL MOL/SYB Format

SYBYL is the file format used by the workstation molecular modeling package of the same name by Tripos Associates. When PCMODEL writes files in this format, the default filename extension given is SYB; when files in the format are read into PCMODEL, the default filename extension expected is MOL. Use of the latter can be confusing, since the Alchemy MOL and MDL MOL file formats have the same filename extension. In the workstation environment, SYBYL uses the four-character MOL2 filename extension.

PCMODEL was used to prepare the input file {HPRLSYBL.MOL} for L-4-hydroxyproline in the SYBYL file format from the file {HPROLINE.PCM}. This input file is shown in Fig. 4-20, and this corresponds to the structure displayed in Fig. 4-18.

This format is quite similar to that for the Alchemy file shown in Fig. 4-17, though somewhat simpler. It still appears to be position-specific. Line 1 gives the header information:

```
18  mol  (L)-4-Hydroxyproline
```

Here the '18' indicates the number of atoms in the molecule, the term 'mol' is a determinative for a molecule file, and this is followed by the title of the molecule.

Lines 2–19 constitute the coordinate section, and six columns of data arc given. Using the example of line 2:

```
1    1   -0.5119  -0.2181   0.4320C
```

the '1' in the first column gives the atom file number; the '1' in the second column gives the atom type (an sp^3 carbon); the numbers -0.5119, -0.2181, and 0.4320 in the third, fourth, and fifth columns are the X-, Y-, and Z-coordinates; the 'C' in the sixth column (not separated from the Z-coordinate) provides the atom label.

Line 20 signals the beginning of the connection table, where the bond information is obtained. The '18' in this line gives the number of bonds to read.

Lines 21–38 characterize the bonds in this molecule. Each contains four columns of data, as shown in the example of line 21:

```
1    2   1     1
```

The '1' in the first column gives the bond file number; the '2' in the second column and the '1' in the third column give the atom file numbers of the two atoms making up the bond described; the '1' in the fourth column gives the multiplicity of the bond, a single bond in this case.

Line 39 is the final line of the structure file, and contains a '0' as a stop token to signal the file parser to stop reading input.

4.5.4. The MacroModel MAC Format

MacroModel is the file format for the workstation and VAX platform molecular modeling program of the same name developed by the research group of

```
 1  | 18 mol (L)-4-Hydroxyproline
 2  |  1  1  -0.5119  -0.2181   0.4320C
 3  |  2  1   0.8661   0.6339   0.5790C
 4  |  3  5  -0.7269  -0.6341  -1.0500N
 5  |  4  2  -1.7489   0.6529   0.9380C
 6  |  5  1   1.7701   0.0209  -0.6270C
 7  |  6  1   0.6381  -0.3491  -1.7350C
 8  |  7  8  -2.5189   0.2099   2.0000O
 9  |  8  9  -2.0319   1.7699   0.3590O
10  |  9  8   2.5251  -1.2091  -0.1790O
11  | 10 13  -0.4599  -1.1471   1.0190H
12  | 11 13   0.7041   1.7279   0.4760H
13  | 12 13   1.3411   0.5069   1.5750H
14  | 13 13  -1.5539  -0.0661  -1.5230H
15  | 14 13   2.5961   0.6559  -0.9780H
16  | 15 13   0.5121   0.4709  -2.4720H
17  | 16 13   0.9361  -1.2331  -2.3370H
18  | 17 13  -2.8659   0.6319   2.2490H
19  | 18 13   3.2901  -0.9311   0.5760H
20  | 18 mol
21  |  1  2  1     1
22  |  2  3  1     1
23  |  3  4  1     1
24  |  4  5  2     1
25  |  5  6  3     1
26  |  6  6  5     1
27  |  7  7  4     1
28  |  8  8  4     2
29  |  9  9  5     1
30  | 10 10  1     1
31  | 11 11  2     1
32  | 12 12  2     1
33  | 13 13  3     1
34  | 14 14  5     1
35  | 15 15  6     1
36  | 16 16  6     1
37  | 17 17  7     1
38  | 18 18  9     1
39  |  0
```

Figure 4-20. The SYBYL MOL Format Input File for L-4-Hydroxyproline, {HPRLSYBL.MOL}.

W. C. Still at Columbia University.[285,286] An input file in this format, {HPROLINE.MAC}, was written in PCMODEL from {HPROLINE.PCM}, and is shown in Fig. 4-21. The structure for this input file corresponds to the atom file numbers seen in Fig. 4-2.

The input file has 25 lines, one line for header information and the rest for atom specification. The data fields appear to be position-specific. Line 1 gives

```
24(L)-4-Hydroxyproline
 1   3  2 1 1    3 1 1    4 1    10 1  0 0  -0.511875  -0.218083   0.432000  0 0
 2   3  1 1 1    5 1 1   11 1    12 1  0 0   0.866125   0.633917   0.579000  0 0
 3   3  1 1 1    6 1 1   13 1    14 1  0 0  -0.726875  -0.634083  -1.050000  0 0
 4  26  1 1 1    8 2 1    7 1 1   0 0  0 0  -1.748875   0.652917   0.938000  0 0
 5   2  1 1 1    9 1 1    6 1 1  15 1  0 0   1.770125   0.020917  -0.627000  0 0
 6   3  2 1 1    5 1 1    6 1 1  17 1  0 0   0.638125  -0.349083  -1.735000  0 0
 7   3  1 1 1    5 1 1   16 1 1   0 0  0 0  -2.518875   0.209917   2.000000  0 0
 8  16  4 1 1   18 1 1   19 1 1   0 0  0 0  -2.031875   1.769917   0.359000  0 0
 9  15  4 2 1   20 1 1   21 1 1   0 0  0 0   2.525125  -1.209083  -0.179000  0 0
10  16  5 1 1   22 1 1   23 1 1  24 1  0 0  -0.459875  -1.147083   1.019000  0 0
11  41  1 1 1    0 0 0    0 0 0   0 0  0 0   0.704125   1.727917   0.476000  0 0
12  41  2 1 1    0 0 0    0 0 0   0 0  0 0   1.341125   0.506917   1.575000  0 0
13  41  2 1 1    0 0 0    0 0 0   0 0  0 0  -1.553875  -0.066083  -1.523000  0 0
14  43  3 1 1    0 0 0    0 0 0   0 0  0 0  -1.020875  -1.701083  -1.141000  0 0
15  41  3 1 1    0 0 0    0 0 0   0 0  0 0   2.596125   0.655917  -0.978000  0 0
16  41  5 1 1    0 0 0    0 0 0   0 0  0 0   0.512125   0.470917  -2.472000  0 0
17  41  6 1 1    0 0 0    0 0 0   0 0  0 0   0.936125  -1.233083  -2.337000  0 0
18  44  7 1 1    0 0 0    0 0 0   0 0  0 0  -2.865875   0.631917   2.249000  0 0
19  63  7 1 1    0 0 0    0 0 0   0 0  0 0  -2.448875  -0.371083   2.133000  0 0
20  63  8 1 1    0 0 0    0 0 0   0 0  0 0  -2.420875   2.195917   0.591000  0 0
21  63  8 1 1    0 0 0    0 0 0   0 0  0 0  -1.724875   2.006917  -0.112000  0 0
22  42  9 1 1    0 0 0    0 0 0   0 0  0 0   3.290125  -0.931083   0.576000  0 0
23  63  9 1 1    0 0 0    0 0 0   0 0  0 0   1.824125  -1.942083   0.273000  0 0
24  63  9 1 1    0 0 0    0 0 0   0 0  0 0   3.030125  -1.682083  -1.046000  0 0
```

Figure 4-21. The MacroModel Format Input File for L-4-Hydroxyproline, {HPROLINE.MAC}.

```
18
 1  C  -0.511875  -0.218083   0.432000  64  2   3   4  10  1  1  0  0  0  0  0
 2  C  -0.866125   0.633917   0.579000  64  1   5  11  12  1  0  0  0  0  0  0
 3  N  -0.726875  -0.634083  -1.050000  73  1   6  13   0  1  0  0  0  0  0  0
 4  C  -1.748875   0.652917   0.938000  63  1   8   7   0  0  1  0  0  0  0  0
 5  C   1.770125   0.020917  -0.627000  64  2   9   6  14  1  1  0  0  0  0  0
 6  C   0.638125  -0.349083  -1.735000  64  3   5  15  16  1  1  0  0  0  0  0
 7  O  -2.518875   0.209917   2.000000  82  4  17   1   0  0  0  0  0  0  0  0
 8  O  -2.031875   1.769917   0.359000  81  5   2   0   0  0  0  0  0  0  0  0
 9  O   2.525125  -1.209083  -0.179000  82  1  18   1   0  0  0  0  0  0  0  0
10  H  -0.459875  -1.147083   1.019000  11  1   0   0   0  0  0  0  0  0  0  0
11  H   0.704125   1.727917   0.476000  11  2   0   0   0  0  0  0  0  0  0  0
12  H   1.341125  -0.506917   1.575000  11  2   0   0   0  0  0  0  0  0  0  0
13  H  -1.553875  -0.066083  -1.523000  11  3   0   0   0  0  0  0  0  0  0  0
14  H   2.596125   0.655917  -0.978000  11  5   0   0   0  0  0  0  0  0  0  0
15  H   0.512125   0.470917  -2.472000  11  6   0   0   0  0  0  0  0  0  0  0
16  H   0.936125  -1.233083  -2.337000  11  6   0   0   0  0  0  0  0  0  0  0
17  H  -2.865875   0.631917   2.249000  11  7   0   0   0  0  0  0  0  0  0  0
18  H   3.290125  -0.931083   0.576000  11  9   0   0   0  0  0  0  0  0  0  0
```

Figure 4-22. The CHEM3D Format Input File for L-4-Hydroxyproline, {HPROLINE.C3D}.

24 for the number of atoms in the molecule, followed immediately by the title, '(L)-4-Hydroxyproline'. The MacroModel format does include pseudoatoms for the lone pairs.

Lines 2–25 constitute a combination of the coordinate section and the connection table. The atom file number is implied in the order of appearance. There are 17 columns of data in each of the lines, which can be assigned by examining line 2 for atom 1:

```
3     2 1    3 1    4 1    10 1    0 0    0 0
              -0.511875    -0.218083    0.432000    0
```

The '3' in the first column gives the atom type. The next 12 columns are arranged in pairs, and constitute the connection table. A maximum of hexavalency can be accommodated; and the first number of a pair is the atom file number of an attached atom, and the second is the multiplicity of the bond. For atoms which are less than hexavalent, values of 0 are entered in the fields for the unused valencies. In this example, our atom 1 has a single bond to atom 2 ('2 1' in the second and third columns), a single bond to atom 3 ('3 1' in the fourth and fifth columns), a single bond to atom 4 ('4 1' in the sixth and seventh columns), and a single bond to atom 10 ('10 1' in the eighth and ninth columns). The tenth through thirteenth columns hold zeros because atom 1 is only tetravalent.

The fourteenth through sixteenth columns hold the values -0.511875, -0.218083, and 0.432000, which are the X-, Y-, and Z-coordinates, respectively. The seventeenth column gives a value of 0 for the charge on atom 1.

4.5.5. The Chem-3D C3D Format

Chem-3D is the file format for the desktop molecular modeling program of the same name developed by CambridgeSoft Corp. An input file in this format, {HPROLINE.C3D}, was written in PCMODEL from {HPROLINE.PCM}, and is shown in Fig. 4-22. The structure of {HPROLINE.C3D} as read into PCMODEL, with atom file number labels, is shown in Fig. 4-23.

The lone pairs have been stripped in this conversion; like the Tripos force field, the Chem3D force field does not employ lone pair pseudoatoms. The data fields appear to be position-specific. The structure in Fig. 4-23 resembles that for the Alchemy MOL format structure seen in Fig. 4-18, except that the conversion routine has produced a C—O single bond for the carbonyl group, between atoms 4 and 8. As has been mentioned elsewhere, it is not uncommon for some structure editing to be necessary after reading a nonnative format into a modeling program.

The input file consists of 19 lines. Line 1 contains only the value 18 for the number of atoms in the molecule. Lines 2–19 contains 14 columns of data, which will be explained using line 2 as an example:

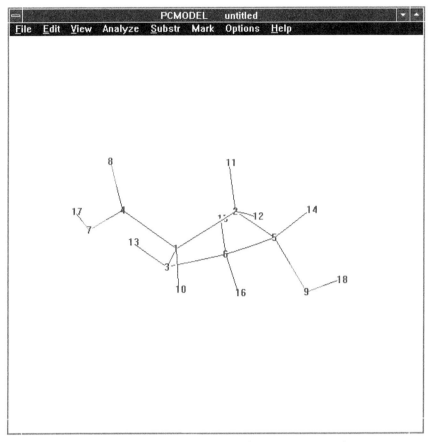

Figure 4-23. L-4-Hydroxyproline Structure from {HPROLINE.C3D}, with Atom File Number Labels.

```
C   1   -0.511875   -0.218083   0.432000
                            64   2   3   4   10   1   0   0   0
```

where the 'C' in the first column gives the atom label. The '1' in the second column is the atom file number. The three numbers −0.511875, −0.218083, and 0.432000 in the third through fifth columns are the X-, Y-, and Z-coordinates. The number 64 in the sixth column gives the atom type. The numbers 2, 3, 4, and 10 in the seventh through tenth columns give the attached atoms, and the '1' in the eleventh column indicates a single bond. The zeros in the remaining three columns are placeholders.

Hond, or HONDO, is an *ab initio* molecular orbital calculation package, developed by M. DuPuis. In a Hond format file, the criterion for drawing single, double, or triple bonds is distance. Since this is a crude estimate, bond

4.5. OTHER FILE FORMATS AND INTERCONVERSION 167

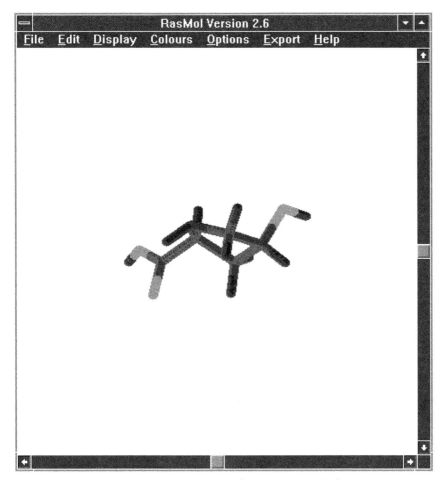

Figure 4-24. L-4-Hydroxyproline from {HPROLINE.PDB} in RASMol.

multiplicity may often be in error. The bonds can be edited with the drawing tools described in Section 3.1.

Gaussian is the file format for the *ab initio* quantum mechanics calculational package developed by Pople *et al.* and available commercially in various forms.

One more file format will be discussed here, the Brookhaven **PDB** (Protein Data Bank) format.[137,280,281] Although PCMODEL does not read such files, the vast majority of structures available on the World Wide Web are found in this format. There are several visualization and conversion programs which will convert PDB files into a format which PCMODEL can read. In this case we have used **RASMol**, a versatile molecular graphics program which was discussed in Section 1.4.4.

168 FILE FORMATS

For the purposes of this discussion, the ALCHEMY-III structure file {HPRLALCH.MOL} for L-4-hydroxyproline was read into RASMOL, and the display option chosen was 'Sticks'. After some rotation, the structure appeared as is shown in Fig. 4-24 (compare Fig. 4-18 for atom numbering), and was written to a PDB format file named {HPROLINE.PDB} and depicted in Fig. 4-25.

The PDB file format was originally devised by researchers at Brookhaven National Laboratory to store the single or multiple structures of proteins and various cofactors and prosthetic groups, along with accompanying archival and physical information. Its use for small molecules is a secondary development, and came about largely because it was the first molecular structure format extensively used to transmit structures across the internet. The PDB files follow Fortran format, and the data is position-specific, which means that it must appear in the correct columns of a line to be read properly. A brief description of this format is available on a Web page,[281] and the explanation here is derived from information given at that source. The most comprehensive information is available from the Protein Data Bank itself.[137,280]

The input file {HPROLINE.PDB} has 21 lines. Line 1 is a header, and contains an Rtyp (record type) descriptor of 'COMPND', followed by the title of the structure. Lines 2–19 contain 10 columns for the atom specifications and Cartesian coordinates for the 18 atoms of the structure (PDB format does not accommodate lone pairs). Take line 2 for an example:

```
ATOM   1   C3   MOL   1   -0.720   -0.484   0.544   1.00   0.00
```

Here the Rtyp field contains the 'ATOM' descriptor, signaling that data for a particular atom is being read. The second column gives a value of 1 for the

```
 1  COMPND   (L)-4-Hydroxyproline
 2  ATOM    1  C3  MOL  1   -0.720  -0.484   0.544  1.00  0.00
 3  ATOM    2  C3  MOL  1    0.652   0.364   0.688  1.00  0.00
 4  ATOM    3  N3  MOL  1   -0.936  -0.900  -0.936  1.00  0.00
 5  ATOM    4  C2  MOL  1   -1.960   0.384   1.048  1.00  0.00
 6  ATOM    5  C3  MOL  1    1.556  -0.248  -0.512  1.00  0.00
 7  ATOM    6  C3  MOL  1    0.424  -0.616  -1.620  1.00  0.00
 8  ATOM    7  O3  MOL  1   -2.728  -0.060   2.112  1.00  0.00
 9  ATOM    8  O2  MOL  1   -2.240   1.500   0.468  1.00  0.00
10  ATOM    9  H3  MOL  1    2.312  -1.476  -0.064  1.00  0.00
11  ATOM   10  H   MOL  1   -0.668  -1.412   1.132  1.00  0.00
12  ATOM   11  H   MOL  1    0.488   1.456   0.584  1.00  0.00
13  ATOM   12  H   MOL  1    1.128   0.236   1.684  1.00  0.00
14  ATOM   13  H   MOL  1   -1.764  -0.332  -1.408  1.00  0.00
15  ATOM   14  H   MOL  1    2.380   0.388  -0.864  1.00  0.00
16  ATOM   15  H   MOL  1    0.300   0.200  -2.360  1.00  0.00
17  ATOM   16  H   MOL  1    0.724  -1.500  -2.224  1.00  0.00
18  ATOM   17  H   MOL  1   -3.076   0.360   2.360  1.00  0.00
19  ATOM   18  H   MOL  1    3.076  -1.196   0.688  1.00  0.00
20  TER    19      MOL  1
21  END
```

Figure 4-25. The PDB Format Structure File for L-4-Hydroxyproline, {HPROLINE.PDB}.

'Num' field, corresponding to the atom file number. The third column gives a value of 'C3' for the Atm field, corresponding to an atom type of sp^3 carbon. The fourth column gives a value of 'MOL' for the 'Res' (residue) field, which here designates that a molecule is being described. The fifth column contains a value of 1 for the 'ResN' (residue number) field, which in this case indicates that residue 1 is being described. The sixth through eighth columns hold the values -0.720, -0.484, and 0.544, which are the X-, Y- and Z-coordinates, respectively. The ninth column gives a value of 1.00 for the Occ (occupancy factor) field, and the tenth column gives a value of 0.00 for the Temp (temperature factor) field. The temperature factor is high for atoms which are disordered in a crystal; it has little relevance for small molecules whose structures are not derived from diffraction patterns.

Line 20 gives an Rtyp of 'TER', a stop token to terminate the parsing of input for this structure. Although values are also given for the Num, Res, and ResN fields, no more specific atom input is given at this point. Line 21 has an Rtyp of 'END'.

4.5.6. Interconversion between File Formats

We have had some experience with interconversion between file formats earlier in this chapter. In common with most of the modern, graphically oriented desktop molecular modeling programs, PCMODEL itself may be used as an input file conversion utility. Currently, structures may be imported to PCMODEL in 12 formats (see Fig. 3-24), and may be written to seven formats (see Fig. 3-25). PCMODEL can also prepare an image file in three graphics formats.

There are also numerous standalone interconversion utility programs, three of which will be described briefly: CONSYSTANT/CHEMELEON by Exographics, BABEL from the University of Arizona, and MOL2MOL from Cherwell Scientific Publishing. In addition, the two World Wide Web structure viewers which were discussed in Section 1.4.4., RASMOL and CHIME™, can also carry out file interconversion. The file formats involved are described in Appendix D, and links to World Wide Web sites with more information are provided in the HTML document "Molecular Modeling Resources on the World Wide Web," which is found on the diskette which accompanies this book.

ExoGraphics offers CONSYSTANT© and CHEMELEON©, which can handle import and export of the structure and graphics formats shown in Table 4-6. CONSYSTANT operates as a standalone interconversion utility program, while CHEMELEON function as a "terminate stay resident" (**TSR**) program performing selected conversions of file formats as they are copied to and from the Windows clipboard. These programs are bundled with the newer versions of MDLI's ISIS suite.

BABEL is a popular new input file interconversion program, which is offered as freeware by an academic group at the University of Arizona, and is available at many FTP sites worldwide. Originally available only for the Unix

TABLE 4-6. Importable and Exportable Structure File Formats for CONSYSTANT

Software (Distributor)	Portability	Native Format
Molecular Presentation Graphics (Hawk Scientific)	I/E	MPG
ISIS (MDL, Inc.)	I/E	SKC
ISIS (MDL, Inc.)	I/E	MOL
ISIS (MDL, Inc.)	I/E	TGF
ISIS (MDL, Inc.)	E	SDF
HYPERCHEM (HyperCube, Inc.)	I/E	HIN
SMILES code (Daylight Chem. Inf. Syst.)	I/E	SMI
ROSDAL (Beilstein)	I/E	ROS
CHEMDRAW (CambridgeSoft Corp.)	I/E	CDR
CHEMWINDOW (SoftShell)	I/E	SCF
ALCHEMY-III MOL (Tripos)	I/E	ALC
SYBYL MOL2 (Tripos)	I/E	SY2
KEKULÉ (PSI International)	I/E	STR
CHEMDRAW CT (CambridgeSoft Corp.)	I/E	CDT
CATALYST (Molecular Simulations, Inc.)	I/E	TPL
PCMODEL (Serena Software)	I/E	PCM
X-ray coordinates	I	XRA
Word Perfect Graphics	E	WPG
Encapsulated Postscript	E	EPS
HP-GL (Hewlett–Packard)	E	HPG
CGM Metafile	E	CGM
Windows Metafile (Microsoft)	E	WMF
Encapsulated Postscript with Windows Metafile	E	EPS

CONSYSTANT™ and CHEMELEON™ are registered trademarks of ExoGraphics. Under Portability, I indicates that CONSYSTANT and CHEMELEON can import files in the respective format, E indicates that CONSYSTANT and CHEMELEON can export files in the repective format, and I/E indicates that CONSYSTANT and CHEMELEON can both import and export in the repective format. Under Native Format is given the filename extension used by CONSYSTANT and CHEMELEON for files in the respective format, which is not necessarily the same filename extension used in the third-party application. Additional information on these applications and formats may be found in Appendix D, and World Wide Web links for most of the software applications and companies are provided in the HTML document "Modeling Resources on the World Wide Web" which accompanies this book.

platform, versions are now available for the DOS Windows and Macintosh environments. BABEL focuses on the structure file formats, and does not currently include graphic file capability. Among current file conversion programs, it handles the broadest range of structure input file formats, which are shown in Table 4-7.

MOL2MOL is another standalone input file interconversion program, and is offered commercially by Cherwell Scientific Publishing. The MOL2MOL program is available for a variety of platforms, and can handle the import and export of the structure and graphics formats shown in Table 4-8.

4.5. OTHER FILE FORMATS AND INTERCONVERSION

TABLE 4-7. Importable and Exportable File Formats for BABEL

Software (Distributor)	Portability	Native Format
ALCHEMY-III (Tripos, Inc.)	I/E	MOL
AMBER (UCSF)	I	
Ball & Stick (Cherwell Scientific Publ.)	I/E	
Biosym CAR	I	CAR
Boogiee	I	
CACAO Cartesian	I/E	
CACHE (Oxford Molecular)	E	
CADPAC (Cambridge Univ.)	I	
CHARMM (Harvard Univ.)	I	
CHEM3D Cartesian 1, Cartesian 2 (CambridgeSoft Corp.)	I/E	
CHEMDRAW CT (CambridgeSoft Corp.)	E	
CSSR (Cambridge Cryst. Data Cent.)	I/E	
FDAT, GSTAT (Cambridge Cryst. Data Cent.)	I	
DIAGNOSTICS	E	
Free Form Fractional	I	
GAMESS Input (Iowa State Univ.)	E	
GAMESS Output (Iowa State Univ.)	I	LOG
GAUSSIAN Z-Matrix (Pople, CMU)	I/E	
GAUSSIAN Output (Pople, CMU)	I	OUT
HYPERCHEM (HyperCube, Inc.)	I/E	HIN
IDATM	E	
MACCS, ISIS (MDLI)	I/E	MOL
MACMOLECULE (Univ. Arizona)	I/E	
MACROMODEL (Columbia Univ.)	I/E	
Micro World	I/E	
MM2 Input, Output (Allinger, QCPE)	I/E	
MM3 (Allinger, QCPE)	I/E	
MMADS	I/E	
MOLIN	I	
MOPAC Cartesian, Internal	I/E	
MOPAC Output	I	
PCMODEL (Serena Software)	I/E	PCM
Protein Data Bank (Brookhaven Natl. Lab.)	I/E	PDB
QUANTA (Molecular Simulations, Inc.)	I	
Report	E	
ShelX	I	
SPARTAN (Wavefunction, Inc.)	I/E	
SYBYL (Tripos)	I/E	MOL, MOL2
XMOL (Minn. Supercomp. Cent.)	I/E	XYZ

Native Format indicates the filename extension used in the DOS or other operating system environment for the files indicated. Under Portability, I indicates that BABEL can import files in the format, E indicates that BABEL can export files in the format, and I/E indicates that BABEL can both import and export in the format. BABEL is capable of creating three types of GAMESS input files: Cartesian coordinates, Gaussian style Z-matrix, and MOPAC style-Z-matrix. QUANTA files are binary, and different systems use different binary representations, so BABEL should be run on the same type of machine which created the QUANTA file. In PDB files, a bond is assigned when the interatomic distance is less than the sum of the atoms' covalent radii.

172 FILE FORMATS

TABLE 4-8. Importable and Exportable Structure File Formats for MOL2MOL

Software Distributor	Portability	Native Format
ALCHEMY III (Tripos, Inc.)	I/E	MOL
AMPAC Z-matrix	I/E	
Beilstein ROSDAL (Softron)	I	
ChemDraft (C_Graph Software)	I/E	
CHEMWINDOW (SoftShell)	I/E	
Desktop Molecular Modeler	I/E	
Free format Cartesian	I*	
Free format X-ray fractional	I*	
Free format Z-matrix	I/E*	
HYPERCHEM (HyperCube Inc)	I/E	HIN
ISIS/MACCS (MDLI)	I/E	MOL
MOBY Cartesian (Springer)	I/E	
MOLIDEA (CheMicro Ltd.)	I/E	
MOPAC Z-matrix	I/E	
PCMODEL (Serena Software)	I/E	PCM
PC MOL	I/E	
Protein Data Bank (Brookhaven Natl. Lab.)	I/E	PDB
X-ray fractional CIF	I	
X-ray fractional, Cambridge FDAT	I	
X-ray fractional, Cambridge MODEL	I	
PLUTO (CSD and PC versions)	E	
PLT (RYLAZ Products)	E	
WIMP (Aldrich Chemical Co.)	E	FTR
Schakal (Univ. Freiburg)	E	

An asterisk indicates that the insertion of a marker or token is necessary for the input file to be parsed properly. Under Portability, I indicates that MOL2MOL can import files in the format, E indicates that MOL2MOL can export files in the format, and I/E indicates that MOL2MOL can both import and export in the format. Native Format indicates the filename extension used in the DOS environment for the files indicated. MOL2MOL makes the translation automatically, but because different programs may use different special atom or bond types (extended atoms, dummy atoms, lone electron pairs, aromatic bonds or atoms, etc.), MOL2MOL will sometimes prompt the user for additional information.

RASMOL was one of the Web browser structure viewers which were discussed in Section 1.4.4. It is available as freeware for a variety of computer environments from the desktop up to mainframes. It functions as a standalone in the Windows 3.1 environment, and it can serve as a Web browser plug-in in Windows 95 and higher level environments. RASMOL can handle the import and export of the structure and graphics formats shown in Table 4-9.

CHIME™ from MDL is the other Web browser structure viewer which was discussed in Section 1.4.4. It is currently available as freeware for the Windows environment, and implementation in the Macintosh and Unix environments is imminent. CHIME functions as a true plug-in to the Netscape NAVIGATOR

TABLE 4-9. Importable and Exportable File Formats for RASMOL

Software (Distributor)	Portability	Native Format
Structure Formats		
ALCHEMY-III (Tripos, Inc.)	I	MOL
CHARMM	I	
ISIS/MACCS (MDL)	I	MOL
Protein Data Bank (Brookhaven Natl. Lab.)	I	PDB
RASMOL Script		SCRIPT
SYBYL (Tripos, Inc.)		MOL2
XMOL (Minn. Supercomp. Cent.)	I	XYZ
Graphics Formats		
PICT (Apple)	E	PICT
Bitmap (Microsoft)	E	BMP
Encapsulated PostScript	E	PS, EPSF
GIF (CompuServe)	E	GIF
MAGE (D. Richardson)	E	KINEMAGE
MOLSCRIPT Input Script (P. Kraulis)	E	MOLSCRIPT
Monochrome Encapsulated PostScript	E	MONOPS
Sun Rasterfile	E	RAS
Portable Pixmap	E	PPM
POSTSCRIPT, raster	E	
POSTSCRIPT, vector	E	

Under Portability, I indicates that RASMOL can import files in the format, and E indicates that RASMOL can export files in the format. Native Format indicates the expected filename extension used in the Web environment for the files indicated.

TABLE 4-10. Importable File Formats for CHIME™

Software (Distributor)	Native Format
Chemical Structure Markup Language	CSM, CSML
GAUSSIAN Input	GAU
EMBL Nucleotide	EMB, EMBL
ISIS/MACCS (MDL)	MOL
MOPAC input	MOP
RASMOL Script	SPT
RxnFile (MDL)	RXN
Protein Data Bank (Brookhaven Natl. Lab.)	PDB, ENT
Transportable Graphics File (MDL)	TGF
XMOL (Minn. Supercomp. Cent.)	XYZ

Native Format indicates the expected filename extension used in the Web environment for the files indicated.

174 FILE FORMATS

2.0+ Web browser, which means that when properly configured it is associated with the filename extension(s) common to a particular structure file format. Activating links to such files in the World Wide Web environment launches the CHIME viewing window from within NETSCAPE. CHIME can handle the import and export of the structure and graphics formats shown in Table 4-10.

4.6. MODIFICATION OF THE INPUT FILE

In Sections 4.1 through 4.5 we have explored different types of structure file format. In large part, these files are created and modified through the graphic user interface, using the appropriate commands to set or clear flags, enter parameters, etc. The other way to modify the input file is by editing the file itself directly. The user must exercise caution to maintain the proper format conventions, especially in those formats in which the data is position-specific.

One should *always* make a working copy of the file to be edited, and carry out modifications only on it, so that the source file remains available in its original state.

A text processor program is the best tool to achieve this goal, since the user will likely be familiar with one or more such programs and be able to take advantage of editing short-cuts. However, these input files are all **ASCII** files, i.e., they may include only the 96 standard keyboard characters: alphabetic, numeric, and symbolic (punctuations and symbols). Control characters or higher order characters will corrupt the file and make it unreadable. Edited files should be saved under their original names, or a close equivalent, in order to be recognizable by the modeling program. They should never be saved with document formatting of any kind, except hard returns, and all spaces should be hard spaces, as opposed to the soft returns and soft spaces used in most document formats. One easy way to ensure compliance with these restrictions is to edit input files only in strike-over (or replace, or overwrite) mode as opposed to insert mode.

The author has found it convenient to prepare a document template specifically for this purpose. All of the file formats we have explored in this chapter employ a maximum of 80 characters per line. It is easier to follow the flow of the file when a nonproportional font is used, since in this case all characters have an identical width, and all columns will be properly aligned. A template file for Microsoft Word for Windows v. 6.0, entitled {ASCII.DOT}, is included in the input file library for those who wish to employ it. It should be copied into the {\TEMPLATE} subdirectory of the Windows directory for proper access. This template employs a True Type nonproportional Courier New 13-point font, and for the portrait mode ($8\frac{1}{2}$ in. wide by 11 in. high), with 0.0-in. margins on the top, sides, and bottom. This works out to 80 characters per line.

4.6. MODIFICATION OF THE INPUT FILE

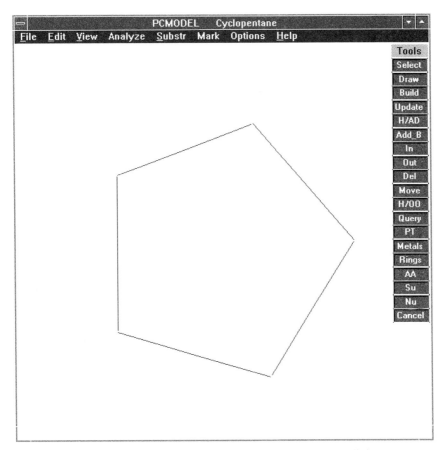

Figure 4-26. First Step in the Preparation of the Magnesium Ethandiolate Structure.

There are a variety of changes which may be easier to do by editing. An obvious one is to edit the title of the file. A new title must conform to the position requirements of the format. Taking the PCM format as an example, the starting token '{' in column 1 and the 'PCM' label in columns 2–4 musn't be disturbed. An empty space must be left after the 'PCM', before beginning the new title, which should not exceed the allowed 60 characters.

The flags for π-atoms, hydrogen bonding, substructure membership, etc. can all be changed by editing, but these changes are made more reliably from within PCMODEL. An exception might be if an appended format file were being modified or created by cut-and-paste cloning of a single input file. In these cases, editing through a text interface can be much simpler and faster.

Another modification that can be made is between atom types. This can include easy changes (such as from carbon to silicon, oxygen to sulfur, etc.), but these can also be made easily from the PCMODEL graphical interface. A

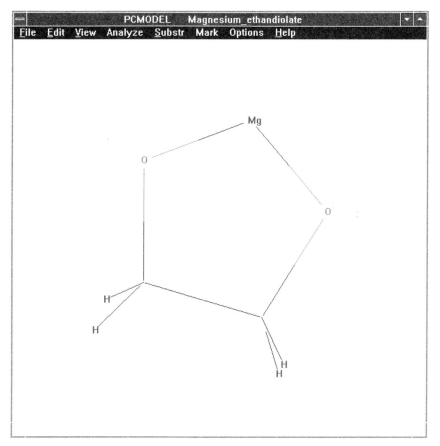

Figure 4-27. Second Step in the Preparation of the Magnesium Ethandiolate Structure.

change that cannot be effected so easily in PCMODEL is between the different types of hydrogen (@5, @21, @23, @24, @28). For example, one application would involve modeling an alcohol which is more acidic than normal (such as a perfluoroalkanol); one may want to use an @28 hydrogen (enol) rather than the standard @21 hydroxyl hydrogen. Such substitutions are always subject to the availability of suitable parameters in the MMX force field. If an "illegal" substitution is made, the parser may reverse the change prior to minimization, or the minimization may fail, although descriptive error messages will be given to help detect what the problem was.

A few other items which can be changed are the flags for the level of output, or the level of output for π-calculations, or the name of a metal atom and its covalent radius may be altered. As was mentioned above, appended format files can be created by cut-and-paste "cloning," and files which are already in appended format can be dissected apart into individual files.

4.6. MODIFICATION OF THE INPUT FILE   **177**

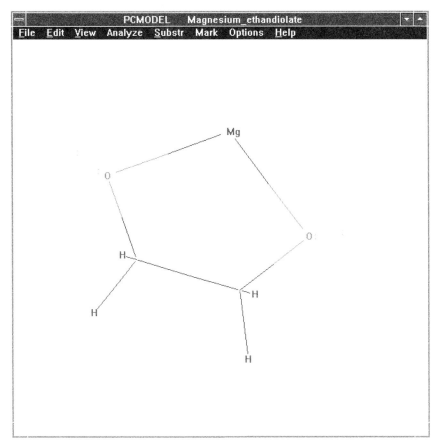

Figure 4-28. The Optimized Structure for Magnesium Ethandiolate, {MGALKOX.PCM}.

Many examples of this will be encountered later in the book, especially in the higher level exercises in Chapter 6, but two examples here will serve to illustrate the method.

Exercise 4-1: Modified Metal Atom Covalent Radii in Magnesium Ethandiolate. Activate the **Rings** tool and import cyclopentane (**C5**) into the PCMODEL work space. Use the **H/AD** tool to remove the hydrogens to give a structure like that in Fig. 4-26.

Activate the **Metals** tool and click on the **Mg** button to load the mouse cursor with this metal, which should be used to convert the top carbon of the cyclopentane structure to a magnesium atom. Next, activate the **PT** tool, click on the **O** button in the 'Periodic Table' menu, and then click on the two atoms connected to the magnesium to convert these to oxygens. Finally, use the

H/AD tool to add the hydrogens and lone pairs to give a structure like that in Fig. 4-27, which may be given a title such as 'magnesium ethandiolate'.

The **Analyze\Minimize** command may be used to run this structure through energy minimization, during the course of which the "magnesium corner" of this ring will be seen to expand to accommodate the metal atom's covalent radius of 1.6 Å. The result should be saved to an input file such as {MGALKOX.PCM}, and will resemble the structure shown in Fig. 4-28. The input file itself is displayed in Fig. 4-29.

The input file {MGALKOX.PCM} consists of seventeen lines. Line 4 describes atom 1 in this molecule, which happens to be the magnesium atom. We see that a value of 'Mg' identifies the atom as an @80 metal with 'Mg' for an atom label, and the last portion of that line, 'R 1.6', gives the metal atom's covalent radius as 1.6 Å. This parameter will be modified by editing as follows.

Close, iconize, or background PCMODEL, then open up a preferred text processing program (such as Microsoft Windows), and import the file {MGALKOX.PCM}. Select the entire file, i.e. lines 1–17, and copy this to the clipboard. Then, open or create a new file, and paste the contents of the clipboard into this new file. Next, move the cursor to the end of line 4, and overwrite the value for the covalent radius from 'R 1.6' to 'R 0.8'. This value is not realistic for magnesium, but it will serve to illustrate the modification process. Save this as a new file with a name such as {MGALKOX2.PCM}, making sure that there are no extraneous spaces or carriage returns in the text. Figure 4-30 shows what this new file should look like, with the modified portion shown in bold.

Close this file and exit the text processor. Bring up PCMODEL and read in the file {MGALKOX2.PCM}. Run this through several cycles of energy minimization by activating the **Analyze\Minimize** command; during the first minimization sequence, watch the "magnesium corner" of the ring shrink down in response to the altered covalent radius parameter. Save the resulting, optimized structure to the same filename, {MGALKOX2.PCM}, overwriting or

```
1    {PCM Magnesium_ethandiolate
2    NA 13
3    FL EINT4 UV1 PIPL1
4    AT 1 Mg     0.469    1.593     0.006 B 2 1 5 1 S 0 R 1.6
5    AT 2 6     -1.380    0.939    -0.499 B 1 1 3 1 6 1 7 1 S 0 C 0.1531343
6    AT 3 1     -0.937   -0.303     0.080 B 2 1 4 1 8 1 9 1 S 0 C 0.1268657
7    AT 4 1      0.623   -0.762    -0.084 B 3 1 5 1 10 1 11 1 S 0 C 0.1268657
8    AT 5 6      1.668    0.037     0.502 B 1 1 4 1 12 1 13 1 S 0 C 0.1531343
9    AT 6 20    -1.485    1.014    -1.074 B 2 1 S 0 C -0.14
10   AT 7 20    -1.801    1.286    -0.276 B 2 1 S 0 C -0.14
11   AT 8 5     -1.159   -0.242     1.173 B 3 1 S 0
12   AT 9 5     -1.592   -1.096    -0.352 B 3 1 S 0
13   AT 10 5     0.845   -0.823    -1.177 B 4 1 S 0
14   AT 11 5     0.743   -1.786     0.341 B 4 1 S 0
15   AT 12 20    2.211    0.102     0.282 B 5 1 S 0 C -0.14
16   AT 13 20    1.795    0.040     1.078 B 5 1 S 0 C -0.14
17   }
```

Figure 4-29. The Structure Input File {MGALKOX.PCM} for Magnesium Ethanediolate.

4.6. MODIFICATION OF THE INPUT FILE   179

```
 1  {PCM Magnesium_ethandiolate
 2  NA 13
 3  FL EINT4 UV1 PIPL1
 4  AT  1 Mg   0.469    1.593    0.006 B 2 1 5 1 S 0 R 0.8
 5  AT  2  6  -1.380    0.939   -0.499 B 1 1 3 1 6 1 7 1 S 0 C 0.1531343
 6  AT  3  1  -0.937   -0.303    0.080 B 2 1 4 1 8 1 9 1 S 0 C 0.1268657
 7  AT  4  1   0.623   -0.762   -0.084 B 3 1 5 1 10 1 11 1 S 0 C 0.1268657
 8  AT  5  6   1.668    0.037    0.502 B 1 1 4 1 12 1 13 1 S 0 C 0.1531343
 9  AT  6 20  -1.485    1.014   -1.074 B 2 1 S 0 C -0.14
10  AT  7 20  -1.801    1.286   -0.276 B 2 1 S 0 C -0.14
11  AT  8  5  -1.159   -0.242    1.173 B 3 1 S 0
12  AT  9  5  -1.592   -1.096   -0.352 B 3 1 S 0
13  AT 10  5   0.845   -0.823   -1.177 B 4 1 S 0
14  AT 11  5   0.743   -1.786    0.341 B 4 1 S 0
15  AT 12 20   2.211    0.102    0.282 B 5 1 S 0 C -0.14
16  AT 13 20   1.795    0.040    1.078 B 5 1 S 0 C -0.14
17  }
```

Figure 4-30. The Modified Structure Input File {MGALKOX2.PCM} for Magnesium Ethanediolate.

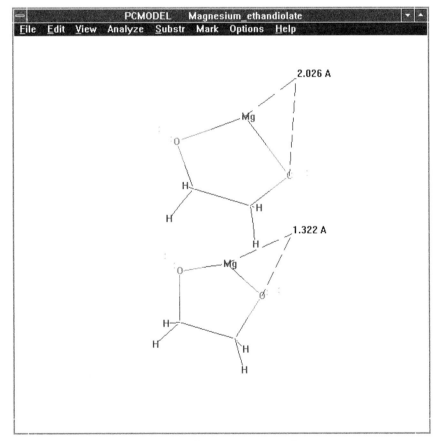

Figure 4-31. Comparison of Original {MGALKOX.PCM} and Modified {MGALKOX2.PCM} Structures for Magnesium Ethandiolate.

replacing its predecessor. The two structures may be compared by reading in {MGALKOX.PCM} as a substructure with the **Substr\Read** command. If the two structures are now repositioned, and the **Query** tool used to give a reading of the O—Mg bond length, the result is shown in Fig. 4-31.

Exercise 4-2. Modified Hydrogen Atom Types in 2-(2'-Hydroxyethyl)pyridine. Activate the **Rings** tool and import benzene (**Ph**) into the PCMODEL work space. Activate the **Build** tool, and then click on the topmost hydrogen atom of the benzene ring to convert it to a methyl group. Click once again on one of the hydrogens of this new methyl group to enlarge this chain by one member. If one of the upper right-hand hydrogens is converted, the result will look like that shown in Fig. 4-32.

Next, delete the hydrogen from the benzene ring position immediately adjacent to the new ethyl chain. Then, activate the **PT** tool, and from the 'Periodic

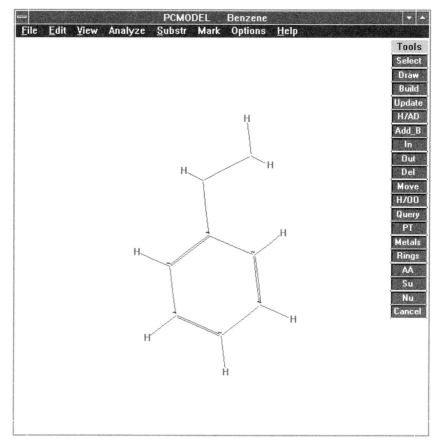

Figure 4-32. First Stage in Preparing the 2-(2'-Hydroxyethyl)pyridine Structure.

Table' menu, select first the 'N' and use this to convert the unsubstituted benzene ring atom to nitrogen. Load the mouse with the **O** atom from the 'Periodic Table' menu, and use it to convert one of the three terminal hydrogens of the ethyl group to oxygen. Finally, activate the **H/AD** tool twice to fill out the proper complement of hydrogens and lone pairs, and activate the **Mark\H-Bonds** command to set the flag for hydrogen bonding. After renaming the structure to '2-(2'-Hydroxyethyl)pyridine', the resulting structure will resemble the one shown in Fig. 4-33.

Carry out several cycles of energy minimization with the **Analyze\Minimize** command. It is possible that the desired hydrogen bond will not be found in this way, so the user may help it along by using the **Select** tool to mark in succession the hydroxyl H, the hydroxyl O, the attached carbon, and then the chain carbon attached to the benzene ring. Activation of the **Edit\Rotate_Bond** command allows the dihedral angle to be varied, and this should be done until

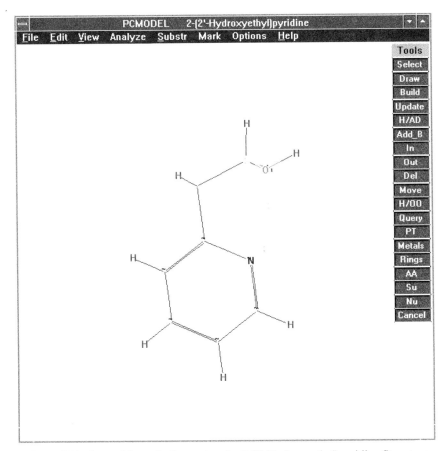

Figure 4-33. Second Stage in Preparing the 2-(2'-Hydroxyethyl)pyridine Structure.

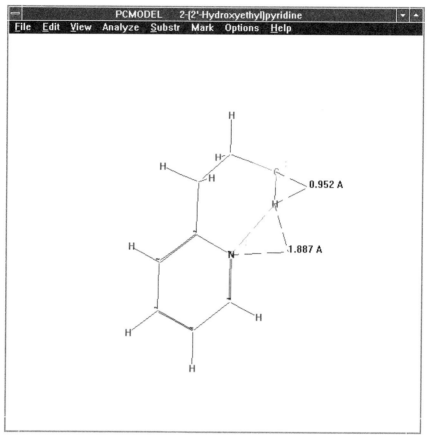

Figure 4-34. Minimized Structure for 2-(2'-Hydroxyethyl)pyridine, With Default @21 Hydroxyl Hydrogen.

a hydrogen bond is seen to form between the hydroxyl hydrogen and the ring nitrogen. Then, a few more cycles of energy minimization will produce the finished structure depicted in Fig. 4-34. In this case, the **Query** tool has been used to measure the O—H bond length and the distance between the hydroxyl hydrogen and the ring nitrogen. This structure should be saved to a file with a name such as {HBONDER.PCM}.

The 25 lines of the PCM-format input file {HBONDER.PCM} for 2-(2'-hydroxyethyl)pyridine are shown at the top of Fig. 4-35. Line 22 carries the identifiers for the hydroxyl hydrogen, atom 19, assigned as the default @21 for a hydroxyl hydrogen. By creating modified versions of this input file with different atom types for the hydroxyl proton, we will be able to alter the character of the model.

Close, iconize, or background PCMODEL, then open up a preferred text processing program (such as Microsoft Windows), and import the file

4.6. MODIFICATION OF THE INPUT FILE

```
 1  {PCM 2-(2'-Hydroxyethyl)pyridine
 2  NA 21
 3  FL EINT4 UV1 PIPL1
 4  AT  1  2   -0.508   -0.069    0.340 B 2 2 6 1 7 1 S 0 P  C  0.2341195
 5  AT  2  2   -1.570   -0.866    0.790 B 1 2 3 1 10 1 S 0 P C -0.09162764
 6  AT  3  2   -1.607   -2.226    0.467 B 2 1 4 2 11 1 S 0 P C  0.002055299
 7  AT  4  2   -0.578   -2.783   -0.305 B 3 2 5 1 12 1 S 0 P  C -0.07038431
 8  AT  5  2    0.472   -1.974   -0.741 B 4 1 6 2 13 1 S 0 P  C  0.156759
 9  AT  6 37    0.494   -0.639   -0.409 B 1 1 5 2 14 1 S 0 P  C -0.2155947
10  AT  7  1   -0.455    1.403    0.661 B 1 1 8 1 15 1 16 1 S 0 C  0.04208417
11  AT  8  1    0.447    2.187   -0.303 B 7 1 9 1 17 1 18 1 S 0 C  0.1268657
12  AT  9  6    1.771    1.695   -0.243 B 8 1 19 1 20 1 21 1 S 0 C -0.1366746
13  AT 10  5   -2.375   -0.422    1.399 B 2 1 S 0 C 0.03814714
14  AT 11  5   -2.440   -2.858    0.818 B 3 1 S 0 C 0.03814714
15  AT 12  5   -0.597   -3.854   -0.567 B 4 1 S 0 C 0.03814714
16  AT 13  5    1.288   -2.400   -1.348 B 5 1 S 0 C 0.03814714
17  AT 14 20    0.972   -0.293   -0.572 B 6 1 S 0 C -0.21
18  AT 15  5   -1.485    1.831    0.622 B 7 1 S 0
19  AT 16  5   -0.078    1.518    1.704 B 7 1 S 0
20  AT 17  5    0.479    3.269   -0.028 B 8 1 S 0
21  AT 18  5    0.089    2.095   -1.355 B 8 1 S 0
22  AT 19 21    1.729    0.783   -0.512 B 9 1 S 0 H  C  0.2898089
23  AT 20 20    1.917    1.639    0.338 B 9 1 S 0 C -0.14
24  AT 21 20    2.084    1.933   -0.698 B 9 1 S 0 C -0.14
25  }
```

```
22*   AT 19 36    1.729    0.783   -0.512 B 9 1 S 0 H  C  0.2898089
```

```
22**  AT 19 28    1.729    0.783   -0.512 B 9 1 S 0 H  C  0.2898089
```

Figure 4-35. Modification of the Input File for 2-(2'-Hydroxyethyl)pyridine.

{HBONDER.PCM}. Select the entire file, i.e. lines 1–25, and copy this to the clipboard. Position the cursor at the very beginning of the file, and paste in the contents of the clipboard *twice* in succession. Next, move the cursor to the equivalent of line 22 within the second structural input portion, and modify the atom type of the hydroxyl hydrogen from @21 to @36 (deuterium), as is shown in line 22* in Fig. 4-35. Then, move the cursor to the equivalent of line 22 within the third structural input portion, and modify the atom type of the hydroxyl hydrogen from @21 to @28 (enolic/phenolic hydrogen), as is shown in line 22** in Fig. 4-35. Save this file in text-only form as {HBONDER.PCM}. This editing has had the effect of converting {HBONDER.PCM} from a single structure file to three different structures contained in the same file in appended format.

After closing out the text processor, return to PCMODEL and with no structures occupying the work space, activate the command **Analyze\Batch**, and choose the file {HBONDER.PCM}. Proceed with the default settings and click on **OK** to begin the batch-mode energy minimization. When the first cycle has finished, exit PCMODEL and first delete {HBONDER.PCM} (or rename it or move it to another directory), then rename {HBONDER.OUT} to {HBONDER.PCM}, and finally delete {HBONDER.SUM}.

Reenter PCMODEL and run through the batch-mode minimization again, as before. When this cycle has finished, repeat the procedures described in the previous paragraph, and run the newest {HBONDER.PCM} through a third

Summary information from batch calculation			
Num	Name	Energy	Hf
1	2-(2'-Hydroxyethyl)pyridine	7.275	-10.06
2	2-(2'-Hydroxyethyl)pyridine	10.630	-5.88
3	2-(2'-Hydroxyethyl)pyridine	8.850	-8.51

Figure 4-36. The Summation File {HBONDER.SUM} for the Energies of the Three Models of 2-(2'-Hydroxyethyl)pyridine.

cycle of energy minimization. When this is finished, carry out the file manipulations described above, with the exception that {HBONDER.SUM} should be kept.

The summation file {HBONDER.SUM}, for the energies of the three models of 2-(2'-hydroxyethyl)pyridine, is shown in Fig. 4-36. At this point it is

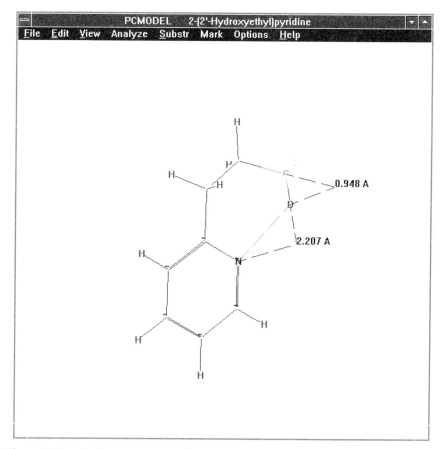

Figure 4-37. 2-(2'-Hydroxyethyl)pyridine Modeled with an @36 (Deuterium) Atom for the Hydroxyl Hydrogen, {HBONDER2.PCM}.

```
{PCM 2-(2'-Hydroxyethyl)pyridine
NA 21
FL EINT4 UV1 PIPL1
AT  1  2    -0.517   -0.103    0.307 B 2 2 6 1 7 1 S 0 P   C  0.2343609
AT  2  2    -1.554   -0.905    0.807 B 1 2 3 1 10 1 S 0 P  C -0.09152441
AT  3  2    -1.595   -2.268    0.497 B 2 1 4 2 11 1 S 0 P  C  0.001981866
AT  4  2    -0.596   -2.823   -0.316 B 3 2 5 1 12 1 S 0 P  C -0.07033198
AT  5  2     0.427   -2.009   -0.802 B 4 1 6 2 13 1 S 0 P  C  0.1565206
AT  6 37     0.453   -0.671   -0.482 B 1 1 5 2 14 1 S 0 P  C -0.2156808
AT  7  1    -0.472    1.372    0.620 B 1 1 8 1 15 1 16 1 S 0 C  0.04208417
AT  8  1     0.440    2.180   -0.317 B 7 1 9 1 17 1 18 1 S 0 C  0.1268657
AT  9  6     1.801    1.822   -0.147 B 8 1 19 1 20 1 21 1 S 0 C -0.1366746
AT 10  5    -2.336   -0.463    1.447 B 2 1 S 0 C 0.03814714
AT 11  5    -2.407   -2.903    0.888 B 3 1 S 0 C 0.03814714
AT 12  5    -0.619   -3.896   -0.569 B 4 1 S 0 C 0.03814714
AT 13  5     1.219   -2.434   -1.441 B 5 1 S 0 C 0.03814714
AT 14 20     0.917   -0.329   -0.668 B 6 1 S 0 C -0.21
AT 15  5    -1.507    1.784    0.551 B 7 1 S 0
AT 16  5    -0.130    1.496    1.674 B 7 1 S 0
AT 17  5     0.363    3.268   -0.071 B 8 1 S 0
AT 18  5     0.147    2.053   -1.388 B 8 1 S 0
AT 19 36     1.958    0.948   -0.481 B 9 1 S 0 H   C  0.2898089
AT 20 20     1.887    1.734    0.441 B 9 1 S 0 C -0.14
AT 21 20     2.124    2.146   -0.538 B 9 1 S 0 C -0.14
}
```

Figure 4-38. 2-(2'-Hydroxyethyl)pyridine Modeled with an @36 (Deuterium) Atom for the Hydroxyl Hydrogen, {HBONDER2.PCM}.

significant only to note that the energies are different; this example and its implications will be considered in greater detail in Section 6.6.

As an additional exercise, the individual files within {HBONDER.PCM} may be dissected out within a text processor and formed into separate input files. All of the structures from this exercise are available in the input file library on diskette which accompanies this book, in the CHAP4 subdirectory. The second structure, with the @36 (deuterium) atom for the hydroxyl hydrogen is {HBONDER2.PCM}, for which the structure is depicted in Fig. 4-37 and the input file in Fig. 4-38.

The third structure, with the @28 (enolic/phenolic hydrogen) atom for the hydroxyl hydrogen, is {HBONDER3.PCM}, for which the structure is depicted in Fig. 4-39 and the input file in Fig. 4-40.

4.7 USING ADDED CONSTANTS

The use of added constants can be very helpful in approaching some modeling problems, but it should be done very cautiously. As is the case with any force field, the parameters used by MMX are the result of years of work, and the body of parameters along with the force field potential equations themselves constitute a carefully balanced whole. Making a significant change in one or a few parameters will necessarily unbalance this system. The effects of such a change must be assessed by applying the modified force field to a representative basis set of molecules, in order to determine whether the model is still accurate. Along these same lines, it is best not to employ parameters obtained

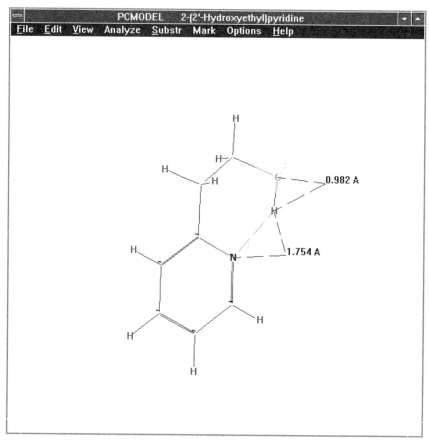

Figure 4-39. 2-(2'-Hydroxyethyl)pyridine Modeled with an @28 (Enolic/Phenolic Hydrogen) Atom for the Hydroxyl Hydrogen, {HBONDER3.PCM}.

from other force fields, which tend to be internally self-consistent but are not always translatable from one force field to another. Parameters are discussed at greater length in Section 5.2.

If a molecule of interest contains structural moieties which are not parametrized in the MMX force field, or if the user wishes to employ parameters which are not part of the standard set, the new parameters must be made available to PCMODEL. This can be accomplished in two ways. The first method is by editing the existing parameter files, {MMXCONST.PAR}, {GENCONST.PAR}, and {PICONST.PAR}. This is tedious because of the size of these files, and one must be sure to keep a secure, unaltered version of these files for future use. The second method is by preparing a constants file, and flagging PCMODEL to use them by activating the **Options\Added Constants** command. When this option is selected, a dialog box entitled 'Added Constants File' (of the "read file" type) pops up, prompting the user to select the

```
{PCM 2-(2'-Hydroxyethyl)pyridine
NA 21
FL EINT4 UV1 PIPL1
AT  1  2   -0.496    -0.068     0.346 B 2 2 6 1 7 1 S 0 P  C  0.2341553
AT  2  2   -1.552    -0.868     0.804 B 1 2 3 1 10 1 S 0 P  C -0.09157829
AT  3  2   -1.594    -2.225     0.468 B 2 1 4 2 11 1 S 0 P  C  0.002072107
AT  4  2   -0.579    -2.773    -0.330 B 3 2 5 1 12 1 S 0 P  C -0.07041161
AT  5  2    0.464    -1.960    -0.773 B 4 1 6 2 13 1 S 0 P  C  0.1567016
AT  6 37    0.494    -0.629    -0.425 B 1 1 5 2 14 1 S 0 P  C -0.2156122
AT  7  1   -0.436     1.400     0.684 B 1 1 8 1 15 1 16 1 S 0 C  0.04208417
AT  8  1    0.450     2.190    -0.289 B 7 1 9 1 17 1 18 1 S 0 C  0.1268657
AT  9  6    1.759     1.637    -0.293 B 8 1 19 1 20 1 21 1 S 0 C -0.06291506
AT 10  5   -2.347    -0.432     1.431 B 2 1 S 0 C  0.03814714
AT 11  5   -2.422    -2.861     0.826 B 3 1 S 0 C  0.03814714
AT 12  5   -0.604    -3.841    -0.603 B 4 1 S 0 C  0.03814714
AT 13  5    1.268    -2.379    -1.400 B 5 1 S 0 C  0.03814714
AT 14 20    0.968    -0.282    -0.593 B 6 1 S 0 C -0.21
AT 15  5   -1.465     1.832     0.673 B 7 1 S 0
AT 16  5   -0.036     1.501     1.720 B 7 1 S 0
AT 17  5    0.519     3.266     0.005 B 8 1 S 0
AT 18  5    0.049     2.124    -1.328 B 8 1 S 0
AT 19 28    1.642     0.692    -0.533 B 9 1 S 0 H  C  0.2160494
AT 20 20    1.895     1.716     0.297 B 9 1 S 0 C -0.14
AT 21 20    2.016     1.965    -0.739 B 9 1 S 0 C -0.14
}
```

Figure 4-40. 2-(2′-Hydroxyethyl)pyridine Modeled with an @28 (Enolic/Phenolic Hydrogen) Atom for the Hydroxyl Hydrogen, {HBONDER3.PCM}.

file which contains the desired constants (see Fig. 4-41). Once selected, the new constants will be in effect until the parameter base is reset by activating the **Options\Standard Constants** command, or by ending the PCMODEL session.

When a constants file is read, the parameter arrays are first set to zero, and then the named constants file is read, followed by the standard constants files. Since they are read first, the added constants will appear first in the parameter arrays accessed by the MMX minimizer, and will be used in preference to the standard constants.

The **Added Constants** procedure does not help the user to write a constants file, which must be prepared separately in the proper format with a text processor. Added constants files are not restricted as to filename, but commonly the filename extensions {*.DAT}, {*.PAR}, and {*.STO} are used so that the nature of the file may be inferred from the name. Added constants files should follow the same format used in the {MMXCONST.PAR} file, and often the simplest way to prepare such files is by copying and pasting from this source. Making the desired changes then involves overwriting the existing data with the new values.

The standard MMX parameter files may contain data for up to ten types of parameters. These will be discussed further in Section 5.2, and they are summarized in Table 4-11 and explained briefly below.

MMX parameter files consist of lines of code, each of which governs a type of interaction for a single unit consisting of between one and four atoms. The type of interaction governed is indicated by a code word in the initial 3–4 spaces of the line: TOR specifies the torsional parameters for a string of four atoms; BND specifies the bond stretching parameters for a pair of atoms;

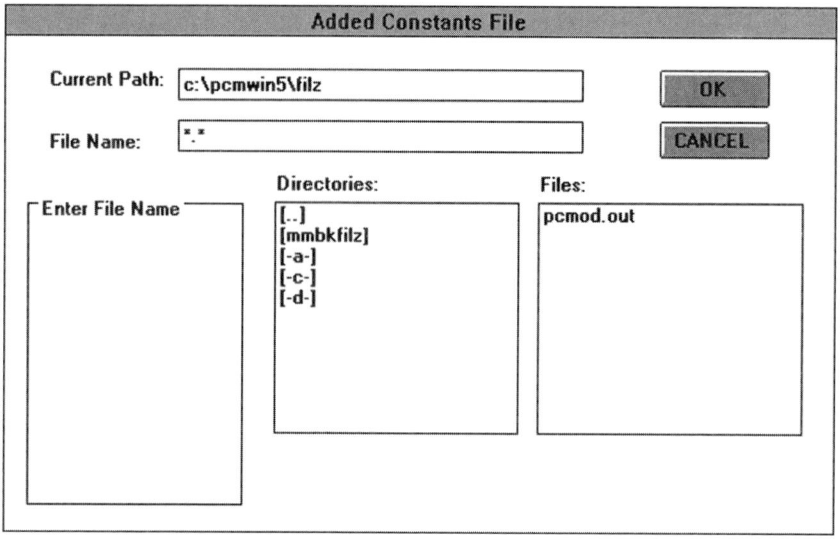

Figure 4-41. The 'Added Constants' Dialog Box.

VDW specifies the parameters for the van der Waals nonbonding interactions for an atom; ANG specifies the angle bending parameters for a string of three atoms; ANG4 specifies the angle bending parameters for a string of three atoms contained in a four-membered ring; ANG3 specifies the angle bending parameters for a string of three atoms contained in a three-membered ring; and OOP specifies the out-of-plane angle bending parameters for a pair of atoms. There are three types of parameters used in the heat of formation calculations: BDHF provides the enthalpy and entropy contributions from a specified bond (two atoms); EVHF provides the heat contribution from a specified structural feature or functional group; and PIHF provides the contributions from a specified bond (two atoms) which is part of a conjugated system.

The sample lines (bold type) contained in Table 4-11 all begin with one of these code words, followed by a *sequence number*. This number is for the user's benefit in keeping track of lines of parameter code; it must be present for the parser to read the line properly, but its value has no effect upon the energy calculations. In Table 4-11, the sequence number is always 1, since these examples are all the first ones in their respective sections of {MMXCONST.PAR}.

The MMX torsional strain potential is discussed in Section 5.1.4. In the example line from Table 4-11, the line is flagged as a torsional strain parameter by the initial code word 'TOR', and there are eight fields of data. The first, '1', is the sequence number. The second through fifth fields contain the four atom types which make up the torsional angle of interest, here '1 1 1 1', signifying four @1 carbon atoms. The sixth, seventh, and eighth fields contain the

TABLE 4-11. Format for MMX Parameter Files

Interaction Type	Parameter Line										
Torsional	TOR	1	1	1	1	1	0.200	0.270	0.093		
Bond stretching	BND	1	1	4.400	1.523		0.000	0.000	0.000	0.000	
van der Waals	VDW	1	CSP3				1.900	0.044	0	0	
Angle bending	ANG	1	1	1	1		0.450	109.470	109.510	109.510	109.500
	ANG4	1	56	56	56		0.340	109.470	109.510	109.510	109.500
	ANG3	1	22	22	22		0.450	60.000	60.000	60.000	60.000
Out-of-plane bending	OOP	1		2	1		0.600				
Heat of formation	BDHF		1	1	C-C SP3 SP3			−0.004	−16.4 00		
	EVHF		1	ISO (ALKANE)				0.078			
	PIHF	1	2	2	C≡C DELOCALIZED			11.200	−40.000	−6.400	250.000

190 FILE FORMATS

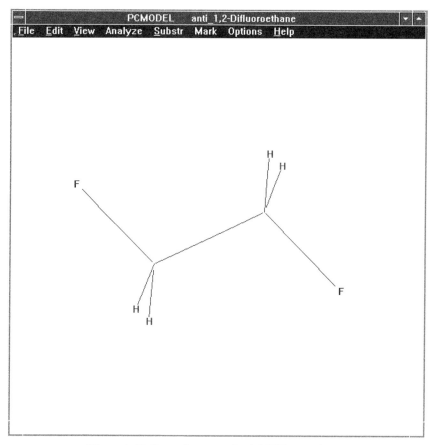

Figure 4-42. Input Structure for 1,2-Difluoroethane.

values for k_{tor1}, k_{tor2}, and k_{tor3}, which are the parameters for the one-, two-, and threefold torsional barriers, respectively, in kilocalories per mole.

The MMX bond stretching potential is discussed in Section 5.1.1. In the example line from Table 4-11, the line is flagged as a bond stretch parameter by the initial code word 'BND', and there are nine fields of data. The first, '1', is the sequence number. The second and third fields contain the two atom types which make up the bond of interest, here '1 1', signifying two @1 carbon atoms. The fourth field contains the value 4.400 for k_{bnd}, the force constant for bond stretching in units of millidynes per angstrom. The fifth field contains the value 1.523 for r_0, the equilibrium or natural bond length, in angstroms. The sixth field contains the value 0.000 for an alternative r_0, the equilibrium or natural bond length in cases where the two constituent atoms bear only one or no hydrogen substituents, in angstroms. In cases where the value of this parameter is 0, the first value for r_0 will be used. The seventh field contains the value 0.000 for the bond moment μ of the bond, in debyes (D). Convention

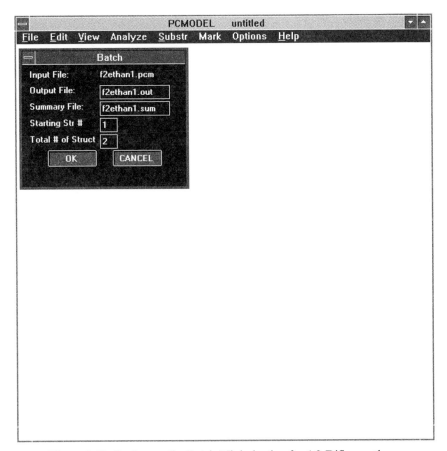

Figure 4-43. Setting up the Batch Minimization for 1,2-Difluoroethane.

dictates that the second of the two atom types be the more electronegative. The eighth field contains the value 0.000 for the force constant correction factor SLPS for cases when the bond is part of a conjugated system, in millidynes per angstrom. The ninth field contains the value 0.000 for the natural bond length correction factor SLPT for cases when the bond is part of a conjugated system, in angstroms.

The MMX van der Waals nonbonding interaction potential is discussed in Section 5.2. In the example line from Table 4-11, the line is flagged as a van der Waals parameter by the initial code word 'VDW', and there are seven fields of data. The value in the first field, '1', is both the sequence number and the atom type, i.e. @1. The second field gives an abbreviated description of the atom type, here 'CSP3', for sp^3-hybridized carbon. The third field gives the value 1.900 for the van der Waals radius in angstroms. The fourth field gives the value 0.040 for the hardness ε in kilocalories per mole. The force constant for van der Waals interaction is a function of the hardness parameter for each

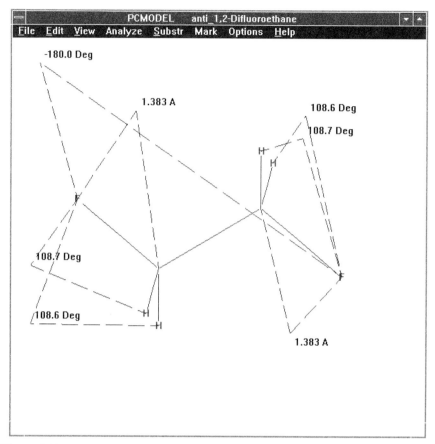

Figure 4.44. Geometric Measures on *anti*-1,2-Difluoroethane, with Standard Parameters.

of the atoms involved. The fifth through seventh fields contain parameters used in calculating the hydrogen bonding potential (see Section 6.6). The fifth field gives the value 0 for the LPD parameter, a measure of the lone pair donor ability in a hydrogen bonding interaction. The sixth field gives the value 0 for the IHTYP parameter, a measure of the acidity of the hydrogen in a hydrogen bonding interaction. The seventh field gives the value 0 for the IHDON parameter, a measure of the ability certain nitrogen atom types to act as hydrogen bond donors.

The MMX angle bending potential is discussed in Section 5.1.2. There are three example lines given in Table 4-11, the lines are flagged respectively as a normal angle bend parameter by the initial code word 'ANG', an angle bend parameter for a four-membered ring by the initial code word 'ANG4', and an angle bend parameter for a three-membered ring by the initial code word 'ANG3'. Each line has eight fields of data, and the data types contained in

4.7. USING ADDED CONSTANTS   193

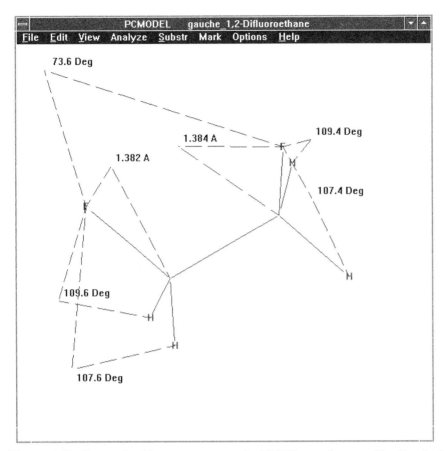

Figure 4-45. Geometric Measures on *gauche*-1,2-Difluoroethane, with Standard Parameters.

each are the same, so the explanation for the first line may be applied to the other two. The value in the first field, '1', is the sequence number. The second through fourth fields contain the three atom types which make up the angle of interest, here '1 1 1', signifying three @1 carbon atoms. The fifth field contains the value 0.450 for k_{ang}, the force constant for angle bending in units of millidynes per degree. The sixth, seventh, and eighth fields contains the values for three equilibrium or natural bond angles, referred to as 'TYPE-1', 'TYPE-2', and 'TYPE-3', in degrees. For some tetravalent atoms the angle bending parameter will depend upon the level of substitution on the atom, and in this example the value is 109.470 when the central @1 carbon is quaternary ('>CR2'), the value is 109.510 when the central @1 carbon is tertiary ('>CRH'), and the value is 109.500 when the central @1 carbon is secondary ('>CH2').

The MMX out-of-plane angle bending potential is discussed in Section

5.1.3. In the example line from Table 4-11, the line is flagged as an out of plane angle bending parameter by the initial code word 'OOP', and there are four fields of data. The first, '1', is the sequence number. The second and third fields give the atom type numbers of the central atom and the reference atom for this interaction, here '2 1' for an @2 carbon and an @1 carbon. The fourth field gives the value 0.600 for k_{oop}, the force constant for out-of-plane angle bending in millidynes per degree.

The final set of parameters are those used in the heat-of-formation calculations. There are three types found in the example line from Table 4-11, and each has a different format.

A parameter line is flagged as providing data for the bond contribution to the heat of formation by the initial code word 'BDHF', and there are six data fields. In the example line from Table 4-11, the first contains the value '1' for the sequence number. The second and third fields give the atom type numbers for the two atoms which make up the bond under consideration, here '1 1' for two @1 carbon atoms. The fourth field contains a brief description of the bond type, here 'C-C SP3 SP3' to indicate a $C(sp^3)$–$C(sp^3)$ bond. The fifth field gives the value of -0.004 for the enthalpy in kilocalories per mole. The sixth field gives the value of -16.400 for the entropy in entropy units.

A parameter line is flagged as providing data for the environmental contribution to the heat of formation by the initial code word 'EVHF', and there are three data fields. In the example line from Table 4-11, the first contains the value '1' for the sequence number. The second field contains a brief description of the structural feature, here 'ISO (ALKANE)', indicating a branched chain. The third field gives the value of 0.078 for the enthalpic contribution, in kilocalories per mole.

A parameter line is flagged as providing data for the resonance contribution to the heat of formation by the initial code word 'PIHF', and there are eight data fields. In the example line from Table 4-11, the first contains the value '1' for the sequence number. The second and third fields give the atom type numbers for the two atoms which make up the bond which is a component of the delocalized system under consideration, here '2 2' for two @2 carbon atoms. The fourth field contains a brief description of the bond type, here 'C = C DELOCALIZED' to indicate a $C(sp^2)$—$C(sp^2)$ bond which is part of a conjugated system. The fifth through eighth fields give 11.200, -40.00, -6.400, 250.000 of the parameters for this contribution.

In order to demonstrate the use of added constants, we will prepare a file which contains several altered parameters, and then employ this in an attempt to improve the MMX model for 1,2-difluoroethane. It is known from electron diffraction[287] and high-level molecular orbital studies[288–295] that the *gauche* conformer of this molecule is more stable than the *anti* conformer by approximately 1–2 kcal/mol. This *gauche effect* is reproduced in the later MM2 fo mulations, MM3, and MMFF94,[296] but it will be seen that MMX does not model it well prior to adding new constants.

4.7. USING ADDED CONSTANTS

Added Constants Exercise: Adjusted Parameters to Account for the Gauche Preference in 1,2-Difluoroethane. In this exercise we will prepare the structure 1,2-difluoroethane, and generate a file which contains both the *anti* and *gauche* conformers. This will be subjected to energy minimization first under the standard parameters and then with an added-parameters file designed to mimic the experimentally observed preference for the *gauche* conformer, and the differences will then be compared. This exercise will focus on the use of the added-parameters file; other aspects of this problem will be covered in Section 6.3.

The similar molecule 1,2-dichloroethane {DICLETHN.PCM} was prepared in Section 3.6.3. The requisite structure for the current exercise may be prepared by modification of that earlier structure (replacing the chlorines with fluorines from the 'Periodic Table' menu), or by following the same drawing procedure, which will be reviewed briefly here. In the PCMODEL window, activate the **Draw** tool, and use it to draw a zigzag of four atoms in a line as a lopsided N-shape. Bring up the 'Periodic Table' menu and click on the fluorine **F**, then use this to convert the two end atoms to fluorines, and then add hydrogens with the **H/AD** tool. The **View\Control_Panel** command can be used to do some rotations and improve the perspective. This will be the *anti* conformer, as shown in Fig. 4-42. Name this molecule as '*anti*_1,2-Difluoroethane,' and write an input file as {F2ETHAN1.PCM}.

Now mark the four heavy atoms in order with the **Select** tool, and activate the **Edit\Rotate_Bond** command. From the 'Rotate_Bond' dialog box, use the scroll bar to adjust the dihedral angle from $\approx 180°$ to $\approx 60°$. Cancel out of the dialog box, name this structure 'gauche_1,2-Difluoroethane', and write it to the same input file created above, {F2ETHAN1.PCM}. In answer to the prompt that the 'File Exists', click on the 'Append' button. This will store both conformations in the same input file. Use the **Edit\Erase** command to clear the screen.

Now use the **Analyze\Batch** command to run the two structures in the appended file through energy minimization. This will use {F2ETHAN1.PCM} as input, and produces a new input file named {F2ETHAN1.OUT} which

From {MMXCONST.PAR}

TOR		180	11	1	1	11	0.190	-1.550	-0.680	
BND	9		1	11	5.100		1.380	0.000	1.820	0.000	0.000
ANG		86		5	1	11	0.490	109.500	0.000	0.000

{F2ETHANU.PAR}

TOR		180	11	1	1	11	-0.150	-2.850	0.750	
BND	9		1	11	5.100		1.389	0.000	1.820	0.000	0.000
ANG		86		5	1	11	0.490	108.000	0.000	0.000

Figure 4-46. The Resident (top) and Modified (bottom) Parameters for 1,2-Difluoroethane.

contains the optimized structures, and a summary containing the final energies of the minimized structures named {F2ETHAN1.SUM} (see Fig. 4-43).

When this is finished, move the original {F2ETHAN1.PCM} and the {F2ETHAN1.SUM} file to another directory or delete them, then rename {F2ETHAN1.OUT} to {F2ETHAN1.PCM}, and run it through the batch minimization again. When this is finished, delete the existing {F2ETHAN1.PCM}, then rename the new {F2ETHAN1.OUT} to {F2ETHAN1.PCM}. When the two minimized structures are viewed in PCMODEL, some geometric features may be measured, and these are shown in Figs. 4-44 and 4-45. These values and the energies will be collected below in Table 4-12.

Thus, the standard MMX model predicts approximately equal energies for the two conformations, the *anti* being slightly lower in energy. This is inconsistent with the *gauche* conformational preference reported from electron diffrac-

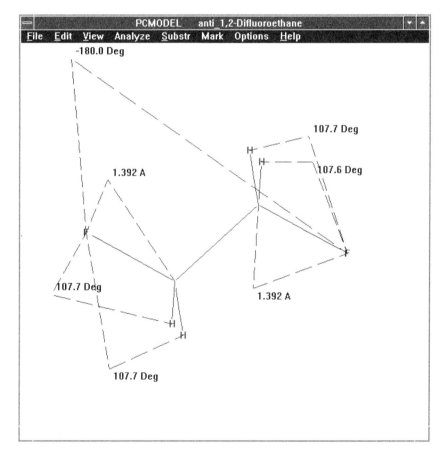

Figure 4-47. Geometric Measures on *anti*-1,2-Difluoroethane, with Modified Parameters from {F2ETHANU.PAR}.

4.7. USING ADDED CONSTANTS   197

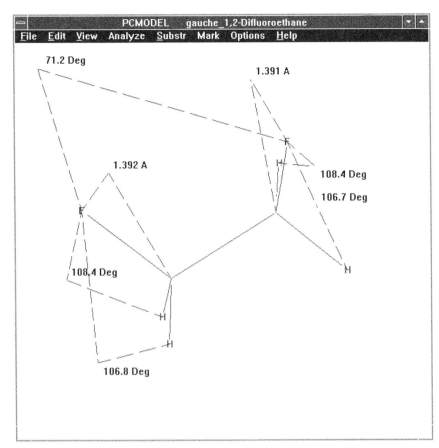

Figure 4-48. Geometric Measures on *gauche*-1,2-Difluoroethane, with Modified Parameters from {F2ETHANU.PAR}.

tion and MO calculations,[288-295] but this weakness of the MMX force field is known.[296] Some geometric measures are slightly off, such as the F—C—H angle and the F—C bond length, but the biggest discrepancy between the MMX model and published data is found in the F—C—C—F torsional angle for the *gauche* conformer.

The standard parameters from {MMXCONST.PAR} for the F—C—C—F torsional angle, the F—C bond, and the F—C—H angle are displayed in the top half of Fig. 4-46. The lower half of this figure shows the modified parameters contained in the added-constants file. Although there are systematic methods for optimizing parameters,[19,297] the values for the changed torsional parameters k_{tor1}, k_{tor2}, and k_{tor3} (sixth through eighth fields in the first line) were obtained by trial and error in this case. The new value for the natural bond length (fifth field in the second line) and for the natural angle span were adapted from the published electron diffraction data.[287] The lines of

TABLE 4-12. Energy and Geometry Values for the Conformers of 1,2-Difluoroethane

Conformer	MMX/E (kcal/mol)	F—C—C—F (deg)	F—C—H (deg)	F—C (Å)
Gauche (MMX)	−8.736	73.6	108.5	1.383
Anti (MMX)	−8.792	−180.0	108.7	1.383
Gauche (MMX*)	−9.958	71.2	107.4	1.392
Anti (MMX*)	−8.725	180.0	107.7	1.392
Gauche (ED/CP)		71.3	107.9	1.389
Anti (ED/CP)		180	107.9	1.389

MMX/E is the steric energy calculated by PCMODEL; the column headed F—C—C—F gives the values for the torsional angle; the column headed F—C—H gives the values for the bond angle; the column headed by F—C gives the values for the bond length; (MMX) indicates values calculated by PCMODEL with the standard parameters set; (MMX*) indicates values calculated by PCMODEL with the new parameters created for this experiment; (ED/CP) indicates values derived from experimental electron diffraction data using a cosine-potential model.[287]

{F2ETHANU.PAR} were prepared by overwriting a copy of the extant parameter lines.

Either prepare an {F2ETHANU.PAR} file using a text processor, or use the file provided in the input file library. In the PCMODEL window, activate the **Options\Added Constants** command, and use the 'Added Constants File' dialog box to select the {F2ETHANU.PAR} file. PCMODEL will now use these

TABLE 4-13. Energy Summary for the Conformers of 1,2-Difluoroethane

Method	ΔH(gauche–anti) (kcal/mol)
MMX	0.056
MMX*	−1.233
ED/CP	−1.76
Raman (g)	−0.81
Raman (l)	−1.98
DFT	−1.76 to −1.96
LCAO–SCF	1.04
RHF	−0.19

ΔH(gauche–anti) is the difference in energy between the *gauche* and *anti* conformers determined by the indicated method (a negative value indicates that the *gauche* conformer is more stable); MMX indicates values calculated by PCMODEL with the standard parameters set; MMX* indicates values calculated by PCMODEL with the new parameters created for this experiment; ED/CP indicates the value derived from experimental electron diffraction data using a cosine-potential model;[287] Raman (g) indicates the value obtained from gas-phase Raman spectroscopy;[293] Raman (l) indicates the value obtained from liquid phase Raman spectroscopy;[298] DFT indicates the range of values calculated from density functional theory, with five basis sets;[294] LCAO–SCF indicates the value obtained from molecular orbital calculations employing the extended 4-31G basis set;[288] RHF indicates the value obtained from restricted Hartree–Fock calculations employing the 6-311++G** basis set.[293]

TABLE 4-14. Comparison of Force Field Parameters

Parameter	MMX	MMX*	MM3	MMFF94
k_{bnd} for C—F (mdyn/Å)	5.100	5.100	5.900	6.011
r_0 for C—F (Å)	1.380	1.389	1.392	1.360
k_{ang} for H—C—F (mdyn · Å/rad²)	0.490	0.490	0.650	0.875
θ_0 for H—C—F (deg)	109.500	108.0	109.0	107.897
k_{tor1} for F—C—C—F (kcal/mol)	0.190	−0.150	0.000	−0.387
k_{tor2} for F—C—C—F (kcal/mol)	−1.550	−2.850	0.000	−0.543
k_{tor3} for F—C—C—F (kcal/mol)	−0.680	0.750	0.2250	1.405

MMX refers to the standard MMX parameter set; MMX* refers to the modified MMX parameters used in this experiment; MM3 refers to the standard MM3 parameter set (see Section 5.4.1.3); MMFF94 refers to the standard MMFF94 parameter set (see Section 5.4.6); k_{bnd} is the bond stretch force constant; r_0 is the equilibrium bond length; k_{ang} is the angle bend force constant; in θ_0 is the equilibrium angle span; and k_{tor1}, k_{tor2}, and k_{tor3} are the one-, two-, and three-fold torsional force constants, respectively.

amended constants in all calculations carried out in the current session. Use the **Analyze\Batch** command to run the {F2ETHAN1.PCM} file through energy minimization. In the 'Batch' dialog window, enter 'F2ETHAN2.PCM' in the 'Output File:' text entry field, and 'F2ETHAN2.SUM' in the 'Summary File:' text entry field. When this is finished, repeat the batch minimization, this time indicating {F2ETHAN2.PCM} as the input file, and accept the defaults in the 'Batch' dialog window. When this is finished, move the original {F2ETHAN2.PCM} file to another directory or delete it, then rename {F2ETHAN2.OUT} to {F2ETHAN2.PCM}.

When the two new minimized structures are viewed in PCMODEL, the same geometric features as before should be measured again, and these are shown in Figs. 4-47 and 4-48.

The geometric measures and the energies for these runs, along with the data from electron diffraction,[287] are found in Table 4-12, and a summary of values for the gauche–anti energy difference from various methods is found in Table 4-13.

A substantial improvement in the MMX model for this substrate was achieved in this experiment. A comparison of the force field parameters used in this experiment with the analogous parameters from MM3 (see Section 5.4.1.3) and MMFF94 (see Section 5.4.6) are found in Table 4-14. The trends are interesting, but it should be kept in mind that each force field has a unique construction, and such parameters should not be transferred from one force field to another without careful consideration.

CHAPTER 5

The Molecular Mechanics Force Field

The origins and history of the molecular mechanics force field model were discussed in Chapter 1. The first section of the present chapter begins with an overview of the components which make up modern force fields. This is followed by discussions of how the MMX component potentials are implemented, and the parametrized atom types used in the MMX force field are introduced and described. After this come descriptions of a number of the other current major force fields and their component potentials, with comparisons with MMX. The subject of parameters and how they are obtained is examined. Electrostatics and solvation are covered next, followed by discussions of energy calculations and minimum-energy structures. Next the subjects of conformation searching and global minimization are covered, and the chapter ends with a discussion of molecular dynamics.

For purposes of clarity, attempts were made to standardize the nomenclature and symbols used in different force fields, to enable more ready comparison between terms and across force fields. Thus, the appearance of the potential forms and equations used in this chapter may differ from those reported by the original authors in their publications. Every attempt has been made to ensure that the substitutions are consistent and accurate.

General discussions of the makeup of force fields, some in much greater detail or with a different focus than that presented here, are available from other sources,[7–8,15,19,24,34,104,105,107,111,112,115,125,299–302] and many of the details specific to the MMX force field are available in the documentation[303] and were published in a book chapter by the Serena Software developers.[304]

An excellent general description of the development of the theoretical basis for molecular mechanics force fields is given by Burkert and Allinger,[34] and the reader is directed to that source for a more in-depth discussion. To summarize, in molecular mechanics the interactions between the atoms of a molecule are represented by macroscopic models governed by the laws of classical physics, i.e. perturbed harmonic oscillators. These interactions are grouped as bonded interactions (different types of vibration such as bond stretching and angle bending) and nonbonded interactions such as steric (van der Waals) and electrostatic.

For the bonded interactions, the total force field is described by a two-dimensional matrix of these vibrations, to take account of the effects of coupling between different types of vibration. A first approximation considers only the diagonal terms in this matrix, which correspond to the uncoupled vibrations: pure bond stretching, pure angle bending, etc. The model composed of only these terms is called a **diagonal force field** or a **valence force field**. An unaugmented valence force field may constitute a poor model, since in reality many of the vibrations are coupled, and the resultant force field may not be transferable from one molecule to another. The judicious treatment of nonbonded interactions (steric and electrostatic) may compensate effectively, as in the AMBER force field, or selected cross terms may be included to achieve a more accurate model, as in the MM2-type force fields. Many molecular mechanics force fields contain off-diagonal or cross terms, but they differ from each other in which ones are employed. A balance is struck between including more cross terms in the model and simplifying the calculations so that accurate structure and property information may be obtained with a reasonably short amount of computer time.

Recently, a new paradigm in the development of molecular mechanics force fields has been formulated the **Class II force field**.[90,92,93] The potential functions in this new formulation are wholly or predominantly from the quantum mechanical potential energy surface, with appropriate scaling to achieve consistency with empirical data (see Sections 5.4.2.4 and 5.4.6).

Performance comparisons across force fields have appeared from time to time,[7,104,305-314] but a recent study is the most comprehensive.[296] This group surveyed the MM2*, MM3*, and AMBER* force fields of MACROMODEL; the "MM2" force field of CHEM3D [modified from Allinger's MM2(87)]; the DREIDING 2.21, UFF 1.01, and MMFF93 force fields of CERIUS²; the MM2(91) force field of MACMIMIC; the TRIPOS force field of Alchemy III; the MMX force field of PCMODEL; the CVFF and CFF(91) force fields in DISCOVER 2.95; and a Macintosh implementation of MM3(92). Each of these modeling systems was applied to eight tasks, where their results could be compared with experimental values: rotational barriers, conformational energies for hydrocarbons, conformational energies of oxygen-containing compounds, conformational energies of nitrogen-containing compounds, conformational energies of cyclohexane derivatives, conformational energies of haloalkanes, conformational energies of halocyclohexanes, and conformational energies of conjugated compounds. The conclusion was that the best overall performance was gotten from the members of the extended MM2/MM3 family (including PCMODEL's MMX), and from MMFF93 (the precursor to the MMFF94 discussed in Section 5.4.6). The strengths and weaknesses of each of the subject force fields were also discussed.

5.1. COMPONENTS OF THE MOLECULAR MECHANICS FORCE FIELD

In general, the components of molecular mechanics force fields fall into three groups: the valence terms, the valence cross terms, and the nonbonding terms.

202 THE MOLECULAR MECHANICS FORCE FIELD

Each type of term will be discussed and illustrated before moving on to the descriptions of the MMX (Section 5.2) and other force fields in use (Section 5.4).

5.1.1. Valence Terms

Among the valence terms are the bond stretching potential V_{bnd}, the angle bending potential V_{ang}, the torsional interaction potential V_{tor}, the out-of-plane angle bending potential V_{oop}, and the improper torsional interaction potential V_{impt}. Generally only one of the last two will be used in a force field, to reinforce planarity or nonplanarity. A generic depiction of each of these interactions is provided below, and these definitions will be referred to in the descriptions of the individual force fields in Sections 5.2–5.4.

5.1.1.1. Bond Stretching Potential[315,316]

A generic molecular moiety to demonstrate the bond stretching potential is shown in Fig. 5-1. The two atoms are referred to as I and J, and the internuclear distance is given as r_{IJ}. As was seen in Chapter 1, the valence force field uses models from classical physics to describe the interatomic forces operating at the molecular level. Atoms are considered to be spheres having a specified hardness, and the bonds connecting them are viewed as springs having a specified elasticity. **Hooke's law** governs the behavior of a harmonic oscillator wherein the restoring force is proportional to the displacement from equilibrium, and this simple potential function gives a parabolic potential curve. The Hooke's law equation has the form

$$V_{bnd} = K_{bnd} \cdot \tfrac{1}{2}k_{bnd}(r - r_0)^2$$

where K_{bnd} is a constant for units conversion; k_{bnd} is the force constant for the bond; r (equivalent to r_{IJ} in Fig. 5-1) is the distance between the two atoms, and r_0 is the equilibrium or natural bond length; and $r - r_0$ gives the displacement from equilibrium. The Hooke's law harmonic potential curve is given by the dotted line in Fig. 5-2.

The Hooke's law harmonic potential approximation works well in the portion of the curve near the minimum-energy point, but elsewhere an anharmonic potential is needed. More sophisticated calculations and the results of

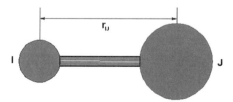

Figure 5-1. Definitions for the Bond Stretching Potential.

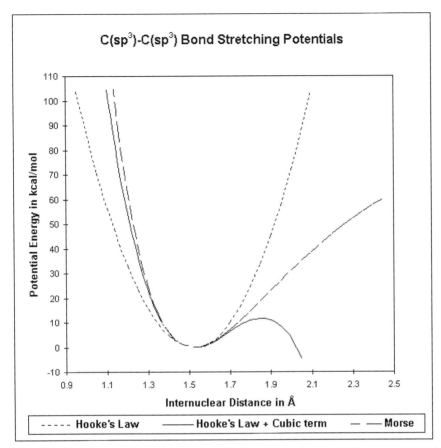

Figure 5-2. $C(sp^3)$—$C(sp^3)$ Bond Stretching Potentials: Simple Hooke's Law, Hooke's Law + Cubic Term, and the Morse Potentials.

experimental measurements indicate that bond compression should meet greater resistance than that given by the simple harmonic potential. Bond stretching must eventually lead to bond breaking, so that the potential energy should approach zero asymptotically as the internuclear distance approaches infinity, and not meet an infinitely large restoring force as predicted by the simple harmonic potential. The anharmonic potential derived from quantum mechanical considerations is modeled fairly well by using the empirical Morse function,[317-320]

$$V_{bnd} = D_{bnd}(1 - e^{-\alpha \delta r})^2$$

where

$$\alpha = (k_{bnd}/2D_{bnd})^{1/2}$$

Here D_{bnd} is the bond dissociation energy in kilocalories per moles, δr is the internuclear distance $r - r_0$, and k_{bnd} is the force constant for bond stretching.

Calculating the exponential-based Morse potential is computationally intensive, and a more facile equation for an anharmonic oscillator is obtained with a power series function of $r - r_0$; specifically the exponential terms raised to an odd power confer the anharmonicity. Truncation of the series at the cubic term provides an improved fit within the normal range of bond lengths while keeping the calculation manageable. The form of this modified equation is shown below:

$$V_{bnd} = (K_{bnd} \cdot \tfrac{1}{2}k_{bnd})(r - r_0)^2 + (K_{bnd} \cdot \tfrac{1}{2}k_{bnd}k_{cub})(r - r_0)^3$$

or

$$V_{bnd} = K_{bnd} \cdot \tfrac{1}{2}k_{bnd}[1 + k_{cub}(r - r_0)](r - r_0)^2$$

where k_{cub} is the force constant for the cubic correction term.

Figure 5-2 shows the three types of stretching potential curves for the $C(sp^3)$—$C(sp^3)$ bond using the Hooke's law potential (short-dashed line), the modified potential (Hooke's law + cubic term, solid line), and the Morse potential function (long-dashed line). In this calculation, a value of 88 kcal/mol is used for the bond dissociation energy D_{bnd} in the Morse potential.

Figure 5-3 shows the convergence of these three potential curves near the equilibrium point ($\pm \approx 0.05$ Å). In either direction outside this range, as in the case molecules with moderate strain or steric crowding, the modified Hooke's law function follows the Morse potential fairly closely, and models the bond stretch potential significantly better than the simple Hooke's law function.

It can be seen in Fig. 5-2 that the curve for the modified Hooke's law potential inverts just before the $C(sp^3)$—$C(sp^3)$ internuclear distance reaches ≈ 1.90 Å, and dips below zero energy as the displacement reaches ≈ 2.00 Å (intercept at 2.023 Å). This is the so-called **cubic stretch catastrophe**,[321] where the stretching potential goes to negative infinity, which would cause the two atoms to repel each other as the bond stretches instead of undergoing a gradual bond cleavage. To avoid this problem in the early stages of energy optimization where the bond stretch may be extensive, the simple Hooke's law function is used by MMX (as with most other modern force field programs) when the longer bond lengths in the input geometry would be modeled poorly. The Hooke's law potential is used to keep the atoms together until after the gross corrections in the bond length have been made, and the more accurate modified Hooke's law function is used over the rest of the bond length range in the later stages of minimization.

A recent report has proposed the use of a quadratic potential proportional to the inverse of the internuclear distance r_{IJ}.[322] This study employed *ab initio* calculations to test various formulations of the bond energy potential.

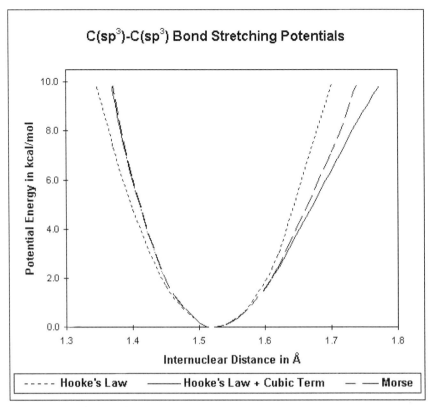

Figure 5-3. $C(sp^3)$—$C(sp^3)$ Bond Stretching Potentials: Simple Hooke's Law, Hooke's Law + Cubic Term, and the Morse Potentials, Near the Equilibrium Point.

5.1.1.2. Angle Bending Potential[315,323] A generic molecular moiety to demonstrate the angle bending potential is shown in Fig. 5-4. The three atoms are referred to as I, J and K, and the angle span is given as θ_{IJK}. Some form of Hooke's law function is generally used to model the contribution to the potential energy from angle bending:

$$V_{\text{ang}} = K_{\text{ang}} \cdot \tfrac{1}{2}k_{\text{ang}}(\theta - \theta_0)^2$$

where K_{ang} is the constant for unit conversion, k_{ang} is the force constant for the three-atom angle; θ (equivalent to θ_{IJK} in Fig. 5-4) gives the angle span across the three atoms, and $\theta - \theta_0$ gives the displacement of the angle span from the equilibrium value.

In some force fields, such as the Allinger MM-type and MMX, the angles θ_{IJK} are placed in one of three groups, depending on the substituents on the

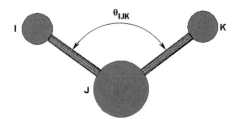

Figure 5-4. Definitions for the Angle Bending Potential.

central atom J besides I and K: type 1 has two nonhydrogen substituents, type 2 has one hydrogen and one nonhydrogen substituent, and type 3 has two hydrogen substituents. These types will often be differently parametrized.

5.1.1.3. Torsional Interaction Potential[324,325]

A generic molecular moiety to demonstrate the torsional interaction potential is shown in Fig. 5-5. The four atoms are referred to as I, J, K, and L, and the torsional angle span is given as ω_{IJKL}.

Pitzer provided early guidance towards using a potential expression which was a function of the cosine of the torsional angle, with a factor for the periodicity.[40] He also described what the torsional potentials of simple hydrocarbons should look like, and what structural factors would affect torsional energy barriers. Modern molecular mechanics force fields generally employ one or more of the parts of the following torsional potential function, which contains terms for one-, two-, or threefold periodicity:

$$V_{\text{tor}} = \tfrac{1}{2}[k_{\text{tor}1}(1 + \cos \omega) + k_{\text{tor}2}(1 - \cos 2\omega) + k_{\text{tor}3}(1 + \cos 3\omega)]$$

where $k_{\text{tor}1}$, $k_{\text{tor}2}$, and $k_{\text{tor}3}$ are the parameters for the one-, two-, and threefold torsional barriers, respectively, in kilocalories per mole and ω (equivalent to ω_{IJKL} in Fig. 5-5) is the torsional angle in radians. More sophisticated force

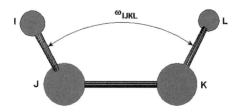

Figure 5-5. Definitions for the Torsional Interaction Potential.

fields (such as MM4) use fourfold and higher terms for certain torsional sets (see Section 5.4.1.4). Many other force fields (such as AMBER) include only those torsional terms whose periodicity is dictated by the geometry and hybridization of the attached atoms.

It is tempting to associate physical significance with each of these terms in the Fourier series for the torsional potential, which one may do with caution. The onefold term could represent dipole–dipole interactions and/or some nonbonded effect. The twofold term could correlate with hyperconjugation in alkanes, or conjugation in unsaturated compounds. The hyperconjugative-type effect should be more pronounced if one of the terminal atoms of the dihedral were electronegative, enabling the interaction between an electron lone pair and an antibonding σ-orbital at the other end of the dihedral. Consistent with this is the fact that the k_{tor1} and k_{tor2} parameters tend to be numerically much larger in dihedrals which contain heteroatoms, where the dipole and hyperconjugative effects would be larger. The threefold term k_{tor3} may be viewed as essentially steric in nature. It has been variously described as either a special case of the van der Waals interaction, or due to repulsions between the electron pairs of the bonds to the terminal atoms of the dihedral.[108,324]

The torsional potential energy component V_{tor} models the barrier(s) to free rotation about a dihedral angle, e.g. the central bond of four contiguous attached atoms. An excellent discussion of the development of this term may be found in Burkert and Allinger,[324] upon which the following summary is based. Earlier force fields lacking a specific torsional term failed to model dihedral rotational barriers acceptably. The V_{vdW} parameters could be modified to take account of the nonbonded interactions between vicinal substituents (i.e. between I and L above) in such a way that torsional barriers were accurately reproduced. However, the use of these parameters seriously degraded the accuracy of the model in the interactions between atoms more distantly linked, so that other properties of the molecule were no longer properly predicted. The earliest molecular mechanics force fields for saturated hydrocarbons employed a torsional potential containing only a threefold term, which is identified with bond–bond repulsion. By comparison, force fields from vibrational spectroscopy employ a Fourier series of terms based upon cosines of multiples of the dihedral angle, containing up to six terms. In later work on the development of molecular mechanics force fields, it was found that torsional potentials containing the first three terms provide significant improvement in the model while minimizing the number of terms considered.

When viewed in isolation, the potential curves for some I—J—K—L substrates appear unnatural. This is because each such curve is a partial torsional potential, and it is the sum of all such partial potentials which is expected to provide a predictive molecular model. As will be seen below in the discussion of the MMX torsional potential in Section 5.2, there are several four-atom moieties which contain the same two central carbon atoms, and the sum of these contributions constitutes the modeled torsional energy barrier in this molecule. In addition, van der Waals interactions and, especially in the case of electronegative elements, electrostatic effects will also affect the overall barrier to dihedral rotation, so the partial contributions are not always representative of the overall torsional potential.

5.1.1.4. Out-of-Plane Angle Bending Potential An early description of the out-of-plane angle bending component was given by Dunitz in his study of distorted-medium-ring cycloalkenes.[326] He pointed to two contributors to the strain energy in such nonplanar olefins: the twisting (torsion) about the double bond, and the out-of-plane distortion at the trigonal atoms. Although not evaluated rigorously, the out-of-plane contribution seemed to be more significant accoding to contemporary estimates.

The generic molecular moiety shown in Fig. 5-6, consisting of a central trigonal atom J and its three attached atoms I, K, and L, demonstrates four conventions for quantifying the out-of-plane angle bending.

The MM2 **out-of-plane angle** in (a), $\theta_{\text{oop/MM2}}$, is given as the angle defined by the bond between the central atom J and one of its attached atoms, e.g. I, and the projection of that bond onto the plane determined by the three attached atoms, I, K, and L. This convention is used in MM2, MM3, and MMX (see Sections 5.2 and 5.4.1).

The CFF out-of-plane angle in (b), $\theta_{\text{oop/CFF}}$, is given as the angle between two planes: the plane defined by the three atoms J, K, and L, and the plane defined by atoms, I, K, and L. This convention is used in CFF and related force fields (see Section 5.4.2).

The distance h in (c) has been referred to as the **pyramid height**,[327] and is defined as the displacement of the central atom J from its projection onto the plane determined by the three attached atoms, I, K, and L. This convention is used in the Tripos force field (see Section 5.4.5).

The **Wilson angle** in (d),[327,328] θ_{Wilson}, is the angle defined by the bond between the central atom J and one of its attached atoms, e.g. I, and the plane determined by the bonds from J to the other two substituents, i.e. J—K and J—L. This convention is used in the DREIDING force field (see Section 5.4.8).

The out-of-plane bending potential acts as a restoring force to keep certain atom types J planar, or to enforce nonplanarity (asymmetry) upon a central atom J, and a simple Hooke's law function is generally used to model this contribution to the potential energy. It takes the same form as the angle bending potential, but since the equilibrium angle is zero (planar arrangement), the equation simplifies to

$$V_{\text{oop}} = K_{\text{oop}} \cdot \tfrac{1}{2} k_{\text{oop}} (\theta_{\text{oop}})^2$$

where K_{oop} is the constant for unit conversion, k_{oop} is the force constant for the out-of-plane angle bending, and θ_{oop} gives the angle span for the pyramidal deformation from planarity.[329]

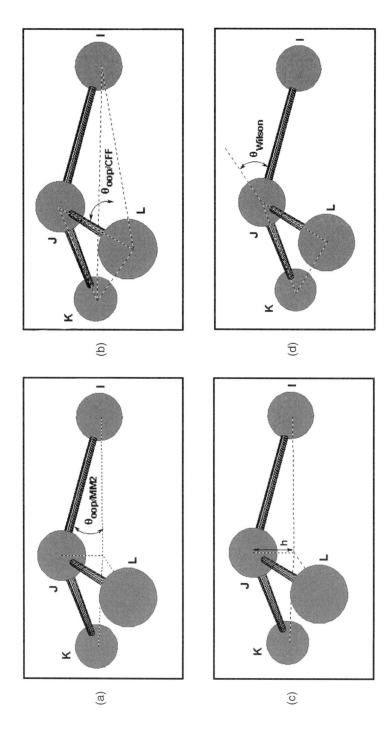

Figure 5-6. Definitions for Measures of Pyramidal Distortion: (a) MM2/MMX Out-of-Plane Angle; (b) CFF Out-of-Plane Angle; (c) Pyramid Height; (d) Wilson Angle.

209

Newer formulations of this potential are being developed.[330] An alternative implementation of this potential is the improper torsion (see below).

5.1.1.5. Improper Torsional Interaction Potential

A generic molecular moiety to demonstrate the improper torsional interaction potential is shown in Fig. 5-7, where the four atoms are referred to as I, J, K, and L. There are two conventions for designating the improper torsional angle. That employed in the CHARMM and MM4 force fields is depicted on the left, where the three improper torsional angles ω are defined by the sequences of atoms J—I—K—L (ω_1), J—K—L—I (ω_2), and J—L—I—K (ω_3). In the convention employed in the AMBER force field, the three improper torsional angles ω are defined by the sequences of atoms: K—L—J—I ($\omega_{1'}$), L—I—J—K ($\omega_{2'}$), and I—K—J—L ($\omega_{3'}$).

The improper torsional interaction potential acts as a restoring force to keep certain atom types J planar, or to enforce tetrahedral asymmetry upon a central atom J when it is being modeled by a trisubstituted united atom type. This potential can take the form

$$V_{\text{impt}} = k_{\text{impt}(JIKL)}(1 - \cos 2\omega_1) + k_{\text{impt}(JKLI)}(1 - \cos 2\omega_2) + k_{\text{impt}(JLIK)}(1 - \cos 2\omega_3)$$

where the k_{impt} factors are the improper torsion force constants, and the ω are the values for the corresponding improper torsional angles as defined above.

An alternative implementation of this potential is out-of-plane angle bending (see above).

5.1.2. Valence Cross Terms

A force field composed solely of valence terms and nonbonded terms can provide a balanced, accurate molecular model, but some applications require the refinement of valence cross terms.[331] These terms allow for the fact that the coupling of some valence term interactions may have a significant effect upon the calculated structure or properties. The inclusion of cross terms is especially important if the force field is designed to reproduce vibrational spectra and rotational barriers for congested molecules.[93]

Sometimes the goals of computational molecular modeling may be addressed either by the addition of one or more cross terms or by the addition or enhancement of nonbonded terms, such as the choice between a bond-stretch–angle-bend valence cross term and a Urey–Bradley nonbonded geminal term. In order to reduce the difficulty and length of calculations it is best to keep the number of cross terms included in a force field to the minimum consistent with an accurate molecular model. One such cross term, the stretch–bend potential $V_{\text{bnd/ang}}$, will be used as an illustration, and other examples of such terms will be mentioned without extensive discussion.

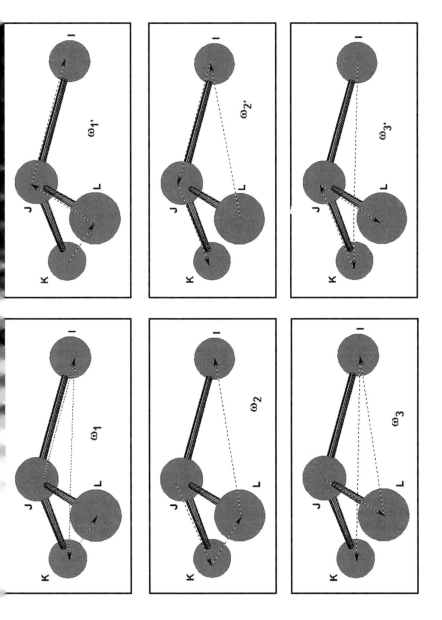

Figure 5-7. Definitions for the Improper Torsional Angle Potential: ω_1, ω_2, and ω_3 are for the CHARMM and MM4 force fields; $\omega_{1'}$, $\omega_{2'}$, and $\omega_{3'}$ are for the AMBER force field.

5.1.2.1. Stretch–Bend Cross-Term Potential A generic molecular moiety to demonstrate the bond-stretch–angle-bend cross term potential is shown in Fig. 5-8. The three atoms are referred to as I, J, and K, the angle span is given as θ_{IJK}, and the internuclear distances for the two bonded pairs of atoms are given as r_{IJ} and r_{JK}. This potential essentially combines the forms of the bond stretch and angle bend potentials, and is given as

$$V_{\text{bnd/ang}} = K_{\text{bnd/ang}} \cdot \tfrac{1}{2} k_{\text{bnd/ang}} [(r' - r'_0) + (r'' - r''_0)](\theta - \theta_0)$$

where $K_{\text{bnd/ang}}$ is a unit cancellation factor, $k_{\text{bnd/ang}}$ is the stretch–bend force constant, the quantities r' and r'' are the I—J and J—K internuclear distances (corresponding to r_{IJ} and r_{JK} in Fig. 5-8, respectively), and the quantities $r' - r'_0$ and $r'' - r''_0$ give the corresponding displacements from the equilibrium values; and θ (equivalent to θ_{IJK} in Fig. 5-8) gives the angle span across the three atoms, while $\theta - \theta_0$ gives the displacement of that angle span from the equilibrium value.

This cross term takes account of the fact that in a three-atom system I—J—K, the bond stretching of the pair of two-atom bond components will be interdependent with the angle bending of the system as a whole. It is difficult to model adequately the two-center nonbonded repulsions (1,3-interactions) between the geminal substituents I and K in an I—J—K system by simple adjustment of the general V_{vdW} parameters without affecting the accuracy of the model for the nonbonded interactions between atoms more distantly linked. As the I—J—K angle opens up (becomes more obtuse), the I—J and J—K bonds may shrink to relieve strain, and likewise as the I—J—K angle is compressed (becomes more acute), the I—J and J—K bonds may stretch to relieve strain. This is what is meant by these vibrations being coupled.

5.1.2.2. Other Cross-Term Potentials In one illustrative example, the recently published MM4 force field contains seven binary cross terms ($V_{\text{ang/ang}}$, $V_{\text{bnd/ang}}$, $V_{\text{tor/bnd}}$, $V_{\text{bnd/bnd}}$, $V_{\text{tor/ang}}$, $V_{\text{tor/tor}}$, $V_{\text{tor/impt}}$) and two ternary cross terms ($V_{\text{ang/tor/ang}}$, $V_{\text{impt/tor/impt}}$). As these encompass nearly all of the cross terms encountered in major force fields, each of these will be discussed here briefly,

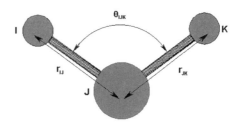

Figure 5-8. Definitions for the Bond-Stretch–Angle-Bend Cross-Term Potential.

5.1. COMPONENTS OF THE MMF FIELD

in generic terms. More details may be found in the sections on individual force fields, in particular MM4 (see Section 5.4.1.4).

The angle-bend–angle-bend term ($V_{\text{ang/ang}}$) addresses the coupling between the angle bending vibrations in the three-atom systems I—J—K and K—J—L which share the common bond J—K. It is a function of the two angles θ_{IJK} and θ_{KJL}, as depicted in Fig. 5-9.

The stretch–bend term ($V_{\text{bnd/ang}}$) was discussed in Section 5.1.2.1.

Two types of torsion–stretch cross term ($V_{\text{tor/bnd}}$) are possible. As defined in Fig. 5-10, a *type 1* torsion–stretch term addresses the coupling between the torsional rotation across I—J—K—L with the stretching of the central bond J—K, and is a function of the torsional angle ω_{IJKL} and the central bond distance r_{JK}.

As defined in Fig. 5-11, a *type 2* torsion-stretch term addresses the coupling between the torsional rotation across I—J—K—L with the stretching of one of the terminal bonds, I—J; it is a function of the torsional angle ω_{IJKL} and the terminal bond distance r_{IJ}.

The stretch–stretch term ($V_{\text{bnd/bnd}}$) addresses the coupling between two bond stretching vibrations (I—J and J—K) in the three-atom system (I—J—K) where a single atom (J) is the terminus of both bonds (see Fig. 5-12). It is a

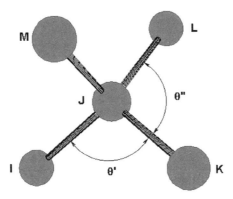

Figure 5-9. Definitions for the Angle-Bend–Angle-Bend Cross-Term Potential.

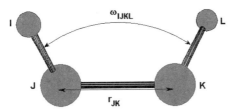

Figure 5-10. Definitions for the Type 1 Torsion–Bond-Stretch Cross Term.

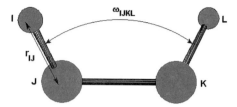

Figure 5-11. Definitions for the Type 2 Torsion–Bond-Stretch Cross Term.

function of the two internuclear distances r_{IJ} and r_{JK}. This is the same molecular moiety addressed by the stretch–bend term (see Section 5.1.2.1), but in this case no angle bending is involved. These vibrations corresponds to the familiar symmetric stretch and asymmetric stretch from vibrational spectroscopy.

The torsion–angle-bend term ($V_{tor/ang}$) addresses the coupling between the torsional interaction in I—J—K—L and the angle bending for one of the two three-atom angles contained in this moiety (I—J—K: see Fig. 5-13). It is a function of the torsional angle ω_{IJKL} and the angle θ_{IJK}.

The torsion–torsion term ($V_{tor/tor}$) addresses the coupling between two four-atom torsional systems (I—J—K—L and J—K—L—M), which share a common three-atom angle (J—K—L), as depicted in Fig. 5-14. It is a function of the two torsional angles ω_{IJKL} and ω_{JKLM}.

The torsion–improper-torsion term ($V_{tor/impt}$) addresses the coupling between the torsional interaction (I—J—K—L) and the improper torsional

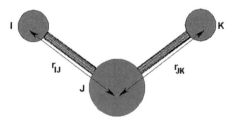

Figure 5-12. Definitions for the Bond-Stretch–Bond-Stretch Cross Term.

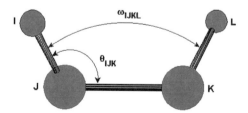

Figure 5-13. Definitions for the Torsion–Angle-Bend Cross Term.

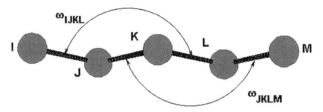

Figure 5-14. Definitions for the Torsion–Torsion Cross Term.

potential at one of the internal atoms, e.g. J. It is a function of the torsional angle ω_{IJKL} and the improper torsional angle ω_{JIMK}, as depicted in Fig. 5-15. It acts to decrease pyramidalization at J when the torsional angle is rotated toward $90°$.[332]

The bend–torsion–bend term ($V_{\text{ang/tor/ang}}$) addresses the coupling between the torsional interaction (I—J—K—L) and the angle bending for the two three-atom angles contained in this moiety (I—J—K and J—K—L). It is a function of the torsional angle ω_{IJKL} and the angles θ_{IJK} and θ_{JKL}, as defined in Fig. 5-16.

The improper-torsion–torsion–improper-torsion term ($V_{\text{impt/tor/impt}}$) addresses the coupling between the torsional interaction (I—J—K—L) and the improper torsional potentials at both of the internal atoms J and K in molecules which have extended conjugation. It is a function of the torsional angle ω_{IJKL} and the improper torsional angles ω_{JIMK} and ω_{KLNJ}, as depicted

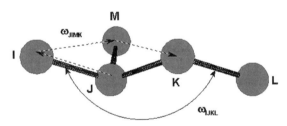

Figure 5-15. Definitions for the Torsion–Improper-Torsion Cross Term.

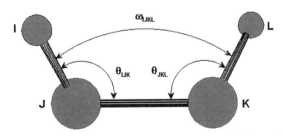

Figure 5-16. Definitions for the Angle-Bend–Torsion–Angle-Bend Cross Term.

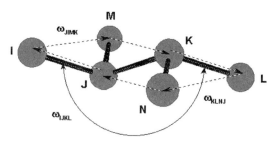

Figure 5-17. Definitions for the Improper-Torsion–Torsion–Improper-Torsion Cross Term.

in Fig. 5-17. This term counters out-of-plane bending in conjugated systems.[332,333]

5.1.3. Nonbonded Interaction Potentials[334]

Nonbonded interactions fall into two groups, those due to nonpermanent charges (steric or van der Waals) and those due to permanent fixed charges (electrostatic). These are described and illustrated below.

5.1.3.1. van der Waals Nonbonded Interactions Potential[334,335]

In the simplest mechanical model of molecules, atoms are hard spheres. In this model, two atoms not bonded together have no effect upon each other until the distance between them narrows to the sum of the radii of the two spheres, at which point an infinitely high repulsive force keeps them apart. It is known from experiment that molecules in the gas phase suffer inelastic collisions (scattering phenomena), and in the solid phase display a range of closest contact points (crystal diffraction data). A model incorporating relative hardness enables much more accurate prediction of properties and behavior. The colligative properties of matter also require an attractive force between nonbonded atoms in the intermediate range of separation between closest contact and distant noninteraction.

In the molecular orbital description of atoms and molecules, the continuum of electron orbital density requires a continuum of the forces which result from the nonbonding interaction of these orbitals: the electron–electron repulsions at close range, and the attraction between complementary pairs of induced transient dipoles at intermediate range. Taken over time, the electron distribution should be homogeneous, but at any particular instant there will be minute variations in the surface electron density. Such local variations represent small temporary dipoles, and these will induce compensating changes in nearby electron cloud surfaces, either from a neighboring molecule or from another portion of the same molecule. Such temporary dipole pairs will experience a distance-dependent attraction for the duration of their short existence.

Many such compensating pairs will exist at any one moment, and they will continually form and disperse, with the net result of a weakly attractive interaction between atomic or molecular surfaces over a short distance, limited by electron–electron repulsions as the intersurface distance approaches zero. This phenomenon is called the van der Waals attractive force, and the minimum distance over which this force will be attractive is van der Waals radius. The development of potential energy functions to describe these interactions over the course of this century has been reviewed,[334] and the expressions employed in the force fields described in this chapter show a large variation.

A generic molecular moiety to define the van der Waals interaction potential is shown in Fig. 5-18. The two atoms are referred to as I and J, and are separated by a distance r_{IJ}. Each atom type in a force field will generally have a characteristic van der Waals radius r_{vdW}, which represents the distance of closest approach based upon close contacts within the molecules of the basis set. An atom type will also have a force constant for its van der Waals potential, k_{vdW}, also known as the *hardness* and traditionally symbolized by ε. Often, a van der Waals interaction potential will consist of two parts which depend upon different exponential powers of the inverse of r_{IJ}, as will be seen in the descriptions of the individual force fields.

The spherical model for atoms works well for larger atoms, but deformation density effects become significant for the first-row elements such as nitrogen, oxygen, fluorine, and especially hydrogen, where this model becomes inadequate. In the case of nitrogen and oxygen, the model in some force fields, such as MM2[10,336] and MMX,[303,304] is amended to allow for these perturbations by including one and two lone pairs of electrons,[337] respectively, as small pseudoatoms. In this way, torsional interactions are reproduced more accurately, and these lone pairs also play an important role in the hydrogen bonding and metal complexation potentials in MMX (see Sections 6.6 and 6.7, respectively). In other force fields (such as MM3 and MM4), the lone pair pseudoatoms are not used, and the effects of the asymmetric electron clouds are modeled with more sophisticated electrostatics and an adjusted torsional potential.

The greatest discrepancy occurs for the smallest element, hydrogen and its isotope deuterium, as was discussed in Section 3.3. At least in organic compounds, hydrogen is always the more electropositive when it is one of the two atoms involved in a bond. This translates into a diminished electron cloud

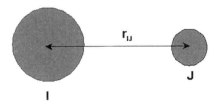

Figure 5-18. Definitions for the van der Waals Nonbonded Interaction Potential.

around the hydrogen nucleus. It is the interactions between these electron clouds which give rise to van der Waals attractions and repulsions, and so a correction is made for this deformation from sphericity in specifying the van der Waals radii for these two atom types.

To account for this effect on the hydrogen in a C—H bond, the standard bond length r may be *reduced* by a factor of r_δ, and the hydrogen assumes a new *virtual position* with different Cartesian coordinates at a distance r_{red} from the carbon atom (see Fig. 5-19). The hydrogen's van der Waals surface is constructed about this virtual position, and this mimics the distorted van der Waals surface about the actual hydrogen position. This may be seen by comparing the van der Waals surface representations for the C–H bond at the lower left and the lower right in Fig. 5-19. All calculations of the van der Waals interactions will be based on the hydrogens' virtual position, which will allow closer approach. Often, a modified force constant ε will be used for these reduced hydrogen interactions.

A recent publication discusses the use of test particle calculations to derive van der Waals parameters.[338]

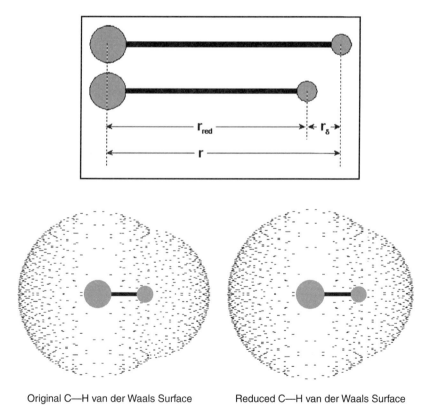

Original C—H van der Waals Surface   Reduced C—H van der Waals Surface

Figure 5-19. Definitions for the Reduced C—H van der Waals Surface.

5.1.3.2. Urey–Bradley Geminal Nonbonded Interaction Potential In many force field treatments, including the Allinger MM series and MMX, steric and electrostatic nonbonded interactions are not evaluated between geminal atoms (1,3-atoms). In the course of developing such fields, it was discovered that the standard van der Waals and electrostatic potentials gave poor results when applied to geminal moieties. The net effects of these geminal interactions are represented by a bond-stretch–angle-bend cross term $V_{bnd/ang}$ and by other aspects of the overall parametrization.

Another option is to employ a specific harmonic stretching potential to handle these geminal interactions. Such a potential function is called a **Urey–Bradley term**, named for the codevelopers of an early force field.[19,33,34] Some versions of the Consistent Force Field (CFF, see Section 5.4.2) and the CHARMM force field (see Section 5.4.4) employ a Urey–Bradley term. A force field including a Urey–Bradley term is mathematically equivalent to a valence force field, in that both represent simplifications of the complete spectroscopic force field, differing only in which terms are retained to provide the working force field algorithm.[339] In one sense, it is a more consistent formulation, since the force field contains only diagonalized, harmonic potentials. It does, however, increase the amount of computation needed.

It was found that force fields which include a Urey–Bradley term still require off-diagonal parametrization (i.e. cross terms) to handle the vibrational frequencies of molecules such as ethylene and aromatic molecules.[339]

A generic molecular moiety to define the Urey–Bradley potential is shown in Fig. 5-20. It is generally represented by a one- or two-part harmonic function of the internuclear distance[33,71,104] r_{IK}, and takes the form

$$V_{UB} = K_{UB1}k_{UB1}(r - r_0) + K_{UB2} \cdot \tfrac{1}{2}k_{UB2}(r - r_0)^2$$

where K_{UB1} is a unit cancellation constant for the linear term; k_{UB1} is the force constant for the linear term; r is the internuclear distance (corresponding to r_{IK} in Fig. 5-20), and $r - r_0$ is its departure from the equilibrium distance; K_{UB2} is a unit cancellation constant for the quadratic term; and k_{UB2} is the force constant for the quadratic term.

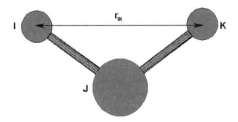

Figure 5-20. Definitions for the Geminal Nonbonded (Urey–Bradley) Potential.

5.1.3.3. Electrostatic Interaction Potentials In all but the simplest, most symmetrical hydrocarbons, electrostatic interactions play a very important role in structure and properties. There are two main models for assessing the contribution of these effects: the charge–charge model, which employs a Coulomb potential, and the dipole–dipole model, which employs a Jeans potential.[340] When force fields which use the dipole–dipole model (such as the Allinger MM series) are applied to structures which bear full permanent charges (zwitterions, charged intermediates, etc.), a composite treatment including dipole–dipole, dipole–charge, and charge–charge interactions must be used. In this section, the two electrostatic models will be described, but the reader is referred to the specific implementations covered in later sections, as well as Section 5.6 on the subject of electrostatics and solvation.

A generic molecular moiety to define the Coulomb potential for the charge–charge electrostatic model is shown in Fig. 5-21. The two atoms I and J may or may not be in the same molecule. The Coulomb potential is calculated from the following general equation:

$$V_{\text{elec/QQ}} = K_{\text{elec/QQ}} \frac{(q_I q_J)}{D r_{IJ}}$$

where $K_{\text{elec/QQ}}$ is a unit cancellation factor; q_I, q_J are the charges on atoms I and J; D is the dielectric constant of the medium; and r_{IJ} is the distance between the two charges, generally taken as the internuclear distance between atoms I and J.

A variation on this is the distance-dependent dielectric model, in which the potential varies as the inverse square of the internuclear distance and takes the form

$$V_{\text{elec/QQ}} = K_{\text{elec/QQ}} \frac{(q_I q_J)}{r_{IJ}^2}$$

A generic molecular moiety to define the Jeans potential for the dipole–dipole electrostatic model is shown in Fig. 5-22. This shows two dipolar bonds, I—J and K—L, with their dipole vectors indicated as μ_a and μ_b. These two bond dipoles may or may not be in the same molecule. The Jeans poten-

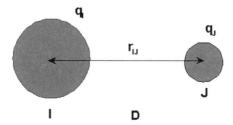

Figure 5-21. Definitions for the Charge-Charge Model Electrostatic Potential.

5.1. COMPONENTS OF THE MMF FIELD

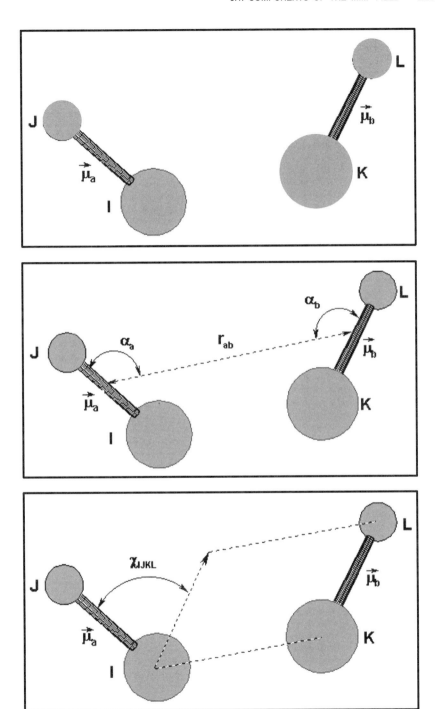

Figure 5-22. Definitions for the Dipole–Dipole Model Electrostatic Potential.

tial for this interaction is calculated from the following general equation:

$$V_{\text{elec/DD}} = K_{\text{elec/DD}} \frac{|\mu_a||\mu_b|}{Dr_{ab}^3} (\cos \chi - 3 \cos \alpha_a \cos \alpha_b)$$

where $K_{\text{elec/DD}}$ is a unit cancellation factor; $|\mu_a|$ and $|\mu_b|$ are the magnitudes of the two dipole moments, in debyes; D is the unitless dielectric constant of the medium; r_{ab} is the distance between the centers of the two dipoles; χ is the flap angle between the two dipole vectors; α_a and α_b are the angles between the vectors μ_a and μ_b and the interdipole distance r_{ab}.

Values for the dielectric constant for common solvents are given in Table A-4 in Appendix A.

5.2. THE MMX FORCE FIELD

The MMX molecular mechanics force field,[303,304] employed by the PCMODEL program, originally derives from Allinger's MM2 force field (see Section 5.4.1.2). The π-VESCF routines come from Allinger's MMP1 program. These routines were modified for open-shell species by J. McKelvey of Kodak, and improvements to the heat of formation calculations were carried out by J. J. Gajewski. Gajewski and Gilbert devised a larger library of atom types for MMX, added the ability to handle transition metals and transition states, and increased the number of parameters.

MMX currently has 70 defined atom types (see Section 5.3 and Appendix C), and the nine component potentials are shown in the following equation:

$$V_{\text{tot}} = V_{\text{bnd}} + V_{\text{ang}} + V_{\text{oop}} + V_{\text{tor}} + V_{\text{bnd/ang}} + V_{\text{vdW}} + V_{\text{elec}} + V_{\text{HB}} + V_{\text{metal}}$$

where V_{tot} is the total potential energy, V_{bnd} is the bond stretch component, V_{ang} is the angle bend component, V_{oop} is the out-of-plane angle bend component, V_{tor} is the torsional angle strain component, $V_{\text{bnd/ang}}$ is the stretch–bend cross term component, V_{vdW} is the component for van der Waals nonbonded interactions, V_{elec} is the component for electrostatic (charge–charge) nonbonded interactions, V_{HB} is the hydrogen bonding component, and V_{metal} is the metal-coordination component. The component potentials are all given in kilocalories per mole.

Bond Stretching Interactions The MMX bond stretching potential (see Fig. 5-1) is a modified Hooke's law potential including a cubic term, and is given by

$$V_{\text{bnd}} = K_{\text{bnd}} \cdot \tfrac{1}{2} k_{\text{bnd}} [1 + k_{\text{bnd3}}(r - r_0)](r - r_0)^2$$

where K_{bnd} is a unit cancellation constant with the value 143.83; k_{bnd} is the force constant for bond stretching, in millidynes per angstrom, k_{bnd3} is the force constant for the cubic correction term, in Å^{-1}; and r is the internuclear distance (corresponding to r_{IJ} in Fig. 5-1), and $r - r_0$ gives its departure from the equilibrium value.

Typical values for the parameters k_{bnd} and r_0 are 4.40 mdyn/Å and 1.523 Å for $I\text{---}J = C(sp^3)\text{---}C(sp^3)$, 9.60 mdyn/Å and 1.337 Å for $I\text{---}J = C(sp^2)\text{---}C(sp^2)$, and 15.60 mdyn/Å and 1.200 Å for $I\text{---}J = C(sp)\text{---}C(sp)$; k_{bnd3} has the value $-2.00\ \text{Å}^{-1}$.

The MMX bond stretch potential curves are illustrated for the case of the sp^3, sp^2, and sp carbon–carbon bonds in Fig. 5-23.

Angle Bending Interactions The MMX angle bending potential (see Fig. 5-4) is a modified Hooke's law potential including quadratic and sextic terms, given by

$$V_{ang} = K_{ang} \cdot \tfrac{1}{2}k_{ang}(\theta - \theta_0)^2[1 + k_{ang6}(\theta - \theta_0)^4]$$

where K_{ang} is the constant for unit cancellation, which takes the value 0.043828; k_{ang} is the force constant for the angle defined by atoms $I\text{---}J\text{---}K$, in mdyn $\cdot$ Å/rad^2; θ gives the value for the angle span (corresponding to θ_{IJK} in Fig. 5-4); and $\theta - \theta_0$ gives the displacement of the angle span from the equilibrium value, in degrees; k_{ang6} is the sextic force constant, in deg^{-4}, which takes the value 7.00×10^{-8} deg^{-4}. For some bond angles, the parameter θ_0 will vary with degree of substitution and will be different for quaternary, tertiary, and secondary atoms.

Typical values for k_{ang} and θ_0 are 0.450 mdyn $\cdot$ Å/rad^2 and 109.47° (type 1), 109.51° (type 2), and 109.50° (type 3) for $I\text{---}J\text{---}K = C(sp^3)\text{---}C(sp^3)\text{---}C(sp^3)$; 0.430 mdyn $\cdot$ Å/rad^2 and 120.0° for $I\text{---}J\text{---}K = C(sp^2)\text{---}C(sp^2)\text{---}C(sp^2)$; and 0.120 mdyn $\cdot$ Å/rad^2 and 180.0° for $I\text{---}J\text{---}K = C(sp)\text{---}C(sp)\text{---}H$. k_{ang6} takes the value 7.00×10^{-8} deg^{-4}. There is a separate parameter list for three- and four-membered rings.

The MMX angle bending potential curves are illustrated in Fig. 5-24 for the three cases given above: type 1 $C(sp^3)\text{---}C(sp^3)\text{---}C(sp^3)$, $C(sp^2)\text{---}C(sp^2)\text{---}C(sp^2)$, and $C(sp)\text{---}C(sp)\text{---}H$.

These inverted bell-shaped curves are similar in shape, differing in height, width, and X-axis intercept (equilibrium bond angle). These potential functions also invert at their extremes; if other force field components (such as the stretch–bend cross term) didn't act to keep the angles within these boundaries, the structures would tend to minimize to severely distorted geometries.

Out-of-Plane Bending Interactions The MMX force field employs an out-of-plane angle bending potential (see Fig. 5-6) to reinforce the planarity of certain atom types: @2 (alkene sp^2 carbon), @3 (carbonyl sp^2 carbon), @6 (oxygen, when invoked as a π-atom), @9 (amide and enamine nitrogen), @26 (trigonal boron), @29 (carbon radical), @30 (carbocation), @37 (imine sp^2 nitrogen), @40 (aromatic carbon), @48 (carbanion, when invoked as a π-atom), and

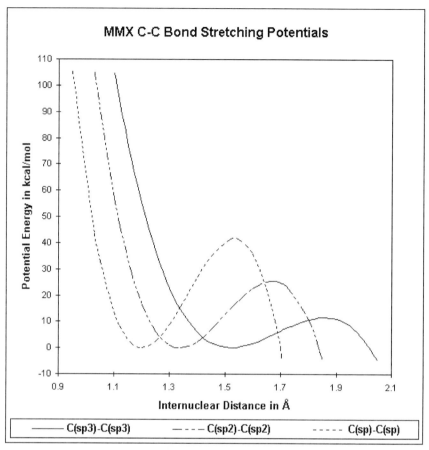

Figure 5-23. MMX Bond Stretching Potentials: Carbon–Carbon Single, Double, and Triple Bonds.

@80, @81, @82 (generalized metal atoms).

This potential is analogous to that of the angle bending potential, but since the equilibrium angle is zero (planar arrangement), the equation simplifies to

$$V_{oop} = K_{oop} \cdot \tfrac{1}{2} k_{oop} \theta_{oop}^2$$

where K_{oop} is the constant for unit cancellation, which takes the value 0.043828; k_{oop} is the force constant for the out-of-plane angle bending, in mdyn · Å/rad^2; and θ_{oop} gives the out-of-plane angle span, as defined in Fig. 5-6.

Some typical values for k_{oop} are 0.600 for alkene $C(sp^2)$, 0.140 for amide and enamine $N(sp^3)$, 0.600 for trigonal boron, and 0.200 for carbon radical C·, all in mdyn · (Å/rad^2).

The out-of-plane bending potential is illustrated in Fig. 5-25 for the case of trimethylboron, where the out-of-plane angle θ_{oop} is given by the angle

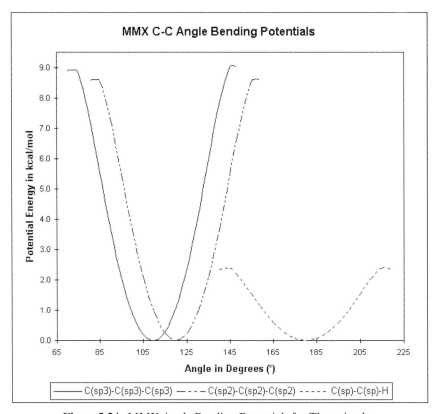

Figure 5-24. MMX Angle Bending Potentials for Three Angles.

between the B—C bond and the projection of that bond onto the plane determined by the three methyl groups (cf. Fig. 5-6). The form of the potential is practically identical to that for simple angle bending.

Torsional Interactions The MMX force field employs a threefold torsional potential function (see Fig. 5-5), which takes the form

$$V_{tor} = \tfrac{1}{2}[k_{tor1}(1 + \cos \omega) + k_{tor2}(1 - \cos 2\omega) + k_{tor3}(1 + \cos 3\omega)]$$

where k_{tor1}, k_{tor2}, and k_{tor3} are the force constants for the one-, two-, and threefold torsional barriers, respectively, in kilocalories per mole, and ω is the dihedral angle (corresponding to ω_{IJKL} in Fig. 5-5) in radians.

Sample parameters for k_{tor1}, k_{tor2}, and k_{tor3} are 0.000, 0.000, and 0.237 kcal/mol for I—J—K—L = H—C(sp^3)—C(sp^3)—H; 0.200, 0.270, and 0.093 kcal/mol for C(sp^3)—C(sp^3)—C(sp^3)—C(sp^3); 0.000, −0.600, and 0.300 kcal/mol for O(sp^3)—C(sp^3)—C(sp^3)—O(sp^3); 0.000, 0.000, and 0.406 kcal/mol for H—C(sp^3)—C(sp^3)—Cl; 0.190, −1.550, and −0.680 kcal/mol for F—C(sp^3)—

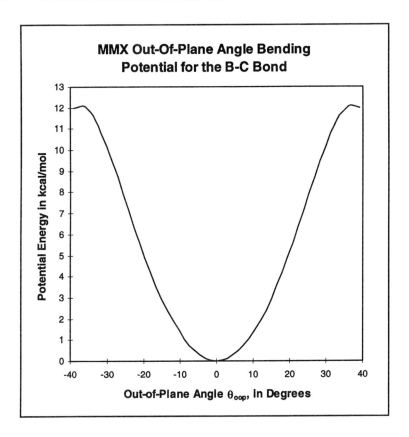

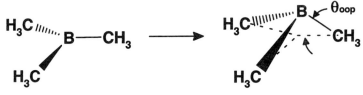

Figure 5-25. MMX Out-of-Plane Angle Bending Potential in Trimethylboron.

$C(sp^3)$—F; −0.240, 0.620, and 0.540 kcal/mol for Cl—$C(sp^3)$—$C(sp^3)$—Cl; and 0.000, 15.000, and 0.000 kcal/mol for H—$C(sp^2)$—$C(sp^2)$—H.

The potential energy curves for the partial torsional contribution of representative four-atom dihedrals, using the parameter values given above, are shown in Fig. 5-26. These four examples all have carbons for the two central atoms. All of the torsional potentials in Fig. 5-26 reach zero energy in the antiperiplanar conformation ($\omega = 180°$ or $-180°$). Two of the dihedrals (H—C—C—H and Cl—C—C—Cl) show the expected trimodal potential, with energy minima at $\approx 60°$, 180° and $\approx 240°$ ($= -60°$). The other two (C—C—

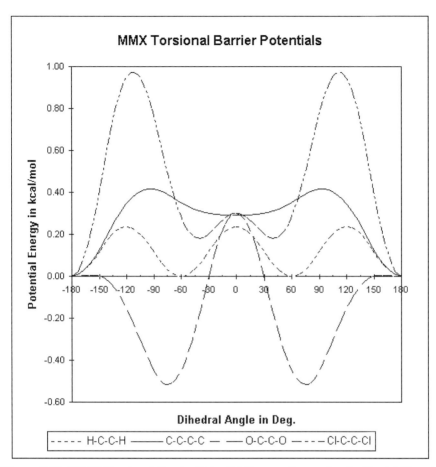

Figure 5-26. MMX Torsional Barrier Potential Energy Curves for the Partial Contribution of Representative Four-Atom Dihedrals.

C—C and O—C—C—O) show a bimodal potential—this is less intuitive, especially the inverted curve in the latter case.

The example of 1,2-dichloroethane will be used below to examine first how the one-, two- and threefold torsional terms contribute to torsional potential for a particular moiety in the molecule, Cl—C—C—Cl, and then how all of the component torsions for this molecule (for I—J—K—L = Cl—C—C—Cl, Cl—C—C—H, and H—C—C—H) are summed to produce the overall modeled "pure" torsional energy barrier in this molecule (cf. Section 5.1.1.3). In addition to these purely torsional interactions, van der Waals interactions, and especially in the case of electronegative elements, electrostatic effects will also affect the overall barrier to dihedral rotation.

With the sample values for the force constants k_{tor1}, k_{tor2} and k_{tor3} given above, Fig. 5-27 shows the contribution of each of these torsional terms, as

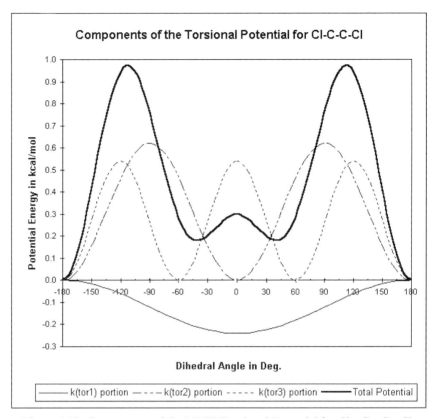

Figure 5-27. Components of the MMX Torsional Potential for Cl—C—C—Cl.

well as the overall torsional potential energy for the Cl—C—C—Cl moiety. The onefold contribution is unimodal and negative, with a relatively shallow minimum at 0°. This "dipole" term may be viewed as a correction to the electrostatic repulsion between the chlorine atoms, which is substantial. This repulsion would otherwise distort the molecule more as the chlorines approached each other, with a maximum at 0° when the chlorines are eclipsed, or synperiplanar.

The twofold contribution is positive and bimodal, with minima at $-180°$, 0°, and 180°. These minima in the *hyperconjugative* term may be attributed to the contribution from hyperconjugative stabilization. Hyperconjugation depends on orbital overlap between the antibonding orbital of one C—Cl bond and the bonding orbital of the other. Thus, the stabilization will be optimal at $-180°$, 0° and 180°, and be minimal at $-90°$ and 90°, which is borne out by the k_{tor2} curve.

The threefold contribution is positive and trimodal, with minima at $-180°$, $-60°$, 60°, and 180°. These energy minima in the *steric* term correspond to the *anti* and *gauche* conformations of the chlorine substituents, which is in accord with intuition.

It was mentioned above that each of these dihedral sets is only a partial contributor to the corresponding overall torsional barrier. The two central atoms of ClCH$_2$CH$_2$Cl are tetravalent, so there are $3 \times 3 = 9$ total interactions to account for. The nine torsional interactions in the 1,2-dichloroethane molecule are detailed in Table 5-1. In Fig. 5-28 are shown the partial and total potential energy curves, and sawhorse and Newman projections for the subject molecule with the atoms labeled for differentiation.

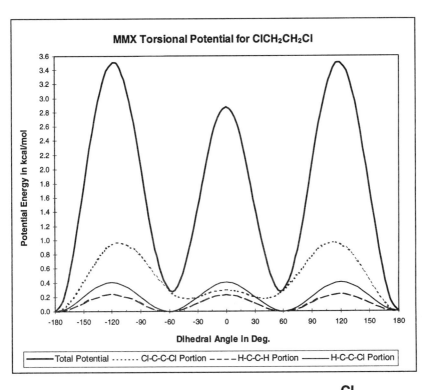

Figure 5-28. Overall Torsional Potential for 1,2-Dichloroethane.

230 THE MOLECULAR MECHANICS FORCE FIELD

TABLE 5-1. The Torsional Interactions in 1,2-Dichloroethane

Atom 1	Atom 2	Range (Clockwise)
Cl_1	Cl_2	-180 to $180°$
H_{a1}	H_{b2}	-180 to $180°$
H_{b1}	H_{a2}	-180 to $180°$
Cl_1	H_{a2}	-60 to $-420°$
H_{a1}	Cl_2	-60 to $-420°$
H_{b1}	H_{b2}	-60 to $-420°$
Cl_1	H_{b2}	60 to $420°$
H_{b1}	Cl_2	60 to $420°$
H_{a1}	H_{a2}	60 to $420°$

Due to equivalence, these nine torsional interactions simplify to three different components: The Cl—C—C—Cl, Cl—C—C—H, and H—C—C—H dihedrals. In the latter two cases, only k_{tor3} has a nonzero coefficient, and this leads to a uniform and symmetrical threefold curve for each. Since the cosine is the same for the arguments $-180°$, $-60°$, and $60°$, the potential energy contributions for all four Cl—C—C—H dihedrals are equal, as they are for all four H—C—C—H dihedrals. Thus, the total potential curve shown in Fig. 5-28 is given by the equation

$$V_{tor}[\text{overall}] = \delta V_{tor}[\text{Cl—C—C—Cl}] + 4\,\delta V_{tor}[\text{Cl—C—C—H}]$$
$$+ 4\,\delta V_{tor}[\text{H—C—C—H}]$$

As was mentioned above, this represents only the torsional strain component of the energy; nonbonded interactions such as the van der Waals and electrostatic effects will also contribute to the overall barrier to rotation calculated by the force field.

Bond-Stretch–Angle-Bend Cross Term The MMX force field models the coupled bond-stretching–angle-bending behavior across three bonded atoms (see Fig. 5-8) with a harmonic potential of the form

$$V_{\text{bnd/ang}} = K_{\text{bnd/ang}} \cdot \tfrac{1}{2}k_{\text{bnd/ang}}[(r' - r'_0) + (r'' - r''_0)](\theta - \theta_0)$$

where $K_{\text{bnd/ang}}$ is a unit cancellation factor which has the value of 5.02236; $k_{\text{bnd/ang}}$ is the stretch–bend force constant (in millidynes per radian); r' and r'' are the internuclear distances in angstroms for the bond pairs I—J and J—K, respectively (corresponding to r_{IJ} and r_{JK} in Fig. 5-8), and the quantities $r' - r'_0$ and $r'' - r''_0$ give their departures from the equilibrium distances; θ is the angle span across the I—J—K system in degrees (equivalent to θ_{IJK} in Fig. 5-8), and the quantity $\theta - \theta_0$ gives its departure from the equilibrium value.

This can be simplified to

$$V_{bnd/ang} = K_{bnd/ang} \cdot \tfrac{1}{2} k_{bnd/ang} (r_{(sum)} - r_{(sum)0})(\theta - \theta_0)$$

where $r_{(sum)} = r' + r''$ and $r_{(sum)0} = r'_0 + r''_0$.

There are only four values for the stretch–bend force constant, which depend whether the central atom of I—J—K is a first- or second-row element, and whether K is a hydrogen or not. If K is a hydrogen, the contribution due to the J—K bond is not added to the potential, including methylene CH_2. Thus $k_{bnd/ang} = 0.12$ mdyn/rad for I—J—K = C—C—C, 0.25 mdyn/rad for C—Si—C, 0.09 mdyn/rad for C—C—H, and -0.40 mdyn/rad for C—S—H. There are separate parameter sets for use in small rings such as cyclopropane and cyclobutane.

As seen above, the equation allows the combination of the bond length differences, so the $V_{bnd/ang}$ potential becomes a function of two independent variables, the bond length difference and the bond angle. An example of the energy surface produced for the $C(sp^3)$—$C(sp^3)$—$C(sp^3)$ system is shown in Fig. 5-29.

It can be seen that this energy surface is saddle-shaped. A trend to lower energy is seen at larger bond stretch coupled with a more acute angle and at

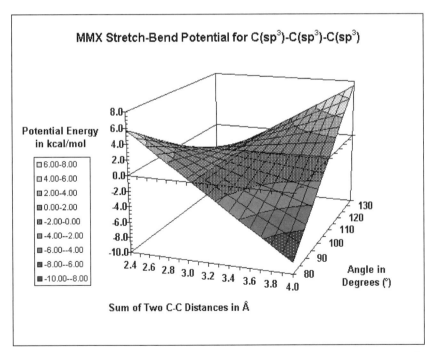

Figure 5-29. The MMX Bond-Stretch–Angle-Bend Potential Surface for the $C(sp^3)$—$C(sp^3)$—$C(sp^3)$ System.

shorter bond compression coupled with a more obtuse angle, while a trend to higher energy is seen at larger bond stretch coupled with a more obtuse angle and at shorter bond compression coupled with a more acute angle. This makes intuitive sense in terms of one of these coupled energy terms being able to ameliorate or accentuate the other.

van der Waals Interactions To model the van der Waals potential which governs the interactions between nonbonded nongeminal atoms (see Fig. 5-18), the MMX force field uses the exact same pair of functions as MM2 (see Section 5.4.1.2), based on the ratio of the ideal to the actual internuclear distance, which combine attractive and repulsive components and which are functions of the adjusted force constant k_{vdW} and the internuclear distance r_{IJ}:

$$V_{vdW} = K_{vdW} k_{vdW} p^6 + K'_{vdW} k_{vdW} e^{-12.50/p} \quad \text{(when } p \leqslant 3.311\text{)}$$

and

$$V_{vdW} = K''_{vdW} k_{vdW} p^2 \quad \text{(when } p > 3.311\text{)}$$

where the scaling and unit cancellation factors take the values $K_{vdW} = -2.25$, $K'_{vdW} = 2.90 \times 10^5$ and $K''_{vdW} = 336.176$; k_{vdW} is the force constant for nonbonded interactions between I and J (see below), in kilocalories per mole; and p is a unitless ratio of the equilibrium distance between I and J to the actual distance.

The force constant k_{vdW} is the geometric mean of the hardness parameters ε for the two atoms (I and J) involved in the van der Waals interaction

$$k_{vdW} = (\varepsilon_I \varepsilon_J)^{1/2}$$

The distance ratio p is given by

$$p = \frac{r_{I(0)} + r_{J(0)}}{r_{IJ}}$$

where $r_{I(0)}$ and $r_{J(0)}$ are the van der Waals radius parameters for the two atoms involved (I and J), and r_{IJ} is the actual internuclear distance.

Typical values for the hardness ε and the van der Waals radius r_0 are found in Table C-1 in Appendix C.

This form of the potential function for nonbonded interactions was adapted by Allinger from the Buckingham potential, and is depicted graphically for the van der Waals interaction of two fluorines in Fig. 5-30.

Considering this composite potential function, the first of the two expressions has two parts: the first component is a repulsive potential, and is an exponential function of the ratio $-12.50/p$, while the second component is an attractive potential, and is porportional to the sixth power of p. The overall potential V_{vdW} is obtained by subtracting the attractive potential from the repulsive potential.

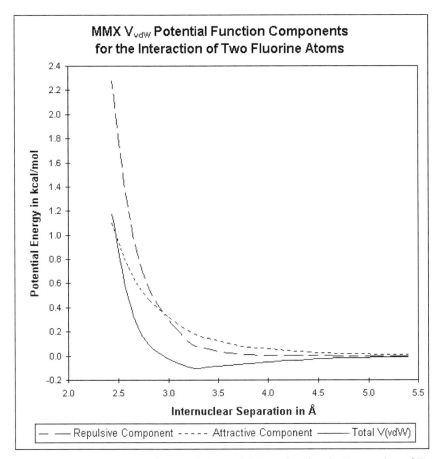

Figure 5-30. The MMX van der Waals Potential Function for the Interaction of Two Fluorines: Repulsive Potential, Attractive Potential, and Total Potential.

The second of the two equations is another repulsive potential, this time proportional to the square of p, and intersects the combination potential from the first equation at $p = 3.311$. It is a correction factor, and the crossover is required because the combination potential inverts at short internuclear distances; this is shown for the van der Waals interaction of two fluorines in Fig. 5-31.

The V_{vdW} nonbonded potential interactions in the MMX force field between some typical atom–atom pairs are depicted in Fig. 5-32.

The discussion in Section 5.1.3.1 covered the use of reduced hydrogen positions as a correction factor for van der Waals interactions involving hydrocarbon C—H (see Fig. 5-19). The MMX force field employs a reduction factor of 8.5% (i.e. $100 \times r_\delta/r = 8.5$, using the notation of Fig. 5-19). Two of the curves

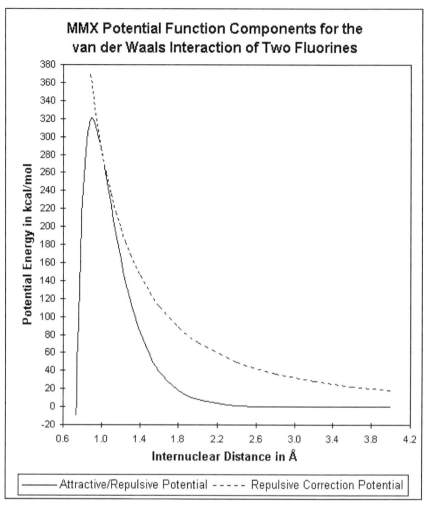

Figure 5-31. The Attractive/Repulsive Potential and the Correction Potential for the van der Waals Interactions of Two Fluorine Atoms.

in Fig. 5-32 involve this reduced hydrogen interaction, and for these a collinear approach to the C—H bond was used (i.e. C—H·····X), and the distance to the hydrogen nucleus along the C—H bond was reduced by 8.5% to correct for the deformation from sphericity of the electron density around the hydrogen atom. This has the net effect of allowing closer approach to the reduced hydrogen, and this is evident from the position of the minimum of the potential wells for the standard vs. the reduced H—H interaction in Fig. 5-32.

A known problem in the MM2 force field was that the benzene face-to-face dimer was calculated to have a 3-kcal/mol energy minimum at a separation of

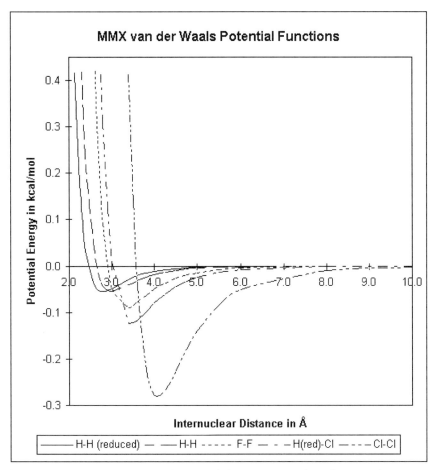

Figure 5-32. MMX van der Waals Potential Energy Functions for Two-Atom Interactions: H—H (reduced), H—H, F—F, H (reduced)—Cl, and Cl—Cl.

approximately 3.5 Å. To redress this excessive attractive interaction, in the MMX force field the attractive portion of the van der Waals potential is ignored for pairs including @2, @4, @40, and @48 carbons. Thus in PCMODEL, where the interactions are modeled on the basis of van der Waals attractive and repulsive forces and simple electrostatic forces, the benzene face-to-face dimer now has no more than a 1-kcal/mol minimum and the perpendicular dimer is about equivalent in energy (see Section 6.1).

Electrostatic Interactions To calculate the potential V_{elec} which governs the electrostatic interactions between nonbonded non-geminal atoms I and J by the charge–charge model (see Fig. 5-21), the MMX force field uses the Coulomb potential:

$$V_{elec} = K_{elec} \frac{q_I q_J}{Dr_{IJ}}$$

where K_{elec} is a constant for scaling and unit cancellation with the value 333.6; q_I and q_J are the charges on the corresponding atoms, in atomic charge units; D is the unitless dielectric constant of the medium (with a default value of 1.5); and r_{IJ} is the internuclear distance between atoms I and J, in angstroms. As in the case of the van der Waals nonbonded interaction for hydrogens bonded to carbon, the reduced hydrogen positions are used to determine r_{IJ} for the electrostatic calculation.

MMX determines the net charge q on an atom I from the bond dipole moments stored in the parameter set for all bonds which include atom I, e.g. for the bonds I—X, I—Y, I—Z, etc. A simplified derivation is provided in Section A.4 for an exercise in Section 5.6. The partial charge contributions from each bond dipole moment, e.g. μ_{IX}, μ_{IY}, μ_{IZ}, are simply summed to give the total charge on atom I, and may be calculated by the expression

$$q_{I(X)} = K_{\mu q} \frac{|\mu_{IX}|}{r_{IX}}$$

where $q_{I(X)}$ is the partial charge contribution on atom I due to the bond dipole with atom X, in units of the elementary charge e; $K_{\mu q}$ is a unit cancellation constant which has a value of $0.208194 \, e \cdot \text{Å} \cdot D^{-1}$; $|\mu_{IX}|$ is the absolute value of the dipole moment for the I—X bond, in debyes; and r_{IX} is the length of the I—X bond, in angstroms. When this calculation has been carried out, a charge of $+q_{I(X)}$ is placed on whichever of atoms I and X is more electropositive, and a charge of $-q_{I(X)}$ is placed on whichever is more electronegative.

Typical values for the bond dipole parameter μ are 0.00 D for $C(sp^3)$—H, 0.850 D for $C(sp^3)$—O, 1.820 D for $C(sp^3)$—F, -0.200 D for $C(sp^2)$—H, 0.000 D for $C(sp^2)$—O, and 1.480 D for $C(sp^2)$—F.

The example of (E)-1,2-difluoroethene (Fig. 5-33) will be used to demonstrate the electrostatic calculation. There are only two nonzero bond dipole moments in this molecule, that for the $C(sp^2)$—H and $C(sp^2)$—F bonds. In the energy-minimized molecule, the bond lengths are found to be $r_{CH} = 1.101$ Å and $r_{CF} = 1.352$ Å. Using the above equation, these charges are

$$q_H = +\left[0.208194 \frac{0.200}{1.101}\right] = +0.0378 e$$

$$q_F = -\left[0.208194 \frac{1.480}{1.352}\right] = -0.2279 e$$

The interatomic distances calculated from the energy-minimized molecule are $r_{FF} = 3.577$ Å, $r_{HH} = 2.553$ Å, and $r_{FH} = 2.929$ Å, where in the latter two cases the reduced hydrogen positions are used. Thus, the overall electrostatic

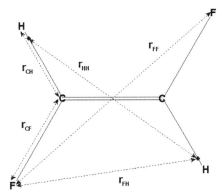

Figure 5-33. Electrostatic Interactions in (E)-1,2-Difluoroethene.

contribution to the potential energy is given by

$$V_{elec} = \frac{K_{elec}}{D}\left(2\frac{q_F q_H}{r_{FH}} + \frac{q_H^2}{r_{HH}} + \frac{q_F^2}{r_{FF}}\right)$$

$$= \frac{333.6}{1.5}\left(2 \times \frac{0.0378 \times (-0.2279)}{2.929} + \frac{(0.0378)^2}{2.553} + \frac{(-0.2279)^2}{3.577}\right)$$

$$= 2(-0.762) - 0.110 + 3.280$$

$$= 1.866 \text{ kcal/mol}$$

The MMX electrostatic model has a correction for 1,3-electrostatic interactions when two geminal atoms possess nonzero charges are attached to sp^3 carbon. This correction is not made with any other central atom, and its absence can be a source of error.

Hydrogen Bonding Interactions The hydrogen bonding potential, V_{Hbnd}, is implemented as a special case of electrostatic interaction, in which a combination of the charge–charge model and the dipole–dipole model is used. This mixed model has been found to give the best replication of experimental data, and takes account of the directionality of the basic-atom–lone-pair group. It will be discussed in Section 6.6.

Metal-Coordination Interactions The organometallic potential V_{metal} is also based upon a combination of the charge–charge model and the dipole–dipole model. The mixed model has been found to give the best replication of experimental data, and takes account of the directionality of the **haptophilic**-atom–lone-pair group. This and other aspects of the implementation of metal atom bonding and complexation are explored in Section 6.7.

5.3. ATOM TYPES IN THE MMX FORCE FIELD

As was discussed earlier, force field models have standard parametrized atom types for representing different elements, for their different valence states, and for some cases in which a particular environment requires a separate atom type, such as an acidic hydrogen or a transition state atom. There are 80 atom types used by PCMODEL, and they are described further in Appendix C. This discussion will consider the atom types briefly, by element or by groups of elements. Following the convention established for this book, the atom type numbers are prefixed with an @. Further information on the atom types which can be included in a π-calculation may be found in Section 6.5; further information on the atom types which can be involved in hydrogen bonding may be found in Section 6.6, and further information on transition state atom types may be found in Section 6.9.

There are eight atom types for hydrogen. The most commonly used hydrogen atom type is @5, for hydrogens attached to carbon, phosphorus, sulfur, selenium, or a transition metal hydride. There are four atom types for more acidic hydrogens: @21 for hydroxyl hydrogen, @23 for amino or imino hydrogen, @24 for the acidic hydrogen in a carboxylic acid or in an ammonium or immonium salt, and @28 for the hydrogen of an enol or phenol. There is the atom type @36 for heavy hydrogen, or deuterium (D).[341,342] There is a special atom type for hydrogens involved in a transition state , @45 (H*), and PCMODEL version 6.00 supports the @70 transition metal bridging hydride.

The attractive potential of hydrogen bonds may be evaluated for any hydrogen attached to nitrogen, oxygen, sulfur, or selenium, including appropriate @5's and @36's, and any @21, @23, @24, or @28 hydrogens.

Beginning the Main Group in the periodic table, there are three atom types for boron. Trigonal, trivalent boron is modeled with @26, and tetrahedral, tetravalent boron is modeled with @27 (B⁻). For transition state boron such as in a hydroboration activated complex, the @43 atom type (B#) is used.

PCMODEL version 6.00 has added some aluminum atom types, with @44 for tricoordinate aluminum and @58 for tetracoordinate aluminum. Aluminum can also be modeled with the @80 metal atom type.

All five members of the carbon family are represented by MMX atom types: there are 17 atom types for carbon, and one each for silicon, germanium, tin, and lead. One may carry out π-calculations or model transition states with only certain of the carbon atom types, and not with any of the other members.

The most commonly used atom types for carbon are @1 for sp^3-hybridized carbon, @2 for the sp^2-hybridized carbon in an alkene, @3 for the sp^2-hybridized carbon in a carbonyl, and @4 for sp-hybridized carbon. A special atom type is used for cyclopropane carbons, @22, and another special type for aromatic carbons, @40. There are three atom types for reactive intermediates: @29 (C ·) for radicals, @30 (C+) for carbocations, and @48 (C−) for carbanions. The @48 carbanion atom bears a lone pair. Six of these atom types may be included in a π-calculation: @2, @3, @4, @29, @30, and @48.

Four atom types are available for modeling different kinds of carbon transition states: @49 (C*), @50 (C#), @51 (trivalent, C$), @52 (pentavalent, C%).

PCMODEL version 6.00 has added four new carbon atom types for organometallics: @61 for a metal carbene, @62 for a metal carbyne, @63 for a metal carbonyl carbon, and @71 for a bridging carbon.

For the higher members of the family, @19 is used for silicon, @31 is used for germanium, @32 is used for tin, and @33 is used for lead. Only van der Waals parameters are resident for @33; all others must be supplied by the user.

In the pnicogen family, there are atom types for only the first two members. Certain of the nitrogen atom types are supported for π-calculations, but phosphorus is not.

Among the seven atom types for nitrogen, the most commonly used are @8 for sp^3-hybridized nitrogen, @9 for the sp^2-hybridized nitrogen in enamines and amides, @37 for imines, and @10 for sp-hybridized nitrogen. Atom types @8, @10, and @37 will bear a lone pair of electrons. Tetravalent cationic ammonium and iminium groups are modeled with the @41 atom type (N$^+$). For transition state nitrogen such as in an aza–Cope rearrangement activated complex, the @55 atom type (N$^\#$) is used. PCMODEL version 6.00 supports an @72 atom type for a bridging nitrogen in organometallic complexes.

Four of these nitrogen atom types may be included in a π-calculation: @9, @10, @37, @41.

There are two phosphorus atom types: @25 is used for trivalent phosphorus, which bears a lone pair of electrons, and @47 is used for pentavalent phosphorus.

There are atom types for four members of the chalcogen family: five atom types for oxygen, five for sulfur, two for selenium, and one for tellurium. One may carry out π-calculations or model transition states with only certain of the oxygen atom types, and not with any of the other members. The first three chalcogens can be hydrogen bond donors, but tellurium cannot.

Among the thirteen atom types for oxygen, the most commonly used are @6 for sp^3-hybridized oxygen and @7 for the sp^2-hybridized oxygen, and both of these bear two lone pairs of electrons. Oxygen anions are modeled with the @42 (O−). Oxonium ions are modeled with the @46 (O+), which is also used for the oxygen in terminal metal carbonyl groups. All four of the former may be included in a π-calculation. For transition state oxygen such as in a Claisen rearrangement or oxynucleophile S$_N$2-activated complex, the @53 atom type (O$^\#$) is used.

The standard divalent sulfur is modeled with @15, which bears two lone pairs of electrons. For a trivalent sulfonium group, @16 (S+) is used. There

TABLE 5-2. Comparison of van der Waals Radii.[344]

Type	Symbol	Name	Radius (Å) PCMODEL	Standard
@1	C	Carbon	1.900	1.85
@5	H	Hydrogen	1.500	1.20
@6	O	Oxygen	1.740	1.40
@8	N	Nitrogen	1.820	1.54
@11	F	Fluorine	1.650	1.35
@12	Cl	Chlorine	2.030	1.81
@13	Br	Promine	2.180	1.95
@14	I	Iodine	2.320	2.15
@15	S	Sulfur	2.110	1.85
@19	Si	Silicon	2.250	2.00
@25	P	Phosphorus	2.050	1.90
@26	B	Boron	1.980	2.08

are three special atom types for oxidized sulfur. The sulfoxide group is modeled with @17, and the sulfone with @18, both of which require explicit double bonds to oxygen. The @38 is used for doubly bonded thiocarbonyl sulfur. No sulfur atom may be included in a π-calculation.

For selenium, there are two atom types: @34 for the bivalent form, which bears two lone pairs of electrons, and @39 for a selenoxide, which requires an explicit double bond to oxygen.

Tellurium is modeled with @35. The limited parameters available for this atom type in PCMODEL include a full van der Waals set, bond parameters for @2—@35, @3—@35, and @35—@35; angle parameters for @2—@35—@35 and @3—@35—@35; and generalized torsional parameters for @0—@35—@35—@0.

There are MMX atom types for four members of the halogen family. Fluorine is modeled with @11, chlorine is modeled with @12, bromine is modeled with @13, and iodine is modeled with @14. There is also a transition state iodine, @54 (I%), for use as a leaving group in modeling S_N2 activated complexes. PCMODEL version 6.00 supports atom types for bridging halides in organometallic complexes: @73 for bridging fluorine, @74 for bridging chlorine, @75 for bridging bromine, and @76 for bridging iodine.

The single atom type @80 is used for most metals not in the Main Group of the periodic table: the alkali metals, the alkaline earths, the transition metals, and selected lanthanides and actinides. Earlier versions of PCMODEL had the capacity to model a maximum of three metals, for which the atom types @44, @56, and @57 were used. These three atom type numbers have been reassigned in version 6.00 of PCMODEL. The limit on the number of metals has been removed in the current MMX force field, and @80 is used for all parametrized metals. When more than one metal is present in a molecule, each

TABLE 5-3. Comparison of Covalent Radii.

Type	Symbol	Name	Radius (Å)		
			PCMODEL	Standard Ref. 344	Ref. 345
@1	C	sp^3-Hybridized carbon	0.762	0.77	0.765
@2	C	sp^2-Hybridized carbon	0.669	0.67	0.661
@4	C	sp-Hybridized carbon	0.60	0.60	0.591
@5	H	Hydrogen	0.35	0.32	0.328
@6	O	sp^3-Hybridized oxygen	0.645	0.66	0.661
@7	O	sp^2-Hybridized oxygen	0.538	0.56	0.549
@46	O	sp-Hybridized oxygen	0.55	0.528	
@8	N	sp^3-Hybridized nitrogen	0.678	0.70	0.704
@9	N	sp^2-Hybridized nitrogen	0.601	0.63	0.618
@10	N	sp-Hybridized nitrogen	0.564	0.56	0.545
@11	F	Fluorine	0.618	0.64	0.621
@12	Cl	Chlorine	1.021	0.99	1.04
@13	Br	Bromine	1.182	1.142	1.20
@14	I	Iodine	1.377	1.333	1.40
@15	S	Sulfur	1.053	1.04	1.05
@19	Si	Silicon	1.118	1.10	1.10
@25	P	Trivalent phosphorus	1.098	1.08	1.09
@26	B	Trigonal boron	0.798	0.79	0.83
@34	Se	Selenium	1.221	1.21	1.21

new one is assigned a different atom type number, with the second metal being typed as @81, the third as @82, etc. These metals are characterized by their atomic symbols, covalent radius, and charge (if any).

There are four atom types which don't fall easily into any group. The electron lone pair is modeled with @20, and is a pseudoatom with a van der Waals radius of 1.2 Å, and a hardness value $\varepsilon = 0.016$ kcal/mol.[343] The two wildcard atoms @58 and @59 are user-definable atoms for which added constants will be needed. The @60 atom is a spherical approximation of the water molecule. Its parameters are: van der Waals radius = 1.53 Å, $\varepsilon = 0.5$ kcal/mol; @60—@60 bond length = 2.8 Å; $k_{bnd} = 1.0$ mdyn/Å; charge = 0.20; @60—@60—@60 angle = 80°.

For purposes of comparing some MMX atomic size parameters with analogous values in general usage, some representative examples are shown in Table 5-2 for van der Waals radii, and in Table 5-3 for the covalent radii.[344,345]

5.4. OTHER FORCE FIELDS

In this section a number of other molecular mechanics force fields will be described. The discussions will highlight the essential aspects of each force field, and where appropriate will compare and contrast it with MMX. This discussion is not meant to be exhaustive, and the interested reader should consult the original literature reports and review accounts for further information on individual force fields and the relationships between them.

It should be remembered that force fields are not all created with the same intended scope of application. There are significant differences between those developed primarily for small molecules and those for macromolecules; for polar and for nonpolar molecules; for organic, for organometallic, and for inorganic molecules; etc. For this reason, force field components, potential functions, and parameters are transferred between different force fields only at great risk, and the validity of such borrowing is subject to extensive testing and validation.

These concerns noted, it is educational to compare the makeup of different force fields and the values of their parameters. These issues are important to consider when selecting the most appropriate force field to model the system of interest. It helps to know that one force field places greater emphasis on cross-term potentials to improve the prediction of vibrational frequencies, while another employs a basic set of valence and cross terms, but uses a sophisticated electrostatic model in order to describe accurately the behavior of charged molecules in a polar medium.

Proportionately more attention in this section is devoted to the Allinger MM series, since they represent the origin of the MMX force field. This uneven coverage is not intended as a value judgment, but rather is driven by cladistics and the idea that knowledge of the ancestors and collateral relatives increases the understanding of the individual.

5.4.1. The Allinger MM Force Fields

The force field development work of the Allinger group over the past thirty years has been the engine which has driven much of the adoption and use of computer molecular modeling by the broader chemical community. It is appropriate to cover this work first among the force fields discussed here, not only because it was the most immediate ancestor of the MMX formulation used in PCMODEL, but also because it has been used and adapted so widely that the names Allinger and MM2 have become synonymous with force field modeling. Even those researchers who find fault with these force field formulations will still find the scientific dialog dominated by them, for MM2 and versions thereof account for more of the published work in force field modeling than any other program.

The following sub-sections cover the development of these force fields from MMI in 1974 to MM4 in 1996.

5.4.1.1. The MMI Force Field In a 1974 paper on the implications of the *gauche*-hydrogen interaction for conformational analysis,[72] Wertz and Allinger published one of the first formulations of what later became know as the 1973 MMI force field. MMI was a hydrocarbon force field, made no provision for electrostatics, and was composed of six component potentials:

$$V_{tot} = V_{vdW} + V_{bnd} + V_{ang} + V_{bnd/ang} + V_{tor} + V_{ang/tor/ang}$$

where V_{tot} is the total potential energy, V_{vdW} is the component for van der Waals nonbonded interactions, V_{bnd} is the bond stretch component, V_{ang} is the angle bend component, $V_{bnd/ang}$ is the stretch–bend cross-term component, V_{tor} is the torsional angle strain component, and $V_{ang/tor/ang}$ is the angle-bend–torsional-angle–angle-bend cross-term component. The potential energy functions are all given in kilocalories per mole.

MMI was parametrized only for two atom types, hydrogen and sp^3-hybridized carbon.

van der Waals Interactions MMI employed a two-part van der Waals potential V_{vdW} which is used in MMX to govern the steric interactions between nonbonded nongeminal atoms. The attractive portion was proportional to the sixth power of the distance, and the repulsive portion was an exponential function. For atoms I and J (see Fig. 5-18), the steric interaction was given by

$$V_{vdW} = K_{vdW} k_{vdW} p^6 + K'_{vdW} k_{vdW} e^{-13.60/p}$$

where K_{vdW} is the unit cancellation constant for the attractive potential with a value of -2.25; k_{vdW} is the force constant for nonbonded interactions between I and J, derived as the geometric mean of the individual hardness parameters, ε as defined in Section 5.2, in kilocalories per mole; p is a unitless ratio of the equilibrium distance between I and J to the actual, as defined in Section 5.2; and K'_{vdW} is the unit cancellation constant for the repulsive potential with a value of 8.28×10^5.

Some typical values for the hardness parameter ε are 0.063 kcal/mol for H and 0.041 kcal/mol for C(sp^3), and for the van der Waals radius r_0 are 1.500 Å for H and 1.750 Å for C(sp^3).

For the steric interaction between a carbon and a carbon-bound hydrogen, the C—H bond distance is reduced by 7.5% (i.e., the hydrogen nucleus is moved closer to the bonded carbon), for reasons discussed in Section 5.1.3.1.

Bond Stretching Interactions MMI employed a harmonic bond stretching potential V_{bnd} with a cubic term, with the bond I—J as defined in Fig. 5-1:

$$V_{bnd} = K_{bnd} k_{bnd} (r - r_0)^2 [1 + k_{bnd3}(r - r_0)]$$

where K_{bnd} is a unit cancellation constant with a value of 71.94; k_{bnd} is the force constant for bond stretching, in millidynes per angstrom; r is the distance between the I and J nuclei (corresponding to r_{IJ} in Fig. 5-1), and $r - r_0$

is its departure from the equilibrium bond length, in angstroms; and k_{bnd3} is the cubic force constant and has the value -2.00 Å^{-1}.

Some typical parameter values are: $k_{bnd} = 4.600$ mdyn/Å for a $C(sp^3)$—H bond and 4.400 mdyn/Å for a $C(sp^3)$—$C(sp^3)$ bond; $r_0 = 1.095$ Å for a $C(sp^3)$—H bond and 1.514 Å for a $C(sp^3)$—$C(sp^3)$ bond.

Angle Bending Interactions MMI used a modified Hooke's law function to model the contribution to the potential energy from angle bending, V_{ang}, for the angle I—J—K as defined in Fig. 5-4. The quadratic and cubic terms from the power series are used in the expression:

$$V_{ang} = K_{ang} k_{ang} (\theta - \theta_0)^2 [1 + k_{ang3}(\theta - \theta_0)]$$

where K_{ang} is the constant for unit conversion, which takes the value 0.021914; k_{ang} is the quadratic force constant for the three-atom angle, in mdyn · Å/rad^2; k_{ang3} is the cubic force constant, which takes the value -0.006 deg^{-1}; θ is the current span across the angle (corresponding to θ_{IJK} in Fig. 5-4), in degrees; and $\theta - \theta_0$ gives the displacement of the angle span from the equilibrium value, in degrees. For some bond angles, the parameter θ_0 will vary with the level of substitution and will be different for quaternary, tertiary, and secondary atoms.

Some typical values for k_{ang} are 0.380 mdyn · Å/rad^2 for $C(sp^3)$—$C(sp^3)$—$C(sp^3)$ and 0.240 mdyn · Å/rad^2 for $C(sp^3)$—$C(sp^3)$—H; and for θ_0 are 110.2° for type 3 (secondary) $C(sp^3)$—$C(sp^3)$—$C(sp^3)$, 110.5° for type 2 (tertiary) $C(sp^3)$—$C(sp^3)$—$C(sp^3)$, and 109.467° for type 1 (quaternary) $C(sp^3)$—$C(sp^3)$—$C(sp^3)$.

Bond-Stretch–Angle-Bend Cross Term MMI modeled the coupling between bond stretching and angle bending with a bond-stretch–angle-bend cross-term potential $V_{bnd/ang}$, applied to the three-atom unit I—J—K (as defined in Fig. 5-8) and taking the form

$$V_{bnd/ang} = K_{bnd/ang} k_{bnd/ang} (\theta - \theta_0)[(r' - r'_0) + (r'' - r''_0)]$$

where $K_{bnd/ang}$ is a unit conversion constant which has the value 2.51124; $k_{bnd/ang}$ is the stretch–bend force constant in millidynes per radian; θ is the current span across the angle (corresponding to θ_{IJK} in Fig. 5-8), in degrees, and $\theta - \theta_0$ gives its displacement from the equilibrium value, in degrees; r' is the distance between the I and J nuclei (corresponding to r_{IJ} in Fig. 5-8), and $r' - r'_0$ is the departure from the equilibrium bond length, in angstroms; and r'' is the distance between the J and K nuclei (corresponding to r_{JK} in Fig. 5-8), and $r'' - r''_0$ is the departure from the equilibrium bond length, in angstroms.

Typical values for the force constant $k_{bnd/ang}$ are 0.12 mdyne/rad for $C(sp^3)$—$C(sp^3)$—$C(sp^3)$ and 0.40 mdyne/rad for $C(sp^3)$—$C(sp^3)$—H. The values used for r'_0, r''_0, and θ_0 are the same as those taken from the corresponding bond stretch and angle bending potentials.

Torsional Interactions MMI employs a simple potential V_{tor} to govern torsional interactions, which contains a single, threefold term based on the cosine of the torsional angle ω, as defined by the atom string *I—J—K—L* (see Fig. 5-5), and takes the form

$$V_{tor} = \tfrac{1}{2}k_{tor3}(1 + \cos 3\omega)$$

where k_{tor3} is the force constant for the torsional interaction in kilocalories per mole, and ω gives the span of the torsional angle (corresponding to ω_{IJKL} in Fig. 5-5), in radians.

A single value of 0.53 kcal/mol is given for k_{tor3} for any hydrocarbon torsional angle.

Angle-Bend–Torsion–Angle-Bend Cross Term MMI modeled the coupling between torsional interactions and angle bending across the four bonded atoms *I—J—K—L* (see Fig. 5-13) with a harmonic potential $V_{tor/ang}$ of the form

$$V_{ang/tor/ang} = k_{ang/tor/ang}(1 + \cos 3\omega)[(\theta' - \theta'_0) + (\theta'' - \theta''_0)]$$

where $k_{ang/tor/ang}$ is the angle-bend–torsion–angle-bend force constant, in kilocalories per mole; ω gives the span of the torsional angle (corresponding to ω_{IJKL} in Fig. 5-16), in radians; θ' is the current span across the first angle (corresponding to θ_{IJK} in Fig. 5-13), in degrees, and $\theta' - \theta'_0$ gives the displacement of the angle span from its equilibrium value, in degrees; and θ'' is the current span across the second angle (corresponding to θ_{JKL} in Fig. 5-16), in degrees, and $\theta'' - \theta''_0$ gives the displacement of the angle span from its equilibrium value, in degrees.

The value of $k_{ang/tor/ang}$ is -0.0110 kcal/mol for *I—J—K—L* = C(sp^3)—C(sp^3)—C(sp^3)—C(sp^3), and is 0.00 kcal/mol otherwise.

5.4.1.2. The MM2 Force Field The MM2 force field was initially published in 1977,[346] and is described best in two standard sources.[10,272] A number of publications have appeared since, which explore various aspects of the MM2 force field such as amides and ureas,[347] anilines and enamines,[348] deuterium compounds,[342] disulfides,[349] electrostatics,[350–352] ketenes,[353] nitro compounds,[354] silanes,[355,356] and siloxanes.[357]

It is generally similar to MMI, although the hydrogen atoms are smaller in size and softer, and the carbon atoms are larger. A three-part torsional potential was employed. The new twofold torsional term accounts for hyperconjugation effects, and its inclusion allows the model to deemphasize the *gauche*-hydrogen repulsions as the main component of the torsional potential, and permits the above-mentioned softening of the hydrogens.[358] MM2 added many more atom types and introduced the electrostatics term. It dropped the angle-bend–torsion–angle-bend cross-term potential and introduced the out-of-plane angle bending term. A companion force field, MMP2, was intended for use in calculations of conjugated systems (π-calculations are discussed in Section 6.6).

MM2 rapidly became the standard in the field, and is still the most widely referenced force field in organic chemistry. In the period of over two decades before MM3 was released (see the following section), many in-line upgrades and enhancements were added to the force field, but the following discussion covers essentially the "1977 MM2 force field" as described in the two sources cited above.

The many similarities between the MM2 and MMX force fields will be apparent. The functional forms are nearly identical, with the major difference being the use of a charge–charge model for electrostatics in MMX, where MM2 employs the dipole–dipole model. Unlike MMX, MM2 does not employ specific hydrogen bonding or metal atom coordination potential terms.

MM2 has 28 parametrized atom types, including five types of hydrogen, five types of carbon, three types of nitrogen, two types of oxygen, halogens, some second row elements, and a pseudoatom representing a lone pair of electrons.

There are seven component potentials in the overall force field, as follows:

$$V_{tot} = V_{vdW} + V_{bnd} + V_{elec} + V_{ang} + V_{bnd/ang} + V_{oop} + V_{tor}$$

where V_{tot} is the total potential energy, V_{vdW} is the component for van der Waals nonbonded interactions, V_{bnd} is the bond stretch component, V_{elec} is the component for electrostatic (dipole–dipole) nonbonded interactions, V_{ang} is the angle bend component, $V_{bnd/ang}$ is the stretch–bend cross-term component, V_{oop} is the component for the out-of-plane bending potential, and V_{tor} is the torsional angle strain component. The potential energy functions are all given in kilocalories per mole.

van der Waals Interactions To model the van der Waals potential which governs the interactions between two nonbonded nongeminal atoms I and J (see Fig. 5-18), MM2 uses a pair of functions based on the ratio of the ideal to the actual internuclear distance. The first part of the MM2 van der Waals nonbonding potential V_{vdW} combines attractive and repulsive components, which are functions of the adjusted force constant k_{vdW} and the internuclear distance r_{IJ}.

The MM2 potential differs from that of MMI in several ways. The most dramatic change is the addition of a hyperbolic potential as a correction for the inversion which takes place at smaller values of r_{IJ} (see Fig. 5-31). Less noticeable at first glance, but still significant, is the decrease in the absolute value of the numerical coefficient of the exponential term from 13.60 to 12.50, and the decrease in the value of the constant K'_{vdw} for the repulsive potential from 8.28×10^5 to 2.90×10^5. The equations for the van der Waals potential are

$$V_{vdW} = k_{vdW}(K_{vdW} p^6 + K'_{vdW} e^{-12.50/p}) \quad \text{(when} \quad p \leqslant 3.311)$$

and

$$V_{vdW} = K''_{vdW} k_{vdW} p^2 \quad \text{(when} \quad p > 3.311)$$

where the scaling and units cancellation factors take the values $K_{vdW} = -2.25$, $K'_{vdW} = 2.90 \times 10^5$, and $K''_{vdW} = 336.176$; k_{vdW} is the adjusted force constant for nonbonded interactions between I and J, derived as the geometric mean of the individual hardness parameters as in MMI, in kilocalories per mole; and p is the unitless ratio of the equilibrium distance between I and J to the actual distance. The factors k_{vdW} and p are defined from the atomic hardness parameters ε_I and ε_J and the van der Waals radii r_I and r_J in the same way as in the MMI formulation.

Some typical values for the hardness parameter ε are 0.047 kcal/mol for hydrogen and 0.044 kcal/mol for C(sp^3), and for the van der Waals radius r_0 are 1.500 Å for hydrogen and 1.900 Å for carbon.

For the steric interaction between a carbon and a carbon-bound hydrogen, special reduced parameters are used for reasons discussed in Section 5.1.3.1. The C—H bond distance is reduced by 8.5% (up from 7.5% in MMI), a reduced force constant k_{vdW} of 0.046 kcal/mol is used, and the sum of the van der Waals radii used is 3.340 Å.

Bond Stretch Interactions MM2 employs the MMI harmonic potential V_{bnd} with a cubic term, for the stretching of bond I—J as defined in Fig. 5-1. The bond stretching potential takes the form

$$V_{bnd} = K_{bnd} \cdot \tfrac{1}{2} k_{bnd}(r - r_0)^2 [1 + k_{bnd3}(r - r_0)]$$

where K_{bnd} is a unit cancellation constant with a value of 143.88 (note that now a factor of $\tfrac{1}{2}$ has been broken out of this constant); k_{bnd} is the force constant for bond stretching in millidynes per angstrom; r is the distance between the I and J nuclei (corresponding to r_{IJ} in Fig. 5-1), and $r - r_0$ is its departure from the equilibrium bond length, in angstroms; and k_{bnd3} is the cubic force constant, which has the value -2.00 Å^{-1}.

Some typical parameter values are: $k_{bnd} = 4.600$ mdyn/Å for a C(sp^3)—H bond and 4.400 mdyn/Å for a C(sp^3)—C(sp^3) bond; $r_0 = 1.113$ Å for a C(sp^3)—H bond and 1.523 Å for a C(sp^3)—C(sp^3) bond.

Electrostatic Interactions The inclusion of heteroatoms makes MM2 qualitatively different from MMI, one result being the requirement for an electrostatic term. MM2, unlike nearly all other major force fields, employs the dipole–dipole model for electrostatic interactions. The potential energy for the electrostatic interaction between two dipole vectors **a** and **b** corresponding to the two bonds *I*—*J* and *K*—*L* (see Fig. 5-22 for illustrations) is given by the Jeans formula:

$$V_{elec} = K_{elec} \frac{|\mu_a||\mu_b|}{Dr_{ab}^3} (\cos \chi - 3 \cos \alpha_a \cos \alpha_b)$$

where K_{elec} is a unit cancellation factor; $|\mu_a|$ and $|\mu_b|$ are the magnitudes of the two dipole moments in debyes; D is the dielectric constant of the medium with a default value of 1.5; r_{ab} is the distance between the two dipoles in angstroms; χ is the flap angle between the two dipole vectors; and α_a and α_b are the angles between the axes of the vectors **a** and **b** (respectively) and the interdipole distance r_{ab}.

Some typical parameter values for the magnitude of the bond moment are 0.300 D for the C(sp^3)—C(sp^2) bond, 0.000 D for the C(sp^3)—H bond, and 1.820 D for the C(sp^3)—F bond.

Angle Bending Interactions MM2 uses a modified Hooke's law function is used to model the potential energy contribution from angle bending V_{ang} for the angle *I*—*J*—*K* as defined in Fig. 5-4. The MMI potential was modified to include the quadratic and sextic terms from the power series:

$$V_{ang} = K_{ang} \cdot \tfrac{1}{2} k_{ang}(\theta - \theta_0)^2 [1 - k_{ang6}(\theta - \theta_0)^4]$$

where K_{ang} is the unit cancellation constant, which takes the value 0.043828; k_{ang} is the quadratic force constant for the three-atom angle, given in mdyn · Å/rad^2; θ is the current span across the angle (corresponding to θ_{IJK} in Fig. 5-4), in degrees, and $\theta - \theta_0$ gives the displacement of the angle span from the equilibrium value, in degrees; and k_{ang6} is the sextic force constant, which takes the value 7.00×10^{-8} deg^{-4}. For some bond angles, the parameter θ_0 will vary with the level of substitution and will be different for quaternary, tertiary, and secondary atoms.

Some typical values for k_{ang} are 0.450 mdyn · Å/rad^2 for C(sp^3)—C(sp^3)—C(sp^3) and 0.360 mdyn · Å/rad^2 for C(sp^3)—C(sp^3)—H; and for θ_0 are 109.470° for type-1 (secondary) C(sp^3)—C(sp^3)—C(sp^3), 109.510° for type 2 (tertiary) C(sp^3)—C(sp^3)—C(sp^3), and 109.500° for type 3 (quaternary) C(sp^3)—C(sp^3)—C(sp^3).

Bond-Stretch–Angle-Bend Cross Term MM2 models the coupling between bond stretching and angle bending with the same bond-stretch–angle-bend cross-term potential $V_{bnd/ang}$ used in MMI, applied to the three-atom unit *I*—*J*—*K* as defined in Fig. 5-8.

Some representative values for the force constant $k_{bnd/ang}$ are 0.12 mdyne/rad when J is a first-row element with two nonhydrogen atoms as I and K; 0.25 mdyne/rad when J is a first-row element with one hydrogen and one nonhydrogen atom as I and K; 0.09 mdyne/rad when J is a second-row element with two nonhydrogen atoms as I and K; and -0.40 mdyne/rad when J is a second-row element with one hydrogen and one nonhydrogen atom as I and K.

The values used for r'_0, r''_0, and θ_0 are the same as those taken from the corresponding bond stretch and angle bending potentials.

Out-of-Plane Angle Bending Interactions MM2 introduced the out-of-plane bending potential V_{oop}, a modification of the angle bending potential V_{ang} discussed above, in order to maintain planarity for some atom types. The specifics of this interaction are defined in Fig. 5-6, and it is governed by the potential equation

$$V_{oop} = K_{ang} \cdot \tfrac{1}{2} k_{oop} \theta^2$$

where K_{ang} is the unit cancellation constant, which takes the value 0.043828; k_{oop} is the quadratic force constant in mdyn $\cdot$ Å/rad^2; and θ is the out-of-plane angle as defined in Section 5.1.3 (corresponding to θ_{oop} in Fig. 5-6), in degrees.

Some typical parameter values for k_{oop} are 0.050 mdyn $\cdot$ Å/rad^2 for J—I = @2 C(sp^2)—C(sp^3), 0.800 mdyn $\cdot$ Å/rad^2 for J—I = @3 C(sp^2)–C(sp^3), and 0.050 mdyn $\cdot$ Å/rad^2 for J—I = @9 N(sp^2)—C(sp^3). The equilibrium value for θ is 0°, which simplifies the potential equation given above compared to that for the angle bending potential.

Torsional Interactions MM2 employs a three-part torsional potential V_{tor} based on the cosine of the torsional angle ω, as defined by the atom string I—J—K—L (see Fig. 5-5), to govern torsional interactions. It has one-, two-, and threefold angle terms, and takes the form

$$V_{tor} = \tfrac{1}{2} [k_{tor1}(1 + \cos \omega) + k_{tor2}(1 - \cos 2\omega) + k_{tor3}(1 + \cos 3\omega)]$$

where k_{tor1}, k_{tor2}, and k_{tor3} are the force constants kilocalories per mole, and ω gives the span of the torsional angle (corresponding to ω_{IJKL} in Fig. 5-5), in radians. A positive value for k_{tor1} and k_{tor3} will destabilize eclipsed conformations, while a positive value for k_{tor2} will lead to energy minima at 0° and 180°. There is a separate set of constants for four-membered rings.

Some typical parameter values for I—J—K—L = C(sp^3)—C(sp^3)—C(sp^3)—C(sp^3) are $k_{tor1} = 0.200$ kcal/mol, $k_{tor2} = 0.270$ kcal/mol, and $k_{tor3} = 0.093$ kcal/mol; and for H—C(sp^3)—C(sp^3)—H are $k_{tor1} = 0.0$ kcal/mol, $k_{tor2} = 0.0$ kcal/mol, and $k_{tor3} = 0.237$ kcal/mol.

5.4.1.3. The MM3 Force Field The MM3 force field[277-279,354,359-368] was designed to incorporate many of the changes which had been introduced into revisions of MM2, within the context of an overall reparametrization. Specific deficiencies which were addressed were the underestimation of high C—C rotational barriers in congested hydrocarbons, overestimation of hydrogen–hydrogen nonbonded repulsions at very short distances, underestimation of the bond length when the bonds are in an eclipsed conformation, the failure to model the stability of the perpendicular benzene dimer vs. the face-to-face dimer, and difficulties with the angle bending potential at large deformations, specifically with three-, four-, and five-membered rings.

Another significant driver for the development of the MM3 force field was the desire to produce more accurate predictions of vibrational spectra.

The new force field contains more elaborate potential expressions for bond stretching and angle bending, a modified van der Waals potential, and several additional cross terms. The parametrization for three-, four-, and five-membered rings has been expanded, and five-membered-ring carbon atom types were introduced. The use of explicit lone pair pseudoatoms was discontinued.[363] Electrostatics for uncharged species is still calculated on the dipole–dipole interaction model. Many individual applications of the MM3 force field have been reported, such as aldehydes and ketones,[365] alkyl iodides,[369] alkynes and nitriles,[370] anilines and enamines,[348] azoxy compounds,[371] β-lactams,[372] carboxylic acids and esters,[366] corannulene derivatives,[373] hydrogen bonding,[374] disulfides,[375] hydroxylamine and methyl derivatives,[376] ketenes,[352] nitro compounds,[354] propargyl alcohol derivatives,[377] sulfides,[378] sulfones,[368] and sulfoxides.[379]

MM3 has 152 parametrized atom types, including for example 9 types of hydrogen, 21 types of carbon, 21 types of nitrogen, 46 types of oxygen (many of these in various acid derivatives), 8 types of sulfur, 5 noble gases, a large number of metals, and a dummy atom. The lone pair of electrons has been removed from the atom list.

There are nine component potentials in the overall force field, as follows:

$$V_{\text{tot}} = V_{\text{bnd}} + V_{\text{ang}} + V_{\text{tor}} + V_{\text{bnd/ang}} V_{\text{tor/bnd}} + V_{\text{ang/ang}} + V_{\text{oop}} + V_{\text{vdW}} + V_{\text{elec}}$$

where V_{tot} is the total potential energy, V_{bnd} is the bond stretch component, V_{ang} is the angle bend component, V_{tor} is the torsional angle strain component, $V_{\text{bnd/ang}}$ is the stretch–bend cross-term component, $V_{\text{tor/bnd}}$ is the torsion–stretch cross-term component, $V_{\text{ang/ang}}$ is the one-center bend–bend cross-term component, V_{oop} is the component for the out-of-plane bending potential, V_{vdW} is the component for van der Waals nonbonded interactions, and V_{elec} is the component for electrostatic (dipole–dipole) nonbonded interactions. The seven terms from the MM2 force field have been retained, and two new cross terms were developed for MM3. The authors of the new force field considered and rejected the idea of adding a torsion–angle-bend cross term.[277] The potential energy functions are all kilocalories per mole.

5.4. OTHER FORCE FIELDS

Bond Stretch Interactions MM3 employs the MM2 Hooke's law harmonic potential V_{bnd} for the stretching of bond I—J as defined in Fig. 5-1, with the addition of a quartic term in order to avoid the **cubic stretch catastrophe**.[321] The bond stretching potential takes the form

$$V_{bnd} = K_{bnd} k_{bnd} (r - r_0)^2 [1 + k_{bnd3}(r - r_0) + k_{bnd4} k_{bnd3}^2 (r - r_0)^2]$$

where K_{bnd} is a unit cancellation constant and has the value 71.94 (note the return to incorporation of the factor of $\frac{1}{2}$ into this constant); k_{bnd} is the bond stretching force constant in millidynes per angstrom; r is the distance between the I and J nuclei (corresponding to r_{IJ} in Fig. 5-1), and $r - r_0$ is its departure from the equilibrium bond length, in angstroms; k_{bnd3} is the cubic stretch constant with a value of -2.55 Å^{-1}; and k_{bnd4} is the quartic stretch constant with a value of $-\frac{7}{12}$ Å^{-2}. There are specialized parameters for use with bonds included in a conjugated system and in a three-, four-, or five-membered ring.

Some typical parameter values are $k_{bnd} = 4.74$ mdyn/Å and $r_0 = 1.112$ Å for an aliphatic C—H bond, and $k_{bnd} = 4.49$ mdyn/Å and $r_0 = 1.112$ Å for a $C(sp^3)$—$C(sp^3)$ bond.

MM3 also introduces an electronegativity correction to the bond length, δr_{elec}, based upon the atom types of the substituents attached to I and J of the bond. In general, an electropositive substituent lengthens the bond, and an electronegative substituent shrinks it. Some typical correction values for I—$J = C(sp^3)$—$C(sp^3)$ are $\delta r_{elec} = -0.0255$ Å for an attached fluorine, 0.0030 Å for an attached selenium, -0.0145 Å for an attached sulfonamide nitrogen, and 0.0090 Å for an attached silicon.

Angle Bending Interactions MM3 employs a modified Hooke's law harmonic potential function to model the contribution from angle bending, V_{ang}, for the angle I—J—K as defined in Fig. 5-4. The MM2 potential was modified to include all of the power series terms from the quadratic up through the sextic:

$$V_{ang} = K_{ang} k_{ang} (\theta - \theta_0)^2 [1 + k_{ang3}(\theta - \theta_0) + k_{ang4}(\theta - \theta_0)^2 + k_{ang5}(\theta - \theta_0)^3 + k_{ang6}(\theta - \theta_0)^4]$$

where K_{ang} is a unit cancellation constant and has the value 0.021914 (here again the factor of $\frac{1}{2}$ was folded back into this constant); k_{ang} is the angle bending force constant in mdyne · Å/rad^2; θ is the current span across the angle (corresponding to θ_{IJK} in Fig. 5-4), in degrees, and $\theta - \theta_0$ gives its displacement from the equilibrium value, in degrees; k_{ang3} is the cubic bend constant with a value of -0.014 deg^{-1}; k_{ang4} is the quartic bend constant with a value of 5.6×10^{-5} deg^{-2}; k_{ang5} is the quintic bend constant with a value of -7.0×10^{-7} deg^{-3}; and k_{ang6} is the sextic bend constant with a value of 9.0×10^{-10} deg^{-4}. There are specialized parameters for use with angles which include a delocalized bond or a three-, four-, or five-membered ring. As was the case in MM2, for some bond angles, the parameter θ_0 will vary with the level of substitution and will be different for quaternary, tertiary, and secondary atoms.

Some typical parameter values are $k_{ang} = 0.59$ mdyn · Å/rad^2 and $\theta_0 = $ 109.8°, 109.3°, and 110.7° for a type 1, type 2, and type 3 C(sp^3)—C(sp^3)—H angle, respectively; and $k_{ang} = 0.67$ mdyn · Å/rad^2 and $\theta_0 = 109.5°$, 110.2°, and 111.0° for a type 1, type 2, and type 3 C(sp^3)—C(sp^3)—C(sp^3) angle, respectively.

Torsional Interactions MM3 employs the same three-part torsional potential V_{tor} as was used in MM2, based on the cosine of the torsional angle ω, as defined by the atom string *I—J—K—L* (see Fig. 5-5). There are specific sets of constants for three-, four-, and five-membered rings.

Some typical parameter values for *I—J—K—L* = C(sp^3)—C(sp^3)—C(sp^3)—C(sp^3) are $k_{tor1} = 0.185$ kcal/mol, $k_{tor2} = 0.170$ kcal/mol, and $k_{tor3} = 0.520$ kcal/mol; and for H—C(sp^3)—C(sp^3)—H are $k_{tor1} = 0.000$ kcal/mol, $k_{tor2} = 0.000$ kcal/mol, and $k_{tor3} = 0.238$ kcal/mol.

Bond-Stretch–Angle-Bend Cross Term MM3 employs the same form as in MMI and MM2 for the bond-stretch–angle-bend cross-term potential $V_{bnd/ang}$, applied to the three-atom unit *I—J—K* (as defined in Fig. 5-8). There are specific sets of constants for three-, four-, and five-membered rings.

Some typical parameter values are $k_{bnd/ang} = 0.130$ mdyn/rad for *I—J—K* = C(sp^3)—C(sp^3)—C(sp^3) and $k_{bnd/ang} = 0.080$ mdyn/rad for *I—J—K* = C(sp^3)—C(sp^3)—H.

Torsion–Bond-Stretch Cross Term MM3 models the coupling between torsional interactions and bond stretching with a torsion–stretch cross-term potential $V_{tor/bnd}$ applied to the torsional angle ω defined by the four-atom unit *I—J—K—L* and the bond corresponding to *J—K* (as defined in Fig. 5-10). The addition of this term causes the central bond to lengthen more realistically when the substituents are eclipsed, a property which was not modeled well in MM2. This cross-term potential takes the form

$$V_{tor/bnd} = K_{tor/bnd} \cdot \tfrac{1}{2} k_{tor/bnd}(r - r_0)(1 + \cos 3\omega)$$

where $K_{tor/bnd}$ is a unit cancellation constant and has the value 11.995; $k_{tor/bnd}$ is the torsion–stretch force constant in kcal/Å · mol; r is the distance between the *J* and *K* nuclei (corresponding to r_{JK} in Fig. 5-10), and $r - r_0$ is its departure from the equilibrium bond length, in angstroms; and ω gives the span of the torsional angle (corresponding to ω_{IJKL} in Fig. 5-10), in radians.

A typical parameter value is $k_{tor/bnd} = 0.059$ kcal/Å · mol when the central bond *J—K* is C(sp^3)—C(sp^3).

Angle-Bend–Angle-Bend Cross Term The angle-bend–angle-bend cross-term potential $V_{\text{ang/ang}}$ is applied to two angles, I—J—K and K—J—L, which share one bond including the central atom, as depicted in Fig. 5-9. The addition of this term improves the prediction of vibrational spectra, while in its absence the calculated frequencies of the two angles are too close together. The force constants for this potential are somewhat generic, and depend only upon whether the attached atoms I and K are heavy atoms or hydrogens. To model this potential, MM3 employs the expression

$$V_{\text{ang/ang}} = K_{\text{ang/ang}} k_{\text{ang/ang}} (\theta' - \theta'_0)(\theta'' - \theta''_0)$$

where $K_{\text{ang/ang}}$ is a unit cancellation constant and has the value -0.021914; $k_{\text{ang/ang}}$ is the angle bending force constant in mdyne · Å/rad^2; θ' gives the span of the angle defined by atoms I—J—K (corresponding to θ_{IJK} in Fig. 5-9), and $\theta' - \theta'_0$ is its departure from the equilibrium angle span, in degrees; and θ'' gives the span of the angle defined by atoms K—J—L (corresponding to θ_{KJL} in Fig. 5-9), and $\theta'' - \theta''_0$ is its departure from the equilibrium angle span, in degrees.

Some typical values for $k_{\text{ang/ang}}$ are 0.30 mdyn · Å/rad^2 when the angle $\theta' = C(sp^3)$—$C(sp^3)$—H, and 0.24 mdyn · Å/rad^2 when $\theta' = C(sp^3)$—$C(sp^3)$—$C(sp^3)$.

Out-of-Plane Angle Bending Interactions MM3 employs the identical expression used in MM2 for the out-of-plane bending potential V_{oop}, which governs the degree of pyramidalization for certain atoms, specifically the departure of an atom J from the plane defined by three attached atoms I, K, and L, as defined in Fig. 5-6.

Typical parameter values for k_{oop} are 0.100 mdyn · Å/rad^2 for J—I = @2 $C(sp^2)$—$C(sp^3)$; 0.590 mdyn · Å/rad^2 for J—I = @3 $C(sp^2)$—$C(sp^3)$; and 0.010 mdyn · Å/rad^2 for J—I = @9 $N(sp^2)$—$C(sp^3)$.

van der Waals Interactions To model the van der Waals potential which governs the interactions between two nonbonded nongeminal atoms I and J (see Fig. 5-18), MM3 uses the two-part function of MM2 based on the ratio of the ideal to the actual internuclear distance. The van der Waals nonbonding potential V_{vdW} combines attractive and repulsive components, which are functions of the adjusted force constant k_{vdW} and the internuclear distance r_{IJ}.

The MM3 potential differs from that of MM2 in several ways. MM3 has dropped the hyperbolic correction function. The constants have been modified again to soften the function; the absolute value of the numerical coefficient of the exponential term has dropped from 12.50 to 12.00, and the value of the constant K'_{vdW} for the repulsive potential has decreased from 2.90×10^5 to 1.84×10^5. The new function takes the form

$$V_{\text{vdW}} = k_{\text{vdW}}(K_{\text{vdW}} p^6 + K'_{\text{vdW}} e^{-12.00/p})$$

where k_{vdW} is the force constant for nonbonded interactions between I and J, derived as the geometric mean of the individual hardness parameters as in MMI and MM2, in kilocalories per mole; the unit cancellation factors take the values $K_{vdW} = -2.25$, $K'_{vdW} = 1.84 \times 10^5$; and p is defined as above.

Some typical values for the hardness parameter ε are 0.020 kcal/mol for hydrogen and 0.027 kcal/mol for carbon, and for the van der Waals radius r_0 are 1.62 Å for hydrogen and 2.04 Å for carbon.

For the van der Waals interaction when I is a carbon and J is a carbon-bound hydrogen or deuterium, special reduced parameters are used (see Section 5.1.3.1).[279] The C–H or C–D bond distance is by 7.7% (down from 8.5% in MM2), and a reduced force constant k_{vdW} of 0.046 kcal/mol is used. The sum of the van der Waals radii used is 3.560 Å for the C······H interaction, and 3.557 Å for C······D.

Electrostatic Interactions For the electrostatic nonbonding potential V_{elec}, MM3 uses the same dipole–dipole model and the identical potential form which was employed in MM2. The potential energy for the electrostatic interaction between two dipole vectors **a** and **b** corresponding to the two bonds I—J and K—L (see Fig. 5-22 for illustrations) is given by the Jeans formula (see Section 5.4.1.2). When net charges are present, these are modeled as charge–dipole and charge–charge interactions.[277,359]

Some typical parameter values for the magnitude of the bond moment are: 0.900 D for the @1—@2 $C(sp^3)$—$C(sp^2)$ bond, 0.000 D for the $C(sp^3)$—H bond, -0.6000 D for the @2—@5 $C(sp^2)$—H bond, and 1.820 D for the $C(sp^3)$—F bond.

5.4.1.4. The MM4 Force Field

The first reports of the newest force field in this series, MM4, have recently been published.[93,332,333,380,381] The information available at this writing is limited to hydrocarbons, so the present analysis will be incomplete. MM4 appears to employ the same library of parametrized atom types introduced in MM3, with a few additions: @112 is now used for a hydrogen bonded to an sp^2 carbon, @122 is now used for an sp^2 carbon in a five-membered ring, and @123 is now used for an sp^3 carbon in a five-membered ring.

MM4 was designed to address several specific deficiencies in MM3, namely, vibrational spectra were not yet calculated with sufficient accuracy, rotational barriers for sterically congested molecules were still too low, heat-of-formation calculations did not explicitly include vibrational energy, and there were still some problems with simple structures. There are fifteen component potentials in the overall force field, as follows

$$V_{tot} = V_{bnd} + V_{ang} + V_{tor} + V_{vdW} + V_{elec} + V_{ang/ang} + V_{bnd/ang} + V_{impt} + V_{tor/bnd}$$
$$+ V_{bnd/bnd} + V_{tor/ang} + V_{ang/tor/ang} + V_{tor/tor} + V_{tor/impt} + V_{impt/tor/impt}$$

where V_{tot} is the total potential energy, V_{bnd} is the bond stretch component, V_{ang} is the angle bend component, V_{tor} is the torsional angle strain component, V_{vdw} is the component for van der Waals nonbonded interactions, V_{elec} is the component for electrostatic (dipole–dipole) nonbonded interactions, $V_{ang/ang}$ is the one-center bend–bend cross-term component, $V_{bnd/ang}$ is the stretch–bend cross-term component, V_{impt} is the improper torsional angle strain component (successor to the out-of-plane angle bending term), $V_{tor/bnd}$ is the torsion–stretch cross-term component, $V_{bnd/bnd}$ is the stretch–stretch cross-term component, $V_{tor/ang}$ is the torsion–bend cross-term component, $V_{ang/tor/ang}$ is the bend–torsion–bend cross-term component, $V_{tor/tor}$ is the torsion–torsion cross-term component, $V_{tor/impt}$ is the torsion–improper-torsion cross-term component, and $V_{impt/tor/impt}$ is the improper-torsion–torsion–improper-torsion cross-term component. The first seven terms have been retained from the MM3 force field, the eighth and ninth are modified from the MM3 force field, and the remaining six are new cross terms developed for MM4. The potential energy functions are all given in kilocalories per mole.

Bond Stretch Interactions   To model the stretching of a bond I—J (defined in Fig. 5-1), MM4 has elaborated upon the quartic potential V_{bnd} used in MM3 by including up to the sextic term in the power series. This new potential takes the form

$$V_{bnd} = K_{bnd} k_{bnd}(r - r_0)^2 [1 + k_{bnd3}(r - r_0) + k_{bnd4} k_{bnd3}^2(r - r_0)^2$$
$$+ k_{bnd5} k_{bnd3}^3(r - r_0)^3 + k_{bnd6} k_{bnd3}^4(r - r_0)^4]$$

where K_{bnd} is a unit cancellation constant and has the value 71.94; k_{bnd} is the bond stretching force constant in millidynes per angstrom; r is the distance between the I and J nuclei (corresponding to r_{IJ} in Fig. 5-1), and $r - r_0$ is its departure from the equilibrium bond length, in angstroms; k_{bnd3} is the cubic stretch constant, with a value of -2.20 Å^{-1} used for bonds to hydrogen, and a value of -3.00 Å^{-1} used for all other bonds; k_{bnd4} is the quartic stretch constant, with a value of $\frac{7}{12}$; k_{bnd5} is the quintic stretch constant, with a value of $\frac{1}{4}$; and k_{bnd6} is the sextic stretch constant, with a value of $\frac{31}{360}$. There are specialized parameters for use with bonds included in a conjugated system and in a three-, four- or five-membered ring.

Some typical parameter values are $k_{bnd} = 4.740$ mdyn/Å and $r_0 = 1.1120$ Å for an aliphatic C—H bond, and $k_{bnd} = 4.550$ mdyn/Å and $r_0 = 1.5270$ Å for a C(sp^3)—C(sp^3) bond. There are specialized parameters for use with bonds included in a conjugated system.

Angle Bending Interactions   MM4 employs the same sextic potential V_{ang} to govern bending for the angle I—J—K (defined in Fig. 5-4) as was used in MM3. There are specialized parameters for use with angles which include a conjugated bond or a three-, four-, or five-membered ring.

Some typical parameter values are $k_{ang} = 0.59$ mdyn · Å/rad^2, and $\theta_0 = 108.9°$, $109.47°$, and $110.8°$ for a type 1, type 2, and type 3 C(sp^3)—C(sp^3)—H angle, respectively; and $k_{ang} = 0.74$ mdyn · Å/rad^2, and $\theta_0 = 109.5°$, $110.4°$, and $111.8°$ for a type 1, type 2, and type 3 C(sp^3)—C(sp^3)—C(sp^3) angle, respectively.

Torsional Interactions MM4 employs the same three-part potential V_{tor} to govern torsional interactions as was used in MM3 and MM2, with two additional terms to be used in certain specific instances. The function is based on the cosine of the torsional angle ω, as defined by the atom string I—J—K—L (see Fig. 5-5), and there are specific sets of constants for three-, four-, and five-membered rings.

A fourfold term is added for cases where both of the two central atoms (i.e. J and K) are C(sp^2).[332] This results in a broader potential well for these systems, and contributes to a more accurate modeling of their vibrational frequencies, as well as better fitting the difference between planar and pyramidal transition states for alkenes. A sixfold term is added to the potential for treating the torsional strain only in the specific case of I—J—K—L = H—C(sp^3)—C(sp^3)—H (@5—@1—@1—@5); this confers a slight improvement in the calculation of torsional frequencies and energy barriers. The complete modified form for the potential is thus

$$V_{tor} = \tfrac{1}{2}[k_{tor1}(1 + \cos \omega) + k_{tor2}(1 - \cos 2\omega) + k_{tor3}(1 + \cos 3\omega) + k_{tor4}(1 - \cos 4\omega) + k_{tor6}(1 - \cos 6\omega)]$$

where k_{tor4} is the force constant for the fourfold component, k_{tor6} is the force constant for the sixfold component, and the other constants and variables are as defined previously.

Some typical parameter values for I—J—K—L = H—C(sp^3)—C(sp^3)—H are $k_{tor1} = 0.000$ kcal/mol, $k_{tor2} = 0.000$ kcal/mol, $k_{tor3} = 0.260$ kcal/mol; $k_{tor6} = 0.008$ kcal/mol (and 0.00 for all other systems); for I—J—K—L = C(sp^3)—C(sp^3)—C(sp^3)—C(sp^3) are $k_{tor1} = 0.239$ kcal/mol, $k_{tor2} = 0.024$ kcal/mol, and $k_{tor3} = 0.637$ kcal/mol; for I—C(sp^2)—C(sp^2)—L, $k_{tor4} = -0.09k_{tor2}$ when neither I nor L is C(sp^2), and $k_{tor4} = -0.06k_{tor2}$ when either I or L or both is C(sp^2).

van der Waals Interactions To model the van der Waals potential which governs the interactions between two nonbonded nongeminal atoms I and J (see Fig. 5-18), MM4 employs the same function used in MM3 with some varied parameters. In MM4, there are different atom types and van der Waals parameters for hydrogen, depending upon whether it is attached to an sp^3 carbon (@5 hydrogen) or to an sp^2 carbon (the new @112 hydrogen), as well as new atom types for carbons contained within five-membered rings: @122 for a C(sp^2) and @122 for a C(sp^3).

As was the case in MM3, special reduced parameters are used for the van der Waals interaction when I is a carbon and J is a carbon-bound hydrogen or deuterium. MM4 alters their positions by invoking a 6.0% reduction in the length of the C—H or C—D bond,[93] down from 7.7% in MM3. For the interaction of an @5 H and an @1 carbon, a force constant k_{vdW} of 0.024 kcal/mol and a combined van der Waals radius r_0 of 3.44 Å is used; for an @5 H and an @2 carbon, the values are $k_{vdW} = 0.048$ kcal/mol and $r_0 = 3.20$ Å; and for an @112 H and an @2 carbon, the values are $k_{vdW} = 0.034$ kcal/mol and $r_0 = 3.58$ Å.

Some typical values for the hardness parameter ε are 0.017 kcal/mol for hydrogen, 0.057 kcal/mol for C(sp^2), and 0.037 kcal/mol for C(sp^3); the van der Waals radius r_0 is 1.640 Å for hydrogen, and 1.960 Å for either sp^2 or sp^3 carbon.

Electrostatic Interactions For the electrostatic nonbonding potential V_{elec}, MM4 uses the same dipole–dipole model which was employed in MM3 and MM2. The potential energy for the electrostatic interaction between two dipole vectors **a** and **b** corresponding to the two bonds I—J and K—L (see Fig. 5-22 for definitions) is given by the Jeans formula (see Section 5.4.1.2). When net charges are present, these are modeled as charge–dipole and charge–charge interactions.

Dipole moments of 0.60 and 0.95 D have been assigned to the C(sp^2)—H and C(sp^2)—C(sp^3) bonds, respectively, with the C(sp^2) atom as the negative end of the dipole in the latter case. The C(sp^3)-H bond bears no net dipole.

Angle-Bend–Angle-Bend Cross Term MM4 employs the same angle-bend–angle-bend cross-term potential $V_{ang/ang}$ which is used in MM3. This term is used to model the coupled vibrations of two angles, I—J—K and K—J—L, which share one bond including the central atom, as depicted in Fig. 5-9.

Some typical values for $k_{ang/ang}$ are 0.350 mdyn · Å/rad² when I—J—K = C(sp^3)—C(sp^3)—H, and 0.204 mdyn · Å/rad² when I—J—K = C(sp^3)—C(sp^3)—C(sp^3).

Bond-Stretch–Angle-Bend Cross Term MM4 employs the same form as MM2 for the bond-stretch–angle-bend cross-term potential $V_{bnd/ang}$, applied to the three-atom unit I—J—K (as defined in Fig. 5-8). There are specific sets of constants for three-, four-, and five-membered rings.

Some typical parameter values are $k_{bnd/ang} = 0.140$ mdyn/rad for I—J—K = C(sp^3)—C(sp^3)—C(sp^3) and $k_{bnd/ang} = 0.100$ mdyn/rad for I—J—K = C(sp^3)—C(sp^3)—H.

Improper Torsional Interactions The improper torsion potential term, V_{impt}, is employed by MM4 to model the departure from planarity of an sp^2-hybridized atom J as a function of the improper torsional angles it forms with its three attached atoms I, K, and L, as defined in Fig. 5-7. MM4 assigns the improper torsional force constants based upon the identity of the first (i.e.

central) and last atom of the improper torsional angle. The equation for this potential takes the form

$$V_{impt} = k_{impt1}(1 - \cos 2\omega_1) + k_{impt2}(1 - \cos 2\omega_2) + k_{impt3}(1 - \cos 2\omega_3)$$

where k_{impt1}, k_{impt2}, and k_{impt3} are the force constants (in kilocalories per mole) for the improper torsional angles ω_1, ω_2, and ω_3, as defined in Fig. 5-7, and would be assigned to the atom pairs J—L, J—I, and J—K, respectively.

Some typical values for k_{impt} are 1.20 kcal/mol for $J = C(sp^2)$ and $L = C(sp^3)$, and 1.60 kcal/mol for $J = C(sp^2)$ and $L = C(sp^2)$.

This potential is simpler in computational terms than the out-of-plane angle bending term V_{oop} used in MM3 and MM2, for which it substitutes.

Torsion–Bond-Stretch Cross Term The MM4 force field contains two types of torsion–bond-stretch cross-terms $V_{tor/bnd}$, known as type 1 and type 2. As defined in Fig. 5-10, the type 1 term addresses the coupling between the torsional rotation across I—J—K—L with the stretching of the central bond J—K, and is a function of the torsional angle ω_{IJKL} and the central bond distance r_{JK}.[332] The addition of this term causes the central bond to lengthen more realistically when the substituents are eclipsed. The MM4 implementation uses an expression which has been modified from that used in MM3 by inclusion of onefold and twofold terms, and which takes the form

$$V_{tor/bnd} = -\tfrac{1}{2}K_{tor/bnd}(r - r_0)[(1 + k_{tor/bnd1}\cos\omega)$$
$$+ (1 - k_{tor/bnd2}\cos 2\omega) + (1 + k_{tor/bnd3}\cos 3\omega)]$$

where $K_{tor/bnd}$ is a unit cancellation constant which takes the value 11.995; r is the distance between the atoms of the central bond J—K (corresponding to r_{JK} in Fig. 5-10), and $r - r_0$ is its departure from the equilibrium bond length, in angstroms; $k_{tor/bnd1}$, $k_{tor/bnd2}$, and $k_{tor/bnd3}$ are the one-, two-, and three-fold torsion–stretch force constants (respectively) in kcal/Å · mol; and ω is the torsional angle (corresponding to ω_{IJKL} in Fig. 5-10).

Typical parameter values for this type 1 term are $k_{tor/bnd1} = 0.0$, $k_{tor/bnd2} = 0.0$, and $k_{tor/bnd3} = 0.660$ kcal/Å · mol when the central bond J-K is $C(sp^3)$—$C(sp^3)$. This term leads to a longer central bond in eclipsed and skew conformations, and a shorter central bond in *gauche* and *anti* conformations.

As defined in Fig. 5-11, the type 2 torsion–bond-stretch term addresses the coupling between the torsional rotation across I—J—K—L with the stretching of one of the terminal bonds, I—J.[381] It is a function of the torsional angle ω_{IJKL} and the terminal bond distance r_{IJ}, and takes the form

$$V_{tor/bnd} = -\tfrac{1}{2}K_{tor/bnd}(r - r_0)[(1 + k_{tor/bnd1}\cos\omega)$$
$$+ (1 - k_{tor/bnd2}\cos 2\omega)]$$

5.4. OTHER FORCE FIELDS

where $K_{\text{tor/bnd}}$ is a unit cancellation constant which has the value 11.995; r is the internuclear distance between atoms I and J (corresponding to r_{IJ} in Fig. 5-11), and $r - r_0$ is its departure from the equilibrium bond length, in angstroms; $k_{\text{tor/bnd1}}$ and $k_{\text{tor/bnd2}}$ are the one- and twofold torsion–stretch force constants (respectively) in units of kcal/Å · mol; and ω is the torsional angle (equivalent to ω_{IJKL} in Fig. 5-11).

Typical parameter values for the type 2 term are $k_{\text{tor/bnd1}} = 0.150$ kcal/Å · mol and $k_{\text{tor/bnd2}} = 0.216$ kcal/Å · mol for I—J—K—L = H—$C(sp^2)$—$C(sp^3)$—$C(sp^3)$.[381]

This type 2 term is important for unsaturated compounds (modeling hyperconjugation effects), or when the constituent atoms bear lone pairs of electrons (modeling the Bohlmann effect). These effects are generally small, and their magnitude approaches the error limits of the method, but inclusion of this potential does lead to better approximations of those structures which are shown by physical methods to have abnormally long bonds.[381]

Bond-Stretch–Bond-Stretch Cross Term The bond-stretch–bond-stretch potential $V_{\text{bnd/bnd}}$ is a new cross term employed by MM4[332] to model the coupling of the stretching vibrations for the two bonds I—J and J—K in the three-atom system I—J—K. A single atom J is the terminus of both bonds. It is a function of the two internuclear distances r_{IJ} and r_{JK}, as defined in Fig. 5-12, and takes the form

$$V_{\text{bnd/bnd}} = K_{\text{bnd/bnd}} k_{\text{bnd/bnd}} (r' - r'_0)(r'' - r''_0)$$

where $K_{\text{bnd/bnd}}$ is a units cancellation constant which takes the value 143.88; $k_{\text{bnd/bnd}}$ is the stretch–stretch force constant in millidynes per angstrom; r' is the internuclear distance between atoms I and J (corresponding to r_{IJ} in Fig. 5-12), and $r' - r'_0$ is the departure from its equilibrium value, in angstroms; and r'' is the internuclear distance between atoms J and K (corresponding to r_{JK} in Fig. 5-12), and $r'' - r''_0$ represents the departure from its equilibrium value, in angstroms. These coupled stretching vibrations correspond to the familiar symmetric stretch and asymmetric stretch from vibrational spectroscopy, which are particularly significant in the model for the unsaturated hydrocarbons isobutene and benzene.[332]

Typical values for the stretch–stretch force constant $k_{\text{bnd/bnd}}$ are 0.4 mdyn/Å for I—J—K = $C(sp^3)$—$C(sp^2)$—$C(sp^3)$ and 0.8 mdyn/Å for I—J—K = $C(sp^2)$—$C(sp^2)$—$C(sp^3)$. In its current level of development, MM4 invokes this term only when the central atom J is an @2 $C(sp^2)$.

Torsion–Angle-Bend Cross Term The torsion–angle-bending potential term, $V_{\text{tor/ang}}$, is a new cross term employed by MM4 to model the coupling between the torsional interaction over the four-atom system I—J—K—L and the angle bending in the two three-atom angles contained in this moiety,

260 THE MOLECULAR MECHANICS FORCE FIELD

$I\text{—}J\text{—}K$ and $J\text{—}K\text{—}L$, as defined in Fig. 5-16. It is a function of the torsional angle ω_{IJKL} and the angles θ_{IJK} and θ_{JKL}, and takes the form:

$$V_{\text{tor/ang}} = K_{\text{tor/ang}}\{[k'_{\text{tor/ang1}}(1 + \cos \omega) + k'_{\text{tor/ang2}}(1 - \cos 2\omega)$$
$$+ k'_{\text{tor/ang3}}(1 + \cos 3\omega)](\theta' - \theta'_0)$$
$$+ [k''_{\text{tor/ang1}}(1 + \cos \omega)$$
$$+ k''_{\text{tor/ang2}}(1 - \cos 2\omega)$$
$$+ k''_{\text{tor/ang3}}(1 + \cos 3\omega)](\theta'' - \theta''_0)\}$$

where $K_{\text{tor/ang}}$ is a unit cancellation constant and has the value 2.51124; $k'_{\text{tor/ang1}}$, $k'_{\text{tor/ang2}}$, and $k'_{\text{tor/ang3}}$ are the one-, two-, and threefold force constants with respect to first angle defined by $I\text{—}J\text{—}K$; ω is the torsional angle (corresponding to ω_{IJKL} in Fig. 5-16); θ' is the current span for the angle $I\text{—}J\text{—}K$ (corresponding to θ_{IJK} in Fig. 5-16), and $\theta' - \theta'_0$ gives the departure of this angle from its equilibrium value; $k''_{\text{tor/ang1}}$, $k''_{\text{tor/ang2}}$, and $k''_{\text{tor/ang3}}$ are the one-, two-, and threefold force constants with respect to second angle defined by $J\text{—}K\text{—}L$; and θ'' is the current span for the angle $J\text{—}K\text{—}L$ (corresponding to θ_{JKL} in Fig. 5-16), and $\theta'' - \theta''_0$ gives the departure of this angle from its equilibrium value.

Typical values for these parameters are: for $I\text{—}J\text{—}K\text{—}L = C(sp^3)\text{—}C(sp^2)\text{—}C(sp^2)\text{—}C(sp^3)$, $k'_{\text{tor/ang1}} = k''_{\text{tor/ang1}} = 0.006$ kcal/mol, and $k'_{\text{tor/ang2}} = k'_{\text{tor/ang3}} = k''_{\text{tor/ang2}} = k''_{\text{tor/ang3}} = 0.0$ kcal/mol.

This function acts to open bond angles as the torsional angle goes from *anti* to *gauche*. This term becomes important when atoms containing a lone pair of electrons are involved (for modeling the Bohlmann effect), and also favors and renders more puckered the boat and tub forms of cycloheptatriene and cyclo-octatetraene.[332]

Angle-Bend–Torsion–Angle-Bend Cross Term The angle-bend–torsion–angle-bend potential term $V_{\text{ang/tor/ang}}$ is a new cross term employed by MM4 to model the in-plane coupling between the torsional interaction over the four-atom system $I\text{—}J\text{—}K\text{—}L$, and the two included three-atom angles for the three-atom subgroups $I\text{—}J\text{—}K$ and $J\text{—}K\text{—}L$, as defined in Fig. 5-16. It is a function of the torsional angle ω_{IJKL} and the angles θ_{IJK} and θ_{JKL}, and takes the form

$$V_{\text{ang/tor/ang}} = K_{\text{ang/tor/ang}} k_{\text{ang/tor/ang}}(\cos \omega)(\theta' - \theta'_0)(\theta'' - \theta''_0)$$

where $K_{\text{ang/tor/ang}}$ is a unit cancellation factor which takes the value 0.043828; $k_{\text{ang/tor/ang}}$ is the force constant for the angle-bend–torsion–angle-bend potential, in kilocalories per mole; ω is the value for the torsional angle (corresponding to ω_{IJKL} in Fig. 5-16) in radians; and θ' and θ'' are the current values for the two angles (corresponding to θ_{IJK} and θ_{JKL}, respectively, in Fig.

5-16) in degrees, and $\theta' - \theta'_0$ and $\theta'' - \theta''_0$ are their departures from the equilibrium angle spans. Inclusion of this cross term improves the C(sp^2)—H in-plane bending frequencies, for example the ethylene B_{1g} and B_{2u} rocking frequencies.[332,333]

Typical values for the angle-bend–torsion–angle-bend force constant $k_{\text{ang/tor/ang}}$ are: -0.08 kcal/mol for I—J—K—L = H—C(sp^3)—C(sp^2)—C(sp^2), and -0.04 kcal/mol for I—J—K—L = C(sp^3)—C(sp^2)—C(sp^2)—C(sp^2).

Torsion–Torsion Cross Term The torsion–torsion potential term $V_{\text{tor/tor}}$ is a new cross term employed by MM4 to model the coupling between the two torsional interactions over the five-atom system I—J—K—L—M, as defined in Fig. 5-14. It is a function of the two torsional angles ω_{IJKL} and ω_{JKLM} and the bond order P of the central bond of each torsional angle, P_{JK} and P_{KL}. It takes the form

$$V_{\text{tor/tor}} = -k'_{\text{tor/tor}}(1 - P')(1 + \cos 3\omega')k''_{\text{tor/tor}}(1 - P'')(1 + \cos 3\omega'')$$

where $k'_{\text{tor/tor}}$ and $k''_{\text{tor/tor}}$ are the force constants for the first and second torsional angles ω' and ω'' (corresponding to ω_{IJKL} and ω_{JKLM}, respectively, in Fig. 5-14); and P' and P'' are the bond order parameters for the central bonds of the two torsional angles, J—K and K—L, respectively (corresponding to P_{JK} and P_{KL}, respectively, in Fig. 5-14). The torsion–torsion cross term is important for large out-of-plane bending frequencies, especially in benzene. It also lowers the energies of more highly aromatic compounds (with more nearly equivalent adjacent bond orders), which leads to more accurate calculation of heats of formation.[332]

A typical value is $k_{\text{tor/tor}} = 0.85$ kcal/mol for I—J—K—L = C(sp^2)—C(sp^2)—C(sp^2)—C(sp^2).

Torsion–Improper-Torsion Cross Term The torsion–improper-torsion potential term, $V_{\text{tor/impt}}$, is a new cross term employed by MM4 to model the in-plane coupling between the torsional interaction for the four-atom unit I—J—K—L and the improper torsional potential at the internal atoms J and K, defined respectively by the torsional angle ω_{IJKL} and the improper torsional angles ω_{JIMK} and ω_{KLNJ}, as depicted in Fig. 5-17. It is a function of the torsional angle ω_{IJKL} and the improper torsional angles ω_{JIMK} and ω_{KLNJ}, and takes the form

$$V_{\text{tor/impt}} = (1 - \cos^2 \omega)[k'_{\text{tor/impt}}(1 - \cos \omega') + k''_{\text{tor/impt}}(1 - \cos \omega'')]$$

where ω is the torsional angle (corresponding to ω_{IJKL} in Fig. 5-17); $k'_{\text{tor/impt}}$ and $k''_{\text{tor/impt}}$ are the respective force constants (in kilocalories per mole) for ω' and ω''; and ω' and ω'' are the two improper torsional angles in radians (corresponding to ω_{JIMK} and ω_{KLNJ} in Fig. 5-17, respectively). MM4 assigns force constants to this potential based upon the identity of the first (i.e. central) and last atom of the improper torsional angle. This term counters out-of-plane bending in conjugated systems.[332,333] It serves to flatten $C(sp^2)$ in ethylene and similar molecules, and acts to decrease pyramidalization at atom J when the torsional angle is rotated toward 90°, which serves to reproduce more closely the rotational barriers found for these molecules in *ab initio* computational studies.[332]

The force constant $k_{\text{tor/impt}}$ typically has a value of 5.0 kcal/mol, e.g. when J—K is $C(sp^2)$—$C(sp^3)$ or $C(sp^2)$—$C(sp^2)$.

Improper-Torsion–Torsion–Improper-Torsion Cross Term The improper-torsion–torsion–improper-torsion potential term, $V_{\text{impt/tor/impt}}$, is a new cross term employed by MM4 to model the out-of-plane coupling between the torsional interaction (I—J—K—L) and the improper torsional potentials at the internal atoms J and K. It is a function of the torsional angle ω_{IJKL} and the improper torsional angles ω_{JIMK} and ω_{KLNJ}, as depicted in Fig. 5-17, and takes the form

$$V_{\text{impt/tor/impt}} = k_{\text{impt/tor/impt}} \cos \omega \cos \omega' \cos \omega''$$

where $k_{\text{impt/tor/impt}}$ is the force constant for the improper-torsion–torsion–improper-torsion potential in kilocalories per mole; ω is the torsional angle (corresponding to ω_{IJKL} in Fig. 5-17), in radians; ω' is the improper torsional angle beginning at atom J (corresponding to ω_{JIMK} in Fig. 5-17), in radians; and ω'' is the improper torsional angle beginning at atom K (corresponding to ω_{KLNJ} in Fig. 5-17), in radians. MM4 assigns force constants to this potential based upon the identity of the first (i.e. central) and last atom of the improper torsional angle. This term counters out-of-plane bending in conjugated systems. For example, it improves the calculation of the $C(sp^2)$—H out-of-plane bending frequencies with B_{1g} and B_{2u} symmetry in ethylene.[332,333]

Typical values for the force constant $k_{\text{impt/tor/impt}}$ are 3.5 kcal/mol when J—K is $C(sp^2)$—$C(sp^3)$, and 3.0 kcal/mol when it is $C(sp^2)$—$C(sp^2)$.

5.4.2. The Consistent Force Field (CFF)

The **Consistent Force Field** (CFF) is a force field formulation which has undergone as extensive an evolutionary process as has the Allinger family of force fields. The CFF is so named because its developers (S. Lifson *et al.*) sought to develop a single set of force field equations which would predict equilibrium conformations, vibrational spectra, strain energies, and vibrational enthalpies with comparable accuracy.[15,71,90,92,300,382–397] The initial sources of parameters were a set of infrared and Raman fundamental frequencies for cyclohexane, the excess enthalpies of the cycloalkanes determined from calorimetry, and the main conformational features of the cyclohexane chair conformation, obtained from the electron diffraction structure.[71] The CFF formulation was originally developed with the ultimate goal of calculating protein conformation.[9]

5.4.2.1. The Urey–Bradley Consistent Force Field (UBCFF) A series of publications near the end of the 1960s may be taken as the introduction of the CFF.[71,300,382,384] Its initial form was that of a hydrocarbon force field, as was MMI. The force field as it was presented consisted of seven component potentials:

$$V_{tot} = V_{bnd} + V_{ang} + V_{tor} + V_{vdW} + V_{UB} + V_{elec} + V_{ang/tor/ang}$$

where V_{tot} is the total potential energy, V_{bnd} is the bond stretch component, V_{ang} is the angle bend component, V_{tor} is the torsional angle strain component, V_{vdW} is the component for van der Waals nonbonded nongeminal interactions, V_{UB} is the Urey–Bradley component for nonbonded geminal interactions, V_{elec} is the component for electrostatic (charge–charge) nonbonded interactions, and $V_{ang/tor/ang}$ is the angle-bend–torsion–angle-bend cross-term component which models the coupling between torsion and bending of the two contained angles. The potential energy functions are all given in kilocalories per mole.

Bond Stretch Interactions UBCFF employed a harmonic bond stretching potential V_{bnd} for a bond I—J as defined in Fig. 5-1:

$$V_{bnd} = k_{bnd}(r - r_0)^2$$

where k_{bnd} is the force constant for bond stretching in kcal/mol · Å² (the value of this constant includes the factor of $\frac{1}{2}$ which normally appears in a harmonic function); and r is the distance between the I and J nuclei (corresponding to r_{IJ} in Fig. 5-1), and $r - r_0$ is its departure from the equilibrium bond length, in angstroms.

Some typical parameter values are $k_{bnd} = 286.9$ kcal/mol · Å² and $r = 1.099$ Å for I—J = C(sp^3)—H, and $k_{bnd} = 111.0$ kcal/mol · Å² and $r = 1.455$ Å for I—J = C(sp^3)—C(sp^3).

Angle Bending Interactions UBCFF used a harmonic angle bending potential V_{ang} for an angle I—J—K as defined in Fig. 5-4:

$$V_{ang} = k_{ang}(\theta - \theta_0)^2$$

where k_{ang} is the force constant for angle bending in mdyn · Å/rad² (the value of this constant includes the factor of $\frac{1}{2}$ which normally appears in a harmonic function); and θ is the current span across the angle (corresponding to θ_{IJK} in Fig. 5-4), and $\theta - \theta_0$ gives its displacement from the equilibrium value, in degrees.

Some typical values for k_{ang} are 22.0 kcal/mol · rad^2 for C(sp^3)—C(sp^3)—C(sp^3) and 26.79 kcal/mol · rad^2 for C(sp^3)—C(sp^3)—H; θ_0 in all cases is taken to be 109.47°.

Torsional Interactions UBCFF employed a simple potential V_{tor} to govern torsional interactions, which contains a single, threefold term based on the cosine of the torsional angle ω, as defined by the atom string *I—J—K—L* (see Fig. 5-5), and takes the form

$$V_{tor} = k_{tor3}(1 + \cos 3\omega)$$

where k_{tor3} is the force constant for the torsional interaction in kilocalories per mole (the value of this constant includes the factor of $\frac{1}{2}$ which normally appears in a harmonic function), and ω gives the span of the torsional angle (corresponding to ω_{IJKL} in Fig. 5-5), in radians.

UBCFF associates the torsional force constant with the two central atoms, which in the case of hydrocarbons will always be C(sp^3)—C(sp^3). The force constant k_{tor3} takes the value 1.418 kcal/mol.

van der Waals Interactions To model the van der Waals potential which governs the interactions between two nonbonded nongeminal atoms *I* and *J* (see Fig. 5-18), UBCFF uses a variation of the Lennard–Jones 6–12 potential, which is a function of the adjusted force constant k_{vdW} and the internuclear distance r_{IJ}. The sixth-power term provides the attractive component and the twelfth power term provides the repulsive component of the potential:

$$V_{vdW} = k_{vdW}(p^{12} - 2p^6)$$

where k_{vdW} is the force constant for nonbonded interactions between *I* and *J*, in kilocalories per mole, and p is a unitless ratio of the equilibrium distance between *I* and *J* to the actual distance, as defined in Section 5.2.

Some typical parameter values are: for the H·····H interaction, $k_{vdW} = 0.0045$ kcal/mol and $r_{IJ(0)} = 2.936$ Å; for the C·····C interaction, $k_{vdW} = 0.0196$ kcal/mol and $r_{IJ(0)} = 4.228$ Å.

Urey–Bradley Geminal Nonbonded Interactions To model the interactions between two nonbonded geminal atoms *I* and *K* (see Fig. 5-20), UBCFF uses a harmonic potential which includes linear and quadratic terms,[71] given by

$$V_{UB} = k_{UB}(r - r_0) + k_{UB2}(r - r_0)^2$$

where k_{UB} is the force constant for first term, in kcal/mol · Å; r is the distance between the *I* and *K* nuclei (corresponding to r_{IK} in Fig. 5-20), and $r - r_0$ is the departure from the equilibrium internuclear separation, in angstroms; and k_{UB2} is the force constant for the quadratic term, in kcal/mol · Å² (the value of this constant includes the factor of $\frac{1}{2}$ which normally appears in a harmonic function).

Some typical parameter values are: for the H······H interaction, $k_{UB} = -0.104$ kcal/mol · Å, $k_{UB2} = 2.900$ kcal/mol · Å², and $r_{IK(0)} = 1.8$ Å; for the C······H interaction, $k_{UB} = -0.746$ kcal/mol · Å, $k_{UB2} = 43.61$ kcal/mol · Å², and $r_{IK(0)} = 2.2$ Å; for the C······C interaction, $k_{UB} = -1.547$ kcal/mol · Å, $k_{UB2} = 37.31$ kcal/mol · Å², and $r_{IK(0)} = 2.5$ Å.

Electrostatic Interactions The Coulomb charge–charge model is applied to the electrostatic interactions between two nonbonded nongeminal atoms I and J (see Fig. 5-21). No correction for the dielectric constant of the medium is employed at this point. The Coulombic potential between two atoms I and J is given by

$$V_{\text{elec}} = \frac{q_I q_J}{r_{IJ}}$$

where q_I and q_J are the charges on the respective atoms I and J, in atomic charge units, and r_{IJ} is the internuclear distance between atoms I and J, in angstroms.

The charges used by UBCFF are $+0.144$ for hydrogen atoms, and a compensating charge of the opposite sign on the carbon atoms.

Angle-Bend–Torsion–Angle-Bend Cross Term An angle-bend–torsion–angle-bend cross-term potential was introduced[382] into the UBCFF for modeling the coupled interactions between two angles I—J—K and J—K—L centered at the adjacent carbons J and K and the encompassing torsion for the four-atom system I—J—K—L, as defined in Fig. 5-16. It is equivalent to the angle-bend–torsion–angle-bend potential term $V_{\text{ang/tor/ang}}$ employed by MM4 (see Section 5.4.1.4), and so is presented here as such. The inclusion of this term enables the more accurate prediction of some vibrational frequencies,[384] and is a function of the torsional angle ω_{IJKL} and the angles θ_{IJK} and θ_{JKL} as

$$V_{\text{ang/tor/ang}} = k_{\text{ang/tor/ang}}(\cos \omega)(\theta' - \theta'_0)(\theta'' - \theta''_0)$$

where $k_{\text{ang/tor/ang}}$ is the force constant for the angle-bend–torsion–angle-bend potential, in units of kilocalories per mole; ω is the value for the torsional angle (corresponding to ω_{IJKL} in Fig. 5-16) in radians; and θ' and θ'' are the current values for the two angles (corresponding to θ_{IJK} and θ_{JKL}, respectively, in Fig. 5-16) in degrees, and $\theta' - \theta'_0$ and $\theta'' - \theta''_0$ are their departures from the equilibrium angle spans.

Typical values for the angle-bend–torsion–angle-bend force constant $k_{ang/tor/ang}$ are -2.3 kcal/mol · rad^2 for I—J—K—L = C(sp^3)—C(sp^3)—C(sp^3)—C(sp^3), and -9.5 kcal/mol · rad^2 for I—J—K—L = H—C(sp^3)—C(sp^3)—H.

The application of the UBCFF treatment beyond the realm of hydrocarbons was reported in a series of papers in the early 1970s,[383,398,399] which described a potential function to model the hydrogen bonding interaction in amides. The amide parameters and the hydrogen bonding potential were superimposed upon the eight components of the existing hydrocarbon force field described above. In almost all cases, the same potential energy equation was used. Some sample parameters are given below.

For the bond stretch potential, some typical parameter values are: $k_{bnd} = 405$ kcal/mol · Å^2 and $r_0 = 0.980$ Å for I—J = N—H; $k_{bnd} = 403$ kcal/mol · Å^2 and $r_0 = 1.278$ Å for I—J = C(sp^2)—N; $k_{bnd} = 595$ kcal/mol · Å^2 and $r_0 = 1.200$ Å for I—J = C(sp^2)=O; and $k_{bnd} = 261$ kcal/mol · Å^2 and $r_0 = 1.457$ Å for I—J = C(sp^3)—N.

For the angle bend potential, some typical parameter values are $k_{ang} = 54.5$ kcal/mol · rad^2 for I—J—K = C(sp^2)—N—C(sp^3), $k_{ang} = 48.5$ kcal/mol · rad^2 for I—J—K = N—C(sp^2)=O, and $k_{ang} = 26.6$ kcal/mol · rad^2 for C(sp^2)—N—H; θ_0 for all trigonal angles is taken to be 120.0°.

For the torsional interaction potential in the trigonal systems, a twofold potential was employed in keeping with symmetry, to give

$$V_{tor} = k_{tor2}(1 + \cos 2\omega)$$

Some typical parameter values are $k_{tor2} = 1.655$ kcal/mol for I—J—K—L = C(sp^3)—C(sp^2)—N—H, and $k_{tor2} = 4.045$ kcal/mol for I—J—K—L = O=C(sp^2)—N—H. Typical parameter values for the threefold terms involving tetrahedral atoms are $k_{tor3} = 0.500$ kcal/mol for I—J—K—L = H—C(sp^3)—C(sp^2)—∗; and $k_{tor3} = 1.500$ kcal/mol for I—J—K—L = H—C(sp^3)—N—∗, where ∗ is a wildcard.

For the Urey–Bradley potential, some typical parameter values are: $k_{UB2} = 55.8$ kcal/mol · Å^2 and $r_{IK(0)} = 2.0$ Å for I—J—K = C(sp^2)—N—H; $k_{UB2} = 180$ kcal/mol · Å^2 and $r_{IK(0)} = 2.186$ Å for I—J—K = N—C(sp^2)=O; $k_{UB2} = 32.4$ kcal/mol · Å^2 and $r_{IK(0)} = 2.400$ Å for I—J—K = C(sp^2)—N—C(sp^3); and $k_{UB2} = 76.8$ kcal/mol · Å^2 and $r_{IK(0)} = 1.975$ Å for C(sp^2)—C(sp^3)—H. The force constant for the linear term (k_{UB}) is related to that for the quadratic term (k_{UB2}) by

$$k_{UB} = -\tfrac{1}{10}k_{UB2}$$

For the out-of-plane angle bending potential in a secondary amide, some typical parameter values are: for deformation at the carbonyl C(sp^2), $k_{oop} = 8.08$ kcal/mol · rad^2, where the out-of-plane angle is defined to be the span between the planes containing C(sp^3)—C(sp^2)—N and O=C(sp^2)—N; and for deformation at the amide N, $k_{oop} = 1.38$ kcal/mol · rad^2, where the out-of-plane angle is defined to be the span between the planes containing C(sp^2)—N—C(sp^3) and C(sp^2)—N—H.

Special attention was paid to the nonbonding interactions, since it was in this context that the hydrogen bonding potential would be formulated. As a point of departure, the authors formulated a four-part nonbonding potential: the attractive and repulsive nonbonding components (i.e. the two parts of a Lennard–Jones potential), an electrostatic component (Coulombic potential), and an explicit hydrogen bonding attractive component, using a Morse potential function.[317-320]

Following optimization, it was found that the hydrogen bonding effects could be modeled quite well without the explicit Morse potential bonding component, and that the standard steric and electrostatic nonbonding components would suffice. Little difference was found between the 6–9 and 6–12 versions of the Lennard–Jones potential used for this interaction—the equation below uses the 6–9 form. Thus, the total nonbonding potential energy is given by

$$V_{nonbonding} = k_{vdW}(2p^9 - 3p^6) + \frac{q_I q_J}{r_{IJ}}$$

where the variables are defined as above.

The hydrogen bonding effect is implemented by suitable modification of the parameters for those atoms involved in the interaction. Specifically, the van der Waals radius r_0 for the "acidic" amide hydrogen, and the van der Waals force constant k_{vdW} for the interaction between the carbonyl oxygen and the amide hydrogen, are set to zero, and the attraction is driven electrostatically without steric repulsion. This formulation of the hydrogen bond potential has reappeared since in many other force fields.

Some typical parameter values for the amide group constituents are: for C(sp^3)-bound hydrogen, $\varepsilon = 0.0025$ kcal/mol, $r_0 = 1.77$ Å, and $q = +0.11$; for "acidic" N-bound hydrogen, $\varepsilon = 0.0$ kcal/mol, $r_0 = 0.0$ Å, and $q = +0.26$ (secondary amide), $+0.41$ (primary amide); for the carbonyl oxygen, $\varepsilon = 0.198$ kcal/mol, $r_0 = 1.825$ Å, and $q = -0.46$; and for amide nitrogen, $\varepsilon = 0.161$ kcal/mol, $r_0 = 2.005$ Å, and $q = -0.26$ (secondary amide), -0.82 (primary amide).

An extension of the UBCFF to conjugated hydrocarbons was presented in 1972,[400] which separated the σ- and π-electron systems. The σ-framework was treated using the UBCFF method, with a modified Buckingham potential for nonbonded interactions instead of the Lennard–Jones 6–9 potential, in order to use the same potential form for the saturated and unsaturated substructures. They found that the electrostatic Coulomb term could be omitted for the systems under study, without introducing significant errors. They employed the Pariser–Parr–Pople molecular orbital approximation, corrected for nearest neighbor overlap (SCF-MO-CI), to treat the π-system.

5.4.2.2. The Valence Consistent Force Field (CVFF) Beginning in 1973, the CFF was reformulated as a **valence force field** through the elimination of the Urey–Bradley geminal interaction term and the introduction of a number of off-diagonal elements or cross terms.[386,389] This led to the existence of two separate versions of this force field: the Valence Consistent Force Field (VCFF, or later CVFF), and the Urey–Bradley Consistent Force Field (UBCFF). Still primarily a hydrocarbon force field, the CVFF consisted of the ten components shown below. At this stage, many more moieties were parametrized, including sp^2 carbons. The electrostatic potential had been dropped, as it had been judged to have an insignificant effect on the overall potential energy of hydrocarbons. The expression for this potential is

$$V_{tot} = V_{bnd} + V_{ang} + V_{oop} + V_{tor} + V_{bnd/bnd}$$
$$+ V_{ang/ang} + V_{ang/tor/ang} + V_{bnd/ang} + V_{oop/oop} + V_{vdW}$$

where V_{tot} is the total potential energy, V_{bnd} is the bond stretch component, V_{ang} is the angle bend component, V_{oop} is the component for out-of-plane angle bending interactions, V_{tor} is the torsional angle strain component, $V_{bnd/bnd}$ is the component for the bond-stretch–bond-stretch cross term, $V_{ang/ang}$ is the component for the angle-bend–angle-bend cross term, $V_{bnd/ang}$ is the component for the bond-stretch–angle-bend cross term, $V_{oop/oop}$ is the component for the out-of-plane-angle-bend–out-of-plane-angle-bend cross term, and V_{vdW} is the component for van der Waals nonbonded nongeminal interactions. The potential energy functions are all given in kilocalories per mole.

Bond Stretch Interactions The 1973 CVFF employed the same harmonic bond stretching potential V_{bnd} as the 1968 version. Some typical parameter values are: $k_{bnd} = 654.0$ kcal/mol · Å^2 and $r = 1.105$ Å for $I—J$ = methylene $C(sp^3)$—H; $k_{bnd} = 1309.9$ kcal/mol · Å^2 and $r = 1.333$ Å for $I—J = C(sp^2)=C(sp^2)$; and $k_{bnd} = 645.3$ kcal/mol · Å^2 and $r = 1.526$ Å for $I—J = C(sp^3)—C(sp^3)$.

Angle Bending Interactions The 1973 CVFF used the quadratic term of the harmonic angle bending potential V_{ang} as the 1968 version, without the first term used for the Urey–Bradley parametrization. Some typical values are $k_{ang2} = 93.2$ kcal/mol · rad^2 and $\theta_0 = 109.47°$ for $I—J—K = C(sp^3)—C(sp^3)—C(sp^3)$ with a quaternary central carbon; $k_{ang2} = 72.4$ kcal/mol · rad^2 and $\theta_0 = 122.3°$ for $I—J—K = C(sp^2)=C(sp^2)—C(sp^3)$; and $k_{ang2} = 79.0$ kcal/mol · rad^2 and $\theta_0 = 109.6°$ for $I—J—K = H—C(sp^3)—H$.

Out-of-Plane Angle Bending Interactions The 1973 CVFF incorporated the out-of-plane bending potential, V_{oop}, which had been introduced in the earlier treatment of the amide group.[383] This potential is a modification of the general angle bending potential V_{ang}, used to maintain planarity for $C(sp^2)$. The specifics of this interaction are depicted in Fig. 5-6, where in CFF the out-of-plane bending angle $\theta_{oop}(JI,KL)$ for atom J is defined as the angle between the plane including atoms J, K, and L and that including I, K, and L.[383] The associated energy is given by the potential equation

$$V_{oop} = \tfrac{1}{2} k_{oop}\, \theta_{oop}^2$$

where k_{oop} is the quadratic force constant in kcal/mol · rad², and θ is the out-of-plane angle as defined for CVFF,[386] in radians.

The force constant k_{oop} had a value of 22.9 kcal/mol · rad² where the central atom J is C(sp^2). The equilibrium value for θ is 0°, which simplifies the potential equation given above compared to that for the angle bending potential.

Torsional Interactions The 1973 CVFF torsional potential V_{tor} contains a single term based on the cosine of the torsional angle ω, as defined by the atom string I—J—K—L (see Fig. 5-5). The periodicity depends upon the multiplicity of the central bond. This potential takes the form

$$V_{tor} = k_{tor}(1 + \cos n\omega)$$

where k_{tor} is the force constant for the torsional interaction in kilocalories per mole (the value of this constant includes the factor of $\tfrac{1}{2}$ which normally appears in a harmonic function); n is the periodicity and takes a value of 3 for C—C single bonds and 2 for C=C double bonds; and ω gives the span of the torsional angle (corresponding to ω_{IJKL} in Fig. 5-5), in radians.

CVFF associates the torsional force constant k_{tor} with the two central atoms, and evaluates the contribution of each I—J—K—L set while dividing the given parameters by the number of sets. For the sets where I—J—K—L is C=C—C—C or C=C—C—H, the cosine term is subtracted rather than added.

Some typical values for the force constant k_{tor} are 32.7 kcal/mol for J-K as a 1,2-disubstituted C(sp^2)=C(sp^2); 2.532 kcal/mol for J—K as a C(sp^2)—C(sp^2); and 2.845 kcal/mol for J—K as a C(sp^3)—C(sp^3).

Bond-Stretch–Bond-Stretch Cross Term The 1973 CVFF introduced a bond-stretch–bond-stretch cross-term potential $V_{bnd/bnd}$ for the system I—J—K as defined in Fig. 5-12:

$$V_{bnd/bnd} = k_{bnd/bnd}(r' - r'_0)(r'' - r''_0)$$

where $k_{bnd/bnd}$ is the bond-stretch–bond-stretch force constant in kcal/mol · Å²; r' is the internuclear distance between atoms I and J (corresponding to r_{IJ} in Fig. 5-12), and $r' - r'_0$ is its departure from the equilibrium value, in angstroms; and r'' is the internuclear distance between atoms J and K

(corresponding to r_{JK} in Fig. 5-12) and $r'' - r''_0$ is its departure from the equilibrium value, in angstroms.

This cross term was parametrized only for I—J—K = C(sp^3)—C(sp^3)—C(sp^3), where the value for $k_{\text{bnd/bnd}}$ is 28.5 kcal/mol · Å^2.

Angle-Bend–Angle-Bend Cross Term The 1973 CVFF introduced an angle-bend–angle-bend cross-term potential $V_{\text{ang/ang}}$ applied to two angles, I—J—K and K—J—L, which share one bond including the central atom, as depicted in Fig. 5-9. The force constants for this potential depend only upon whether the two atoms of the shared bond J—K, are C—H or C—C. To model this potential, CVFF employs the expression:

$$V_{\text{ang/ang}} = k_{\text{ang/ang}}(\theta' - \theta'_0)(\theta'' - \theta''_0)$$

where $k_{\text{ang/ang}}$ is the angle bending force constant in kcal/mol · deg^2; θ' gives the span of the angle defined by atoms I—J—K (corresponding to θ_{IJK} in Fig. 5-9), and $\theta' - \theta'_0$ is its departure from the equilibrium angle span for this first angle, in degrees; and θ'' gives the span of the angle defined by atoms K—J—L (corresponding to θ_{KJL} in Fig. 5-9), and $\theta'' - \theta''_0$ is its departure from the equilibrium angle span for this second angle, in degrees.

A typical value for $k_{\text{ang/ang}}$ is -7.9 kcal/mol · deg^2 when J—K = C(sp^3)—C(sp^3).

Angle-Bend–Torsion–Angle-Bend Cross Term Although treated now as a special case of the angle-bend–angle-bend cross-term potential, the 1973 CVFF incorporated an angle-bend–torsion–angle-bend cross term $V_{\text{ang/tor/ang}}$ identical in form to that used in the earlier force field.

Typical values for the angle-bend–torsion–angle-bend force constant $k_{\text{ang/tor/ang}}$ are -10.5 kcal/mol · deg^2 for I—J—K—L = H—C(sp^2)—C(sp^2)—C(sp^3), and -10.0 kcal/mol for I—J—K—L = H—C(sp^2)—C(sp^2)—H. It was determined that the contribution of this cross term in the case of I—J—K—L = C(sp^3)—C(sp^2)—C(sp^2)—C(sp^3) was negligible.

Bond-Stretch–Angle-Bend Cross Term The 1973 CVFF introduced a bond-stretch–angle-bend cross-term potential $V_{\text{bnd/ang}}$, applied to the three-atom unit I—J—K (as defined in Fig. 5-8) and taking the form

$$V_{\text{bnd/ang}} = k_{\text{bnd/ang}}(\theta - \theta_0)(r - r_0)$$

where $k_{\text{bnd/ang}}$ is the stretch–bend force constant in kcal/mol · deg · Å; θ is the current span across the angle (corresponding to θ_{IJK} in Fig. 5-8), in degrees, and $\theta - \theta_0$ gives the displacement of the angle span from its equilibrium value, in degrees; r is the distance between the I and J nuclei (corresponding to r_{IJ} in Fig. 5-8), and $r - r_0$ is the departure from the equilibrium bond length, in angstroms.

Some typical values for the force constant $k_{\text{bnd/ang}}$ are 60.2 kcal/mol · deg · Å when $I—J—K = C(sp^3)—C(sp^3)—C(sp^3)$, and 38.4 kcal/mol · deg · Å when $I—J—K = C(sp^3)—C(sp^3)—H$. The values used for r_0, and θ_0 are the same as those taken from the corresponding bond stretch and angle bending potentials.

Out-of-Plane–Out-of-Plane Cross Term The 1973 CVFF introduced the out-of-plane–out-of-plane cross-term potential $V_{\text{oop/oop}}$, to model the coupling between out-of-plane angle bending at two adjacent atoms. This is roughly equivalent to the improper-torsion–torsion–improper-torsion cross term used in MM4 (see Section 5.4.1.4), as defined in Fig. 5-17 for the molecular moiety $I—J(—M)—K(—N)—L$, although in CVFF the out-of-plane angle θ is defined differently.[386] The associated potential energy is given by the equation

$$V_{\text{oop/oop}} = k_{\text{oop/oop}} \theta' \theta''$$

where $k_{\text{oop/oop}}$ is the force constant in kcal/mol · deg^2, θ' is the CVFF out-of-plane angle for the central atom J, in degrees, and θ'' is the CVFF out-of-plane angle for the central atom K, in degrees. This cross term has a profound effect upon the two out-of-plane bending modes of ethylene, but is less important for cycloalkenes.[386] A similar observation has been made about the improper-torsion–torsion–improper-torsion cross term in the MM4 force field.[332,333]

This interaction is defined only where the two central atoms are C(sp^2), and the force constant $k_{\text{oop/oop}}$ has a value of 3.31 kcal/mol · rad^2 in that case.

van der Waals Interactions The 1973 CVFF employed a modified van der Waals potential to govern the interactions between two nonbonded non-geminal atoms I and J (see Fig. 5-18). The Lennard–Jones 6–12 potential used earlier is now joined by an alternative, the 6–9 potential, which is still a function of the adjusted force constant k_{vdW} and the internuclear distance r_{IJ}. The sixth power term still provides the attractive component, while the repulsive component of the potential now varies as the ninth power:

$$V_{\text{vdW}} = k_{\text{vdW}}(2p^9 - 3p^6)$$

where k_{vdW} and p are defined as before. Hydrogen was made a bit larger, and carbon a bit smaller.

Some typical parameter values are: for the H······H interaction, $k_{\text{vdW}} = 0.0641$ kcal/mol, and r_0 for hydrogen is 1.816 Å for the C······C interaction, $k_{\text{vdW}} = 0.4072$ kcal/mol, and r_0 for carbon is 1.759 Å.

Beginning in 1979, the hydrogen bonding treatment was updated to include carboxylic acids.[390,401–403] Consideration of experimental data from a large basis set, and the results of *ab initio* studies of the charge distributions in these molecules, several conclusions were reached: the amide parameters could be transferred to the carboxyl carbonyl, the carbonyl and hydroxyl oxygens in the functional group could be treated as equivalent, the COOH group could be treated as overall neutral, and the van der Waals radius of the carboxyl proton could be neglected.

Thus, the amide parameters from the Lennard–Jones 6–9 potential function listed earlier could now be augmented with values for the carboxylic acid. For the "acidic" O-bound hydrogen, $\varepsilon = 0.0$ kcal/mol, $r_0 = 0.0$ Å, and $q = +0.41$; for either of the two carboxyl oxygens, $\varepsilon = 0.198$ kcal/mol, $r_0 = 1.825$ Å and $q = -0.46$, with the remaining charge balanced by the carboxyl carbon.

5.4.2.3. The Lyngby CFF In 1970, the Rasmussen group at Lyngby (Denmark) began work on an alternative formulation of the CFF, which is documented in a 1977 monograph.[388] The Lyngby CFF was derived from the UBCFF, but a later version drops the explicit Urey–Bradley term. As this group had a serious interest in modeling metal coordination compounds, appropriate parametrization was included. There were twelve atoms types, including a generic halide (@3) and a generic metal (@10).[404] Niketic and Rasmussen also surveyed the reports on CFF parameter sets and potential functions, including hydrocarbons, crystalline hydrocarbons, amides and lactams, amino acids, metal coordination compounds, and carbohydrates.[405]

In the most general formulation, the potential energy is portrayed as being derived from six component potentials:

$$V_{tot} = V_{bnd} + V_{ang} + V_{tor} + V_{vdW} + V_{UB} + V_{elec}$$

where V_{tot} is the total potential energy, V_{bnd} is the bond stretch component, V_{ang} is the angle bend component, V_{tor} is the torsional angle strain component, V_{vdW} is the component for van der Waals nonbonded nongeminal interactions, V_{UB} is the component for the Urey–Bradley nonbonded geminal interaction, and V_{elec} is the component for electrostatic (charge–charge) nonbonded interactions. The force field may be augmented with additional terms to describe specific interactions, such as a hydrogen bonding potential. The potential energy functions are all given in kilocalories per mole.[105]

Bond Stretch Interactions The Lyngby CFF now accommodates several variants of the bond stretching potential V_{bnd} for a bond I—J as defined in Fig. 5-1. The most general formulation is given by

$$V_{bnd} = k_{bnd1}(r - r_0) + \tfrac{1}{2}k_{bnd2}(r - r_0)^2 + \tfrac{1}{6}k_{bnd3}(r - r_0)^3$$

where k_{bnd1} is the linear force constant for bond stretching in kcal/mol · Å; r is the distance between the I and J nuclei (corresponding to r_{IJ} in Fig. 5-1), and $r - r_0$ is its departure from the equilibrium bond length, in angstroms; k_{bnd2} is the quadratic force constant for bond stretching in kcal/mol · Å^2; and k_{bnd3} is the cubic force constant for bond stretching in kcal/mol · Å³. For conformational calculations, only the quadratic term is used.

A Morse potential[317-320] for bond stretching is also available, which takes the form

$$V_{\text{bnd}} = D_{IJ}[(e^{-\alpha(\delta r)} - 1)^2 - 1]$$

where

$$\alpha = (k_{\text{bnd}}/2D_{IJ})^{1/2}$$

Here D_{IJ} is the bond dissociation energy for the bond I—J in kilocalories per mole, δr is the internuclear distance $r - r_0$ in angstroms, and k_{bnd} is the quadratic force constant for bond stretching in kcal/mol · Å².

Angle Bending Interactions For modeling the angle bending potential V_{ang} for an angle I—J—K (as defined in Fig. 5-4), the Lyngby CFF uses the generalized form

$$V_{\text{ang}} = k_{\text{ang}}(\theta - \theta_0) + \tfrac{1}{2}k_{\text{ang}2}(\theta - \theta_0)^2 + \tfrac{1}{6}k_{\text{ang}3}(\theta - \theta_0)^3$$

where k_{ang} is the linear force constant for angle bending in mdyn · Å/rad; θ is the current span across the angle (corresponding to θ_{IJK} in Fig. 5-4), in radians, and $\theta - \theta_0$ gives the displacement of the angle span from the equilibrium value, in radians; $k_{\text{ang}2}$ is the quadratic force constant for angle bending, in mdyn · Å/rad²; and $k_{\text{ang}3}$ is the cubic force constant for angle bending, in mdyn · Å/rad³.

Torsional Interactions There is some flexibility in the Lyngby CFF treatment of torsional interactions, and the actual potential V_{tor} will contain selected terms from the following general expression, which is based on the cosine of the torsional angle ω:

$$V_{\text{tor}} = k_{\text{tor}1}(1 + \cos \omega) + k_{\text{tor}2}(1 + \cos 2\omega) + k_{\text{tor}3}(1 + \cos 3\omega) + \cdots$$

where $k_{\text{tor}1}$ is the force constant for the onefold torsional barrier, in kilocalories per mole; ω gives the span of the torsional angle defined by the atom string I—J—K—L (corresponding to ω_{IJKL} in Fig. 5-5), in radians; $k_{\text{tor}2}$ is the force constant for the twofold torsional barrier; $k_{\text{tor}3}$ is the force constant for the threefold torsional barrier; etc.

The threefold term is the one used most in hydrocarbon modeling, since the tetrahedral carbon will have at least approximate C_{3v} symmetry. The twofold term is useful for modeling torsions about double bonds, and a sixfold term has been employed for torsions between two atoms which possess C_{3v} and C_{2v} symmetry, respectively. Higher periodicity terms would be necessary to treat species having higher valency, such as metal complexes, although the torsional effects in these systems are often small.[105]

van der Waals Interactions The Lyngby CFF can model van der Waals interactions with either one of two potentials: the Lennard–Jones 6–12 (or 6–9) potential, or the modified Buckingham potential (akin to that used in MMX and in the Allinger MM force fields).

The most common formulation of the Lennard–Jones potential is

$$V_{vdW} = k_{vdW}(p^{12} - 2p^{-6})$$

or

$$V_{vdW} = k_{vdW}(\tfrac{2}{3}p^9 - p^{-6})$$

where k_{vdW} is the adjsted force constant for nonbonded interactions, and p is the ratio of the equilibrium internuclear distance to the actual separation, as defined in Section 5.2.

The modified Buckingham potential is shown below:

$$V_{vdW} = \frac{k_{vdW}}{1 - \dfrac{6}{\alpha}} \left[\frac{6}{\alpha} \exp\left(\alpha \bigg/ \left(1 - \frac{1}{p}\right)\right) - p^6 \right]$$

where k_{vdW} is the adjusted force constant for the interaction, α is an additional unitless parameter which governs the steepness of the potential curve, and p is defined as above.

Urey–Bradley Interactions To model the interactions between two non-bonded geminal atoms I and K (see Fig. 5-20), the Lyngby CFF uses a two-part harmonic potential consisting of a linear term and a quadratic term, as described earlier in this section.

Electrostatic Interactions The Coulomb charge–charge model is used for electrostatic interactions between two nonbonded nongeminal atoms I and J (see Fig. 5-21), now with a correction for the dielectric constant of the medium is employed at this point. The Coulombic potential between two atoms I and J is given by

$$V_{elec} = \frac{q_I q_J}{D r_{IJ}}$$

where q_I and q_J are the charges on the respective atoms I and J, in elementary atomic charge units; D is the dielectric constant of the medium, with a default value of 2; and r_{IJ} is the internuclear distance between atoms I and J, in angstroms.

Over the next decade and a half, the Lyngby CFF formulation continued to evolve. An updated version was documented in a second monograph,[406,407] with details of the application to organometallic complexes,[408] saccharides,[409] and some miscellaneous compounds.[410] A report in 1987 described a parametrization for imines and oximes.[391] More recently, two updated general treatments of the Lyngby CFF version have appeared,[392,395] including a description of a graphic user interface program called GOPT (for graphics optimization tool), and a step-by-step description of the use of the programs for molecular modeling. A series of four papers from 1994–1995 present the updated formulation of the Lyngby CFF program, known as PEF91L.[393,394,396,397] In summary, the Morse potential described above has become the standard treatment for modeling the bond stretch vibration, the angle bending and torsional interactions have remained largely unchanged, and the out-of-plane angle bending differs only in having a nonzero value for the equilibrium angle (see below). The electrostatic treatment is identical to that described above, and the van der Waals nonbonded interactions are modeled with the Lennard–Jones 6–12 potential.

For simplicity, no cross-term potentials are employed. Geminal nonbonded interactions are not explicitly treated. The parameters are still derived from experimental data with necessary refinements, and no *ab initio* results are incorporated. Currently, parameter sets are available for alkanes,[394] aldehydes and ketones,[396] and aliphatic and alicyclic ethers (including anomeric carbons).[397] Some representative parameters are discussed below.

For bond stretching modeled with the Morse potential shown above, some typical parameters are: $D_{IJ} = 425.1$ kJ/mol, $\alpha = 0.018139$ pm^{-1}, and $r_0 = 109.50$ pm for I—$J = C(sp^3)$—H; $D_{IJ} = 369.0$ kJ/mol, $\alpha = 0.022545$ pm^{-1}, and $r_0 = 151.85$ pm for I—$J = C(sp^3)$—$C(sp^3)$; $D_{IJ} = 376.6$ kJ/mol, $\alpha = 0.021558$ pm^{-1}, and $r_0 = 138.99$ pm for I—$J = C_{ano}(sp^3)$—$O(sp^3)$; and $D_{IJ} = 525.9$ kJ/mol, $\alpha = 0.023751$ pm^{-1}, and $r_0 = 121.18$ pm for I—$J = C(sp^2)$=$O(sp^2)$. Here, $C_{ano}(sp^3)$ indicates an anomeric carbon atom.

For angle stretching modeled with the single-term quadratic potential shown above, some typical parameters are: $k_{ang} = 110.8551$ J/mol · deg^2 and $\theta_0 = 109.5°$ for $C(sp^3)$—$C(sp^3)$—$C(sp^3)$; $k_{ang} = 111.0844$ J/mol · deg^2 and $\theta_0 = 109.5°$ for $C(sp^3)$—$C(sp^3)$—H; $k_{ang} = 109.5392$ J/mol · deg^2 and $\theta_0 = 105.02°$ for $C(sp^3)$—$O(sp^3)$—$C_{ano}(sp^3)$; and $k_{ang} = 167.219$ J/mol · deg^2 and $\theta_0 = 106.60°$ for $C(sp^3)$—$C(sp^2)$=$O(sp^2)$. Here again, $C_{ano}(sp^3)$ indicates an anomeric carbon atom.

For the torsional interactions, the bond torsion formulation is used and the terms of the appropriate periodicity (i.e. twofold, threefold, etc.) are selected according to the symmetry of the moiety of interest. With respect to the generalized torsional potential equation shown above, some typical parameters are: $k_{tor3} = 0.9272$ kJ/mol for H—$C(sp^3)$—$C(sp^3)$—H; $k_{tor3} = 6.7224$ kJ/mol for $C(sp^3)$—$C(sp^3)$—$C(sp^3)$—$C(sp^3)$; $k_{tor6} = 0.0372$ kJ/mol for $O(sp^2)$=$C(sp^2)$—

C(sp^3)—C(sp^3); k_{tor3} = 1.3795 kJ/mol for C(sp^3)—O(sp^3)—C$_{ano}$(sp^3)—O(sp^3); and k_{tor3} = 19.6744 kJ/mol for O(sp^3)—C(sp^3)—C(sp^3)—O(sp^3). Here again, C$_{ano}$(sp^3) indicates an anomeric carbon atom.

An out-of-plane angle bending potential is employed for sp^2-hybridized carbon in carbonyl compounds. The modern form is close to that described in earlier CFF versions, save in having a nonzero value for the equilibrium angle. It is given by

$$V_{oop} = \tfrac{1}{2} k_{oop} (\theta - \theta_0)^2$$

where k_{oop} is the quadratic force constant in J/mol · deg^2; and θ is the out-of-plane angle as defined for CFF,[386] and $\theta - \theta_0$ is the displacement of the out-of-plane angle from its equilibrium value θ_0, in degrees. For an example such as the sp^2-hybridized carbon in acetone, referring to the depiction in Fig. 5-6, the CFF PEF91L out-of-plane bending angle θ_{oop}O(sp^2)[C(sp^2)—C(sp^3)—C(sp^3)] (corresponding to θ_{oop} $I[J$—K—$L]$) for the C(sp^2) atom J is defined as the angle between two planes: that including atoms J, K, and L, and that including I, K, and L. For this example, k_{oop} = 127.5 J/mol · deg^2 and θ_0 = 0.01°.

Nonbonded interactions are handled with a combined potential equation including the Lennard–Jones 6–12 potential function for the van der Waals interactions and the Coulomb potential for the electrostatic interactions:

$$V_{nonbond} = A r_{IJ}^{-12} - B r_{IJ}^{-6} + \frac{q_I q_J}{D r_{IJ}}$$

Some typical parameters are: A = 4742.736 × 10^{-12} (J · pm^{12}/mol)$^{0.5}$, B = 153.495 × 10^{-6} (J · pm^6/mol)$^{0.5}$, and q = 0.1627 for H; A = 71174.935 × 10^{-12} (J · pm^{12}/mol)$^{0.5}$, B = 2076.96 × 10^{-12} (J · pm^6/mol)$^{0.5}$, and q = 0.0934 for C(sp^3); A = 65130.26 × 10^{-12} (J · pm^{12}/mol)$^{0.5}$, B = 1801.22 × 10^{-6} (J · pm^6/mol)$^{0.5}$, and q = 0.0347 for C(sp^2); A = 33430.81 × 10^{-12} (J · pm^{12}/mol)$^{0.5}$, B = 2010.14 × 10^{-6} (J · pm^6/mol)$^{0.5}$, and q = −0.6200 for O(sp^3); and A = 31687.69 × 10^{-12} (J · pm^{12}/mol)$^{0.5}$, B = 1260.67 × 10^{-6} (J · pm^6/mol)$^{0.5}$, and q = −0.2881 for O(sp^2).

5.4.2.4. The Quantum Mechanical Force Field (QMFF)

Two members of the other extant branch of the CFF family tree, QMFF and CFF93, will be discussed in this and the following sections. QMFF is an acronym for quantum mechanical force field. It was introduced in a series of publications from the Hagler group at Biosym Technologies (now MSI, Inc.),[90–92,94,327] and is based on the derivation and parametrization of analytic representations of the Hartree–Fock *ab initio* potential energy surface of a wide variety of alkane molecules. CFF93 is derived by scaling the QMFF formulation to be consistent with experimental data.[91,92]

The QMFF formulation differs in several significant respects from the Lyngby version, in that quantum mechanical calculations played a significant role in the parametrization, and extensive use is made of off-diagonal (cross-term) components in the force field equations. The potential energy is composed of ten component potentials:

$$V_{tot} = V_{bnd} + V_{ang} + V_{tor} + V_{nonbond} + V_{bnd/bnd} + V_{bnd/ang}$$
$$+ V_{ang/ang} + V_{ang/ang/tor} + V_{bnd/tor} + V_{ang/tor}$$

where V_{tot} is the total potential energy; V_{bnd} is the bond stretch component; V_{ang} is the angle bend component; V_{tor} is the torsional angle strain component; $V_{nonbond}$ is the component for nonbonded nongeminal interactions, which includes the van der Waals potential (V_{vdW}) and the electrostatic potential (V_{elec}); $V_{bnd/bnd}$ is the stretch–stretch cross-term component; $V_{bnd/ang}$ is the stretch–bend cross-term component; $V_{ang/ang}$ is the one-center bend–bend cross-term component; $V_{ang/ang/tor}$ is the bend–bend–torsion cross-term component; and $V_{bnd/tor}$ is the stretch–torsion cross-term component; and $V_{ang/tor}$ is the bend–torsion cross-term component.

Bond Stretch Interactions To model the stretching of a bond I—J (defined in Fig. 5-1), QMFF has elaborated upon the cubic potential V_{bnd} used in earlier versions of CFF by adding a quartic term. The new potential takes the form

$$V_{bnd} = k_{bnd2}(r - r_0)^2 + k_{bnd3}(r - r_0)^3 + k_{bnd4}(r - r_0)^4$$

where k_{bnd2} is the quadratic bond stretching force constant in kcal/mol · Å^2; r is the distance between the I and J nuclei (corresponding to r_{IJ} in Fig. 5-1), and $r - r_0$ is its departure from the equilibrium bond length, in angstroms; k_{bnd3} is the cubic stretch constant in kcal/mol · Å^3; and k_{bnd4} is the quartic stretch constant in kcal/mol · Å^4.

Some typical parameter values are $k_{bnd2} = 417.4$ kcal/mol · Å^2, $k_{bnd3} = -851.3$ kcal/mol · Å^3, $k_{bnd4} = 1040.0$ kcal/mol · Å^4, and $r_0 = 1.0845$ Å for an aliphatic C(sp^3)—H bond; and $k_{bnd2} = 340.2$ kcal/mol · Å^2, $k_{bnd3} = -586.1$ kcal/mol · Å^3, $k_{bnd4} = 758.0$ kcal/mol · Å^4, and $r_0 = 1.5297$ Å for an aliphatic C(sp^3)—C(sp^3) bond.

Angle Bending Interactions QMFF employs a three-part potential function to model the contribution from angle bending, V_{ang}, for the angle I—J—K as defined in Fig. 5-4:

$$V_{ang} = k_{ang2}(\theta - \theta_0)^2 + k_{ang3}(\theta - \theta_0)^3 + k_{ang4}(\theta - \theta_0)^4$$

where k_{ang2} is the quadratic angle bending force constant in kcal/mol · rad^2; θ is the current span across the angle (corresponding to θ_{IJK} in Fig. 5-4), and $\theta - \theta_0$ gives the displacement of the angle span from its equilibrium value, in degrees; k_{ang3} is the cubic bend constant in kcal/mol · rad^3; and k_{ang4} is the quartic bend constant in kcal/mol · rad^4.

Some typical parameter values are $k_{ang2} = 52.7$ kcal/mol · rad^2, $k_{ang3} = -10.9$ kcal/mol · rad^3, $k_{ang4} = -11.3$ kcal/mol · rad^4, and $\theta_0 = 110.8°$ for a C(sp^3)—C(sp^3)—H angle; and $k_{ang2} = 52.2$ kcal/mol · rad^2, $k_{ang3} = -12.1$ kcal/mol · rad^3, $k_{ang4} = -11.3$ kcal/mol · rad^4, and $\theta_0 = 112.9°$ for a C(sp^3)—C(sp^3)—C(sp^3) angle.

Torsional Interactions QMFF employs a three-part potential V_{tor}, based on the cosine of the torsional angle ω (as defined by the atom string *I—J—K—L* shown in Fig. 5-5), to govern torsional interactions. The equation for this potential is given by

$$V_{tor} = k_{tor1}(1 - \cos \omega) + k_{tor2}(1 - \cos 2\omega) + k_{tor3}(1 - \cos 3\omega)$$

where k_{tor1}, k_{tor2}, and k_{tor3} are the force constants for the onefold, twofold, and threefold torsional barriers, respectively, in kilocalories per mole; and ω gives the span of the torsional angle (corresponding to ω_{IJKL} in Fig. 5-5) radians.

Some typical parameter values for *I—J—K—L* = C(sp^3)—C(sp^3)—C(sp^3)—C(sp^3) are $k_{tor1} = -1.152$ kcal/mol, $k_{tor2} = 0.012$ kcal/mol, and $k_{tor3} = -0.179$ kcal/mol; and for *I—J—K—L* = H—C(sp^3)—C(sp^3)—H are $k_{tor1} = -1.152$ kcal/mol, $k_{tor2} = 0.012$ kcal/mol, and $k_{tor3} = -0.179$ kcal/mol.

Nonbonded Interactions The equation for the QMFF nonbonded interaction potential combines a Coulombic potential to model the electrostatic interactions with a Lennard–Jones 6–9 potential to model the van der Waals interactions between atoms *I* and *J* (see Figs. 5-18 and 5-21):

$$V_{nonbond} = \frac{q_I q_J}{r_{IJ}} + k_{vdW}(2p^9 - 3p^6)$$

where q_I and q_J are the charges on atoms *I* and *J* respectively, in elementary atomic charge units; r_{IJ} is the actual internuclear distance between atoms *I* and *J* in angstroms; k_{vdW} is the force constant for the steric interaction between atoms *I* and *J* as defined below, in kilocalories per mole; and p is the unitless ratio of the equilibrium internuclear distance to the actual distance, as defined in Section 5.2. New combination rules shown below are used to derive the force constant k_{vdW} and the equilibrium van der Waals separation $r_{IJ(0)}$ from the corresponding atomic parameters. These newer forms were found by the force field developers to work better than the standard arithmetic and geometric means.[90]

The atomic charges q_I and q_J are calculated by summing the bond increment contributions δ_{IK} from each of the bonds containing the atom of interest. By convention, the first atom in the subscript (i.e. atom *I*) is the charge acceptor, and the second atom in the subscript (i.e. atom *K*) is the charge donor.

The force constant for the steric interaction, k_{vdW}, is derived from the hardness parameters ε from the individual atoms by the following relationship:

$$k_{vdW} = \frac{(\varepsilon_I \varepsilon_J)^{0.5} \times 2(r_{I(0)} r_{J(0)})^3}{r_{I(0)}^6 + r_{J(0)}^6}$$

where ε_I and ε_J are the hardness parameters for the corresponding atoms I and J, and $r_{I(0)}$ and $r_{J(0)}$ are their van der Waals radii.

The factor p is defined as in Section 5.2, except that the equilibrium internuclear separation between atoms I and J, $r_{IJ(0)}$, is derived from the individual van der Waals radius parameters by

$$r_{IJ(0)} = \left(\frac{r_{I(0)}^6 + t_{J(0)}^6}{2}\right)^{1/6}$$

Some typical values for the bond increment factor δ_{IK} are 0.053 elementary charge units for the H—$C(sp^3)$ bond and 0.00 for the $C(sp^3)$—$C(sp^3)$ bond; for the hardness parameter ε are 0.020 kcal/mol for hydrogen and 0.054 kcal/mol for $C(sp^3)$; and for the van der Waals radius r_0 are 2.995 Å for hydrogen and 4.010 Å for $C(sp^3)$.

Bond-Stretch–Bond-Stretch Cross Term The QMFF bond-stretch–bond-stretch cross-term potential, $V_{bnd/bnd}$, is the same as that employed in the CVFF formulation[386,389] to model the coupling of the stretching vibrations for the two bonds I—J and J—K in the three-atom system I—J—K (as defined in Fig. 5-12).

Some typical parameter values are $k_{bnd/bnd} = 11.769$ kcal/mol · Å² for I—J—K = H—$C(sp^3)$—H, $k_{bnd/bnd} = 11.310$ kcal/mol · Å² for I—J—K = $C(sp^3)$—$C(sp^3)$—H, and $k_{bnd/bnd} = 10.062$ kcal/mol · Å² for I—J—K = $C(sp^3)$—$C(sp^3)$—$C(sp^3)$.

Bond-Stretch–Angle-Bend Cross Term The QMFF bond-stretch–angle-bend cross-term potential, $V_{bnd/ang}$, is the same as that employed in the CVFF formulation[386,389] to model the coupling between the stretching of the bond I—J (or J—K) with the angle spanning the three-atom system I—J—K (as defined in Fig. 5-8).

Some typical parameter values are: $k_{bnd/ang} = 22.7$ kcal/mol · Å · rad for I—J—K = H—$C(sp^3)$—H, $k_{bnd/ang} = 34.5$ kcal/mol · Å · rad for I—J—K = $C(sp^3)$—$C(sp^3)$—H, and $k_{bnd/ang} = 18.3$ kcal/mol · Å · rad for I—J—K = $C(sp^3)$—$C(sp^3)$—$C(sp^3)$.

Angle-Bend–Angle-Bend Cross Term The QMFF angle-bend–angle-bend cross-term potential $V_{ang/ang}$ is the same as that employed in the CVFF formulation[386,389] to model the coupling of the bending vibrations of two angles, I—J—K and K—J—L, which share one bond including the central atom, as depicted in Fig. 5-9.

Some typical values are: $k_{ang/ang} = 0.930$ kcal/mol · rad² for I—J—K, K—J—L = H—C(sp^3)—H, H—C(sp^3)—H; $k_{ang/ang} = -2.18$ kcal/mol · rad² when I—J—K, K—J—L = H—C(sp^3)—C(sp^3), C(sp^3)—C(sp^3)—H; and $k_{ang/ang} = -8.97$ kcal/mol · rad² when I—J—K, K—J—L = C(sp^3)—C(sp^3)—C(sp^3), C(sp^3)—C(sp^3)—C(sp^3).

Angle-Bend–Torsion–Angle-Bend Cross Term In the QMFF formulation, the angle-bend–torsion–angle-bend cross-term potential $V_{ang/tor/ang}$ takes the same form as that employed by the earliest CFF formulation.[382] It models the coupling between the torsional interaction over the four-atom system I—J—K—L, and the bending of the two included three-atom angles for the three-atom subgroups I—J—K and J—K—L, as defined in Fig. 5-16.

Typical values for the angle-bend–torsion–angle-bend force constant $k_{ang/tor/ang}$ are -35.6 kcal/mol · rad² for I—J—K—L = C(sp^3)—C(sp^3)—C(sp^3)—C(sp^3), and -14.8 kcal/mol · rad² for I—J—K—L = H—C(sp^3)—C(sp^3)—H.

Torsion–Bond-Stretch Cross Term The QMFF formulation employs a torsion–bond-stretch cross term, $V_{tor/bnd}$, which corrects for the coupling between the torsional rotation across I—J—K—L and the stretching of an included bond. These torsion–bond-stretch interactions come in two different arrangements. The type 1 term addresses the coupling of the torsional rotation across the atoms I—J—K—L with the stretching of the central bond J—K, and is a function of the torsional angle ω_{IJKL} and the central bond distance r_{JK}, as defined in Fig. 5-10. The type 2 term addresses the coupling between the torsional rotation across I—J—K—L and the stretching of a terminal bond, I—J, and is a function of the torsional angle ω_{IJKL} and the terminal bond distance r_{IJ}, as defined in Fig. 5-11. Unlike in the MM4 force field (see Section 5.4.1.4), QMFF employs the same potential form for both types 1 and 2. It is a function of the bond length r and the cosine of the torsional angle ω:

$$V_{tor/bnd} = (r - r_0)(k_{tor/bnd1} \cos \omega + k_{tor/bnd2} \cos 2\omega + k_{tor/bnd3} \cos 3\omega)$$

where r is the internuclear distance between the two atoms of the bond of interest (J and K in type 1, corresponding to r_{JK} in Fig. 5-10, and I and J in type 2, corresponding to r_{IJ} in Fig. 5-11), and $r - r_0$ is its departure from the equilibrium bond length, in angstroms: $k_{tor/bnd1}$, $k_{tor/bnd2}$, and $k_{tor/bnd3}$ are the one-, two-, and threefold torsion–stretch force constants (respectively) in kcal/mol · Å; and ω is the torsional angle (corresponding to ω_{IJKL} in Fig. 5-11), in radians.

Some typical parameter values are given here. For I—J—K—L = C(sp^3)—C(sp^3)—C(sp^3)—C(sp^3), the type 1 parameters are $k_{tor/bnd1}$, $k_{tor/bnd2}$, and $k_{tor/bnd3} = -47.281, -5.641$, and 0.315 kcal/mol · Å, respectively, and the type

5.4. OTHER FORCE FIELDS

2 parameters are $k_{tor/bnd1}$, $k_{tor/bnd2}$, and $k_{tor/bnd3} = 2.294$, 0.765, and 0.357 kcal/mol · Å, respectively. For I—J—K—L = H—C(sp^3)—C(sp^3)—H, the type 1 parameters are $k_{tor/bnd1}$, $k_{tor/bnd2}$, and $k_{tor/bnd3} = -52.239$, -0.784, and -0.901 kcal/mol · Å, respectively, and the type 2 parameters are $k_{tor/bnd1}$, $k_{tor/bnd2}$, and $k_{tor/bnd3} = 0.778, 0.476$, and 0.079 kcal/mol · Å, respectively.

Torsion–Angle-Bend Cross Term   The QMFF formulation employs a torsion–angle bending cross-term potential, $V_{tor/ang}$, to correct for the coupling between the torsional rotation across the four-atom system I—J—K—L and the bending of a three-atom angle contained in this moiety, I—J—K; as defined in Fig. 5-13. It is a function of the torsional angle ω_{IJKL} and the angle θ_{IJK}, and takes the form

$$V_{tor/ang} = (\theta - \theta_0)(k_{tor/ang1} \cos \omega + k_{tor/ang2} \cos 2\omega + k_{tor/ang3} \cos 3\omega)$$

where θ is the current span for the angle I—J—K (corresponding to θ_{IJK} in Fig. 5-13), in radians, and $\theta - \theta_0$ gives the departure of this angle from its equilibrium value; $k_{tor/ang1}$, $k_{tor/ang2}$, and $k_{tor/ang3}$ are the one-, two-, and three-fold force constants with respect to first angle defined by I—J—K, in kcal/mol · rad; and ω is the torsional angle (corresponding to ω_{IJKL} in Fig. 5-13), in radians.

Some typical parameter values are given here: for I—J—K—L = C(sp^3)—C(sp^3)—C(sp^3)—C(sp^3), the parameters $k_{tor/ang1}$, $k_{tor/ang2}$, and $k_{tor/ang3}$ are -0.030, -0.015, and -0.003 kcal/mol · rad, respectively; for I—J—K—L = H—C(sp^3)—C(sp^3)—H, the parameters $k_{tor/ang1}$, $k_{tor/ang2}$, and $k_{tor/ang3}$ are $-1.622, 0.739$, and -0.449 kcal/mol · rad, respectively.

5.4.2.5. CFF93

As was discussed in Section 5.4.2.4, the CFF93 formulation is a recent member in the Consistent Force Field family. It is derived from the Quantum Mechanical Force Field (QMFF) formulation (see Section 5.4.2.4), by the application of scaling factors to those among the analytic representations of the Hartree–Fock *ab initio* potential energy surface which make up the valence portion of the QMFF.[90–92,94] Five separate scaling factors were developed for the different types of potential function. For bond stretching, there is the factor S_b, which takes the value 0.88 for C—C bonds and 0.83 for C—H bonds. For angle bending, there is the factor S_θ, which takes the value 0.81. For the torsional barrier potential, there is the factor S_ω, which takes the value 0.84. For out-of-plane angle bending, there is the factor S_χ, which is not yet assigned. For cross-term potentials, there is the factor S_c, which takes the value 0.87.

CFF93 is an example of a **class II force field**, defined by the Hagler group as a force field which contains anharmonic potentials through the addition of cubic and higher terms, and utilizes explicit off-diagonal terms from the force constant matrix.[90,92,93] Class II force fields may be applied with equivalent success to isolated small molecules, to condensed phases, and to macromolecular systems; it fits highly strained molecules including small rings with the same parameters; it employs a Morse or quartic bond stretching potential and a quartic angle bending potential; it has well-characterized one-, two-, and threefold torsional terms, contains cross terms, and employs an exponential or a ninth power nonbonded repulsion.

The CFF93 potential energy is composed of eleven component potentials:

$$V_{tot} = V_{bnd} + V_{ang} + V_{tor} + V_{oop} + V_{nonbond} + V_{bnd/bnd}$$
$$+ V_{ang/ang} + V_{bnd/ang} + V_{bnd/tor} + V_{ang/tor} + V_{ang/ang/tor}$$

where V_{tot} is the total potential energy; V_{bnd} is the bond stretch component; V_{ang} is the angle bend component; V_{tor} is the torsional angle strain component; V_{oop} is the out-of-plane angle bend component; $V_{nonbond}$ is the component for nonbonded nongeminal interactions, which includes the van der Waals potential (V_{vdW}) and the electrostatic potential (V_{elec}); $V_{bnd/bnd}$ is the stretch–stretch cross-term component; $V_{ang/ang}$ is the one-center bend–bend cross-term component; $V_{bnd/ang}$ is the stretch–bend cross-term component; $V_{bnd/tor}$ is the stretch–torsion cross-term component; $V_{ang/tor}$ is the bend–torsion cross-term component; and $V_{ang/ang/tor}$ is the bend–bend–torsion cross-term component.

The forms of the potential function equations are identical to the corresponding ones in QMFF. The force constants for the valence terms of CFF93 may be calculated as the product of the corresponding QMFF parameter and the scaling factor mentioned above. The equilibrium bond lengths are slightly different in CFF93: r_0 for the C—H bond is 1.111 Å, and for the C—C bond is 1.535 Å. The van der Waals and bond increment parameters are identical between the two formulations.

The initial presentations of CFF93 gave only the results for saturated hydrocarbons, although reports documenting further parametrization have appeared. In a study of polycarbonates, out-of-plane angle bending parameters for carbonyl carbon and aromatic carbon are given.[411] A CVFF-related force field for urea– and melamine–formaldehyde resins has been developed.[412]

More recently, the Hagler group has completed a more extensive parametrization of this force field, including many organic functional groups and coverage of biological small molecules and macromolecules such as carbohydrates, lipids, nucleic acids, and proteins. This updated version, known as CFF95, is currently available in the MSI Inc. Discover modeling package.[413]

5.4.3. The AMBER Force Field

The AMBER (assisted model building with energy refinement) molecular modeling program[414] and its attendant force field[415,416] were developed initially by the Kollman research group; it has continued to evolve primarily under the guidance of the Kollman group,[313,415-420] but with contributions from others as well.[421-430] AMBER was designed primarily to model biomacromolecules such as proteins and nucleic acid polymers (e.g. DNA), but it can handle a wide variety of molecules. The authors credit several earlier programs with contributing to the makeup of AMBER, including MMI (see Section 5.4.1.1), QCFF/PI,[431] UNICEPP,[69,70] CAMSEQ, and CHARMM (see Section 5.4.4). The AMBER program and force field are supported by a World Wide Web site.[419] Given the emphasis on biomacromolecules, AMBER is a diagonal force field, modeling bonds and angles with simple harmonic potentials with no cross terms. On the other hand, the treatment of electrostatic interactions is more sophisticated than would be found in a force field designed primarily for small, relatively nonpolar molecules.

Each subject molecule is defined as a protein (*P*), nucleotide (*N*), or other (*O*). In the first two cases, this identification will initiate access to an internal library of amino acid or nucleotide fragments, respectively. AMBER uses a "tree" representation of a molecule. This includes main chain atoms, such as the backbone of a protein; side chain atoms, which are straight segments of branches off the main chain, and as such are attached to two other atoms; branch atoms, which form the branching points along the side chains, and as such are attached to three other atoms; and end atoms, which are the final atoms in a side chain, and as such are attached to one other atom. There are descriptors which can define cyclic structures which don't fit the strict definition of a tree.

A modified conjugate gradient minimizer is used in the energy optimization. The user can choose between a gas phase and a solution potential; in the latter charged interactions are greatly reduced and the van der Waals interactions of hydrophobic atoms are modified.

In these first formulations of AMBER, the **united atom** version had 40 parametrized atom types, including 18 carbon types, eight nitrogen types, four oxygen types, six hydrogen types, two sulfurs, a phosphorus, and a lone pair pseudoatom.[415] The **all atom** version added five more sp^2 carbon types, in order to address nonbonded interactions between aromatic groups.[416] There are no sp-hybridized atom types.

In its early presentation, the AMBER force field consisted of five component potentials,[415,416] and had the form

$$V_{tot} = V_{bnd} + V_{ang} + V_{tor} + V_{nonbond} + V_{HB}$$

where V_{tot} is the total potential energy; V_{bnd} is the bond stretch component; V_{ang} is the angle bend component; V_{tor} is the torsional angle strain component, and the same potential equation is used for improper torsions; $V_{nonbond}$ is the component potential governing nonbonding interactions, which combines a van der Waals steric interaction term, V_{vdW} with a Coulombic term V_{elec} to model the electrostatic interactions; and V_{HB} is a hydrogen bonding potential. The potential energy functions are all given in kilocalories per mole.

Bond Stretch Interactions AMBER employs a harmonic bond stretching potential V_{bnd} for a bond I—J as defined in Fig. 5-1:

$$V_{bnd} = k_{bnd}(r - r_0)^2$$

where k_{bnd} is the force constant for bond stretching in kcal/mol · Å^2; and r is the distance between the I and J nuclei (corresponding to r_{IJ} in Fig. 5-1), and $r - r_0$ is its departure from the equilibrium bond length, in angstroms.

Some typical values for the AMBER bond stretch potential parameters follow. For C(sp^3)—C(sp^3) (i.e. —CH$_2$—CH$_2$—), $k_{bnd} = 260$ kcal/mol · Å^2 (united atom) and 310 kcal/mol · Å^2 (all atom), with $r_0 = 1.526$ Å; for C(sp^3)—H, $k_{bnd} = 331$ kcal/mol · Å^2 and $r_0 = 1.09$ Å; and for C(sp^2)=O(sp^2), $k_{bnd} = 570$ kcal/mol · Å^2 and $r_0 = 1.229$ Å.

Angle Bending Interactions AMBER uses a harmonic angle bending potential V_{ang} for an angle I—J—K as defined in Fig. 5-4:

$$V_{ang} = k_{ang}(\theta - \theta_0)^2$$

where k_{ang} is the force constant for angle bending; and θ is the current span across the angle (corresponding to θ_{IJK} in Fig. 5-4), and $\theta - \theta_0$ gives the displacement of the angle span from the equilibrium value, in radians.

Some typical values for the AMBER angle bending potential parameters follow. For C(sp^3)—C(sp^3)—C(sp^3) (i.e. —CH$_2$—CH$_2$—CH$_2$—), $k_{bnd} = 63.0$ kcal/mol · rad^2 and $\theta_0 = 112.4°$ (united atom), and $k_{bnd} = 40.0$ kcal/mol · rad^2 and $\theta_0 = 109.5°$ (all atom); for C(sp^3)—C(sp^2)=O(sp^2), $k_{bnd} = 80.0$ kcal/mol · rad^2 and $\theta_0 = 120.4°$; and for N(sp^2)—C(sp^2)—N(sp^2) (i.e. heteroaromatic), $k_{bnd} = 70.0$ kcal/mol · rad^2 and $\theta_0 = 129.1°$.

Torsional Interactions AMBER employs a simple potential V_{tor} to govern torsional interactions, which has an n-fold term based on the cosine of the torsional angle ω, as defined by the atom string I—J—K—L (see Fig. 5-5) with a reference angle ω_0 (denoted by γ in the AMBER publications). The same potential is used to determine the contribution from improper torsions, and takes the form

$$V_{tor} = \tfrac{1}{2}k_{tor(n)}[1 + \cos(n\omega - \omega_0)]$$

where $k_{tor(n)}$ is the force constant for the n-periodic torsional interaction, in kilocalories per mole; n is the periodicity of the torsional barrier; ω gives the span of the torsional angle corresponding to ω_{IJKL} in Fig. 5-5 for standard torsions, and to $\omega_{1'}$ (ω_{KLJI}), $\omega_{2'}$ (ω_{LIJK}), or $\omega_{3'}$ (ω_{IKJL}) in Fig. 5-7 for improper torsions, in degrees; and ω_0 is the reference or phase offset angle for the given atom string, in degrees. Most torsional interactions are governed by a single term, but some require two terms.

Three types of torsional interaction are parametrized in AMBER: a general torsion, where only the central atoms (J and K) are specified; a specific torsion, where all four atoms (I—J—K—L) are specified; and the improper torsion, in which a twofold barrier is used to enforce planarity at trigonal atoms, and a threefold barrier is used to enforce asymmetry upon the trivalent united atom representation of a tetrahedral carbon.

Some typical values for AMBER torsional potential parameters follow. Among the general torsional parameters, for X—$C(sp^3)$—$C(sp^3)$—X (i.e. X—CH_2—CH_2—X), $\frac{1}{2}k_{tor3} = 2.0$ kcal/mol (united atom) and 1.6 kcal/mol (all atom) with $\omega_0 = 180°$; and for X—$C(sp^2)$—$N(sp^2)$—X (i.e. amide C—N), $\frac{1}{2}k_{tor2} = 10.0$ kcal/mol with $\omega_0 = 180°$.

Most of the specific torsions are governed by a two-part potential. For the united atom glycol unit $O(sp^3)$—$C(sp^3)$—$C(sp^3)$—$O(sp^3)$, where $\omega_0 = 0°$, one has $\frac{1}{2}k_{tor2} = 0.5$ kcal/mol and $\frac{1}{2}k_{tor3} = 2.0$ kcal/mol. For the united atom disulfide group $C(sp^3)$—S—S—$C(sp^3)$, where $\omega_0 = 0°$, one has $\frac{1}{2}k_{tor2} = 3.5$ kcal/mol and $\frac{1}{2}k_{tor3} = 0.6$ kcal/mol.

The improper torsions can be general or specific in format, while the reference angle ω_0 is always 180°. For maintaining asymmetry at a methine $C^*(sp^3)$, in the general united atom format X—$C(sp^3)$—$C^*(sp^3)$—X, $\frac{1}{2}k_{tor3} = 14.0$ kcal/mol. For maintaining planarity at a carbonyl carbon in the general united atom format X—X—$C(sp^2)$=$O(sp^2)$, $\frac{1}{2}k_{tor2} = 10.5$ kcal/mol.

Nonbonded Interactions   The AMBER nonbonded potential has two parts: a Lennard–Jones 6–12 potential to model the van der Waals steric interactions between two nonbonded nongeminal atoms I and J (see Fig. 5-18), and a Coulombic potential to model the electrostatic nonbonded interactions due to charges on the two atoms I and J (see Fig. 5-21). This nonbonded interaction potential takes the form

$$V_{nonbond} = \frac{A_{IJ}}{r_{IJ}^{12}} - \frac{B_{IJ}}{r_{IJ}^{6}} + \frac{q_I q_J}{D r_{IJ}}$$

where A_{IJ} and B_{IJ} are parameters for the repulsive and attractive portions of the steric potential, respectively; r_{IJ} is the distance between atoms I and J; q_I and q_J are the charges on the respective atoms I and J, in atomic charge units; and D (denoted by ε in the AMBER publications) is the dielectric constant for the medium. In some cases, the electrostatic term is modified to a distance-dependent dielectric model where D is replaced by r_{IJ} (i.e. an inverse square dependence upon intercharge distance).

The derivation of the parameters A_{IJ} and B_{IJ} from the more familiar van der Waals parameters ε and $r_{IJ(0)}$ is as follows:

$$A_{IJ} = k_{vdW} r_{IJ(0)}^{12}$$

and

$$B_{IJ} = 2k_{vdW} r_{IJ(0)}^6$$

Atom-centered monopoles (i.e. point charges) are used in the electrostatic calculation; the exception is the sulfur atom, for which the model is improved when two lone pair pseudoatoms are affixed to it. The charges q for the atoms of a particular molecule are obtained from quantum mechanically derived electrostatic potentials (ESPs) fitted to a point charge model,[432,433] as discussed further in Section 5.6.

Nonbonding interactions are calculated only for 1,4-interactions or higher, and typically nonbonded interactions are evaluated only within a cutoff radius of ≈ 9.0 Å.

Some typical values for the AMBER van der Waals potential parameters are: for united atom methylene $C(sp^3)$, $\varepsilon = 0.12$ kcal/mol and $r_0 = 1.925$ Å; for united atom methyl group $C(sp^3)$, $\varepsilon = 0.15$ kcal/mol and $r_0 = 2.00$ Å; for all atom $C(sp^3)$, $\varepsilon = 0.06$ kcal/mol and $r_0 = 1.80$ Å; for aliphatic H, $\varepsilon = 0.038$ kcal/mol and $r_0 = 1.375$ Å; for amide H, $\varepsilon = 0.02$ kcal/mol and $r_0 = 1.00$ Å; for the lone pair pseudoatom, $\varepsilon = 0.016$ kcal/mol and $r_0 = 1.20$ Å; for ether $O(sp^3)$, $\varepsilon = 0.15$ kcal/mol and $r_0 = 1.65$ Å.

Hydrogen Bonding Interactions AMBER uses the hydrogen bonding potential from the CHARMM program, which has components proportional to the 10th and 12th powers of r_{IJ} when the atoms I and J are a hydrogen bond acceptor and donor. The form of the potential is

$$V_{HB} = \frac{C_{IJ}}{r_{IJ}^{12}} - \frac{D_{IJ}}{r_{IJ}^{10}}$$

where C_{IJ} and D_{IJ} are parameters for the repulsive and attractive portions of the hydrogen bonding potential, respectively, and r_{IJ} is the distance between atoms I and J.

Some typical values for the AMBER hydrogen bonding potential parameters are as follows. For H·····$N(sp^2)$ (amide NH, heterocyclic N) and for H·····$O(sp^2)$ (hydroxyl OH, carbonyl O), $C = 7557$ kcal · Å^{12}/mol and $D = 2385$ kcal · Å^{10}/mol. For H·····$N(sp^2)$ (H in heterocyclic NH$_2$, basic N, in five-membered ring heterocycle) and for H·····$O(\sim sp^2)$ (amide NH, carboxyl or phosphate nonbonded O), $C = 4019$ kcal · Å^{12}/mol and $D = 1409$ kcal · Å^{10}/mol. For H·····S (amide NH, disulfide S; or hydroxyl OH, disulfide S), $C = 265,720$ kcal · Å^{12}/mol and $D = 35,429$ kcal · Å^{10}/mol.

AMBER parameters for some nucleosides were developed by an independent group.[424]

Further work on the components of the AMBER force field by the Kollman group led to the finding that the discrete Lennard–Jones 10–12 hydrogen bonding potential may be folded into the standard 6–12 nonbonding potential by using a curve-fitting procedure to generate the new parameters from the old.[417] For example, in the hydrogen bonding pairs H······N(sp^2) (amide NH, heterocyclic N) and H······O(sp^2) (hydroxyl OH, carbonyl O) described above, the parameters $C = 7557$ kcal·Å^{12}/mol and $D = 2385$ kcal·Å^{10}/mol become $A = 1694$ kcal·Å^{12}/mol and $B = 55$ kcal·Å^6/mol. This greatly simplifies the functional form of the force field.

The Kollman group also developed a new electrostatic treatment called the two-stage Restrained Electrostatic Potential (or RESP) fitted charge model,[418,434] to address the problem that the ESP charges had a conformational dependence, and the observation that these charges were often too large when compared with experimental results. The RESP model is discussed further in Section 5.6.

The second generation AMBER force field incorporated both of these upgrades, the 6–12 hydrogen bonding model and the RESP derived charges.[420] Among its other aspects, it was constructed to enable its use with explicit solvation models such as the **OPLS** formulation of the Jorgensen group (see Section 5.4.7).

There are 40 atom types: thirteen carbons, twelve hydrogens, seven nitrogens, five oxygens, two sulfurs, and a phosphorus. The big increase in the hydrogen menu is due to the addition of six atom types appropriate for attachment to carbons which bear an electronegative substituent. There is a new oxygen type and a new hydrogen type for use in the **TIP3P** water solvent model. The united atom carbons and the lone pair pseudoatom have been dropped. In a comparative study of the cyclopentane molecule by molecular dynamics methods,[313] AMBER compared well with MM3 as a predictor of structure and properties. The major discrepancy was the failure to predict different lengths for the axial and the equatorial C—H bonds.

This new force field has only four components (besides the specific hydrogen bonding potential term):

$$V_{tot} = V_{bnd} + V_{ang} + V_{tor} + V_{nonbond}$$

where the terms are defined as above. The potential forms are essentially identical as well.

The Kollman group reoptimized many of the V_{bnd}, V_{ang}, and V_{tor} parameters from the earlier set. With respect to the sample parameters given above for typical values for the AMBER bond stretch potential parameters, the only change is for C(sp^3)—H, where now $k_{bnd} = 340$ kcal/mol·Å^2. The values for the new AMBER angle bending potential parameters are identical with those given above. Of the two general AMBER torsional potential parameters given above, only one value is different: for the threefold torsional force constant in X—C(sp^3)—C(sp^3)—X, $\frac{1}{2}k_{tor3} = 1.4$ kcal/mol. The general torsional potential parameter for two aromatic carbons, i.e. k_{tor2} for X—CA—CA—X, has been increased from 5.30 to 14.5 kcal/mol, in order to reduce the excessive flexibility of benzene rings in the earlier AMBER. There are several changes in the specific AMBER torsional potential parameters given above. For O(sp^3)—

C(sp^3)—C(sp^3)—O(sp^3), where $\omega_0 = 0°$, one has for $n = 2$, $\frac{1}{2}k_{tor2} = 1.0$ kcal/mol, and for $n = -3$, $\frac{1}{2}k_{tor3} = 0.144$ kcal/mol.

New torsional parameters were developed for the torsional angles ϕ and ψ torsional angles in peptide backbones, and for the torsional angle χ in nucleosides. The new improper torsion is not parametrized for maintaining asymmetry in united atoms, but the same parameters are used as earlier for maintaining planarity at a carbonyl carbon.

The $V_{nonbond}$ potential is still evaluated for all interactions separated by three or more bonds, and scaled for the 1,4-interactions. New van der Waals parameters had been developed in connection with the RESP work,[418] and others were adopted from the OPLS formulation.[435] Some typical values for the AMBER van der Waals potential parameters are: for C(sp^3), $\varepsilon = 0.1094$ kcal/mol and $r_0 = 1.908$ Å; for C(sp^2), $\varepsilon = 0.0860$ kcal/mol and $r_0 = 1.908$ Å; for amide H, $\varepsilon = 0.0157$ and $r_0 = 0.60$ Å; for the TIP3P water hydrogen, $\varepsilon = 0.00$ kcal/mol and $r_0 = 0.00$ Å; and for ether O(sp^3), $\varepsilon = 0.17$ kcal/mol and $r_0 = 1.6837$ Å.

Note that the amide H is smaller ($r_0 = 0.60$ Å), and both the alcoholic/acidic hydrogen and the TIP3P hydrogen have both ε and r_0 set to zero, in order to facilitate hydrogen bonding. The new force field also shrinks the size of aliphatic hydrogens when more electron withdrawing groups are placed on the carbon. Thus, $\varepsilon = 0.0157$ kcal/mol for aliphatic H, and $r_0 = 1.487, 1.387, 1.287$, and 1.187 Å for H—C, H—CX, H—CX_2, and H—CX_3, respectively.

As noted above, the RESP model for deriving atomic charges[418,434] is used to handle the electrostatic potential in this new version of AMBER.

A number of add-on parameter sets have been developed for the AMBER force field, including a force field for metalloproteins,[421] a set for modeling platinum complexes of guanine derivatives,[427] a set for sulfates and sulfamates,[428] a set for monosaccharides and (1 → 4) linked polysaccharides,[426] and the GLYCAM_93 set for oligosaccharides and glycoproteins.[429] The GLYCAM_93 parameter set can be downloaded from the World Wide Web.[436] AMBER* sets for the peptide backbone[422] for pyranoses[430] have been published. There is also a World Wide Web-based utility for calculating AMBER force constants for C—C and C—N bonds with the bond length or the π—bond order as input.[437]

5.4.4. The CHARMM Force Field

CHARMM is an acronym for "Chemistry at Harvard Macromolecular Mechanics," a molecular modeling program and attendant force field developed by M. Karplus and coworkers for carrying out static energy and dynamics calculations on proteins, nucleic acids, prosthetic groups and substrates, polypeptides, and other related macromolecules.[438–445] The program can be used to study static energy, molecular dynamics, solvation, electrostatics, crystal packing, vibrational analysis, free energy perturbation, combined quantum mechanics and molecular mechanics, stochastic dynamics, etc.

5.4. OTHER FORCE FIELDS

The CHARMM World Wide Web site[445] is a good source of current information on this force field.

More recently, as the academic and commercial development of this program have diverged, the academic version has been referred to as CHARMM, and the commercial version through MSI as CHARMm.

CHARMM has three methods to model hydrogen atoms: the **all atom** model, where all hydrogens are treated as discrete atoms; the **extended atom** model (corresponding to the united atom model for AMBER), which uses atom types for a combined heavy atom plus its attached hydrogens; and an intermediate model which uses extended atoms for hydrocarbon units which are not part of a functional group, but applies the all atom treatment to hydrogen bonding moieties.

An early report by Gelin and Karplus may be considered the genesis of CHARMM, although it wasn't termed that at the time.[438] In their conformational analysis of peptide side chains, they developed a five-component potential. This extended atom formulation had no explicit hydrogens, and had 12 atom types including carbon, oxygen, nitrogen, and sulfur. The potential had the following form:

$$V_{tot} = V_{bnd} + V_{ang} + V_{tor} + V_{nonbond} + V_{HB}$$

where V_{tot} is the total potential energy, V_{bnd} is the bond stretch component, V_{ang} is the angle bend component, V_{tor} is the torsional angle strain component, $V_{nonbond}$ is the component potential governing nonbonding interactions (which combines a van der Waals steric interaction term V_{vdW} with a Coulombic term V_{elec} to model the electrostatic interactions), and V_{HB} is a hydrogen bonding potential. The potential energy functions are all given in kilocalories per mole.

Bond Stretch Interactions The Gelin–Karplus treatment employs a harmonic bond stretching potential V_{bnd} for a bond I—J as defined in Fig. 5-1:

$$V_{bnd} = \tfrac{1}{2} k_{bnd}(r - r_0)^2$$

where k_{bnd} is the force constant for bond stretching in kcal/mol · Å²; and r is the distance between the I and J nuclei (corresponding to r_{IJ} in Fig. 5-1), and $r - r_0$ is its departure from the equilibrium bond length, in angstroms.

Some typical values for the Gelin–Karplus bond stretch potential parameters follow. For $C(sp^3)$—$C(sp^3)$ (i.e. —CH_2—CH_2—), $\tfrac{1}{2}k_{bnd} = 400$ kcal/mol · Å², with $r_0 = 1.518$ Å; and for $C(sp^2)$=$O(sp^2)$, $\tfrac{1}{2}k_{bnd} = 600$ kcal/mol · Å² and $r_0 = 1.234$ Å.

Angle Bending Interactions The Gelin–Karplus treatment uses a harmonic angle bending potential V_{ang} for an angle I—J—K as defined in Fig. 5-4:

$$V_{ang} = \tfrac{1}{2} k_{ang}(\theta - \theta_0)^2$$

where k_{ang} is the force constant for angle bending; and θ is the current span across the angle (corresponding to θ_{IJK} in Fig. 5-4), and $\theta - \theta_0$ gives the displacement of the angle span from its equilibrium value, in radians.

Some typical values for the Gelin–Karplus angle bending potential parameters follow. For $C(sp^3)$—$C(sp^3)$—$C(sp^3)$ (i.e. —CH_2—CH_2—CH_2—), $\frac{1}{2}k_{bnd} = 30.0$ kcal/mol · rad^2 and $\theta_0 = 109.5°$; for $C(sp^3)$—$C(sp^2)$=$O(sp^2)$ (i.e. —CH—C=O), $\frac{1}{2}k_{bnd} = 50.0$ kcal/mol · rad^2 and $\theta_0 = 122.3°$; and for $N(sp^2)$—$C(sp^2)$=$N(sp^2)$ (i.e. amidine), $\frac{1}{2}k_{bnd} = 60.0$ kcal/mol · rad^2 and $\theta_0 = 120.3°$.

Torsional Interactions The Gelin–Karplus treatment employed a simple potential V_{tor} to govern torsional interactions, which had an n-fold term based on the cosine of the torsional angle ω (called ϕ by Gelin and Karplus), as defined by the atom string I—J—K—L (see Fig. 5-5) with a reference angle ω_0 (called δ by Gelin and Karplus), and takes the form

$$V_{tor} = \tfrac{1}{2}k_{tor(n)}[1 + \cos(n\omega - \omega_0)]$$

where $k_{tor(n)}$ is the force constant for the n-periodic torsional interaction, in kilocalories per mole; n is the periodicity of the torsional barrier; ω gives the span of the torsional angle corresponding to ω_{IJKL} in Fig. 5-5, in degrees; and ω_0 is the reference or phase offset angle for the given atom string, in degrees. Only general torsions were parametrized, i.e. of the type X—J—K—Y characterized by the central atoms J and K.

Some typical values for Gelin–Karplus are: for X—$C(sp^3)$—$C(sp^3)$—X (i.e. X—CH_2—CH_2—Y), ($\tfrac{1}{2}k_{tor3} = 0.5$ kcal/mol, with $\omega_0 = 0°$; for X—$C(sp^2)$—$N(sp^2)$—Y (i.e. amide C—N), $\tfrac{1}{2}k_{tor2} = 7.0$ kcal/mol, with $\omega_0 = 180°$; and for X—S—S—Y, $\tfrac{1}{2}k_{tor2} = 4.0$ kcal/mol, with $\omega_0 = 0°$.

Nonbonded Interactions The Gelin–Karplus nonbonded potential has two parts: a Lennard–Jones 6–12 potential to model the van der Waals steric interactions between two nonbonded nongeminal atoms I and J (see Fig. 5-18), and a Coulombic potential to model the electrostatic nonbonded interactions due to charges on the two atoms I and J (see Fig. 5-21). This nonbonded interaction potential takes the form

$$V_{nonbond} = \frac{A_{IJ}}{r_{IJ}^{12}} - \frac{B_{IJ}}{r_{IJ}^{6}} + \frac{q_I q_J}{D r_{IJ}}$$

where A_{IJ} and B_{IJ} are parameters for the repulsive and attractive portions of the steric potential, respectively; r_{IJ} is the distance between atoms I and J; q_I and q_J are the charges on the respective atoms I and J, in atomic charge units; and D is the unitless dielectric constant for the medium. In some cases, the electrostatic term is modified to a distance-dependent dielectric model where D is replaced by r_{IJ} (i.e. an inverse square dependence upon intercharge distance).

In this potential, the parameters A_{IJ} and B_{IJ} (C in the Gelin–Karplus report) are derived from the polarizabilities, α_I and α_J, the effective numbers of outer shell electrons, N_I and N_J, and the van der Waals radii, $r_{0(I)}$ and $r_{0(J)}$, of the two interacting atoms I and J. The algorithm for deriving the charges q was not provided in this report.

Some typical values for the Gelin–Karplus van der Waals potential parameters are: for united atom methylene (CH$_2$), $\alpha = 1.77$ Å^3, $N = 7$, and $r_0 = 1.90$ Å; for united atom methyl group (CH$_3$), $\alpha = 2.17$ Å^3, $N = 8$, and $r_0 = 1.95$ Å; for carbonyl C(sp^2), $\alpha = 1.65$ Å^3, $N = 5$, and $r_0 = 1.80$ Å; for amide NH, $\alpha = 1.4$ Å^3, $N = 7$, and $r_0 = 1.65$ Å; and for carboxyl oxygen, $\alpha = 2.14$ Å^3, $N = 6$, and $r_0 = 1.60$ Å.

Hydrogen Bonding Interactions The Gelin–Karplus hydrogen bonding potential (also used in the AMBER force field; see Section 5.4.3) has components proportional to the 10th and 12th powers of r_{IJ} when the atoms I and J are the extended atom hydrogen bond donor and acceptor. The form of the potential is

$$V_{HB} = \frac{C_{IJ}}{r_{IJ}^{12}} - \frac{D_{IJ}}{r_{IJ}^{10}}$$

where C_{IJ} (A' in the Gelin–Karplus report) and D_{IJ} (C' in the Gelin–Karplus report) are parameters for the repulsive and attractive portions of the hydrogen bonding potential, respectively, and r_{IJ} is the distance between atoms I and J. The parametrization is in terms of an E_{min}, in kilocalories per mole, and an R_{min}, in angstroms, which are defined as

$$E_{min} = -0.067 D_{IJ}^6 / C_{IJ}^5$$

$$R_{min} = (1.2 C_{IJ}/D_{IJ})^{0.5}$$

Some typical values for the Gelin–Karplus hydrogen bonding potential parameters are as follows: for NH······O(sp^2) (amine NH, carbonyl oxygen), $E_{min} = -2.5$ kcal/mol and $R_{min} = 2.87$ Å; for NH······O(sp^2) (peptide NH, carbonyl oxygen), $E_{min} = -3.0$ kcal/mol and $R_{min} = 2.95$ Å; and for OH······O(sp^2) (hydroxyl OH, carbonyl oxygen) $E_{min} = -3.5$ kcal/mol and $R_{min} = 2.80$ Å.

The more fully developed CHARMM was reported a few years later,[439,440] equipped for both statics and dynamics calculations. This version permits the choice between all atom and extended atom representations of hydrogens. There are 29 atom types, including three hydrogen types, six carbon types, 11 nitrogen types, six oxygen types, two sulfurs, and an iron atom for modeling hemes. Nonbonded interactions were now broken out into van der Waals and electrostatic terms, and five electrostatics models were now available. This version of CHARMM had a seven-component potential, with two additional constraint potentials for use in dynamics calculations:

292 THE MOLECULAR MECHANICS FORCE FIELD

$$V_{tot} = V_{bnd} + V_{and} + V_{tor} + V_{impt} + (V_{vdW} + V_{elec}) + V_{HB}(V_{c/r} + V_{c/tor})$$

where V_{tot} is the total potential energy, V_{bnd} is the bond stretch component, V_{ang} is the angle bend component, V_{tor} is the torsional angle strain component, V_{impt} is the improper torsional angle strain component, V_{vdW} is the van der Waals nonbonding potential component, V_{elec} is the nonbonding electrostatic potential component, V_{HB} is a hydrogen bonding potential, $V_{c/r}$ is a potential for constraining specified atom–atom distances in dynamics calculations, and $V_{c/tor}$ is a potential for constraining specified torsional angles in dynamics calculations. The potential energy functions are all given in kilocalories per mole.

Bond Stretch Interactions CHARMM employs the same harmonic bond stretching potential described above, with the factor of $\frac{1}{2}$ folded into the force constant. Some typical values for the CHARMM bond stretch potential parameters follow. For $C(sp^3)$—$C(sp^3)$ (i.e. —CH_2—CH_2—), $k_{bnd} = 225$ kcal/mol · Å^2 with $r_0 = 1.52$ Å (extended atom), and $k_{bnd} = 200$ kcal/mol · Å^2 with $r_0 = 1.53$ Å (all atom); for $C(sp^3)$—H, $k_{bnd} = 300$ kcal/mol · Å^2 with $r_0 = 1.08$ Å; and for $C(sp^2)$=$O(sp^2)$, $k_{bnd} = 580$ kcal/mol · Å^2 with $r_0 = 1.23$ Å.

Angle Bending Interactions CHARMM employs the same harmonic angle bending potential described above, with the factor of $\frac{1}{2}$ folded into the force constant. Some typical values for the CHARMM angle bending potential parameters follow. For $C(sp^3)$—$C(sp^3)$—$C(sp^3)$ (i.e. —CH_2—CH_2—CH_2—), $k_{bnd} = 45.0$ kcal/mol · rad^2, with $\theta_0 = 110.0°$ (extended atom) and $\theta_0 = 111.0°$ (all atom); for $C(sp^3)$—$C(sp^2)$=$O(sp^2)$ (i.e. —CH—C=O), $k_{bnd} = 85.0$ kcal/mol · rad^2 with $\theta_0 = 121.5°$ (both extended and all atom); and for $N(sp^2)$—$C(sp^2)$—$N(sp^2)$ (i.e. a five-membered heterocycle), $k_{bnd} = 70.0$ kcal/mol · rad^2 with $\theta_0 = 109.0°$ (all atom).

Torsional Interactions CHARMM employed a modified expression for the torsional interaction potential V_{tor}, still an *n*-fold term based on the cosine of the torsional angle ω (called ϕ by Karplus *et al.*), as defined by the atom string *I*—*J*—*K*—*L* (see Fig. 5-5). The factor of $\frac{1}{2}$ has been folded into the force constant, and the potential takes the form

$$V_{tor} = |k_{tor(n)}| - k_{tor(n)} \cos n\omega$$

where $k_{tor(n)}$ is the force constant for the *n*-periodic torsional interaction, in kilocalories per mole; *n* is the periodicity of the torsional barrier; and ω gives the span of the torsional angle corresponding to ω_{IJKL} in Fig. 5-5, in radians. Several specific torsions were included in the parametrization, but most are still general torsions.

An example of a CHARMM specific torsion parameter is for $C(sp^3)$—$C(sp^2)$—$N(sp^2)$—$C(sp^3)$ (i.e. the peptide ω-angle), $k_{tor2} = 10.0$ kcal/mol. Some typical values for CHARMM general torsions are: for X—$C(sp^3)$—$C(sp^3)$—X (i.e. X—CH_2—CH_2—Y), $k_{tor3} = -1.6$ kcal/mol (both extended atom and all atom); for X—$C(sp^2)$—$N(sp^2)$—Y (i.e. generic peptide ω-angle), $k_{tor2} = 8.2$ kcal/mol; for X—$C(sp^3)$—$N(sp^2)$—Y (i.e. generic peptide ϕ-angle), $k_{tor3} = -0.3$ kcal/mol (both extended atom and all atom); for X—$C(sp^3)$—$C(sp^2)$—Y (i.e. generic peptide ψ-angle), $k_{tor3} = 0.0$ kcal/mol (both extended atom and all atom); and for X—S—S—Y, $k_{tor2} = -4.0$ kcal/mol.

Improper Torsional Interactions CHARMM employs a simple harmonic potential, V_{impt}, to model improper torsional interactions. This potential is used to inhibit the departure from planarity of an sp^2-hybridized atom, or to enforce asymmetry upon the trivalent united atom representation of a tetrahedral carbon. This is modeled as a function of the improper torsional angle formed between the atom J and its three attached atoms I, K, and L, as defined in Fig. 5-7. CHARMM assigns the improper torsional force constants based upon the identity of the first (i.e. central) and last atom of the improper torsional angle. The equation for this potential takes the form

$$V_{impt} = k_{impt}(\omega - \omega_0)^2$$

where k_{impt} is the force constant for the improper torsional, in kcal/mol · rad^2; ω is the improper torsional angle (either ω_1, ω_2, or ω_3, as defined in Fig. 5-7), in degrees, and ω_0 is the reference angle for the interaction, in degrees. In cases where planarity is being enforced, $\omega_0 = 0.0°$, and where asymmetry is being enforced, $\omega_0 = 35.26439°$.

CHARMM parametrizes some general and some specific improper torsional angles. Some typical general values are: for $C(sp^2)$—X—X—$O(sp^2)$ (i.e. for a carbonyl carbon), k_{impt} is 100 kcal/mol · rad^2; for $C(sp^3)$—X—X—$C(sp^3)$ (i.e. for an extended atom methine), k_{impt} is 55 kcal/mol · rad^2; for $N(sp^2)$—X—X—$C(sp^3)$ (i.e. for an alkylated amide nitrogen), k_{impt} is 45 kcal/mol · rad^2. A typical specific value is for $C(sp^2)$—$O(\sim sp^2)$—$O(\sim sp^2)$—$C(sp^3)$ (i.e. for a carboxyl carbonyl carbon): k_{impt} is 100 kcal/mol · rad^2.

van der Waals Nonbonded Interactions CHARMM retains the Lennard–Jones 6–12 potential to model the van der Waals steric interactions, but has added a switching function to give an ≈ 8-Å cutoff. In this potential, the parameters A_{IJ} and B_{IJ} are derived from the polarizabilities, α_I and α_J, the effective number of outer shell electrons, N_I and N_J, and the van der Waals radii, $r_{0(I)}$ and $r_{0(J)}$, of the two interacting atoms I and J.

The values given earlier[438] for these parameters in the extended atom types are still in use. Some typical values for the new all atom types are: for $C(sp^3)$, $\alpha = 0.980$ Å^3, $N = 5$, and $r_0 = 1.80$ Å; for aliphatic H, $\alpha = 0.100$ Å^3, $N = 1$, and $r_0 = 1.468$ Å; for a hydrogen-bonding H, $\alpha = 0.044$ Å^3, $N = 1$, and $r_0 = 0.800$ Å; for a secondary amide N (i.e. peptide N with one H), $\alpha = 1.100$ Å^3,

$N = 6$, and $r_0 = 1.600$ Å; and for hydroxyl oxygen, $\alpha = 0.840$ Å^3, $N = 6$, and $r_0 = 1.60$ Å.

Electrostatic Nonbonded Interactions Many more options for the treatment electrostatic nonbonded interactions are presented in this version of CHARMM. In addition to the original Coulomb law potential, a distance-dependent potential is also available. These potentials take the form:

$$V_{\text{elec}} = \begin{cases} q_I q_J / D r_{IJ} & \text{(Coulomb)} \\ q_I q_J / r_{IJ}^2 & \text{(distance-dependent)} \end{cases}$$

Applying a simple distance cutoff to electrostatic effects leads to anomalous results because charges still affect each other from long distances, unlike van der Waals forces. CHARMM has three additional electrostatic treatments to deal with the cutoff issue: the shifted dielectric, different electrostatic treatments according to functional group, and a multipole expansion which is a function of the dipoles and quadrupoles. A discussion of these models in detail is beyond the scope of the book, and the interested reader is directed to the original literature. The derivation of the charges q was not provided in this report.

Hydrogen Bonding Interactions CHARMM has a more sophisticated hydrogen bonding potential. The 10–12 potential has been modified for the all atom formulation, and is scaled with two cosine factors to include an angle dependence for the geometry of the hydrogen bond. This hydrogen bonding potential operates in addition to the electrostatic interactions. The form of the potential is

$$V_{\text{HB}} = \left(\frac{C_{IJ}}{r_{IJ}^{12}} - \frac{D_{IJ}}{r_{IJ}^{10}} \right) \cos^m \theta_{I-H-J} \cos^n \theta_{H-J-J^*}$$

where C_{IJ} (A' in reference 439) and D_{IJ} (B' in reference 439) are parameters for the repulsive and attractive portions of the hydrogen bonding potential, respectively, and are defined as before; r_{IJ} is the distance between the hydrogen bond donor heavy atom I and the acceptor atom J; m and n are exponents chosen from a topology file; θ_{I-H-J} is the angle between the hydrogen bond donor heavy atom I, the acidic hydrogen atom H, and the acceptor atom J, and θ_{H-J-J^*} is the angle between the acidic hydrogen atom H, the acceptor atom J, and the acceptor antecedent atom J^* (i.e. the other atom attached to J). There are also two switching functions for turning this potential on and off.

The CHARMM hydrogen bonding parameters are given for donor-atom–acceptor-atom pairs, in terms of $E_{\min}$ and $R_{\min}$ as defined previously. For

N–N, $E_{min} = -3.0$ kcal/mol and $R_{min} = 3.0$ Å; for N–O, $E_{min} = -3.5$ kcal/mol and $R_{min} = 2.9$ Å; for O–N, $E_{min} = -4.0$ kcal/mol and $R_{min} = 2.85$ Å; and for O–O, $E_{min} = -4.25$ kcal/mol and $R_{min} = 2.75$ Å.

The two constraint functions mentioned earlier for use in molecular dynamics simulations are simple harmonic Hooke's law potentials which operate upon the interatomic distance r_{IJ} or the torsional angle ω, which is to be constrained. There is also a water solvation potential for explicit inclusion of solvent interactions.

The second of these two reports[440] expanded the parameter set to cover nucleic acids. Among other changes, the torsional interaction potential once again contains the reference angle ω_0.

More recently, the Karplus group has extended the application of the CHARMM force field to conformational analysis in small hydrocarbon molecules, with satisfactory results.[441] Not surprisingly, the electrostatic and hydrogen bonding terms were not employed for these studies. At this time, the newly named **CHARMm**, force field for modeling small molecules debuted in the QUANTA® modeling package by Polygen Corp (now MSI).[446]

A new formulation of CHARMM was reported in 1995, in which a number of changes had been made.[445] The new CHARMM is compatible with current solvation models. The explicit hydrogen bonding potential has been dropped, and the hydrogen bonding is modeled with judicious van der Waals and electrostatic parametrization for the donor and acceptor atoms. A Urey–Bradley term has been added for geminal interactions. The new CHARMM is an all atom force field, with parameters derived both from empirical data and from potential surfaces obtained through *ab initio* calculations, and is represented by a seven-component potential:

$$V_{tot} = V_{bnd} + V_{ang} + V_{UB} + V_{tor} + V_{impt} + (V_{bdW} + V_{elec})$$

where V_{tot} is the total potential energy, V_{bnd} is the bond stretch component, V_{ang} is the angle bend component, V_{UB} is a Urey–Bradley geminal interaction potential, V_{tor} is the torsional angle strain component, V_{impt} is the improper torsional angle strain component, V_{vdW} is the van der Waals nonbonding potential component, and V_{elec} is the nonbonding electrostatic potential component. The forms of the individual component potentials follow closely on the previous models. The Urey–Bradley potential is a simple harmonic potential based upon the internuclear distance between the geminal atoms (see Section 5.1.3.2).

The Harvard group has developed new formulations for the torsional and improper torsional components which have computational advantages.[444] Grootenhuis, Haasnoot, and Kouwijzer have developed a companion carbohydrate force field, **CHEAT**,[447–449] which employs extended atom hydroxyl groups. Huige and Altona have derived parameters for sulfates and sulfamates for the CHARMm force field.[428] A collaborative group has reported a

CHARMM parameter set for NAD coenzymes and nucleotide pyrophosphates.[450] CHARMM documentation and parameter sets, as well as other related information, are available from the CHARMM World Wide Web site.[445]

5.4.5. The Tripos Force Field

The Tripos force field was developed for use in the commercial molecular modeling programs offered by Tripos, Inc., such as SYBYL™ for the workstation platform and ALCHEMY™ for the desktop platform. The force field and its parameters have been developed over time from earlier literature reports[271,451] and internal investigations at Tripos, Inc.[306] A large number of studies have been published which employ the Tripos force field, and applications of it such as the CoMFA electrostatic surface analysis.

The Tripos 5.2 force field has 31 parametrized atom types, including for example one type of hydrogen, four types of carbon, seven types of nitrogen, two types of oxygen, halogens, some second-row elements and metal atoms, a pseudoatom representing a lone pair of electrons, and a dummy atom. Hydrogen bond donor/acceptor atoms include fluorine, nitrogen, and oxygen.

There are six component potentials in the overall force field, as follows:

$$V_{tot} = V_{bnd} + V_{ang} + V_{tor} + V_{vdW} + V_{oop} + V_{elec}$$

where V_{tot} is the total potential energy, V_{bnd} is the bond stretch component, V_{ang} is the angle bend component, V_{tor} is the torsional angle strain component, V_{vdW} is the component for van der Waals nonbonded interactions, V_{oop} is the component for the out-of-plane bending potential, and V_{elec} is the component for electrostatic (charge–charge) nonbonded interactions. The potential energy functions are all given in kilocalories per mole. There is no explicit hydrogen bonding potential included, but acidic hydrogens are given a van der Waals radius r_0 of 0.00 Å and a hardness parameter ε of 0.00 kcal/mol to facilitate these interactions passively.

Bond Stretch Interactions Tripos employs a simple Hooke's law harmonic potential V_{bnd} for the stretching of bond I—J as defined in Fig. 5-1, which takes the form

$$V_{bnd} = \tfrac{1}{2}k_{bnd}(r - r_0)^2$$

where k_{bnd} is the force constant for bond stretching in kcal/mol · Å^2; and r is the distance between the I and J nuclei (corresponding to r_{IJ} in Fig. 5-1), and $r - r_0$ is its departure from the equilibrium bond length, in angstroms.

Some typical parameter values are: $k_{bnd} = 662.4$ kcal/mol · Å^2 for a C(sp^3)—H bond and 633.6 kcal/mol · Å^2 for a C(sp^3)—C(sp^3) bond; $r_0 = 1.100$ Å for a C(sp^3)—H bond, and 1.540 Å for a C(sp^3)—C(sp^3) bond. If no bond parameters are found, the input geometry value of r is taken for r_0, and $k_{bnd} = 600.0$ kcal/mol · Å^2.

Angle Bending Interactions Tripos uses a simple Hooke's law function to model the potential energy contribution from angle bending, V_{ang}, for the angle I—J—K as defined in Fig. 5-4, which takes the form

$$V_{ang} = \tfrac{1}{2} k_{ang} (\theta - \theta_0)^2$$

where k_{ang} is the force constant for the three-atom angle given in kcal/mol · deg^2; and θ is the current span across the angle (corresponding to θ_{IJK} in Fig. 5-4), and $\theta - \theta_0$ gives the displacement of the angle span from its equilibrium value, in degrees.

Some typical values for k_{ang} are 0.024 kcal/mol · deg^2 for C(sp^3)—C(sp^3)—C(sp^3) and 0.016 kcal/mol · deg^2 for C(sp^3)—C(sp^3)—H (generic parameter); and for θ_0 are 109.5° for C(sp^3)—C(sp^3)—C(sp^3) and 109.5° for C(sp^3)—C(sp^3)—H (generic parameter). If no angle parameters are found, the input geometry value of θ is taken for θ_0, and $k_{ang} = 0.02$ kcal/mol · deg^2.

Torsional Interactions Tripos employs a one-part torsional potential V_{tor} based on the cosine of the torsional angle ω, as defined by the atom string I—J—K—L (see Fig. 5-5), to govern torsional interactions. The periodicity of the force constant is a property of the atom string I—J—K—L, and most torsions are characterized semigenerically by the central atoms J—K. The torsional potential takes the form

$$V_{tor} = \tfrac{1}{2} k_{tor} (1 \pm \cos |f| \omega)$$

where k_{tor} is the force constant in kilocalories per mole; f is the periodicity of the given torsion, the algebraic sign of which dictates the choice of $+$ or $-$ in $\pm$; and ω gives the span of the torsional angle (corresponding to ω_{IJKL} in Fig. 5-5) in degrees.

Some typical parameter values for I—J—K—L are: for C(sp^3)—C(sp^3)—C(sp^3)—C(sp^3), $k_{tor} = 0.500$ kcal/mol and $f = +3$; for $*$—C(sp^3)—C(sp^3)—H ($*$ is a wildcard), $k_{tor} = 0.32$ kcal/mol and $f = +3$; for $*$—C(ar)—C(ar)—$*$ (ar is an aromatic atom: aromatic central bond), $k_{tor} = 2.0$ kcal/mol and $f = -2$; for $*$—C(sp^2)=C(sp^2)—$*$ (central bond is double), $k_{tor} = 12.5$ kcal/mol and $f = -2$. If no match is found for the two central atoms, a message is entered into the output file and default values of $k_{tor} = 0.200$ kcal/mol and $f = +3$ are used.

van der Waals Interactions To model the van der Waals potential V_{vdW} which governs the interactions between two nonbonded nongeminal atoms I and J (see Fig. 5-18), Tripos uses a standard Lennard–Jones 6–12 potential, which is a function of the adjusted force constant k_{vdW} and the ratio p of the ideal to the actual internuclear distance r_{IJ}. The sixth power term provides the attractive component and the twelfth power term provides the repulsive component of the potential:

$$V_{vdW} = k_{vdW}(p^{12} - 2p^6)$$

where k_{vdW} is the adjusted force constant for nonbonded interactions between I and J, derived as the geometric mean of the individual hardness parameters ε, in kilocalories per mole, and p is a unitless ratio of the equilibrium distance between I and J to the actual distance r_{IJ}, and defined in Section 5.2.

Some typical parameter values are: for hydrogen, $\varepsilon = 0.042$ kcal/mol and $r_0 = 1.5$ Å; for carbon, $\varepsilon = 0.107$ kcal/mol and $r_0 = 1.7$ Å; for fluorine, $\varepsilon = 0.109$ kcal/mol and $r_0 = 1.47$ Å; and for nitrogen, $\varepsilon = 0.095$ kcal/mol and $r_0 = 1.55$ Å. Hydrogens attached to hydrogen bond donors have $\varepsilon = 0.00$ kcal/mol for interactions with hydrogen bond acceptor atoms.

Out-of-Plane Angle Bending Interactions The Tripos force field employs a simple out-of-plane bending potential V_{oop} which is a function of the distance h over which the atom in question has been displaced from the projection of that atom upon the plane formed by its three attached atoms, as defined in Fig. 5-6. It is given by the potential equation

$$V_{oop} = k_{oop} h^2$$

where k_{oop} is the force constant in kcal/mol · Å^2; and h is the displacement from the plane formed by the three attached substituents (see Fig. 5-6), in angstroms. In the Tripos force field, the force constant k_{oop} is a property only of the central atom.

Some typical parameter values for k_{oop} are 480 kcal/mol · Å^2 for carbon; and 120 kcal/mol · Å^2 for nitrogen.

Electrostatic Interactions To calculate the potential V_{elec} which governs the electrostatic interactions between nonbonded nongeminal atoms I and J by the charge–charge model (see Fig. 5-21), the Tripos force field uses the Coulomb potential:

$$V_{elec} = K_{elec} \frac{q_I q_J}{D r_{IJ}}$$

where K_{elec} is a constant for scaling and unit cancellation with the value 332.17; q_I and q_J are the charges on the corresponding atoms, in units of atomic charge; D is either the unitless dielectric constant of the medium, or a distance-dependent dielectric function; and r_{IJ} is the internuclear distance between atoms I and J, in angstroms.

The method of Gasteiger and Marsili[452] is used to calculate the charges for these atoms.

5.4.6. The MMFF94 Force Field

MMFF94 stands for Merck Molecular Force Field 1994. This force field is described in publications by the Merck group headed by T. A. Halgren,[314,321,453-455] and the present description has been adapted from those publications. The currently published form is intended for use in molecular dynamics studies, while the form designed for energy minimization is still under development. MMFF94 was developed from **MM2X**,[456] and both have affinities with the **MM2, MM3** series by Allinger and coworkers (see Section 5.4.1). MMFF94 is currently implemented in **OPTIMOL**,[456] the molecular mechanics platform for which it was developed, as well as in **CHARMm** and **MacroModel** BatchMin. The full parameter file is available as supplementary material from the *J. Comput. Chem.*, and may be downloaded from their World Wide Web site.[457]

The current version has several advantages over earlier ones, and over other force fields. It employs a unique functional form for the van der Waals potential. The core parametrization is based upon the results of *ab initio* calculations and crystal diffraction structures. The basis set for developing the parametrization includes a wide variety of molecules at the outset. The nonbonded (i.e. van der Waals and electrostatic) parametrization has been carried out in a way that ensures a proper balance between solvent–solvent, solvent–solute, and solute–solute interactions in aqueous simulations. Acknowledged drawbacks are limitations in the electrostatic model, the use of only electrostatic and van der Waals components in the organometallic portion of the potential, and the absence of certain cross terms such as the one linking bond length to degree of conjugation.[321]

MMFF94 has 99 parametrized atom types, including several aromatic-type carbons and nitrogens, halide ions, and a number of specific metal cations. There are seven component potentials in the overall force field, as follows:

$$V_{tot} = V_{bnd} + V_{ang} + V_{bnd/ang} + V_{oop} + V_{tor} + V_{vdW} + V_{elec}$$

where V_{tot} is the total potential energy, V_{bnd} is the bond stretch component, V_{ang} is the angle bend component, $V_{bnd/ang}$ is the stretch–bend cross-term component, V_{oop} is the out-of-plane torsional angle strain component, V_{tor} is the torsional angle strain component, V_{vdW} is the component for van der Waals nonbonded interactions, and V_{elec} is the component for electrostatic nonbonded interactions. The potential energy functions are all given in kilocalories per mole. The values employed for the force constants were obtained using the PROBE program, which uses a least squares fitting routine to extract the constants from the data obtained from *ab initio* calculations on the basis set molecules.

Bond Stretch Interactions The potential V_{bnd} which governs bond stretching in MMFF94[321,453] uses a quartic function:

$$V_{bnd} = K_{bnd} \cdot \tfrac{1}{2}k_{bnd}(r - r_0)^2[1 + k_{bnd/cub}(r - r_0)$$
$$+ \tfrac{7}{12}k_{bnd/cub}^2(r - r_0)^2]$$

where K_{bnd} is a unit cancellation constant and has the value 143.9325; k_{bnd} is the bond stretching force constant in millidynes per angstrom, $r - r_0$ is the difference between the actual and the equilibrium bond length in angstroms, and $k_{bnd/cub}$ is the cubic stretch constant with a value of -2 Å^{-1}. There are specialized parameters for use with bonds included in a conjugated system. Some typical parameter values are $k_{bnd} = 4.766$ mdyn/Å and $r_0 = 1.093$ Å for an aliphatic C—H bond, and $k_{bnd} = 4.258$ mdyn/Å and $r_0 = 1.508$ Å for a C(sp^3)—C(sp^3) bond.

Angle Bend Interactions The potential V_{ang} which governs angle bending in MMFF94[321,453] uses a cubic function for most angles:

$$V_{ang} = K_{ang} \cdot \tfrac{1}{2}k_{ang}(\theta_0 - \theta)^2[1 + k_{ang/cub}(\theta_0 - \theta)]$$

where K_{ang} is a unit cancellation constant and has the value 0.043844, k_{ang} is the angle bending force constant in mdyn·Å/rad^2, $\theta_0 - \theta$ is the difference between the actual and the equilibrium angle span in degrees, and $k_{ang/cub}$ is the cubic bend constant with a value of -0.007 deg^{-1} ($= -0.4$ rad^{-1}). There are specialized parameters for use with angles including a delocalized bond or a small ring. Some typical parameter values are $k_{ang} = 0.636$ mdyn·Å/rad^2 and $\theta_0 = 110.549°$ for an aliphatic C(sp^3)—C(sp^3)—H bond, and $k_{ang} = 0.851$ mdyn·Å/rad^2 and $\theta_0 = 109.608°$ for a C(sp^3)—C(sp^3)—C(sp^3) angle.

Angles which approach 180° are treated with a simplified potential expression $V_{ang/lin}$ given by

$$V_{ang/lin} = 143.9325 k_{ang}(1 - \cos \theta)$$

This form is also used in the generic force fields DREIDING and UFF.

Bond-Stretch–Angle Bend Cross Term For an angle composed of the three atoms I—J—K, the potential $V_{bnd/ang}$ in MMFF94[321,453] corresponding to the stretch–bend cross term, which couples the bond stretch with the angle bend, is given by

$$V_{bnd/ang} = K_{bnd/ang}[k_{bnd/ang(IJK)}(r_{IJ} - r_{IJ(0)})$$
$$+ k_{bnd/ang(KJI)}(r_{KJ} - r_{KJ(0)})](\theta - \theta_0)$$

where $K_{bnd/ang}$ is a unit cancellation constant and has the value 2.51210; $k_{bnd/ang(IJK)}$ and $k_{bnd/ang(KJI)}$ are the stretch–bend force constants which couple the I—J and the K—J bond stretch, respectively, with the angle bend for I—J—K, in millidynes per radian; $r_{IJ} - r_{IJ(0)}$ and $r_{KJ} - r_{KJ(0)}$ are the displacements from the equilibrium bond length for the I—J and K—J bonds (respectively); and θ-θ_0 is the difference between the actual and the equilibrium angle span in degrees, as above. The stretch–bend term is omitted for angles approaching 180°.

Some typical parameter values are $k_{\text{bnd/ang}(IJK)} = 0.227$ mdyn/rad, $k_{\text{bnd/ang}(KJI)} = 0.070$ mdyn/rad for an aliphatic $C(sp^3)$—$C(sp^3)$—H bond, and $k_{\text{bnd/ang}(IJK)} = k_{\text{bnd/ang}(KJI)} = 0.206$ mdyn/rad for a $C(sp^3)$—$C(sp^3)$—$C(sp^3)$ angle.

Out-of-Plane Interactions For an appropriate central atom J bonded to three other atoms I, K, and L, MMFF94[321,453] treats out-of-plane bending interactions with a potential V_{oop} given by

$$V_{\text{oop}} = K_{\text{oop}} \cdot \tfrac{1}{2} k_{\text{oop}} \theta_{\text{oop}}^2$$

where K_{oop} is a unit cancellation constant and has the value 0.043844; k_{oop} is the out-of-plane bending force constant in mdyn · Å/rad², and θ_{oop} is defined as in MM2, as the angle between the bond J—L and its projection onto the plane determined by the atoms I—J—K (see Fig. 5-6).

A typical parameter value for I—J—K[—L] = $C(sp^3)$—$C(sp^2)$—$C(sp^3)$[= $C(sp^2)$] is $k_{\text{oop}} = 0.030$ mdyn · Å/rad², and for $C(sp^3)$—$C(sp^2)$—H[=O(sp^2)] is $k_{\text{oop}} = 0.122$ mdyn · Å/rad².

Torsional Interactions The potential V_{tor} which governs torsional angle interactions in MMFF94[321,454] uses a standard threefold function based on the cosine of the torsional angle ω defined by the atoms I, J, K, and L:

$$V_{\text{tor}} = \tfrac{1}{2}[k_{\text{tor}1}(1 + \cos \omega) + k_{\text{tor}2}(1 - \cos 2\omega) + k_{\text{tor}3}(1 + \cos 3\omega)]$$

where the force constants $k_{\text{tor}1}$, $k_{\text{tor}2}$, and $k_{\text{tor}3}$ depend upon the four atoms which make up the torsional angle. These terms were derived to fit data for gas phase conformations obtained from *ab initio* calculations. There are specific sets of constants for four-membered rings and for saturated five-membered rings, and also for those cases in which any of the three constituent bonds is part of a multiple bond or part of a conjugated system.

Some typical parameter values for I—J—K—L = $C(sp^3)$—$C(sp^3)$—$C(sp^3)$—$C(sp^3)$ are $k_{\text{tor}1} = 0.103$ kcal/mol, $k_{\text{tor}2} = 0.681$ kcal/mol, and $k_{\text{tor}3} = 0.332$ kcal/mol; and for H—$C(sp^3)$—$C(sp^3)$—H are $k_{\text{tor}1} = 0.284$ kcal/mol, $k_{\text{tor}2} = -1.386$ kcal/mol, and $k_{\text{tor}3} = 0.314$ kcal/mol.

van der Waals Nonbonding Interactions The MMFF94 potential V_{vdW} which governs van der Waals interactions has a novel "buffered 14–7" form—this is one of the most distinguishing features of this force field.[314,321] This potential is calculated for any two atoms I and J which belong to different molecules or are separated by three or more bonds, and takes the form

$$V_{\text{vdW}} = k_{\text{vdW}} \left(\frac{1.07 r_{IJ(0)}}{r_{IJ} + 0.07 r_{IJ(0)}}\right)^7$$
$$\times \left(\frac{1.12 r_{IJ(0)}^7}{r_{IJ}^7 + 0.12 r_{IJ(0)}^7} - 2\right)$$

where the force constant k_{vdW} is a measure of the potential well depth or hardness, r_{IJ} is the internuclear distance for the two atoms in angstroms, and $r_{IJ(0)}$ is a buffering constant equivalent to the equilibrium internuclear distance between atoms I and J.

The force constant k_{vdW}, corresponding to the hardness factor ε_{IJ}, is given by the expression

$$k_{\text{vdW}} = \frac{181.16 G_I G_J \alpha_I \alpha_J}{(\alpha_I/N_I)^{0.5} + (\alpha_J/N_J)^{0.5}} \cdot \frac{1}{r_{IJ(0)}^6}$$

In this expression, G_I and G_J are scale factors for the two atoms I and J, α_I and α_J are the assigned atomic polarizabilities for the two atoms I and J, and N_I and N_J are the Slater–Kirkwood effective numbers of valence electrons. For example, in the case of simple aliphatic carbon, α has the value 1.050 Å^3, G has the value 1.282, and N has the value 2.490, and in the case of simple hydrogen, α has the value 0.250 Å^3, G has the value 1.209, and N has the value 0.800.

The buffering constant $r_{IJ(0)}$ is given by the expression

$$r_{IJ(0)} = 0.5(r_{II(0)} + r_{JJ(0)})\{1 + 0.2[1 - \exp(-12\gamma_{IJ}^2)]\}$$

where $r_{II(0)}$ and $r_{JJ(0)}$ are the minimum-energy separations for same-atom pairs $I\cdots\cdots I$ and $J\cdots\cdots J$ respectively, and are obtained from the assigned atomic polarizabilities by the relationship

$$r_{II(0)} = A_I \alpha_I^{0.25} \quad \text{and} \quad r_{JJ(0)} = A_J \alpha_J^{0.25}$$

where A_I and A_J are scale factors which are usually invariant across a row of the periodic table. For example, in the case of simple aliphatic carbon, A has the value 3.890 Å$^{0.25}$, and in the case of simple hydrogen, A has the value 4.200 Å$^{0.25}$.

The exponential factor γ_{IJ} is a function of the two minimum-energy separations for same-atom pairs $I\cdots\cdots I$ and $J\cdots\cdots J$ as

$$\gamma_{IJ} = \frac{r_{II(0)} - r_{JJ(0)}}{r_{II(0)} + r_{JJ(0)}}$$

Hydrogen bonding is modeled as a special case of the V_{vdW} potential function. This treatment was derived from *ab initio* calculations on hydrogen-bonded water. It has the same form, but uses modified values of $r_{IJ(0)}$ and k_{vdW}. These changes have the net effect of reducing the size of the hydrogen-bonding hydrogen in the context of this interaction, while maintaining its steric size for modeling the nonspecific van der Waals interactions. The hardness parameter k_{vdW}, calculated with the standard value $r_{IJ(0)}$, is scaled by a factor of 0.5 for a hydrogen bonding interaction. $r_{IJ(0)HB}$ is given by the simple arithmetic mean, equivalent to the general expression when the exponential factor goes to zero, and is scaled by a factor of 0.8 when I–J is a donor–acceptor interaction:

$$r_{IJ(0)HB} = 0.8[0.5(r_{II(0)} + r_{JJ(0)})]$$

MMFF94 does not currently employ a reduced bond length for C—H bonds for purposes of calculating nonbonded interactions, in the interest of computational speed.

Electrostatic Nonbonding Interactions Electrostatic interactions between atoms I and J in MMFF94 are modeled with a buffered Coulomb potential,[314,321,455] V_{elec}, for any two atoms I and J which belong to different molecules or are separated by three or more bonds. For two atoms I and J which are separated by three bonds, the interactions are scaled by a factor of 0.75. The potential takes the form

$$V_{elec} = \frac{K_{elec} q_I q_J}{D(r_{IJ} + \delta)^n}$$

where K_{elec} is a unit cancellation constant and has the value 332.0716; q_I and q_J are partial atomic charges calculated by adding the contributions from the charge increments of the bonds to the respective atom, in elementary charge units; r_{IJ} is the internuclear distance for the two atoms, in angstroms; D is the unitless dielectric constant of the medium, with a default value of 1, δ is the electrostatic buffering constant, which takes a value of 0.05 Å, and n is a factor which allows toggling between a constant ($n = 1$, the default) and a distance-dependent ($n = 2$) dielectric model.

The default values for D and for n were set as above because MMFF94 was designed to support application to condensed phases, especially for ones with water as an explicit solvent.

The bond charge increments to q_I reflect the polarity of the bonds to atom I from all attached atoms. For example, a hydrogen attached to an olefinic or aromatic carbon bears a charge of $+0.15$, the hydrogen in water bears a charge of $+0.43$, and a carbonyl oxygen bears a charge of -0.57.

5.4.7. The OPLS Force Field

The discussion here of the **OPLS** force field (Optimized Potentials for Liquid Simulation) covers the work by the Jorgensen group to establish potentials and parameters for modeling solvation. OPLS is different from the other force fields examined in this chapter in that it deals with molecular dynamics simulations of the bulk medium (many solvent molecules) rather than a single molecule. However, the use of solvation systems like OPLS in running molecular dynamics simulations of biomacromolecules is quite common, and so it was selected for discussion as representative of this class.

Building upon previous work, Jorgensen presented his early formulations for water, methanol, ethanol, and ethers in 1981.[458-460] These TIPs (Transferable Intermolecular Potential functions) employed standard bond lengths and angles, and a nonbonding potential was used to model the interactions between solvent molecules. Hydrogen bonding interactions are modeled through the electrostatic term, and the Lennard–Jones parameters for acidic hydrogens are both set to zero. United atom representations worked just as well as all atom representations for the alkyl groups in these simulations. The charges on the united atom components of the alkyl groups were adjusted to achieve neutrality with the charges on oxygen and hydrogen shown below. It was determined that the inclusion of lone pair pseudoatoms did not confer an advantage which justified the additional computation. This potential was summed over all interacting atom pairs I and J, and combined a Lennard–Jones 6-12 potential with a scaled Coulomb potential:

$$V_{\text{TIP}} = q_I q_J \frac{e^2}{r_{IJ}} + \frac{A_I A_J}{r_{IJ}^{12}} - \frac{B_I B_J}{r_{IJ}^6}$$

where q_I and q_J are the charges on the respective atoms I and J, in atomic charge units; e^2 is an electrostatic force constant with the value 332.17752 kcal · Å/mol; r_{IJ} is the distance between atoms I and J, in angstroms; A_I and A_J are steric repulsion parameters for atoms I and J, in [kcal · Å^{12}/mol]$^{1/2}$; and B_I and B_J (C in the Jorgensen report) are steric attraction parameters for atoms I and J, in [kcal · Å^6/mol]$^{1/2}$.

Some typical parameters are: for water oxygen O, $q = -0.80$, $10^{-3}A^2 = 580$ kcal · Å^{12}/mol, and $B^2 = 525$ kcal · Å^6/mol; for alcohol oxygen O, $q = -0.685$, $10^{-3}A^2 = 515$ kcal · Å^{12}/mol, and $B^2 = 600$ kcal · Å^6/mol; for ether oxygen O, $q = -0.58$, $10^{-3}A^2 = 500$ kcal · Å^{12}/mol, and $B^2 = 625$ kcal · Å^6/mol; for water and alcohol hydrogen H, $q = 0.40$, $10^{-3}A^2 = 0.0$ kcal · Å^{12}/mol, and $B^2 = 0.0$ kcal · Å^6/mol; for CH$_3$, $10^{-3}A^2 = 7950$ kcal · Å^{12}/mol and $B^2 = 2400$ kcal · Å^6/mol; and for CH$_2$, $10^{-3}A^2 = 7290$ kcal · Å^{12}/mol and $B^2 = 1825$ kcal · Å^6/mol.

In the case of ethanol, a torsional potential was included. This had a standard three-part expression for one-, two-, and threefold barriers (see Section 5.1.1.3), and the parameters were fitted from a potential energy surface from an *ab initio* calculation.[460]

The results for alcohols and ethers were satisfactory, but limitations were observed in this model for the structure of liquid water,[458] and the Purdue group continued to refine their model.

Two years later, Jorgensen presented a comparison of six liquid water formulations, including two new ones developed in his group, TIP3P and TIP4P.[461] TIP3P refers to a Transferable Intermolecular Potential, three-point model, and TIP4P is the corresponding four-point model. The three-point model refers to the oxygen and two hydrogens as the interactive points. The four-point model displaces the charge due to the oxygen atom away from the nuclear center, to a virtual point M located 0.15 Å toward the hydrogens along a line bisecting the H—O—H angle.[462,463] The TIP4P formulation, a refinement of the earlier TIPS2 water model, was found to reproduce physical and thermodynamic properties most faithfully. This analysis also produced formulations for simulating liquid hydrocarbons, with the addition of an intramolecular torsional potential for the appropriate molecules:

$$V_{tor} = k_{tor0} + \tfrac{1}{2}[k_{tor1}(1 + \cos \omega) + k_{tor2}(1 - \cos 2\omega)$$
$$+ k_{tor3}(1 + \cos 3\omega)]$$

where the force constants k_{tor0}, k_{tor1}, k_{tor2}, and k_{tor3} (in kilocalories per mole) are fitted from the MM2-calculated torsional potential for the subject molecules. The new hydrocarbon potentials and the TIP4P model are collectively called OPLS models.

The parameters for TIP4P, where the two interacting atoms are always hydrogen and oxygen, are given by $10^{-3} A_{OH} = 600$ kcal·Å^{12}/mol, $B_{OH} = 610$ kcal·Å^6/mol; for O, $q = 0.00$, for H, $q = 0.52$, and for the virtual point M, $q = -1.04$.

The hydrocarbon parametrization uses united atom units, and covers 12 types. Electrostatic effects are neglected. The Lennard–Jones portion of the hydrocarbon potential uses A_{IJ} and B_{IJ} values which are the geometric mean of the corresponding parameters for the two interacting groups. The individual parameters are defined in terms of the parameters ε (hardness) and σ (van der Waals internuclear radius) as

$$A_{II} = 4\varepsilon_I \sigma_I^{12} \quad \text{and} \quad B_{II} = 4\varepsilon_I \sigma_I^6$$

Some typical hydrocarbon values are: for CH$_3$—C(2°) (i.e. methyl attached to a secondary carbon), $\sigma = 3.905$ Å, $\varepsilon = 0.175$ kcal/mol; for CH$_3$—C(4°) (i.e. methyl attached to a quaternary carbon), $\sigma = 3.960$ Å, $\varepsilon = 0.145$ kcal/mol; for CH$_2$ (i.e. methylene carbon), $\sigma = 3.905$ Å, $\varepsilon = 0.118$ kcal/mol; for CH$_2(sp^2)$ (i.e. vinylidene), $\sigma = 3.850$ Å, $\varepsilon = 0.140$ kcal/mol; for CH(sp^2) (i.e. aromatic), $\sigma = 3.750$ Å, $\varepsilon = 0.110$ kcal/mol; for C(4°) (i.e. quaternary carbon), $\sigma = 3.800$ Å, $\varepsilon = 0.050$ kcal/mol.

This work has been followed by reports of for OPLS formulations for amides (formamide, N-methylacetamide, N,N-dimethylacetamide, and extension to peptides),[464] further work on alcohols,[465] and thiols and thioethers.[466] This culminated in a report of the adaptation of OPLS for proteins, through parametrization of the standard amino acid pendant groups.[435] The adaptation to these more complex systems came about through amalgamation with the AMBER force field, from which came the terms for intramolecular interactions (bond stretch, angle bend, torsional interactions), to produce the so-called AMBER/OPLS for simulations of biomacromolecules.

More recent efforts by the Jorgensen group have focused on calculating partition coefficients,[467] the structure of benzene dimer,[468] nucleic acid interactions,[469,470] parametrization for acetic acid and methyl acetate,[471] specific examples of biomacromolecule modeling,[472,473] and aspects of molecular recognition.[474,475]

Recently, a modified AMBER/OPLS has been reported.[476] This is an all atom formulation for small organic molecules and peptides, and is termed OPLS-AA. It incorporates the bond stretch and angle bend potentials from AMBER, torsional potentials which have been obtained from *ab initio* calculations, and the OPLS nonbonding potential terms (i.e. Coulomb electrostatics and Lennard–Jones 6–12 sterics). Thus, OPLS has become a full-fledged force field.

5.4.8. Miscellaneous Force Fields

In this section, a number of other force fields will be discussed more briefly than the foregoing. The selection of force fields to describe here was not an easy one, and the depth of coverage is not intended as a judgment of quality or importance. Rather, the choice of formulations to be afforded more detailed coverage over the preceding seven sections was dictated by their similarity to MMX, as well as some historical factors, the availability of published documentation, and the author's familiarity. Coverage of the older force fields is minimal, and the interested reader is referred to other sources for a discussion of these treatments.[7–10,13,15,19]

The force fields and modeling programs described here briefly are **MacroModel, ECEPP, EAS, DREIDING, UFF, EFF, PIMM,** and **YETI.**

5.4.8.1 MacroModel. MACROMODEL®, with its predecessors RINGMAKER and MODEL,[477,478] played a very important role in the transfer of molecular modeling technology to the general chemistry community. Its release in 1983 marks a turning point in the use of computer molecular modeling by the organic synthesis community.[22] The program is easy to use and has a friendly graphic user interface. It also played a role in the development of PCMODEL, through collaborations between the two development groups. It would not be unreasonable to observe that MACROMODEL has had some effect on the look and feel of all of the major commercial modeling packages.

MACROMODEL is not a force field per se, but a modeling environment.[285,286] It is a workstation and mainframe platform molecular modeling program for biomacromolecules as well as small molecules, developed by the research group of W. C. Still at Columbia University. It can host or implement several force fields as slightly modified versions, usually indicated by an asterisk postfix on the name of the modified field, for example, MM2*, MM3*, AMBER*, OPLSA*. It can also employ the force fields AMBER94 and MMFF in their native form. It has 64 atom types, with a thesaurus of atom type equivalences for the corresponding types in MM3, CHARMM, and AMBER. MACROMODEL enables structure creation and manipulation, energy minimizations, docking, and molecular dynamics simulations. It incorporates the analytical GB/SA continuum treatment for solvation (see Section 5.6), and algorithms for conformation searching (see Section 5.8), comparisons of structures, predictions of NMR coupling constants, and construction of molecular surfaces. More information on MACROMODEL may be obtained from its World Wide Web site.[286]

5.4.8.2. The ECEPP and UNICEPP Force Fields ECEPP stands for empirical conformational energy program for peptides, and UNICEPP stands for united atom conformational energy program for peptides. These force field programs were developed by the research group of H. A. Scheraga at Cornell University, and the source code is available from QCPE as program 286 (ECEPP),[479] program 361 (UNICEPP),[69] and program 454 (ECEPP/2 or ECEPP83).[480]

ECEPP and ECEPP/2 are implementations of an empirical potential energy algorithm for the conformational analysis of polypeptides, which utilizes previously described empirical parameters.[68,70,307,30,481-484] UNICEPP is a united atom version of ECEPP.

There are 28 atom types for these force fields, including five hydrogens, six carbons, three nitrogens, three oxygens, two sulfurs, and a chlorine. UNICEPP adds the **united atom** atom types for CH_3, CH_2, aliphatic CH, and aromatic CH groups. There is a library of 26 amino acid residues and 20 end group substructures to use for constructing peptides. ECEPP/2 also has a supplementary set of 10 additional amino acid residues. The residues (including end groups) are numbered consecutively from the N-terminal end. This force field model was parametrized for linear polypeptides and for those containing one or more cystine disulfide bridges. It does not adequately address cyclic polypeptides without additional parametrization.

These programs read an input data set containing standard residue information for 26 amino acid residues and 8 end groups. The initial structural input includes: the desired amino acid sequence, including end groups and sites of cystine disulfide bridges; the stereochemistry (D or L) of each amino acid residue; the initial conformation, characterized by dihedral angles, the number of conformations to be examined; and which dihedral angles will be treated as variables for the conformational search, and the new values for these variables.

Bond lengths and bond angles are kept at ideal values, supplied from a standard set, and the potential energy equation for these force fields has the following form:

$$V_{tot} = V_{tor} + V_{vdW} + V_{elec} + V_{HB}$$

where V_{tot} is the total potential energy; V_{tor} is the torsional angle strain component, which sums cosine and sine functions of the dihedral angle; V_{vdW} is the component potential governing nonbonding steric interactions, with a Lennard–Jones 6–12 potential; V_{elec} models the nonbonding electrostatic interactions with a Coulomb potential; and V_{HB} is a hydrogen bonding potential, which uses a 10–12 potential function. There is an algorithm for driving the closure of a cystine disulfide bond, and also a separate potential for the torsions about the disulfide bond in a linked cystine.

Subroutines in ECEPP operate upon the input structure to generate all possible conformations about the specified torsional angles, and each of these is subjected to energy minimization and later comparison. Nonbonded steric interactions are scaled to one-half for 1,4-interactions.

ECEPP83 was reported as an updated version of the force field,[485,486] followed later by ECEPP/3.[487]

Comparisons of the models produced for peptides by CHARMM, AMBER, and ECEPP have been reported,[307,308] and a comparison between the ECEPP, HSEA, MM2, MM2CARB, and PCILO models of methyl β-xylobioside found ECEPP to give the best results.[488]

5.4.8.3. The EAS Force Field Building upon earlier work from the Schleyer group on the STRAIN program,[73,489] this hydrocarbon force field was given an abbreviation from the names of the authors of the main report, Engler, Andose, and Schleyer.[7] The form of the potential was relatively simple, and consisted of four components:

$$V_{tot} = V_{bnd} + V_{ang} + V_{tor} + V_{vdW}$$

where V_{tot} is the total potential energy; V_{bnd} modeled the bond stretch interactions with a Hooke's law harmonic potential; V_{ang} modeled the angle bend interactions with a modified Hooke's law potential including a cubic term; V_{tor} is the torsional angle strain component, which employed a one-part potential with a threefold periodicity; and V_{vdW} is the component potential governing nonbonding steric interactions, which employed a modified Buckingham potential devised. All energy values were given in units of 10^{-11} ergs/molecule (≈ 144.0 kcal/mol).

Agreement between with results of molecular structure calculations with the EAS force field and with Allinger's MMI force field were generally good. The individual components and especially the parametrization were different, but both force fields had been optimized against physical data and thus both produced good models.

In 1978, Schleyer, Mislow, and coworkers collaborated to produce the program BIGSTRN (QCPE 348),[490] which could employ either the EAS or

MMI force field. BIGSTRN could calculate structures of up to 100 atoms. The EAS force field could calculate the structure of hydrocarbons and carbocations, but for other kinds of compounds the more broadly parametrized MMI was needed. The updated BIGSTRN-2 (QCPE 348) was produced in 1981,[491] and this program could now host Allinger's MMI and MM2 force fields, the EAS force field, and the MUB-2 force field of Bartell et al.[78,79]

Allinger's MMI force field, and especially its successor MM2, became increasingly popular among practicing chemists, but the EAS force field did not achieve such wide acceptance. In his later work with empirical force field modeling, for example on the transannular interactions in cage compounds[492] or strained olefins,[493] Schleyer carried out calculations either with both the EAS and Allinger force fields, or with the Allinger force field alone. Recent reports from the Schleyer group discuss modified geometry parameters for Allinger's MMP2 force field program, used to achieve good results in modeling localized and delocalized carbocations.[494,495]

5.4.8.4. The DREIDING Force Field One of the generic force fields is DREIDING, developed by Goddard et al.[496] The emphasis is on the use of general force constants and geometry parameters based on simple hybridization considerations, in order to simplify the overall form of the force field. Parameters are obtained from extant structural data, or extrapolated from the data for closely related systems.

The initial report gave 37 atom types, including elements from the first four rows of the Main Group of the periodic table, plus the metals sodium, calcium, iron, and zinc. Atom types are denoted by a five-character code: the first two spaces are for the elemental symbol, the third space for a number or letter indicating the hybridization or geometry, the fourth space the number of implicit hydrogens, and the fifth space for a indicator for special characteristics such as formal oxidation state.

The potential energy equation for this force field has the following form:

$$V_{tot} = V_{bnd} + V_{ang} + V_{tor} + V_{oop} + V_{vdW} + V_{elec} + V_{HB}$$

where V_{tot} is the total potential energy; V_{bnd} is the bond stretch component, modeled with either a simple harmonic potential or a Morse potential; V_{ang} is the angle bend component, described below; V_{tor} is the torsional angle strain component, which takes a form very close to that found in AMBER (see Section 5.4.3) and CHARMM (see Section 5.4.4); V_{oop} is the out-of-plane component, described below; V_{vdW} is the component potential governing nonbonding steric interactions, and can employ either a Lennard–Jones 6–12 or a modified Buckingham potential; V_{elec} models the nonbonding electrostatic interactions with a standard Coulomb potential; and V_{HB} is the hydrogen bonding potential, which employs a form very close to the CHARMM version[439] (see Section 5.4.4). The potential energy functions are all given in kilocalories per mole.

Angle Bending Interactions DREIDING uses a harmonic angle bending potential V_{ang}, a function of the cosine of the angle *I—J—K* as defined in Fig. 5-4

$$V_{ang} = \tfrac{1}{2}k_{ang}(\cos\theta - \cos\theta_0)^2$$

where k_{ang} is the force constant for angle bending, in kilocalories per mole; θ is the current span across the angle (corresponding to θ_{IJK} in Fig. 5-4); and θ_0 is the equilibrium value for the angle.

Out-of-Plane Potential DREIDING employs an out-of-plane angle bending potential to enforce planarity in certain trigonal atoms (e.g. the sp^2 carbon in alkenes), and to enforce nonplanarity in others (e.g. the sp^3 nitrogen in ammonia). This component, termed the inversion potential in the report,[496] is a function of the Wilson angle (see Fig. 5-6) and takes the form

$$V_{oop} = \frac{1}{2}\frac{k_{oop}}{\sin^2\theta_{Wilson}}\cos\theta_{Wilson} - \cos\theta_{Wilson(0)})^2$$

where k_{oop} is the force constant for the out-of-plane bending, in kilocalories per mole; θ_{Wilson} is the current span across the Wilson angle (corresponding to θ_{Wilson} in Fig. 5-6); and $\theta_{Wilson(0)}$ gives the equilibrium value for this angle.

The authors compared the geometry of structures resulting from DREIDING force field energy minimization with the coordinates registered in the Cambridge Crystallographic Database. For 76 molecules, the overall root mean square (RMS) deviation of relative atomic positions was 0.235 Å. Other comparison measures were acceptably good, including the treatment of diborane (B_2H_6). The authors consider this force field to be a method for the rapid examination of novel structures, which may be followed by the application of more specialized force fields to enable more accurate predictions of particular properties.

The DREIDING force field has been used to model the atomic structure of an HIV glycoprotein,[497] a monolayer of organic molecules on a metal surface,[498] and the structure of polymers[499–501] and polymerization catalysts,[502] and in other studies.[503,504] The DREIDING and the newer DREIDING II force fields have been implemented in the Polygraf and Cerius modeling products from MSI.

5.4.8.5. The UFF Formulation Another of the generic force fields is UFF, for Universal Force Field, developed by Rappé et al.[26,505–507] The emphasis is again on the use of general force constants and geometry parameters based on the element, its hybridization, and its connectivity, in order to simplify the overall form of the force field. Parameters are obtained from available structural data, or extrapolated from the data for closely related systems.

The initial reports gave 126 atom types, covering the entire periodic table. Atom types are denoted by a five-character code: the first two spaces are for the element symbol, the third space for a number or letter indicating the hybridization or geometry, and the fourth and fifth spaces are used for alternative parameters such as formal oxidation state.

The potential energy equation for this force field has the following form:

$$V_{tot} = V_{bnd} + V_{ang} + V_{tor} + V_{oop} + V_{vdW} + V_{elec}$$

where V_{tot} is the total potential energy; V_{bnd} is the bond stretch component, modeled with either a simple harmonic potential or a Morse potential; V_{ang} is the angle bend component, described below; V_{tor} is the torsional angle strain component, described below; V_{oop} is the out-of-plane component, described below; V_{vdW} is the component potential governing nonbonding steric interactions, and employs a Lennard–Jones 6–12 potential; and V_{elec} models the nonbonding electrostatic interactions with a standard Coulomb potential. The potential energy functions are all given in kilocalories per mole.

Natural bond lengths are determined as the sum of the contributions from each of the two atoms, plus a bond order correction, plus an electronegativity correction. As in the DREIDING force field above, a single value is used for the bond dissociation energy component of the Morse potential, multiplied for multiple bonds. The bond stretch force constants are derived from the generalized Badger's rules,[508–511] as a function of the effective charges on each atom.

Angle Bending Interactions For the angle I—J—K as defined in Fig. 5-4, UFF uses a Fourier expansion in θ, which simplifies in most cases to the angle bending potential shown below:

$$V_{ang} = \frac{k_{ang}}{n^2}(1 - \cos n\theta)$$

where k_{ang} is the force constant for angle bending, in kilocalories per mole; n is an indicator of the geometry of the central atom J (1 for linear, 3 for trigonal, etc.); and θ is the current span across the angle (corresponding to θ_{IJK} in Fig. 5-4). The force constants are a function of the effective charges on the two attached atoms, the bond lengths, and the cosine of the equilibrium value for the angle, θ_0.

Torsional Interactions For the torsional angle ω (called ϕ by Rappé et al.) as defined by the atom string I—J—K—L (see Fig. 5-5), the UFF treatment employs a simple potential V_{tor} to govern torsional interactions. This equation has an n-fold term based on the product of the cosines of ω and of the reference angle ω_0 (called ϕ_0 by Rappé et al.), and takes the form

$$V_{tor} = \tfrac{1}{2}k_{tor(n)}(1 - \cos n\omega \cos n\omega_0)$$

where $k_{tor(n)}$ is the force constant for the n-periodic torsional interaction, in kilocalories per mole; n is the periodicity of the torsional barrier; ω gives the span of the torsional angle corresponding to ω_{IJKL} in Fig. 5-5; and ω_0 is the reference or phase offset angle for the given atom string. Only general torsions are parametrized, i.e. of the type $(X-J-K-Y)$ characterized by the central atoms J and K. There is a separate expression to derive torsional barriers between bonds of multiplicity other than 1. The force constants are composed from the geometric mean of atom-type-specific parameters.

Out-of-Plane Potential UFF employs an out-of-plane angle bending potential to enforce planarity in certain trigonal atoms (e.g. the sp^2 carbon in alkenes) and to enforce nonplanarity in others (e.g. the sp^3 nitrogen in ammonia). This component, termed the inversion potential in the reports,[505-507] is a function of the Wilson angle (see Fig. 5-6) and takes the form

$$V_{oop} = k_{oop}(K_{oop0} + K_{oop1} \cos \theta + K_{oop2} \cos 2\theta)$$

where k_{oop} is the force constant for the out-of-plane bending, in kilocalories per mole; K_{oop0}, K_{oop1}, and K_{oop2} are constants which depend upon the atom types involved; and θ is the current span across the Wilson angle (equivalent to θ_{Wilson} in Fig. 5-6). This potential is evaluated for each of the three Wilson angle combination of the central atom J with the attached atoms I, K, and L, and each makes a one-third contribution to the out-of-plane potential.

Applications of the UFF to the structure of organic molecules,[506] main group molecules,[507] metal complexes,[512] and polymers[513] have been reported.

5.4.8.6. The EFF Formulation Dillen has reported on two generations of a hydrocarbon force field formulation termed EFF for Empirical Force Field.[514-516] It was developed to address some of the deficiencies of MM2, many of which have also been remedied in the newer Allinger force field MM3 (see Section 5.4.1). Consistent with the limitation to hydrocarbon systems, there is no electrostatic potential term. The potential energy equation for this force field, parametrized for both the gas phase and crystalline media, has the following form:

$$V_{tot} = V_{bnd} + V_{ang} + V_{tor} + V_{vdW}$$
$$+ V_{bnd/ang} + V_{bnd/tor} + V_{bnd/bnd} + V_{ang/ang} + V_{ang/tor/ang}$$

where V_{tot} is the total potential energy; V_{bnd} is the bond stretch component, which employs a modified Hooke's law potential with cubic and quartic terms much as in MM3; V_{ang} is the angle bend component, which employs a modified Hooke's law potential with a cubic term; V_{tor} is the torsional angle strain component, which employs a four-part potential similar to that in MM4 with terms for one-, two-, three-, and sixfold periodicity; V_{vdW} is the component potential governing nonbonding steric interactions, and employs a modified

Buckingham potential; $V_{bnd/ang}$ is the cross-term potential which models the coupling between the bond stretch and angle bend vibrations, with an MM2-type potential; $V_{bnd/tor}$ is the cross-term potential which models the coupling between the bond stretch and torsional interactions, with an MM3-type potential; $V_{bnd/bnd}$ is the cross-term potential which models the coupling between two bond stretch vibrations, with an MM4-type potential; $V_{ang/ang}$ is the cross-term potential which models the coupling between two angle bend vibrations, with an MM3-type potential; and $V_{ang/tor/ang}$ is the angle-bend–torsion–angle-bend cross-term component, and employs an MM4-type potential. The potential energy functions are all given in kilocalories per mole.

5.4.8.7. The PIMM Force Field An early force field which combined a π-SCF–LCAO–MO treatment of conjugated systems with a molecular mechanics treatment of the σ-framework was the **PIMM** force field (an abbreviation for "pi molecular mechanics") developed by Lindner.[517,518]

In its earlier form,[517] it combined the π-SCF method of Dewar and Harget with a hydrocarbon force field patterned closely on that of Westheimer, and the main application was to conjugated or aromatic systems. The force field portion had the form

$$V_{tot} = V_{bnd} + V_{ang} + V_{tor} + V_{oop} + V_{vdW}$$

where V_{tot} is the total potential energy; V_{bnd} is the bond stretch component, modeled with a simple harmonic potential; V_{ang} is the angle bend component, modeled with a simple harmonic potential; V_{tor} is the torsional angle strain component, evaluated only for π-bonds; V_{oop} is the out-of-plane component, modeled with a simple harmonic potential; V_{vdW} is the component potential governing nonbonding steric interactions, and employed a Lennard–Jones 6–12 potential for C······C nonbonded interactions and a modified Buckingham potential for H······H and C······H nonbonded interactions.

A new version was reported in 1991,[518] which combined the Pople–Pariser–Parr π-SCF treatment of conjugated systems and a Marsili–Gasteiger **PEOE** (partial equalization of orbital electronegativity) estimation of charges for the σ-framework, together with an updated molecular mechanics treatment for organic molecules and metal complexes. The force field is documented in a doctoral dissertation (M. Kroeker, 1994, Technische Hochscule Darmstadt) which is mounted on the World Wide Web.[519]

There are 10 atom types: C(sp^3), C(sp^2), N(sp^3, pyrrole), N(sp^2, pyridine), O(sp^3, ether), O(sp^2, carbonyl), S(sp^3, thioether), Cl, F, and H; and 14 metal cation types.

The PIMM91 potential takes the form

$$V_{tot} = V_{bnd} + V_{ang} + V_{tor} + V_{oop} + V_{vdW} + V_{elec}$$

where V_{tot} is the total potential energy; V_{bnd} is the bond stretch component, modeled with a simple harmonic potential; V_{ang} is the angle bend component, modeled with a simple harmonic potential; V_{tor} is the torsional angle strain component, discussed below; V_{oop} is the out-of-plane component, modeled with a simple harmonic potential; V_{vdW} is the component potential governing nonbonding steric interactions, modeled with a modified Buckingham potential; and V_{elec} is the nonbonding electrostatic component, which employed a standard Coulomb potential where the effective atomic charges are the sum of the σ- and π-components. Metal coordination and hydrogen bonding are not represented by discrete potentials, but rather by appropriate parametrization of the electrostatic and van der Waals potentials for the affected atoms, as has been done in other force fields.

The use of the PIMM91 force field in NMR shift reagent studies on flexible molecules has been reported recently.[520]

5.4.8.8. The YETI Force Field

The YETI force field formulation from Vedani and Dunitz was first reported as the combination of a new hydrogen bonding potential (see Section 6.6) with the other force field components borrowed from AMBER and CHARMM (see Sections 5.4.3 and 5.4.4).[521] This new potential was derived from data on hydrogen bonding environments culled from the Cambridge Crystallographic Structure Database. A similar approach was used to examine the coordination environments of zinc atoms to develop a metal coordination potential (see Section 6.7), which was then used to model the binding between the enzyme human carbonic anhydrase II and various small molecules.[522] In this study, the YETI steric nonbonding model was given as a Lennard–Jones 6–12 potential.

A broader description appeared two years later.[523] YETI emerges as largely a force field for nonvalence interactions between biomacromolecules and small-molecule substrates, where generally the conformations of substructures (i.e. valence interactions such as bond stretch, angle bend, and torsional strain) are kept constant. In cases where conformational energy must be optimized, the AMBER force field is employed. There are 17 atom types, and most carbons are in united atom format. YETI can handle structures up to 3500 atoms.

Further improvements were provided in a more recent report.[524] The metal coordination potential has been reoptimized with newer structural data, and cobalt is now included. The hydrogen bonding potential has been calibrated against more empirical data as well as *ab initio* level calculations. The torsional potential is now explicitly included in YETI, while bond lengths and angles are still optimized within AMBER and then frozen. The YETI force field potential is thus

$$V_{nonbond} = V_{tor} + V_{vdW} + V_{elec} + V_{HB} + V_{coord}$$

where V_{tot} is the total potential energy; V_{tor} is the torsional strain component, modeled with an AMBER/CHARMM-like potential; V_{vdW} is the component potential governing nonbonding steric interactions, modeled with a 6–12

Lennard–Jones potential; V_{elec} is the nonbonding electrostatic component, modeled with a distance-dependent dielectric potential; V_{HB} is the hydrogen bonding component, modeled with a 10–12 potential modified to account for the geometry; and V_{coord} is the metal-atom-coordination component, modeled with a 10–12 potential modified to allow for the geometry. For metal–ligand interactions, the electrostatic component is scaled to one-half. Four metal coordination geometries were supported: tetrahedral, square pyramidal, trigonal bipyramidal, and octahedral. The van der Waals interactions for atoms in a 1,4-relationship are scaled to one-half.

More recently, an algorithm has been added to YETI for determining the systematic solvation of proteins based upon the directionality of the hydrogen bonds near the solvent interface.[525] A related application was developed for creating a *virtual* **pharmacophore** or *pseudoreceptor*, that is, a representation of the structural complement to putative binding interactions which are culled from the structural characteristics of a number of known inhibitors. This application is termed **YAK**,[168] and utilizes mainly hydrogen extension vectors (**HEV**s), lone pair vectors (**LPV**s), and hydrophobicity vectors (**HPV**s).

YETI is beginning to find use in the modeling of biomacromolecule–small-molecule interactions.[526,527]

5.5. PARAMETERS—THEIR SOURCES AND UTILITY

Constructing a set of equations for the potential functions of a force field is the first step in developing a system for molecular modeling. The critical second step is the selection and optimization of the parameters and constants to be used with these equations. Generally, an initial set of parameters will have been used during the formation of the force field, but these must be refined, and it behooves the developers to formulate additional parameter sets which will extend the scope of the force field and make it more useful to a broader clientele. Indeed, that clientele may be the biggest source for new parameters, which they will develop in order to model the systems of interest to themselves.

The subject of the derivation and optimization of force field parameters has been addressed in a number of sources.[19,24,78,111,113,306,339,528–531] In earlier days, the initial parameters were obtained from empirical data such as vibrational spectroscopy, crystal diffraction structures, NMR **NOE** measurements, and enthalpy measurements. Parameter optimization was done by iterations of inspection and manual change. While these parameters are derived from the properties of real molecules, after the optimization process they can no longer be considered real outside of their contribution to a refined model. This was clarified early on by Hill, in connection with his early force field model: "The force constants are hypothetical force constants. They are not the force constants which might be determined experimentally, but rather those which would exist in the absence of steric effects. Hence, they must be estimated by analogy, guessed, or calculated backwards from experimental data."[35]

More recently, force field development has been carried out by using a least squares optimization procedure.[297,454] For many of the newer force fields, such as MM4, CFF93, and MMFF94 (see Sections 5.4.1, 5.4.2, and 5.4.6), the parameters are derived in whole or in part from *ab initio* level calculations, and scaled against a test set of structures.

Below are found some sources of parameters and how they are related. The use and limitations of some of these types of data are discussed in Section 3.3.

X-ray diffraction—in solid phase, measures orbital electron distributions (or *electron clouds*), gives three-dimensional picture of nuclei in molecule.

NMR spectroscopy—in liquid or solution phase, measures spin coupling and enhancement by energy transfer from nearby nuclei (NOE), gives bonded and nonbonded interatomic distances.

Electron diffraction—in gas phase, measures positions of nuclei.

Infrared and Raman spectroscopy—in liquid phase, solution or gas phase, gives interatomic distances, bond stretch, and angle bend force constants.

Microwave spectroscopy—in gas phase, measures moments of inertia, from which internuclear distances can be determined.

Ab initio calculation—theoretical, derives all aspects of molecular structure from first principles.

However they are obtained, parameters must be optimized over as large a basis set as possible, in order to achieve transferability, or applicability to the most members of the most classes of structures.

New parameters for existing force fields are often reported, either by the development group, or by other researchers who require an application outside the original basis set. Published force field parameters have been collected recently in a review publication,[532] and a tutorial on the subject is available within the AMBER World Wide Web site.[533]

5.6. ELECTROSTATICS AND SOLVATION

The two areas of electrostatics models and solvation models are quite closely intertwined in molecular modeling, which justifies their being discussed together. The earliest force field modeling formulations had provision for neither electrostatics nor solvation. As force fields have moved away from nonpolar hydrocarbons in the gas phase to molecules containing heteroatoms and molecules being simulated in solution or some solvated state, models for these effects have evolved.

Electrostatic effects have been introduced in Section 5.1.3.3, and the specific implementations in various force fields were covered in Sections 5.2 and 5.4.

5.6. ELECTROSTATICS AND SOLVATION

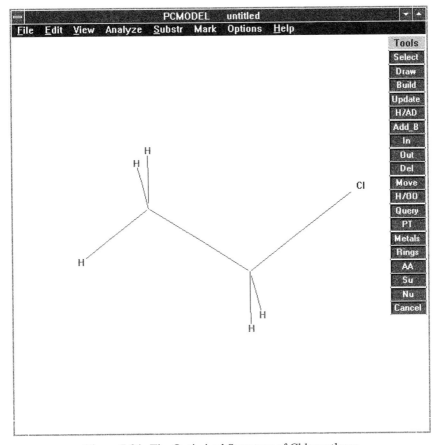

Figure 5-34. The Optimized Structure of Chloroethane.

Solvation phenomena were mentioned in the discussion of the OPLS formulation in Section 5.4.7. The treatment here will be general and will involve experiments to demonstrate these models.

5.6.1. Electrostatics

There are a number of general discussions of electrostatics available.[108,117,120,334,340,350,352,534-540] We will consider here two models for treating electrostatic interactions. The charge–charge model (i.e. $V_{\text{elec/QQ}}$) uses the Coulomb potential between the charges on two atoms I and J separated by an internuclear distance r_{IJ} (see Fig. 5-21)—either the classical form with the dielectric constant of the medium, D, in the denominator, or the distance-dependent dielectric form with an inverse square dependence on r_{IJ}:

$$V_{\text{elec/QQ}} = \begin{cases} q_I q_J / D r_{IJ} & \text{(classical Coulomb)} \\ q_I q_J / r_{IJ}^2 & \text{(distance-dependent Coulomb)} \end{cases}$$

318 THE MOLECULAR MECHANICS FORCE FIELD

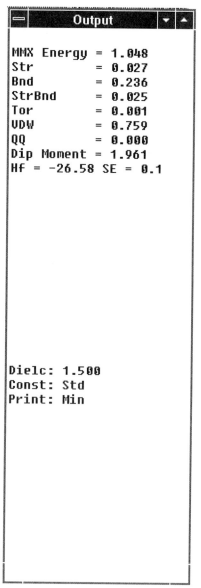

Figure 5-35. The 'Output' Window for the Optimized Structure of Chloroethane.

5.6. ELECTROSTATICS AND SOLVATION

In MMX, for the most part, these charges are derived from the bond dipole parameters (see Section A.4), although heteroatom π-systems will also contribute charges.

The dipole–dipole model (i.e. $V_{\text{elec/DD}}$) uses the Jeans potential between two bond dipoles **a** and **b**, whose midpoints are separated by a distance r_{ab} and whose relative geometry is described as is shown in Fig. 5-22. The dipole–dipole model is used in the Allinger force fields:

$$V_{\text{elec/DD}} = K_{\text{elec/DD}} \frac{|\mu_a||\mu_b|}{Dr_{ab}^3} (\cos \chi - 3 \cos \alpha_a \cos \alpha_b)$$

In the dipole–dipole model, if there are discrete charges present, the model must be expanded to include dipole–charge and charge–charge interactions.

Electrostatics Exercise 1: The Dipole Moment of Chloroethane. In this experiment, we will prepare chloroethane structure, carry it through energy minimization, and compare the calculated dipole moment with an empirical value. We will also examine how the charges are assigned by PCMODEL.

In the PCMODEL window, activate the **Draw** tool, and sketch in three atoms in a wide V-shape. Activate the **PT** tool and click first on the 'Cl' button and then on one of the terminal atoms in the V-shaped structure. Then, activate the **H/AD** tool to add the necessary five hydrogens, and use the **Analyze\Minimize** command to carry out an energy minimization. Rerun this command several times to ensure a thorough minimization. The final structure should resemble Fig. 5-34, and the outcome of the energy minimization is reported in the 'Output' child window, shown in Fig. 5-35.

Note the value given for the 'Dip Moment' (dipole moment), 1.961 D. Compare this with the literature value of 2.05 D,[541] which is fairly close agreement.

With this molecule still on screen, we will query the charges assigned to the atoms by PCMODEL. Activate the **Query** tool, then click first on the chlorine atom and then on an empty space to call out a report. This report line will resemble the following:

```
xyz:2.00  0.62  0.22  Q:-0.23
```

where 'xyz' refers to the Cartesian axes, the numbers are the corresponding coordinates, and the 'Q' denotes charge, with the value being -0.23.

Repeat this reporting procedure with the carbon atom to which chlorine is attached, and to one of the hydrogen atoms on each carbon. Also, click in turn on the chlorine, its attached carbon, and an empty space to produce a report of the bond length. The resulting screen should resemble the depiction in Fig. 5-36.

Only the chlorine and the carbon to which it is attached bear charges, equal and opposite. PCMODEL derives these charges from the value for the bond

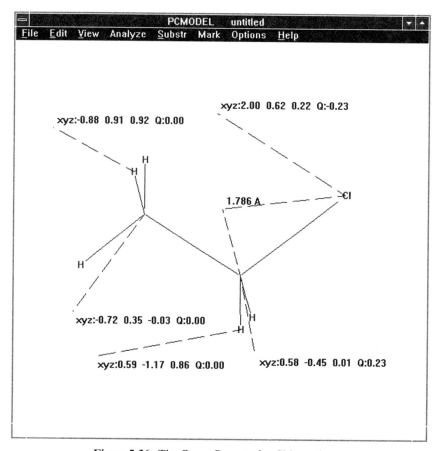

Figure 5-36. The **Query** Reports for Chloroethane.

moment parameter of the @1—@12 [(sp^3)carbon–chlorine] bond, 1.940 D; the length of the bond, 1.786 Å; and the appropriate conversion units. A detailed derivation is given in Section A.4, and the result is a charge of 0.23 on each of the atoms, with the algebraic sign depending upon the relative electronegativity of the atom.

The chloroethane structure may be found as {CLETHAN.PCM} on the input file library diskette in the CHAP5 directory.

> Some way of estimating the electrostatic interactions both within and between large molecules is of the utmost importance in any quantitative modeling package, and as virtually all molecular mechanics calculations involve atom-centered steric interactions this almost of necessity imposes the same constraints on the calculation of the electrostatic interactions, i.e. that they result from atom-centred partial charges.[542]

5.6. ELECTROSTATICS AND SOLVATION

That the assignment of charges is a critical part of any electrostatics model is borne out by literature reports.[117,418,432–434,452,542–552] A few of the currently used methods are described here briefly.

One early method is the Mulliken population analysis,[117] in which a molecular orbital calculation is carried out, the electron populations of the orbitals on an atom are summed, and the electron density in bonding orbitals is divided equally between the two atoms involved. This method was not designed for producing point charges, which are often too large for force field electrostatics models. Figure 5-37 shows the chloroethane molecule with atom labels, and the charges calculated from a Mulliken population analysis of the results of an SCF/STO-3G *ab initio* calculation[553] are given in Table 5-4.

In the del Re method[543] and later modifications,[556] the charge on an atom is found as the sum of contributions from each of the bonds to that atom. The

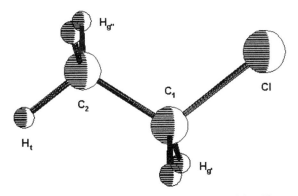

Figure 5-37. Chloroethane, with Atoms Marked for Charges.

TABLE 5-4. Charges Calculated for the Chloroethane Molecule by Various Methods

Method	Cl	C_1	$H_{g'}$	C_2	$H_{g''}$	H_t
MMX	−0.23	+0.23	0.0	0.0	0.0	0.0
Mulliken	−0.176	−0.060	+0.092	−0.172	+0.076	+0.072
Del Re	−0.177	+0.001	+0.068	−0.089	+0.043	+0.043
CNDO/2	−0.1452	+0.0595	+0.014	+0.0055	+0.014	+0.014
Smith–Eyring	−0.157	−0.0502	+0.0808	−0.156	+0.0675	+0.0675

See Fig. 5-37. Cl gives the charge on the chlorine atom; C_1 gives the charge on carbon 1; $H_{g'}$ gives the charge on the *gauche* hydrogens attached to C_1; C_2 gives the charge on carbon 2; $H_{g''}$ gives the charge on the *gauche* hydrogens attached to C_2; and H_t gives the charge on the *trans*-situated hydrogen attached to C_2. The MMX charges are those calculated in PCMODEL; the Mulliken charges are those derived from an SCF/STO-3G *ab initio* calculation[553]; the Del Re charges are those derived from the MO–LCAO method[543]; the CNDO/2 charges are those derived from this type of calculation, including the 3*d* orbitals[554]; the Smith–Eyring charges are those derived from the corresponding analysis.[555]

bond-specific contributions are derived from a semiempirical MO–LCAO calculation, and results were compared with data from empirical dipole moment data. The del Re charges for chloroethane are given in Table 5-4. For additional comparisons, the charges assigned to chloroethane from the results of CNDO/2 calculations (including 3d orbitals)[554] and from the Smith–Eyring method[555,557,558] are also given in Table 5-4. A modified Smith–Eyring (MSE) method was developed by Allinger to calculate dipole moments and predict the conformational equilibria in compounds containing electronegative atoms.[351,559,560]

Gasteiger and Marsili developed an algorithm to calculate atomic charges called *partial equalization of orbital electronegativity* (PEOE).[452] In this method, an equation expresses the electronegativity of a bonding orbital on an atom as a quadratic polynomial function of the total charge on that atom, with coefficients which depend upon the valence state of the atom. These derived orbital electronegativities are then used to generate partial charges on the two bonded atoms. The partial charges on an atom are then summed to provide a new total charge. A new iteration then begins in which the orbital electronegativities are recalculated from the new charges, etc., and this sequence is repeated until the charge distribution becomes self-consistent. These iterations tend to equalize the orbital electronegativities partially, but not totally, because of the damping effect of the local electrostatic field generated when charge is transferred from one atom to another. The resulting charges correlate well with available empirical data on electron binding energies and acidity constants. In Fig. 5-38 is found the structure of fluoroethane with atom labels, and Table 5-5 gives the MMX-derived charges and the Gasteiger–Marsili charges for this molecule.[452,561]

The ESP and RESP electrostatic models used in the AMBER force field were introduced in Section 5.4.3.[418,432,433,434] In the ESP model, an *ab initio* calculation using the 6-31G** basis set is first carried out on the molecule of interest. Then, the Connolly surface algorithm (see Section 6.2) is used to con-

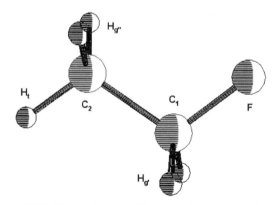

Figure 5-38. Fluoroethane, with Atoms Marked for Charges.

TABLE 5-5. Charges Calculated for the Fluoroethane Molecule by the MMX and Gasteiger–Marsili Methods

Method	F	C_1	$H_{g'}$	C_2	$H_{g''}$	H_t
MMX	−0.28	+0.28	0.0	0.0	0.0	0.0
Gasteiger–Marsili	−0.1777	+0.0874	+0.0254	−0.0367	+0.0254	+0.0254

See Fig. 5-38. F gives the charge on the fluorine atom; C_1 gives the charge on carbon 1; $H_{g'}$ gives the charge on the *gauche* hydrogens attached to C_1; C_2 gives the charge on carbon 2; $H_{g''}$ gives the charge on the *gauche* hydrogens attached to C_2; and H_t gives the charge on the *trans*-situated hydrogen attached to C_2. The MMX charges are those calculated in PCMODEL; the Gasteiger–Marsili charges are those derived from the PEOE method.[452,561]

struct a molecular surface at a selected radius, and the electrostatic potential from the quantum calculation is superimposed upon this surface. An initial set of atom-centered partial charges is arranged within this electrostatic surface, and these charges or their locations are varied in a least squares optimization to achieve the best fit with the surrounding electrostatic surface.

The two-stage Restrained Electrostatic Potential (RESP) fitted charge model was developed to address the problem that the ESP charges had a conformational dependence (compare H_g vs. H_t in Table 5-6) and the observation that these charges were often too large when compared with experimental results. Briefly, in the first stage, ESP charges are calculated with no restraints on hydrogen atoms and weak restraints on nonhydrogen atoms, except that structurally equivalent atoms (e.g. the two oxygens in $HOCH_2CH_2OH$) are constrained to have equivalent charges. In the second stage, the charges are frozen on all nonhydrogen atoms which are not part of a methyl or methylene group; charges on the methyl and methylene group are reoptimized with strong restraints on the carbons and with the constraint of equivalence on the hydrogens of each such group. Application of a 1,4-electrostatic scale factor also improved the model.

In Fig. 5-39 is found the structure of methanol with atom labels, and Table 5-6 gives the MMX-derived charges, a Mulliken analysis of the SCF/6-31G* results, the Del Re charges,[543] the ESP[433] and two-stage RESP[418,434] charges based on the same SCF/6-31G* results, the QEq data from the charge equilibration algorithm of Rappé and Goddard,[549] the QEq-PD data from the algorithm of Bakowies and Thiel[552] based on charge equilibration and potential-derived charges, and the charges derived by the CHARGE2 program of Abraham.[542] Note in particular the conformational dependence for the charge on hydrogens (compare H_g vs. H_t in Table 5-6) for the Mulliken, ESP, and QEq charge distributions.

The inclusion of electrostatic terms for the hydrocarbon portions of molecules (i.e. for hydrogen and carbon), instead of solely for the heteroatom substituents, has been proposed as a simplification which would standardize the treatment of nonbonded repulsions across different force fields.[562] This

324 THE MOLECULAR MECHANICS FORCE FIELD

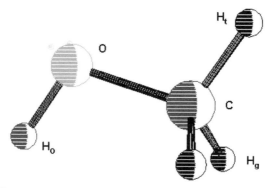

Figure 5-39. Methanol, with Atoms Marked for Charges.

approach has met with some failures,[563] and the apparent conclusion is that saturated hydrocarbons are better modeled without an electrostatics component in the force field. The inclusion of charges for (sp^2)C—H does improve the nonbonding interaction model, however, especially in aromatic moieties, and this is implemented in the MMX force field (see Section 5.2).

More sophisticated electrostatics models would include polarizability factors and the contributions from higher atomic multipoles, but this is not generally done.[314] The CHARMM force field (see Section 5.4.4) has an

TABLE 5-6. Charges Calculated for the Methanol Molecule by Various Methods

Method	H_o	LP	O	C	H_g	H_t
MMX	+0.29	−0.14	−0.14	+0.13	0.0	0.0
Mulliken	+0.4383		−0.7261	−0.1747	+0.1447	+0.1732
Del Re	+0.318		−0.472	−0.011	+0.055	+0.055
ESP	+0.4233		−0.6680	+0.1955	−0.0038	+0.0568
RESP	+0.4215		−0.6498	+0.1166	+0.0372	+0.0372
QEq	+0.36		−0.66	−0.15	+0.14	+0.18
QEq-PD	+0.400		−0.613	+0.011	+0.068	+0.068
CHARGE2	+0.329		−0.458	−0.036	+0.055	+0.055

See Fig. 5-39. H_O gives the charge on the hydroxyl hydrogen atom; LP gives the charge on the lone pair pseudoatom; O gives the charge on the oxygen atom; C gives the charge on the carbon atom; $H_{g'}$ gives the charge on the *gauche* hydrogen atoms; and H_t gives the charge on the *trans*-situated hydrogen atom. The MMX charges are those calculated in PCMODEL; the Mulliken charges are those derived from an analysis of the SCF/6-31G* calculations[418,434]; the Del Re charges are derived by the corresponding method[543]; the ESP charges are those derived from the SCF/6-31G* calculations, as described in the text[433]; the RESP charges are those derived from the SCF/6-31G* calculations, as described in the text[418,434]; the QEq data are from the charge equilibration algorithm of Rappé and Goddard[549]; the QEq-PD data are from the charge equilibration-potential derived charges algorithm of Bakowies and Thiel[552]; and the CHARGE2 charges are those derived by the method of Abraham.[542]

extended electrostatics option which does employ electric quadrupole moments.[439] Most force field formulations have eliminated the use of lone pair pseudoatoms, although MMX retains them for modeling the torsional interactions with oxygen and nitrogen, and they are essential components of the hydrogen bonding potential (see Section 6.6) and the metal-atom-coordination potential (see Section 6.7).

There are two PCMODEL tools and commands which deal directly with electrostatics, Options\Dielc and Options\DP_DP. In addition, Mark\Reset has an 'Atomic Charge' option which allows applied charges to be removed, and the two surface commands, View\CPK_Surface and View\Dot_Surface, have options to build an electrostatic surface (see Sections 3.6 and 6.2).

The Options\Dielc command allows the user to change the value used in the force field calculation for the dielectric constant of the medium to a value other than the default of 1.5. Activating this command brings up a simple dialog box for changing the dielectric constant, as shown in Fig. 5-40. Entering a numerical value in the text input box will cause the electrostatic calculation to use that value for the dielectric constant.

It is rare for increases in the dielectric constant to reproduce accurately the properties of molecules in higher dielectric solvents, so this option should be used with care. If explicit solvent molecules are employed, Dielc should be set to 1.0.

The Options\DP_DP command allows the user to change the model employed by the MMX force field to calculate the electrostatic contribution to the energy from the charge–charge (Coulomb potential, default) model to the bond-dipole–bond-dipole model (Jeans formula). Activating this command brings up a simple dialog box for changing the electrostatic model, as shown in Fig. 5-41. Selecting one or the other option box will invoke the corresponding electrostatic model for ensuing energy calculations.

Electrostatics Exercise 2: The Effect of the Electrostatics Model on the Conformational Energy of 1,2-Dichloroethane. In this experiment, we will examine the difference in energy between the two conformers of 1,2-dichloroethane,

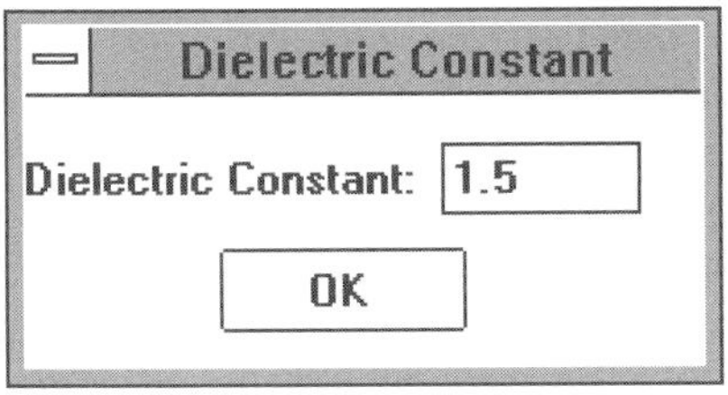

Figure 5-40. The 'Dielectric Constant' Dialog Box.

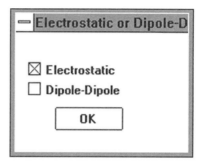

Figure 5-41. The 'Electrostatic or Dipole-Dipole' Dialog Box.

{DICLETHN.PCM}, as a function of the dielectric constant of the medium and the electrostatic model employed, charge–charge vs. dipole–dipole.

Open the input file {DICLETHN.PCM}, found on the input file library diskette in the CHAP3\SECT6 subdirectory, and select the *anti* conformer. Run the structure through a minimization sequence by activating **Analyze\Minimize**, and repeat the sequence a few times. Note the results given in the 'Output' window, which will resemble that depicted on the left side (a) of Fig. 5-42: the 'MMX Energy' is given as -4.636 kcal/mol, the electrostatic component ('QQ') is given as -5.822 kcal/mol, and the dipole moment ('Dip Moment') is given as 0.014 D.

Repeat this procedure for the *gauche* isomer contained in {DICLETHN.PCM}. The results given in the 'Output' window will resemble that depicted on the right side (b) of Fig. 5-42: the MMX energy is given as -2.925 kcal/mol, the electrostatic component (QQ) is given as -4.966 kcal/mol, and the dipole moment is given as 3.044 D. These data will be collected in Table 5-7.

Now, open the original *anti* conformer of {DICLETHN.PCM} again, and then activate the **Options\DP_DP** command and select the 'Dipole–Dipole' choice. Run through several courses of energy minimization, and note the values given in the 'Output' window for the 'MMX Energy', 'QQ', and 'Dip Moment'. Carry out the same calculations on the *gauche* conformer of {DICLETHN.PCM} while the 'Dipole–Dipole' model is still in effect.

These four calculations—the *anti* conformer and the *gauche* conformer with each of the two electrostatic models, the default charge–charge and the dipole–dipole—should also be carried out with other values for the dielectric constant. For example, the *anti* conformer of {DICLETHN.PCM} should be opened again. Then, the **Options\Dielc** command should be activated, and a value of 5 substituted for the default value of 1.5. Next, the energy minimization should be carried out, and the values for the 'MMX Energy', 'QQ', and 'Dip Moment' should be noted. Then, the same procedure should be used for the *gauche* conformer, and then for the *anti* and *gauche* conformers under the dipole–dipole model.

5.6. ELECTROSTATICS AND SOLVATION

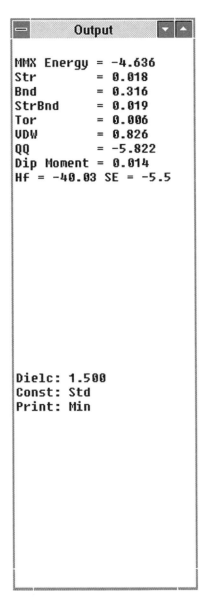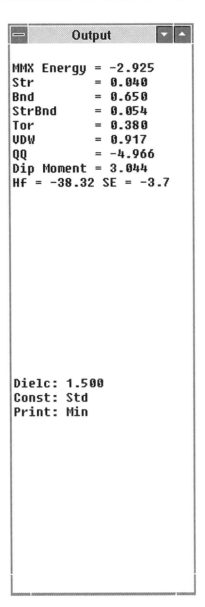

Figure 5-42. Output Summaries for the Energy Minimization of 1,2-Dichloroethane: (a) *anti* Isomer; (b) *gauche* Isomer.

Finally, all four calculations should be carried out with a dielectric constant of 10. The data from all twelve of these runs are collected in Table 5-7, along with the ΔE values showing the difference between the *anti* and *gauche* conformers. Among the trends which come out, the *anti* conformer is always the more stable of the two, although the difference is slightly less under the dipole–dipole model. The difference becomes smaller, and the values for the two electrostatic models approach each other, with a higher effective dielectric constant. A similar trend is seen when only the electrostatic energy contributions are considered.

The most surprising trend is in the dipole moments for each of the conformers, for which the two models make quite different predictions. The charge–charge model predicts a negligible dipole moment for the *anti* con-

TABLE 5-7. Energy Differences Between *anti* and *gauche* Conformations of 1,2-Dichloroethane

Dielectric constant D	Electrostatic model	$E_{tot/anti}$	$E_{tot/gauche}$	ΔE_{tot}	ΔE_{obs}		
1.5	QQ	−4.636	−2.952	−1.711	−1.20		
	DD	3.253	4.190	−1.657	−1.20		
5.0	QQ	−0.567	0.538	−1.105	−0.42		
	DD	1.797	2.896	−1.099	−0.42		
10.0	QQ	0.303	1.281	−0.978	−0.31		
	DD	1.485	2.460	−0.975	−0.31		

Dielectric constant D	Electrostatic model	$E_{elec/anti}$	$E_{elec/gauche}$	ΔE_{tot}	Dipole moment (D)	
					gauche	anti
1.5	QQ	−5.822	−4.966	−0.856	3.044	0.014
	DD	2.080	2.869	−0.789	0.882	1.612
5.0	QQ	−1.740	−1.486	−0.254	3.069	0.00
	DD	0.624	0.873	−0.249	0.903	1.616
10.0	QQ	−0.870	−0.743	−0.127	3.074	0.00
	DD	0.312	0.437	−0.125	0.908	1.612

The two electrostatic models are QQ for charge–charge and DD for dipole–dipole. The dielectric constant D is varied from the default value of 1.5 up to 10. $E_{tot/anti}$ is the MMX energy for the *anti* conformer; $E_{tot/gauche}$ is the MMX energy for the *gauche* conformer; ΔE_{tot} is the difference in MMX energy, $E_{anti} - E_{gauche}$. ΔE_{obs} is the experimentally observed energy difference in solutions of the indicated dielectric constant, for comparison.[564] $E_{elec/anti}$ is the electrostatic component of the MMX energy for the *anti* conformer; $E_{elec/gauche}$ is the electrostatic component of the MMX energy for the *gauche* conformer; ΔE_{elec} is the difference in the electrostatic component of the MMX energy, $E_{anti} - E_{gauche}$. All energy values are in kilocalories per mole. The dipole moments are in debyes.

former, consistent with theory,[564] while the dipole–dipole model predicts a higher dipole moment for the *anti* than for the *gauche* conformer. This is clearly a problem with the dipole–dipole model in MMX.

5.6.2. Solvation

There are a number of published general discussions of solvation available.[115,117,120,474,535,565–567] Classical molecular mechanics calculations model the molecule in the gas phase, which is not always a good predictor for solution or solid phase behavior. In MMX, only a crude solvation treatment is possible by alteration of the value used for the dielectric constant of the medium (D), which would model behavior in a different solvent which has the same dielectric constant. One concern is that the bulk dielectric constant is the one available from measurements, but it is the effective dielectric constant which is appropriate. There can be large discrepancies between the bulk and effective dielectric constants. In addition, the alteration of the dielectric constant is valid only when the distance between the atomic charges or bond dipoles of interest is large compared to the distance between neighboring charges or the magnitude of the dipole. This is only rarely the case, because the medium in between these two charges or dipoles of interest is most often the rest of the molecule, not solvent molecules. The default value for D used in PCMODEL is 1.5, in order to model a "generic" hydrocarbon skeleton. It is rare for increases in the dielectric constant to reproduce accurately the properties of molecules in higher dielectric solvents, so they should be used with great care.

The modeling of solvation effects has evolved to an advanced level, particularly for aqueous solutions with the obvious application to biological systems. A brief discussion here will serve to introduce the topic, but a more complete treatment is beyond the scope of this book. For the sake of simplicity, we will consider two approaches: the explicit solvent model and solvent continuum model. A single example has been chosen arbitrarily for each.

5.6.2.1. Explicit Solvent Model In this solvation model, explicit solvent molecules are employed to form an ensemble with the solute(s) of interest. The system is generally run through a molecular dynamics or Monte Carlo simulation (see Section 5.9) in order to mimic the behavior in solution. Due to the computational complexities in such multibody systems, this type of solvation model is generally limited to the force field level of theory. The transferable intermolecular potential functions (TIPS; see Section 5.4.7) of the Jorgensen group will be used to exemplify this model.

Evaluating the energy in an explicit solvent model simulation depends upon the number of interatomic interactions (a function of the internuclear distances). For this reason, the united atom format is used as much as possible for the solvent molecules, such as the carbons plus attached hydrogens of alkyl groups. As was seen in Section 5.4.7, a nonbonding potential consisting of the

6–12 Lennard–Jones components and a Coulomb component is used for these solvent interactions:

$$V_{\text{TIP}} = q_I q_J \frac{e^2}{r_{IJ}} + \frac{A_I A_J}{r_{IJ}^{12}} - \frac{B_I B_J}{r_{IJ}^6}$$

where q_I and q_J are the charges on the respective atoms I and J, in units of atomic charge; e^2 is an electrostatic force constant with the value 332.17752 kcal · Å/mol; r_{IJ} is the distance between atoms I and J, in angstroms; A_I and A_J are steric repulsion parameters for atoms I and J, in [kcal · Å^{12}/mol]$^{1/2}$; and B_I and B_J are steric attraction parameters for atoms I and J, in [kcal · Å^6/mol]$^{1/2}$.

Thus, for each "atom" (i.e. interaction point) in the solvent molecule, there will be three parameters: the Lennard–Jones repulsive parameter A, the Lennard–Jones attractive parameter B, and the charge q. The Lennard–Jones parameters for acidic hydrogens are set to 0.0 for reasons discussed previously. All of the parameters underwent extensive optimization against empirically obtained data and the results of *ab initio* level calculations for solvent dimers.

Generally for these simulations, 125–250 solvent molecules are employed, depending upon the size of the solute.

5.6.2.2. Solvent Continuum Model In this solvation model, the medium is calculated as a statistical continuum, without the use of explicit solvent molecules. This continuum may be used in molecular dynamics or Monte Carlo simulations. The **GB/SA** (generalized Born/surface area) solvent continuum treatment developed by Still and coworkers[568] will be used to exemplify this model.

The solvation free energy (G_{solv}) is decomposed into three components: the solvent–solvent cavity term (G_{cav}), a solute–solvent van der Waals term (G_{vdW}), and a solute–solvent electrostatic polarization term (G_{pol}):

$$G_{\text{solv}} = G_{\text{cav}} + G_{\text{vdW}} + G_{\text{pol}}$$

The solvent–solvent cavity term describes the energy it takes to create a cavity within the bulk solvent to be occupied by the solute. The solute–solvent van der Waals term covers the steric interactions between the solute and the solvent, and the solute–solvent electrostatic polarization term covers the electrostatic interactions between the solute and the solvent.

The quantity $G_{\text{cav}} + G_{\text{vdW}}$ may be determined as the sum over all atom types I of the terms $\sigma_I \text{SA}_I$, where σ_I is an empirical atomic solvation parameter for atom type I, and SA_I is the total solvent-accessible surface area (whence the SA part of the model's name) for atom type I.

The solute–solvent electrostatic polarization term G_{pol} contains a Born equation contribution and a Coulomb contribution. This Coulomb component accounts for the effect of the dielectric medium on the pairwise interactions of charged particles, less the *in vacuo* Coulomb contribution. Combining the solution phase Coulomb term with the Born term gives rise to the *generalized* Born equation (the GB part of the name):

$$G_{\text{pol}} = -166\left(1 - \frac{1}{\varepsilon}\right) \sum_{I=1}^{n} \sum_{J=1}^{n} \frac{q_I q_J}{f_{\text{GB}}}$$

where ε is the dielectric constant of the medium, q_I and q_J are the charges on the interacting atoms I and J, f_{GB} is a function of the internuclear distance r_{IJ} between atoms I and J, and α_I and α_J are the Born radii for the respective atoms:

$$f_{\text{GB}} = (r_{IJ}^2 + \alpha_{IJ}^2 \, e^{-D})^{0.5}$$

$$\alpha_{IJ} = (\alpha_I \alpha_J)^{0.5}$$

$$D = r_{IJ}^2/(2\alpha_{IJ})^2$$

The Born radius α_I for an atom I is calculated by solving the Born equation for that atom alone, evaluating G_{Born} by the finite difference method, and assuming that all other atoms are neutral and displace the dielectric:

$$G_{\text{Born}} = -166\left(1 - \frac{1}{\varepsilon}\right)\frac{q_I^2}{\alpha_I}$$

The results of the GB/SA continuum solvation model were compared by the developers with the corresponding results from the free energy perturbation method and with experimental measurements, and have been shown to give good predictions of the properties of polar molecules in solution.

Other studies have been published in which the GB/SA continuum solvation model provides results consistent with those of explicit solvation calculations and with experiment, at much reduced computation time. For example, a conformational study of 1,2-dimethoxyethane[569] and a method for determining cyclic peptide conformation[570] have demonstrated the advantage of this solvation model. On the other hand, drawbacks of this methodology have also been pointed out.[571]

Two of the terms in the equation for the GB/SA solvation energy equation involve the solvent-accessible molecular surface. Molecular surfaces are discussed in Section 6.2, and some examples were provided in Section 3.6. Several of these molecular surface formats are important in the context of electrostatics and solvation, such as the solvent-accessible surface and the electrostatically coded surface. These surface display formats are available under the commands **View\CPK_Surface** and **View\Dot_Surface**. Electrostatic surfaces are available in both formats (CPK and dot surface) by selecting the radio button titled 'Charge' in the appropriate dialog box, and a solvent-accessible dot surface is available by selecting the radio button titled 'Water'. For the electrostatic surfaces, the color code is like the one for standard pH paper: dark blue is very electronegative, light blue and green is less so, and white is neutral; yellow is slightly positive, light red is more so, and dark red is very electropositive. If the molecule has no or few bond dipoles, the electrostatic surface will look the same as the dot cloud surface. The electrostatic CPK surface will be demonstrated with a brief exercise.

Electrostatics Exercise #3: Electrostatic Surface for Formamide. In this exercise, an electrostatic surface will be generated for the formamide molecule.

In the PCMODEL window, activate the **Draw** tool, and with the mouse cursor draw in the heavy atoms of formamide, like a broad V with one leg as a double bond. Activate the **PT** tool, and replace the terminus of the double bond with an oxygen and the terminus of the single bond with a nitrogen. The result should resemble the display shown in Fig. 5-43.

Click on the **H/AD** tool to add hydrogens and lone pairs. Run the resulting molecule through several cycles of energy minimization with the

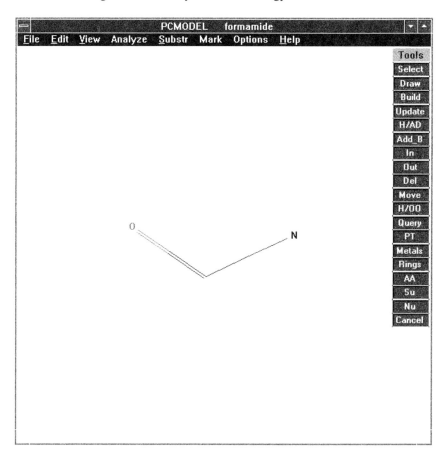

Figure 5-43. Drawing the Formamide Molecule.

5.6. ELECTROSTATICS AND SOLVATION 333

Analyze\Minimize command. If desired, the structure may be given a descriptive title, such as 'formamide', with the **Edit\Structure Name** command. Orient the molecule for easy viewing by activating the **Select** tool and then marking in succession the carbonyl carbon, the nitrogen, and the nitrogen-bound hydrogen which is the closer to oxygen. Next, activate the **Edit\Orient_XY_Plane** command, then double click on **Select** to remove the atom marker dots. Since the molecule has been resized to fill the work space, it will be helpful to shrink the scale a bit with the **View\Control_Panel** command, 'Scale' scroll bar.

To get an idea of the molecular surface of formamide, use the **View\Dot_Surface** command to build a default van der Waals surface (the 'VdW' option), the result of which is depicted in Fig. 5-44. On a color monitor, the surface dots will be color-coded according to their parent atom.

Now, remove the dot cloud surface (**View\Dot_Surface**, 'Cancel'), and build the electrostatic CPK surface by activating **View\CPK_Surface** and selecting

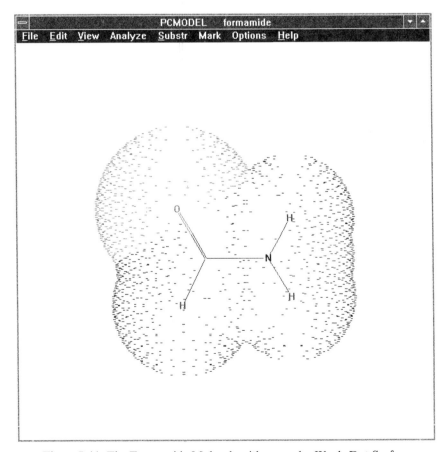

Figure 5-44. The Formamide Molecule with a van der Waals Dot Surface.

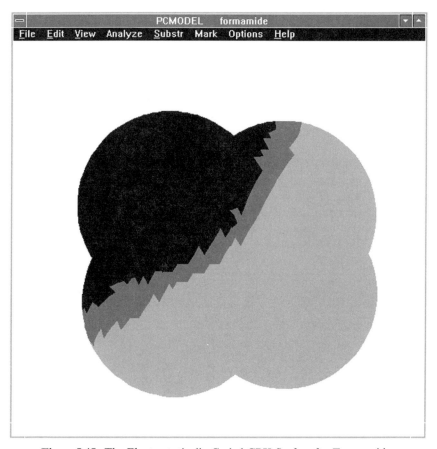

Figure 5.45. The Electrostatically Coded CPK Surface for Formamide.

the 'Charge' option. Figure 5-45 gives a rendering of the result in gray tones, but the actual display should follow the pH paper color code mentioned above, with the basic, electronegative carbonyl oxygen appearing deep blue, the acidic, electropositive amide hydrogens appearing deep red, and intermediate colors in the transition zone.

5.7. ENERGY CALCULATIONS

The immediate objectives of any molecular modeling experiment are to obtain an optimum geometry for the molecule of interest and to calculate its energy. These two goals are linked, in that we generally wish to have the energy for the optimum geometry. The energy calculation is a function of the force field employed, since each of the force field equations contributes to the overall energy.

Almost any molecular geometry will yield a value for the energy. With the results of an initial energy calculation available, one of many minimization engines (or minimizers) may be employed to drive the geometry toward an optimum and the energy toward a minimum value. In this section, we will discuss first the single-point energy, that is, the energy calculated on the basis of an input geometry with no attempt at optimization. Next, energy minimization and the operation of different minimizers will be covered. Lastly, the topic of minimum-energy structures is examined.

Energy calculations[11,104,111–113,115,118,157,178,304,530,572–575] are central to molecular mechanics modeling, and the topic has been reviewed on various levels of detail.

5.7.1. The PCMODEL Output Window

In PCMODEL, when either the **Single Point Energy** or the **Minimize** command is invoked, a child window entitled 'Output' is spawned which monitors the progress of the energy calculation and presents a summary of the results at the end. The 'Output' window has been discussed before, in Sections 3.2.6 and 5.6.

Examples of the use of the 'Output' window will be provided in the next subsections, and some simple guidelines will be provided here. The 'Output' menu window may be moved about the screen; it is not limited by the confines of the PCMODEL window, and it may be resized to an extent by dragging its edge. However, resizing the 'Output' window should be done cautiously, since if it is downsized beyond a critical minimum it cannot communicate with the parent window. In this case the Windows environment will freeze and the only recourse will be a reboot. This can also happen if the 'Output' window is iconified by clicking on its Windows 'Minimize' button. When the PCMODEL program is iconified, the 'Output' window follows faithfully. The 'Output' window can be closed at any time when a calculation is not in progress. It is not necessary to close the 'Output' window between minimization sequences, as it will refresh with each new sequence. The 'Output' window gives the results from the commands **Analyze\Minimize**, **Analyze\Single Point E**, and **View\Compare**. The contents of the 'Output' window may be printed with or without the structure window (i.e. the work space) by making the appropriate radio button selections from the 'Print' dialog box (see Fig. 3-48). These options are titled 'Energy Window' and 'Compare Window'.

It is also important to ensure that there are no lost fragments or atoms present during an energy calculation, a result of incomplete cleanup after editing the structure(s) of interest. Such lost fragments will affect the energy calculations and complicate their interpretation.

If the total energy is labeled 'MMX ENERGY', then no generalized parameters are in use. If however it is labeled 'MM ENERGY', then at least one generalized parameter was used. A report will also be seen at the bottom of the energy summary above, 'Gnrl Param'. The {PCMOD.OUT} file will indicate those parameters as 'GENERALIZED'.

It is generally a good idea to run through at least two or three cycles of energy minimization, in order to give the system a chance to wander away from an energy saddle point, and to ensure a good local energy minimum.

5.7.2. Single-Point Energy

The single-point energy (**Single Point E**) calculation takes the input geometry and directly applies the force field equations to produce a value for the energy, without any geometry optimization or energy minimization. While not very useful in its own right, this type of calculation is generally the first step in an energy minimization, to establish the initial state from which the minimizer can proceed. Either a **Single Point E** or a **Minimize** calculation is a prerequisite to launching such operations as **Rot_E** and **Dynam** which need the energy values as input. **Single Point E** is also useful for generating energy values for previously minimized structures more rapidly than running them through a full energy minimization sequence, or for carrying out an energy calculation on a geometry obtained from some other source (such as a crystal diffraction structure) for comparison purposes.

A variant of **Single Point E** is **Rot_E**, or rotational energy. This command performs a single-point energy calculation at a number of steps of rotation of a specified bond. This command is discussed in Section 6.3.1.

With a structure in the work space, activating the **Analyze\Single Point E** command initiates the energy calculation. The input file is first read and analyzed in terms of the molecular interactions for which the force field equations will be used to make the energy calculation. Next, the standard parameter files are read, along with any added parameter files whose presence has been flagged (see Section 4.7). If there are any interactions which are not covered by specific or general parameters in the available files, an error message is generated and the calculation is aborted. Finally, the energy components of all of the appropriate molecular interactions are calculated, and the results are displayed in a standard 'Output' window. This will be exemplified in a simple exercise.

Single-Point Energy Calculation Exercise: In this exercise, an input structure is prepared for di-t-butyl ether, and the energy is calculated.

In the PCMODEL window, activate the **Draw** tool, and sketch in a broad, inverted V. Bring up the 'Periodic Table' menu, pick up an oxygen, and place this on the central atom of the inverted 'V.' Click on the **H/AD** tool to add hydrogens and lone pairs. Use **View\Control_Panel** to readjust the perspective for better viewing if necessary. Next, activate the **Build** tool, and click on each of the six methyl hydrogens of the dimethyl ether molecule, which converts each to a methyl group itself. The structure may be titled with a name such as 'di-t-Butyl_ether' if desired. This structure should resemble that shown in Fig. 5-46.

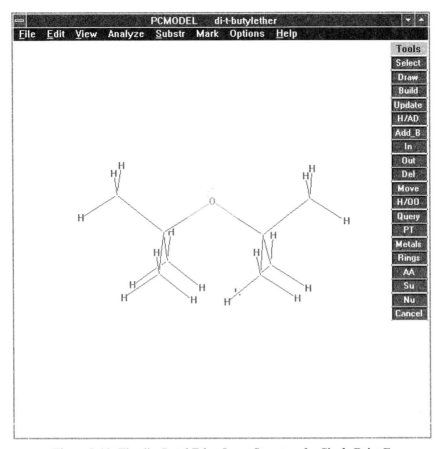

Figure 5-46. The di-*t*-Butyl Ether Input Structure for **Single Point E.**

Now, activate the **Analyze\Single Point E** command. The 'Output' child window pops up and provides the energy summary, as in Fig. 5-47.

This structure should be saved to a file with some descriptive name such as {TBU2O.PCM}, for use in the next exercise.

The values reported in the 'Output' window have the following meanings. The 'MMX Energy' is the total energy calculated from the MMX force field equations. The 'Str' is the component of the total energy due to bond stretching (see Section 5.1.1.1). The 'Bnd' is the component of the total energy due to angle bending (see Section 5.1.1.2). The 'StrBnd' is the component of the total energy due to stretch–bend cross terms (see Section 5.1.2.1). The 'Tor' is the component of the total energy due to torsional interactions (see Section 5.1.1.3). The 'VDW' is the component of the total energy due to nonbonding van der Waals interactions (see Section 5.1.3.1). The 'QQ' is the component of

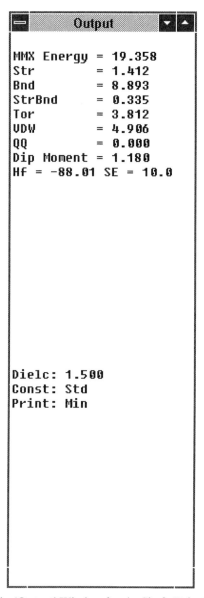

Figure 5-47. The 'Output' Window for the **Single Point E** Calculation.

the total energy due to nonbonding electrostatic interactions (see Section 5.1.3.3). The 'Dip Moment' is the value calculated for the dipole moment (see Section 5.6). The 'Hf' is the value calculated for ΔH_{form} by summing the energy, the bond enthalpy, and the partition function contribution (a conformational population increment plus a torsional contribution plus a translation/rotation term). The 'SE' is the strain energy (energy plus environmental corrections). The 'Dielc:' is the value used for the dielectric constant of the medium. The 'Const:' refers to the constants or parameters used, in this case 'Std' or standard. The 'Print:' gives the level of output printout requested, in this case 'Min' or minimum.

Note the relatively high value for the MMX energy of 265.952 kcal/mol, most of which is due to 'VDW', 252.55 kcal/mol. This is because of the unrelieved crowding of the two *t*-butyl groups.

5.7.3. Energy Minimization

The energy minimization (**Analyze\Minimize**) sequence begins with the input geometry and the initial energy calculation, and then iteratively modifies the Cartesian coordinates of the atoms of the molecule in such a way that the energy of the system decreases, until a constant minimum energy is reached. The resulting molecular geometry has been at least partially optimized, but is not necessarily the global minimum geometry. There are a variety of minimization algorithms available, which will be discussed briefly below. As mentioned above, either a **Single Point E** or a **Minimize** calculation is a prerequisite to launching such operations as **Rot_E** and **Dynam** which need the energy values as input.

There are several commands available which are related to **Minimize**. One is **Analyze\Dihedral Driver**, which performs an energy minimization at a number of steps of rotation around one or two specified bonds. The **Dihedral Driver** command is discussed in Section 6.3.2. Another variant is the **Analyze\Batch** command, which will carry out energy minimizations consecutively on all or some of the components of an appended format input file. A related command is **Substr\Don't Minimize**, which exempts a designated substructure from undergoing energy minimization. This command was discussed in Section 3.2.6.

When the **Analyze\Minimize** command is activated, the PCMODEL energy minimization engine begins to operate on the contents of the work space, whether it is one or more molecules. The input file is first read and analyzed in terms of the molecular interactions which will be subjected to energy minimization. PCMODEL uses the block-diagonal Newton–Raphson minimization algorithm, which is described below. Next, the standard parameter files are read, along with any added parameter files whose presence has been flagged (see Section 4.7). If there are any interactions which are not covered by specific or general parameters in the available files, an error message is generated and the calculation is aborted. Then, the energy components of all of the appropriate molecular interactions are summed for the initial energy calculation, and the results are displayed in the 'Output' window. The progress of the energy minimization will be reported in this window, and its final results appear at the conclusion of the process. The minimization may be interrupted by hitting the ⟨Esc⟩ key on the keyboard.

Two new files are created whenever the **Minimize** command is used. One is {PCMOD.BAK}, a PCM-format structure file which contains the final or most recent structure for which the energy has been calculated. The second is {PCMOD.OUT}, the output file which contains a record of the progress of the minimization, energy data, etc. The user can select between a "minimum" and a "full" level of detail in this output (see Section 3.5). If more than one molecule is processed in a PCMODEL session, the {PCMOD.BAK} file will overwritten with each new structure, while the {PCMOD.OUT} file will contain all of the minimization records or logs for each structure processed, each succeeding entry appended to the existing {PCMOD.OUT} file. If the user wishes to keep these files, they should be renamed before further energy calculations are carried out.

The use of the **Minimize** command will be exemplified in a simple exercise.

Energy Minimization Exercise. In this exercise, an input structure is prepared for di-t-butyl ether, and a energy structure is arrived at by calculation.

In the PCMODEL window, read in the file {TBU2O.PCM} prepared in the previous exercise. Activate the **Analyze\Minimize** command. The 'Output' child window pops up; near the top of the window is the energy breakdown described in the previous experiment, and below this it provides updates on the progress of the minimization every five iterations, with report lines that look something like

```
Iter   Move   Energy
  5    150    50.000
 10    100    25.000
 "  "   "  "   "  "  "  "
```

where 'Iter' gives the number of iterations processed, 'Move' gives the average atom movement in units of 10^{-15} m (femtometers, or fm), and 'Energy' gives the total MMX energy for the last of the five most recent iterations. If the average atom movement is greater than 999 fm, a value of 999 will be displayed, and if the calculated energy is greater than 999 kcal/mol, a string of asterisks will be displayed in the 'Energy' field. The structure in the work space will be seen to change after each update. After a short time, the minimization will finish. The iteration update reports will cease, and the final energy summary report will remain in the 'Output' window.

The minimum-energy structure should resemble that shown in Fig. 5-48, and the energy summary will be similar to that depicted in Fig. 5-49.

The final structure may be written to the same filename used earlier, {TBU2O.PCM}, by overwriting that earlier version.

Four of the algorithms used in molecular mechanics minimization engines will now be discussed: steepest descent, conjugate gradient, Newton–Raphson, and block-diagonal Newton–Raphson. More details on these techniques may be found in the sources cited at the beginning of this section. In particular, Comba and Hambley provide helpful diagrams illustrating the different minimum-energy search methods.[575]

5.7.3.1. Steepest Descent Algorithm This mathematical method was first applied to molecular mechanics for discovering a minimum-energy molecular structure by K. Wiberg.[57] Following an initial energy calculation based upon the input geometry, the algorithm calculates the derivative of the change in energy with respect to the motion of each of the atoms in the molecule along its Cartesian coordinates. It does this by changing each coordinate by a small amount and evaluating the change in energy which results. Then, that coordinate is returned to its original value, and the next coordinate is changed and the change in energy evaluated. This is continued until all of the coordinates have been examined in this way (one iteration).

From this data is determined the vector direction of steepest descent of the energy for each atom. The atoms are then moved together along their respective vectors of steepest descent, and this process is reiterated as long as the energy continues to decrease. If further travel along a particular vector begins to increase the energy, the movement of that atom along that vector is reversed.

When the change in energy between iterations drops below a threshold level, the resulting structure is judged to occupy an energy minimum. The steepest descent method converges rapidly when far from the equilibrium position, but the approach slows as the atoms draw closer to their equilibrium positions.

5.7.3.2. Conjugate Gradient Algorithm This mathematical method was first applied to molecular mechanics strain energy minimization by Engler, Andose, and Schleyer.[7]

As in the steepest descent, the conjugate gradient algorithm changes each coordinate by a small amount and evaluates the change in energy which results. The difference is that the coordinate remains at its new, more favorable value while the next coordinate is examined. In this way, the conjugate gradient method stores the history of the gradients, which has the effect of pointing the system more precisely toward the minimum energy geometry. Thus,

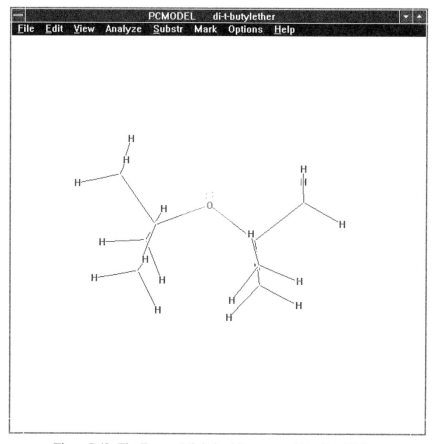

Figure 5-48. The Energy-Minimized Structure of di-*t*-Butyl Ether.

the movement along one vector does not accidentally undo the progress towards the minimum established along another vector. Consequently, the conjugate gradient method converges more rapidly than does the simple steepest descent method.

It is also known as the pattern search method.[104,572]

5.7.3.3. Newton–Raphson Algorithm This method was first applied to molecular mechanics strain energy minimization by Boyd,[58] who employed numerical differentiation. Analytical formulas are used today. This method applies both the first and second derivative of the change in energy in order to guide the system to equilibrium.

For n atoms, the movements of each of which have components along the X-, Y-, and Z-axes, the Newton–Raphson method involves calculating a matrix of second derivatives of dimensions $3n$ by $3n$. This is referred to as the Hessian matrix. In a geometry optimization, only $3n-6$ degrees of freedom

5.7. ENERGY CALCULATIONS 343

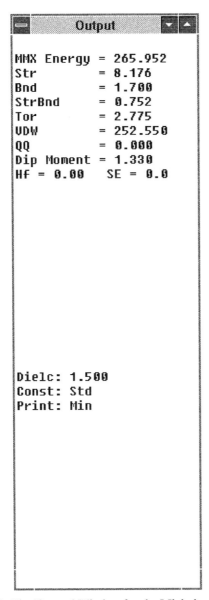

Figure 5-49. The 'Output' Window for the **Minimize** Calculation.

need to be considered, since three translations and three rotations will not affect the energy of the system. After these nonproductive degrees of freedom are subtracted, the matrix is inverted to obtain a lower energy geometry. This process is repeated until convergence to a local minimum.

The Newton–Raphson algorithm performs best on harmonic energy surfaces near the equilibrium point, and does poorly when the system is distant from its equilibrium point. The second derivatives are also computationally more intensive.

The method is also known as the valence force minimization.[104]

5.7.3.4. Block-Diagonal Newton–Raphson Algorithm This method was first applied to molecular mechanics strain energy minimization by Allinger and coworkers.[572,576] It is a simplification of the Newton–Raphson algorithm, which avoids the computational load of calculating the full matrix by operating only upon the 3 by 3 portion of the matrix for each atom. These submatrices are 3 by 3 blocks which lie along the diagonal portion of the full matrix, which gives rise to the "block-diagonal" portion of the name. For a molecule of n atoms, this method requires only $9n$ elements. In comparison with the full matrix Newton–Raphson method, the computer time per iteration will be much less with the block-diagonal. However, the minimization process may require more iterations with the block-diagonal method, because the geometry makes less progress toward equilibrium during each iteration.

This method is also known as nonsimultaneous local energy minimization.[104]

5.7.4. Minimum Energy Structures

There are some important caveats to remember regarding the minimum-energy structures produced in any molecular modeling experiment. Energy minimization is an iterative geometry optimization which seeks a local minimum. Unless there is only a single minimum-energy conformation for a molecule, the final conformation arrived at depends on the geometry of the input structure. The result will be the minimum-energy conformation closest, in some sense, to the input structure. The problems of multiple minima and the finding the global minimum have been addressed in a number of other sources.[304,577]

If the input geometry is flawed—for example, having a planar carbon, overlapping or superimposed atoms, hypervalent atoms, inappropriate placement of hydrogen atoms, etc.—the geometry optimization process may fail to reach a realistic final structure. These flaws in input geometry are not necessarily remedied by the standard minimization engines. The user may notice unexpectedly large values for the energy, in which case the molecular structure should be inspected visually for flaws. Deleting and re-adding the hydrogens (two applications of the **H/AD** tool in PCMODEL) can sometimes help. Running the structure through two or three cycles of energy minimization can also address these problems.

This problem is least troublesome with highly rigid structures, or those with only a handful of obvious conformations. In this case, the appropriate input geometries can be prepared and optimized, and the output can be interpreted with confidence. However, the more degrees of internal rotational freedom exist, the less reliable becomes this simple approach to conformation sampling.

The potential functions of all but the simplest molecules are multidimensional, but it is easier to visualize the problem as a three-dimensional model (energy vs. two spatial coordinates. The potential surface of a molecule over various conformations will have ridges and valleys, hills and wells. The potential well into which the molecule will slide (minimize) is affected by upon which side of which hill it is introduced to the surface.

Ideally, the modeler would like to be aware of all of such wells or energy minima. Some will be local minima, the lowest energy attainable without rolling up over another hill. Some may even be saddle points, metastable positions from which the conformation could minimize further by varying different components. For many applications it is sufficient to sample a reasonable number of manually generated conformations. This approach should involve a judicious analysis of the family of low-energy conformations obtained to determine whether any pertinent conformational space has not been sampled. The chance of being stuck on a saddle point is lessened if each conformation minimization is run twice or three times.

There are no certain ways of arriving at the global minimum; all known methods give an increasing probability with an increasing number of trials, and the time expenditure for a trial increases dramatically with increasing complexity of the molecule. The following section deals with methods of sampling conformational space and the methods used to search for the structure with global energy minimum.

5.8. CONFORMATION SEARCHING AND GLOBAL MINIMIZATION

Conformation searching, or conformational analysis, is probably the single most common goal of computer-aided structure modeling experiments. This technique is used to determine which arrangements in space a molecule will adopt in a relaxed, or low-energy, state. A conformational analysis evaluates all, or more commonly a representative subset of all of the theoretically possible conformations. The lowest energy conformation is referred to as the ground state, or global minimum, and there may be a number of other conformations of reasonably low energy which may also help to rationalize the molecule's properties. A conformation with global energy minimum conformation can be elusive, and is not always self-evident even when in hand—there is no good proof that a particular conformation is at the global minimum. One can only increase the level of confidence by making a good faith effort to locate any lower energy conformations. Careful application of the techniques to be described below will increase the likelihood of defining the scope of the conformational space for a molecule, if not the global minimum.[573]

Understanding the conformational behavior of nonrigid molecules is essential to gaining insight into structure–reactivity and structure–(bio)activity relationships. We will see an example of this in the experiment on NMR in Section 6.4, where a population analysis of 1-chloro-2-butanone will be used to predict a ^{1}H-NMR coupling constant.

Conformational analysis is an area where molecular mechanics has a clear advantage over other calculation methods. The CPU time required is minimal compared to semiempirical and *ab initio* methods. The conformational geometries discovered and the relative energies among conformations determined by molecular mechanics correlate very closely with the *ab initio* results. However, serious discrepancies arise with semiempirical methods such as AM1, which are believed due to an excessive attractive component to the H·····H nonbonded interaction, which underestimates the strain in crowded conformations such as boat cyclohexane.[312]

5.8.1. Strategies and Methods

It is safe to state that search methods for the conformational analysis of flexible systems constitute one of the most popular topics in computational chemistry publications. Many good reviews are available,[115,264,374,573,574,578–583] and a number of experiments have been described which are amenable to use in computer modeling courses.[191,197,198,201,205,221,584,585] In particular, the use of conformation generation and conformational search techniques has become a popular way to search 2D or 3D structure databases for ligands in conformations which satisfy a set of constraints (see Section 3.4).[586,587] This has become an important research strategy for the discovery of novel bioactive substances, and has applications in material science as well.

The report of a collaborative study between three research groups to evaluate methods for conformational searching on cycloheptadecane provides an excellent comparison between the extant methods.[588] The objective was to discover all unique conformations within a 3-kcal/mol envelope. The methods used were the Cartesian stochastic search of Saunders,[89,310,589,590] the torsional tree methods of Houk *et al.* and of Still and Lipton,[591] the torsional Monte Carlo method of Still *et al.*,[592] the distance geometry method (**DGEOM** program) of Blaney, Crippen, Dearing, and Dixon to generate test structures,[593] and molecular dynamics searches with C—H bonds constrained by **SHAKE**[594,595] (see Section 5.9). All of the methods except distance geometry located the same global minimum, and no single method located all 262 conformations, although the Monte Carlo torsional method seemed to locate the most conformations most rapidly.

Several other reports have appeared in which various conformational searching methods are compared.[264,586,587,596,597]

5.8. CONFORMATION SEARCHING AND GLOBAL MINIMIZATION

PCMODEL offers several techniques for conformation searching. As will be discussed in Section 6.3, the **Rot-E** and **Dihedral Driver** operations permit a search over one or two torsional angles. A standalone, DOS environment program called GMMX (Global MMX) is the conformational search module for PCMODEL, and this will be discussed in Section 5.8.2.

We briefly describe here some of the strategies and methods (and their variants) for conformational searching: the systematic search, the stochastic search (torsions or coordinates), molecular dynamics, genetic algorithms, distance geometry, and directed tweak.

5.8.1.1. Systematic Search

The systematic search, or torsional tree method, or brute force method involves the generation of multiple conformations by varying each rotatable torsional angle by a set amount. These conformations are all subjected to minimization, and generally only one from a set of duplicate, equivalent, or mirror image structures is saved. The group of distinct conformations resulting is taken as representing the conformational space for the particular molecule.[591,598–600]

A modification of this method would include the use of geometric constraints on two or more atoms to reduce the number of initial files considered. Examples would be a requirement for ring closure in a RINGMAKER-type search,[477,478] or distance constraints from NMR studies.

The systematic approach was used by the Scheraga group to carry out conformational analysis of 20 amino acid residues, for use in protein modeling.[481]

The Still group has introduced the Systematic Unbounded Multiple Minimum (SUMM) search algorithm.[601] With this method, all rotatable bonds are searched first at low resolution (120°), and when this stage is complete the resolution is incremented and the search repeated. The user has control over how many stages of resolution are explored, and what criteria to use for closing the search.

The program **WIZARD** applies the techniques of artificial intelligence and expert systems to conformational searching.[270,602,603] A more directed approach to the systematic search technique, called the bond-by-bond iterative method, has been reported.[604]

5.8.1.2. Stochastic Search

The **stochastic**, or **Monte Carlo**, search comes in two versions, depending on whether the algorithm operates upon torsional space or coordinates space.

The Monte Carlo torsional search has similarities with the systematic torsional search, except that the torsional arrays and the angle variations are chosen randomly or semirandomly. A random subset of the rotatable bonds is selected, and a random rotational increment is applied. The resulting conformation is subjected to some preliminary tests to screen out severe steric strain or violation of any imposed constraints. If it passes, the structure is subjected to energy minimization, and a determination is made of whether the structure falls within the desired energy envelope, and whether it is a duplicate or equivalent of any existing conformation. Structures which surmount all of these hurdles are kept for final analysis.[570,592] This method works well with acyclic structures or substructures, or with cyclics if the ring is first broken for the application of the torsional displacement, and then a ring-making criterion is employed for acceptance of the resulting conformation.

In the Monte Carlo Cartesian coordinate search, each of the constituent atoms suffers a displacement of randomly determined direction and (within user-defined limits) magnitude, the so-called *random kick*.[89,310,589,590,605] This new arrangement of atoms is subjected to energy minimization, and the resulting conformation then measured against the criteria mentioned above to determine whether it is discarded or kept. This method works particularly well with cyclic or polycyclic structures or substructures.

A variation on the stochastic search methods is the use of the **simulated annealing** approach, which employs the so-called **Metropolis algorithm**, or Metropolis Monte Carlo algorithm.[42,573,606-616] In this method the molecule is placed in a very hot *bath* in which a high degree of conformational flexibility is possible. The conformations generated by random displacements of torsions or coordinates are then sampled and analyzed as above, while the temperature of the bath is decreased. Towards the end of the cooling cycle, the molecule will settle into a conformation of local minimum energy, which is evaluated and either discarded or kept. Then, this conformation is placed in a very hot bath, and the process is repeated as often as the user wishes. The displacements are generated either by a simple random number generator, or by a temperature-dependent Cauchy generating function in which their amplitude decreases gradually as the bath is cooled. The heating cycles allow the structure to visit new conformations by overcoming the energetic barrier that would prevent a transition under normal conditions.

The simulated annealing algorithm is also used in the **Dock** command; see Section 6.8.

5.8.1.3. Molecular Dynamics

The **molecular dynamics** technique is discussed in the following section, but briefly it involves simulating the response of a structure to the laws of Newtonian physics, which determines a *molecular trajectory*, or history of animated movement. These movements allow the molecule to visit different areas of conformational space, the more so the longer the dynamics simulation is run. Sample conformations are pulled periodically, subjected to energy minimization, and evaluated as above.[617]

While there are some similarities to the stochastic search methods, molecular dynamics is deterministic by contrast, in that the same molecule in the same conformation subjected to a dynamics simulation under the same parameters will cover the same molecule trajectory.

There have been attempts to link molecular dynamics to Monte Carlo simulations. The DMC (dynamics Monte Carlo) program generates trial configurations with short bursts of high-temperature dynamics. Constant temperature and simulated annealing search protocols are then applied to work through the conformations generated.

5.8.1.4. Genetic Algorithms Genetic algorithms randomly build *individuals* (sets of solutions) from some encoded forms known as *chromosomes* or *genomes*. The genetic algorithm requires the specification of three operations on objects, each typically probabilistic, called *strings*, which may be real-valued vectors: (1) reproduction, or combining strings in the population to create a new string or offspring; (2) mutation, or spontaneous alteration of characters in a string; and (3) crossover, or combining strings to exchange values, creating new strings in their place. Thus, the genomes of individuals are combined or mutated to breed new individuals, influenced by an ongoing process of competition, or natural selection within populations.

The application to conformation searching requires a molecular graph representation for conformations. Several schemes have been presented for the use of genetic algorithms for geometry optimization and searching for the global minimum.[586,597,618] The individual conformations may be optimized with respect to energy, or to maximizing attractive intermolecular interactions in a docking experiment.

5.8.1.5. Distance Geometry Distance geometry is a method which uses the N by N internuclear distance matrix for the set of N atoms constituting a molecule to generate sets of Cartesian coordinates which satisfy the distance constraints. The constraints may be either specific or generic. The upper right triangle of the matrix is used for the upper bounds, and the lower left triangle of the matrix is used for the lower bounds.

Specific constraints may be derived from experimental data such as NMR NOE studies, or from the requirements of a hypothetical biophore. For the generic constraints, the upper bound for a pair of atoms will be the length of a fully extended chain connecting the two atoms, and the lower bound will be the sum of the van der Waals radii, or, hydrogen bonding distances for polar interactions.

A set of intramolecular distances is then chosen at random, and eigenvalue techniques are used to find the best three-dimensional fit to those distances. Energy minimization of the molecule, either alone or interacting with a second molecule, is then employed to rank the found conformation.[264,593,619]

5.8.1.6. Directed Tweak The **Directed Tweak** is a three-dimensional structure searching technique designed by Hurst to facilitate the discovery of matches between a 3D structure query and accessible conformations of flexible molecules contained in a structure database.[620]

In directed tweak, the rotatable bonds of a conformationally flexible structure are adjusted at search time with a torsional space minimizer, to produce a conformation which most closely matches the 3D structure query. The tolerances of the query, and an assessment of conformational energy based upon the sum of van der Waals interactions, are used to score search hits.

5.8.1.7. Other Methods Among several less widely used techniques for conformational analysis is the mapping of large rings onto the diamond lattice.[621]

An additional technique for ensuring uniform coverage in a conformational search is **poling**, developed by a research group at MSI. Poling may be used with any conformational search routine, and associates a penalty function, or poling function, with each found conformation. The poling function modifies the potential surface by raising the energy drastically in the vicinity of a known conformation in order to avoid discovering another close by on the potential surface. Only those new conformations outside the "penalized conformational space" surrounding known conformations will have sufficiently low energy to be kept.[622–624]

5.8.2. The GMMX Program

The conformational search module for PCMODEL is available as a standalone program for the DOS environment, GMMX for global MMX. GMMX was developed from the work of Still, Saunders, and Houk,[588] and incorporates two Monte Carlo (stochastic) algorithms for searching conformational space and seeking a global energy minimum. The first method, used for an acyclic structure or substructure, carries out random rotations of all or a subset of indicated rotatable bonds. The second, used for a cyclic structure or substructure, delivers random kicks to the Cartesian coordinates of the ring atoms. These two algorithms can be used singly or in combination.

The use of GMMX will be illustrated with two examples: *n*-pentane as a model acyclic molecule, and 2-methyl-1,3-dioxolane as a model cyclic molecule. The preparation of the structure files will be presented very briefly.

Conformation Search Exercise. Building the Test Structures. In the PCMODEL window, activate the **Draw** tool, and from left to right draw a five-atom zigzag pattern. Use the **H/AD** tool to add hydrogens, and then run the resulting structure through several cycles of **Analyze\Minimize**. The final structure should look like that depicted in Fig. 5-50. Write the structure to a file for use in GMMX. This structure file is available on the structure file library diskette in the CHAP5 directory as {PENTANE.PCM}.

In the PCMODEL window, bring up the 'Rings' menu (**Rings** tool) and click on 'C5' to place the cyclopentane structure in the workspace. Use the **H/AD** tool to delete the hydrogens, and use the 'Labels' dialog box (**View\Labels** command) to change the label format to 'Atom Numbers'. Then, bring up the 'Periodic Table' menu (**PT** tool), click on 'O', and convert atoms 1 and 3 in the working structure to oxygen. The label format may then be reset to the default 'Hydrogens and Lone Pairs'. Next, add hydrogens and lone pairs to the structure with the **H/AD** tool, and use the **Build** tool to convert the pseudoequatorial hydrogen on C2 (between the oxygens) to a methyl group. The structure may be rotated for better viewing with **View\Control_Panel**.

5.8. CONFORMATION SEARCHING AND GLOBAL MINIMIZATION 351

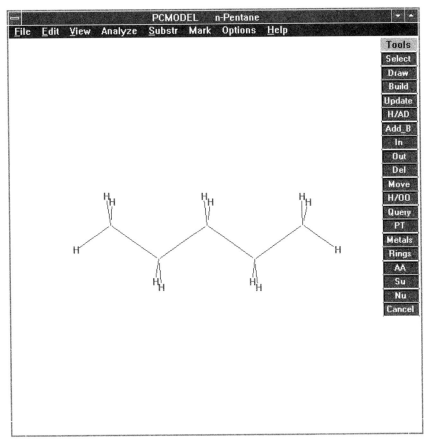

Figure 5-50. The Pentane Molecule.

Run this structure through several cycles of energy minimization. The final structure should look like that depicted in Fig. 5-51. Write the structure to a file for use in GMMX. This structure file is available on the structure file library diskette in the CHAP5 directory as {MEDO.PCM}.

Prior to preparing the GMMX input command files, the atom numbering should be confirmed for each of the structures by changing the label format to 'Atom Numbers'. The numbering should be as shown in Fig. 5-52.

5.8.2.1. The Input File with GMMXINP This section contains a discussion of the operation of GMMXINP, the DOS environment program which prepares an input command file for GMMX from the structure file together with user-supplied parameters.

Once the structure file of interest has been copied to the directory which contains GMMX, the user will need to open a DOS window, or leave the

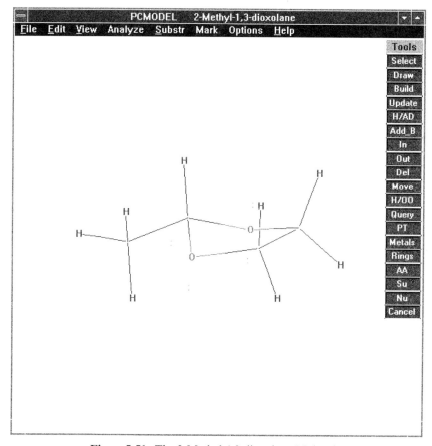

Figure 5-51. The 2-Methyl-1,3-dioxolane Molecule.

Windows environment and return to the DOS environment. The input command file for GMMX is prepared with the GMMXINP program by linking the structure file with a combination of parameters and options selected by the user through an interactive session.

The prompts and explanatory material provided during the course of the GMMXINP process are reproduced and discussed below, for the two cases of {PENTANE.PCM} and {MEDO.PCM}. The data is read in Fortran format, which means that numerical values should be given without spaces, and two or more values given in the same line should be separated by commas. This format requirement is generally provided within the text of the prompts. In the presentation below, the program queries are provided in monospace, left-justified. The user responses for this exercise are also provided in monospace, right-justified, with either the text or numerical data requested, or a ⟨CR⟩ to indicate that the default has been accepted.

5.8. CONFORMATION SEARCHING AND GLOBAL MINIMIZATION 353

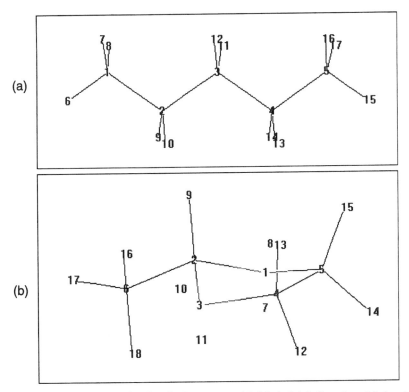

Figure 5-52. Atom Numbering for the Pentane and 2-Methyl-1,3-dioxolane Molecules.

Conformation Search Exercise. Preparing the Input Files: From the DOS prompt in the directory which contains GMMX and associated files, call GMMXINP. This will first bring up a message regarding the use of a **DOS extender**, which enables the large, overlaid programs GMMX and GMMXINP to be run under DOS (see Fig. 5-53). With this extender, each of these programs requires only 184 kB to run.

Next comes the prompt for the name of the structure file, and our response:

```
Enter PCMODEL filename:_
                                                PENTANE.PCM
```

```
32-bit Power for Lahey Computer Systems
Phar Lap's 386|DOS-Extender(tm) Version 5.1
Copyright (C) 1986-93 Phar Lab Software, Inc.
      Available Memory = 12880 Kb
```

Figure 5-53. Screen Image for Phar Lap DOS Extender.

This leads to a message about the preparations for input, and an offer of help for the unfamiliar user:

```
... setting up arrays
Do you need help with the conformational search? [n/y]_
```

An answer of 'y' makes available two screens of information, which are found in Fig. 5-54. These explain how the program and its Monte Carlo algorithm operate, and how to make an informed selection between either or both of the two conformational searching modes available, the random torsional angle search and the Cartesian coordinates random kick search.

After these help screens, or directly after the help prompt when answered with a 'n', a new prompt comes up where the user indicates which search mode(s) to use: the random torsional angle search for an acyclic structure

```
Conformational space can be searched by the statistical method

statistical searching can be done on bonds, coordinates, or
both. The bond method will select at random 25% of all the
rotatable bonds and rotate each one by an angle chosen at
random from the resolution set. To ensure adequate scanning,
if the structure has only 4 or less rotatable bonds, all
of them will be rotated and only the angle will be randomized.
If more than 4 but less than 8, 50% will be selected. The
coordinates method will select at random 25% of all the atoms
and vary their coordinates. (Use this method with rigid structures
and if there are side chains use the both methods option.)

All conformers generated will be processed only if they meet
with the criteria (ie bond distances etc) imposed.
                      Press <cr> to continue
```
```
This subroutine will allow you to generate multiple
conformations of the current (minimized) structure using mmsearch.

For chains, you need to know the atom numbers of
the rotatable bonds and for rings, all the ring atoms.

For conformational searches, the 1st and last
atoms of the contiguous set will define the closure
bond. (These two atoms may be multiply bonded.)

You may specify distance constraints....
in init mode chose fix, then two atoms...
set the distance, ie 2.0 (in a.) and instead of
a constant, give the allowed deviation ie 0.5.

For ring conformational searches, the 1st and last
atoms of the contiguous set will define the closure
bond. (These two atoms should be unsubstituted.)
```

Figure 5-54. The Two Initial Help Screens for GMMXINP.

5.8. CONFORMATION SEARCHING AND GLOBAL MINIMIZATION 355

(bonds), the coordinate random kick search for a cyclic structure (coordinates), or both (the default):

```
Search on bonds, coordinates or both (m) (m,b,c) ?_
                                              b
```

In the case of PENTANE.PCM, we elect the bonds option by entering a 'b'. Next comes a prompt for a seed value to initialize the random number generator for use in the Monte Carlo algorithms:

```
Use a seed value of (large odd i5 integer) (71277) ?_
                                              ⟨CR⟩
```

The default value will generally be selected in this case. After this comes a prompt about the energy envelope within which to retain discovered structures, with a default of 3 kcal/mol, and for the maximum number of iterations ('it/s') for the GMMX run, with a default of 990. If the earlier help option was selected, the prompt is preceded by

```
If only the global minimum conformer is sought,
use a delta E of 1 kcal and an it/s of 990
```

which provides guidance for selecting an energy envelope of 1 kcal/mol if only the global energy minimum structure is desired. The prompt is

```
Delta E and it/s value are to be (3,990) :_
                                              ⟨CR⟩
```

Answering this query brings up another prompt which present the default criteria for ending the search, i.e., the same minimum-energy conformer being found a certain number of times after a certain minimum number of conformers has been evaluated:

```
End the search when the same eminim conformer
has been found 5 times after a minimum of 25
conformers have been minimized or when a duplicate
conformer has been found   12 times in a row
after a minimum of   25 conformers minimized. (y,n) ?_
                                              ⟨CR⟩
```

The numbers used will vary with the complexity of the input structure, so that while 5 is always used for the number of repeats of the minimum-energy structure, the value for the minimum number of conformers is generally five for every heavy atom in the structure file used, and the value for the number of duplicate conformers is half the minimum number of conformers.

The next three prompts ask for the number of rings to be searched, the number of rotatable bonds to be searched, and the default resolution for the latter, with a default of 120°. For the PENTANE.PCM case, we indicate no rings and two rotatable bonds, and accept the default resolution:

```
How many rings do you wish to search on (0,max 4) ?_
                                                         0
How many chain rotatable bonds
    (max 121, 0 to skip, -1 to abort) ?_
                                                         2
Enter default resolution in degrees (120)_
                                                       ⟨CR⟩
```

Now will come the prompts to identify each of the two rotatable bonds by the atom numbers (see Fig. 5-50), and to accept the default value of 120° for the torsional rotation, or supply a new value for the resolution:

```
bond no. 1
Enter atom numbers for bond to be rotated
    (neg value to abort) (2i5) :_
                                                       2,3
Use the default resolution of 120(y,n)
                                                         y
bond no. 2
Enter atom numbers for bond to be rotated
    (neg value to abort) (2i5) :_
                                                       3,4
Use the default resolution of 120(y,n)
                                                         y
```

A prompt then comes up to quantify the minimum acceptable contact distance, with a default of $\frac{1}{4}r_I r_J$, where r_I and r_J are the corresponding van der Waals radii of the two interaction atoms.

```
Enter minimum contact distance
    [0=1/4 sum of vdws] (f5.2) :_
                                                         0
```

GMMXINP then reports the theoretical number of conformers to be screened based upon the information supplied, in this case 10, and prompts an answer for whether to screen for van der Waals 1,5-interactions. Answering 'y' to this prompt only makes sense if there is a chain of at least five-heavy-atoms present in the molecule, which is the case with *n*-pentane.

```
The theoretical number of conformers is: 1.000E+01
```

5.8. CONFORMATION SEARCHING AND GLOBAL MINIMIZATION 357

```
Do you wish to screen for bad 1,5 C-C
vdw interactions (y,n) ?_
                                                y
```

The next prompt comes up with the full screen shown in Fig. 5-55, which address how GMMX will determine which structures are duplicates. There are six options to choose from, which incorporate varying levels of recognition of symmetry and equivalence between the atoms in the subject structures.

Option 0, the default, is to compare the root mean square (RMS) deviations between the positions of the atoms in each structure which are mapped to each other by the atom file numbers. This option recognizes no equivalence or symmetry. It may be modified slightly depending upon the answer to a later prompt on conformational enantiomers (see below).

Option 1 enables GMMX to check for equivalent atoms in the determination of duplicates. Selecting this option brings up the explanatory screen for the **ATEQ** mode shown in Fig. 5-56, with an option to continue or not. The example given is that of treating as equivalent the two oxygens of the anion of propionic acid. Choosing 'y' to continue brings up additional prompts on how many sets of equivalent atoms are to be compared, how many pairs of equivalent heavy atoms are in each set, and the atom file numbers for the equivalent atoms.

Option 2 enables GMMX to check for specific sequences of atoms in the determination of duplicates. Selecting this option brings up the explanatory

```
Conformers are checked for duplication via a super-
imposition routine that will classify as identical
any two structures in which every non-volatile atoms
superimposes to within 0.25 a of its counterpart.

The default checking relies on identical numbering
and atom typing. however, due to complex symmetry
arrangements (cyclic unsubstituted hydrocarbons),
additional modes of checking must be used to
distinguish symmetrical or equivalent isomers from
being typed as unique. These modes are as follows:

compare all non-volatile atoms (default)...........(0)
in addition, check for equivalent atoms............(1)
in addition, check for specific sequences of atoms.(2)
in addition, check for numerical isomers...........(3)
in addition, check for reflection isomers..........(4)
compare only specified non-volatile atoms..........(5)

Compare using option (0)?_
```

Figure 5-55. Configuring the Duplication Check Scheme in GGMXINP.

```
The ATEQ command will function as follows;

                O-
               /4
        CH3-CH2-C
         1   2   3\\
                   O
                   5

By defining atoms 4 and 5 as one equiv pair.
the structure will automatically be compared
against atoms 1,2,3,4,5 and also 1,2,3,5,4

Do you wish to continue with this option? [y/n]_
```
```
How many sets of equiv atoms are to be compared    (max25):_
```
```
How many pairs of equiv heavy atoms are            in set 1:_
```
```
Give the    1 pairs of #'s separating     by commas(max20/line) for
set 1._
```

Figure 5-56. The Description of the Logic Used in Checking for Equivalent Atoms, and Prompts for Designating the Equivalent Atoms.

```
The NSEQ command will function as follows

                 O-CH2-CH2
                /4  5    6  \
     CH2-CH2-C           7  CH2
      1   2   3\           /
                 O-CH2-CH2
                 8  9   10

Although the ATER command may also be used here,
by defining the sequence 1,2,3,4,5,6,7,8,9, and 10
to be the same as 1,2,3,8,9,10,7,4,5, and 6, the
structure will be automatically compared against
atoms 1,2,3,4,5,6,7,8,9, and 10 and 1,2,3,8,9,10,7,4,5
and 6

Do you wish to continue with this option? [y/n]_
```
```
How many sets of heavy atoms are to be compared    (max5):_
```
```
How many main str heavy atoms are in set    1:_
```
```
Give the  **#s for the main seq separating
by commas (max 20/line) for set 1._
```
```
Give the  **#s of the compared sequence._
```

Figure 5-57. The Description of the Logic Used in Checking for Specific Sequences of Atoms.

screen for the **NSEQ** mode shown in Fig. 5-57 with an option to continue or not. The example given is that of treating as equivalent the two different sequences for mapping onto 2-ethyl-1,3-dioxocane. Choosing 'y' to continue brings up additional prompts on how many sets of equivalent atoms are to be compared and how many pairs of equivalent heavy atoms are in each set, then supplying the sequence of atom file numbers for the main sequence and also for the comparison sequence.

Option 3 enables GMMX to check for numerical isomers in the determination of duplicates, specifically in cyclic structures. Selecting this option brings up the explanatory screen for the **NSRO** mode shown in Fig. 5-58. The example given is that of treating as equivalent the eight different sequences for mapping onto the ring atoms of 1,3-dioxocane, as well as the reverse of these mappings and the enantiomeric versions of each of these mappings. There are no additional prompts for this option. If the atom types do not match, the numerical isomer sequence will not be processed.

Option 4 enables GMMX to check for reflection isomers in the determination of duplicates. Selecting this option brings up the explanatory screen for the **NSRF** mode shown in Fig. 5-59. The example given is that of treating as equivalent the two mirror image mappings of dimethoxymethane. If the atom types do not match, the reflection sequence will not be processed.

Option 5 enables GMMX to compare only specified nonvolatile atoms (heavy atoms) in the determination of duplicates. Selecting this option brings up the

```
The NSRO command will function as follows

            O-CH2-CH2
           /2   3    4  \
        1 CH2       5 CH2
           \            /
            O-CH2-CH2
            8 7   6

The structure will be automatically compared
against atoms 1,2,3,4,5,6,7,8 and then against
atoms 2,3,4,5,6,7,8,1 and 3,4,5,6,7,8,1,2 and
4,5,6,7,8,1,2,3 and 5,6,7,8,1,2,3,4 and 6,7,8,1
2,3,4,5 and 7,8,1,2,3,4,5,6 and 8,1,2,3,4,5,6,7,
as well as the reverse order of numbering and
against the enantiomeric relationships for a
total of 32 possibilities each time. If the atom
types do not match, that sequence will not be
processed.

Do you wish to continue with this option? [y/n]_
```

Figure 5-58. The Description of the Logic Used in Checking for Numerical Isomers.

```
The NSRF command will function as follows;

         O-CH2-O
        /2  3  4\
      CH3        CH3
       1          5

The structure will be automatically compared
against atoms 1,2,3,4,5 and then against atoms
5,4,3,2,1. if the atom types do not match, the
reflection sequence will not be processed.

Do you wish to continue with this option? [y/n]_
```

Figure 5-59. The Description of the Logic Used in Checking for Reflection Isomers.

explanatory screen for the **COMP** mode shown in Fig. 5-60. There are additional prompts for how many heavy atoms are to be compared, and to identify the atoms by entering their atom file numbers. By modifying the values put into this option (negative numbers), atoms may be removed selectively from the comparison set.

For this phase of the exercise, we select option 2 to designate equivalent sequences of atom file numbers for the atom–atom mapping. Thus, the additional queries are answered as follows:

```
How many sets of heavy atoms are to be compared
    (max5) :_
                                                              1
```

```
The COMP command should be rarely used and only
with good reason

By defining which atoms are to be used in the
compare routine only these atoms will be compare.

If atoms 1,5,10,11,12,15 and 20 are specified.
then the structure will be compared against atoms,
1,5,10,11,12,15 and 20 only.

if a negative number is given in answer to how many
non-volatile atoms are to be compared, then the
numbers given will be REMOVED from the compare list

Do you wish to continue with this option? [y/n]_

How many heavy atoms are to be compared (min 4), (neg. numb. removes):_

Give the numbers of the ** atoms, separating        by commas (max 20/line)._
```

Figure 5-60. The Description of the Logic Used in Checking for Specific Atom Comparisons.

5.8. CONFORMATION SEARCHING AND GLOBAL MINIMIZATION   361

```
It is possible to include a file of conformers
previously found by other searches or programs

However, the numbering sequence of the non-volatile
atoms in the include file, must be exactly the
same as the input structure. Also, if in the
include file, hydrogens and lone pairs are already
in place, each structure must be made compatible
with the force field to be used. This is very
important if switching between AMBER and MM2.

MDLFTS can be used to automatically process
and convert a MODEL multiple conformer file
derived from BAKMDL, MacroModel, etc. for
compatibility

Include filename (<CR>, exits this option):_
```

Figure 5-61. The GMMXINP Prompt for Including a File of Comparison Structures.

How many main str heavy atoms are in set 1:_
                                                                5
Give the   5#s for the main seq separating
   by commas (max 20/line) for set 1._
                                                          1,2,3,4,5
Give the   5#s of the compared sequence._
                                                          5,4,3,2,1

When this is finished, GMMXINP prompts the user for the filenames of the output files it will produce. The first will contain the final, minimized conformers of the subject structure, and the second will contain the second cycle output, if any. The defaults are to use the filename of the subject structure and append a 1 or a 2, respectively. This is only workable if the original filename is seven characters long or shorter. The user may supply alternative names for these files. In this case, we accept the defaults:

Output filename, no extensions, return for: pentane1
                                                                ⟨CR⟩
2nd cycle output filename
   (no extensions, return for) : pentane2
                                                                ⟨CR⟩

Next comes a screen, seen in Fig. 5-61, explaining the procedure for including a file of reference structures against which to compare the new conformers discovered in the GMMX process. In our case, we answer with a ⟨CR⟩ to decline this option.

Following this, another screen comes up as seen in Fig. 5-62. An explanation is given of the differences between two type of RMS comparison: operating either upon the torsional angles within a 15° window (the structure RMS option), or upon the atomic positions within 0.25 Å (the atom RMS option). GMMXINP suggests two stages of comparison, a first cycle using the faster structure RMS and a second cycle using the more exact atom RMS. It also offers the option of using a faster analytical evaluation or the more accurate explicit evaluation.

In our case, we answer the series of prompts with a ⟨CR⟩, a '1' to employ the structure RMS comparison in the first cycle, a '1' to employ the atom RMS comparison in the second cycle, and a '0' to employ the analytical evaluation.

The next prompt line inquires about the energy window within which two structures should be presumed identical and subjected to comparison. The

---

Internal coordinates of non-volatile atoms will
be used to screen isomers within a 15 deg window,
or by a superimposition of each non-volatile atom
to atom with an RMS value of 0.25 Angstroms. If
you turn off the internal coordinate check (set
window to -1 deg) then for certain structures,
the 0.25 Angs criteria may be too strict and
cause many closely overlapping conformers to be
unnecessarily kept.
To avoid this, you may wish to relax the checking
to a structure RMS compare value instead. A fast
analytical comparison routine is defaulted to.
With structures that use the ATEQ, NSEQ, NSRO,
or NSRF commands, some duplicates may be kept.
If this becomes a problem, use the MUCH SLOWER
but MORE ACCURATE explicit method in a third cycle
rather than conducting the search directly with
the explicit option.
Begin comparing after conformer (0):_

For first cycle
Compare based on Atom RMS (0) or Structure RMS (1):_

For second cycle
Compare based on Structure RMS (0) or Atom RMS (1):_

Compare Analytically (0) or Explicitly (1):_

Figure 5-62. GMMXINP Prompt for Setting the Sequence of Structural Comparison Criteria, and the Analytic vs. Explicit Mode for the Comparison Calculation.

default value is 0.25 kcal/mol, and a value of 100 kcal/mol would be used in a molecular dynamics run. We accept the default:

```
Use an Energy window of (0.250), 100.0 for Dynamics_
```
⟨CR⟩

Another prompt line inquires about the value for the RMS internuclear distance to be used as the criterion of equivalence for two conformers in the atom RMS comparison. We accept the default of 0.25 Å:

```
Use an RMS distance of (0.250):_
```
⟨CR⟩

The value for the torsional angle used to establish equivalence in the structure RMS comparison is the subject of the next prompt. We accept the default of 15°:

```
Screen internal coordinates using a window of (15 deg):
A window of -1 deg will turn this option off._
```
⟨CR⟩

As will be seen below, GMMX prepares a Boltzmann population analysis (see Section A.3) on the structures and energies which result from this conformational search. The Boltzmann calculation depends upon the temperature, and GMMXINP prompts the user for this. We accept the default of 300 K (24°C):

```
Use a Temperature of (300.0) deg for
    the Boltzmann distribution?_
```
⟨CR⟩

The discovered conformers will be arranged in the order of increasing energy. The user has the option of using either the MMX energy or ΔH_{form} (see Section 5.7). We choose the MMX energy:

```
Sort by MMX-E (M) or Heat of Formation (H)?_
                                            M
```

Next comes a prompt for whether π-calculations should be included in the energy minimizations during the conformational search. This makes the conformational evaluation of conjugated systems more accurate, but will slow down the overall process. We answer with an 'N':

```
If your system contains PI atoms, do you wish to
    do a PI calculation? [Y/N]_
                                            N
```

An additional parameter for the duplicates evaluation process is prompted next, when GMMXINP asks whether conformational enantiomers (mirror image conformations) should be kept or not (see Fig. 5-63). For the current search we answer with an 'N', since for a first pass we wish to remove duplicates of any kind. GMMXINP responds with a query about whether the first structure should be used as the benchmark for determining conformational enantiomers. For this search we answer with a 'Y'.

As the conformational search proceeds, it is possible to store the types of geometric reports generated by the **Query** tool (see Section 3.6.2), such as predicted ^{1}H-NMR coupling constants, or internuclear distances, bond angles or torsions. We answer this prompt with an 'N':

```
Do you wish to Query any coupling constants
   distances, angles or torsions (N/Y) ?_
                                          N
```

The final prompt of this sequence has to do with the sliding energy window. The screen, shown in Fig. 5-64, proposes a progressive decrease in the maximum energy used as the knockout factor for discarding high-energy conformations. This is calculated from the sum of the lowest energy and the highest energy for the conformers retained up to that point in the search process, plus an amount which decrements every 25 iterations. This additive amount becomes zero at 200 iterations, and has the effect of speeding up the search. It rests on the assumption that as the search progresses, the overall energy envelope will be better defined, and it is less productive to look far outside the current envelope.

The last screen message from GMMXINP signals the end of the input file preparation:

```
The command file pentane.inp
   has been generated:
```

```
Do you wish to keep CONFORMATIONAL enantiomers? [Y/N]_

The sense of chirality for stereogenic C,S,and N
atoms are checked against the same as determined
from the first structure of a multiple structure
file. For non-related structures, the relationship
is maintained according to each structure read in.

Check chirality against first structure (Y,N):_
```

Figure 5-63. The GMMXINP prompt for Evaluating Conformational Enantiomers.

```
The following sliding energy window will be used

iters     energy window (+5 kcal if H bonding)
 50       eminim + elimit + 15.0 kcal
 75       eminim + elimit + 12.5 kcal
100       eminim + elimit + 10.0 kcal
125       eminim + elimit +  7.5 kcal
150       eminim + elimit +  5.0 kcal
175       eminim + elimit +  2.5 kcal
200+      eminim + elimit +  0.0 kcal
Apply sliding energy window (y,n):_
```

Figure 5-64. The GMMXINP Prompt for the Sliding Energy Window.

This file, {PENTANE.INP}, is available on the input file library diskette in the CHAP5 directory, and is depicted in Fig. 5-65.

For comparison we will also prepare an input file for n-pentane which takes no account of symmetry and equivalence. To avoid confusion, the user should prepare a copy of {PENTANE.INP} as {PENTANX.INP} to use for this second part of the exercise. In this second part, the GMMXINP sequence is followed much as it was above, until the prompt for options in the duplicates check (Fig. 5-55). Here we elect option 0, the default. The other difference is electing to retain conformational enantiomers, that is, answering 'Y' to the prompts displayed in Fig. 5-63. This new file, {PENTANX.INP}, is available on the input file library diskette in the CHAP5 directory, and is depicted in Fig. 5-66.

The preparation for the third part of the exercise involves setting up a command file for the cyclic substrate 2-methyl-1,3-dioxolane, the structure file

```
pentane.pcm
pentane1
BGIN   0   0
CRMS   0   0   1   0   0.000    0.000    15.000
NOPI   0   0
CHIG   0   0
NSEQ   1   0
       5
       1   5   2   4   3   3   4   2   5   1
MCLO   3990
       0   2   0   0.00000   2   071277   1   0   0 3.00
       2   3   3
       3   4   3
ISTOP  0   25  5   12
ENDC   0   0
```

Figure 5-65. The {PENTANE.INP} Command File from GMMXINP.

```
pentanx.pcm
pentanx1
BGIN  0    0
CRMS  0    0    1    0    0.000    0.000    15.000
NANT  0    0
NOPI  0    0
CHIG  0    0
MCLO  3990
      0    2    0    0.00000    2    071277    1    0    0 3.00
      2    3    3
      3    4    3
ISTOP 0    25   5    12
ENDC  0    0
```

Figure 5-66. The {PENTANX.INP} Command File from GMMXINP.

{MEDO.PCM} for which was prepared earlier in this section. The GMMXINP program is run again as before, the structure filename MEDO.PCM is supplied at the appropriate point, and the prompt for the type of search is answered with a 'c' for the Cartesian coordinate random kick search. This brings up the screen shown in Fig. 5-67, which allows the user to randomize only selected atoms. We answer this prompt with an 'n'.

Next come prompts for the number of rings to be searched over, for the ring size, for the atom file numbers for the ring atoms, for the default torsional angle resolution, and for designation of the individual dihedral angles and the angle resolution for each (if different from the default), which we answer as shown:

```
How many rings do you wish to search on (0, max 4)?_
                                                                    1
Enter ringsize (min 5—max 50, 0 to abort)
    of the 1st ring._
                                                                    5
Enter atom numbers of the 5 contiguous
    atoms in the 1st ring (spaced by commas).
A neg value will abort and 0 will restart._
                                                            1,2,3,4,5
Enter default resolution in degrees (30)_
                                                                 <CR>
```

```
By default, all unconstrained, non-terminal, non-volatile
atoms are eligible for random selection during the
coordinates (cartesians) part for atom movements.

Do you wish to limit the selection to specific
unconstrained, non-terminal, non-volatile atoms (n,y)?_
```

Figure 5-67. The GMMXINP Prompt for Specific Atoms to Randomize.

5.8. CONFORMATION SEARCHING AND GLOBAL MINIMIZATION

For the dihedral angle 1, defined by the bond
between atoms 2-3
Use the default resolution of 30(y, n)._
                                           y

The remainder of the GMMXINP sequence mirrors that already described. The final command file, {MEDO.INP}, is available on the input file library diskette in the CHAP5 directory, and is shown in Fig. 5-68.

5.8.2.2. The Conformation Search with GMMX In this section, we will demonstrate the operation of the GMMX program by running the three input command files prepared in the previous section: {PENTANE.INP} and {PENTANX.INP} for the pentane molecule, and {MEDO.INP} for 2-methyl-1,3-dioxolane. When each has been run through the conformational search, the resulting structures and Boltzmann population analysis will be presented and discussed.

Since GMMX utilizes a stochastic search process, the same search run at different times may proceed differently, but the final results should be quite comparable.

Conformation Search Exercise. Carrying Out the Conformation Searches: Like the program GMMXINP for the preparation of the input command files, the GMMX program which carries out the actual conformational search operates in the DOS environment, and the user will need to open a DOS window, or leave the Windows environment and operate within the DOS environment. At the DOS system prompt, run GMMX. This brings up the notice for the DOS extender (see Fig. 5-53), and then some messages indicating that the parameter files {MMXCONST.PAR} and {GENCONST.PAR} are being read. Next, GMMX prompts for the name of the input command file, which we answer:

ENTER COMMAND FILE NAME (.INP Extension is assumed)_
                                                   pentane

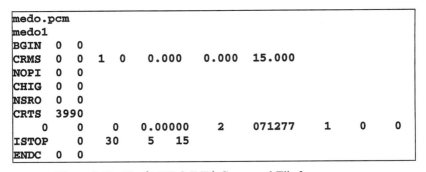

Figure 5-68. The {MEDO.INP} Command File from GMMXINP.

```
CONF    1 E =    2.82. Unique structure    1. Iters =    15. KEPT.
CONF    2 E =    2.82 Identical to    1 found  1 times before. Rejected. 15B
FOUND      1/      2. EMINIM =    2.82 saved as
STR No.    1 found    1 times.    0 Duplcs in a row. High =    0. AcR = 50 %
E of crnt st =    2.82. Iters =   15 fnd    1 tms before. Lst st kept =      1
CPUTIME =    0.0 min (  0.00 h). Delta CPUT =  2.0 sec. AvCPUT =   1.0 sec.

CONF    3 E =    3.72. Unique structure    2. Iters =    15. KEPT.B
FOUND      2/      3. EMINIM =    2.82 saved as
STR No.    1 found    1 times.    0 Duplcs in a row. High =    0. AcR = 66 %
E of crnt st =    3.72. Iters =   15 fnd    0 tms before. Lst st kept =      3
CPUTIME =    0.1 min (  0.00 h). Delta CPUT =  1.0 sec. AvCPUT =   1.0 sec.

CONF    4 E =    6.04. Unique structure    3. Iters =    80. KEPT.B
FOUND      3/      4. EMINIM =    2.82 saved as
STR No.    1 found    1 times.    0 Duplcs in a row. High =    0. AcR = 75 %
E of crnt st =    6.04. Iters =   80 fnd    0 tms before. Lst st kept =      4
CPUTIME =    0.1 min (  0.00 h). Delta CPUT =  4.6 sec. AvCPUT =   1.9 sec.

CONF    5 E =    3.72 Identical to    2 found  1 times before. Rejected. 55B
FOUND      3/      5. EMINIM =    2.82 saved as
STR No.    1 found    1 times.    1 Duplcs in a row. High =    1. AcR = 60 %
E of crnt st =    3.72. Iters =   55 fnd    1 tms before. Lst st kept =      4
CPUTIME =    0.2 min (  0.00 h). Delta CPUT =  3.1 sec. AvCPUT =   2.1 sec.

CONF    6 E =    6.04. Unique structure    4. Iters =    30. KEPT.B
CONF    6 was then rejected as a sequential isomer of    3.
FOUND      3/      6. EMINIM =    2.82 saved as
STR No.    1 found    1 times.    1 Duplcs in a row. High =    1. AcR = 50 %
E of crnt st =    6.04. Iters =   30 fnd    0 tms before. Lst st kept =      4
CPUTIME =    0.2 min (  0.00 h). Delta CPUT =  1.8 sec. AvCPUT =   2.1 sec.

CONF    7 E =    3.72 Identical to    2 found  2 times before. Rejected. 55B
FOUND      3/      7. EMINIM =    2.82 saved as
STR No.    1 found    1 times.    2 Duplcs in a row. High =    2. AcR = 42 %
E of crnt st =    3.72. Iters =   55 fnd    2 tms before. Lst st kept =      4
CPUTIME =    0.3 min (  0.00 h). Delta CPUT =  3.1 sec. AvCPUT =   2.2 sec.

CONF    8 E =    3.72 Identical to    2 found  3 times before. Rejected. 50B
FOUND      3/      8. EMINIM =    2.82 saved as
STR No.    1 found    1 times.    2 Duplcs in a row. High =    2. AcR = 37 %
E of crnt st =    3.72. Iters =   50 fnd    3 tms before. Lst st kept =      4
CPUTIME =    0.3 min (  0.01 h). Delta CPUT =  2.9 sec. AvCPUT =   2.3 sec.

CONF    9 E =    4.45. Unique structure    5. Iters =    30. KEPT.B
FOUND      4/      9. EMINIM =    2.82 saved as
STR No.    1 found    1 times.    0 Duplcs in a row. High =    2. AcR = 44 %
E of crnt st =    4.45. Iters =   30 fnd    0 tms before. Lst st kept =      9
CPUTIME =    0.3 min (  0.01 h). Delta CPUT =  1.8 sec. AvCPUT =   2.3 sec.

CONF   26 E =    4.45 Identical to    5 found  5 times before. Rejected. 25B
EMINIM found    1 times.   14 duplicates in a row. SEARCH ENDED

Finished
pentane.pcm
```

Figure 5-69. Excerpts from the GMMX Readout in the {PENTANE.INP} Search.

```
pentane1
pentane2
BGIN  0    0
NSEQ  1    0  0  0  0.000  0.000   0.000
      5
      1    5  2  4    3      3     4    2  5  1
CRMS  0    0  0  0  0.000  0.000  15.000
CHIG  0    0  0  0  0.000  0.000   0.000
NOPI  0    0  0  0  0.000  0.000   0.000
HREM  0    0  0  0  0.000  0.000   0.000
HADD  0    0  0  0  0.000  0.000   0.000
MINI  3990 0  0  0.000  0.000   0.000
ENDC  0    0
```

Figure 5-70. The {PENTANE1.INP} Command File from GMMXINP.

The conformational search then begins, and a continuous readout on the screen allows the progress to be monitored. An excerpt from the readout in the search over the pentane molecule, which shows the reports of the first nine and the last of the conformations processed, is shown in Fig. 5-69. Brief descriptive remarks are provided below which should help to make the meaning of this output clearer.

The first entry covers conformation 1 ('CONF 1'), which is the input geometry, for which the energy is calculated to be 2.82 kcal/mol ('E = 2.82'). It is the first unique structure ('Unique structure 1') uncovered after 15 iterations ('Iters = 15'), and this conformation was 'KEPT'.

The second entry covers conformation 2, whose energy is calculated to be 2.82 kcal/mol, and which is found to be identical to conformation 1, which has been found once before ('Identical to 1 found 1 times before'). This conformation was 'Rejected.' Other information given in this entry is as follows: one of the two structures evaluated so far is being kept ('FOUND 1/ 2'), some identifiers of the current minimum-energy structure are repeated, the current number and highest number of duplicates to this point are none ('0 Duplcs in a row. High = 0'), the acceptance rate for keeping structures is currently 50% ('AcR = 50%'), the energy of the current structure is repeated, ('E of crnt st = 2.82'), it was found after 15 iterations, it was found once before ('fnd 1 tms before.'), and the last structure kept was conformation 1 ('Lst st kept = 1). The last line gives the total CPU time expended ('CPUTIME = 0.0 min (0.00 h)'), the change in CPU time expended between the previous calculation and the current one ('Delta CPUT = 2.0 sec'), and the average CPU time expended per conformation ('AvCPUT = 1.0 sec').

Both conformations 3 and 4 are unique, and are kept. Conformation 5 is rejected as a duplicate of 3. Conformation 6 is kept initially, but rejected in the duplicate check process as a sequential isomer of 3.

370 THE MOLECULAR MECHANICS FORCE FIELD

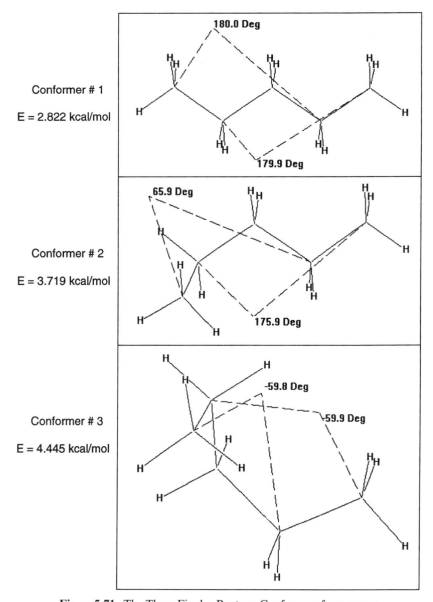

Figure 5-71. The Three Final *n*-Pentane Conformers from GMMX.

Finally, conformation 26 is found to be identical with unique conformation 5, and, this being the fourteenth duplicate in a row, the conformational search is ended. GMMX builds three output files for this exercise:

The first file is {PENTANE1.}, which is an MMX file in appended format containing the heavy atom structure input for the four final conformations.

```
            *** MMX BAKMOD ***

INPUT  = pentane1
OUTPUT = pentane2

OPERATION = BGIN   0   0   0   0   0.000   0.000   0.000
OPERATION = NSEQ   1   0   0   0   0.000   0.000   0.000
5
1   5   2   4   3   3   4   2   5   1
OPERATION = CRMS   0   0   0   0   0.000   0.000  15.000
OPERATION = CHIG   0   0   0   0   0.000   0.000   0.000
OPERATION = NOPI   0   0   0   0   0.000   0.000   0.000
OPERATION = HREM   0   0   0   0   0.000   0.000   0.000
OPERATION = HADD   0   0   0   0   0.000   0.000   0.000
OPERATION = MINI   3 990   0   0   0.000   0.000   0.000
OPERATION = ENDC   0   0   0   0   0.000   0.000   0.000
```

Figure 5-72. The MMX BakMod Portion of the {PENTANE2.PKM} File.

The second file is {PENTANE1.INP}, which is the new input command file for resubmission to GMMX to complete the duplicate check and energy screening process. It is shown in Fig. 5-70.

The third file is {PENTANE1.PKM}, which contains the data and explanation of the Boltzmann population analysis on the four discovered conformations for this first stage.

In order to complete the second stage of this analysis, run GMMX again, and submit {PENTANE1.INP} as the input command file. This gives rise to a much shorter program run, since only the four discovered structures are subjected to the additional analysis. In this case, the fourth (highest energy) conformer discovered in the first stage is discarded as being too high in energy. GMMX builds two output files:

The first file is {PENTANE2}, which contains the all atom versions of the three final conformations, arranged in order of energy. These are depicted in Fig. 5-71, with **Query** reports for the torsional angles. With regard to the two torsional angles of *n*-pentane, these three structures correspond to the *anti,anti*, *gauche,anti*, and *gauche,gauche* conformers.

The second file is {PENTANE2.PKM}, which is a log file of the command operations carried out and includes an explanation of the Boltzmann population analysis on the three final conformations of the *n*-pentane molecule. This file consists of two parts, which are shown in Figs. 5-72 and 5-73. The first part, MMX BakMod (shown in Fig. 5-72), contains the log of the command file operations. The second part, Final Output Conformations (shown in Fig. 5-73), documents the three final conformations and presents the Boltzmann population analysis.

Run the alternate input command file {PENTANX.INP} through GMMX as above. The program creates the expected files {PENTANX1.},

```
        *** FINAL OUTPUT CONFORMATIONS ***


    0 INITIAL STRUCTURES GAVE RISE TO THE FOLLOWING UNIQUE CONFORMATIONS:

    CONFORMATION    1 ENERGY =    2.82 KCAL/MOLE EXCESS STERIC =   0.00 KCAL/MOLE
    DERIVED FROM INPUT STRUCTURE NUMBER    1
    OUTPUT NAME MMX #     1    2.82 HF=   -36.27
    NECESSARY ITERATIONS =   15

    CONFORMATION    2 ENERGY =    3.72 KCAL/MOLE EXCESS STERIC =   0.90 KCAL/MOLE
    DERIVED FROM INPUT STRUCTURE NUMBER    2
    OUTPUT NAME MMX #     2    3.72 HF=   -35.38
    NECESSARY ITERATIONS =   20

    CONFORMATION    3 ENERGY =    4.45 KCAL/MOLE EXCESS STERIC =   1.62 KCAL/MOLE
    DERIVED FROM INPUT STRUCTURE NUMBER    3
    OUTPUT NAME MMX #     3    4.45 HF=   -34.65
    NECESSARY ITERATIONS =   20


    3 CONFORMATIONS FOUND HAVING EXCESS STERIC LESS THAN    3.0 KCAL/MOLE.

The Temperature is:   300.00000 Degrees K
The Boltzmann average energy is:      3.05946 Kcal/mol
The Boltzmann average Heat of formation is:    -36.03655 Kcal/mol
The Boltzmann average Dipole Moment is:      0.00000
The Boltzmann Dipole Moment squared is:      0.00000
Pop conf    1 = 77.65%  E=   2.822,DE=  0.000 HF=  -36.274,DHF= 0.000 DPM= 0.00
Pop conf    2 = 17.24%  E=   3.719,DE=  0.898 HF=  -35.377,DHF= 0.898 DPM= 0.00
Pop conf    3 =  5.11%  E=   4.445,DE=  1.623 HF=  -34.651,DHF= 1.623 DPM= 0.00

The entropy of mixing is:    1.294 cal/mol-K

At  300.0 deg K, TdS =    0.388 Kcal/mol


    OUTPUT FILENAME = pentane2

    CPUTIME =     0.1 MIN.
```

Figure 5-73. The Final Output Conformations Portion of {PENTANE2.PKM}.

{PENTANX1.INP}, and {PENTANX1.PKM}. Inspection of these files reveals a collection of 11 conformations, which represent the same coverage of conformational space as the four conformations found in the first run, but with a higher degeneracy due to the absence of symmetry and equivalence corrections.

Run the {PENTANX1.INP} file through a second stage of GMMX analysis, which creates the new files {PENTANX2.}, and {PENTANX2.PKM}. Inspection of these files reveals the final conformer count to be seven, since the quadruply degenerate highest energy conformer has been discarded based upon the energy criterion. In the final collection, there is a single *anti,anti* conformer, four *gauche,anti* conformers corresponding to two enantiomers each of two numbering sequences, and two enantiomeric *gauche,gauche* conformers. The

5.8. CONFORMATION SEARCHING AND GLOBAL MINIMIZATION

```
7 CONFORMATIONS FOUND HAVING EXCESS STERIC LESS THAN   3.0 KCAL/MOLE.

The Temperature is:   300.00000 Degrees K
The Boltzmann average energy is:       3.32221 Kcal/mol
The Boltzmann average Heat of formation is:   -35.77379 Kcal/mol
The Boltzmann average Dipole Moment is:    0.00000
The Boltzmann Dipole Moment squared is:    0.00000
Pop conf   1 = 49.52%  E=   2.822,DE=  0.000 HF= -36.274,DHF= 0.000 DPM= 0.00
Pop conf   2 = 11.00%  E=   3.719,DE=  0.897 HF= -35.377,DHF= 0.897 DPM= 0.00
Pop conf   3 = 10.99%  E=   3.719,DE=  0.898 HF= -35.377,DHF= 0.898 DPM= 0.00
Pop conf   4 = 10.99%  E=   3.719,DE=  0.898 HF= -35.377,DHF= 0.898 DPM= 0.00
Pop conf   5 = 10.99%  E=   3.720,DE=  0.898 HF= -35.376,DHF= 0.898 DPM= 0.00
Pop conf   6 =  3.26%  E=   4.445,DE=  1.623 HF= -34.651,DHF= 1.623 DPM= 0.00
Pop conf   7 =  3.25%  E=   4.445,DE=  1.624 HF= -34.651,DHF= 1.624 DPM= 0.00

The entropy of mixing is:     3.064 cal/mol-K

At  300.0 deg K, TdS =     0.920 Kcal/mol


    OUTPUT FILENAME = pentane2

    CPUTIME =     0.3 MIN.
```

Figure 5-74. The Boltzmann Population Analysis Portion of {PENTANX2.PKM}.

```
            *** MMX BAKMOD ***

    INPUT  = medo.pcm
    OUTPUT = medo1

    OPERATION = BGIN   0   0   0   0   0.000   0.000    0.000
    OPERATION = CRMS   0   0   1   0   0.000   0.000   15.000
    OPERATION = NOPI   0   0   0   0   0.000   0.000    0.000
    OPERATION = CHIG   0   0   0   0   0.000   0.000    0.000
    OPERATION = NSRO   0   0   0   0   0.000   0.000    0.000
    OPERATION = CRTS   3 990   0   0   0.000   0.000    0.000
         0    0    0   0.00000    2   071277    1    0    0 0.00    0
   KSTOP    0   30    5   15
    OPERATION = ENDC   0   0   0   0   0.000   0.000    0.000

            *** MMX BAKMOD ***

    INPUT  = medo1
    OUTPUT = medo2

    OPERATION = BGIN   0   0   0   0   0.000   0.000    0.000
    OPERATION = NSRO   0   0   0   0   0.000   0.000    0.000
    OPERATION = CRMS   0   0   0   0   0.000   0.000   15.000
    OPERATION = CHIG   0   0   0   0   0.000   0.000    0.000
    OPERATION = NOPI   0   0   0   0   0.000   0.000    0.000
    OPERATION = HREM   0   0   0   0   0.000   0.000    0.000
    OPERATION = HADD   0   0   0   0   0.000   0.000    0.000
    OPERATION = MINI   3 990   0   0   0.000   0.000    0.000
    OPERATION = ENDC   0   0   0   0   0.000   0.000    0.000
```

Figure 5-75. The MMX BakMod Portions of {MEDO1.PKM} (above) and {MEDO2.PKM} (below).

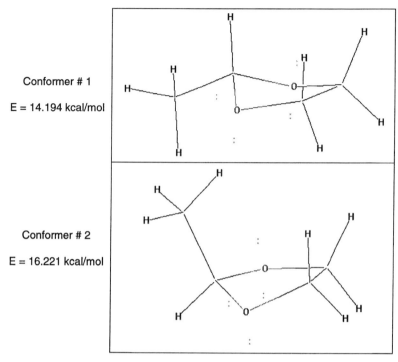

Figure 5-76. Two Conformers of 2-Methyl-1,3-dioxolane, and the Boltzmann Analysis.

portion of {PENTANX2.PKM} which contains the Boltzmann population analysis is depicted in Fig. 5-74.

At this point, it is obvious that the degenerate occupancy of these conformational states has a significant effect on the outcome of the Boltzmann analysis. Ignoring the degeneracy, the conformational profile for *n*-pentane is 77.65% *anti,anti*, 17.24% *gauche,anti*, and 5.11% *gauche,gauche*. Allowing for the degeneracy of the conformational energy states, the profile for *n*-pentane becomes 49.52% *anti,anti*, 43.97% *gauche,anti*, and 6.51% *gauche,gauche*.

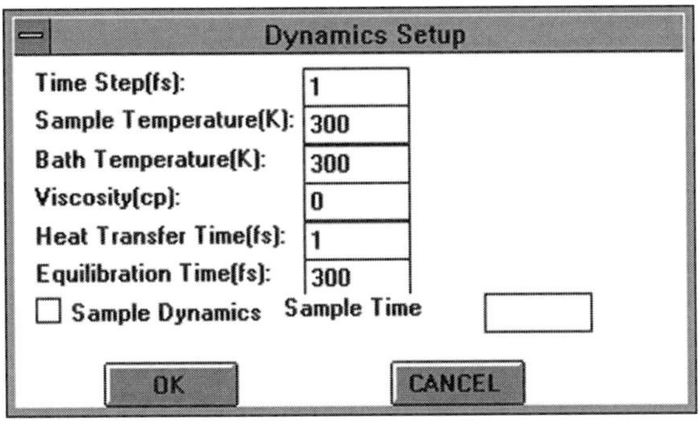

5.8. CONFORMATION SEARCHING AND GLOBAL MINIMIZATION 375

Figure 5-77. The 'Dynamics Setup' Dialog Box.

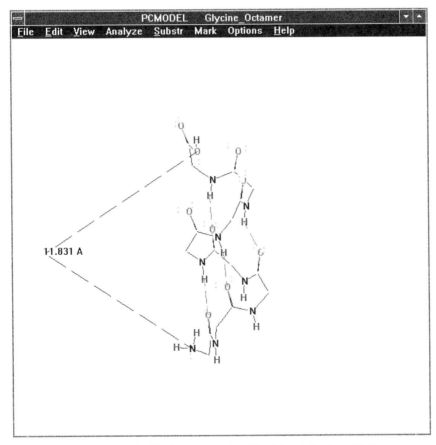

Figure 5-78. The H_2N—$(Gly)_8$—CO_2H Molecule in the α-Helix.

376 THE MOLECULAR MECHANICS FORCE FIELD

Time	Temp	Energy
0	0.013	-56.133
10	6.208	-55.757
20	168.460	-47.578
30	272.018	-23.069
40	282.421	-7.564
50	287.047	7.332
60	300.975	11.337
70	297.930	10.275
80	284.858	18.866
90	293.777	24.306
100	291.059	30.125
110	311.088	26.964
120	321.170	10.345
130	307.552	1.513
140	294.282	5.135
150	298.002	6.046
160	296.617	9.049
170	297.682	11.209
180	304.572	5.837
190	303.472	0.826
200	288.204	8.683

Figure 5-79. Excerpt from the 'Output' Window Display for the Molecular Dynamics Run.

For the last part of this exercise, run the input command file {MEDO.INP} through GMMX. This gives rise to the expected three files {MEDO1.}, {MEDO1.INP}, and {MEDO1.PKM}. Inspection of these files show that two conformers were discovered for 2-methyl-1,3-dioxolane, corresponding to the structure with a pseudoequatorial methyl group and the one with a pseudoaxial methyl group.

Run the {MEDO1.INP} input command through GMMX, which gives rise to the files {MEDO2.}, and {MEDO2.PKM}. Figure 5-75 shows the MMX BakMod portions of {MEDO1.PKM} and {MEDO2.PKM}. This analysis does not allow for degeneracy.

The two final conformations of 2-methyl-1,3-dioxolane are depicted in Fig. 5-76, along with the portion of the {MEDO2.PKM} file which details the results of the Boltzmann population analysis.

5.9. MOLECULAR DYNAMICS

Molecular dynamics describes a family of techniques which model the dynamic, time-dependent behavior of molecules by applying the Newtonian

0	5e-016	-56.13	0.00	-56.13	189.6
10	5.5e-015	-55.76	1.13	-54.63	191.9
20	1.05e-014	-47.58	35.70	-11.87	251.6
30	1.55e-014	-23.07	46.78	23.71	289.7
40	2.05e-014	-7.56	94.70	87.13	293.5
50	2.55e-014	7.33	26.55	33.88	295.2
60	3.05e-014	11.34	1.13	12.46	300.4
70	3.55e-014	10.28	29.49	39.76	299.2
80	4.05e-014	18.87	110.93	129.79	294.4
90	4.55e-014	24.31	118.59	142.90	297.7
100	5.05e-014	30.13	94.13	124.25	296.7
110	5.55e-014	26.96	16.91	43.87	304.1
120	6.05e-014	10.35	7.53	17.88	307.8
130	6.55e-014	1.51	53.46	54.97	302.8
140	7.05e-014	5.13	31.26	36.39	297.9
150	7.55e-014	6.05	81.20	87.24	299.3
160	8.05e-014	9.05	19.92	28.96	298.8
170	8.55e-014	11.21	3.01	14.22	299.1
180	9.05e-014	5.84	69.34	75.17	301.7
190	9.55e-014	0.83	194.43	195.26	301.3
200	1.005e-013	8.68	113.89	122.57	295.7

Figure 5-80. Excerpt from the {PCMOD.OUT} Log File for the Molecular Dynamics Run.

laws of motion to individual atoms and tracking the effect over time in order to simulate the gross movement of the molecule. There are many methods of carrying out molecular dynamics simulations, and there are a number of reviews[88,112,113,118,121,157,161,178,264,573,577,582,615,625] available on the subject, of which perhaps the most informative is by van Gunsteren and Berendsen.[581] For further study, the interested reader is directed to these sources, as well as to tutorials and resources available through the links provided in the HTML document on the diskette which accompanies this book, "Molecular Modeling Resources on the World Wide Web."

In a molecular dynamics experiment, a static molecular structure (generally a minimum-energy structure derived from force field calculations) is animated in a series of small steps (on the order of 10^{-15} s, or 1 fs) by the application of a force to each atom of the molecule. The acceleration of each atom will be a function of this force and of the mass of the atom, and the velocity of each

average kinetic energy:	39.4397	std dev:	30.1573
average potential energy:	20.7041	std dev:	5.75553
average total energy:	60.1438	std dev:	28.108807
average sample temp:	284.284	std dev:	16.6297

Figure 5-81. The Dynamics Run Summary for 800 Iterations on Octaglycine.

atom is calculated from its position and acceleration by the dynamics algorithm. The force in one dimension on an individual atom 'I' is calculated from the following equation:

$$F_I = m_I a_I = -\frac{\delta V}{\delta r_I} = m_I \frac{\delta^2 r_I}{\delta t^2}$$

where the first two members constitute Newton's force law, with F_I being the force on atom I, m_I being the mass of atom I; and a_I being the acceleration experienced by atom I. The third member is the ratio of the change in potential energy, δV, to the change in position, δr_I, i.e. the distance that the atom moves, and the fourth member approximates the second derivative of the position with respect to time. In practice, these quantities are evaluated in three spatial dimensions.

Each succeeding dynamic step takes the current static structure through another set of atomic displacements, the direction and magnitude of which is a function of the forces acting upon that atom at that time. For the dynamic simulation to represent physically meaningful motions of particles, the time step must be an order of magnitude shorter than the fastest vibrational mode in the molecule. The fastest vibrations would be bond stretch, which are typically on the order of $\leqslant 10$ fs; this dictates a time step on the order of 0.5–1 fs for most dynamics simulations. Larger time steps may be used if the bond stretch and angle bend vibrations are frozen, and this is often done with biomacromolecules. Carbon–hydrogen bond stretch vibrations are among the fastest, so that fixing these bond lengths, or employing a united atom or heavy-atom-only model, would allow slightly longer steps, 1.0–1.5 fs. Constraining all bonds would allow an ≈ 2.0-fs time step.

In most dynamics simulations, the system is allowed to equilibrate for a period of from 10^4–10^6 time steps, and then the behavior is monitored by saving sample conformations at timed intervals. These stored conformations may be subjected to energy minimization, and later presented in sequence as a frame-by-frame "movie" of the simulated molecular movement. This sequence is referred to as the molecular dynamics trajectory, and constitutes a way to visit conformational space which would be difficult to access by searching methods based upon the static structure.

Molecular dynamics simulations have a number of applications: to correlate molecular structure with experimental data, such as diffraction crystallography or NMR experiments;[626,627] as a conformational search method;[119,264,313,579,588,628] to study the behavior of molecules in solution (see Section 5.6);[629,630] to simulate intermolecular interactions such as docking and molecular recognition;[173,497] and to simulate macroscopic processes and properties such as melting, diffusion,[631] and the mechanical properties and conformational transitions of macromolecules,[440,442,501,632] among many others. Some force fields, such as MMFF94 (see Section 5.4.6), UFF (see

Section 5.4.8), the QM/MM potential of Karplus et al.,[633] have been specifically developed for use in molecular dynamics as opposed to static calculations.

Molecular dynamics simulations are potentially a powerful tool for studying molecular structure and behavior; their main limitations are the need for high computational capacity and copious memory for storage of the data generated. Both of these limitations put the desktop computer platform at a disadvantage. Still, some simple dynamics calculations may be carried out at this level, as will be seen below in the exercise. The molecular dynamics approach to the conformational analysis of n-pentane as a computer experiment has been published.[205]

In PCMODEL, molecular dynamics simulations on the molecule(s) in the work space are available by using the **Analyze\Dynam** command. Prior to a dynamics run, the structure to be examined must be energy-minimized if the simulation is to succeed. Activation of this command brings up the 'Dynamics Setup' dialog box with the parameters for the dynamics run, as shown in Fig. 5-77.

The value in the box labeled 'Time Step(fs):' sets the interval between calculated dynamic states, given in femtoseconds, or 10^{-15} s. The value in the box labeled 'Sample Temperature(K):' sets the effective temperature of the sample, with a default value of 300 K, equivalent to 27°C. The value in the box labeled 'Bath Temperature(K):' sets the effective temperature of the medium containing the treated molecule, which will determine how much kinetic energy (or heat) will be available to transfer to the subject molecule. The default bath temperature of 300 K is equivalent to 27°C. The value in the box labeled 'Viscosity(cp):' sets the viscosity of the medium, in centipoise (cp), which will determine how much the movement of atoms within the molecule will be impeded. The higher the viscosity, the more sluggish the movement. The default value of 0 cp means unimpeded movement (see Appendix A for the definitions of units and some examples of solvent viscosities). The value in the box labeled 'Heat Transfer Time(fs):' corresponds to the time over which heat transfer from the bath to the sample will occur, given in femtoseconds. The default time period of 1 fs will give heat transfer on the same time scale as the default time step. The larger the value entered here, the slower the net heat transfer to the sample, and values longer than the total time of the dynamics simulation will result in no net heat transfer. The value in the box labeled 'Equilibration Time(fs):' corresponds to the time over which equilibration takes place between the energy sampling points. Selecting the 'Sample Dynamics' check box initiates a sampling protocol on the scale set in the sample time, and making this selection brings up the entitled 'Write File for Batch, Dock and Dihedral Driver' dialog box (shown in Fig. B-2), for designating the desired file.

Care must be taken in increasing the value for the time step, particularly with small molecules. For example, dynamics can be run on boat cyclohexane with up to a 1.75-fs step, but with a 2.0-fs step the atoms of the molecule will fly apart, and the program will crash. Another problem which may arise is that lone pairs may wander away from their parent heteroatoms. If this happens, the dynamics run should be stopped, and an energy minimization should be run to realign the constituents.

PCMODEL employs the *leapfrog* scheme for the integration of Newton's equations of motions. The leapfrog algorithm derives its name from a mathematical device used in the calculation. To determine the position $r(t_n)$ of an atom at time t_n, it integrates the force F over a time interval Δt from $t_n - \Delta t/2$ to $t_n + \Delta t/2$. This gives the velocities at these two points in time, from which the desired position $r(t_n)$ of an atom at time t_n may be derived, i.e. it "leaps" from a point just before t_n to a point just afterward in order to determine the position at t_n. This scheme is ". . . one of the most accurate, stable, and yet simple and efficient algorithms available for molecular dynamics of fluidlike systems."[581] The atomic velocity occurs explicitly in this algorithm, in contrast to the expression used in the Verlet algorithm for dynamics,[634] and this allows the system to be coupled to a heat bath by velocity scaling, enabling control over the temperature.

The gas phase calculation used here corresponds to a vacuum boundary condition, which is known to introduce some distortion.[581] The PCMODEL implementation of molecular dynamics is limited by the capabilities of the desktop computing environment, so that calculations on a system with explicit solvation or a multibody system are not easily carried out.

Molecular Dynamics Exercise: Relaxation of α-Helical Octaglycine. In this exercise the molecule octaglycine will be prepared in the α-helical conformation, and a molecular dynamics simulation will be carried out on it. This simulation is deterministic in nature, and so starting from the same initial structure at the same initial energy with the same dynamics parameters should produce the same molecular trajectory, and the same conformations along the way. However, the objective for this exercise is not to obtain a specified final structure, but more to observe the process of the simulation an gain an appreciation of what can be obtained from this method.

Building the test molecule will follow closely the procedure detailed in Template Drawing Exercise 1 in Section 3.2.2. Bring up the 'Amino Acids' template menu by clicking on the **AA** tool. Clicking in the 'Connect' choice box also selects the 'a-helix' option as the default mode. Next, the mouse cursor is used to click once on the **Gly** button, then seven more times on the **Gly** button to complete the octapeptide H_2N–Gly–Gly–Gly–Gly–Gly–Gly–Gly–Gly–CO_2H. Cancel the 'Amino Acids' template menu, and use **View\Control Panel** to give the octapeptide a better perspective for viewing, with the axis of the α-helix roughly parallel to the Y-axis. Some of the confusing detail can be removed by hiding the volatile hydrogens with the **H/OO** tool. This structure should be run through several cycles of energy minimization to give an MMX energy value in the neighborhood of -56.146 kcal/mol. If the initial energy is different from this value, the course of the dynamics run will differ from that described here. The resulting structure is depicted in Fig. 5-78, where the end-to-end distance, from N-terminal nitrogen to the C-terminal carboxylic oxygen, is indicated to be 11.831 Å. This measure is a crude but convenient way to compare the conformations which will be produced in this exercise. This structure is available as {GLY8MER.PCM} on the input file library diskette, in the CHAP5 directory.

Activate the **Analyze\Dynam** command. Change the 'Time Step' and the 'Heat Transfer Time' to 0.5, and accept the remaining default values and begin the dynamics runs by clicking on the 'OK' button. As the run begins, the 'Output' window will begin to present numerical updates on the progress of the dynamics simulation. The columns of data will be identified once the 'Output' window screen has refreshed once, and Fig. 5-79 shows the data which would be displayed on screen in a typical simulation. The value given under 'Time' is actually the number of iterations, which is proportional to the time by the time step factor. The value given under 'Temp' is the temperature of the bath, and the 'Energy' is the MMX energy of the structure currently displayed. Note that the temperature of the bath starts out close to absolute zero ($T = 0.013$ K at time 0), and then increases over the next 60 iterations to reach ≈ 300 K, about which it will oscillate for the remainder of the dynamics run. The energy of the structure also increases, although it is important to remember that this is not a minimum-energy structure. It can also be seen that the energy of the structure continues to increase as long as the bath temperature is below 300 K and rising, and this energy begins to decrease when the bath temperature rises above 300 K and subsequently falls. This is because in the former case the bath is being heated and transfers energy to the structure, while in the latter case the bath is being cooled and absorbs energy from the structure.

During the dynamics run, the {PCMOD.OUT} file logs the progress of the dynamics run with a series of status lines which look like those shown in Fig. 5-80. There are six columns of data, which have the following meaning: the first (from the left) gives the number of iterations, the second gives the elapsed time in seconds, the third gives the average potential energy (corresponding to the MMX energy), the fourth gives the average kinetic energy, the fifth gives the average total energy, and the sixth gives the temperature of the structure.

Considering the data in both Figs. 5-79 and 5-80, note that at the zero iteration stage, a time interval of 0.5 fs is indicated, the potential and total energy (third and fifth columns) are the same, and the kinetic energy (fourth column) is zero (prior to first "leap"); the temperature of the bath is 0.013 K, and that of the structure is 189.6 K. As the bath is heated, the kinetic energy at each interval rises rapidly and then falls as the temperature of the bath (300.975 K) and sample (300.4 K) plateau at 60 iterations. During this time, the hydrogen bonds between every four glycine residues, which maintain the α-helix conformation, will be broken.

Then, as the dynamics run continues, these temperatures cycle up and down, with the kinetic energy increasing or decreasing as the bath is heated or cooled, respectively, resulting in a feedback loop to keep the sample temperature nominally near 300 K. During this time, hydrogen bonds will form and break often between adjacent glycine residues, and the 1,4-hydrogen bonds will form and break occasionally.

As indicated in the 'Dynamics Setup' window earlier, the first 300 iterations are the equilibration phase of the run. During this time, a relatively large interval (or leap) of 0.5 fs is used in the calculations. After the initial equilibration period is over, a smaller leap of 0.001 fs will be used, and the temperature fluctuations will tend to be less pronounced. The system will occasionally reenter an equilibration cycle over the course of the run. The data for the period just before and just after the end of the initial equilibration are shown in Table 5-8. Note the stabilization of both the energy values and the temperatures just after 300 iterations.

At 600 iterations, and for every 200 iterations after that, a status line will be printed in the {PCMOD.OUT} file giving the values averaged over those 200 iterations:

```
Ave: 1    35.49    25.39    60.88    292.86
```

where the 1 is a counter, 35.49 is the potential energy in kcal/mol, 25.39 is the kinetic energy in kcal/mol, 60.88 is the total energy in kcal/mol, and 292.86 is the temperature of the structure in kelvin.

The dynamics run may be interrupted at any time by hitting the ⟨Esc⟩ key. The current structure will remain on screen, and a summary will be printed in the {PCMOD.OUT}, as shown in Fig. 5-81 for a dynamics run terminated at

TABLE 5-8. Excerpt from the Dynamics Run on Octaglycine

Iters.	Time (fs)	E_{pot} (kcal/mol)	E_{kin} (kcal/mol)	E_{tot} (kcal/mol)	T_{struct} (K)	T_{bath} (K)
250	1.255E-O13	37.69	18.73	56.42	300.7	301.9
260	1.305E-O13	25.98	34.02	60.00	306.0	316.2
270	1.355E-O13	21.08	22.03	43.11	301.4	303.9
280	1.405E-O13	28.86	40.65	69.51	296.1	289.5
290	1.455E-O13	35.61	140.95	176.56	296.1	289.5
300	1.505E-O13	39.68	208.05	247.73	300.4	301.1
310	1.55499E-O13	32.28	70.31	102.59	305.0	313.6
320	1.60499E-O13	23.72	24.27	47.99	302.8	307.5
330	1.65499E-O13	24.38	25.17	49.55	297.7	293.8
340	1.70499E-O13	35.02	22.35	57.37	297.5	293.3
350	1.75499E-O13	43.27	16.64	59.92	298.0	294.7

The iteration numbers (I.) are found in the first column of both the on-screen 'Output' window dynamics reports and the {PCMOD.OUT} log file. The time, E_{pot}, E_{kin}, E_{tot}, and T_{struct} are from the corresponding columns of the {PCMOD.OUT} log file, and the T_{bath} is from the on-screen 'Output' window dynamics reports.

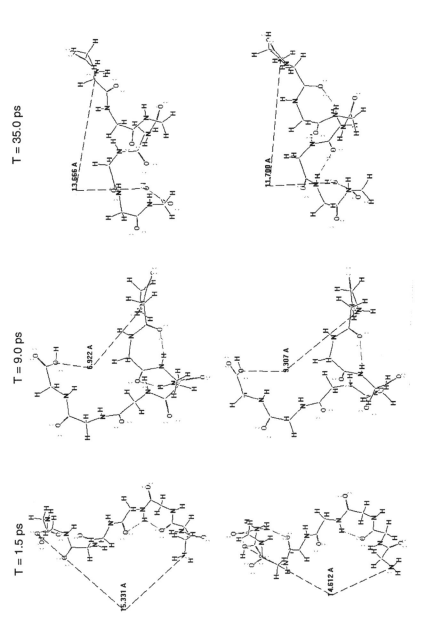

Figure 5-82. Conformations from the Dynamics Simulation of {GLY8MER.PCM}. Top, from dynamics run; bottom, energy-minimized.

800 iterations. It gives the averages of the kinetic energy, potential energy, total energy, and sample temperature over the course of the dynamics run, along with their standard deviations.

Structures were extracted from the dynamics run at three different times, and these are shown in Fig. 5-82. At the top of the figure are found the structures as they were pulled from the dynamics run, with single-point energies of 25.665, 18.959, and 26.978 kcal/mol, and at the bottom are found the energy-minimized structures, which had final energies of -47.833, -44.720, and -50.316 kcal/mol. These are available from the structure input file library in the CHAP5 directory as the appended format file {GLY8MD.PCM}. The conformation at 1.5 ps, "just after" the breakdown of the α-helical structure, is relatively elongated. The conformation at 9.0 ps has collapsed inward a bit, while that at 35.0 ps has begun to extend again.

We don't expect the molecule to resume the α-helical structure, since dynamics isn't an global energy search, but rather a method to explore conformational space. The absence of any constraints or solvent molecules in this simulation greatly hastened the denaturation. Under the conditions used in this dynamics run for the octaglycine molecule, 1 ps of simulation takes about 35 min on a desktop computer with a 486 chip operating at 50 MHz, which translates to about 41 ps per day of continuous operation.

CHAPTER 6
Applications

> When we mean to build,
> We first survey the plot, then draw the model;
> And when we see the figure of the house,
> Then must we rate the cost of the erection;
> Which if we find outweighs ability,
> What do we then but draw anew the model
> In fewer offices, or at last desist
> To build at all?
> —Shakespeare, *King Henry IV*, Part 2, Act 1, Scene 3.

This chapter covers a wide variety of applications of molecular modeling, and it is in this chapter that most of the exercises with PCMODEL are found. Each section addresses a particular area of modeling or type of experiment. The pattern followed is that the concepts and manipulations are discussed first, after which one or more exercises are presented which exemplify the topic in an application. The order of presentation of the topics is somewhat arbitrary, and the reader may choose to follow an order of their own.

The discussion in each section has been organized so that it is presented generally in order of increasing complexity or sophistication. This is done to make the material more useful to the general audience by allowing the user to gauge their own point of entry. The initial material in a section assumes little beyond the basics of force field modeling discussed in Chapter 5, after which the discussion begins to draw upon the concepts and techniques covered in other sections of Chapter 6.

There will be two distinct methods by which the reader may work through the discussions and exercises in Chapter 6. In the *survey method* one would cover the introductory treatments in each of the sections first, returning afterwards to finish each section. Alternatively, in the *in-depth method* one would select a section and work through it to the end before moving on to the next.

6.1. MODELING WITH GEOMETRIC CONSTRAINTS

The most straightforward approach to molecular mechanics modeling is to prepare a "reasonable" input geometry, and then to let the minimization algorithm work within the force field to arrive at a final geometry in which the total energy of the structure has been optimized. However, there are many cases where it is desirable to freeze out one or more degrees of freedom, to fix some aspects of the geometry while allowing the others to minimize normally. In this section, we will learn the ways in which this can be achieved, and see examples in which this technique provides information of value.

PCMODEL has four options to impose geometric constraints upon a molecule: the **Fix_Atoms** command allows the user to fix the position of one or more atoms in a structure; the **Fix_Distance** command allows the user to fix the relative position of two atoms in a structure; the **Fix_Torsions** command allows the user to fix the value for a torsional angle in a structure; and the **Don't Minimize** command exempts a designated substructure from the energy minimization process. The operation of these constraint options will be explored first, followed by exercises to exemplify each type.

The **Fix_** commands for introducing geometric constraints will be introduced now, followed by a simple application for each. These options are all found under the **Mark** command on the menu bar, and all require that the affected atoms be marked using the **Select** tool. These constraints will remain in effect until explicitly removed with the **Mark\Reset** command, or until the current session is finished. When an input file is written from such a constrained structure, the constraints are embedded in that file. There is some overlap in the effect of these options; for example, using the **Fix_Atoms** command on two atoms will effectively fix the distance between them.

Fixing the coordinates in this way permits one to impose upon the molecule a geometry obtained from some other source, such as from physical or spectroscopic data, or a minimum-energy structure obtained through optimization within another force field. Then, the calculated energies with and without the constraint may be compared in order to evaluate the consistency of the currently used force field with empirical results, or with those from another modeling program.

The **Fix_Atoms** command allows the user to fix any one or all three of the Cartesian coordinates X, Y, and Z which has the effect of freezing the motion of these selected atoms along the Cartesian axes during minimization. In addition to the uses mentioned above, this option may also be used to freeze the orientation, so that it will not change during a minimization sequence.

After first marking the atom(s) to be fixed with the **Select** tool, the **Mark\Fix_Atoms** command is activated, which pops up a dialog box entitled 'Fix Atoms', as shown in Fig. 6-1. Here the user may use the radio buttons to fix 'All' three Cartesians, or only one from among 'X', 'Y', or 'Z'.

Geometric Constraint Exercise 1: X-Ray Structure of a Giberrellic Acid Synthesis Intermediate. To demonstrate one use of the **Fix_Atoms** command, we will read in the 'X-Ray' format input file for an intermediate in an approach to the synthesis of giberrellic acid.[635] The heavy atoms in this structure will then be fixed, after which hydrogens and lone pairs will be added with the **H/AD** tool, and a minimization carried out. After noting the energy values for the optimized structure, the constraints will be lifted, and another minimization sequence will be carried out, in which the geometry of the entire structure is

6.1. MODELING WITH GEOMETRIC CONSTRAINTS

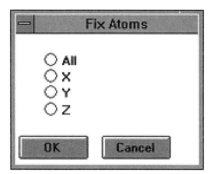

Figure 6-1. The 'Fix Atoms' Dialog Box.

free to change. Finally, the energy values obtained in the two sequences will be compared.

From the input file library diskette, in the CHAP6\SECT1 subdirectory, read in the 'X-Ray' format file {GIBRELN.XRA}. This file was assembled from data reported in the literature.[636] Use **View\Control_Panel** to alter the perspective for a better presentation. Next, activate the **Select** tool and use the mouse cursor to mark all twenty-two atoms in this structure. Then, use the **Mark\Fix_Atom** command, and select the 'All' option, to fix the atoms in all three Cartesian coordinates. The result should resemble Fig. 6-2.

Now, activate the **H/AD** tool to add hydrogens and lone pairs. Then, activate the **Analyze\Minimize** command to initiate a structure optimization sequence. Note that the relative positions of the heavy atoms will not change, only the positions of the hydrogens and lone pairs will be optimized. After repeating the **Minimize** command several times, note the energy values shown in the 'Output' child window. The resulting structure should be like that depicted in Fig. 6-3, and may be found in CHAP6\SECT1 of the molecule input file library as {GIBRELN1.PCM}.

Next, activate the **Mark\Reset** command and select the 'All' check box to remove the constraints. Run the unconstrained structure through several cycles of energy minimization, and then note the energy values shown in the 'Output' child window. The resulting structure should be like that depicted in Fig. 6-4, and may be found on the input file library diskette in the CHAP6\SECT1 of the molecule input file library as {GIBRELN2.PCM}.

On visual inspection the two structures don't appear very dissimilar, but there is a difference of just over 9 kcal/mol in their calculated MMX energies.

The **Fix_Distance** command allows the user to fix the relative position of two atoms in a structure, or of one atom in each of two structures. **Fix_Distance** may be used to freeze two atoms at a relative distance obtained by some physical measure, or to set a specific distance between two structures

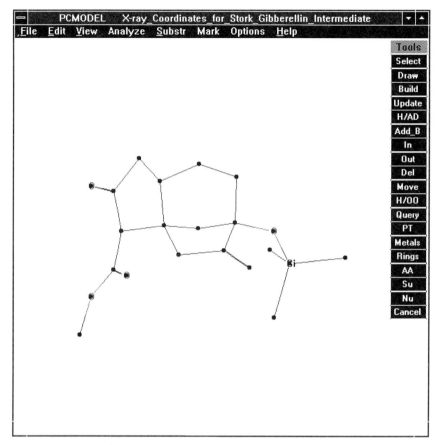

Figure 6-2. {GIBRELN.XRA} with all Atoms Selected for **Fix_Atom**.

using an atom in each as "handles." Up to ten pairs of atoms may be fixed in this way, and the relative geometry is maintained by imposing a force constant of 5.0 mdyne/Å upon the virtual bond between the two atoms.

After first marking the two atoms being constrained with the **Select** tool, the **Mark\Fix_Distance** command is activated, which pops up a dialog box entitled 'Fix Distance,' as shown in Fig. 6-5. Four lines of parameter fields are presented: the first line, entitled 'Atoms:', shows two small data boxes which give the atom file numbers of the two atoms to be constrained, in this example atoms 1 and 4; the second line, entitled 'Current Distance:', shows a data box in which is given the value of the initial distance between the selected atoms, in this case 2.79989 Å; the third line, entitled 'Fix At:', provides a data entry box where the user may enter the desired value for the fixed internuclear distance; the fourth line, entitled 'Force Constant:', provides a data entry box containing the default value of 5.0 mdyn/Å for this constraint, where the user may overwrite it with a larger or smaller value if desired.

6.1. MODELING WITH GEOMETRIC CONSTRAINTS   389

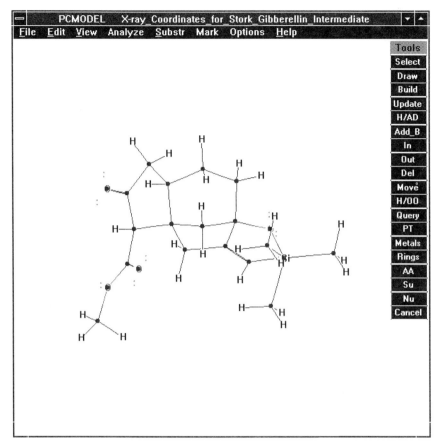

Figure 6-3. Fixed Heavy Atom Skeleton From X-Ray Structure, with Hydrogens and Lone Pairs Added, after Minimization {GIBRELN1.PCM}.

Geometric Constraint Exercise 2: Stacking Interaction of Benzene Dimer. In this exercise a benzene structure will be prepared in which the @40 aromatic carbon is used, and two of these molecules will be brought into the work space. After they are oriented along parallel planes, distance constraints will be placed between atoms on each of the rings. These will serve to hold the rings in the desired relative conformation, while the ensemble is subjected to energy minimization. Then, the constraints will be reset at increasingly shorter distance, and the change in energy will be monitored.

From the **Tools** bar, activate the **Rings** menu and select **Ph** to enter a benzene ring in the work space. This benzene ring is flagged for π-calculations, so activate the **Mark\Reset** command and select the 'Pi Atoms' to remove these flags. Next, activate the **PT** tool and select the **CA** atom, the @40 aromatic carbon. Thus loaded, use the mouse cursor to click on all six of the

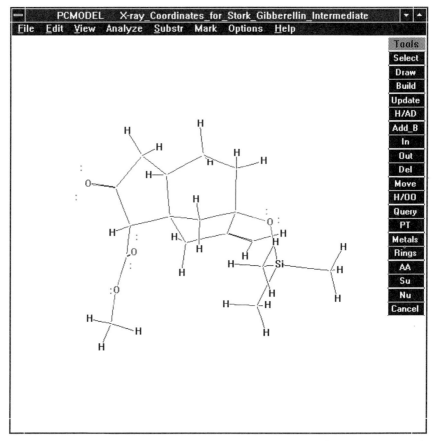

Figure 6-4. Final Structure from Unconstrained Minimization, {GIBRELN2.PCM}.

benzene ring carbon atoms. Activate the **Analyze\Minimize** command to run this structure through an energy minimization sequence, and when the sequence has finished, repeat the minimization a few times. The structure may be given a new title, such as 'Benzene_with_Aromatic_Carbons'; then use the

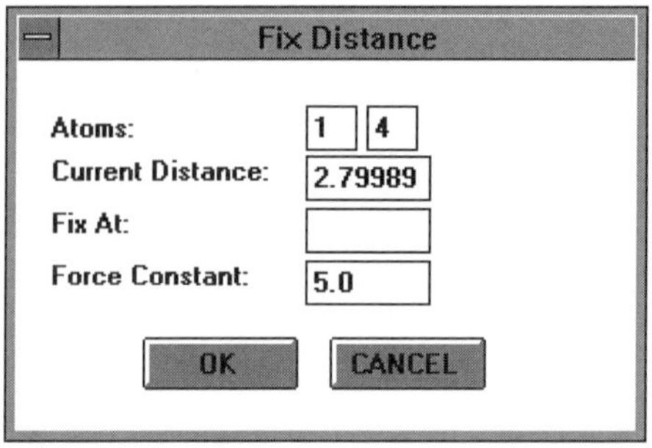

Figure 6-5. 'Fix Distance' Pop-up Dialog Box.

File\Save command to write it to a file with a name like {BENZAROM.PCM}. This file is available from the input file library diskette in the CHAP6\SECT1 subdirectory under this filename, and the result should resemble Fig. 6-6.

With this structure still in the work space, activate the **Substr\Read** command, and read in {BENZAROM.PCM}; the two structures will be placed side by side. Activate the **View\Control_Panel** command to bring up the 'Dials' dialog box, and use the left-hand scroll button of the 'Rotate' 'X' scroll bar to tilt both structures until they are almost perpendicular to the plane of the screen, where the value of X will be about 100. The result after the 'Dials' dialog box has been canceled is depicted in Fig. 6-7.

Now, we will move the right-hand structure into a parallel planes position above the left-hand structure. Activate the **Substr\Move** command, and select the second substructure. In the new 'Dials' dialog box, use the right-hand

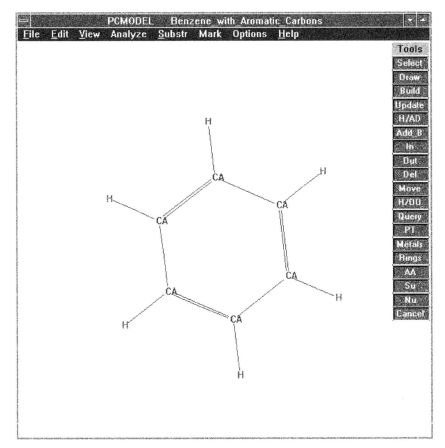

Figure 6-6. Benzene Molecule with 'CA' Aromatic Carbon Atoms, {BENZAROM.PCM}.

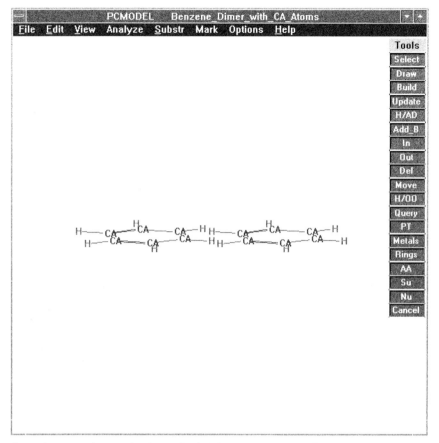

Figure 6-7. Two {BENZAROM.PCM} Substructures Rotated Close to the *YZ* Plane.

scroll button of the 'Translate' 'Y' scroll bar to move the right-hand structure upwards a bit farther than its height, and then the left-hand scroll button of the 'Translate' 'X' scroll bar to move the structure into position directly above the first structure. When this is finished, use the **Select** tool to mark the lower front 'CA' atom of each ring, and then activate the **Mark\Fix_Distance** command. When the 'Fix Distance' dialog box comes up, the result should be like that shown in Fig. 6-8.

In the 'Fix At' line of the dialog box, click in the data box to introduce a data entry cursor, type in 5.5 to fix the distance at 5.5 Å, and then click on the 'OK' button to complete the introduction of the distance constraint flag. A dotted line will appear between these two atoms. Double-click on the **Select** tool to clear the existing selections, and then mark the *same* 'CA' atom in the upper molecule, and an *adjacent* 'CA' atom in the lower molecule. Introduce a distance constraint of 5.5 Å as above. The constraints are set up to form a *virtual* three-membered ring in this way in order to give greater stability to the

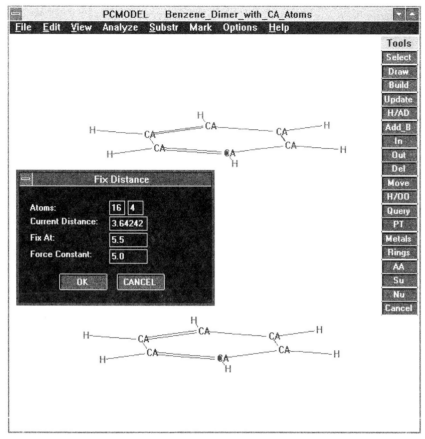

Figure 6-8. Setting the Distance Constraint between Atoms in the {BENZAROM.PCM} Substructures, Stacked in an Orientation Parallel in the YZ Plane.

ensemble, and to lead to a more stable, staggered arrangement of the two rings. Repeat this procedure to create a virtual three-membered ring between the upper rear 'CA' atom of the upper ring and the two 'CA' atoms opposite those marked on the lower ring. Complete the introduction of distance constraints of 5.5 Å for each of these two virtual bonds, and the result should look like Fig. 6-9.

Before running through the energy minimization, a brief digression on the geometry of this ensemble will clarify the results to be obtained. Because of the staggered arrangement of the two rings, the value of the distance constraint does not give the distance between the two rings. The latter is slightly lower, and some triangulation is necessary to arrive at the normal distance between the planes of the two rings. The details of this relationship will be discussed

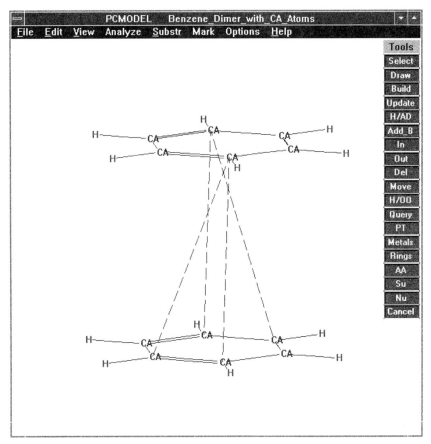

Figure 6-9. Two {BENZAROM.PCM} Substructures with Four Distance Constraints Set.

later in this section, but for the purposes of this exercise one should realize that we are dealing with rough distance measures.

Use the **Analyze\Minimize** command to carry out an energy minimization under the imposed distance constraints. The ensemble should be run through several cycles of minimization to achieve the best geometry. The rings may not end up in exactly parallel planes at this point, but the geometry will straighten out as the rings become closer. The geometry resulting from a distance constraint of 5.5 Å should resemble that seen in Fig. 6-10.

When the minimization is finished, note the value for the energy given in the 'Output' child window, and save the result in an input file with a name like {BENZDIMR.PCM}. Then, activate the **Mark\Reset** command and select the 'Fixed Distances' check box to remove the imposed constraints. *It is necessary to remove the first set of constraints before applying the second; otherwise the effective constraint will be the average of the two.*

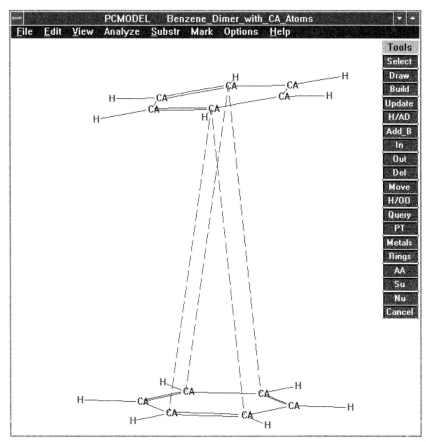

Figure 6-10. Two Stacked {BENZAROM.PCM} Substructures Constrained at 5.5 Å, after Energy Minimization.

For the next stage in this experiment, repeat the procedures detailed above for marking atoms and imposing a distance constraint, this time using a value of 5.0 Å. Continue this up through the energy minimizations, note the value for the energy given in the 'Output' child window, and save the result in the {BENZDIMR.PCM} input file, selecting the 'Append' option when the 'File Exists' dialog box pops up. In this way, this structure file will concatenate the individual files for the benzene dimer ensemble within a single input file.

Repeat these procedures for distance constraints with values of 4.5, 4.0, 3.5, and 3.0 Å, noting the values given for the energy after minimization at each stage. If the rings had been tilted slightly with respect to each other early, they should assume rigid parallel planarity by the time the distance constraint is 3.5 Å. After energy minimization with a distance constraint of 3.0 Å, the ensemble should be as is shown in Fig. 6-11.

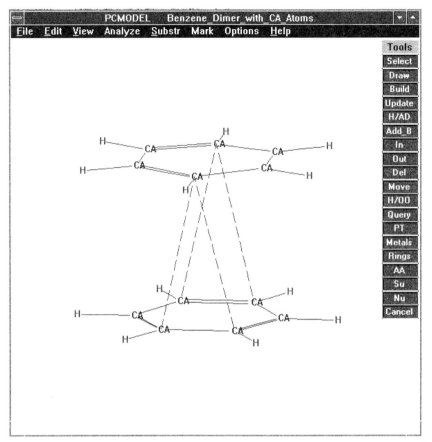

Figure 6-11. Two Stacked {BENZAROM.PCM} Substructures Constrained at 3.0 Å, after Energy Minimization.

Table 6-1 gives an example of the results for this experiment. Note that the energy approaches a minimum when the constraint is around 4.0 Å, and rises sharply at shorter distances. In the later discussion, we will carry out a more precise assessment of this stacking interaction.

The **Fix_Torsions** command allows the user to fix the value for one or two torsional angles in a structure. The constrained angle(s) must be defined by contiguous atoms; placing a **Fix_Torsions** constraint on a flap angle leads to unusual results when an energy minimization is carried out. The algorithms used for fixing a torsional angle are the same as those used for the dihedral driver. A **Fix_Torsions** operation is equivalent to one step of a dihedral driver calculation, which is discussed in greater detail in Section 6.3.

The four atoms comprising the dihedral angle to be constrained first must be marked using the **Select** tool. Then, when the **Fix_Torsions** command is

TABLE 6-1. Results for Benzene Dimer Stacking Experiment

Constraint (Å)	MMX Energy (kcal/mol)
5.5	16.337
5.0	16.170
4.5	15.922
4.0	15.728
3.5	16.428
3.0	22.798

activated, a dialog box entitled 'Fix Torsions' appears as in Fig. 6-12. Except for the title, this dialog box closely resembles the one provided for setting up a dihedral driver operation (see Fig. 6-70). A total of two torsional angles may be fixed in this procedure, and these are labeled 'Angle 1' and 'Angle 2'. These angles may be acyclic or contained within rings.

There are three lines of parameters for each angle: the first line, entitled 'Atom Numbers:', contains four data fields, one for each of the atom file numbers of the four atoms defining the torsional angle; the second line contains a field entitled 'Current Angle:', which gives the initial value in degrees for the defined angle; the third line contains a field entitled 'Fix at:', into which the user may enter a new value at which the angle is to be fixed. There is also a button entitled 'Get Second Angle', which would exit the dialog box and permit the user to mark a new set of four atoms and address the fixation of that second angle.

When finished entering parameters, click on the 'OK' button to confirm the **Fix_Torsions** setup. Each of the four atoms of the dihedral will be overprinted with a 'd1' for the first constrained dihedral (and a 'd2' if a second dihedral constraint has been set). The molecule is now ready for energy minimization. If an angle is set to be fixed at a value other than the current one, the leg

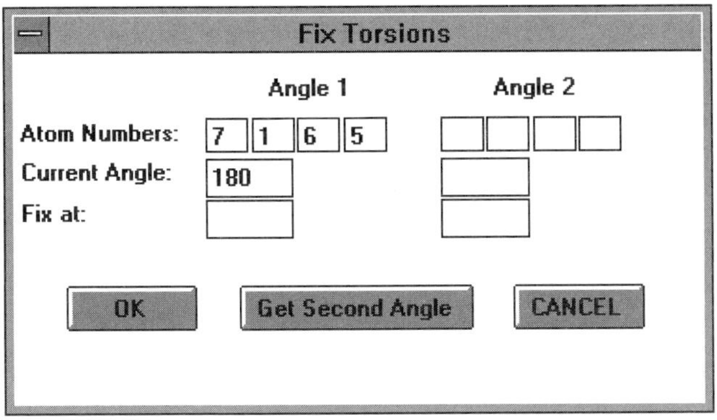

Figure 6-12. The 'Fix Torsions' Dialog Box.

defined by the first two marked atoms will be moved while the leg defined by the last two will remain still.

The uses of the **Fix_Torsions** option would include fine tuning of a dihedral driver sequence, to determine a barrier to rotation or conformational interconversion, or enforcing one or two torsional angles in a flexible molecule to enable the construction of a model having a specific conformation. One important case is that of peptide chains, where one may wish to build a model having a specified tertiary structure. In PCMODEL, it is possible to do this during the building process from the 'Amino Acids' menu window. This is described in Section 3.2.2, where one may choose from the options of an α-helix, a β-sheet, and a type 1 or type 2 β-turn.

If the central atoms are acyclic, any value can be used for the new angle; however, with an endocyclic dihedral a large angle might distort the ring so that the structure might not recover upon minimization. It might be necessary to proceed in steps with a smaller alteration of the angle, fixing the dihedral, doing an energy minimization, and this sequence continued until the desired dihedral angle is reached. This is equivalent to a manual dihedral driver option.

Geometric Constraint Exercise 3: Reinforced Conformation of 2-(2'-Phenolylmethyl)phenol. In this exercise we will assemble the title molecule from phenyl substructures and using the building tools, and then use the **Fix_Torsions** command to reinforce a conformation which will be stabilized by hydrogen bonding. The target torsional angles are obtained from the structure of the related cyclic tetramer calix[4]arene.

Use **File\Open** to open the {BENZAROM.PCM} structure file which was prepared in Exercise 2. Then, use **Substr\Read** to open the {BENZAROM.PCM} structure file again as a substructure. The presentation of these structures may be improved by activating the **Substr\Move** command, and selecting the first (left-hand) substructure first. Use the left-hand scroll button for 'Translate' 'X' in the 'Dials' dialog box to translate the left-hand benzene ring further to the left, and then use the 'Rotate' 'Z' scroll bar to orient the long axis of the ring vertically. After exiting the 'Dials' dialog box, reactivate the **Substr\Move** command, and selecting now the second (right-hand) substructure. Use the 'Rotate' 'Z' scroll bar to orient the long axis of this ring vertically, then exit the 'Dials' dialog box.

Activate the **PT** tool to bring up the 'Periodic Table' menu, and select **O** for oxygen. Use the loaded mouse cursor to convert the uppermost hydrogen on each benzene ring to oxygen, and then activate the **H/AD** tool twice to add hydrogens and lone pairs to these phenolic oxygens. Activate the **Build** tool, and then click on the upper left hydrogen of the right-hand benzene ring to install a methyl group. The result will look like Fig. 6-13.

These two substructures will next be joined together by activating the **Select** tool and marking both the upper right hydrogen of the left hand benzene ring and the lower hydrogen of the new methyl group, followed by activating the

6.1. MODELING WITH GEOMETRIC CONSTRAINTS   399

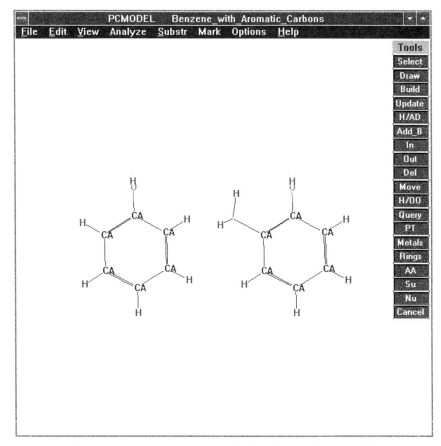

Figure 6-13. Two Substructures for the 'Fix Torsions' Exercise.

Substr\Connect command. Use Edit\Structure Name to name the new molecule '2-(2'-Phenolylmethyl)phenol'. The reconstituted structure should resemble Fig. 6-14.

With the structure in hand, we will prepare for the energy minimization. Activate the Mark\H-Bonds command to set the flag for hydrogen bonding interactions (see Section 6.6). Now, activate the Select tool again, and define the first dihedral by marking in the following order the carbon atom in the right-hand ring which is attached to the phenolic oxygen, the ortho carbon of this ring, the bridging methylene carbon, and the ortho carbon of the left-hand ring. When this is done, activate the Mark\Fix_Torsions command to bring up the 'Fix Torsions' dialog box. In the example, the first angle has an initial value of -120.0. Based upon our calix[4]arene model (see Section 3.4.4), this angle should be $-95°$, and so we will fix it by entering that value in the data entry box beside the 'Fix at:' label (see Fig. 6-15).

Next we select the option to 'Get Second Angle', which dismisses the dialog box and returns us to the structure in the work space, where now the four

400 APPLICATIONS

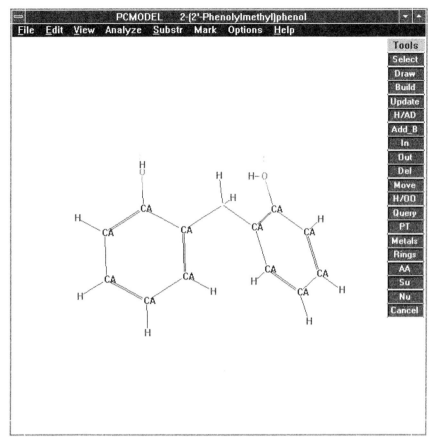

Figure 6-14. Test Structures 2-(2'-Phenolylmethyl)phenol for the 'Fix Torsions' Exercise.

atoms of the first designated dihedral are overprinted with a small 'd1'. We use the same procedure to designate the second dihedral, this time marking in order the ortho carbon of the right-hand ring, the bridging methylene carbon, the ortho carbon of the left-hand ring, and the carbon atom in the left-hand ring which is attached to the phenolic oxygen. The **Mark\Fix_Torsions** command is activated again, and we see that the second angle has an initial value of 149.36. From the calix[4]arene model this angle should be $+95°$, and so this value is entered in the data entry box beside the 'Fix at:' label, as in Fig. 6-16. Clicking on the 'OK' button then confirms the selection of constraints.

After clicking on the **Select** tool once again to unmark the atoms, we initiate an energy optimization by activating the **Analyze\Minimize** command. If desired, the effect of the constraints may be monitored by using the **Query** tool to set up callout report lines for the two dihedral angles. If this is done, it may

6.1. MODELING WITH GEOMETRIC CONSTRAINTS 401

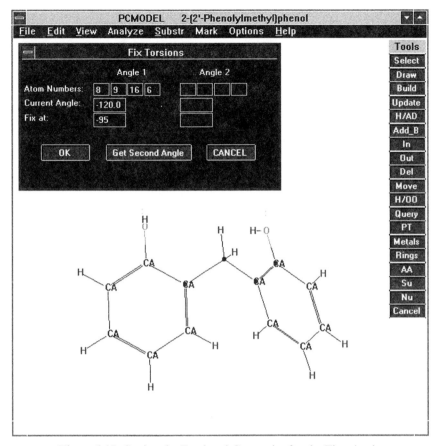

Figure 6-15. Setting the Torsional Constraint for the First Angle.

be seen that these constraints are already realized by the first screen refresh after five iterations. The hydrogen bond should take effect about halfway through the energy minimization, and will be visible as a dotted line between one hydrogen and the other phenolic oxygen. After several cycles of minimization, the resulting structure will be like that found in Fig. 6-17.

Finally, we remove the constraints by activating **Mark\Reset** and selecting the 'Fixed Torsions' check box, and then carry out a final series of energy minimizations. It will be seen that the angle which includes the hydrogen bond donor shrinks, down to approximately 81°, while the other angle expands a bit, up to approximately 98°, as shown in Fig. 6-18. This final structure is available from the input file library diskette in the CHAP6\SECT1 subdirectory as the file {FNOLFNOL.PCM}.

The **Don't Minimize** command exempts a designated substructure from the geometry optimization process, although its energy is still calculated on a

402 APPLICATIONS

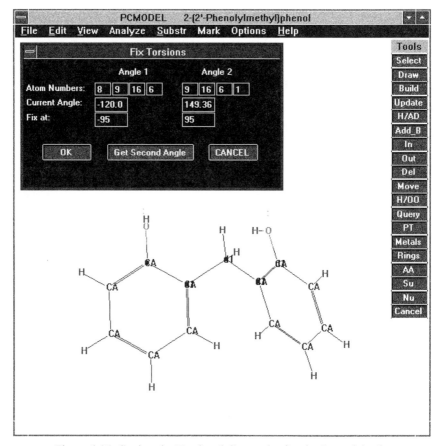

Figure 6-16. Setting the Torsional Constraint for the Second Angle.

static basis. It is equivalent to applying **Fix_Atoms** to all of the atoms in that substructure. This command allows the user to carry out geometry optimization selectively on one of several structures, or on a portion of one structure in the work space. Applications include maintaining the geometry of a structure or substructure while actively subjecting a companion structure or substructure to MMX force field minimization. This will be desirable if the exempted structure has been obtained from experimental methods (e.g. X-ray or neutron diffraction), or from other calculational methods (e.g. another force field, or from quantum mechanical methods). The energy values reported after energy minimization will reflect any strain inherent in the unoptimized geometry of the exempted substructure. The substructure marked with **Don't Minimize** must still be represented in the MMX parameter set, or the energy minimization will fail.

The substructure must be defined prior to invoking this exemption. The **Substr\Don't Minimize** command operates as a toggle, and activating it a

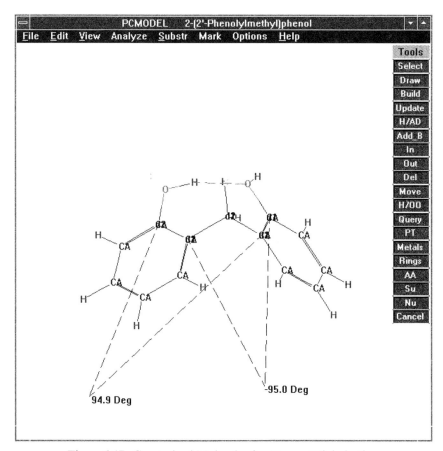

Figure 6-17. Constrained Molecule after Energy Minimization.

second time and selecting the same substructure cancels the exemption. The flag will also be removed if the substructure status is revoked, such as with **Mark\Reset** and checking the 'Substructure Membership' option.

Geometric Constraint Exercise 4: X-ray Structure vs. Geometry Optimized Structure of 2-(2'-Phenolylmethyl)phenol. In this exercise the heavy atom structure of the title compound is read in X-ray format, and this structure is modified by adding hydrogens and other cosmetic changes, and is written to a PCMODEL format file. This latter file is then opened and also read in as a substructure. After repositioning, this substructure is flagged with the **Don't Minimize** command so that no geometry optimization will be carried out on it. The **Query** tool is used to set up a callout report on one of the dihedral angles in each structure, so that the relative effects of an energy minimization sequence on each of the structures may be followed.

404 APPLICATIONS

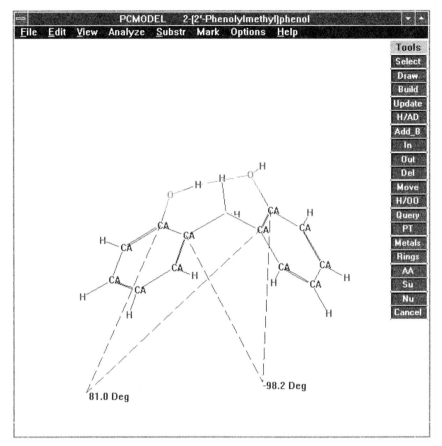

Figure 6-18. Unconstrained Molecule after Energy Minimization.

In a study published on some acetone clathrates of the parent calix-[4]arene, the authors presented partial coordinates for an X-ray crystal structure of the (1:1) clathrate.[637] From this data, an XRA format input file has been prepared for the partial structure corresponding to 2-(2'-phenolylmethyl)-phenol, which is available from the input file library diskette in the CHAP6\SECT1 subdirectory as {CALXCLTH.XRA}. Read this structure file in with **File\Open**, at which point it will appear as in Fig. 6-19.

Note that the structure has some cumulated double bonds in the benzene ring. This is an artifact of the **Read** algorithm for XRA format structures, which uses internuclear distance as the criterion for bond multiplicity. Use the **Del** and **Add_B** tools to remedy this, so that each ring has alternating single and double bonds. Then, use the **PT** tool to bring up the 'Periodic Table' menu, select the aromatic carbon **CA**, and use this to convert all 12 of the ring carbons. Use the **H/AD** tool to add hydrogens and lone pairs, which should produce a structure as shown in Fig. 6-20. Careful inspection of the structure

6.1. MODELING WITH GEOMETRIC CONSTRAINTS

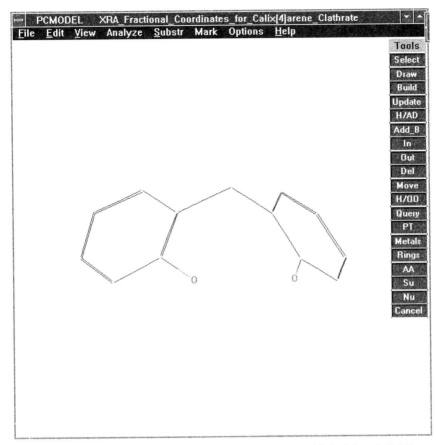

Figure 6-19. Structure from the Fractional Coordinates Published for Calix[4]arene Acetone Clathrate (1:1).[637]

with added hydrogens provides a check on whether the bond multiplicities have been properly adjusted. Write this structure to an input file with a name like {CALXCLTH.PCM}, a sample of which is available from the input file library diskette.

Erase the working structure, open the new structure file {CALXCLTH.PCM}, and then use the **Substr\Read** command to import {CALXCLTH.PCM} as a substructure. Use the **Substr\Move** command to reposition this substructure directly above the initial structure. Activate the **Mark\H-Bonds** command to ensure that these interactions will be included in the energy calculations. Next, activate the **Substr\Don't Minimize** command, and select the second of the two named substructures in order to exempt it from the geometry optimization. Then, use the **Query** tool to set up callout report lines for one of the torsional angles between the two rings, selecting the same angle in each substructure, so that the result is like Fig. 6-21.

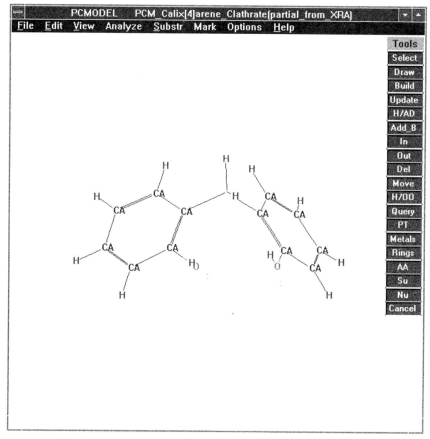

Figure 6-20. Modified Structure for Calix[4]arene Fragment.

Now activate the **Analyze\Minimize** command to begin the energy minimization. It will be seen that the value for the reported torsional angle in the upper structure remains the same, while that in the lower structure changes during the course of the optimization. Several cycles of energy minimization may be carried out to reach a final state for the ensemble. In the case depicted in Fig. 6-22, the OH of the right-hand ring of the lower structure has become the hydrogen bond donor, and the final value for the queried angle is 97.2°, while that for the upper structure has not changed.

No higher level concepts will be dealt with in the context of geometric constraints. Here we will revisit the experiment on the stacking interaction of benzene in a more quantitative way, and examine how PCMODEL handles the input file flags for the **Fix_Distance** command.

Before proceeding with this, some brief comments will be made on the subject of stacking interactions of aromatic rings. For a long time, an assump-

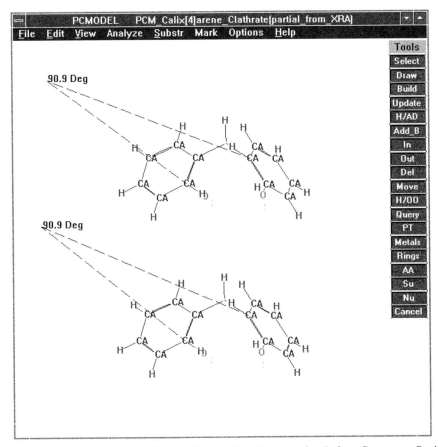

Figure 6-21. Structure Ensemble for **Don't Minimize** Exercise, before Geometry Optimization.

tion of structural chemistry was that the face-to-face, or parallel plane arrangement of aryl rings was the most stable, and this assumption has played a major role in thinking about mechanism in general, and for biological molecules and ensembles in particular.[638] In cases where a charge transfer component is present the face-to-face dimer would benefit from better orbital overlap, but in the absence of such strong electrostatic effects computational studies have repeatedly pointed to an edge-on or T-shaped perpendicular arrangement of the aromatic rings as being more stable.

The *ab initio* calculations show an $\approx$2-kcal/mol potential well for the perpendicular geometry and an $\approx$0.5-kcal/mol potential well for the stacked geometry, but the use of an empirical potential produced equivalent potential wells of $\approx$2 kcal/mol for both arrangements.[639] A comprehensive study employing an empirical potential with Monte Carlo optimizations pointed toward perpendicular geometries being 0.4–0.8 kcal/mol more stable than the

408 APPLICATIONS

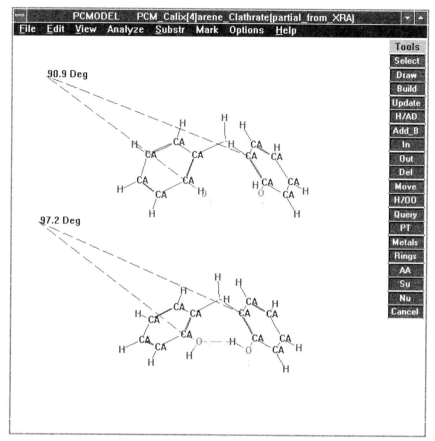

Figure 6-22. Structure Ensemble for **Don't Minimize** Exercise, after Geometry Optimization.

staggered stacked arrangement.[468,640] The structural predictions derived from *ab initio* calculations on benzene-containing molecular structures have been compared with empirical results in a review.[641]

Molecular beam studies, which approximate the gas phase environment which is a basic tenet of empirical force field calculations, have shown that benzene dimer is a polar species.[642,643] A perpendicular arrangement is the only one which can accommodate a dipole moment. The investigators note that a perpendicular relative geometry has been shown in the liquid and solid phases, and they calculate a distance between ring centroids of approximately 5 Å.

In the following exercise, benzene dimer interactions will be examined in two relative geometries, and the interaction potentials are obtained by using a constrained distance driver with intermediate structure optimizations. A known problem in the MM2 force field was that the benzene face-to-face dimer was calculated to have a 3-kcal/mol energy minimum at a separation of approximately 3.5 Å. To redress this excessive attractive interaction, in the MMX force field the attractive portion of the van der Waals potential is ignored for pairs including @2, @4, @40, and @48 carbons. Thus in PCMODEL, where the interactions will be modeled on the basis of van der Waals attractive and repulsive forces and simple electrostatic forces, the benzene face-to-face dimer now has no more than a 1-kcal/mol minimum, and the perpendicular dimer is about equivalent in energy. It should also be noted that an exhaustive search over relative geometries is not used here, nor are any techniques employed for stochastic optimization of geometry (such as constrained molecular dynamics).

Geometric Constraint Exercise 5: Comparison of Benzene Dimer Interactions. In this experiment, we will examine the steric interaction potential of two benzene rings in two different relative geometries, the *staggered stacked* arrangement and the *leading corner perpendicular* arrangement. Appended format input files will be set up for each relative geometry, and edited to provide the best description of the steric interaction potential guided by a series of decreasing distance constraints.

We will begin with an examination of the format of the file {BENZDIMR.PCM} prepared in Exercise 2, which is seen in Fig. 6-23.

For a general explanation of the PCMODEL file format, see Section 4.1. One item of note is the equal and opposite charges on each of the hydrogens and carbons:

for carbon: C −0.03814714
for hydrogen: C 0.03814714

The portion of this file most relevant to the experiment consists of lines 28–32. These five lines, all of which begin with the prefix 'FIX DIS', provide the parameters for the distance constraints. Immediately following the prefix in each of the first four of these lines come the two atom file numbers for those atoms to be connected by the virtual bond. After this we see an 'R', identifying the field for the constraint distance, followed by the value 5.500, in angstroms. The last portion of each line has a 'K' to flag the force constant, followed by its value of 5.000, in millidynes per angstrom. In the fifth line, all of the numerical values are given as 0 or 0.000. This is the signal for PCMODEL to finish processing constraint parameters. As will be seen later, when input files containing constraint commands are present in an appended format file and processed in **Batch** mode, or if a single file is subjected to repeated energy minimization cycles, PCMODEL will insert another line with null values, in order to clear the arrays before processing a new constraint set.

The input file for a study such as the one we will do now may be generated in two ways. One would be to prepare each constituent structure file separately, and writing them consecutively to an appended format file. The second,

APPLICATIONS

```
1    {(PCM Benzene_Dimer_with_CA_Atoms
2    NA 24
3    FL EINT4 UV1 PIPL1
4    AT 1  40   -0.592   -2.513    1.412 B 2 2 6 1 7 1 S 0 C -0.03814714
5    AT 2  40   -1.445   -2.686    0.311 B 1 2 3 1 8 1 S 0 C -0.03814714
6    AT 3  40   -0.906   -2.895   -0.969 B 2 1 4 2 9 1 S 0 C -0.03814714
7    AT 4  40    0.486   -2.934   -1.147 B 3 2 5 1 10 1 S 0 C -0.03814714
8    AT 5  40    1.339   -2.758   -0.046 B 4 1 6 2 11 1 S 0 C -0.03814714
9    AT 6  40    0.801   -2.546    1.233 B 1 1 5 2 12 1 S 0 C -0.03814714
10   AT 7  5    -1.015   -2.349    2.417 B 1 1 S 0 C 0.03814714
11   AT 8  5    -2.539   -2.657    0.451 B 2 1 S 0 C 0.03814714
12   AT 9  5    -1.577   -3.031   -1.834 B 3 1 S 0 C 0.03814714
13   AT 10 5     0.909   -3.100   -2.152 B 4 1 S 0 C 0.03814714
14   AT 11 5     2.433   -2.785   -0.186 B 5 1 S 0 C 0.03814714
15   AT 12 5     1.472   -2.407    2.098 B 6 1 S 0 C 0.03814714
16   AT 13 40    0.213    2.924    1.248 B 14 2 18 1 19 1 S 0 C -0.03814714
17   AT 14 40   -1.067    2.706    0.714 B 13 2 15 1 20 1 S 0 C -0.03814714
18   AT 15 40   -1.228    2.504   -0.666 B 14 1 16 2 21 1 S 0 C -0.03814714
19   AT 16 40   -0.108    2.521   -1.513 B 15 2 17 1 22 1 S 0 C -0.03814714
20   AT 17 40    1.172    2.738   -0.979 B 16 1 18 2 23 1 S 0 C -0.03814714
21   AT 18 40    1.333    2.939    0.401 B 13 1 17 2 24 1 S 0 C -0.03814714
22   AT 19 5     0.340    3.082    2.332 B 13 1 S 0 C 0.03814714
23   AT 20 5    -1.947    2.693    1.379 B 14 1 S 0 C 0.03814714
24   AT 21 5    -2.234    2.333   -1.085 B 15 1 S 0 C 0.03814714
25   AT 22 5    -0.234    2.362   -2.597 B 16 1 S 0 C 0.03814714
26   AT 23 5     2.052    2.750   -1.644 B 17 1 S 0 C 0.03814714
27   AT 24 5     2.339    3.110    0.820 B 18 1 S 0 C 0.03814714
28   FIX DIS 16 4 R   5.500 K   5.000
29   FIX DIS 16 3 R   5.500 K   5.000
30   FIX DIS 13 1 R   5.500 K   5.000
31   FIX DIS 13 6 R   5.500 K   5.000
32   FIX DIS 0  0 R   0.000 K   0.000
33   }
```

Figure 6-23. The First Portion of the Input File {BENZDIMR.PCM}, with Line Numbers in the Left-Hand Column.

and the one we will employ, is to modify an existing file such as {BENZDIMR.PCM} with a text processor to include all of the necessary input files with the distance constraint parameters entered. Then, the input file may be subjected to energy minimization in **Batch** mode, through one or several cycles, and the relative energies recovered from the {*.SUM} log file produced.

In order to get the data for a descriptive potential curve, the values for the distance constraint at different points must be chosen carefully, usually with some trial and error. We recall from Table 6-1 that a minimum energy value was obtained for a constraint of 4.0 Å. Several values of the constraint bracketing 4.0 will help to define the minimum potential well, several values will be needed at much greater distance where the potential should approach that for no interaction between the molecules, and several values will be needed near the minimum distance to help describe the rapidly increasing repulsive portion of the curve. The reader may choose to experiment on their own with this aspect of the problem, but the presentation here will deal with an appended format file containing 18 values for the distance constraint: 7.5, 6.5, 5.5, 5.25, 5.0, 4.75, 4.5, 4.3, 4.2, 4.1, 4.0, 3.9, 3.8, 3.7, 3.6, 3.4, 3.2, and 3.15 Å.

Using a text processor (such as Microsoft Word for Windows™), open the file {BENZDIMR.PCM}, and, once open, save it to a new filename (such as

{BENZDMRF.PCM}) in an ASCII text or non-document mode, to avoid introducing higher order control characters which will derail the parser. Copy to the Windows™ clipboard the entire 33 lines (from start token to stop token) of the first structure's input file. With the cursor to the left of the first start token, paste this file in twice. In lines 28–31, overwrite the distance constraints of '5.500' to '7.500' as is indicated in Fig. 6-24. Then, scroll down to the second inserted file portion, and overwrite the distance constraints of '5.500' to '6.500' analogously to the above.

When this file is run in **Batch** mode, PCMODEL will use the structure from the coordinates, but will enforce the encoded distance constraints prior to energy minimization. These minimizations will go more smoothly if the initial geometry is close to the one required by the constraints, so we will iterate through the equivalent of the initial six files in {BENZDIMR.PCM}, and insert the additional files with the new constraint values by modifying copies of those portions of the original file with an initial geometry closest to the intended new one, until we have built up the complete new file. A copy of the completed, energy-minimized file {BENZDMRF.PCM} is available from the input file library diskette in the CHAP6\SECT1 subdirectory.

The completed file should be subjected to energy minimization by activating the **Analyze\Batch** command, and supplying the appropriate name in the 'Write File for Batch, Dock and Dihedral Driver' dialog box. Then, the 'Batch' dialog box will appear, and offer the default of beginning with the first of 18 structures, and of writing a new input file entitled {BENZDMRF.OUT} and a log file entitled {BENZDMRF.SUM}, which will list the final energies. The simplest procedure is to accept these defaults. When the batch run is finished, the old {BENZDMRF.PCM} may be deleted and the new {BENZDMRF.OUT} should be renamed to {BENZDMRF.PCM}, and again run through a batch energy minimization. After three or four runs of this type, the final {BENZDMRF.OUT} may be renamed to {BENZDMRF.PCM} and saved, and the {BENZDMRF.SUM} file will give the energies of each stage of this constrained interaction.

In order to compare the data just obtained with that for the second portion of this study, it is convenient to convert the values for the distance constraint to the distance between the planes of the benzene rings, equivalent in this case to the distance between the benzene ring **centroids**. This involves triangulation, and the full derivation is found in Section A.4, Appendix A. For the purposes of this exercise, the relevant data are found in Table 6-2, and this data will be presented in graphical form together with the data from the perpendicular-approach portion of this experiment.

28	FIX DIS 16 4 R	5.500 K	5.000	➔	FIX DIS 16 4 R	7.500 K	5.000
29	FIX DIS 16 3 R	5.500 K	5.000		FIX DIS 16 3 R	7.500 K	5.000
30	FIX DIS 13 1 R	5.500 K	5.000		FIX DIS 13 1 R	7.500 K	5.000
31	FIX DIS 13 6 R	5.500 K	5.000		FIX DIS 13 6 R	7.500 K	5.000

Figure 6-24. Editing the Distance Constraints in the {BENZDMRF.PCM} Input File.

TABLE 6-2. Distance Constraints and Steric Energy for the Face-to-Face Benzene Dimer Interaction

Distance Constraint (Å)	Distance between Centroids (Å)	Steric Energy (kcal/mol)
7.50	7.46	16.557
6.50	6.46	16.497
5.50	5.45	16.336
5.25	5.20	16.263
5.00	4.95	16.169
4.75	4.69	16.055
4.50	4.44	15.920
4.30	4.24	15.815
4.20	4.14	15.770
4.10	4.04	15.738
4.00	3.93	15.727
3.90	3.83	15.742
3.80	3.73	15.796
3.70	3.63	15.912
3.60	3.53	16.120
3.40	3.32	16.950
3.20	3.12	18.867
3.15	3.07	19.607

To help visualize the interactions between the two benzene rings, the structure files from three stages of the approach have been chosen for dot-cloud display of the van der Waals surface (see Section 6.2). The ensemble is shown in Figs. 6-25 to 6-27 at ring centroid separations of 7.46 Å, 3.93 Å (minimum energy), and 3.07 Å, respectively.

To begin the second portion of this study, clear the work space and open the file {BENZAROM.PCM}. Activate the **View\Control_Panel** command to bring up the 'Dials' dialog box; ensure that the long axis of the ring is vertical, using the 'Rotate' 'Z' scroll bar to rotate the structure. Next, use the 'Rotate' 'X' scroll bar to tilt the structure until it is almost perpendicular to the plane of the screen, then cancel the 'Dials' box. With this structure still in the work space, activate the **Substr\Read** command, and read in {BENZAROM.PCM} again. Activate the **Substr\Move** command and select the second substructure. Ensure that the long axis of this ring is also vertical, using the 'Rotate' 'Z' scroll bar to rotate the structure. Then, use the 'Translate' 'Y' and 'Translate' 'X' scroll bars to move this second structure into position directly above the first ring. When the structures move too close to the borders, activate the **Update** tool to recenter them. Using the **Query** tool, set up a callout report

6.1. MODELING WITH GEOMETRIC CONSTRAINTS   413

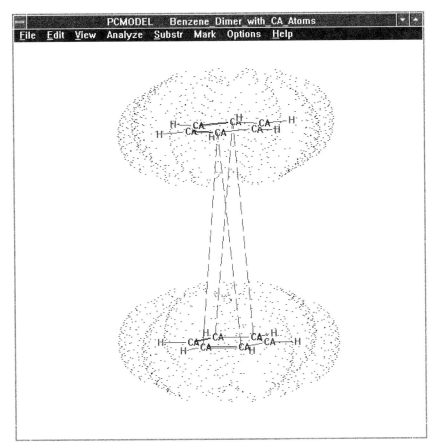

Figure 6-25. Dot-Cloud van der Waals Surface for the Benzene Face-to-Face Dimer at Ring Centroid Separation of 7.46 Å.

line for the distance between the bottom hydrogen of the upper ring and any ring carbon of the lower ring. Use 'Translate' 'Y' to fine-tune the distance to 6.5 Å. Cancel the substructure identities by activating the **Mark\Reset** command and selecting the 'Substructure Membership' choice box. The result after the 'Dials' dialog box has been canceled is depicted in Fig. 6-28.

The next step is to set up distance constraints between the bottom hydrogen of the upper ring and the equivalent of C1, C3, and C5 of the lower ring. If necessary for better viewing, use **View\Control_Panel** to tilt the ensemble so that the individual carbons of the lower ring are clearly visible. Use the **Select** tool to mark the hydrogen and one of the lower ring carbons, then activate the **Mark\Fix_Distance** command as in the earlier exercise. In the 'Fix At:' data entry box in the 'Fix Distance' dialog box, enter the value 6.5. Repeat this

414 APPLICATIONS

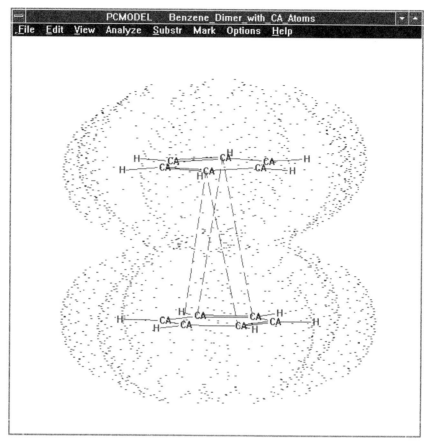

Figure 6-26. Dot-Cloud van der Waals Surface for the Benzene Face-to-Face Dimer at Minimum Energy, Ring Centroid Separation of 3.93 Å.

procedure to establish virtual bonds between the hydrogen and each of C1, C3, and C5 of the lower ring. The result should resemble Fig. 6-29.

Prior to this first minimization, we'd like to ensure a good initial geometry between the two rings. Use **View\Control_Panel** to tilt the ensemble ('Rotate' 'X') so that the viewer is looking down the long axis of the upper ring, and the lower ring is in the plane of the screen. The ensemble should look like Fig. 6-30.

If it is not centered, or if the plane of the upper ring does not bisect the short axis of the lower ring as in Fig. 6-30, designate the upper ring as a substructure (use **Select** to mark one of its attached hydrogens), and then use **Substr\Move** to improve the geometry. It is important to cancel the substructure membership after carrying out such a correction (see above), and then to realign the ensemble to the aspect of Fig. 6-29. Otherwise, proceed to run the ensemble through several cycles of energy minimization. When finished, give

6.1. MODELING WITH GEOMETRIC CONSTRAINTS

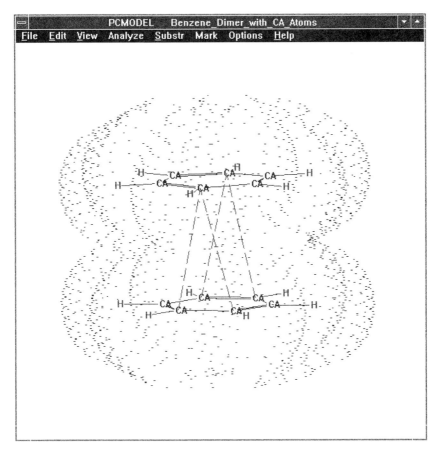

Figure 6-27. Dot-Cloud van der Waals Surface for the Benzene Face-to-Face Dimer at Ring Centroid Separation of 3.07 Å.

the ensemble a descriptive name such as 'Benzene_T-Dimer,' and write it to an input file with a name such as {BENZDMRT.PCM}.

One among several ways to continue this portion of the exercise is to prepare a half dozen energy-minimized benchmark input structures and write these to the same input file in appended format. Next, this input file may be modified with a text processor as before, interpolating enough data points to generate a smooth potential curve. Thus, with the {BENZDMRT.PCM} structure still in the work space, use the **Mark\Reset** command and select the 'Fix Distances' check box to remove the flagged constraints, then reset the same constraints for a value of 5.5, and run through several cycles of minimization, appending the final structure to the {BENZDMRT.PCM} file. Continue this procedure, setting the constraints at 4.5, 3.5, 3.3, and 3.0 in the 'Fix At:' data entry box of the 'Fix Distance' dialog box. Then this appended

416 APPLICATIONS

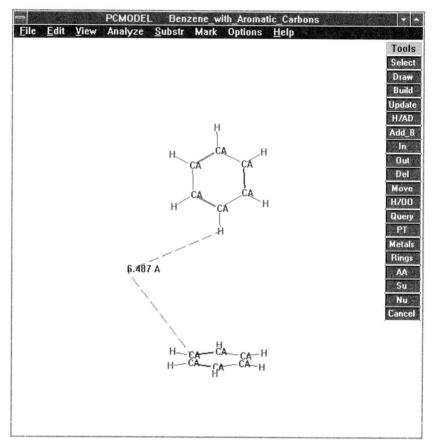

Figure 6-28. Setting Up the Ensemble for the T-Shaped Benzene Dimer; Orienting the {BENZAROM.PCM} Substructures.

format file should be run through a **Batch** mode minimization, as described above.

A useful potential curve may be obtained with a total of 16 constraint values (see Table 6-3). These will be introduced as above, by pasting in modified portions of the input file to generate the remaining 12 desired structure files. For example, the user's current {BENZDMRT.PCM} file contains six structure files. The first three structure files will be fine for their corresponding portions of the potential curve, but between the third (constraint of 4.5 Å) and fourth (constraint of 3.5 Å), we desire two files with intermediate values for the constraint, namely 4.0 and 3.75 Å.

The third structure file (constraint of 4.5 Å) is reproduced in Fig. 6-31. Note in particular line 28, a 'FIX DIS' parameter line filled with null values. While processing multiple files containing constraints in **Batch** mode, PCMODEL introduces a null value line for the constraint portion of each file to zero out any

6.1. MODELING WITH GEOMETRIC CONSTRAINTS   417

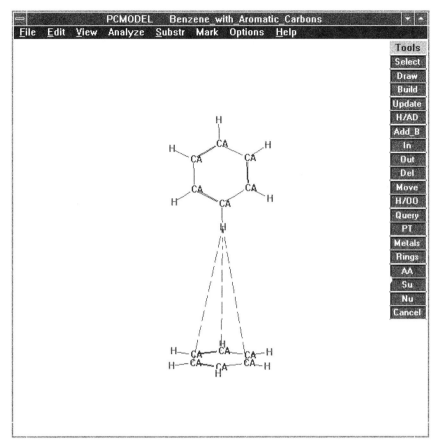

Figure 6-29. Setting Up the Distance Constraints for the T-Shaped Benzene Dimer.

existing parameters prior to reading in the new parameters. An additional line of this type will be introduced for each **Batch** cycle. All but one of these null lines may be removed by text editing without ill effect. Note also lines 29–31, which contain the parameters for a constraint of 4.5, and line 32, which contains another null-value statement to zero out the current constraints.

Using a text processor, copy lines 1–33 to the clipboard, and then paste them in twice, either at the beginning of this portion (i.e. at line 1) or at the beginning of the next portion (i.e. the equivalent of line 34, or line 1 of the following portion). Leaving the first of these three identical portions as is, modify lines 29–31 of the second of these portions as shown at the top of Fig. 6-32, and modify lines 29–31 of the third of these portions as shown at the bottom of Fig. 6-32.

Continue with this sequence to modify two copies of the portion containing a constraint value of 3.5, inserting new values of 3.4 and 3.35; three copies of the portion containing a constraint value of 3.3, inserting new values of 3.25,

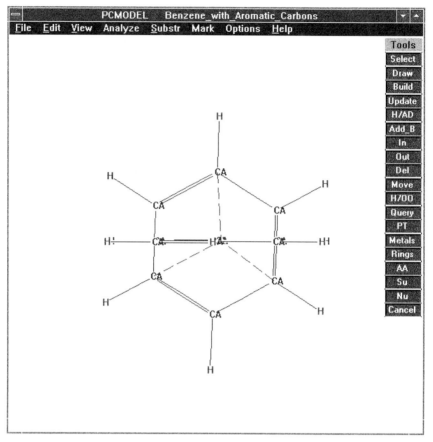

Figure 6-30. Setting Up the Ensemble for the T-Shaped Benzene Dimer; Fine-Tuning the Orientation of the {BENZAROM.PCM} Substructures.

3.2, and 3.1; and three copies of the portion containing a constraint value of 3.0, inserting new values of 2.8, 2.6, and 2.4.

The completed file should be subjected to energy minimization by activating the **Analyze\Batch** command and supplying the appropriate name in the 'Write File for Batch, Dock and Dihedral Driver' dialog box. Then, the 'Batch' dialog box will appear and offer the default: beginning with the first of 18 structures and writing a new input file entitled {BENZDMRT.OUT} and a log file entitled {BENZDMRT.SUM}, which will list the final energies. The simplest procedure is to accept these defaults. When the batch run is finished, the old {BENZDMRT.PCM} may be deleted, and the new {BENZDMRT.OUT} should be renamed to {BENZDMRT.PCM} and again run through a batch energy minimization. After three or four runs of this type, the final {BENZDMRT.OUT} may be renamed to {BENZDMRT.PCM} and

```
1   {PCM Benzene_T-Dimer
2   NA 24
3   FL EINT4 UV1 PIPL1
4   AT 1  40    0.666  -3.045  -1.665 B 2  2  6  1  7  1 S 0 C -0.03814714
5   AT 2  40   -0.708  -3.239  -1.448 B 1  2  3  1  8  1 S 0 C -0.03814714
6   AT 3  40   -1.180  -3.536  -0.160 B 2  1  4  2  9  1 S 0 C -0.03814714
7   AT 4  40   -0.279  -3.638   0.912 B 3  2  5  1 10  1 S 0 C -0.03814714
8   AT 5  40    1.094  -3.444   0.696 B 4  1  6  2 11  1 S 0 C -0.03814714
9   AT 6  40    1.567  -3.148  -0.593 B 1  1  5  2 12  1 S 0 C -0.03814714
10  AT 7  5     1.037  -2.811  -2.677 B 1  1  S 0 C 0.03814714
11  AT 8  5    -1.416  -3.158  -2.290 B 2  1  S 0 C 0.03814714
12  AT 9  5    -2.259  -3.688   0.012 B 3  1  S 0 C 0.03814714
13  AT 10 5    -0.650  -3.871   1.924 B 4  1  S 0 C 0.03814714
14  AT 11 5     1.803  -3.525   1.537 B 5  1  S 0 C 0.03814714
15  AT 12 5     2.646  -2.995  -0.763 B 6  1  S 0 C 0.03814714
16  AT 13 40   -0.149   4.742   0.352 B 14 2 18  1 19  1 S 0 C -0.03814714
17  AT 14 40   -1.246   4.065  -0.202 B 13 2 15  1 20  1 S 0 C -0.03814714
18  AT 15 40   -1.291   2.662  -0.180 B 14 1 16  2 21  1 S 0 C -0.03814714
19  AT 16 40   -0.237   1.936   0.397 B 15 2 17  1 22  1 S 0 C -0.03814714
20  AT 17 40    0.861   2.613   0.951 B 16 1 18  2 23  1 S 0 C -0.03814714
21  AT 18 40    0.906   4.016   0.928 B 14 1 17  2 24  1 S 0 C -0.03814714
22  AT 19 5    -0.114   5.845   0.334 B 13 1 S 0 C 0.03814714
23  AT 20 5    -2.075   4.636  -0.655 B 14 1 S 0 C 0.03814714
24  AT 21 5    -2.153   2.130  -0.615 B 15 1 S 0 C 0.03814714
25  AT 22 5    -0.271   0.834   0.415 B 16 1 S 0 C 0.03814714
26  AT 23 5     1.689   2.042   1.404 B 17 1 S 0 C 0.03814714
27  AT 24 5     1.768   4.548   1.364 B 18 1 S 0 C 0.03814714
28  FIX DIS 0 0 R   0.000 K   0.000
29  FIX DIS 22 2 R  4.500 K   5.000
30  FIX DIS 22 6 R  4.500 K   5.000
31  FIX DIS 22 4 R  4.500 K   5.000
32  FIX DIS 0 0 R   0.000 K   0.000
33  }
```

Figure 6-31. The Third Portion of the Input File {BENZDMRT.PCM}, with Line Numbers in the Left-Hand Column.

saved, and the {BENZDMRT.SUM} file will give the energies of each stage of this constrained interaction.

To help visualize the interactions between the two benzene rings, the structure files from three stages of the approach have been chosen for dot-cloud display of the van der Waals surface (see Section 6.2). The ensemble is shown in Figs. 6-33 to 6-35 at ring centroid separations of 8.85 Å, 5.38 Å (minimum energy), and 4.46 Å, respectively.

In order to compare the data just obtained with that for the first portion of this study, it is convenient to convert the values for the distance constraint to the distance between the benzene ring centroids. This involves triangulation,

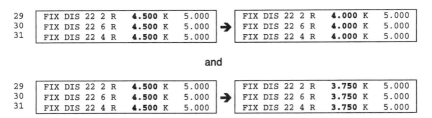

Figure 6-32. Editing the Distance Constraints in the {BENZDMRT.PCM} Input File.

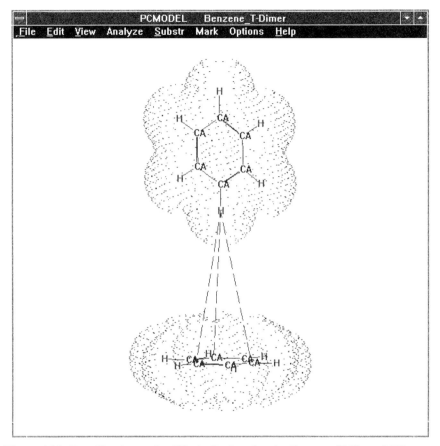

Figure 6-33. Dot-Cloud van der Waals Surface for the Benzene T-Shaped Dimer at Ring Centroid Separation of 8.85 Å.

and the full derivation is found in Section A.4 of Appendix A. For the purposes of this exercise, the relevant data are collected in Table 6-3, and the data for the two portions of this exercise are presented in graph form in Fig. 6-36. It can be seen that the potentials for the two orientations are of similar shape, differing in the distance between ring centroids at minimum energy.

The face-to-face dimer at the PCMODEL minimum-energy geometry has a separation between ring centroids of 3.93 Å, compared to literature values of ≈3.5 Å (empirical),[639] ≈3.8 Å (ab initio),[639] and 3.77 Å (empirical).[468] The perpendicular T-shaped dimer at the PCMODEL minimum-energy geometry has a separation between ring centroids of 5.38 Å, compared to literature values of ≈4.9 Å (empirical and ab initio)[639] and 5.19 Å (empirical).[468] A separation of 4.97 Å was determined for crystalline benzene pairs.[642]

The potential well for both geometries here is approximately 0.8 kcal/mol. Values for the potential well for the face-to-face dimer have been reported in

6.2. SURFACES AND VOLUMES

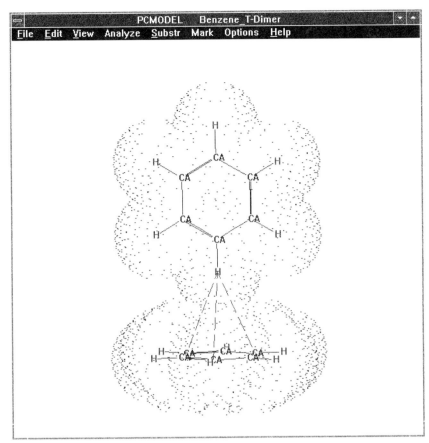

Figure 6-34. Dot-Cloud van der Waals Surface for the Benzene T-Shaped Dimer at Minimum Energy: Ring Centroid Separation of 5.38 Å.

the literature as ≈ 0.6 kcal/mol (*ab initio*),[639] ≈ 2.1 kcal/mol (empirical),[639] and 1.7 kcal/mol (empirical).[468] Values for the potential well for the perpendicular dimer have been reported in the literature as ≈ 2.6 kcal/mol (empirical and *ab initio*)[639] and 2.15 kcal/mol (empirical).[468]

6.2. SURFACES AND VOLUMES

Molecular surface and molecular volume are physical concepts which are arbitrarily defined and calculable, but impractical to determine by direct empirical measurement. Their usefulness depends upon an understanding of these definitions and an appreciation of their implicit limitations. At a basic level, molecular surfaces can aid in visualizing the shape of molecules and can provide a qualitative feel for how two or more molecules might interact with one

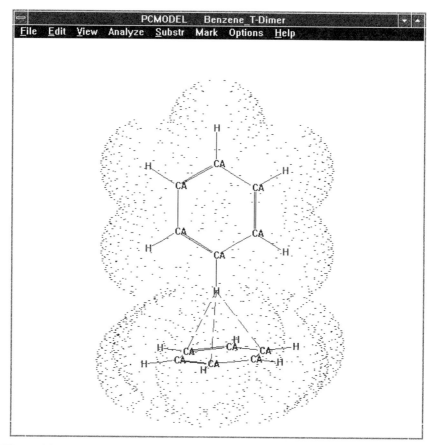

Figure 6-35. Dot-Cloud van der Waals Surface for the Benzene T-Shaped Dimer at Ring Centroid Separation of 4.46 Å.

another. Surfaces which are coded on some scale, such as charge $(+/-)$ or lipophilicity/hydrophilicity, can provide high information density in a single graphic image.

On a more advanced level, molecular surfaces and volumes may be used quantitatively to explore the tertiary structure of macromolecules, solvation phenomena, chemical reactivity, and molecular recognition or docking. One measure of the importance of molecular surfaces and volumes is the number of review articles available on their various aspects.[18,25,113,121,644–650]

In this section, we begin with an explanation of the PCMODEL treatment of molecular surfaces and molecular volume, and the exercises cover simple uses of options available using the amino acid histidine as a substrate. In the later discussion, the construction of molecular surfaces is covered in greater detail and additional exercises are provided. Finally, some more sophisticated appli-

TABLE 6-3. Distance Constraints and Steric Energy for the T-Shaped Benzene Dimer Interaction

Distance Constraint (Å)	Distance between Centroids (Å)	Steric Energy (kcal/mol)
6.50	8.85	16.562
5.50	7.82	16.508
4.50	6.78	16.336
4.00	6.25	16.150
3.75	5.98	16.020
3.50	5.71	15.875
3.40	5.60	15.821
3.35	5.55	15.798
3.30	5.49	15.779
3.25	5.44	15.764
3.20	5.38	15.756
3.10	5.27	15.767
3.00	5.16	15.831
2.80	4.93	16.245
2.60	4.70	17.380
2.40	4.46	19.926

cations of molecular surfaces and volumes are discussed and exemplified with two exercises.

There are four commands in PCMODEL which are concerned with molecular surfaces and volumes: **View\CPK_Surface**, **View\Dot_Surface**, **Analyze\Surface Area**, and **Analyze\Volume**. Each of these will be discussed in turn, with an example to illustrate its usage.

The **View\CPK_Surface** command gives a rendering of a solid molecular surface as the sum of the overlapping spheres corresponding to the constituent atoms. The graphic representation produced by **CPK_Surface** approximates the physical CPK models (see Section 1.1). Activating this command brings up a dialog check and option box entitled 'CPK Models', as shown in Fig. 6-37. This allows the user control over the parameters of the surface construction. The first choice, titled 'Atoms:', governs which portions of the molecule(s) to include: 'All' (all atoms), 'Heavy' (nonhydrogen atoms), or 'Select' (user-selected atoms). The second choice, titled 'Radius:', is what value to use for the radius of the spheroids; here the user may click on the appropriate radio button for 1.0, 0.65, or 0.20 (the fraction of the van der Waals radius). The third choice, titled 'Color By:', is how to color the surface; here the user may click on the appropriate radio button for 'Type' (element or atom type), 'Charge' (electrostatic polarization), or 'Substructure' (substructure membership).

The surface is painted on the screen in layers along the Z-axis, and depending on the hardware configuration may take ten seconds to a minute or so to

424 APPLICATIONS

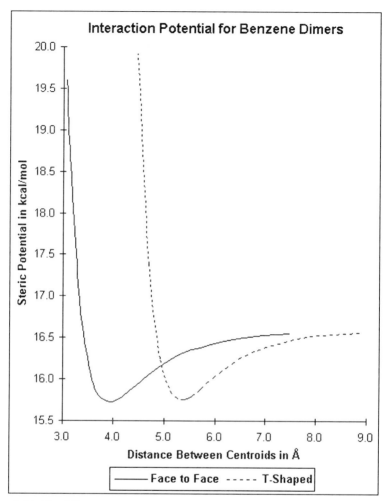

Figure 6-36. The MMX Interaction Potential for the Face-to-Face and the T-Shaped Benzene Dimers.

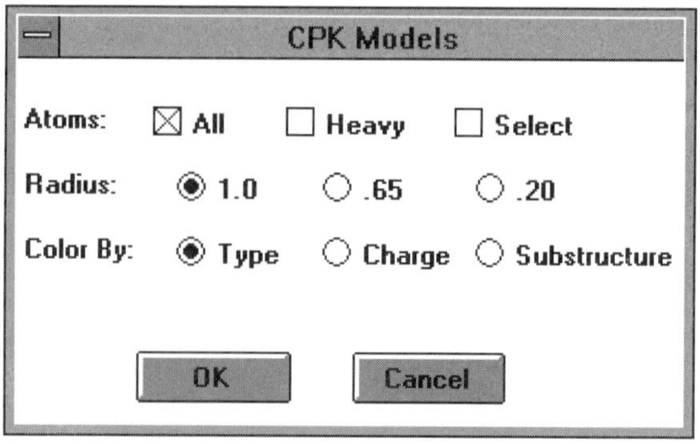

Figure 6-37. The 'CPK Models' Dialog Box.

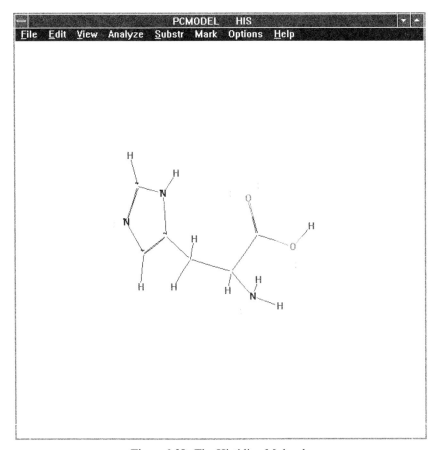

Figure 6-38. The Histidine Molecule.

be complete. It may be removed by again activating the **CPK_Surface** command and then clicking on the 'Cancel' button in the dialog box, or by selecting another option within the **View** menu on the Menu Bar. The **Query** command may be used while a CPK surface is active. After the desired atoms have been selected, PCMODEL will redraw the CPK surface and then produce the requested **Query** report line. The display produced with the 'Radius:' option set at '.20' will roughly approximate the ball-and-stick model discussed above.

The Lee–Richards algorithm[80] is used to calculate the van der Waals molecular surface area, by presenting the aggregate of those portions of the individual atomic van der Waals spheroidal surfaces which are not contained by the spheroidal surfaces of any other atoms in the molecule. The surface spheroidal sections are colored according to atom type unless the user selects one of the other two options.

Exercise 1: The CPK Surface of Histidine. In the work space, activate the **AA** tool, and from this pop up menu click on the **His** button to import the histidine template. It may be reoriented for better viewing, to give a perspective such as is seen in Fig. 6-38.

Activate the **View\CPK_Surface** command. When the 'CPK Models' dialog box pops up, click on the 'OK' button to accept the default settings. The atom color-coded surface is drawn in triangular sections along the Z-axis, so that the surface is drawn first for the part of the molecule "farthest into the screen" from the viewer, gradually working closer. The image should resemble that seen in Fig. 6-39.

If the final CPK surface structure appears too large for proper viewing, the structure of the molecule may be reduced in size by activating the **View\Control_Panel** command and adjusting the 'Scale' scroll bar. For this surface construct, some jagged lines are seen along the seams where the atoms join. This is because the developers chose to optimize the time necessary to

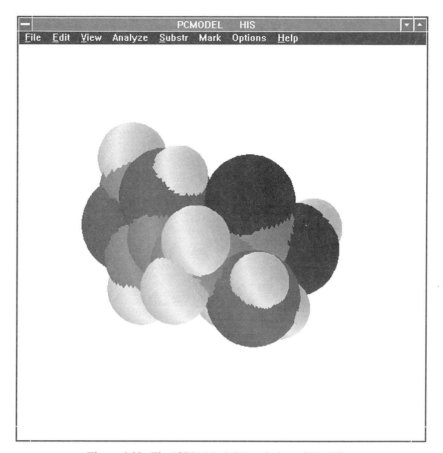

Figure 6-39. The 'CPK Model' Rendering of Histidine.

paint this surface rather than the fine detail of the graphic image. Depending upon the graphics hardware configuration of the desktop platform in use, the user may see a lighter aspect on one side of the structure, which is a depth cuing device.

The **View\Dot_Surface** command draws a diaphanous surface around the structure in the work area to depict the molecular skeleton and a selected surface simultaneously. Activating this command will bring up a dialog options box entitled 'Dot Surface Display', as shown in Fig. 6-40.

Here the user may choose values for three parameters to be used in constructing the surface. The first is the 'Surface Type:' to be drawn: 'VdW', a van der Waals surface colored to show atom types; 'Water', a water accessible surface; or 'Charge', a van der Waals surface colored to show electrostatic affinity. The second option is the type of 'Display:' desired—whether 'All' atoms or only 'Selected' ones will have a surface drawn around them. The last option is the 'Dot Spacing:' to be used in the surface—that is, the distance between dots, which will determine the density or opacity of the surface. The smaller the distance the more opaque the surface.

When the desired parameters have been selected, clicking on the 'OK' button initiates the construction of the surface, which proceeds along the Y-axis from bottom to top. The surface is diaphanous, and the molecular skeleton can be seen within.

Exercise 2: The van der Waals Dot Surface of Histidine. In the work space, activate the **AA** tool, and from its pop-up menu click on the **His** button to import the histidine template. It may be reoriented for better viewing, to give a perspective such as was seen in Fig. 6-38 above.

Activate the **View\Dot_Surface** command. When the 'Dot Surface Display' dialog box pops up, click on the 'OK' button to accept the default settings. The atom color-coded dot surface is drawn in along the Y-axis, so that the

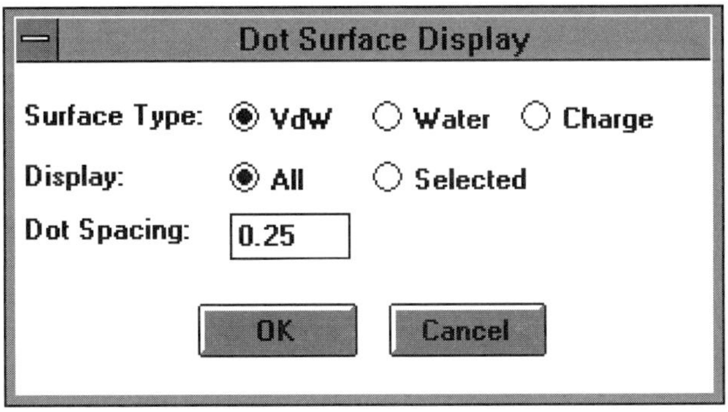

Figure 6-40. The 'Dot Surface Display' Dialog Box.

surface is begun at the bottom part of the molecule, and proceeds up to the top. The final image should resemble that seen in Fig. 6-41.

The **Analyze\Surface Area** command computes the van der Waals molecular surface area of the molecule(s) in the work space. Activating this command brings up the 'Dot Surface Calculation' dialog window shown in Fig. 6-42.

Briefly, the algorithm for calculating the surface area randomly disperses *dots*, or points, on the surfaces of all constituent atoms, eliminates those dots which find themselves inside the surface of any atom, and uses the remaining dots to provide an estimate of the surface area.[25,81,644,651] A related algorithm may be used for the evaluation of molecular volume, as will be seen below.[652]

The default calculation uses 100 dots and repeats the calculation 20 times. The repetition is to provide more confidence in this stochastic calculation. The user may change the resolution (more dots meaning higher resolution) and

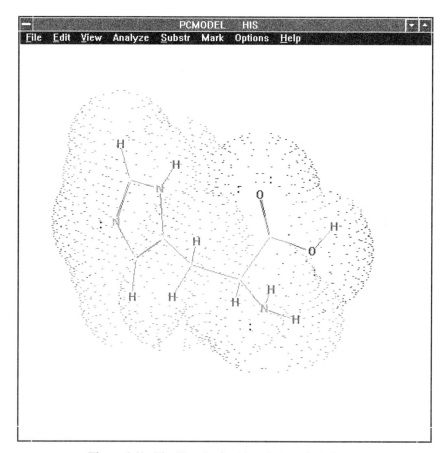

Figure 6-41. The 'Dot Surface' Rendering of Histidine.

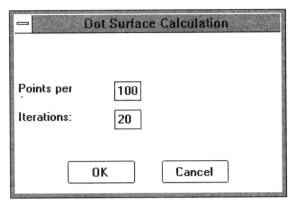

Figure 6-42. The 'Dot Surface' Calculation' Dialog Box.

may select how many iterations of the calculation will be carried out. The unit of area is the square angstrom ($Å^2$).

When the calculation is complete, the program reports out with an information window entitled 'Surface Area Results'. This report gives values for the total area, saturated area, unsaturated area, and polar area. This will be explained in the subsequent discussion.

Exercise 3: The van der Waals Surface Area Calculation for Histidine. With the histidine molecule appearing in the work space, activate the **Analyze\Surface Area** command, and click on the 'OK' button to accept the default parameters for the calculation. The result is given in Fig. 6-43.

The **Analyze\Volume** command computes the volume of the molecule in the work space. The algorithm for making this calculation involves first prepares a molecular surface, as described above: it randomly disperses dots in the area in and around the molecule. The proportion of these dots which fall within the molecular surface is used to generate the volume measurement.

There are no user-selectable parameters for this calculation.

Exercise 4: The Molecular Volume Calculation for Histidine. Import the histidine molecule as above. Activate the **Analyze\Volume** command. After a short time, the results are reported in an information window entitled 'Volume Results', as shown in Fig. 6-44. The molecular volume is given in cubic angstroms, and the molar volume is given in cubic centimeters (cm^3).

The concept of a molecular surface transposes a construct from classical mechanics, the discrete hard physical surface, into a realm governed by quantum mechanics. At the molecular level, the **Heisenberg uncertainty** relation guarantees that the exact positions of subatomic particles are effectively unknowable. The amount of uncertainty is relative, though. Assumptions about the positions of atomic nuclei are fairly (relatively) safe in this context.

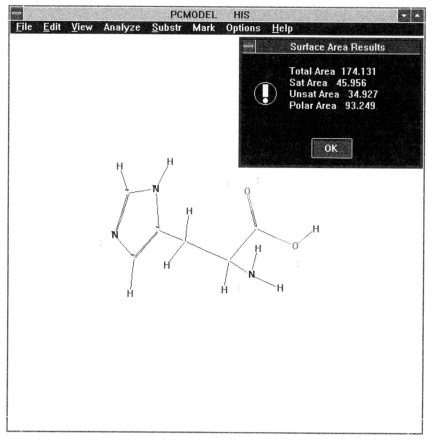

Figure 6-43. The Surface Area Calculation for Histidine.

However, assumptions about the positions of electrons are not justified, and "... the fuzzy, cloud-like electronic distribution of a molecule is very different from a macroscopic body, ... even for a formal, fixed nuclear configuration the electronic cloud does not have a true boundary surface, and no true molecular surface exists."[645]

These caveats having been stated, the attractiveness of the concept of a molecular surface is compelling because it enables the treatment of such a wide variety of intermolecular phenomena. It is also far easier for the modeler to visualize these phenomena in more classical terms, so we recognize the limitations and move forward. Most often, the molecular surface is seen as equivalent to an outer boundary, or probability limit, of the movements of the electrons associated with those atoms located on the periphery of the molecule.

An arbitrarily chosen value of the electron cloud density as it decreases with increasing distance from the associated nuclei may be taken as such a

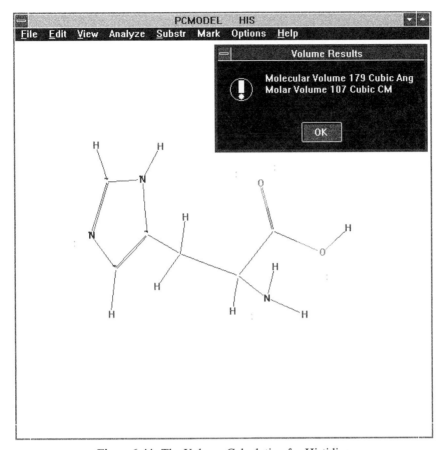

Figure 6-44. The Volume Calculation for Histidine.

limit. One convenient place to draw the boundary is at that distance from the nucleus at which the nonbonded interaction potential has an energy minimum, i.e. where the London dispersion force or van der Waals attraction is maximized, at the van der Waals radius. When two or more nuclei are bonded to each other in a molecule, then that molecule as a whole will possess a surface at which nonbonded attraction will be maximized, the van der Waals surface. This surface consists of the exposed portions of the van der Waals spheroids of the individual constituent atoms.

In their important first paper on this subject,[80] Lee and Richards presented the now classic algorithm for preparing a van der Waals surface. They assigned van der Waals radii to each atom or group of atoms, employing a **united atom** approximation in which nonpolar hydrogens were considered together as a unit with their backbone atom. The molecule, as a collection of spheroids determined by the van der Waals radius, was visualized by taking parallel plane sections separated by a specific spacing. In such a plane, each atomic

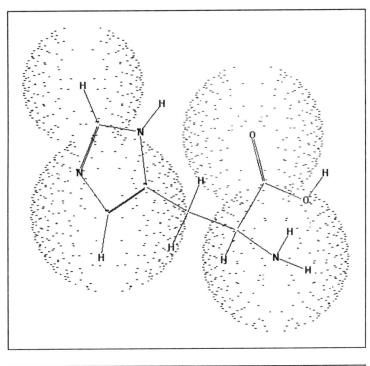
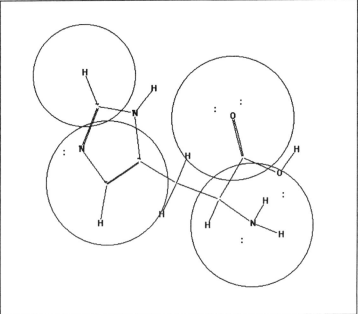

Figure 6-45. Sample van der Waals Spheroids for the Histidine Molecule. Clockwise from lower left: @2 carbon, @5 hydrogen, @7 oxygen, and @8 nitrogen.

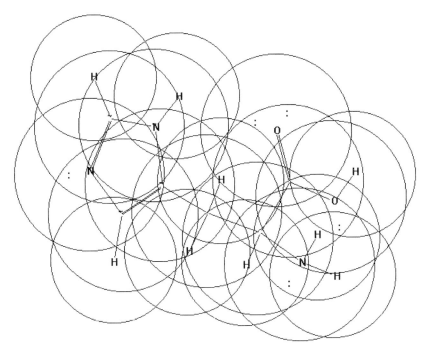

Figure 6-46. All van der Waals Spheroids for the Histidine Molecule.

spheroid projects a circular cross section. For each plane, only the exterior portions of these circles are kept: those portions which are intersected by other circular sections are eliminated. The surface area is then computed as that fraction of the summed atomic spheroidal surface determined by the ratio of the exposed arc sections to the total of circular sections.

In the original method, the arc sections surrounding polar atoms such as oxygen and nitrogen were drawn as dotted lines, and those sections surrounding nonpolar atoms such as carbon and sulfur were drawn as solid lines. With the advent of computer color graphics, programs such as PCMODEL can draw all of the sections as dotted lines, and color them as coded by atom type or by charge density. In the surface area calculation report given (see Fig. 6-43), the total area is reported (in square angtroms), and is further broken down into three types. These designations expand upon the polar–nonpolar distinction made by Richards.[80,644] The 'Sat Area' (saturated area) corresponds to portion of surface ascribed to @5 hydrogen and @1 carbon). The 'Unsat Area' (unsaturated area) corresponds to multiply bonded @2, @3 (bonded to no more than one heteroatom), @4, @22, @29, and @30 carbon, but not @40, @48, @56, or @57 carbon. The 'Polar Area' corresponds to @3 carbon when it is bonded to two or more heteroatoms; @6, @7 oxygen; @8, @9, @10 nitrogen; and attached @21, @23, @24, @28 acidic hydrogen. If the molecule(s) in the work space contain some atoms which are not parametrized

434 APPLICATIONS

for this calculation, the result will be incomplete or will fail, with '0.000' or '-nan' given in the information window report.

The construction of this surface will be illustrated in a two-dimensional cross section in the example of the amino acid histidine, which was seen before in Fig. 6-38. In Fig. 6-45, the structure of histidine is shown with van der Waals spheroids around four of the atoms, above as dot surface representations and below as circles having a radius equal to the van der Waals radius. In this case, the van der Waals radii are as follows: for the sp^2 @2 carbon, 1.94 Å; for the @5 hydrogen, 1.50 Å; for the sp^2 @7 carbonyl oxygen, 1.74 Å; for the sp^3 @8 amino nitrogen, 1.82 Å.

When all of the atomic spheroids have been drawn in, the result is quite busy, as seen in Fig. 6-46. Removing the portions of the atomic spheroids which are contained within other portions yields a much simpler *van der Waals silhouette* of histidine, as shown in Fig. 6-47. In Fig. 6-48, the **View\Dot_Surface** command was employed with the 'Dot Spacing:' parameter reset to 0.10 to prepare a higher density dot surface for this molecule.

Two other types of surface have been defined which have been found to be useful constructs in understanding intermolecular interactions. Unfortunately, in each case there is more than one name attached to the definition in the literature and in common use, making for some confusion. Here, we will associate surface names with the published definitions, and then proceed with what may seem arbitrarily chosen labels for further discussions.

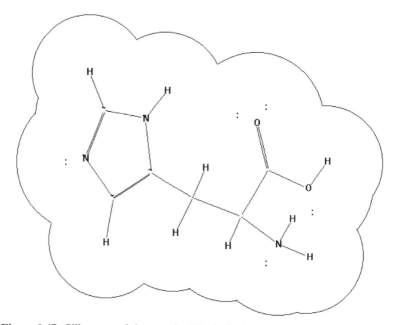

Figure 6-47. Silhouette of the van der Waals Surface for the Histidine Molecule.

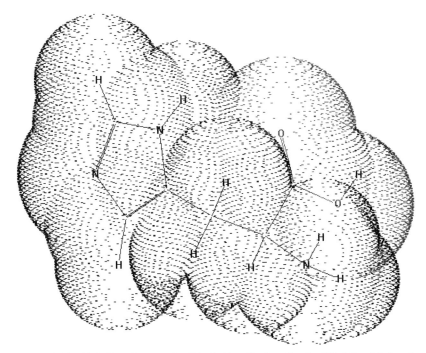

Figure 6-48. High-Density van der Waals Dot Surface for the Histidine Molecule.

Lee and Richards also introduced the concept of a **solvent-accessible surface**, which is related to the collected portions of the molecule which can be brought into van der Waals contact with an external solvent molecule.[80] They conceived of a **probe sphere** of radius 1.4 Å as a convenient approximation of a water solvent molecule, and defined the solvent-accessible surface as the area traced out by the center of this sphere as it rolls over the surface of the molecule. In practice, this was achieved by adding the radius of the probe sphere to the van der Waals radius of the atom concerned when building the atomic spheroids, and then preparing the surface as detailed above. A two-dimensional representation of this is given in Fig. 6-49, where for clarity the probe sphere is smaller than proportionality would dictate. Because this surface is formed by the interaction of a probe sphere with an agglomeration of other spheres, it tends to simplify the complex creases at the junctions of three or more atomic spheroids. These smoothed sections were termed *reentrant surface* by Richards.[644] Several related algorithms have been published.[81,653-655]

In Fig. 6-50 is seen a high-dot-density ('Dot Spacing:' parameter reset to 0.10) water-accessible surface of histidine as constructed by selecting the proper options in **View\Dot_Surface**.

Another type of surface was defined by Richards as the solvent-accessible surface in later work.[644] In this case, it is the sum of the contacts made by the

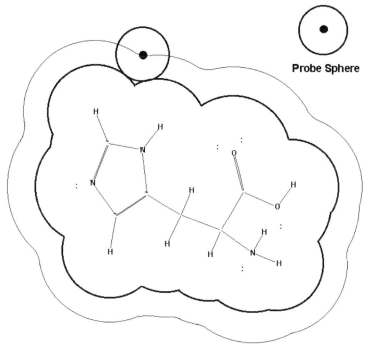

Figure 6-49. Determining the Solvent-Accessible Surface for the Histidine Molecule.

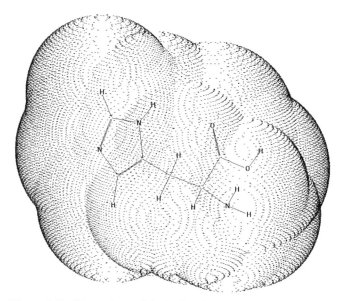

Figure 6-50. Water-Accessible Surface for the Histidine Molecule.

probe sphere with the van der Waals spheroids of the subject molecule which constitute the surface, instead of using the center of the probe sphere, as shown in Fig. 6-51.

Connolly adopted this latter definition of the solvent accessible surface in his presentation of the dot surface and the analytical algorithms for surface calculations.[87,651] For the purposes of these discussions, we use the term *van der Waals surface* to mean the surface determined by the original Lee–Richards algorithm,[80] and the *solvent-accessible surface* also as described in that work; the newer construct due to Richards[644] and Connolly[87,651] will be referred to as the *solvent-excluded surface*.

Two exercises follow which compare analytical approximations of volume with that derived from the dot-surface calculation.

Exercise 5: An Analytical Approximation vs. the Dot-Surface Calculation of Volume for Benzene. In this exercise we will compare the results of a crude analytical surface calculation for benzene with that for the PCMODEL dot-surface calculation.

Begin by importing the 'Ph' benzene structure from the 'Rings' template, and use the 'Dials' dialog box to tilt the ring toward the YZ plane (i.e. edge-on to the viewer). The van der Waals dot surface may be built around the structure to help with visualization, by activating the **View\Dot_Surface** command, and the surface area calculation can be called with the **Analyze\Surface Area**

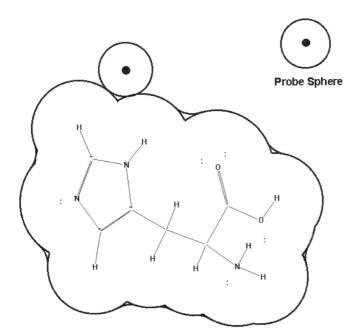

Figure 6-51. The Solvent-Excluded Surface for Histidine.

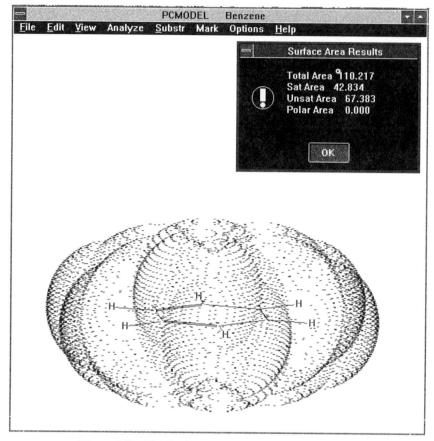

Figure 6-52. The Surface Area Calculation for Benzene.

command, and the result will look like Fig. 6-52. The calculated surface area is 110.217 Å², compared with 107.4 Å² by the molecular free surface algorithm of Gavezzotti.[656]

Now we will approximate the benzene ring as a cylindrical section, and calculate its volume. Using the **Query** tool, the distance from C1 to C4 in this benzene structure will be found to be 2.8 Å, and the van der Waals radius for carbon is 1.94 Å. Ignoring the hydrogens, and adding the carbon-to-carbon distance to twice the carbon van der Waals radius gives a diameter of 6.68 Å, or a radius of 3.34 Å. The height of the cylinder can be given by twice the van der Waals radius of the @2 carbons which make up this ring, 2 × 1.94 Å, or 3.88 Å. A sketch of this cylindrical section is given in Fig. 6-53.

The area of each face of this solid is given by

$$a_{face} = \pi r^2 = \pi(3.34 \text{ Å})^2 = 35.046 \text{ Å}^2$$

Cylindrical Section Approximating Benzene

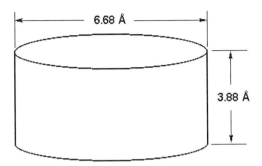

Figure 6-53. The Cylindrical Section Approximation of Benzene.

The area of the side band will be given by the product of the height and the circumference:

$$a_{side} = \pi \, dh = \pi(6.68 \text{ Å})(3.88 \text{ Å}) = 81.425 \text{ Å}^2$$

The total area by this model will be

$$a_{tot} = 2a_{face} + a_{side} = 2 \times (35.046 \text{ Å}^2) + 81.425 \text{ Å}^2 = 151.52 \text{ Å}^2$$

The value produced from this crude approximation is about 37% larger than that calculated by the dot surface algorithm.

Exercise 6: *An Analytical Approximation vs. the Dot Surface Calculation of Surface for Buckminsterfullerene.* In this exercise we will compare the results of a crude analytical surface calculation for buckminsterfullerene with that for the PCMODEL dot surface calculation using different numbers of points and iterations.

Begin by importing the 'C60' buckminsterfullerene structure from the 'Rings' template menu. The van der Waals dot surface may be built around the structure to help with visualization, by activating the **View\Dot_Surface command**, and the surface area calculation can be called with the **Analyze\Surface Area** command; the result will look like Fig. 6-54. The calculated surface area is given as 404.978 Å^2, which we will take as 405 Å^2.

Try additional runs of this calculation, using 100 points and several different values for the number of iterations, and also using several different values for the number of points per atom and 20 iterations. Some representative values are given in Table 6-4, where the values have been rounded off. The average value for this set of determinations is 404.0 Å^2, with a standard deviation of 1.5 Å^2.

440 APPLICATIONS

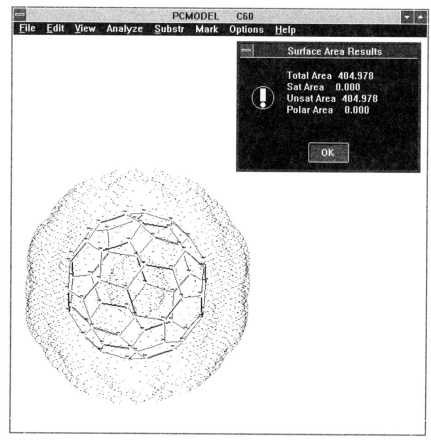

Figure 6-54. The Surface Area Calculation for Buckminsterfullerene.

TABLE 6-4. Surface Area Calculations for Buckminsterfullerene

Points/Atom	Iterations	Surface Area ($Å^2$)
100	10	401.4
100	20	405.0
100	30	403.1
100	40	403.9
50	20	406.0
92	20	402.3
200	20	405.6
300	20	404.2
400	20	404.4

For the second half of this exercise, we will derive the surface area of buckminsterfullerene assuming it to be a perfect sphere. This method depends critically upon the value chosen for the radius, as shown by the formula for a sphere's surface:

$$A = 4\pi r^2$$
$$= 4\pi(5.486)^2$$
$$= 378.2 \text{ Å}^2$$

where the value 5.486 Å for the radius was derived from half the largest carbon–carbon distance ($\frac{1}{2} \times 7.092$ Å) plus the van der Waals radius of @2 carbon (1.940 Å). The crude analytical value of 378.2 Å^2 differs from the average value from the stochastic determinations above (404.0 Å^2) by 6.4%. This may be accounted for by considering the curvature of the molecular surface vs. the ideal sphere.

At this point we will discuss surface analysis and surface comparisons, stochastic vs. analytic methods for determining the surface area of molecules, and the calculation of molecular volume. Molecular surfaces and volumes have been correlated with aspects of biological activity,[178,648,649,657–660] and a computer experiment has been published which correlates molecular size and surface area with odor.[184] The exercises will include probing the inner surface and calculating the contained volume of the cage molecule dodecahedrane, and the calculation of molecular volume of cyclohexane, methanol, and ketene.

6.2.1. Surface Analysis and Surface Comparisons

In the case of a van der Waals surface, given a set of nuclear positions and van der Waals radii for each atom, the representation of surface is straightforward, and analysis of or comparisons between such surfaces of different molecules is likewise straightforward. When the surface at hand cannot be so simply expressed on a mathematical basis, as in the case of the **molecular electrostatic potential** (MEP) surface, analysis and comparison are carried out with topological methods for features (or domains) such as convexity, concavity, or saddle points.[645] In this view, similarity is based upon comparing topologically equivalent domains rather than equivalent geometric surface descriptions. Several newer methods of surface analysis and surface comparisons have been reported,[286,660–664] and the subject has been covered in general discussions.[123,157,158,177,573,645,648,649,665,666]

There are many programs currently being used to carry out molecular surface comparisons; two of the more popular ones will be described here briefly, **Dock** and **Molecular Skins**.

The Dock suite of programs, developed by I. D. Kuntz of UCSF, is used to produce a negative image or cast of the binding pockets (inside surface) of a

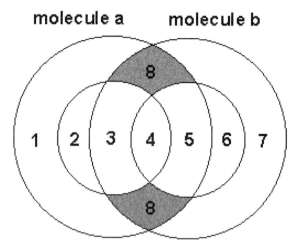

Figure 6-55. Topological Representations for the Molecular Skins Analysis.

biomacromolecule.[667–669] This cast consists of a set of overlapping spheres which complement the molecular surface of the binding site, and can be matched to positive images of putative ligands of known 3D structure. The program has been applied to protein–small-molecule docking, protein–protein docking, and assaying large groups of small-molecule ligands against a protein binding site in order to propose new lead structures for pharmaceutical development.

Dock starts with the crystal coordinates of target receptor, and then generates molecular surface for the active site using the Connolly algorithm.[651] The next step is to use the shapes of cavities in the receptor surface to define spheroids which are generated to fill the active site. The centers of these spheres become potential locations for ligand atoms, and the next step is to match these sphere centers to the ligand atoms, to determine possible orientations for the ligand. Typically on the order of tens of thousands of orientations are generated for each ligand molecule. Each oriented molecule is then scored for fit by one of three schemes: (1) shape scoring, which uses a loose approximation to the Lennard–Jones potential; (2) electrostatic scoring, which uses the program DELPHI to calculate electrostatic potential; or (3) force field scoring, which uses the AMBER potential (see Section 5.4.3). The top scoring orientation for each molecule is then saved, this is compared with the scores of other molecules, and the final results are ordered by score.

Molecular Skins deals with molecular surfaces which have thickness, and was developed as a more reliable way to identify surface similarities or complementarities between molecules of differing sizes. It was developed by B. Masek and coworkers at Zeneca.[670,671]

The method begins with the construction of a "standard" molecular surface, and the authors have used both the van der Waals surface and the solvent-

6.2. SURFACES AND VOLUMES

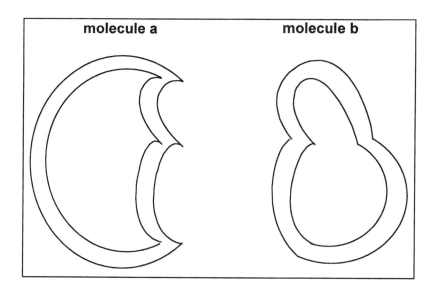

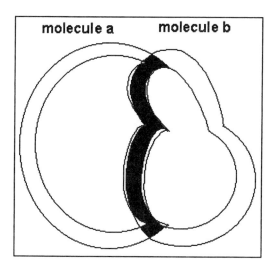

Figure 6-56. Molecular Skins Overlap for Two Irregularly Shaped Molecules.

accessible surface, with both the all-atom model (explicit hydrogens) and the united atom model (implicit hydrogens). This constitutes the *inner surface*, and an *outer surface* is generated by adding 0.4 Å to the inner surface. This is depicted for two spherical surfaces ("molecule a" and "molecule b") in Fig. 6-55.

Topologically, there are eight definable volumes, as indicated in the figure. It is the intersection of the molecular skin volumes, the shaded area 8, which is of interest in determining shape similarity.[670] The Connolly algorithm for

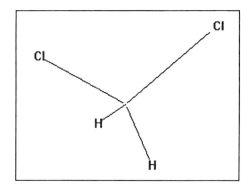

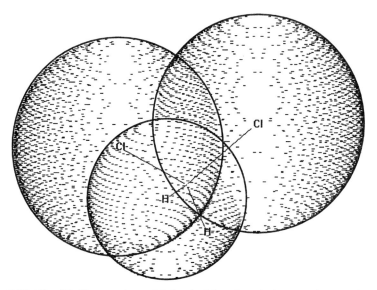

Figure 6-57. The Dichloromethane Molecule (above) and the van der Waals Surface Spheroids for the Two Chlorines and the Nearest Hydrogen (below).

union volumes is used to determine the intersection volume analytically.[652] The alignment is important to this comparison, and alignments may be assigned based on skeletal overlap, or from diffraction structures of bound ligands. These may be used as starting geometries, which are then optimized with respect to the overlap of the respective molecular skins by carrying out translations and rotations.

This is best visualized with an irregularly shaped surface, such as with the two molecules in Fig. 6-56, where the intersection volume of the merged skins is shown shaded.

6.2.2. Numerical vs. Analytical Calculations of Molecular Surfaces and Volumes

Two main methods have been developed for the calculation of molecular surfaces and volumes. The first is referred to as the *numerical*, or *stochastic*, or the *Monte Carlo* method. One such method, described earlier in this section, uses the random dispersion of points or dots upon the exposed surface to calculate surface area, or within the surface to calculate volume, and the number of dots employed is taken as proportional to the surface or volume. A number of other numerical approaches[80,81,653-655,657,658,663,672-684] have been reported.

The other is referred to as the *analytical* method: geometric theorems are employed to calculate the surface area and the volume exactly.[25,87,651,652,659,685-688] This method is far more sophisticated than our attempts in Exercises 5 and 6 in this section. Connolly adapted the Gauss–Bonnet theorem to the analytical calculation of molecular surface and volume.[87,652,689] His approach considered a molecular surface, as defined by contacts with a probe sphere, in three parts: those portions in which the probe sphere contacted only one atomic sphere, and those areas of intersections or overlap in which two or three atomic spheres were contacted. The boundaries between these portions of the surface are edges and vertices, and these portions will constitute convex, toroidal, and concave sections. These are illustrated for a part of the dichloromethane molecule in Figs. 6-57 and 6-58.

In Fig. 6-57 is shown the dichloromethane molecule in perspective (above) and with dot surface spheroids drawn about the two chlorine atoms and the nearest hydrogen. Circles representing these three atomic spheroids are shown in part (a) of Fig. 6-58. Part (b) shows some of the shaded convex portion of one of the chlorine atom spheroids which may be contacted by the probe sphere exclusive of the other atomic spheroids. Part (c) shows some of the shaded toroidal portion generated by the probe sphere as it contacts both chlorine atom spheroids. Part (d) shows the triangular concave portion generated by the probe sphere as it contacts both chlorine atom spheroids and the hydrogen atom spheroid, partly obscured by the latter.

Although these surfaces are all curved, the analogy is drawn to a polyhedron with number of faces equal to the number of curved surface sections. Only those portions of these surfaces which are not contained within any other surface fragments are used in the calculation. A similar procedure leads to the analytical calculation of molecular volume.[652]

More recent work has drawn attention to the need for consideration of higher order overlaps than three.[690] Specifically, in a comparison set of 63 structures it was found that significant contributions to the analytical calculation are made by fourth-, fifth-, and often sixth-order overlaps, and occasional contributions were made by seventh-order interactions, especially in the surface area calculation. Compared with the results of the Connolly method, in which only third-order interactions are evaluated, discrepancies

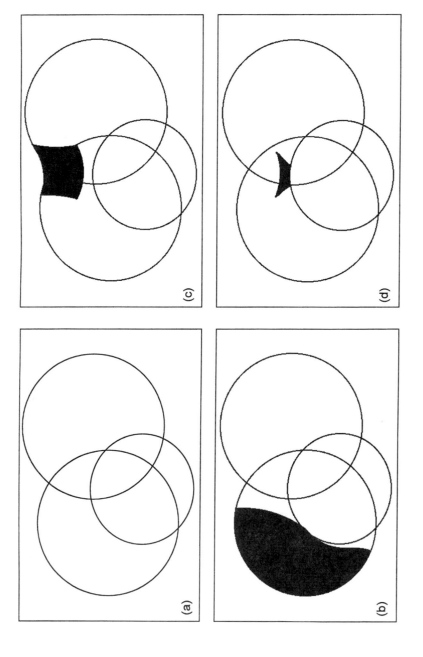

Figure 6-58. Components of an Analytical Surface Area Calculation: (a) Three Atomic Spheroids; (b) the Concave Portion for One-Atom Contact; (c) the Toroidal Portion for Two-Atom Contact; (d) the Concave Section for Three-Atom Contact.

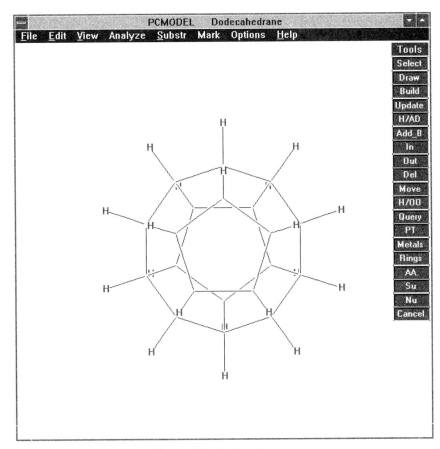

Figure 6-59. Dodecahedrane.

were found of up to 30% in surface area calculations and 5% in volume calculations. When the values obtained from the analytical method are compared with those from the Monte Carlo method, discrepancies were at most 0.1% for either type of calculation.

The final two exercises in this section involve calculating the contained volume of a cage molecule (dodecahedrane) by using a probe atom, which involves aspects of both surface calculation and volume calculation, and calculating the volume for the cyclohexane, methanol and ketene molecules, with comparisons with values obtained from other methods.

Exercise 7: A Calculation of the Contained Volume in the Molecule Dodecahedrane. In this exercise we will calculate the contained volume of the dodecahedrane molecule by enclosing a probe atom within it and gradually growing the

```
 1  {PCM Dodecahedrane
 2  NA 40
 3  FL EINT4 UV1 PIPL1
 4  AT  1  1    1.238   -0.402   -1.705 B  2  1  6  1  7  1 21  1 S 0
 5  AT  2  1    2.004   -0.651   -0.403 B  1  1  3  1  5  1 22  1 S 0
 6  AT  3  1    2.004    0.651    0.402 B  2  1  4  1 10  1 23  1 S 0
 7  AT  4  1    1.239    1.705   -0.402 B  3  1  6  1 16  1 24  1 S 0
 8  AT  5  1    1.239   -1.705    0.402 B  2  1  8  1  9  1 25  1 S 0
 9  AT  6  1    0.765    1.054   -1.705 B  1  1  4  1 20  1 26  1 S 0
10  AT  7  1    0.000   -1.302   -1.705 B  1  1  8  1 19  1 27  1 S 0
11  AT  8  1    0.000   -2.108   -0.403 B  5  1  7  1 11  1 28  1 S 0
12  AT  9  1    0.766   -1.054    1.705 B  5  1 12  1 10  1 29  1 S 0
13  AT 10  1    1.239    0.402    1.705 B  3  1  9  1 13  1 30  1 S 0
14  AT 11  1   -1.239   -1.705    0.403 B  8  1 12  1 15  1 31  1 S 0
15  AT 12  1   -0.765   -1.054    1.705 B  9  1 11  1 14  1 32  1 S 0
16  AT 13  1    0.000    1.302    1.705 B 10  1 14  1 16  1 33  1 S 0
17  AT 14  1   -1.239    0.402    1.705 B 12  1 13  1 17  1 34  1 S 0
18  AT 15  1   -2.004   -0.651   -0.402 B 11  1 17  1 19  1 35  1 S 0
19  AT 16  1    0.000    2.107    0.403 B  4  1 13  1 18  1 36  1 S 0
20  AT 17  1   -2.004    0.651    0.403 B 14  1 15  1 18  1 37  1 S 0
21  AT 18  1   -1.239    1.705   -0.402 B 16  1 17  1 20  1 38  1 S 0
22  AT 19  1   -1.239   -0.402   -1.705 B  7  1 15  1 20  1 39  1 S 0
23  AT 20  1   -0.766    1.054   -1.705 B  6  1 18  1 19  1 40  1 S 0
24  AT 21  5    1.883   -0.611   -2.593 B  1  1 S 0
25  AT 22  5    3.047   -0.990   -0.613 B  2  1 S 0
26  AT 23  5    3.047    0.990    0.612 B  3  1 S 0
27  AT 24  5    1.883    2.592   -0.612 B  4  1 S 0
28  AT 25  5    1.883   -2.592    0.611 B  5  1 S 0
29  AT 26  5    1.163    1.603   -2.592 B  6  1 S 0
30  AT 27  5    0.000   -1.980   -2.592 B  7  1 S 0
31  AT 28  5    0.000   -3.204   -0.612 B  8  1 S 0
32  AT 29  5    1.165   -1.603    2.592 B  9  1 S 0
33  AT 30  5    1.884    0.611    2.592 B 10  1 S 0
34  AT 31  5   -1.883   -2.592    0.612 B 11  1 S 0
35  AT 32  5   -1.163   -1.603    2.592 B 12  1 S 0
36  AT 33  5    0.000    1.980    2.593 B 13  1 S 0
37  AT 34  5   -1.883    0.611    2.593 B 14  1 S 0
38  AT 35  5   -3.047   -0.990   -0.611 B 15  1 S 0
39  AT 36  5    0.000    3.204    0.613 B 16  1 S 0
40  AT 37  5   -3.047    0.990    0.613 B 17  1 S 0
41  AT 38  5   -1.884    2.592   -0.611 B 18  1 S 0
42  AT 39  5   -1.884   -0.612   -2.592 B 19  1 S 0
43  AT 40  5   -1.165    1.602   -2.591 B 20  1 S 0
44  AT 41 Li    0.000    0.000    0.000 B S 0 R 1.52
45  }
```

Figure 6.60. The {DODKAHED2.PCM} Structure Input File before Modification.

probe atom in size until we reach a limit where van der Waals repulsions become significant. Due to dodecahedrane's roughly spherical symmetry, the size of the probe sphere will approximate the inner volume. Dodecahedrane had long been of theoretical interest[7,691–694] before being first prepared by the Paquette group in 1982[695,696] after two decades of synthetic work by several groups.[697] Later approaches to this molecule by the Prinzbach group relied heavily upon molecular modeling to guide the synthetic strategy.[22]

An input file for the subject molecule may either be prepared by the user, or the structure file {DODKAHD1.PCM} may be read in from the input file

6.2. SURFACES AND VOLUMES   449

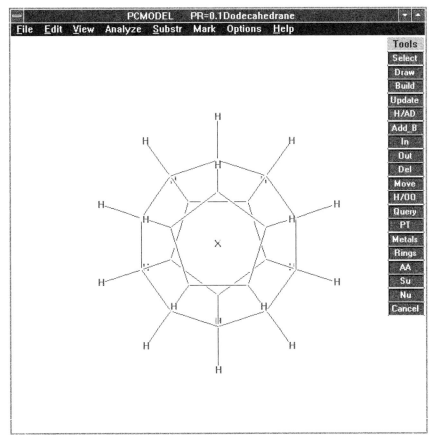

Figure 6-61. Dodecahedrane Containing the 'X' Probe Atom.

library diskette, in the CHAP6\SECT2 subdirectory. With this structure in the PCMODEL window, aligned for easy viewing as shown in Fig. 6-59, activate the **Draw** tool and click once at the center of the structure to place a carbon atom there. Next, bring up the 'Metals' menu and load the mouse cursor with a metal such as lithium. Use the mouse cursor to replace the carbon with the metal atom. Run the resulting structure through energy minimization, and write the result to an input file such as {DODKAHD2.PCM}.

Leave PCMODEL, and bring up the input file {DODKAHD2.PCM} in a text processor to carry out some modifications. This file is shown in Fig. 6-60.

We will need to modify line 44, which describes the lithium atom. Its initial form is

AT 41 Li 0.000 0.000 0.000 B S 0 R 1.52

but the atom label should be changed to 'X' (to avoid confusion) and the van der Waals radius should be changed from 1.52 to 0.05, so that line 44 will look like:

```
AT 41 X   0.000   0.000   0.000 B S 0 R 0.05
```

We start out with a very small probe, of diameter 0.05 Å. In order to examine the limits of the inner cavity, we will need to increase the size of this probe atom. This can be accomplished most easily by text modification of the structure file. Copy the 45 lines of the input file to the clipboard, and insert 10 copies of them at the end of the input file. The value for the radius of the probe atom should be changed in each succeeding structure file unit to 0.1, 0.2, 0.3, 0.4, 0.5, 0.6, 0.7, 0.8, and 0.9 Å.

Now, activate **Analyze\Batch**, and enter the name of the {DODKAHD2.PCM} file in the 'Write File for Batch, Dock, and Dihedral Driver' dialog box. The resulting 'Batch' dialog box will suggest default names for the output files, and will indicate that 10 separate files will be processed. When these conditions are accepted, the batch minimization begins. The structure should resemble that shown in Fig. 6-61.

When the batch minimization is finished, the new {DODKAHD2.OUT} file can be renamed as {DODKAHD2.PCM} and the energy minimization batch process repeated. The results of these calculations may be viewed in the file {DODKAHD2.SUM}, and these values of the probe radius and the MMX energies of the associated probe–molecule complex are found in Table 6-5 and are displayed graphically in Fig. 6-62.

From this graph of the energy vs. size of the probe sphere, we can see that starting from an initially very small size the energy dips slightly as the van der

TABLE 6-5. Results from the Dodecahedrane Contained-Volume Measurement Exercise

r_{probe} (Å)	MMX-E (kcal/mole)	D_{trans} (Å)
0.05	79.813	4.290
0.1	79.58	4.290
0.2	79.235	4.292
0.3	79.206	4.292
0.4	79.732	4.293
0.5	81.127	4.296
0.6	83.772	4.299
0.7	88.118	4.305
0.8	94.665	4.313
0.9	103.95	4.322

The probe sphere radius r_{probe} is the independent variable; the potential energy MMX-E is from the **Batch** energy minimization run; the transannular diameter D_{trans} is measured with the **Query** tool from C9 to C20 in the corresponding {DODKAHD2.PCM} file structures.

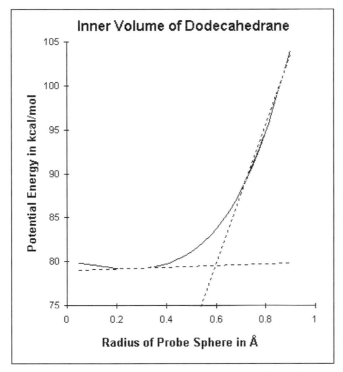

Figure 6-62. Inner Volume Determination for Dodecahedrane.

Waals attractive forces are maximized when the radius is between 0.2 and 0.3 Å. When it reaches 0.4 Å, repulsions begin to be significant. Extrapolations from the portion of the curve at the energy minimum and from the portion where the energy is increasing rapidly point to an idealized transition point ≈0.6 Å, while the energy begins to increase significantly once the probe sphere has grown larger than 0.4 Å.

Taking a conservative value, the inner volume of dodecahedrane would look like that contained by a dot surface around the probe sphere of a 0.4 Å radius, as shown in Fig. 6-63. This leads to a volume measurement of:

$$V = \tfrac{4}{3}\pi r^3 = 0.27 \text{ Å}^3$$

As a check on this, the transannular C------C distance can be measured in {DODKAHD1.PCM} as 4.291 Å with the **Query** tool. Using a value of 1.750 Å for the van der Waals radius of carbon,[698] the difference between the transannular internuclear distance and twice the van der Waals radius of carbon gives us a value of 0.691 Å for the diameter of the inner volume. The corresponding radius, ≈0.35 Å, is a bit less than the number we arrived at. An

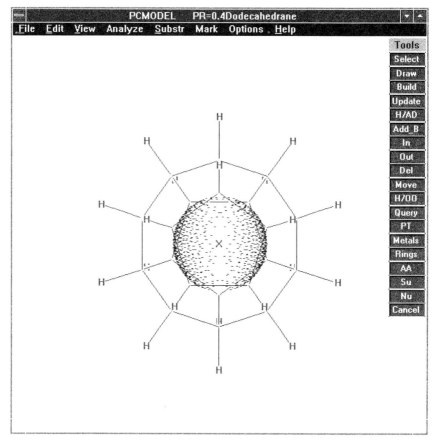

Figure 6-63. The Volume Contained Inside Dodecahedrane.

estimate of 0.91–0.93 Å for the diameter of this enclosed space was given in the literature.[699] In practical terms, the space is too small to enclose any but the smallest ions.

As is the case for molecular surfaces, there is high interest in algorithms which generate a value for the molecular volume.[25,113,177,644,646,647,652,657,663,672,675,679,685,690,698,700–702] The final exercise in this section involves volume calculations on cyclohexane, methanol, and ketene, molecules which have been used as part of a standard basis set for evaluating molecular volume calculation algorithms.

Exercise 8: A Calculation of the Volumes of Cyclohexane, Methanol, and Ketene. In this exercise we employ the PCMODEL **Analyze\Volume** command to calculate the volume of cyclohexane, methanol, and ketene, and these values are compared with results from other methods.

6.2. SURFACES AND VOLUMES

In the PCMODEL window, activate the **Rings** tool, and from the menu click on the **C6** button to bring up the chair cyclohexane structure. **View\Control_Panel** may be used to adjust the perspective. Carry out an energy minimization on the structure to ensure that it has an optimized geometry. Next, activate the **Analyze\Volume** command to calculate the volume for this structure. The resulting screen will look like the depiction in Fig. 6-64.

For the second molecule, clear the screen with the **Edit\Erase** command, activate the **Draw** tool, and draw three atoms making a V-shape. Then, activate the **PT** tool, and pick up the appropriate atoms to convert the vertex of the V into an oxygen, and one terminal atom into a hydrogen. Cancel the 'Periodic Table' menu, click on the **H/AD** tool to add hydrogens and lone pairs, and run this structure through a few cycles of energy minimization. An input file for the resulting methanol molecule may be found on the input file library diskette in the CHAP6\SECT2 subdirectory as {MEOH.PCM}.

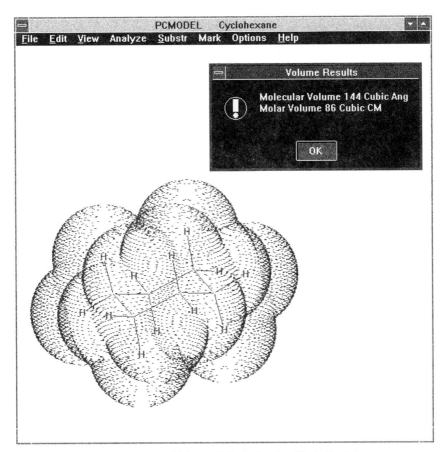

Figure 6-64. PCMODEL Volume Calculation for Chair Cyclohexane.

Activate the **Analyze\Volume** command to calculate the volume for this structure. The resulting screen will look like the depiction in Fig. 6-65.

Table 6-6 compares the values for the PCMODEL molecular volume calculated in this exercise with those from other methods as well as an empirical value. The van der Waals radii from this experiment are almost 50% larger than those obtained from the Gavezzotti[656,698] and Bohacek–Guida[169] methods. Much of this may be attributed to the larger van der Waals radii used by PCMODEL than by the other methods (1.90 Å vs. 1.75 Å), a difference which is amplified in the volume determination. Interestingly, the PCMODEL values for the molal volumes are very close to those derived from the *ab initio* Monte Carlo method, and both are smaller than that obtained from the molecular connectivity method or the value derived from the measured density.[703] The empirical value will include some intermolecular "dead space."

Petitjean also employed his analytical method for determining molecular volume, although his basis set emphasized the branched compounds for which

Figure 6-65. PCMODEL Volume Calculation for Methanol.

TABLE 6-6. Molecular Volumes of Cyclohexane and Methanol

Cyclohexane		Methanol		
V_{vdW} (Å)	V_{molal} (cm³/mol)	V_{vdW} (Å)	V_{molal} (cm³/mol)	Method
144	86	51	30	PCMODEL
			43.9	Molecular connectivity[657]
99.1		34.89		Probe point proximity[698]
101.3		37.6		Numerical bit map[675]
	84.7		30.51	Ab initio Monte Carlo calculation[679]
	108.11		40.49	Empirical density measurements[703]

V_{vdW} is the van der Waals volume; V_{molal} is the molal volume. The PCMODEL values are obtained by employing the Analyze\Volume command; the molecular connectivity value is from a numerical method developed by Hall and Kier[657]; the probe point proximity method of Gavezzotti determines the volume from the proportion of probe points in a volume envelope which are closer than the van der Waals radius to any of the atoms of the molecule[698]; the numerical bit map value is obtained by the method of Bohacek and Guida, where the elements of a bit-mapped volume space are tested to determine whether they are insider or outside the molecular surface[675]; the ab initio Monte Carlo value is derived from a Monte Carlo numerical evaluation of the volume contained within the molecular surface defined as the 0.001-atomic-unit electron density envelope[679]; the values from empirical density measurements are obtained by dividing the empirically obtained density by the molecular weight.[703]

larger contrasts were evident.[690] We select the ketene molecule as a point of comparison between these methods.

For the third molecule, clear the screen with the Edit\Erase command, activate the Draw tool, and once again draw three atoms making a V-shape. Then, convert one terminal atom into an oxygen, cancel the 'Periodic Table' menu, activate the Add_B tool, and click on each of the bonds. Next, click on the H/AD tool to add hydrogens and lone pairs, and run this structure through a few cycles of energy minimization. An input file for the resulting ketene molecule may be found on the input file library diskette in the CHAP6\SECT2 subdirectory as {KETENE.PCM}. Activate the Analyze\Volume command to calculate the volume for this structure. The resulting screen will look like the depiction in Fig. 6-66. Petitjean calculated a molecular volume of 38.522 Å³ for ketene with his algorithm,[690] compared to 59 Å³ calculated with PCMODEL. As above, part of this difference may be attributed to the different van der Waals radii employed (1.90 Å in PCMODEL vs. 1.75 Å in Petitjean).

6.3. DIHEDRAL DRIVER CALCULATIONS

A dihedral driver calculation allows the user to observe the effect upon the potential energy of rotation of a specific bond within a molecule. It may be

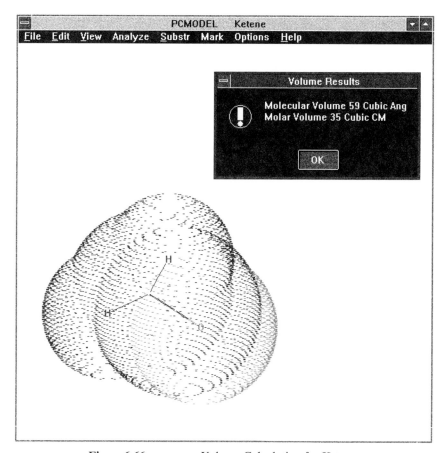

Figure 6-66. PCMODEL Volume Calculation for Ketene.

used to predict a value for a torsional or rotational barrier, or it may be part of a broader conformation search.

The early publications on force field modeling from the Westheimer group were concerned with the analysis of torsional barriers: the barrier to racemization of biphenyl derivatives.[37,38] The first report of a molecular mechanics experiment in which a torsional angle was driven with energy minimization at each stage was from Wiberg and Boyd.[82] This technique was applied early on to macromolecular conformations, as described by Boyd and Breitling who analyzed the conformational properties of polyisobutylene by carrying out dihedral driver calculations on representative small-molecule model compounds.[704] A computer experiment has been published which describes the creation of a potential energy surface for the molecule methyl 2-hydroxyacetoacetate using PCMODEL, where both the **Rot_E** rigid rotor and dihedral driver operation are employed in the analysis.[197] A method was recently reported in which a dihedral driver experiment may be carried out on a metal coordinated with a dihapto ligand.[705]

There are several options within PCMODEL which can be used to carry out dihedral driver calculations. The simplest is the rigid rotor experiment, which is realized in PCMODEL with the **Analyze\Rot_E** command, discussed in Section 6.3.1. This command calculates a crude potential energy curve by a series of fixed rotations about a selected bond, with no energy minimization at each step. A more accurate experiment may be carried out with the **Analyze\Dihedral Driver** command, discussed in Section 6.3.2. In this case, a series of fixed rotations are examined, while a full energy minimization is carried out at each step to optimize the structure. In addition to these, related experiments may be set up manually by using **Edit\Rotate_Bond** to achieve the desired dihedral angle, and then carrying out energy minimization either with or without constraints on the angle.

6.3.1. Rigid Rotor

The rigid rotor operation in PCMODEL, available through the **Analyze\Rot_E** command, gives an imprecise but representative picture of the torsional energy profile. It shows which torsional angles are associated with an energy minimum or maximum, but it greatly exaggerates the rotational energy barrier since no energy minimization takes place. The energy maxima discovered in this way are generally too high, but this operation is helpful in that a large number of steps can be taken. It works on acyclic bonds only, and application to a ring will fail. This operation should only be carried out on a structure which has been run through an energy minimization.

Rot_E (the name is abbreviation for "rotational energy") carries out a rigid rotation of a specified torsional angle over a specified range, with no intermediate energy minimization. The two atoms of the bond to be rotated must first be marked in sequence using the **Select** tool. Activating the **Analyze\Rot_E** command then brings up a new, blank child window entitled 'One Angle Plot', superimposed upon which is a dialog box entitled 'Rot_E' where the user selects the rotation step size (the data entry box labeled 'Step Size:' with a default value of $10°$) and the range (the data entry box labeled 'Total Rotation:' with a default value of $400°$). The 'Rot_E' dialog box is shown in Fig. 6-67.

The user may enter new values for these parameters or accept the defaults. Clicking on the 'OK' button cancels the 'Rot_E' dialog box and initiates the experiment, as will be seen below. When the calculation is finished, a plot of the rotation barrier is given in the 'One Angle Plot' window, with angle (in degrees) on the X-axis, and energy (in kilocalories per mole) on the Y-axis. At the top of the plot is given the structure name of the structure file, followed by the phrase

```
Rotation about atoms   # - # - # - #
```

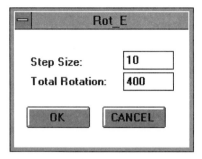

Figure 6-67. The 'Rot_E' Dialog Box.

where the #'s are the atom file numbers of the four atoms defining the torsional angle of interest. The menu bar in the 'One Angle Plot' window has only a single command, **File**, under which are found a **Print** command and an **Exit** command. The operation of the **Rot_E** command will be illustrated with the example of the 1,2-dichloroethane molecule, which we have dealt with before in Structure Comparisons Exercise 1 in Section 3.6, in the discussions of torsional interactions in the MMX force field in Section 5.2, and in Electrostatics Exercise 2 in Section 5.6.

Rigid Rotor Exercise: The Rigid Rotation Energy Profile of 1,2-Dichloroethane. In this experiment, we will carry out a rigid rotation on 1,2-dichloroethane, and view a plot of the energy profile.

Open the input file {DICLETHN.PCM}, found on the input file library diskette in the CHAP3\SECT6 subdirectory, and select the *anti* conformer. Run the structure through a minimization sequence by activating **Analyze\Minimize**, and repeat the sequence a few times. Use the **Select** tool to mark the two carbon atoms. The resulting structure should look like the depiction in Fig. 6-68.

Activate the **Analyze\Rot_E** command, which brings up the 'One Angle Plot' window and the 'Rot_E' dialog box. Accept the default values by clicking on the 'OK' button. Presently, a plot of the rotational energy profile will appear, as shown in Fig. 6-69.

The rotational barriers for calculated by **Rot_E** for 1,2-dichloroethane (approximately 5 and 12 kcal/mol) are too high, as expected. The predicted energy difference between *anti* and *gauche* conformers is ≈ 2.2 kcal/mol, as compared with the value of 1.711 kcal/mol calculated in Exercise 2 of Section 5.6 (Table 5-7). As is indicated at the top of the 'One Angle Plot' window, PCMODEL is tracking the torsional angle with atoms 6, 2, 3, and 4. Energy minima are seen at 60°, 180°, and 300° (corresponding to Cl—C—C—Cl angles of 180°, 300° = −60°, and 60°), and energy maxima are seen at 120°, 240°, and 360° (corresponding to Cl—C—C—Cl angles of 240° = −120°, 0°,

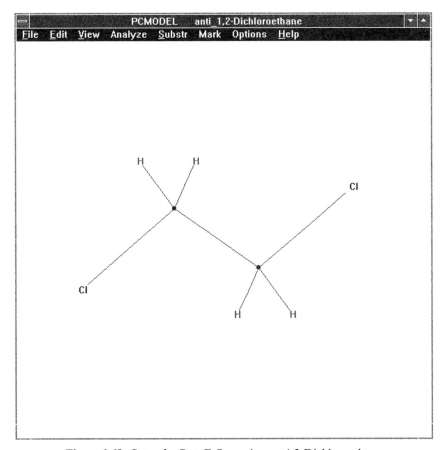

Figure 6-68. Setup for Rot_E Operation on 1,2-Dichloroethane.

and 120°), which fits with expectation. In the next section, a minimized-step dihedral driver operation will be used to analyze this same molecule, for purposes of comparison.

6.3.2. Minimized-Step Dihedral Driver

The minimized-step dihedral driver operation in PCMODEL, available through the **Analyze\Dihedral Driver** command, gives a more precise picture of the torsional energy profile. Aside from the caveats mentioned at the beginning of this section, this option calculates a more realistic rotational energy barrier, since an energy minimization is carried out at each step of the rotation. A drawback is that this process is limited in the number of steps that can be taken.

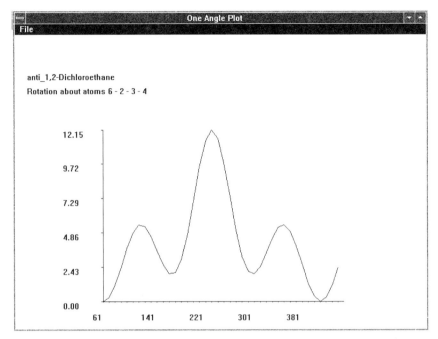

Figure 6-69. Energy Profile Plot for the Results of **Rot_E** on 1,2-Dichloroethane.

The **Dihedral Driver** command may be used on either acyclic or cyclic bonds, but in the latter case the step size should be kept small, on the order of 5–10°. This operation also works best when carried out on structure which has been run through an energy minimization.

The **Dihedral Driver** command sets up the input file to do a dihedral driver calculation on one or two dihedral angles of the molecule in the work space. Prior to activating the **Dihedral Driver** command, the user must employ the **Select** tool to mark consecutively the four contiguous atoms defining a dihedral angle in the molecule of interest. If this has not been done, an error message to that effect will pop up. When marking the atoms of the dihedral, the bond containing the first and second marked atoms will be rotated clockwise with respect to the bond containing the third and fourth marked atoms (which will be held still) along an axis coincident with the bond between the second and third atom. If the user prefers one or the other part to be rotated, then the atoms should be marked in the appropriate order. If the torsional angle to be driven is contained in a ring, it should be confined to the range 0 to 180° and 0 to $-180°$, and for best results increments of no more than 5–10° should be used.

The use of the **Dihedral Driver** command will be exemplified in the exercise, in which the rotational barriers of 1,2-dichloroethane are evaluated with greater precision than in the previous example.

6.3. DIHEDRAL DRIVER CALCULATIONS

Dihedral Driver Exercise: The Minimized Step Rotation Energy Profile of 1,2-Dichloroethane. In this experiment, we will carry out a minimized step rotation on 1,2-dichloroethane, and view a plot of the energy profile.

Open the input file {DICLETHN.PCM}, found on the input file library diskette in the CHAP3\SECT6 subdirectory, and select the *anti* conformer. Run the structure through a minimization sequence by activating **Analyze\Minimize**, and repeat the sequence a few times. Use the **Select** tool to mark the four heavy atoms in sequence, Cl—C—C—Cl. When the **Analyze\Dihedral Driver** command is activated, a dialog box entitled 'Dihedral Driver Setup' appears (see top of Fig. 6-70) which enables the user to set the parameters for this operation.

There is provision for two angles to be driven, and these are labeled 'Angle 1' and 'Angle 2'. In the first line beneath each of these headings are four boxes with the title 'Atom Numbers:'; these are the input file numbers for the selected atoms in the order selected, in this case '4 3 2 1'. That these numbers are correct may be verified by clicking on the 'Cancel' button and using the **Labels** command to have the atom file numbers displayed on the structure itself. In the second line, 'Current Angle:', is given the value of the current torsional angle between the selected atoms. The values for the initial and terminal angles of the dihedral driver procedure are entered by the user at the third line, 'Starting Angle:', and the fourth line, 'Final Angle:', and the torsional angle increment is input at the fifth line, 'Step Size'. The two radio buttons enable the user to choose which structure will display on the screen when the dihedral driver procedure is finished, the final one ('Last Structure') or the initial one ('Original Structure'). The 'Calculate' button will then initiate the dihedral driver calculations.

The button entitled 'Get Second Angle' would enable the user to add the parameters for including a second angle in the calculations without starting over completely. In this case, angle 1 will be incremented first, and then angle 2 will be driven throughout its range, after which angle 1 will be moved to its second increment and angle 2 again driven throughout its range, and so on until all assigned angles have been covered. The 'Cancel' button ends the setup procedure. Initial and final drive angles and the increment angle cannot be changed unless **Reset** is used and the process is begun again.

For this exercise, we select a starting angle of $-180°$, a final angle of $180°$, and a step size of $10°$, as shown in the bottom part of Fig. 6-70. Then, the 'Calculate' button should be activated, and a 'Write File for Batch, Dock and Dihedral Driver' dialog box (as shown in Appendix B, Fig. B-2) comes up prompting the user to supply a name for the output file, which generally has the {*.OUT} format, for example {CL2C2DDR.OUT}. Once this name is supplied, the dihedral driver process will begin when the 'OK' button is clicked. An 'Output' child window appears to the right of the PCMODEL window, which allows the user to monitor the progress of the operation. During this process, the molecule will appear on screen with a 'd1' over each of the four marked atoms, and the title bar of the PCMODEL window will display a status report

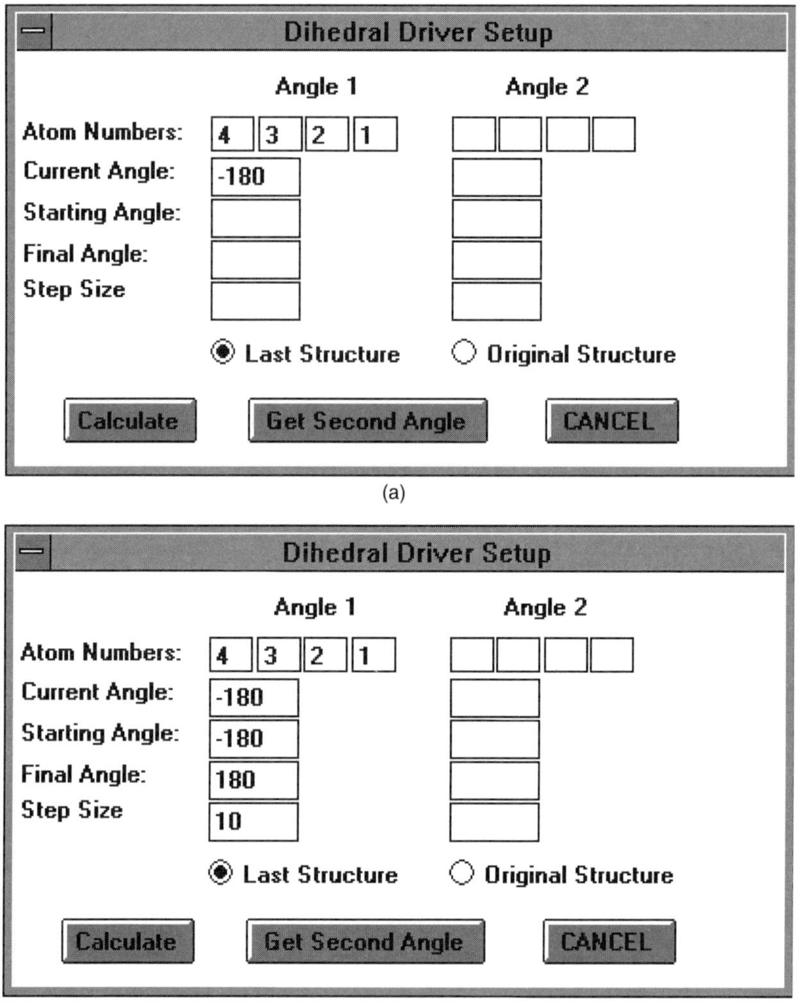

Figure 6-70. The 'Dihedral Driver Setup' Dialog Box before Parameter Entry (above) and after Parameter Entry (below).

line to indicate which structure is being processed. The final structure is depicted in Fig. 6-71.

The status report line contains the string

```
#dihd  36   -4.635813   180   0
```

which indicates that this is the 36th dihedral processed, -4.635813 kcal/mol is the minimized energy for that structure the dihedral angle for angle 1 is 180°,

6.3. DIHEDRAL DRIVER CALCULATIONS

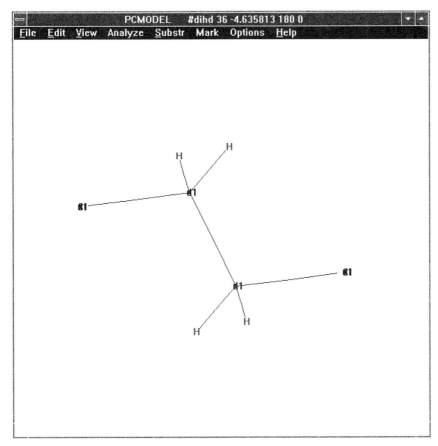

Figure 6-71. Final **Dihedral Driver** Structure for 1,2-Dichloroethane.

and the dihedral angle for angle 2 is 0° (since a second angle was not examined in this experiment).

Once the angle has been driven over the desired range, a 'One Angle Plot' child window opens up to display a graph of energy vs. angle, which is shown in Fig. 6-72. This graph may be reproduced at any time by activating the **View\Dihed Map** command and then reading in the {*.OUT} file from the earlier process. The rotational barriers for calculated by **Dihedral Driver** for 1,2-dichloroethane are 5.18 and 7.55 kcal/mol. The predicted energy difference between *anti* and *gauche* conformers is 1.8 kcal/mol, not far off from the value of 1.711 kcal/mol calculated in Exercise 2 of Section 5.6 (Table 5-7). The value for the barriers could be refined further by fixing the torsion at the respective angle (see Section 6.1) and carrying out several additional cycles of energy minimization.

Close the 'One Angle Plot' window, and clear the 1,2-dichloroethane structure from the work space (**Edit\Erase**). Verify that we can recreate the energy

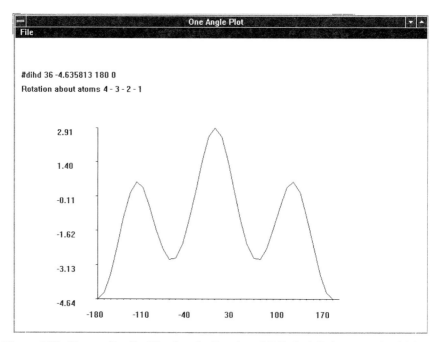

Figure 6-72. Energy Profile Plot for the Results of **Dihedral Driver** on 1,2-Dichloroethane.

vs. angle graph by activating the **View\Dihed Map** command. This brings up a 'Read Dihedral Driver File' dialog box, where we can enter the filename {CL2C2DDR.OUT}. The result should still look like Fig. 6-72.

If two angles had been driven, reading the corresponding {*.OUT} file into **View\Dihed Map** would produce a contour map where the X- and Y-axes would correspond to the values of each of the angles, and the energy would be indicated by color. The highest 10% of energy values would be printed in white, and a spectrum of colors used indicate different energy ranges, with the lowest 10% in red.

A caveat to keep in mind with the use of the dihedral driver method of determining rotational barriers is the *lagging* described by several authors.[706–709] This concerns the observation that the rotational energy surface predicted by this method may yield higher values for the energy, and a discontinuous jump as the substituents on the rotating bond at first snag on each other and then snap past each other.

6.4. NMR CORRELATIONS

One of the most common applications of PCMODEL reported in the literature is to corroborate structural assignments by ^{1}H-NMR measurements, and in particular by comparing actual spin–spin coupling constants with those calculated for minimum-energy conformations. Predictions of most three-bond (H—C—C—H) coupling constants are available through the **Query** tool in PCMODEL. This function uses published algorithms to generate values for most three-bond coupling constants with the H—C—C—H torsional angle as input, and will generate a prediction for vicinal hydrogens where the hybridization for the two parent carbons is sp^3–sp^3 or sp^3–sp^2. In the former case the electronegativity of attached substituents is also factored in.

In the case of a rigid structure, molecular mechanics minimization, can provide a single stable conformation in most cases, and it is straightforward to use these derived geometric relationships between hydrogens on adjacent carbons to generate a predicted coupling constant, or J-value. When the molecule is moderately flexible, one can calculate a small group of conformations ranked by energy, and a simple Boltzmann population analysis can be applied to obtain a predicted coupling constant in these cases.

The prediction of ^{1}H and ^{13}C chemical shifts is less straightforward. The chemical shift phenomenon depends upon the interplay of a number of factors, only a few of which are addressed in the force field model. Still, a moderately successful relationship between π-electron density and chemical shift can be demonstrated, as will be seen later in this section.

Four exercises will be used to exemplify NMR correlations. The first is the prediction of coupling constants for the rigid bicyclic structure bicyclo[2.2.2]octadienone. The second uses 4-methyl-2-methoxytetrahydropyran to show how the coupling constant correlation can be used to assign stereochemistry. In the third, a simple Boltzmann population analysis is applied to the calculated conformers of the conformationally mobile molecule 1-chloro-3-butanone, to enable a prediction of the observed coupling constant. In the last exercise, a series of nitrogen heterocycles is used as a basis set to derive relationships between the calculated VESCF π-electron density and the ^{13}C and ^{1}H chemical shifts; then these relationships are used to predict chemical shifts for a test set of nitrogen heterocycles. Computer molecular modeling experiments have been published which involve ^{1}H-NMR coupling constant analysis,[193,211] and the application of NOE enhancement measurements to conformational analysis.[209]

For readers unfamiliar with the area of NMR spectroscopy, access to a basic text[710–712] or more advanced treatments[713–717] will provide more background than the introduction provided here.

At this stage, we will discuss the use of PCMODEL for NMR correlations and carry out a modeling experiment on a rigid molecule for the purpose of predicting coupling constants. The **Query** tool in PCMODEL can be used to estimate ^{1}H–^{1}H NMR coupling constants in an energy-minimized structure. Two literature algorithms are employed to generate these predicted values, depending on the hybridization of the intervening carbon atoms: the correlation of Haasnoot *et al.*[718] is used for H—C(sp^3)—C(sp^3)—H moieties, and the correlation of Garbisch[719] is used for H—C(sp^2)—C(sp^3)—H moieties. The physical

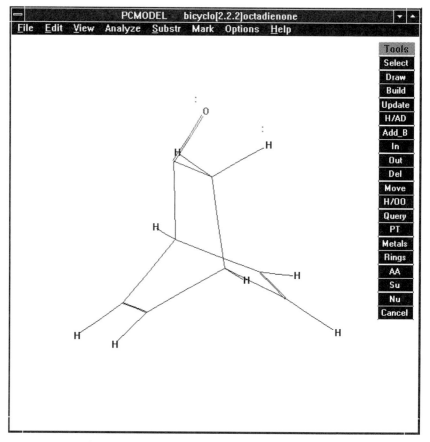

Figure 6-73. The Bicyclo[2.2.2]octadienone Structure.

rationale behind the empirical correlations used here is that the spin of one proton, H_a, pairs with the spins of bonding electrons in the three intervening bonds, through to the second proton, H_b. The ability of these spins to pair will be related to the orbital overlap, which depends on the dihedral angle defined by H_a—C—C—H_b.

The prediction of J-values is limited to the cases described above, and so includes no capability with geminal coupling (i.e. two-bond coupling), with allylic coupling (i.e. four-bond coupling), with hydrogens on two vicinal sp^2 carbons, or when one of the vicinal carbons is sp-hybridized.

The first example demonstrates of the algorithms for predicting coupling constants in both saturated [H—C(sp^3)—C(sp^3)—H] and unsaturated [H—C(sp^3)—C(sp^2)—H] systems in the rigid bicyclic molecule bicyclo[2.2.2]octadienone, for which there is only one conformation.

NMR Exercise 1: Prediction of Coupling Constants: Bicyclo[2.2.2]octadienone. In the PCMODEL window, click on the **Rings** tool, and from the 'Rings' pop-up menu click on the **B222** button. This will bring up the bicyclo[2.2.2]octane skeleton in the work space, which may be rotated for better viewing using **View\Control Panel**. The carbonyl oxygen will be introduced by replacing one of the bridge hydrogens. Click on the **PT** tool, then on the oxygen, **O**, in the 'Periodic Table' pop-up menu, and then on one of the bridge hydrogens. This will result in that hydrogen atom being converted to an oxygen. Next, the **H/AD** tool button should be clicked, which removes all of the hydrogens. The **Add-B** tool is then activated, and the mouse cursor is positioned in turn on the C—O bond and each of the C—C bonds composing the other two bridges of the structure. Clicking again on the **H/AD** tool button will add hydrogens and lone pairs, and will update the structure. If desired, the structure name in the Title Bar may be modified to 'bicyclo[2.2.2]octadienone' using **Edit\Structure Name**. The result should look like Fig. 6-73.

Geometry optimization should then be carried out with the **Analyze\Minimize** command. Some reorienting of the final display may be necessary for better viewing, and then the **Query** tool should be used to generate J-value predictions by clicking on the C4 bridgehead proton and its neighbors (for numbering see Table 6-7): the methylene bridge hydrogens at C3 adjacent to the carbonyl ($J_{4,3}$), and the adjacent vinylic protons at C5 and C8 ($J_{4,5} = J_{4,8}$). The result should look like Fig. 6-74, and the predicted values are found in Table 6-7. The agreement with literature models is quite good. This structure is available on the input file library diskette in the CHAP6\SECT4 subdirectory under the name of {BC222ODN.PCM}.

TABLE 6-7. Predicted Coupling Constants for Bicyclo[2.2.2]octadienone

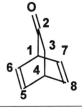

X, Y	$J_{x,y}$ (Hz)	
	Calc.	Lit.
4, 3	2.94	2.6
4, 5; 4, 8	6.58	6.5

Under X, Y are found the atom numbers for the two coupled nuclei. For literature values, $J_{4,3}$ was modeled on dibenzobicyclo[2.2.2]octadienone,[720] and $J_{4,5}$ and $J_{4,8}$ were modeled on bicyclo[2.2.2]octa-2,5-diene.[721]

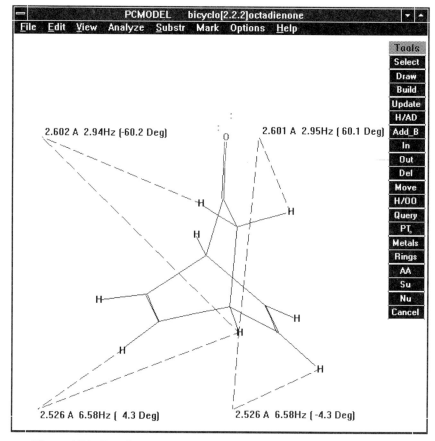

Figure 6-74. Coupling Constant Predictions for Bicyclo[2.2.2]octadienone.

At this point, we will discuss the algorithms used by PCMODEL to generate predictions of coupling constants, and we will carry out two modeling experiments. The first correlates the coupling constant with stereochemistry, and the second addresses the prediction of coupling constants in a simple, conformationally mobile system.

The first correlation for vicinal ^{1}H–^{1}H NMR coupling constants to gain wide acceptance was that of Karplus,[722] who derived the relationship from a valence-bond treatment of the contact electron spin interaction utilizing the corresponding wave functions. Since the nuclear spin–spin coupling is promulgated via interactions between the intermediary electronic spins, this simplified correlation, dependent only upon the dihedral angle, is intellectually satisfying in a qualitative way in that the overlap integral for sp^3-hybridized orbitals is related to a cosine function of the dihedral angle between the axes of the orbitals. With certain assumptions and simplifications, he showed that the coupling constant, J_{ab}, in hertz, between two protons H_a and H_b attached to

sp^3-hybridized carbons related by a dihedral angle ϕ, in radians, can be given by

$$J_{ab} = \begin{cases} 8.5 \cos^2 \phi - 0.28 & \text{when} \quad 0° \leq \phi \leq 90° \\ 9.5 \cos^2 \phi - 0.28 & \text{when} \quad 90° \leq \phi \leq 180° \end{cases}$$

This curve is depicted in Fig. 6-75.

The history of the Karplus correlation and more recent modifications has been reviewed.[723] Unfortunately, this simple expression fails to give accurate predictions in many cases. Current theory sees the vicinal coupling constant for a system H—C—C(X)—H as dependent on the dihedral angle for H—C—C—H, the C—C bond length, the two H—C—C bond angles, and the electronegativity of attached X groups.[711]

The Karplus equations were reexamined by Altona et al. with the goal of correcting them for the effects of geminal and vicinal polar substituents on the coupling constant J_{ab}.[718] A modified correlation was derived which depends upon the electronegativity and orientation of these substituents. The data basis set from which they worked consisted of 315 coupling constants from 109 compounds. The torsional angles were taken from energy-minimized structures produced with the Allinger MMI force field for the heavy atom skeleton, and fixing standardized methylene and methine proton positions: methylene

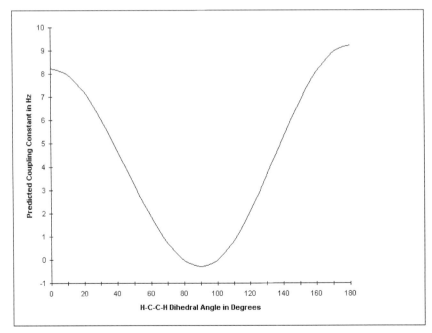

Figure 6-75. The Karplus Correlation for H—C(sp^3)—C(sp^3)—H Coupling Constants.[722]

groups retain C_{2v} symmetry with an H—C—H angle of 107.6°, and methine protons have bond angles equal to those of the heavy atoms. The electronegative atom substituents used in the data set were mostly oxygen, with some nitrogen, sulfur, halogens, silicon, and selenium. An iterative least squares minimization procedure was used to optimize the coefficients, and the correlation equation is as follows:

$$J_{ab} = 13.70 \cos^2 \phi - 0.73 \cos \phi$$
$$+ \sum \Delta\chi_i[0.56 - 2.47 \cos^2(\xi_i \phi + 16.9° \times |\Delta\chi_i|]$$

where ϕ is the H—C—C—H dihedral angle in radians, $\Delta\chi_i$ is the difference in electronegativity between the substituent and hydrogen ($\chi_{subst} - \chi_H$), and ξ_i is a toggle of +1 or −1 according to the orientation of the substituent. The use of the absolute value $|\Delta\chi_i|$ in the last term ensures symmetry of this factor about $\Delta\chi_i = 0$. There is also correction term for vicinal electronegative substituents which improves the accuracy of the predictions. This is the algorithm used by PCMODEL to calculate H_a—$C(sp^3)$—$C(sp^3)$—H_b coupling constants.

Three-bond [i.e. H—$C(sp^2)$—$C(sp^3)$—H] and four-bond [i.e. H—$C(sp^2)$—$C(sp^2)$—$C(sp^3)$—H] vinyl–allylic coupling was analyzed by Garbisch.[719] He proposed both σ- and π-orbital component contributions to the magnitude and algebraic sign of the coupling constant. He modified the Karplus equations by including a term proportional to $\sin^2 \phi$, to take account of the π-orbital contribution. By estimating the necessary parameters, and using the same definitions of J_{ab} and ϕ, he derived the following pair of equations for the three-bond coupling:

$$J_{ab} = \begin{cases} 6.6 \cos^2 \phi + 2.6 \sin^2 \phi & \text{when } 0° \leq \phi \leq 90° \\ 11.6 \cos^2 \phi + 2.6 \sin^2 \phi & \text{when } 90° \leq \phi \leq 180° \end{cases}$$

This curve is depicted in Fig. 6-76. The Garbisch algorithm has been demonstrated to give accurate predictions over a variety of alkenes, and it is used by PCMODEL to predict H_a—$C(sp^2)$—$C(sp^3)$—H_b coupling constants. The four-bond coupling constant calculation is not enabled in PCMODEL.

There are several comprehensive treatments of theoretical correlations with NMR data available.[724]

NMR Exercise 2: Assignment of Stereochemistry from Coupling Constants: 2-Methoxy-4-methyltetrahydropyran. For this exercise, assume that we have obtained a stereoisomer of 2-methoxy-4-methyltetrahydropyran for which we have obtained the ^{1}H-NMR spectrum, and we have determined that the coupling constants $J_{1,2}$ are 2.2 and 8.3 Hz. We wish to model the possible stereoisomers, predict the coupling constants $J_{2,3}$ for each, and compare these predictions with the experimentally obtained ones to assign the relative stereochemistry at C2 and C4.

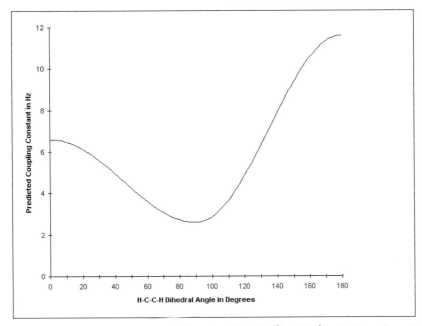

Figure 6-76. The Garbisch Correlation for H—C(sp^2)—C(sp^3)—H Coupling Constants.[719]

In the PCMODEL window, click on the **Rings** tool, and from the 'Rings' pop-up menu click on the **C6** button. This will bring up the chair cyclohexane skeleton in the work space, viewed from a point orthogonal to the plane of the ring. Activate the **Build** tool, and then click on the equatorial hydrogen at the bottom of the structure and on the one at the upper right, and then click again on the upward-pointing hydrogen on this most recently introduced methyl group. Next, remove all hydrogens by activating the **H/AD** tool, then select oxygen from the 'Periodic Table' pop-up menu (**PT** tool). Use the mouse cursor to convert the top ring carbon, and the carbon of the ethyl group which is attached to the ring, to oxygens. Restore the hydrogens and add the appropriate lone pairs by clicking on the **H/AD** tool again. If desired, the structure name in the title bar may be modified to cis-2-methoxy-4-methyltetrahydropyran' using **Edit\Structure Name**, and the result should look like Fig. 6-77.

A viewing perspective more along the plane of the ring will be helpful in detailing the J-value predictions, and this can be achieved by activating the **View\Control Panel** command and effecting the necessary rotations. Then, use the **Analyze\Minimize** command to carry out several cycles of energy minimization on this structure. At this stage it would be prudent to write the structure to an input file with the **File\Save** command. Such an input file is available on the input file library diskette in the CHAP6\SECT4 subdirectory

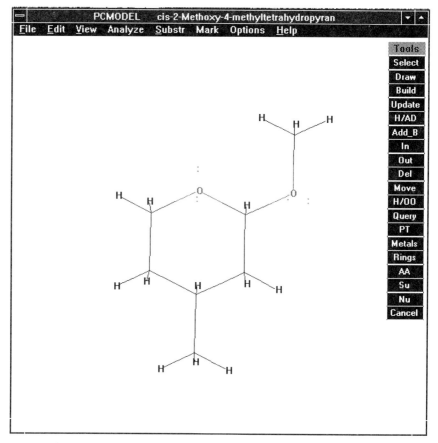

Figure 6-77. The *cis*-2-Methoxy-4-methyltetrahydropyran Structure.

under the name of {MEOMETHP.PCM}. This is an appended-format file, and also contains the two other structures which will be generated later in the course of this exercise; this first structure's name begins, '*cis*-2-Methoxy . . .'. Use the **Query** tool to generate reports of the predicted *J*-values between the hydrogen at C2 and the axial and equatorial hydrogens at C3. The screen containing the structure and *J*-value reports should look like Fig. 6-78 (see Table 6-8 for numbering).

We will now examine the *trans* isomer of this molecule by epimerizing the *cis* isomer. As the *trans* structure will have one equatorial and one axial substituent, there will be two possible chair conformations, and we will examine both in the comparison of predicted vs. literature values for the coupling constants. With the *cis* structure still on screen, use the **Select** tool to mark first C2, then the two attached groups, hydrogen and methoxyl. Next, activate the **Edit\Epimer** command, which interchanges the positions of the two substit-

6.4. NMR CORRELATIONS   473

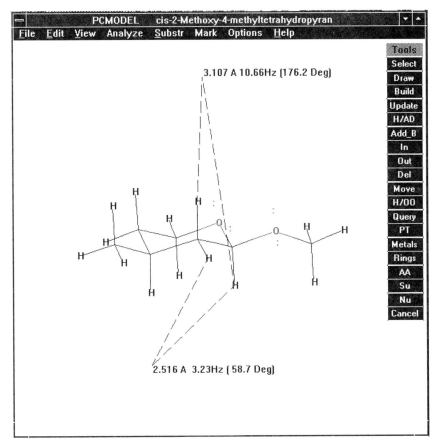

Figure 6-78. Coupling Constant Predictions for *cis*-2-Methoxy-4-methyltetrahydropyran.

uents. To tell the isomers apart, we may amend the name of this structure to '*trans* (a)-2-Methoxy . . .', as described above. Now, carry out several cycles of geometry optimization as above to generate a low-energy conformation for this *trans* isomer. Generate reports of the predicted *J*-values as before, to give a screen which should resemble Fig. 6-79.

The other conformation of the *trans* isomer will have an equatorial methoxyl and an axial methyl group. This next structure can be generated by importing the initial *cis* isomer, by using **File\Open** with either the user's own input file or {MEOMETHP.PCM} from the input file library. This time, the epimerization sequence will be carried out at C4 instead of at C2, and the resulting structure should be run through several cycles of geometry optimization. Finally, the reports of predicted *J*-values should be generated as above, and this screen should resemble Fig. 6-80.

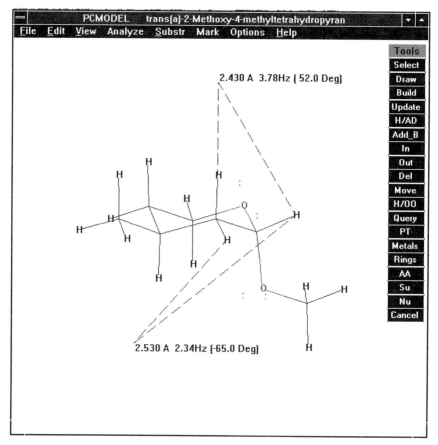

Figure 6-79. Coupling Constant Predictions for *trans*(a)-2-Methoxy-4-methyltetrahydropyran.

The results are summarized in Table 6-8. From rapid inspection, it seems that calculated coupling constants of either the *cis* or *trans* (b) isomer should correlate well with the experimental values. However, of the two *trans* conformers the *trans* (b) isomer is just over 3 kcal/mol higher in energy than the *trans* (a) isomer, ensuring that this latter will predominate at room temperature. Thus the choice comes down to *cis* vs. *trans* (a), and the *cis* stereochemistry is assigned on the basis of the best fit. The calculated and empirical values for the coupling constants in the *cis* isomer do differ somewhat, but in a semiquantitative sense, this analysis of the ^{1}H NMR *J*-values for the C2 methine proton in the *cis* isomer vs. the *trans* isomer enables the user to assign the *cis* stereochemistry with confidence. Being able to quantify the relative stability of the two conformations of the *trans* isomer simplified this analysis by justifying the exclusion of the *trans* (b) conformation from consideration.

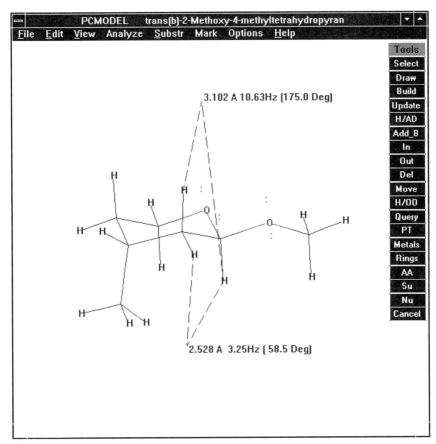

Figure 6-80. Coupling Constant Predictions for *trans*(b)-2-Methoxy-4-methyltetrahydropyran.

In conformationally mobile systems, where the rate of interconversion between conformations is rapid with respect to the time scale of ^{1}H NMR spectroscopy, the spectrum obtained will reflect an average of the absorptions for the individual conformers.[726] When the conformers are of unequal energy, the averaged NMR spectrum can best be modeled with the contributions from each conformation weighted according to a Boltzmann population analysis (see Appendix A). This will be true for chemical shift as well as coupling constants.

The more conformations there are, the more complex such an analysis will be to carry out, and a first-order analysis ignores the contributions from vibrational states. However, in some simple systems with two or only a few conformations, a reasonable approximation may result. The next exercise provides a

TABLE 6-8. Predicted and Empirical Coupling Constants for 2-Methoxy-4-methyltetrahydropyran. Observed couplings: $J = 2.2, 8.3$ Hz

	PCMODEL Calculated Values		
	Coupling Constants		
Isomer	$J_{2,3(cis)}$ (Hz)	$J_{2,3(trans)}$ (Hz)	Relative Energy (kcal/mol)
cis	3.23	10.66	1.35
trans (a)	3.78	2.34	0.00
trans (b)	3.25	10.63	3.10

The observed values are taken from the literature.[725]

simple application of this analysis to the prediction of vicinal ^{1}H coupling constants in a simple acyclic compound which is conformationally mobile.

NMR Exercise 3: Population Analysis for Coupling Constant Prediction: 1-Chlorobutan-3-one. In this example, we will carry out a conformational analysis of 1-chlorobutan-3-one. The calculated energy differences between the conformations can be used to weight the contribution of the predicted coupling constant for each, allowing us to predict an observable coupling constant.

1-Chlorobutan-3-one fits the general formula $X—CH_2—CH_2—Y$. There will be two important conformations, the *anti* and the *gauche*.[726,727] In this simple system, the *anti* conformation is plane-symmetric, and the two *gauche* conformations are mirror images of each other. We will build the molecule, generate the conformations, record the energy and predicted coupling constants for each, and then calculate a predicted observable coupling constant as a weighted average using a Boltzmann population analysis.

While it would not be difficult to draw, we will employ a shortcut in preparing the input structure, by using a portion of chair cyclohexane as a template. In the PCMODEL window, click on the **Rings** tool, and from the 'Rings' pop-up menu click on the **C6** button. This will bring up the chair cyclohexane skeleton in the work space, viewed from a point orthogonal to the plane of the ring. Remove the hydrogens with the **H/AD** tool, and then use the **Del** tool to remove the lower left-hand carbon from the ring. Bring up the 'Periodic Table' pop-up menu, and first pick up a 'Cl' atom and deposit it on the bottom carbon of the five-atom chain, then pick up an 'O' atom and deposit it on the atom at the other end of the chain. Finish the carbonyl group by activating the **Add_B** tool and clicking on the O—C bond. Clicking on the **H/AD** tool again will fill out the structure with hydrogens and lone pairs. We finish the structure by activating the **Build** tool and clicking on the aldehydic hydrogen to convert the aldehyde to our desired methylketone. The structure name may be amended at this point, and the result should resemble Fig. 6-81.

The **View\Control Panel** command may be used to place the molecule in a perspective where the user can sight down the C1–C2 bond. Such a simple system is expected to occupy only *anti* and *gauche* conformations; we will take the shortcut of improving the initial geometry to the *anti* conformation prior to energy minimization. Use the **Select** tool to mark, in order, the Cl—C—C—C(=O)—CH$_3$ atoms, and then activate the **Edit\Rotate Bond** command. From the 'Rotate_bond' pop-up menu, adjust the dihedral angle to near 180° using the scroll bar. When this is finished, cancel the pop-up menu and run the structure through several cycles of energy minimization. When

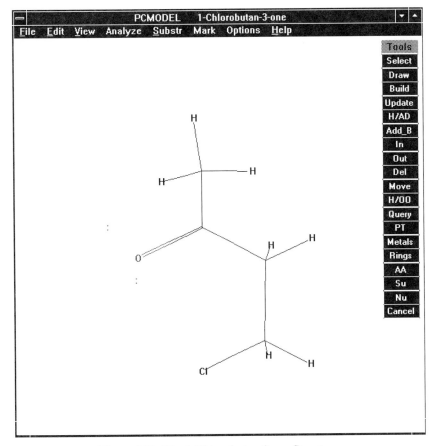

Figure 6-81. The 1-Chlorobutan-3-one Structure.

this is finished, we see that the MMX energy for this *anti* conformation is ≈ −0.62 kcal/mol. Next, activate the **Query** tool and click successively on the four C1H—C2H pairs of hydrogens in order to generate reports on screen of the predicted coupling constants. The result should look like Fig. 6-82.

The user may wish to save this structure; it is available from the input file library diskette in the CHAP6\SECT4 subdirectory as the first structure ('anti 1-Chlorobutan-3-one') in an appended-format input file entitled {CLC4ONE.PCM}. Note that, due to averaging in this conformationally mobile molecule, the apparent coupling constant for this conformer will be the average of the individual values, $J_{H1/H2}(anti) = (3.68 + 12.44 + 3.69 + 12.43)/4 = 8.06$ Hz.

The next step is to mark the **Cl—C—C—C(=O)—CH$_3$** atoms with the **Select** tool again, and to rotate the bond as above to give a value approximating 60° (or −60°). Send this *gauche* conformation through several cycles of energy minimization. When it is finished, we see that the MMX energy for this conformation is ≈ +0.40 kcal/mol. Adjust the perspective if necessary,

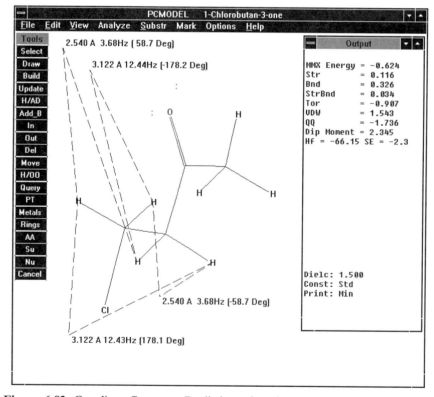

Figure 6-82. Coupling Constant Predictions for the *anti* Conformation of 1-Chlorobutan-3-one.

and probe the calculated coupling constants, as above. The result should look like Fig. 6-83.

The user may wish to save this second structure; it is available from the input file library as the second structure ('gauche 1-Chlorobutan-3-one') in the input file entitled {CLC4ONE.PCM}. The apparent coupling constant for this conformer will be the average of the individual values, $J_{H1/H2}(gauche) = (1.63 + 12.39 + 2.98 + 3.18)/4 = 5.05$ Hz.

We now have average coupling constants for the two important conformations; the energy difference between these two conformations is $0.40 - (-0.62) = 1.02$ kcal/mol. The assumption that these conformations are freely interconverting is a good one, which allows us to use the relationship between equilibrium partitioning and the energy differences (see Section A.3 in Appendix A). Using the values in the table for 20°C and rounding off the energy difference between the two conformational states $(E_{gauche} - E_{anti})$ to 1.00 kcal/mol, we find a population ratio of 85:15, or 1.0:0.18 after conversion to mole fraction. Keeping in mind that there are two degenerate *gauche* states

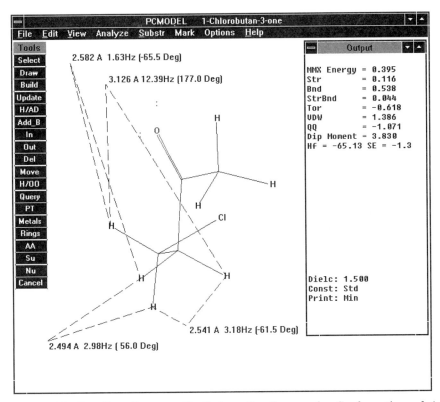

Figure 6-83. Coupling Constant Predictions for the *gauche* Conformation of 1-Chlorobutan-3-one.

(enantiomers of each other), we can make the following simple calculation:

$$J_{H1/H2}(\text{obs.}) = J_{H1/H2}(gauche) \times (\text{weighting factor})$$
$$+ J_{H1/H2}(anti) \times (\text{weighting factor})$$
$$= (5.05 \text{ Hz}) \frac{2(0.18)}{1 + 2(0.18)} + (8.06 \text{ Hz}) \frac{1}{1 + 2(0.18)}$$
$$= 7.3 \text{ Hz}$$

The experimental value[728] is 7.5 Hz, in good agreement with this prediction.

We have seen that molecular mechanics modeling provides good predictions of coupling constants for vicinal protons. Additionally, many attempts have been made to develop modeling correlations for the prediction of chemical shifts.[724,729–733] At the same time, tables of empirically derived additive substituent factors have also been developed for use in generating predictions of chemical shifts.[710,712,715,716] The fact that such empirical methods produce relatively accurate predictions should impel molecular modelers to improve the reliability of their calculational methods.

Thus, the task is to model effectively the chemical environment of the magnetic nucleus. Nuclei experience diamagnetic shielding by nearby bonding electrons, so the geometry and hybridization, as well as bond and angle strain, are important. Through-bond inductive effects caused by electronegative or electropositive substituents play a role. Nonbonding interactions, such as van der Waals repulsion and attraction and Coulombic repulsion, are also crucial. Many of these factors which determine the chemical environment are the same ones which are modeled by the components of a force field.

There are many inherent differences between the ^{1}H and ^{13}C = nuclei, beyond even the obvious one of atomic complexity. In organic molecules the ^{13}C nuclei make up the "skeleton," while the ^{1}H nuclei make up the "skin"; van der Waals contacts and neighboring group anisotropy will be more important for the latter,[733] while hybridization, inductive effects, hyperconjugative effects, mesomeric or resonance effects, and heavy halogen substituent effects will be more important for the former.[732] Conformational effects on proton NMR have been known for a long time. Their importance in ^{13}C spectra is becoming more appreciated now, especially the γ-shielding,[715,716] leading to the development of a powerful tool for acyclic conformational analysis.[734] The interested reader may find comprehensive treatments of ^{13}C NMR spectroscopy in the literature.[717]

In the area of aromatic and heteroaromatic compounds, conformational effects are relatively minor, and the substituent inductive and mesomeric effects on the electron density have been assigned the most important role in determining chemical shift.[730,731,735–739] Steric factors will play a role in crowded molecules, and in heterocycles the ring heteroatoms themselves exert strong inductive effects. Early modeling efforts focused on the π-electron density, which is easy to obtain by **Hückel calculation**. More sophisticated analyses which addressed the total charge density (π-system plus σ-framework) by an Extended Hückel Theory (EHT) treatment,[730,736–738] or by a semiempirical MO treatment,[731,735,738] obtained better correlations.

VESCF Hückel calculations have been employed in an attempt to remedy the neglect of the electronic effect of the σ-framework, since this method allows for the variable electronegativity of the substituents and ring heteroatoms; but this approach has been criticized as inadequate.[735,736] Still, such an analysis meets with some success, and the exercise involves the development of correlations using π-electron density to predict chemical shifts in heteroaromatic compounds. This exercise requires extensive use of VESCF π-electron orbital calculations and the manipulation of the associated output. Those unfamiliar with these operations should first work through the discussions on π-calculations in Section 6.5.

NMR Exercise 4: Predicting ^{1}H- and ^{13}C-NMR Chemical Shifts in Nitrogen Heterocycles. This exercise will involve establishing correlations between π-electron density and the ^{1}H- and ^{13}C-NMR chemical shifts using a basis set of six-membered nitrogen heterocycles, and then using the derived correlations to predict chemical shifts for molecules outside the basis set. Each of the molecules in the basis set will be built and then subjected to MMX energy optimization with concomitant VESCF π-calculations. The π-electron densities for each carbon atom will be tabulated along with literature values for the chemical shifts, to derive the relationships. Finally, the calculated π-electron densities for a selection of six molecules outside the basis set will be fed into these correlations to generate predicted chemical shifts, which will be compared with experimental values.

The six molecules selected for the basis set are depicted in Fig. 6-84. We will work through in detail setting up the input file, running the energy minimization, and analyzing the VESCF calculation output for the first two examples. Afterward, the user should be able to process the remaining four examples without a detailed recipe. If one wishes to skip quickly to the analysis phase of the exercise, the basis set molecules are available from the input file library diskette in the CHAP6\SECT4 subdirectory under the names {PYRIDINE.PCM}, {PYRMIDIN.PCM}, {PYRAZINE.PCM}, {STRIAZIN.PCM}, {2MEPYRDN.PCM}, and {4BRPYRDN.PCM}, respectively.

To prepare the input file for pyridine, in the PCMODEL window click on the **Rings** tool, and from the 'Rings' pop-up menu click on the **Ph** button. Note that the ring carbons are already flagged for π-calculations. This will bring up benzene in the work space, viewed from a point orthogonal to the plane of the ring. The numbering will be important in correlating π-electron density with a particular atom, so activate the **View\Labels** command, and in the 'Labels'

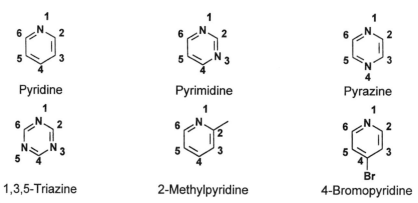

Figure 6-84. Nitrogen Heterocycles Forming the Basis Set for the VESCF π-Electron-Density–Chemical-Shift Correlation.

dialog box click on the radio button for 'Atom Numbers.' Then activate the **H/AD** tool to remove the hydrogens, and from the 'Periodic Table' pop-up menu pick an 'N' atom, and use the mouse cursor to replace the C1 with nitrogen. Since the numbering of the ring in the figures will always run clockwise, one may need to do some rotating to get the molecule on screen oriented in this way. Replace the hydrogens, with a lone pair now on N1, and run the molecule through several cycles of energy minimization. During the times when status readouts in the 'Output' window halt, the minimization engine is carrying out the VESCF calculations.

When the energy of the pyridine structure has been minimized, save the input file if desired, and then exit PCMODEL. Exiting is necessary so that the program will finish writing the output file {PCMOD.OUT}. Using an appropriate text viewer, open the file {PCMOD.OUT} and jump to the end of the file. Hitting the 'Page Up' key on the keyboard several times backs up to the VESCF calculation matrix output. Locate a 6 by 6 matrix entitled 'Electron Densit- Bond Order Matrix', which has been reproduced in Fig. 6-85.

In the actual PCMOD.OUT file, there is no border around the matrix, and in this figure the matrix diagonal cells have been emphasized by enclosing them in boxes. Starting at the upper left, cell 1,1 gives the electron density

```
Electron Densit- Bond Order Matrix
  1.1700   0.6554  -0.0245  -0.3329  -0.0245   0.6554
  0.6554   0.9198   0.6714   0.0580  -0.3281  -0.0475
 -0.0245   0.6714   1.0149   0.6637  -0.0179  -0.3281
 -0.3329   0.0580   0.6637   0.9607   0.6637   0.0580
 -0.0245  -0.3281  -0.0179   0.6637   1.0149   0.6714
  0.6554  -0.0475  -0.3281   0.0580   0.6714   0.9198
```

Figure 6-85. VESCF Electron-Density–Bond-Order Matrix for Pyridine.

6.4. NMR CORRELATIONS

at atom 1 (nitrogen); cell 2,2 gives the electron density at atom 2, etc. It is these values which we will tabulate a bit further on to formulate our correlation. The pyridine molecule has a plane of symmetry running through N1 and C4, and this is manifested in the symmetrical electron distribution, where the π-electron densities at C2 and C6 are equal, as are those at C3 and C5.

To build the pyrimidine molecule, we can use the pyridine molecule as a template. In the PCMODEL window, use the **File\Open** command to read in the user-saved input file. If the file was not saved, it should still be available as the backup file {PCMOD.BAK}. Activate the **View\Labels** command, and select 'Atom Numbers' from the pop-up menu, to ensure proper numbering in the input file. Use the procedure described above to convert C3 to a nitrogen. The atom labels may be returned to the default now, and the resulting molecule should be run through several cycles of energy minimization. Save the input file if desired, exit PCMODEL, and then access the {PCMOD.OUT} file as before to locate the electron density matrix for pyrimidine. This has been reproduced in Fig. 6-86.

Note that the pyrimidine molecule also has an axis of symmetry running through C2 and C5, so that equal π-electron densities are found for the pairs N1, N3 and C4, C6.

The procedures for preparing molecule input files, running through energy minimization, and capturing the matrix diagonal values should be repeated for the other four examples in the basis set: pyrazine, 1,3,5-triazine, 2-methylpyridine, and 4-bromopyridine. The user may notice elements of symmetry in the outputs for these molecules, for example the two axes of symmetry for pyrazine and the threefold symmetry of 1,3,5-triazine, in each case leading to equivalence among the nitrogens and among the carbons. The results of these calculations may be found in Table 6-9, along with ^{13}C-NMR and ^{1}H-NMR chemical shift values taken from the literature.[712,715,740]

These data may be entered into any standard spreadsheet statistical application such as Microsoft® Excel. A least squares fit should be obtained between the π-electron densities and the ^{13}C chemical shifts, and another between the electron densities and the ^{1}H chemical shifts. For the purposes of graphing the data from this exercise, it is most convenient to place the VESCF π-electron densities on the x-axis. For each of the two correlations, this produces an equation of the type $y = ax + b$, where a is the slope and b is the

Electron	Densit-	Bond Order Matrix			
1.1811	0.6613	-0.0478	-0.3255	-0.0424	0.6480
0.6613	0.8411	0.6613	0.0109	-0.3160	0.0110
-0.0478	0.6613	1.1811	0.6479	-0.0424	-0.3255
-0.3255	0.0109	0.6479	0.8845	0.6693	0.1134
-0.0424	-0.3160	-0.0424	0.6693	1.0277	0.6692
0.6480	0.0110	-0.3255	0.1134	0.6692	0.8844

Figure 6-86. VESCF Electron-Density–Bond-Order Matrix for Pyrimidine.

APPLICATIONS

TABLE 6-9. VESCF π-Electron Densities and Literature Chemical Shifts for Some Nitrogen Heterocycles

Molecule	Atom No.	Electron Density	Chemical Shift (ppm)	
			^{13}C	^{1}H
Pyridine	2,6	0.9198	149.5	8.60
	3,5	1.0149	125.6	7.25
	4	0.9607	138.7	7.64
Pyrimidine	2	0.8411	158.4	9.26
	4,6	0.8845	156.9	8.78
	5	1.0277	121.9	7.36
Pyrazine	2,3,5,6	0.9331	145.9	8.63
1,3,5-Triazine	2,4,6	0.8102	166.1	9.25
2-Methylpyridine	2	0.8373	159.9	—
	3	1.0504	123.4	7.27
	4	0.9574	137.2	7.74
	5	1.0297	122.0	7.22
	6	0.9175	149.8	8.67
4-Bromopyridine	2,6	0.9218	152.6	8.68
	3,5	0.9939	127.6	7.73
	4	1.0065	133.2	—

VESCF π-electron densities are obtained from the electron density matrix in the PCMODEL output files for the corresponding molecule. Chemical shift values are relative to internal tetramethylsilane, and are taken from the literature.[712,715,740]

intercept. These results are depicted in Fig. 6-87. Here, the data points are taken from Table 6-9 for the basis set molecules shown in Fig. 6-84. The VESCF π-electron densities are obtained from the PCMODEL output files. The chemical shift values are in parts per million relative to internal tetramethylsilane, and are taken from literature sources.[712,715] In these cases, the least square fits are not too bad, with correlation coefficients of 0.95 and 0.91 for the ^{13}C data and ^{1}H data, respectively.

The next step in this exercise is to test these correlations with some molecules which are outside the basis set, to see whether the correlations we derived have any predictive value. Then, for each molecule, we will prepare the input file, carry out the VESCF π-calculations, and capture the VESCF π-electron densities. These values will then be used to generate predictions of the ^{13}C and ^{1}H chemical shifts, which in turn will be compared with literature values. The set of six molecules selected for the test set are depicted in Fig. 6-88. The input files for these molecules are available from the input file library diskette in the CHAP6\SECT4 subdirectory under the names {UTRIAZIN.PCM}, {ZTRIAZIN.PCM}, {METRIAZN.PCM}, {3CNPYRDN.PCM}, {QUINOLIN.PCM}, and {26NFTHYR.PCM}.

For each of the test set molecules, the input file should be set up with careful attention to the atom numbering, the structure should be run through

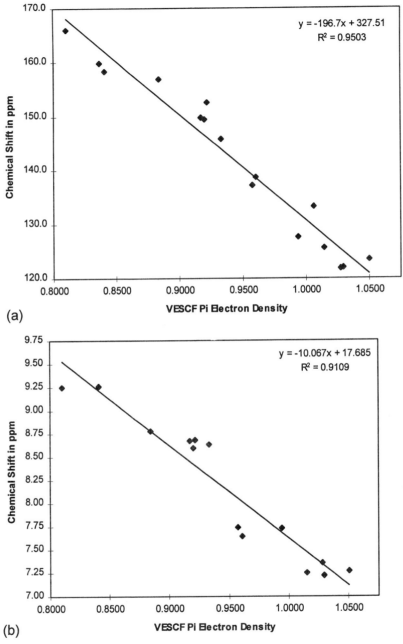

Figure 6-87. Correlations between VESCF π-Electron Density and NMR Chemical Shift: (a) Correlation for ^{13}C-NMR; (b) Correlation for ^{1}H-NMR.

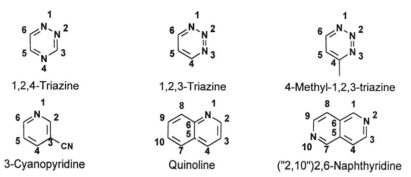

Figure 6-88. Nitrogen Heterocycles in the Test Set for the VESCF π-Electron-Density–Chemical-Shift Correlation.

energy minimization, and the electron density matrix in the output file {PCMOD.OUT} should be analyzed as was done for the molecules of the basis set. In order to generate the predicted chemical shifts, the matrix diagonal values for the electron density at each carbon atom may be entered into any standard spreadsheet statistical application such as Microsoft® Excel, and a second and third column should be defined by the equations given in Fig. 6-87.

Table 6-10 contains the results for this portion of the exercise. There are columns for molecule name, atom number, and electron density, and for each of ^{13}C and ^{1}H both the calculated and a literature value for the chemical shift are provided. For these predictions, the correlation coefficients are 0.91 for the ^{13}C and 0.93 for the ^{1}H shifts. There are several large discrepancies: in the ^{13}C for C5 of 1,2,3-triazine and 1-methyl-1,2,3-triazine, C4 of 3-cyanopyridine, C3 and C6 of quinoline, and C4/C8 of 2,6-naphthyridine; in the ^{1}H for C3 and C6 of 1,2,4-triazine, C2 of quinoline, and C1/C7 of 2,6-naphthyridine. The reasons for these discrepancies have been discussed earlier, but briefly could be due to inadequate modeling of the σ-inductive effect in the case of the ^{13}C shifts, and inadequate modeling of the through-space Coulombic effect in the case of the ^{1}H shifts.

Another application of force field modeling for the interpretation of NMR results has been correlations with lanthanide shift reagent studies. This type of experiment involves the titration of an NMR sample with a soluble lanthanide complex (the shift reagent), and monitoring the changes in chemical shift of certain of the absorptions with increasing concentration of the shift reagent. Many functional groups in organic molecules, such as nitriles, ketones, etc., can act as ligands in metal complexes. A coordinatively unsaturated lanthanide complex, such as a europium(III) tris(β-diketonate), will accept the organic molecule reversibly as a ligand. In the case of ^{1}H-NMR, the protons in the organic molecule which are close to the functional group which is

TABLE 6-10. Predicted Chemical Shifts for Some Nitrogen Heterocycles

Molecule	Atom No.	Electron Density	^{13}C Calc.	^{13}C Lit.	^{1}H Calc.	^{1}H Lit.
1,2,4-Triazine	3	0.8617	158.0	158.1	9.01	9.88
	5	0.8821	154.0	149.6	8.80	8.84
	6	0.9338	143.8	150.8	8.28	9.48
1,2,3-Triazine	4,6	0.9032	149.9	149.7	8.59	9.06
	5	0.9905	132.7	117.9	7.71	7.45
4-Methyl-1,2,3-triazine	4	0.8223	165.8	159.7	—	—
	5	1.0256	125.8	117.8	7.36	7.33
	6	0.9019	150.1	148.8	8.61	8.92
3-Cyanopyridine	2	0.8708	156.2	153.7	8.92	9.22
	3	1.0931	112.5	111.2	—	—
	4	0.9151	147.5	140.5	8.47	8.47
	5	1.0142	128.0	124.7	7.48	7.81
	6	0.8947	151.5	153.7	8.68	9.09
Quinoline	2	0.9258	145.4	150.3	8.36	8.81
	3	1.0081	129.2	121.0	7.54	7.26
	4	0.9627	138.1	136.0	7.99	8.00
	5	1.0220	126.5	128.3	—	—
	6	0.9762	135.5	148.3	—	—
	7	0.9929	132.2	129.4	7.69	8.05
	8	0.9944	131.9	129.4	7.67	7.61
	9	1.0016	130.5	126.5	7.60	7.43
	10	0.9958	131.6	127.7	7.66	7.68
2,6-Naphthyridine	1,7	0.8800	154.4	152.6	8.83	9.27
	3,9	0.9372	143.2	145.5	8.25	8.65
	4,8	0.9953	131.7	119.9	7.67	7.69
	5,6	1.0318	124.6	131.0	—	—

VESCF π-electron densities are obtained from the electron density matrix in the PCMODEL output files for the corresponding molecule. Chemical shift values are relative to internal tetramethylsilane; reference values are taken from the literature.[712,715,740,741]

serving as a ligand will experience a distance-dependent field-moderating effect due to the lanthanide. Generally, these complexes will be paramagnetic and induce a downfield shift. Since the complexation is reversible and rapid on the NMR time scale, an averaged absorption will be recorded. The magnitude of the shift will depend in part upon the proximity of the affected protons to the functional group acting as a ligand, and upon the equilibrium concentration of the substrate–shift-reagent complex, which in turn depends upon the concentration of the shift reagent.

Force field modeling can help rationalize the results of shift reagent studies by enabling a conformational analysis of the predicted complex. A series of reports by Raber et al. provide a good example of this type of study.[742-744] The Allinger MMI force field was used to generate minimum-energy conformers, and a metal atom was then coordinated to the donor ligand atom at a distance and angle dictated by examples from X-ray diffraction structures of similar complexes. Metal-centered angle bend interactions were not evaluated, but geminal nonbonded interactions were allowed to guide the geometry to an optimum conformation. Distances could then be calculated easily between the paramagnetic metal and the affected hydrogens, and these results were correlated with experiment. Modeling of metal-ligand complexes is discussed in Section 6.7.

The PIMM91 force field has been used in NMR shift reagent studies on flexible molecules.[520]

6.5. π-CALCULATIONS

In this section we will discuss the inclusion of π-calculations in force field modeling. First, the implementation of these calculations in PCMODEL will be discussed, along with the tools for carrying them out, and will be exemplified with exercises involving the benzene and 1,3-butadiene molecules. Next, the development of these molecular orbital π-calculations in the molecular mechanics context will be discussed, with emphasis on the **Hückel calculation**, the modified Pariser–Parr method, and the Variable-Electronegativity Self-Consistent Field (**VESCF**) formulation. Some intermediate level exercises are then presented which involve structure calculations on small aromatic heterocycles.

Earlier in the book, it was stated that the basis of force field modeling requires the Born–Oppenheimer approximation, the assumption that the motions of electrons and those of nuclei may be considered separately. This approximation allows us to treat the electrons as a dependent feature of the nuclei, so that we may carry out structural calculations by operating upon and parametrizing only these latter. This approximation breaks down in systems which experience **conjugation**, and the standard force field method will not provide good predictions of structure in these cases. Explicit molecular orbital calculations of the π-electron distributions must be carried out, and these results must be reflected in the bond lengths, bond angles, and torsional angles in order to describe such structures accurately. In the case of heteroconjugated systems, the charges on the constituent heteroatoms must also be factored in. PCMODEL addresses π-conjugation effects in two ways: with a special carbon atom type, and by enabling VESCF π-calculations.

The @40 atom type was developed for the aromatic, benzene-like carbon atom (see Section 5.3 and Appendix C). It gives a reasonable structure for benzene and a good approximation for other simple, symmetrical aromatic systems. However, the @40 atom does not adapt well to structural variation, since the bond order, bond length, and force constants are invariant. The @40 atom type may be assigned by picking the **CA** button from the 'Periodic Table' menu as (see Fig. 2-11).

The π-calculation option involves carrying out VESCF calculations on the π-electrons of the conjugated system in series with the energy minimization of the parent structure. Since the results of each of these operations will have an effect upon the other, each is allowed to proceed alternatively until both have achieved convergence. This method provides a more accurate approximation to structures in conjugated systems, especially in nonsymmetrical and heteroaromatic systems, as well as for conjugated reactive intermediates. After each cycle of π-calculations, the electron densities are used to calculated a bond order between each pair of atoms in the conjugated system, and this value is used to scale the bond length and force constants for bond stretching and for the twofold term of the torsional potential.

The **Mark\Piatoms** command is used to flag the atoms of a structure in the work space for inclusion in a π-calculation. If only a subset of the eligible atoms is to be included in the π-atom array for these calculations, those atoms must first be marked with the **Select** tool. The Windows version of PCMODEL can carry out Hückel molecular orbital calculations on π-systems of up to 100 atoms. The acceptable atoms for π-calculations are the carbons @2, @3, @4, @29 (radical), @30 (carbocation), and @48 (carbanion); nitrogens @9, @10, @37, and @41 (immonium); and oxygens @6, @7, and @42 (oxyanion) (see Appendix C). No other atoms, (sulfur, phosphorus, silicon, the halogens, etc.) may be included in a π-calculation.

Activating the **Piatoms** command pops up the 'Pi Atom Selection' dialog box depicted in Fig. 6-89, in which the user may opt to include all appropriate π-atoms ('All') or only a marked subset ('Selected') by clicking on the respective radio button.

After clicking on the 'OK' button, the structure will be redrawn with the flagged π-atoms indicated by a tilde, '~'. If the 'All' option is chosen, all

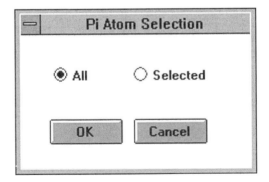

Figure 6-89. The 'Pi Atom Selection' Dialog Box.

unsaturated or multiply bonded atoms (i.e. sp^2- or sp-hybridized) and attached heteroatoms will be so marked, including those for which no VESCF parameters are resident, such as sulfur. However, once an energy minimization sequence is engaged, these inappropriate atoms will be removed automatically from the π-atom array.

The user can set up some of the parameters to be used in the π-calculations with the **Options\MMX_PI_Calc** command after the π-atoms have been marked. Activating this command brings up a dialog box entitled 'MMX Pi Calc Options', as shown in Fig. 6-90, in which several options may be set. The user may select either the doublet closed shell (**RHF**) or singlet open shell (**UHF**) calculation mode; this will be described in greater detail below. Beside the 'Multiplicity' heading is a data entry box in which the multiplicity of the electronic state may be entered. Beside the heading 'Start Calculation With:' are two radio buttons which give the user the choice of whether the force field geometry optimization will be started after the initial Hückel calculation, or after a full cycle of VESCF calculations on the conjugated system. Finally, beside the heading 'Total PI System:' are two radio buttons which the user may select either to allow the conjugated system to deviate from planarity (the default) or to enforce planarity.

π-Calculation Exercise 1: Modeling the Benzene Molecule. In this exercise, benzene will be modeled in three ways: first with full π-calculations, then without π-calculations, and finally using the specially parametrized aromatic atom, @40.

Click on the **Rings** tool to bring up the 'Rings' menu, and then click on the **Ph** button to bring up the benzene structure in the work space. Note that all

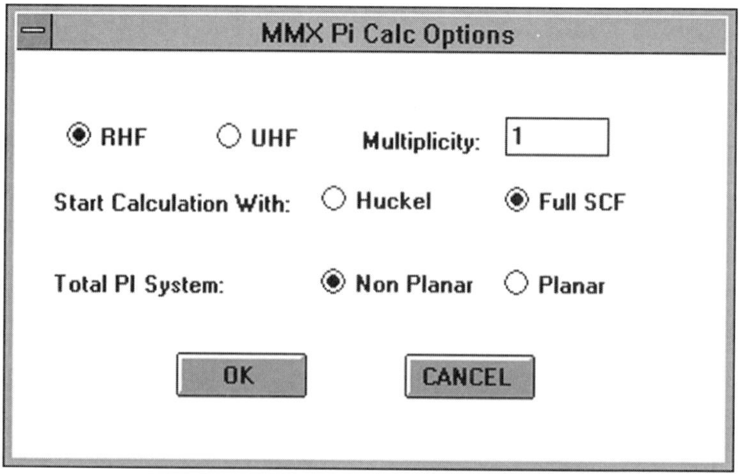

Figure 6-90. The 'MMX Pi Calc Options' Dialog Box.

six carbon atoms are already marked as π-atoms with a small tilde (~). Run the structure through a few cycles of energy minimization, then use **Edit\Structure Name** to give this structure a title such as 'Benzene_with_pi_calcs', and save it to a PCM format file named {BENZENE.PCM} Next, activate the **Mark\Reset** command and select the 'Pi Atoms' check box to remove these atoms from the π-atom array. Use **Edit\Structure Name** to give this new structure a title such as 'Benzene_with_no_pi_calcs,' and run this new structure through a few cycles of energy minimization, then save it to {BENZENE.PCM} in the appended format. Finally, click on the **PT** tool in order to bring up the 'Periodic Table' menu, pick up the 'CA' @40 atom, and use it to replace all six of the benzene ring atoms. Use **Edit\Structure Name** to give this third structure a title such as 'Benzene_with_@40_carbons', and run this structure through a few cycles of energy minimization, then save it to {BENZENE.PCM} in the appended format.

Activate the **Analyze\Batch** command to run the {BENZENE.PCM} file through energy minimization in batch mode. The file {BENZENE.OUT} will contain the final structures, so it may be renamed to {BENZENE.PCM} if desired. This file may be found on the input file library diskette in the CHAP6\SECT5 subdirectory.

Use the **Query** tool to examine bond lengths and bond angles in each of these three models; some sample values are displayed in Fig. 6-91 for each. In all three, the C—H bond length is fairly uniform at 1.103–1.104 Å. In the first example (a), the C—C bond lengths are uniformly 1.400–1.401 Å, and the C—C—H and C—C—C angles are 120°. The model without π-calculations, (b), has quite different bond lengths for the single (1.483 Å) and double (1.340 Å) bonds, and the angles also vary. The model with the @40 atom types, (c), returns to a uniform bond length of 1.404 Å for all C—C bonds, and uniform bond angles of 120°.

π-Calculation Exercise 2: Modeling the 1,3-Butadiene Molecule. In this exercise, the 1,3-butadiene molecule will be examined using the same three models as in the benzene exercise: with full π-calculations, without π-calculations, and with the @40 aromatic atom.

In the PCMODEL window, activate the **Draw** tool and sketch out four atoms connected in a linear zigzag shape. Activate the **Add_B** tool and click on both of the terminal bonds to double them, and then click on the **H/AD** tool button to add hydrogens and to effect an update. Activate the **Mark\Piatoms** command, and click on the 'OK' button in the 'Pi Atom Selection' dialog box. This action will add all four atoms to the π-atom array, and they should be marked with a tilde (~). Run the structure through a few cycles of energy minimization, then use **Edit\Structure Name** to give this structure a title such as '1,3-Butadiene_with_pi_calcs', and save it to a PCM format file named {BUTADIEN.PCM}. Next, activate the **Mark\Reset** command and select the 'Pi Atoms' check box to remove these atoms from the π-atom array. Use **Edit\Structure Name** to give this new structure a title such as '1,3-

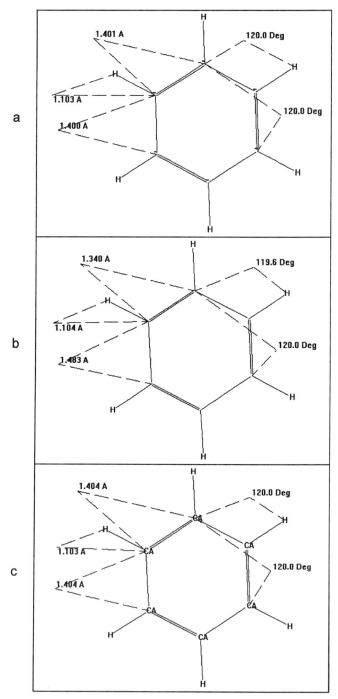

Figure 6-91. Models for the Benzene Molecule with (a) π-Calculations; (b) No π-Calculations; (c) the @40 Aromatic Carbon Atom.

6.5. π-CALCULATIONS   493

Butadiene_with_no_pi_calcs', and run this new structure through a few cycles of energy minimization; then save it to {BUTADIEN.PCM} in the appended format. Finally, click on the **PT** tool in order to bring up the 'Periodic Table' menu, pick up the 'CA' @40 atom, and use it to replace all four of the butadiene carbon atoms. Use **Edit\Structure Name** to give this third structure a title such as '1,3-Butadiene_with_@40_carbons', and run this structure through a few cycles of energy minimization; then save it to {BUTADIEN.PCM} in the appended format.

Activate the **Analyze\Batch** command to run the {BUTADIEN.PCM} file through energy minimization in batch mode. The file {BUTADIEN.OUT} will contain the final structures, so it may be renamed to {BUTADIEN.PCM} if desired. This file may be found on the input file library diskette in the CHAP6\SECT5 subdirectory.

Use the **Query** tool to examine bond lengths and bond angles in each of these three models; some sample values are displayed in Fig. 6-92 for each. In all three, the C—H bond length is 1.104 Å, and the dihedral angle is 180°. The first example, (a), has bond lengths of 1.349 and 1.471 Å for the terminal and internal C—C bonds, respectively. The model without π-calculations, (b), has bond lengths of 1.341 and 1.484 Å for the terminal and internal C—C bonds. The model with the @40 atom types, (c), has virtually identical bond lengths of 1.404–1.405 Å for all C—C bonds.

The algorithm for π-calculations in PCMODEL employs Hückel Molecular Orbital (HMO) theory, a treatment in which only the electrons in π-orbitals are included.[745] The following discussion explains the PCMODEL implementation briefly, and the interested reader should consult the cited works for further details. Book chapters and monographs on the subject of HMO theory may be cited,[746–750] and the subject of π-calculations in conjunction with force field calculations has been described in primary reports[352,362,380,400,431,517,518,751–758] and reviews[113,117,119] as well as reports of comparisons across methods.[759] Some of the aspects of π-calculations in conjunction with force field calculations were discussed briefly in Section 5.4.

The PCMODEL algorithm for π-calculations comes from the implementation of the Pople–Pariser–Parr (PPP) method used by Allinger and coworkers in the MMP1 program,[760] which has evolved over time.[62,761–772] The MMP1 π-routines are linked to the MM2-like PCMODEL MMX force field. Lipkowitz, Naylor and Melchior had reported a similar computational system.[773] The PCMODEL algorithm adopts the use of the VESCF method of Brown and Heffernan,[774] which Allinger et al. used to modify the atomic valence state ionization potential of the π-atoms based upon the electronegativity of the substituents on that atom.[765–767,775] The VESCF treatment leads to much more accurate transition energies for many unsaturated hydrocarbons, heterocycles,[768,772] and unsaturated carbonyl compounds.[766,769] Kao provided an updated treatment which used the results of the HMO π-calculations to scale

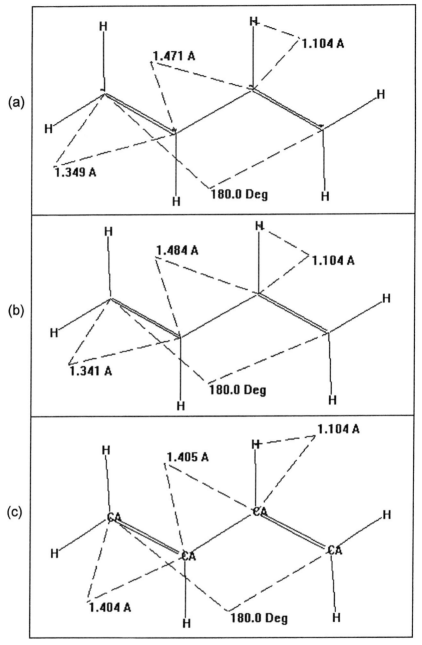

Figure 6-92. Models for the Butadiene Molecule with (a) π-Calculations; (b) No π-Calculations; (c) the @40 Aromatic Carbon Atom.

the bond order (and hence bond length), stretching force constants, and torsional force constants.[776] For the heat-of-formation calculations, PCMODEL follows the Allinger formulation,[770] with modifications introduced by Lo and Whitehead.[71,777]

Certain functional groups within a molecule, such as amides, esters, carboxylic acids, or simple vinyl ethers and enamines, may be modeled without an explicit π-calculation, since the resident MMX parameters already take account of the **delocalization**. If a group of this type is part of a larger conjugated system, a π-calculation should be run to take account of the conjugation effect. Moreover, any diene or structure with four or more electrons in a conjugated system, and any olefin or other unsaturated group which is strained, should be subjected to a π-calculation.

Almost any conjugated system may be modeled with the @40 aromatic carbon atoms, especially in the early stages of geometry optimization, but in the final analysis a π-calculation must be performed in order to obtain reasonable bond lengths and torsion angles. In some charged functional groups the anionic atom should be left out. For example, in a carboxylate the C=O bond should be included in the π-system, but not the C—O$^-$ bond; in a nitro group the $^+$N=O bond should be included in the π-system, but not $^+$N—O$^-$ bond.

In the HMO method, the molecular orbitals (MOs) are obtained as approximate eigenfunctions of a one-electron Hamiltonian. This treatment in its unmodified form fails to allow for interactions between the electrons, but it is not possible to address this deficiency rigorously with a collection of one-electron operators. A set of approximations and assumptions developed by Pople[778,779] and Pariser and Parr[780] permit electron–electron repulsions to be addressed by considering the effect of the mean field of the other electrons upon each electron in turn. The energy of a particular molecular orbital in the π-system may then be expressed as:

$$V_\pi = \alpha + k\beta$$

where α is the Coulomb integral, β is the resonance integral, both in electron volts (eV), and k is a coefficient for the energy of the molecular orbital of interest. The resonance integral β between a given pair of atom types bonded to each other with a given internuclear distance is taken to be constant, independent of their environment. It is a function of the atomic valence state ionization potential of each atom, the overlap integral, and an empirical scaling constant dependent on the type of overlap ($\pi\pi$ vs. $\sigma\sigma$). The value of the resonance integral between nonbonded atoms falls off very rapidly with distance, and earlier Hückel implementations ignored it, but more recent versions include all integrals. The Coulomb integral α incorporates both repulsive and attractive components.

Through an iterative process the coefficients of the orbitals are modified from cycle to cycle until the electronic energy reaches a constant minimum value and the orbitals no longer change. This SCF method overestimates electron–electron repulsion because it doesn't allow electrons to avoid each other. These discrepancies tend to average out in comparisons between structures.

In the earlier discussion of the **Options\MMX_PI_Calc** command (see Fig. 6-90), it was seen that two mutually exclusive options available are 'RHF', for **restricted Hartree–Fock**, and 'UHF', for **unrestricted Hartree–Fock**. In the Hartree–Fock approach to molecular orbital calculations, the energy of a set of MOs can be derived from the basis set functions used to define each orbital, and a set of adjustable coefficients which are used to minimize the total energy of the system.[121,781] Thus, in Hartree's original method the wave function for a number of electrons N was found as the product of N one-electron wave functions. Fock modified the method to use Slater determinants (or matrices) to combine the one-electron wave functions instead of using their products. This method took account of most of the effects of electron spin. Essentially, the Hartree–Fock method employs antisymmetric Slater determinants for combining the one-electron wave functions, and follows the SCF procedure of optimizing the coefficients of the orbitals until the electronic energy achieves a constant minimum value.

The RHF calculation is used for a closed-shell system. It carries the *restriction* that every orbital is either doubly occupied or unoccupied, and that all electrons are spin-paired. The RHF calculation is appropriate for neutral systems, anions, or cations. The UHF calculation is for an open-shell system, where at least one orbital will be singly occupied. It calculates two sets of molecular orbitals, one for each type of spin, named *alpha* and *beta*. *This is distinct from the α and β integrals discussed above.* The UHF approach thus considers each doubly occupied MO to consist of two spin-paired singly occupied MOs. The energies of these latter are not necessarily equal, as would be required in the RHF formulation. The UHF calculation accounts for the perturbation effect of an unpaired spin upon the formally paired spins, and generates more realistic spin densities. The UHF calculation is appropriate for radicals or diradicals. A radical (doublet) will have one more alpha-type electron than beta-type, and a diradical (triplet) will have an excess of two alpha-type electrons.

The UHF calculation causes the electronic energy of the system to be lower than does the RHF calculation for the same system, so that the results from the two types of calculation cannot be compared with each other, but comparisons of calculations made with the same method are valid.

Several computer experiments have been published which employ HMO methods,[185,196,200] and there are several World Wide Web sites which offer information and tutorials on this method,[782–784] and provide standalone utilities for carrying out HMO calculations.[785,786]

For the remaining exercises in this section, first the two examples 1,3-butadiene and benzene will be used to explain the numerical output from the PCMODEL π-calculations. Finally, the energy-minimized structures will be calculated for the two simple heterocycles pyrazole and imidazole under two sets of conditions: with π-calculations, and without π-calculations using the @40 aromatic carbon atoms for the ring. Afterwards we will compare the geometries of each with each other and with literature values, and we will compare the calculated π-electron densities with literature and with known chemical reactivity patterns.

π-Calculation Exercise 3: Numerical Output for 1,3-Butadiene and Benzene. In this exercise we examine and explain the results of the VESCF π-calculations for the two title molecules. Open the {BUTADIEN.PCM} file, and select the first 1,3-butadiene structure from the list (with π-atoms) into the work space. Activate the **Options\MMX_PI_Calc** command to ensure that the defaults are all active (see Fig. 6-90). Then, run this structure through several cycles of energy minimization. Now, exit the PCMODEL program to trigger completion of the {PCMOD.OUT} file. Then, use a text-processing program to open the ASCII-format {PCMOD.OUT} file, and scroll toward the end until the final set of matrices are visible, which appear directly after the statement 'Minimization Done'. They should resemble the 20 lines of output displayed in Fig. 6-93.

The title 'F-matrix' appears in line 2, and the matrix itself occupies lines 3–6. This 4 by 4 matrix represents the overlap integrals for the bonds in this system, where the value gives the overlap between the atoms corresponding to the matrix coordinates of the cell of interest. So, for example, the overlap integral between atoms 1 and 2 is -6.2696, and that between 1 and 4 is 0.5103.

The title 'EIGENVALUES-Most to least stable' appears in line 8, and in line 9 are given the four eigenvalues calculated for this system: -14.5496,

```
1   Minimization Done
2     F Matrix
3     -5.7771   -6.2696   -0.3068    0.5103
4     -6.2696   -6.6647   -3.2102   -0.3071
5     -0.3068   -3.2102   -6.6648   -6.2697
6      0.5103   -0.3071   -6.2697   -5.7773
7
8   EIGENVALUES-Most to least stable
9    -14.5496  -11.0037   -0.6144    1.2471
10    Molecular Orbitals
11     0.4091    0.5772    0.5767    0.4085
12     0.5546    0.4380   -0.4387   -0.5551
13     0.5768   -0.4086   -0.4090    0.5771
14     0.4385   -0.5550    0.5547   -0.4381
15
16    Electron Densit- Bond Order Matrix
17     0.9500    0.9581   -0.0147   -0.2815
18     0.9581    1.0501    0.2815   -0.0147
19    -0.0147    0.2815    1.0500    0.9581
20    -0.2815   -0.0147    0.9581    0.9499
```

Figure 6-93. The VESCF Output Matrices for 1,3-Butadiene.

498 APPLICATIONS

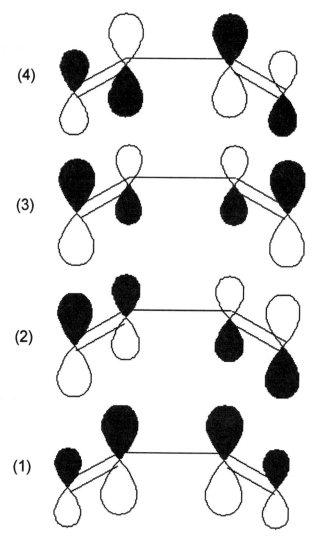

Figure 6-94. The Molecular Orbitals of 1,3-Butadiene, from Lowest Energy (1) to Highest Energy (4).

-11.0037, -0.6144, and 1.2471 eV. These represent the energy of the MOs, with the first two being the bonding orbitals occupied by the two electron pairs of the 1,3-butadiene molecule.

The title 'Molecular Orbitals' appears in line 10, and this 4 by 4 matrix occupies lines 11–14. The values given here are the coefficients for the p-orbitals of the π-system, with the row in line 11 being the first molecular orbital, the row in line 12 being the second molecular orbital, etc. The change in algebraic sign indicates the presence of a node between adjacent orbitals.

6.5. π-CALCULATIONS 499

These coefficients have been used to scale the cartoon representations of the orbitals presented in Fig. 6-94.

The title 'Electron Densit- Bond Order Matrix' appears in line 16, and this 4 by 4 matrix occupies lines 17–20. The values given here along the diagonal are the electron densities for the corresponding atom, and the off-diagonal values give the bond order for the bond between the atoms corresponding to the matrix coordinates of the cell of interest. For example, the electron density on atom 1 is given as 0.9500, and that on atom 2 is given as 1.0501. The π-bond order between atoms 1 and 2 is given as 0.9581, and that between atoms 2 and 3 is given as 0.2815.

Now, open the {BENZENE.PCM} file and select the first benzene structure from the list (with π-atoms) into the work space. Activate the **Options\MMX_PI_Calc** command to ensure that the defaults are all active (see Fig. 6-90). Then, run this structure through several cycles of energy minimization. Now, exit the PCMODEL program to trigger completion of the {PCMOD.OUT} file. Use the text-processing program to open the ASCII-format {PCMOD.OUT} file, and scroll toward the end until the final set of matrices are visible, which appear directly after the statement 'Minimization Done'. They should resemble the 26 lines of output displayed in Fig. 6-95.

In this portion of the output from the VESCF calculations on the benzene molecule, the title 'F-matrix' appears in line 2, and the 6 by 6 matrix itself

```
1  Minimization Done
2    F Matrix
3    -6.8015   -4.9527   -0.3644    0.6418   -0.3645   -4.9542
4    -4.9527   -6.8012   -4.9537   -0.3641    0.6416   -0.3643
5    -0.3644   -4.9537   -6.8007   -4.9522   -0.3643    0.6416
6     0.6418   -0.3641   -4.9522   -6.8007   -4.9539   -0.3639
7    -0.3645    0.6416   -0.3643   -4.9539   -6.8010   -4.9527
8    -4.9542   -0.3643    0.6416   -0.3639   -4.9527   -6.8012
9
10 EIGENVALUES-Most to least stable
11   -16.7922 -12.0315 -12.0299  -0.8422  -0.8417   1.7350
12   Molecular Orbitals
13    0.4080    0.4079    0.4083    0.4085    0.4086    0.4083
14    0.0421   -0.4777   -0.5199   -0.0423    0.4774    0.5196
15   -0.5760   -0.3246    0.2511    0.5756    0.3243   -0.2516
16   -0.5758    0.3257    0.2499   -0.5756    0.3256    0.2502
17    0.0437    0.4768   -0.5205    0.0436    0.4767   -0.5203
18    0.4081   -0.4081    0.4082   -0.4084    0.4084   -0.4083
19
20   Electron Densit- Bond Order Matrix
21    0.9999    0.6665    0.0001   -0.3334    0.0000    0.6668
22    0.6665    1.0000    0.6668    0.0000   -0.3333    0.0000
23    0.0001    0.6668    1.0000    0.6665    0.0000   -0.3333
24   -0.3334    0.0000    0.6665    1.0000    0.6668   -0.0001
25    0.0000   -0.3333    0.0000    0.6668    1.0000    0.6666
26    0.6668    0.0000   -0.3333   -0.0001    0.6666    1.0000
```

Figure 6-95. The VESCF Output Matrices for Benzene.

occupies lines 3–8. Analogously to the butadiene case discussed above, the present F-matrix represents the overlap integrals for the bonds in the benzene system; the value gives the overlap between the atoms corresponding to the matrix coordinates of the cell of interest. So, for example, the overlap integral between atoms 1 and 2 is -4.9527, and that between 1 and 5 is -0.3645.

The title 'EIGENVALUES-Most to least stable' appears in line 10, and in line 11 are given the six eigenvalues calculated for the benzene system: -16.7922, -12.0315, -12.0299, -0.8422, -0.8417, and 1.7350 eV. These represent the energy of the molecular orbitals, with the first three being the bonding orbitals occupied by the three electron pairs of the benzene molecule.

The title 'Molecular Orbitals' appears in line 12, and this 6 by 6 matrix occupies lines 13–18. The values given here are the coefficients for the p-orbitals of the π-system, with the row in line 13 being the first molecular orbital, the row in line 14 being the second molecular orbital, etc. The change in algebraic sign indicates the presence of a node between adjacent orbitals. These coefficients have been used to scale the cartoon representations of the molecular orbitals for benzene, presented in Fig. 6-96.

The title 'Electron Densit- Bond Order Matrix' appears in line 20, and this 6 by 6 matrix occupies lines 21–26. The values given here along the diagonal are the electron densities for the corresponding atom, and the off-diagonal values give the bond order for the bond between the atoms corresponding to the matrix coordinates of the cell of interest. In this highly symmetrical example, the electron density on each of the atoms varies between 0.9999 and 1.0000. The π-bond order between atoms 1 and 2 is given as 0.6665, that between atoms 1 and 3 is given as 0.0001, and that between atoms 1 and 4 is given as -0.3334.

HMO theory needs modification when applied to heteroconjugated systems. Recall the equation for the π-energy:

$$V_\pi = \alpha + k\beta$$

where the factors are defined as above, the Coulomb integral α must be modified by an additive factor proportional to β, and a suitable choice of k will yield the appropriate value for the resonance integral:

$$\alpha_{Het} = \alpha + h_{Het}\beta$$
$$\beta_{Het} = k_{Het}\beta$$

The constants h_{Het} and k_{Het} are chosen to provide the most accurate calculations for a basis set of molecules, and then used for the Coulomb integral α_{Het}, for the heteroatom and for the resonance integral β_{Het} between a heteroatom and carbon or between two heteroatoms. The number of electrons contributed by the heteroatom must also be taken into account. The heteroatom

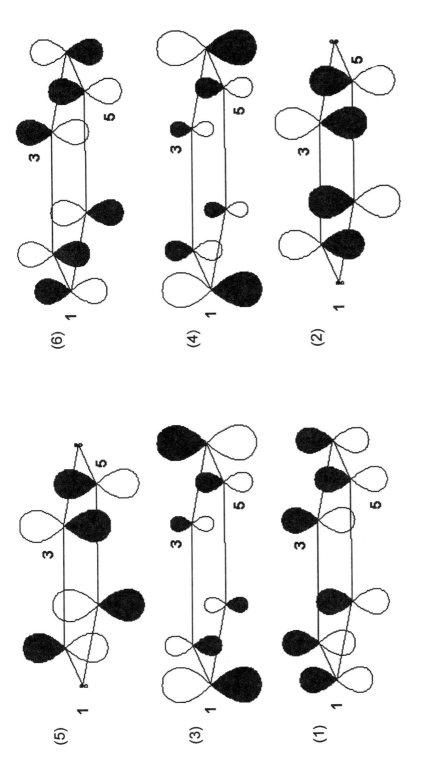

Figure 6-96. The Molecular Orbitals of Benzene, from Lowest Energy (1) to Highest Energy (6).

501

502 APPLICATIONS

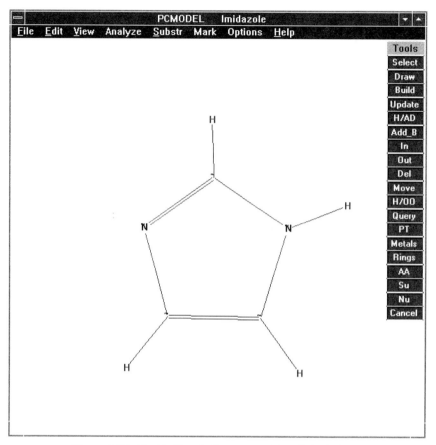

Figure 6-97. The Imidazole Structure with π-Calculations.

in molecules such as pyridine or acetophenone will contribute one electron, while the heteroatom in molecules such as pyrrole or anisole will contribute two electrons.[784]

π-Calculation Exercise 4: Modeling the Imidazole and Pyrazole Molecules. In this exercise model the structures of the two title heteroaromatic molecules, and compare the results with literature values. In the PCMODEL window, activate the **Rings** tool and select 'C5'. The cyclopentane structure which comes up is then modified by removing the hydrogens with the **H/AD** tool, converting two of the ring atoms with a 1,3-relationship into nitrogens (such as the upper left and upper right corners of the pentagon) with the **PT** tool, adding two double bonds to the ring with the **Add_B** tool, and adding hydrogens and lone pairs with the **H/AD** tool. Finally, the **Mark\Piatoms** command is used to flag all of the ring atoms for π-calculations. After assignment of the name 'Imidazole', the result should resemble Fig. 6-97.

Run the resulting structure through several cycles of energy minimization. Name this structure to an input file such as {IMIDAZOL.PCM}; an equivalent file may be found in the input file library diskette in the CHAP6\SECT5 subdirectory.

Use the resulting structure, activate the **Mark\Reset** command and select the 'Pi Atoms' option to unflag the π-atoms. Then, from the 'Periodic Table' menu, load the mouse cursor with the @40 aromatic carbon atom, 'CA', and use it to replace the three carbon atoms in the imidazole molecule. Run this structure through several cycles of energy minimization, and write it to the input file {IMIDAZOL.PCM} in appended format. The optimized structure should look like that depicted in Fig. 6-98.

To build the pyrazole structure, read in the appended format file {IMIDAZOL.PCM}, and select the first of the two structures. Remove the

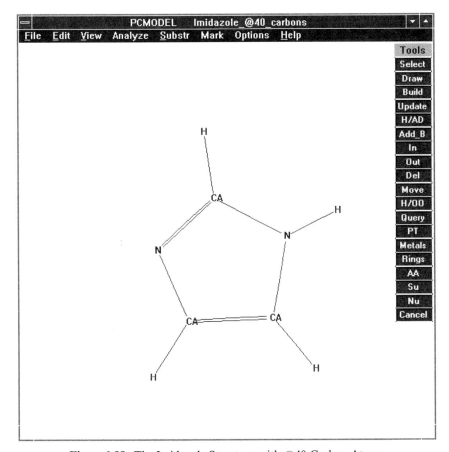

Figure 6-98. The Imidazole Structure with @40 Carbon Atoms.

504 APPLICATIONS

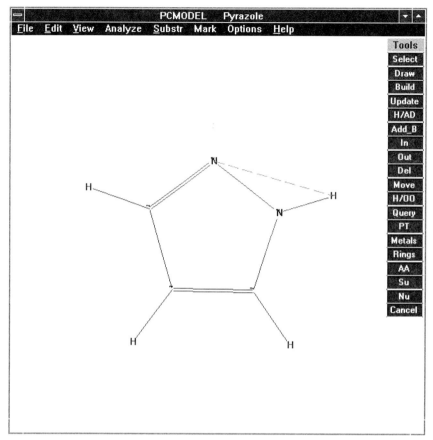

Figure 6-99. The Pyrazole Structure with π-Calculations.

hydrogens and lone pair with **H/AD** and then convert the unprotonated nitrogen to a carbon, and the carbon adjacent to the N—H to another nitrogen. Use the **H/AD** tool to add back the hydrogens and the lone pair; then run this structure through several cycles of energy minimization. Write it to an input file such as {PYRAZOL.PCM}; an equivalent file may be found in the input file library diskette in the CHAP6\SECT5 subdirectory. The optimized structure should look like that shown in Fig. 6-99.

Just as was done above, activate the **Mark\Reset** command and select the 'Pi Atoms' option to unflag the π-atoms in this pyrazole structure. Then, from the 'Periodic Table' menu, load the mouse cursor with the @40 aromatic carbon atom, 'CA', and use it to replace the three carbon atoms in the molecule. Run this structure through several cycles of energy minimization, and write it to the input file {PYRAZOL.PCM} in appended format. The optimized structure should look like that depicted in Fig. 6-100.

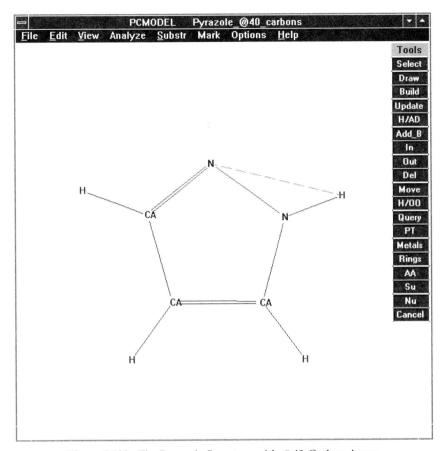

Figure 6-100. The Pyrazole Structure with @40 Carbon Atoms.

Run each of the two files, {IMIDAZOL.PCM} and {PYRAZOL.PCM}, through minimization in **Batch** mode. The minimization process may be repeated by deleting the initial {IMIDAZOL.PCM} and {PYRAZOL.PCM} files, renaming the new {IMIDAZOL.OUT} and {PYRAZOL.OUT} files to their respective {*.PCM} forms, and running them again through the **Batch** minimization. The **Query** tool may then be used to examine the geometric features of the optimized structures, which have been collected in Table 6-11, together with comparable measures obtained from X-ray and neutron diffraction experiments and microwave spectroscopy.

While overall there is good agreement between the results of our calculations and the experimental data, there are some discrepancies of note. In the imidazole structure, the use of π-calculations models the angles 1–2–3 and 5–1–2 better than the use of the @40 atom. Just the opposite result is found

TABLE 6-11. Calculated vs. Experimental Geometric Measures for Imidazole and Pyrazole:

Imidazole

Pyrazole

Structure	Bond angles (deg)					Bond lengths (Å)					Reference
	1–2–3	2–3–4	3–4–5	4–5–1	5–1–2	1–2	2–3	3–4	4–5	5–1	
Imidazole/1	106.1	109.3	107.6	106.1	110.9	1.356	1.397	1.374	1.397	1.388	This exercise
Imidazole/2	107.8	108.9	107.2	107.0	109.1	1.277	1.275	1.403	1.361	1.366	This exercise
Imidazole[a]	105.1	109.2	106.7	106.1	112.8	1.320	1.343	1.375	1.350	1.329	248
Imidazole[b]	105.1	109.8	106.1	107.0	111.9	1.316	1.368	1.358	1.362	1.337	787
Imidazole[c]	105.4	109.8	106.3	107.2	111.3	1.326	1.378	1.358	1.369	1.349	788
Pyrazole/1	109.8	107.0	107.8	108.6	106.7	1.360	1.403	1.389	1.386	1.412	This exercise
Pyrazole/2	111.5	106.5	105.2	109.9	106.9	1.274	1.407	1.407	1.365	1.420	This exercise
Pyrazole[a]	111.6	105.0	107.5	111.5	104.3	1.336	1.375	1.362	1.335	1.352	789
Pyrazole[b]	111.8	104.0	107.5	111.7	104.9	1.323	1.391	1.370	1.339	1.338	790
Pyrazole[c]	111.9	104.5	106.4	113.1	104.1	1.349	1.416	1.373	1.359	1.349	788

Atom numbering follows that arising from the use of the PCMODEL ring template, as in the structures depicted above the table, and is not systematic. Structures marked [a] come from the referenced X-ray diffraction data, those marked [b] are from neutron diffraction data, and those marked [c] come from microwave spectral data.

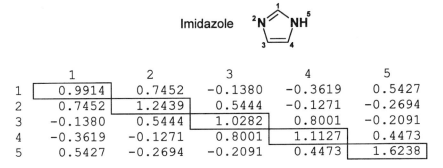

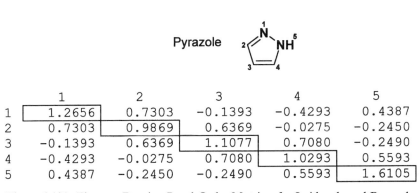

Figure 6-101. Electron-Density–Bond-Order Matrices for Imidazole and Pyrazole.

for modeling the bonds 4–5 and 5–1. In the pyrazole structure, the use of the @40 atom better models the angles 1–2–3, 2–3–4, and 4–5–1, and the bond 4–5. The use of π-calculations gives the better agreement for the bonds 1–2 and 3–4. Both calculational methods produce structures with distorted 5–1–2 angles and 5–1 bonds in pyrazole, suggesting the model for the N—N bond as the source of error. Other sources of error arising from limitations in the transferability of structural parameters were discussed in Sections 3.3 and 5.5.

The {PCMOD.OUT} output file for these energy minimizations will contain the electron density matrices, as described earlier in this section. These may be viewed with the use of any text window utility, and are depicted in Fig. 6-101.

The diagonal values in each of the matrices (enclosed in boxes) correspond to the electron densities at the correspondingly numbered atoms. A comparison between our calculated values for the π-electron densities and values from literature Hückel calculations is found in Table 6-12.

The agreement here is fairly good. With regard to reactivity patterns, imidazole is reactive to nucleophiles at the 2-position (i.e. Atom 1), and this is the carbon atom with the least electron density in our calculation. Imidazole is reactive to electrophiles at the 4-position (Atoms 3 or 4), while our calculations

TABLE 6-12. Calculated vs. Experimental Electron Densities for Imidazole and Pyrazole:

Imidazole: ring with N at position 2, NH at position 5, positions 3 and 4 on bottom, position 1 at top.

Pyrazole: ring with N at position 1, N at position 2, NH at position 5, positions 3 and 4 on bottom.

Structure	Electron Density					Reference
	Atom 1	Atom 2	Atom 3	Atom 4	Atom 5	
Imidazole	0.9915	1.2440	1.0280	1.1129	1.6237	This exercise
Imidazole	1.100	1.101	1.037	1.112	1.650	788
Imidazole	0.9479	1.2308	1.0753	0.9798	1.7662	248
Imidazole[a]	0.9546	1.2626	1.0799	0.9934	1.6993	248
Imidazole	0.894	1.325	1.049	1.054	1.679	730
Pyrazole	1.2658	0.9869	1.1077	1.0291	1.6106	This exercise
Pyrazole	1.162	1.031	1.086	1.109	1.656	788
Pyrazole	1.255	1.034	1.08	1.000	1.625	730

Atom numbering follows that arising from the use of the PCMODEL ring template, as shown above the table, and is not systematic. Literature values are from Hückel calculations. The protonated nitrogen, NH, is taken as atom 1. [a] Hückel calculations on the hydrogen-bonded pentamer, N–H terminal residue. Electron density values are from the unitless ratio of the calculated value over that for a uniform distribution of one electron per atom.

show Atom 4 to have the highest electron density on carbon. For pyrazole, our calculation shows the carbon with the highest electron density is the 4-position (Atom 3), and this is also the preferred site of attack on electrophiles.

6.6. HYDROGEN BONDING

The **hydrogen bond** is arguably the most important type of non covalent interaction. According to the putative "father of the hydrogen bond," M. L. Huggins, ". . . except for electron-pair bonding between atoms and Coulombic attractions and repulsions between ions, the most important structural principle in chemistry and biology is that of hydrogen bonding."[791]

The hydrogen bond may be defined as an attractive interaction between a functional group A—H (where A is generally an electronegative atom) and an atom or group of atoms B in the same or a different molecule. The hydrogen bond is a special case of the electrostatic interaction between two fixed dipoles. A hydrogen atom bonded to a small, highly electronegative atom A forms one of the dipoles, and the other is defined by a lone or nonbonding electron pair associated with another such electronegative atom B. In most cases, the atoms A and B are nitrogen, oxygen, fluorine, or sulfur. Hydrogen bonds generally range in strength from 3 to 8 kcal/mol.

The first description of a hydrogen bond to appear in the scientific literature is generally believed to be that of Latimer and Rodebush. Using Lewis dot structures, they drew diagrams depicting the hydrogen bond between two water molecules and between water and ammonia, and stated that their "explanation amounts to saying that the hydrogen nucleus held between two octets constitutes a weak 'bond'."[792] A footnote in this report alludes to a thesis submitted by Huggins at Berkeley the previous year, in which he introduced "the idea of a hydrogen kernel held between two atoms as a theory in regard to certain organic compounds." Thirty-three years later, the hydrogen bond formed the linchpin of the model of the DNA double helix proposed by Watson and Crick.[41] The theory of the hydrogen bond has continued to evolve through the spectroscopic and diffraction analysis of molecular structure and computational methods.[403,791,793–803]

The development of potential functions to represent the hydrogen bonding interaction[804–807] was covered to some extent in the discussions of individual empirical force fields in Section 5.4, and will be reviewed later in this section in connection with the description of the MMX implementation. In the PCMODEL program, the hydrogen bonding interaction may be turned on with the **Mark\H-Bonds** command, and may be turned off with the **Mark\Reset** command. The default is for the hydrogen bonding interaction to be turned on. Hydrogen bonds between a donor hydrogen and an appropriate acceptor atom are displayed on screen as a red dashed line when the distance between the two atoms is ⩽2.25 Å, but there will be no other visible indication on the affected atoms of the structure.

When the **H-Bonds** command is activated, all hydrogens on appropriate donor atoms will be flagged in the input file as hydrogen bond donors. There are six hydrogen atom types which will act as a hydrogen bond donor: the generic @5 hydrogen, when bonded to sulfur or selenium; the @21 hydrogen when bonded to oxygen; the @23 hydrogen when bonded to aliphatic or amide nitrogen; the @28 hydrogen when bonded to the oxygen of an enol or phenol; the @24 hydrogen when bonded to the oxygen of a carboxylic acid group, or to the O^+ of an oxonium, or to the N^+ of an ammonium or immonium compound; and the @36 deuterium atom when bonded to nitrogen, oxygen, sulfur, or selenium (see Appendix C, and Section 5.3). These are all parametrized for levels of acidity and donor ability.

The MMX atom types which will serve as hydrogen bond acceptors are the @6, @7, and @42 oxygens, the @8, @10, and @37 nitrogens, the @15 and @38 sulfur; and the @25 phosphorus. An acceptor atom must possess an electron lone pair, which it will be seen later is an essential factor in the potential equation. In the case that the hydrogen bonding flag has not been set, or in the case of relatively weak hydrogen atom acceptors such as selenium, an attractive interaction will be modeled solely through the Coulombic electrostatic term. The distance between the donor hydrogen and the acceptor atom may be greater than 2.25 Å, especially in the case of second-row elements as acceptors, and these interactions will not be displayed on screen with the red

dashed line. In PCMODEL, the hydrogen bonding interaction cannot be flagged for a subset of appropriate donor–acceptor pairs—it is either all or none.
The hydrogen bonding interaction will be demonstrated in an exercise which evaluates the methanol dimer in this context.

Hydrogen Bonding Exercise 1: The Methanol Dimer. This exercise shows how the hydrogen bonding interaction mediates the formation of a methanol dimer complex.

A PCM format input file for methanol, {MEOH.PCM}, was prepared in Section 6.2 on volumes and surfaces, and a copy of this file may be found on the input file library diskette in the CHAP6\SECT6 subdirectory. In the PCMODEL window, read in this structure. Then, use the **Substr\Read** command to read in a second copy of {MEOH.PCM} as a substructure. The result should look like Fig. 6-102.

An energy minimization should then be carried out by activating the **Analyze\Minimize** command. During the course of the minimization, the two

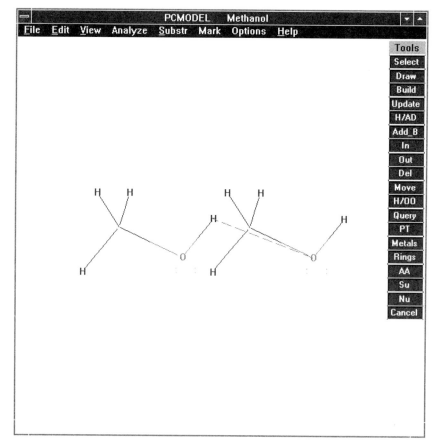

Figure 6-102. Starting Point for the Methanol Dimer Experiment.

methanol molecules will reorient to form a hydrogen-bonded dimer—if not, ensure that the hydrogen bonding flag is set by activating the **Mark\H-Bonds** command. This dimer structure should be sent through several cycles of energy minimization to arrive at a reasonably optimized structure, which will resemble that depicted in Fig. 6-103. This system may be given a name such as 'Methanol Dimer' with the **Edit\Structure Name** command, and an input file for the depicted dimer may be found as the first of four structures in the {MEOHDIMR.PCM} file.

Use the **Query** tool to examine the structure of the methanol dimer. The results from this experiment are comparable with those from other methods, given in Table 6-13.

The PCM-format input file for the methanol dimer ensemble prepared in the first exercise is shown in Fig. 6-104. The PCM input file format is discussed in greater detail in Section 4.1. Note here first that the entire ensemble contains sixteen atoms (line 2), eight for each of the methanol molecules

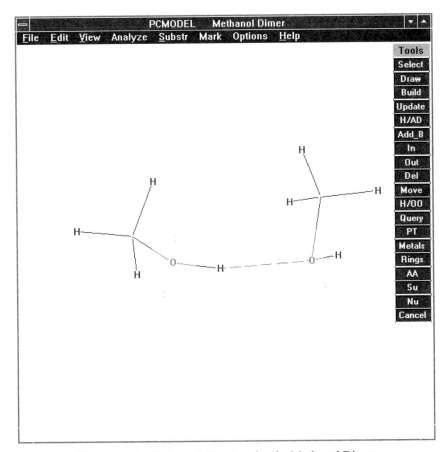

Figure 6-103. Optimized Structure for the Methanol Dimer.

TABLE 6-13. Methanol Linear Dimer Results

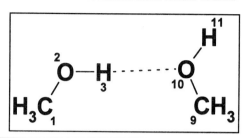

r(3–10) (Å)	r(2–10) (Å)	θ(2–3–10) (deg)	θ(3–10–11) (deg)	Method	Reference
1.799	2.737	169.0	105.2	MMX	This exercise
1.8	2.81	180	110	TIPS	459
2.004	2.957			MM3(94)	374
2.011	2.957			6-31G**	374

For each method, the r's are distances and the values for θ's are angles. Under "Method," MMX, TIPS, and MM3(94) indicate the corresponding force fields (see Chapter 5), and 6-31G** refers to an *ab initio* calculation at the Hartree–Fock level of theory.

(CH_4O plus two lone pairs). Lines 4–11 represent the incipient main structure, as indicated by an 'S 0' in the data string just prior to the charge parameter, and lines 12–19 represent the incipient substructure (see Section 3.2.6), as indicated by an 'S' in the data string. The donor hydrogens are atoms 6 and 14, and the hydrogen bonding flag is indicated by an 'H' in the data string for donor hydrogens (lines 6 and 14), between the substructure and charge fields.

```
1  {PCM Methanol Dimer
2  NA 16
3  FL EINT4 UV1 PIPL1
4  AT 1  1   -1.641    0.140   -0.332 B 2 1 4 1 5 1 6 1 S 0 C  0.1268657
5  AT 2  6   -0.873   -0.354    0.744 B 1 1 3 1 7 1 8 1 S 0 C -0.1366746
6  AT 3 21    0.014   -0.464    0.424 B 2 1 S 0 H   C  0.2898089
7  AT 4  5   -2.700    0.232    0.001 B 1 1 S 0
8  AT 5  5   -1.242    1.137   -0.630 B 1 1 S 0
9  AT 6  5   -1.563   -0.573   -1.185 B 1 1 S 0
10 AT 7 20   -1.028   -0.927    0.840 B 2 1 S 0 C -0.14
11 AT 8 20   -0.792    0.093    1.139 B 2 1 S 0 C -0.14
12 AT 9  1    1.914    0.838   -0.887 B 10 1 12 1 13 1 14 1 S   C  0.1268657
13 AT 10 6    1.723   -0.335   -0.124 B 9 1 11 1 15 1 16 1 S   C -0.1366746
14 AT 11 21   2.231   -0.243    0.668 B 10 1 S  H   C  0.2898089
15 AT 12 5    1.325    0.745   -1.829 B 9 1 S
16 AT 13 5    1.559    1.712   -0.295 B 9 1 S
17 AT 14 5    2.999    0.945   -1.118 B 9 1 S
18 AT 15 20   2.020   -0.783   -0.391 B 10 1 S  C -0.14
19 AT 16 20   1.152   -0.332    0.100 B 10 1 S  C -0.14
20 }
```

Figure 6-104. The PCM Input File for the Methanol Dimer, {MEOHDIMR. PCM}.

The MMX force field models the hydrogen bond by combining the standard charge–charge electrostatic and van der Waals treatment (see Section 5.2) with the modified van der Waals hydrogen bonding potential shown below. There is an attractive term proportional to the inverse square of the hydrogen–lone-pair distance, which is scaled according to the atom donating the lone pair and the acidity of the hydrogen. An additional angle dependence factor is introduced which is proportional to the cube of the difference between the distance from the hydrogen donor atom to the lone pair donor atom and the distance from the hydrogen atom to the lone pair. If the lone pair donor heavy atom is a vinyl ether oxygen which is conjugated and subject to a π-calculation, the difference in π-electron density on an isolated vinyl ether (1.95) and the calculated electron density adds a factor to the hydrogen donating ability. (This discussion is adapted from the account published by the developers of the MMX force field.[807] The MMX hydrogen bonding potential function takes the form

$$V_{HB} = K_{HB} k_{HB} \, e^{-12.50/\hat{p}} - k_{HB} \hat{k}_{HB} \hat{p}^2$$

where K_{HB} is a scaling and units cancellation factor taking the value 2.90×10^5; k_{HB} is the force constant for the hydrogen bonding interaction between the hydrogen and the lone pair, in kilocalories per mole; $\hat{p}$ is a unitless ratio of the equilibrium distance between the hydrogen atom and the lone pair to the actual distance; and $\hat{k}_{HB}$ is a unitless scaling constant for the donor–acceptor interaction.

The force constant k_{HB} is the geometric mean of the hardness parameters ε for the lone pair (0.016 kcal/mol) and for the specific hydrogen atom type involved in the hydrogen bond in question:

$$k_{HB} = (\varepsilon_H \varepsilon_{lp})^{1/2}$$

The distance ratio $\hat{p}$ is given by

$$\hat{p} = \frac{r_H + r_{lp}}{r_{H-lp}}$$

where r_H is the van der Waals radius of the specific hydrogen atom type involved in the hydrogen bond in question, r_{lp} is the van der Waals radius of the lone pair (0.200 Å), and r_{H-lp} is the actual distance between the donor hydrogen and the acceptor lone pair, as depicted for the methanol dimer in Fig. 6-105.

The constant $\hat{k}_{HB}$ is a scaling factor for angle dependence, defined as

$$\hat{k}_{HB} = \frac{(\hat{k}_D + \hat{k}_A)(r_{D-A} - r_{H-lp})^3}{4}$$

514 APPLICATIONS

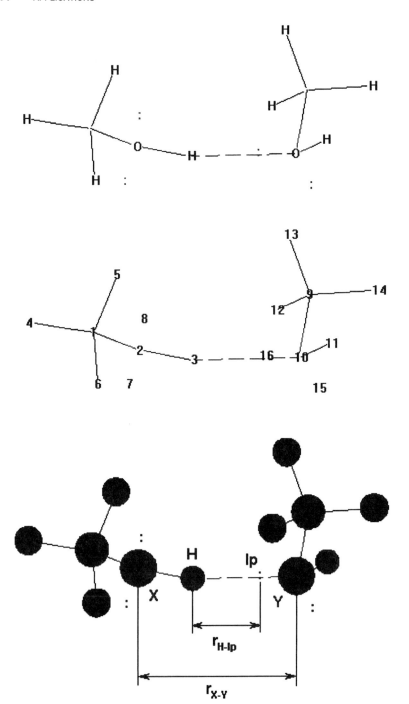

Figure 6-105. The Hydrogen Bonding Interaction in a Linear Methanol Dimer.

where $\hat{k}_D$ is a parameter dependent upon the type of hydrogen bond donor moiety (see Table 6-14), $\hat{k}_A$ is a parameter dependent upon the type of hydrogen bond acceptor moiety (see Table 6-15); r_{D-A} is the distance between the donor moiety heavy atom and the acceptor moiety heavy atom (i.e. r_{X-Y} in Fig. 6-105), and r_{H-lp} is the distance between the donor hydrogen and the acceptor lone pair.

It should be obvious that the factor reflecting the difference in distances between the parts of the donor and acceptor moieties will reach a maximum value when the parts of the X—H—lp—Y system are most collinear. Likewise, higher values of $\hat{k}_D$ and $\hat{k}_A$ will also contribute to a stronger hydrogen bonding interaction.

TABLE 6-14. MMX $\hat{k}_D$-Parameters for Hydrogen Donor Atoms

Atom	Type	H-type	ε_H	r_H (Å)	$\hat{k}_D$	Moiety
C	@1	@5	0.047	1.500	20	Hydrocarbon
O	@6	@21	0.036	1.100	40	Alcohol
N	@9	@23	0.034	1.125	40	Amide
O	@6	@23	0.015	0.900	60	Phenol or enol
N	@37	@24	0.015	0.800	60	Imine
N	@41	@24	0.015	0.800	80	Ammonium
O	@6	@24	0.015	0.800	120	Carboxylic acid

Atom type refers to the parametrization of atoms in the MMX force field; H-type refers to the parametrization of donor hydrogen in the MMX force field; ε_H is the hardness parameter for the specific hydrogen; r_H is the van der Waals radius for the specific hydrogen; $\hat{k}_D$ is a unitless parameter which defines the strength of the hydrogen bonding potential due to the particular donor moiety; moiety refers to the specific functional group which bears the donor hydrogen. These values are taken from reference 807.

TABLE 6-15. MMX $\hat{k}_A$-Parameters for Hydrogen Bond Acceptor Atoms

Atom	Type	$\hat{k}_A$	Moiety
O	@7	5	Carbonyl oxygen
O	@7	15	Sulfoxide oxygen
N	@10	20	Nitrile nitrogen
S	@15	50	Sulfide
S	@38	50	Thione sulfur
P	@25	60	Phosphine
O	@6	90	Ether or alcohol oxygen
N	@37	130	Imine nitrogen
N	@8	200	Amine nitrogen
O	@42	300	Oxygen anion

Atom type refers to the parametrization of atoms in the MMX force field; $\hat{k}_A$ is a unitless parameter which defines the strength of the hydrogen bonding potential due to the particular acceptor moiety; moiety refers to the specific functional group which bears a lone pair of electrons and acts as the hydrogen bond acceptor. These values are taken from reference 807.

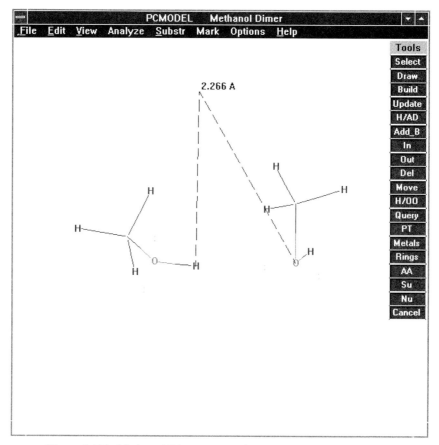

Figure 6-106. Methanol Dimer Reassociated after the Dynamics Run.

Several studies of hydrogen bonding modeled with the MMX force field,[808,809] including one experiment for undergraduate use,[197] have been published.

In a second exercise on the methanol dimer, we will examine the hydrogen bonding interaction under the influence of molecular dynamics.

Hydrogen Bonding Exercise 2: Dynamic Perturbation of the Methanol Dimer. This exercise shows how the methanol dimer complex behaves under molecular dynamics perturbation (see Section 5.9). Read in the previously prepared methanol dimer ensemble, or import it as the first structure in the {MEOHDIMR.PCM} input file. Run the structure through at least one cycle of energy minimization. Then, use the **Query** tool to set up a report of the H······O distance for the hydrogen bond, which should read about 1.798–1.800 Å at the outset. Next, activate the **Analyze\Dynam** command, and change the value of the 'Time Step' in the 'Dynamics Setup' dialog box from 1 to 0.2 fs.

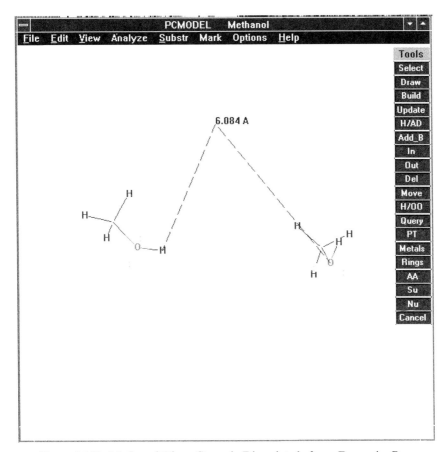

Figure 6-107. Methanol Dimer Strongly Dissociated after a Dynamics Run.

Click on the 'OK' button to initiate the dynamics run with the new time step and the default values for the other parameters. Depending upon the clock speed of the desktop platform used, the two methanol molecules will slowly wander away from each other. Use the ⟨Esc⟩ key to interrupt the dynamics run just as the red dashed line of the hydrogen bond disappears, when the distance is about 2.25 Å. The screen should resemble the depiction in Fig. 6-106.

Now, reactivate the **Analyze\Minimize** command, which will cause the dimer to reform driven by the hydrogen bonding potential interaction. The reassociated dimer complex will resemble Fig. 6-103 again, and the input file for this reconstituted dimer may be found as the second of four structures in the {MEOHDIMR.PCM} file.

The limiting distance over which the dimer can reassociate is approximately 6 Å, which we can check as follows. Read in the first of the four structures in the {MEOHDIMR.PCM} file, bring up the 'Dynamics Setup' dialog box again, and enter a value of 0.75 fs for the time step. Start the dynamics run again, and wait until the H······O distance has reached ≈ 6 Å, at which point the dynamics run should be interrupted. The resulting ensemble is shown in Fig. 6-107.

Reactivate the **Analyze\Minimize** command, which should cause the dimer to re-form as before, although the return will take longer. The result should look like Fig. 6-108, and the input file for this reconstituted dimer may be found as the third of four structures in the {MEOHDIMR.PCM} file.

The user should try running the dynamics for other lengths of time, to determine empirically where the "point of no return" is. The last part of the exercise is to carry out an energy minimization in which the hydrogen bonding flag is turned off, for comparison. Read in again the first of the four structures in the {MEOHDIMR.PCM} file. Activate the **Mark\Reset** command, and select the 'Hydrogen Bonds' option, to remove the flag for this interaction. Then, run the energy minimization. The two methanol molecules will be seen

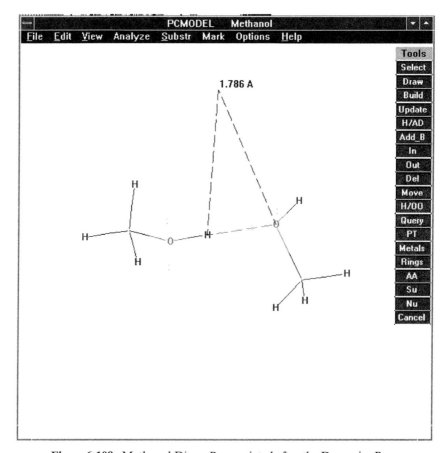

Figure 6-108. Methanol Dimer Reassociated after the Dynamics Run.

to push each other apart, and come to rest with an H······O distance of ≈2.3 Å, as seen in Fig. 6-109.
Comparing the final energy reports (see Fig. 6-110), the energy difference due to the hydrogen bonding interaction in the linear methanol dimer works out to 3.535 kcal/mol, which can be seen to have a van der Waals component ('Vdw', 1.97 kcal/mol) and a charge-charge component ('Q-Q', 1.66 kcal/mol).

The potential functions used to model the hydrogen bonding interaction have evolved a great deal over the history of empirical force field modeling. Many contemporary force fields employ a combination of the van der Waals and electrostatic nonbonding potentials. These are generally tailored for hydrogen bonding with special parameters for the donor hydrogens and acceptor heavy atoms, which permit a closer approach and allow the attraction to be driven electrostatically.

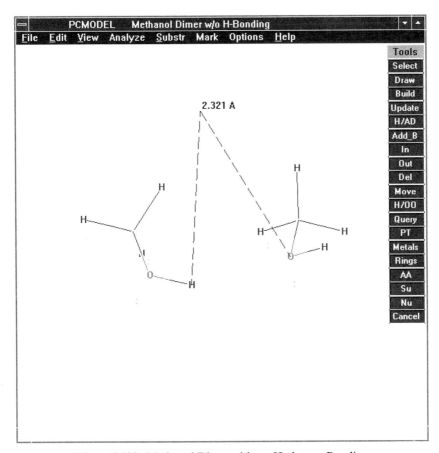

Figure 6-109. Methanol Dimer without Hydrogen Bonding.

```
         Heat of Formation (hf0) = energy + be + pfc  :      -103.934

         Strain Energy (energy+environment corrs.)=   :        -6.566
(a)      MMX Energy = -4.146
         Str = 0.112      Bnd = 0.479
         S-B = 0.034      Tor = 0.003
         Vdw = -1.500     Q-Q = -3.275
         Dip Mom = 2.352
```

```
         Heat of Formation (hf0) = energy + be + pfc  :      -100.402

         Strain Energy (energy+environment corrs.)=   :        -3.034
(b)      MMX Energy = -0.614
         Str = 0.026      Bnd = 0.471
         S-B = 0.031      Tor = 0.003
         Vdw = 0.470      Q-Q = -1.615
         Dip Mom = 3.111
```

Figure 6-110. Calculated Energies for the Methanol Linear Dimer with (a) and without (b) Including the Hydrogen Bonding Potential.

A more sophisticated treatment of hydrogen bonding must take account of the directionality of this interaction.[374,801,803,810–812] The MMX force field does this by including a scaling factor which is dependent upon the difference between the distance from donor to acceptor heavy atoms and the distance from donor hydrogen to acceptor lone electron pair. Other force fields incorporate scaling factors which account for directionality in different ways, and these will be discussed below.

From the theoretical perspective, directional hydrogen bonding between two molecules may be viewed through the Fukui formalism in terms of the interactions of **frontier molecular orbitals**,[813,814] specifically the **LUMO** of the acceptor molecule and the **HOMO** of the donor molecule.[374,810,811] For those for whom visual representations of molecular orbitals are far more enlightening than numerical descriptions, reference should be made to one of the several good collections of such orbital pictures, such as the monograph by Jorgensen and Salem.[815] Considering our example of the hydrogen-bonded methanol dimer,[816] the acceptor HOMO would correspond to one of the two nonbonding MOs, which essentially correspond to what we call the two lone pairs of electrons. These nonbonding MOs have a pronounced directionality with stereochemical consequences,[337] which cannot be adequately modeled as an isotropic electrostatic interaction. The donor LUMO corresponds to the σ^*_{C-O} antibonding MO, which has a prominent lobe in the vicinity of the donor hydrogen. The tendency toward linearity of the X—H—lp—Y moiety in otherwise unstrained ensembles may be seen as a consequence of the symmetry of this HOMO–LUMO interaction.[811]

The hydrogen bonding interaction has also been analyzed with different molecular orbital formalisms such as localized molecular orbitals (LMO),[803] and electrostatic treatments such as the localized charge distribution (LCD)[803] and the distributed multipole analysis (DMA).[801,812] These different approaches give results which are qualitatively much the same. In the discussion which follows, the treatments of hydrogen bonding in a number of force fields are summarized.

Of the Allinger family of force fields (see Section 5.4.1), MMI was a hydrocarbon force field, and thus did not encompass hydrogen bonding. MM2 did not employ a specific hydrogen bonding potential, but relied instead upon electrostatic (dipole–dipole) and van der Waals contributions to model this effect.[805]

A similar treatment was used for amines and alcohols early on in MM3,[363,364] except that specialized van der Waals parameters were used when the two nonbonding atoms were involved in hydrogen bonding. Note that at this stage, the lone pair pseudoatoms were no longer employed. The value used for the sum of the van der Waals radii for the donor hydrogen and the acceptor heavy atom was lowered, and the nonbonding force constant ε for these pairs was made softer to allow for closer approach. By comparison, MM2 underestimates the hydrogen bonding interaction by about 20%.[364] The MM3 hydrogen bonding potential was modified a few years later to account more for the directionality.[374] In the resulting expression,

$$V_{\text{vdW-HB}} = k_{\text{vdW-HB}}(K_{\text{vdW}} K_{\text{HB}} p^6 + K'_{\text{vdW}} e^{-12.00/p})$$

the sixth power term of the van der Waals potential expression now bears a scaling factor, K_{HB}, in contrast with the standard van der Waals expression given in Section 5.4.1.3. This factor is defined as shown below, and the appropriate geometrical features are displayed in Fig. 6-111:

$$K_{\text{HB}} = (\cos \beta) \frac{r_{X-H}}{r^0_{X-H}}$$

where β is the Y—X—H angle r_{X-H} is the actual distance between the donor heavy atom and the donor hydrogen, and r^0_{X-H} is the equilibrium value for this quantity. The factor p is given by

$$p = \frac{r_H + r_Y}{r_{H-Y}}$$

where the sum $r_H + r_Y$ is parametrized to allow for a closer approach for the hydrogen bonding pair H and Y. The force constant $k_{\text{vdW-HB}}$ is derived from the hardness parameters ε for the donor hydrogen and the heavy atom acceptor, as described above.

Indications are that the MM4 implementation of the hydrogen bonding potential will be similar to the above.

In the CFF family of force fields, a potential function for the hydrogen bonding interaction was developed for amides as a model for the peptide

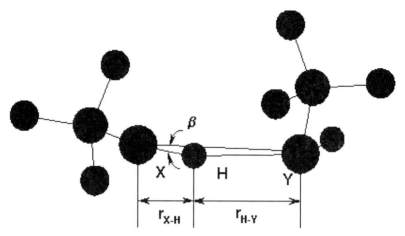

Figure 6-111. Geometrical Features of the MM3 Hydrogen Bonding Potential.

moiety. These parameters and the hydrogen bonding potential were superimposed upon the eight components of the existing UBCFF hydrocarbon force field (see Section 5.4.2.1).[383,398,399,804] The hydrogen bonding interaction was formulated at first as a four-part nonbonding potential including the attractive and repulsive components of the Lennard–Jones potential, a Coulombic potential electrostatic component, and an explicit hydrogen bonding attractive component, using a Morse potential function. After some optimization, it was found that the hydrogen bonding could be modeled quite well without the explicit Morse component. The 6-9 and 6-12 versions of the Lennard—Jones potential appeared to work equally well for this interaction—the equation below uses the 6-9 form. Thus, the hydrogen bonding potential energy is given by

$$V_{HB} = k_{vdW-HB}(2p^9 - 3p^6) + \frac{q_H q_Y}{r_{H-Y}}$$

where these variables are defined as above.

This potential differs from the standard van der Waals system by virtue of suitably modified parameters for those atoms involved in the hydrogen bonding interaction. Specifically, the van der Waals radius r_H^0 for the donor amide hydrogen and the van der Waals force constant k_{vdW-HB} for the interaction between the carbonyl oxygen and the amide hydrogen are set to zero, so that the attraction is driven electrostatically without steric repulsion. A similar approach was used when the hydrogen bonding treatment was later updated to include carboxylic acids.[390,401–403]

The precursor of the CHARMM force field (see Section 5.4.4) introduced a hydrogen bonding potential with components proportional to the 10th and 12th powers of r_{XY} when the atoms X and Y are the extended atom hydrogen bond donor and acceptor.[438] This potential operates in addition to the electrostatic interactions, and takes the form

$$V_{HB} = \frac{C_{XY}}{r_{XY}^{12}} - \frac{D_{XY}}{r_{XY}^{10}}$$

where C_{XY} and D_{XY} are parameters for the repulsive and attractive portions of the hydrogen bonding potential, respectively, and are defined in Section 5.4.4; and r_{XY} is the distance between the hydrogen bond donor heavy atom X and the acceptor atom Y.

The version of CHARMM reported a few years later had a more sophisticated hydrogen bonding potential,[439,440] which had been modified for the all atom formulation, and was scaled with two cosine terms to take of angle dependence of the hydrogen bond. The form of this potential is

$$V_{HB} = \left(\frac{C_{XY}}{r_{XY}^{12}} - \frac{D_{XY}}{r_{XY}^{10}}\right) \cos^m \theta_{X-H-Y} \cos^n \theta_{H-Y-Y^*}$$

where m and n are exponents chosen from a topology file; θ_{X-H-Y} is the angle between the hydrogen bond donor heavy atom X, the acidic hydrogen atom H, and the acceptor atom Y; and θ_{H-Y-Y^*} is the angle between the acidic hydrogen atom H, the acceptor heavy atom Y, and the acceptor antecedent atom Y^* (i.e. the other heavy atom attached to Y). These relationships are displayed in Fig. 6-112.

A new formulation of CHARMM was reported in 1995,[445] in which the explicit hydrogen bonding potential has been dropped, and hydrogen bonding is modeled through a combination of the van der Waals and electrostatic potentials with appropriate parameters for the donor and acceptor atoms.

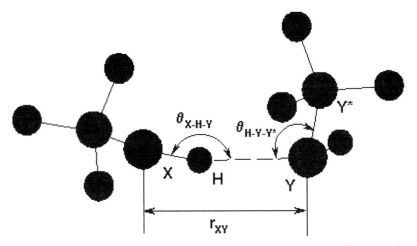

Figure 6-112. Geometric Features of the CHARMM Hydrogen Bonding Potential.

In its early version, the AMBER force field (see Section 5.4.3) employed the CHARMM-style 10–12 potential for modeling the hydrogen bonding interaction.[415,416] Further work on the components of the AMBER force field by the Kollman group led to the finding that the discrete Lennard–Jones 10–12 hydrogen bonding potential may be folded into the standard 6–12 nonbonding potential by using a curve-fitting procedure to generate the new parameters from the old.[417] The "second generation" AMBER force field incorporates both the 6–12 hydrogen bonding model and an improved electrostatic model.[420] For example, the amide H was made smaller, and the alcoholic/acidic hydrogen and the OPLS-style TIP3P hydrogen have both ε and r_0 set to zero, in order to facilitate hydrogen bonding.

Another sophisticated geometric treatment of hydrogen bonding is available in the YETI force field (see Section 5.4.8),[521,523] in which the directionality was parametrized based upon many examples from the Cambridge Crystallographic Structure Database (see Section 4.4).[800] The YETI hydrogen bonding potential resembles the early CHARMM 10–12 form, but incorporates a factor which scores the displacement of the donor hydrogen from a plane defined by acceptor heavy atom and two substituents. This potential takes the form

$$V_{HB} = \left(\frac{C_{HY}}{r_{HY}^{12}} - \frac{D_{HY}}{r_{HY}^{10}}\right) \cos^k \theta_{X-H-Y}$$
$$\times \cos^m(\theta_{H-Y-Y^*} - \theta^0) \cos^n(\omega_{H-Y^*-Y^{**}} - \omega^0)$$

where the exponents k, m and n depend upon the characteristics of the donor, the acceptor, and the lone pair directionality, respectively; θ_{X-H-Y} is the angle between the hydrogen bond donor heavy atom X, the donor hydrogen atom H, and the acceptor atom Y; θ_{H-Y-Y^*} is the angle between the donor hydrogen atom H, the acceptor heavy atom Y, and a substituent atom Y^* attached to Y; θ^0 is a reference angle which depends on the acceptor type; $\omega_{H-Y-Y^*-Y^{**}}$ is an improper torsion between the donor hydrogen atom H, the acceptor heavy atom Y, and two substituent atoms Y^* and Y^{**} attached to Y; and ω^0 is a reference angle which depends on the lone pair directionality for the acceptor. These relationships are displayed in Fig. 6-113.

This example shows the methanol dimer with the methyl hydrogens removed for clarity. The donor oxygen is marked X, the donor hydrogen is marked H, the acceptor oxygen is marked Y, the methyl carbon of the acceptor molecule is marked Y^*, and the hydrogen attached to the acceptor oxygen is marked Y^{**}. The improper torsion angle $\omega_{H-Y-Y^*-Y^{**}}$ gives the departure of the donor hydrogen from the plane determined by the carbon, oxygen, and hydrogen of the acceptor molecule.

The DREIDING force field, developed by Goddard et al.,[496] employs a hydrogen bonding potential of a form very close to the CHARMM version (see Section 5.4.8).

6.6 HYDROGEN BONDING   525

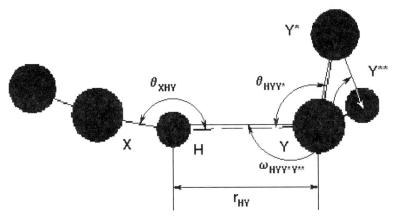

Figure 6-113. Geometric Features of the YETI Hydrogen Bonding Potential. (Hydrocarbon hydrogens removed for clarity.)

In the TRIPOS force field (see Section 5.4.5) and OPLS force field (see Section 5.4.7), there is no explicit hydrogen bonding potential included, but acidic hydrogens are given a van der Waals radius r_0 of 0.00 Å and a hardness parameter ε of 0.00 kcal/mol to facilitate these interactions passively.[271,306,451]

In the MMFF94 force field (see Section 5.4.6), hydrogen bonding is modeled as a special case of the van der Waals potential, but uses modified values for the equilibrium internuclear distance and the force constant which were derived from *ab initio* calculations on hydrogen-bonded water. These changes have the net effect of reducing the size of the hydrogen-bonding hydrogen in the context of this interaction, while maintaining its steric size for modeling the nonspecific van der Waals interactions. For a hydrogen bonding interaction, the hardness parameter is scaled by a factor of 0.5, and the equilibrium internuclear distance is scaled by a factor of 0.8.

Many other force field potentials have been developed for modeling the hydrogen bond, and a comprehensive description of all is beyond the scope of this book. Examples of these treatments include the following force fields: the early report of Taylor regarding a potential for carbohydrates composed of van der Waals, electrostatic, and Morse stretching components[817]; that of Goodford et al. for modeling biomacromolecule–ligand complexes[818]; that of Jeffrey for modeling carbohydrates[819]; that of Dauchez et al., which is parametrized for peptides and other biological molecules[820]; that of Machida et al. for modeling carboxylic acids[821]; and that of Masella et al. for modeling clusters of water and alcohols.[822,823]

An application of hydrogen bonding which has not received much attention in the force field modeling community is the case where weak carbon acids act as the hydrogen bond donors, e.g. the C—H———O interaction.[824,825] Evidence from crystallographic structures indicates that such interactions may be significant in biochemical systems, especially of enzyme-bound inhibitor complexes; but a workable molecular mechanics implementation has not yet appeared.

The hydrogen-bonded methanol dimer system will be examined again in Section 6.8 on the docking of substructures.

6.7. METAL ATOM BONDING AND COORDINATION

This section describes the modeling of metal-containing molecules with empirical force fields. We include in this area the modeling of compounds containing Main Group elements outside the normal organic arena, such as boron, aluminum, germanium, tin, lead, tellurium, etc. After a brief history of organometallic force field modeling, the MMX implementation and options are discussed in general terms. Later, the organometallic portion of the MMX force field is described in greater detail, and the challenges of organometallic modeling in contrast with strictly organic systems are covered. A brief overview of other force field descriptions of organometallic compounds and complexes is also given.

Organometallic and inorganic modeling is the topic of a number of review articles and book chapters,[11,17,20,24,114,116,148,230,408,826,827] and a highly informative monograph.[131] There have also been several reports on the use of inorganic modeling in undergraduate education.[194,207]

The history of organometallic and inorganic modeling has proceeded in parallel with that of hydrocarbon or organic modeling. As was discussed in Chapter 1, even though some of the pivotal early published work in computational modeling involved organometallic complexes,[43] there has only a moderate amount of synergy between the development organometallic force fields and the more purely organic ones. Much of this early work was done by adapting existing force fields to include coordination compounds of metals with Lewis basic donor ligands, both on small complexes[14,59-61,77,306,828-837] and on biomacromolecules.[421,427,439,440,512,827,838,839] Brubaker and Johnson have reviewed much of the earlier published applications to the amine complexes of the late first-row transition metals cobalt, nickel, and copper, encompassing the use of bi-, tri- and tetradentate ligands.[11] Only later were force field models developed for π-complexes.[827,838,840-847] Force fields specifically developed for organometallic compounds have been reported only within the past decade.[309,496,512,522,524,848-852] The molecular mechanics technique has also been applied to structural studies of zeolites[853] and of reactive intermediates and transition states for organometallic reactions.[854-856]

A number of factors make the modeling of organometallics more challenging than that of carbon compounds, and these have been summarized.[24,131,309,848] These factors include the greater variety of hybridizations to parametrize as metal atom types, with several oxidation states, valences, and coordination numbers; the breakdown in the assumption of connectivity, or metal-atom–donor-atom mapping; a greater ambiguity regarding the magnitude and location of charges; large variation of bond angles and the lack of a unique equilibrium bond angle; inadequacy of the standard harmonic angle deformation potentials to model organometallics accurately; and effects unique to organometallic molecules such as the **Jahn–Teller effect** and the **trans influence**. In most cases, the MMX force field treatment of organometallic bonding provides a very reasonable model for these compounds.

In general, the bond between a metal and a ligand is best described in terms of orbital overlap, but the MMX force field includes parameters and algorithms to model organometallic bonds as either **covalent bonds** (classical valence bonds) or as **coordination bonds** (Lewis acid–base complexes or coordination complexes). The user should be aware that the designations "covalent" and "coordination" in this context need have no intrinsic relationship with theoretical models of organometallic bonding—here they are merely convenient metaphors to differentiate alternative force field treatments of bonds to metals. Familiarity with these algorithms coupled with a basic knowledge of the structural chemistry of inorganic or organometallic compounds can often provide a suggestion for which mode of bonding is most appropriate in a particular case.

The force field treatment of organometallic molecules is less well developed than in the standard substituted alkane model. The best practice is to calculate one or more model systems for which comparisons with empirical structural data can be made, and then to optimize the modeling parameters before moving to novel examples. Inorganic and organometallic molecules or complexes can often be modeled several different ways, by varying the values used for the metal atom charge and covalent radius, the bond multiplicity, and the atom type of the ligand(s), and by using either the covalent or the coordination bonding protocols. In some cases, the metal may be covalently bonded to some ligands and coordinatively bonded to others. In the discussion which follows, the term "donor atom" will be used for atoms bonded to a metal in either a covalent or a coordinative sense.

As with the purely organic structures in PCMODEL, a model of a metal-containing compound can be prepared de novo by drawing it on screen with the mouse, or by modifying an existing structure or template. In the MMX model, covalent bonds are created by drawing explicit metal–nonmetal or metal–metal bonds on screen, using the **Draw** and/or **Build** tools. They are treated in the much the same way as are covalent bonds between nonmetals, with some differences which will be discussed below. Such bonds will appear in the connected- and attached-atom lists of the input file. There are parameters resident in PCMODEL for the covalent radii of most of the metals, and the user may wish to substitute an alternative value for the covalent radius in particular cases. These values are summarized in Table 6-16.

TABLE 6-16. PCMODEL-Resident Parameters and Additional Values for Metallic Atom Covalent and Atomic Radii

													Al		
Li	Be												1.900		
1.52	1.11												1.25		
1.23	0.89												1.43		
1.52	1.11														

Na	Mg
1.86	1.60
1.57	1.36
1.86	1.60

K	Ca	Sc	Ti	V	Cr	Mn	Fe	Co	Ni	Cu	Zn	Ga	As
2.31	1.97	1.60	1.46	1.31	1.25	1.29	1.26	1.26	1.26	1.26	1.355	....	....
2.03	1.74	1.44	1.32	1.22	1.19	1.18	1.16	1.16	1.15	1.17	1.25	1.25	1.21
2.27	1.97	1.61	1.45	1.31	1.25	1.37	1.24	1.25	1.25	1.28	1.33	1.22	1.25

Rb	Sr	Y	Zr	Nb	Mo	Tc	Ru	Rh	Pd	Ag	Cd	In	Sb
2.44	2.15	1.85	1.60	1.455	1.39	1.355	1.355	1.355	1.355	1.455	1.51	....	....
2.16	1.91	1.62	1.45	1.34	1.30	1.27	1.25	1.25	1.28	1.34	1.41	1.50	1.41
2.48	2.15	1.78	1.59	1.43	1.36	1.35	1.33	1.35	1.38	1.45	1.49	1.63	1.45

Cs	Ba	→	Hf	Ta	W	Re	Os	Ir	Pt	Au	Hg	Tl	Bi	Po	At
2.62	2.15		1.60	1.455	1.39	1.355	1.355	1.355	1.39	1.455	1.51	....	....	....	....
2.53	1.98		1.44	1.34	1.30	1.28	1.26	1.27	1.30	1.34	1.44	1.55	1.46	1.46	1.45
2.66	2.17		1.56	1.43	1.37	1.37	1.34	1.36	1.37	1.44	1.50	1.70	1.55	1.67	

La	Ce	Pr	Nd	Pm	Sm	Eu	Gd	Tb	Dy	Ho	Er	Tm	Yb	Lu
....	1.65	1.64	1.64	....	1.62	1.85	....	....	....	....	....	....	....	....
1.69	1.83	1.82	1.81	1.63	(1.81)	2.00	1.61	1.59	1.59	1.58	1.57	1.56	1.74	1.56
1.87							1.79	1.76	1.75	1.74	1.73	1.72	1.94	1.72

Ac	Th	Pa	U	Np	Pu
....	1.65	....	1.43	....	....
	1.80	1.56	1.39	1.30	1.51
1.88					

```
Me      = Atomic symbol of metal
0.00    = PCMODEL resident value for covalent radius (Å)
0.00    = Covalent radius (Å), WebElements, version 1.1.1
0.00    = Atomic radius (Å), WebElements, version 1.1.1
```

Table format taken from the periodic table of the elements, with key given above. Vacant positions and missing rows or columns are nonmetallic elements, elements which PCMODEL treats as nonmetals, or elements which are not modeled (e.g. the noble gases). Most reference values for the covalent radius and the atomic radius are taken from the World Wide Web resource WebElements.[857] Atomic radius for Sm comes from the Sargent Welch *Table of Periodic Properties of the Elements*.

A coordination bond is invoked by marking the appropriate metal atom and the associated ligand donor atoms, using the **Select** tool, and then activating the **Metal_Coord** command, which brings up a dialog box for assigning the desired parameters (see below). Eligible donor atoms are those which bear nonbonding lone electron pairs or π-bonding pairs of electrons. Coordination bonding between the metal acceptor and one or more donor atoms is treated in a fashion analogous to that used in the treatment of hydrogen bonds (see Section 6.6).

A menu of metal atoms for which there are parameters resident in PCMODEL is brought up by activating the **Metals** command on the tool bar (see Fig. 2-12). Individual metallic elements can be selected from the 'Metals' menu, and many metallic elements can be used in a structure simultaneously. In the input file these are assigned atom types @80, @81, @82, ..., in the order chosen. All generic parameters for metal atoms are found in the parameter lists under the atom type @80. The two modes of organometallic bonding used in the MMX force field will now be discussed in more detail.

6.7.1. Covalent Bonding

The types of covalent bonds to metals which are recognized in the MMX force field are depicted in Fig. 6-114, where the described bonds are grouped according to the atom bonded to the metal. The graphic depicts the bonding arrangement, where M stands for the metal atom. The specific atom type used for the bond partner is given beneath the graphic, followed by the atomic symbol or group designation such as halogen (see Section 5.3 and Appendix C). Covalent bonds to metals are built in the same fashion as for those bonds involving only the Main Group nonmetal atoms. A bond can be drawn on the screen with the mouse between two points, and the identity of one (or both) of the attached atoms is changed from default carbon to the desired metallic element. Alternatively, one can operate from a template structure wherein an existing lone pair, hydrogen, or heavy atom bonded to a donor atom can be replaced on screen with a defined metal atom, by clicking on its symbol on the 'Metals' menu and clicking again at the desired position in the work space. In most cases, a charged atom type isn't used for the donor atom, even though that atom may appear to be hypervalent by standard valence-bond rules. The exceptions where a donor atom bears an explicit charge are depicted as such in Fig. 6-114, and are explained below.

A specific metal is selected with the mouse cursor from the 40 elements available in the 'Metals' menu (Fig. 2-12), and is then deposited at the appropriate position in the molecule. If the desired metal is not available on the menu, a similar one should be selected, and then prior to energy minimization the input file should be modified to include the covalent radius (see below); and other necessary parameters may be included in a separate constants file (see Section 4.7). The default values in MMX for covalent radii are close or equal to the atomic radii given in standard tables (see Table 6-16); the

530 APPLICATIONS

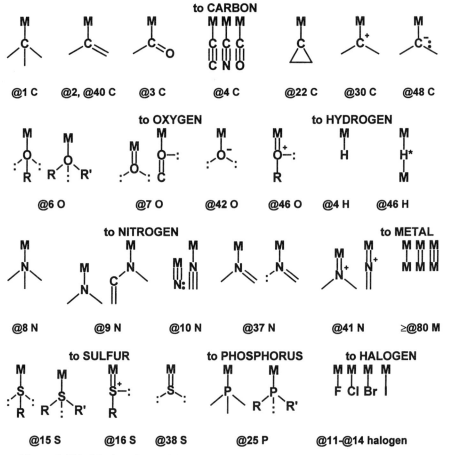

Figure 6-114. Modes of Metal Atom Covalent Bonding in the MMX Force Field.

program developers have found that these values have generally given the best correlation with empirical observations. If the user wishes to use a different value for the covalent radius, the best method is to change its value by editing the input file (see Section 4.6), as will be illustrated in the exercises. Covalently bonded metals receive a default null charge, but this may also be changed by editing the input file to include charges on metals. However, the application of charge to a metal should be done judiciously, in order not to degrade the quality of the model. Putting a charge on the metal may adversely perturb the model by triggering unexpected electrostatic interactions with other atoms.

The **Mark\Metal_Coord** command (see below) may be used to add charge to a metal atom, and to invoke modified equilibrium bond lengths for square planar complexes or for the organometallic compounds of the lanthanides and actinides which have more than eighteen electrons. This may be done even when the molecule involves only "covalent" bonds from the metal to its substituents. To make these changes, only the metal atom is marked with the **Select** tool. Then, the **Mark\Metal_Coord** command is activated, and the appropriate option is selected.

6.7.2. Coordination Bonding

The types of coordination bonds to metals which are recognized in the MMX force field are depicted in Fig. 6-115, where the described bonds are grouped according to the atom coordinated to the metal. The graphic depicts the bonding arrangement, where M stands for the metal atom. The atom type used for the atom coordinated to metal is given below, followed by the atomic symbol or group designation such as halogen (see Section 5.3 and Appendix C). Coordination bonds to metals are not drawn in the same way that covalent bonds are. A donor atom or group is first drawn in or read in as a substructure. Then, it is moved into the vicinity of the metal atom, and an

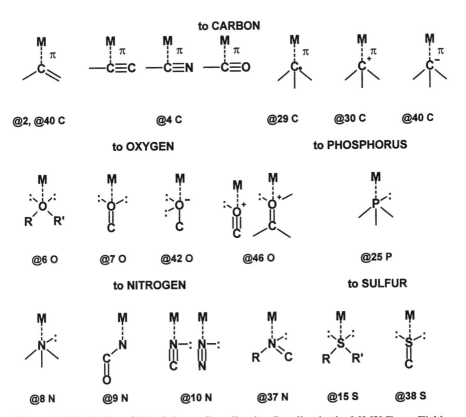

Figure 6-115. Modes of Metal-Atom-Coordination Bonding in the MMX Force Field.

attractive potential for the metal–donor pair is triggered by activating the **Mark\Metal_Coord** command. As with hydrogen bonding, a dotted line will appear on screen between the metal and donor ligand atoms to represent the coordination bond.

The **Mark\Metal_Coord** command sets up the flag for coordination between a lone pair or p-orbital donor ligand and a metal atom. Before it is invoked, the metal acceptor of interest and the corresponding donor ligand(s) must already be present in the work area, and must be marked by using the **Select** tool, in any order. Unless the donor atoms bear their associated lone pairs, **Metal_Coord** will not work properly, so all member substructures of the metal complex should have their full complement of hydrogens and lone pairs attached before this command is invoked. When the **Metal_Coord** command is activated, a dialog box entitled 'Metal Information' appears as shown in Fig. 6-116.

This dialog box gives the elemental symbol of the metal selected in a text entry box titled 'Metal', which can be changed if desired. There is also a text entry box titled 'Charge' which allows the user to place a charge on the metal. In general, the caveat regarding the use of atomic charges discussed above is appropriate for the adding of charge to metals involved in coordination bonding. Unless there is clear reason to place charge on the metal, the uncharged metal will model most coordination complexes adequately. The organometallic portion of the MMX force field was designed to be used with a net charge on the metal only in the presence of a charged substituent or ligand. A nonzero charge will increase the electrostatic bonding to the metal. In some cases this is essential in order to balance one or more charged ligands

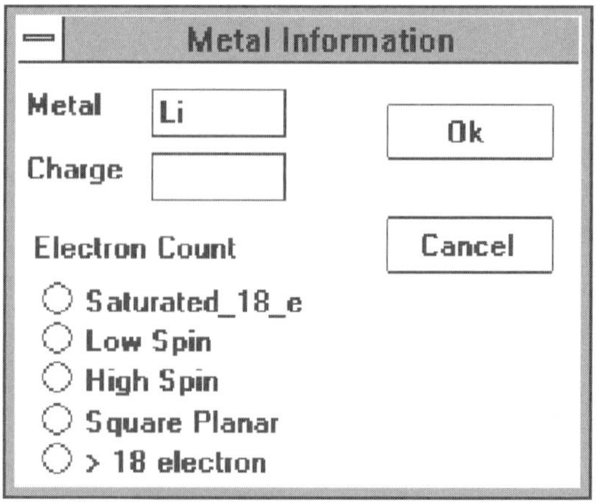

Figure 6-116. The 'Metal Information' Dialog Box.

when a donor atom or substituent bears an explicit charge, as in the case of cyclopentadienyl complexes. An atomic charge can be placed on a metal by selecting it, invoking the **Metal_Coord** command, and entering the desired charge in the 'Metal Information' dialog box which comes up. The charge may be removed by activating the **Mark\Reset** command from the menu bar and selecting 'Atomic Charge' in the 'Reset' option box.

There are then five options available under the heading 'Electron Count', including 'Saturated_18_e', 'Low Spin', 'High Spin', 'Square Planar', and '>18 electron'. These options are used to modify the metal–ligand equilibrium bond lengths. The details of these modifications are provided later in the section. When these selections are complete, the 'OK' button should be clicked.

The dialog box closes, and the complex now has red lines linking the metal atom with the donor atoms. Clicking again on the **Select** button will remove the gray marking dots over the selected atoms. This ensemble will continue to act as a metal complex until the flags are removed with the **Reset** command under **Mark** on the menu bar. Under **Reset**, the 'Coordinated Atoms' option should be checked off to remove only the **Metal_Coord** flags.

In coordination bonding, any lone pairs normally assigned to the donor atom in the **H/AD** process are allowed to remain, and in fact are necessary for the metal–donor electrostatic attraction to be properly evaluated in the MMX force field.

In many cases, the donor ligand is read into the work space as a substructure. The time needed to optimize the geometry during energy minimization may be decreased by adjusting manually the relative positions and orientation of the ligand(s). This can be done by any of the standard PCMODEL methods of substructure manipulation (see Section 3.2.3): translating atoms with **Move** on the tool bar, or translating or rotating substructures with **Substr\Move** on the command bar, or with **View\Control_Panel** on the command bar on substructures marked with the Select tool.

If the user wishes to use the Dock option (see Section 6.8 and Appendix B) to do all of the geometry optimization, the ligands must remain as substructures. The Dock procedure has the apparent advantage of eliminating the prejudice of the user, in that random approaches (the simulated annealing algorithm) are used to optimize the geometry of the complex. In practice, the random nature of the Dock optimization procedure can lead to strange geometries on occasion, and the prejudice of the user will still be necessary to eliminate unreasonable results.

The use of both the covalent and coordination modes of bonding is demonstrated in the exercises of this section, and working through these applications should give the reader a good feel for their use and how they differ from each other.

Organometallic Modeling Exercise 1: Chromyl Chloride. As a basic level exercise with covalent metallic compounds, we will model chromyl chloride,

534 APPLICATIONS

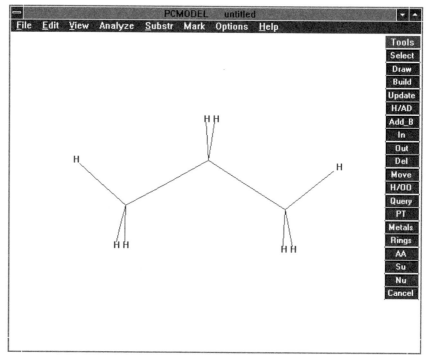

Figure 6-117. Propane Template for Chromyl Chloride Exercise.

CrO_2Cl_2, and compare our calculated geometric values with literature values. The MMX force field provides a moderately accurate model of this inorganic molecule. This exercise demonstrates that generating the structure input for metallic compounds is just as easy as for organic ones. It also shows the process in which the structural input can be modified in order to improve the model.

The **Draw** tool could be used to prepare a starting geometry for this structure directly from its constituent atoms. In that case, all of the bond connections would be drawn in explicitly. The oxygen and chlorine atoms would be assigned by picking them up from the 'Periodic Table' menu (**PT** tool) and using the mouse cursor to drop them on the appropriate atom sites, and the central chromium atom would be picked up from the 'Metals' menu (**Metals** tool). At the outset, all five atoms would lie in the same plane, but the tetrahedral geometry could be approximated by using the **In** and **Out** tools on two of the attached atoms.

However, a shortcut to the target structure will be demonstrated. The sequence is briefly to draw propane, convert this to 2,2-dichloropropane, and then retype the three carbon atoms to make CrO_2Cl_2. The initial, planar three-atom propane structure will be populated with hydrogens by using the

H/AD tool, which will give an approximately correct 3D structure as the starting point for minimization.

In the PCMODEL opening window, activate the **Draw** tool and use the mouse to draw three atoms to form an angle slightly more obtuse than 90°, and then click on the **H/AD** button to add hydrogens. After using the **View\Control_Panel** to rotate a bit about the Y-axis, the result should look like Fig. 6-117.

Next activate the **PT** tool, pick the **Cl** from the 'Periodic Table', and drop it on the two hydrogens of the central carbon with the mouse cursor, to convert them into chlorines. Then click on the **H/AD** tool again to remove the hydrogens from the terminal carbons, and pick **O** from the 'Periodic Table' to replace the two terminal carbons with oxygens. To introduce the chromium center, activate the **Metals** tool, select **Cr** from the 'Metals' menu, and then drop it on the central carbon atom. The **Add_B** tool is used next to create the Cr=O double bonds, and after application of the **H/AD** tool and an **Update** the result should look like Fig. 6-118.

With the input structure ready, we activate the **Analyze\Minimize** command from the menu bar to initiate the energy minimization process. When the first round is finished, the structure should be sent through an energy minimization

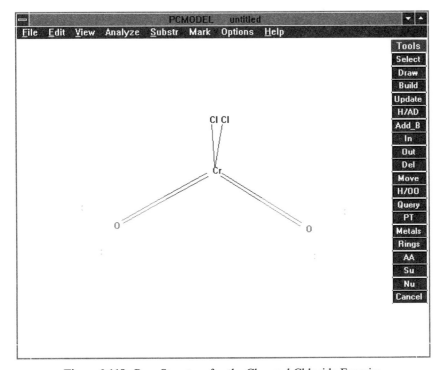

Figure 6-118. Base Structure for the Chromyl Chloride Exercise.

once or twice more, to make sure that a stable minimum has been reached. A copy of this input file is available on the structure library diskette in the CHAP6\SECT7 directory under the name of CRO2CL2.PCM. When the energy minimization is finished, the **Query** tool can be used to display the values for the length of the Cr—O and Cr—Cl bonds, and the O—Cr—O, Cl—Cr—Cl and O—Cr—Cl bond angles. A screen with these values displayed is depicted in Fig. 6-119. All of the geometric values calculated in this exercise are shown in Table 6-17, together with literature values taken from electron diffraction studies.[858]

We can experiment with improving the model for this molecule by changing the input parameters in various ways and examining the effect on the final geometry. The covalent radius for chromium can be altered from the default value of 1.25 Å, the bond multiplicity between chromium and oxygen can be either two or three, and we could use either the @7 O or the @46 O$^+$ for the metal–oxo oxygens.

The bond multiplicity is quite easy to change: first use **H/AD** to remove the lone pairs, then use **Add_B** on both Cr—O bonds, and finally use **H/AD** again to add one lone pair to each O. The result should look like Fig. 6-120 after energy minimization.

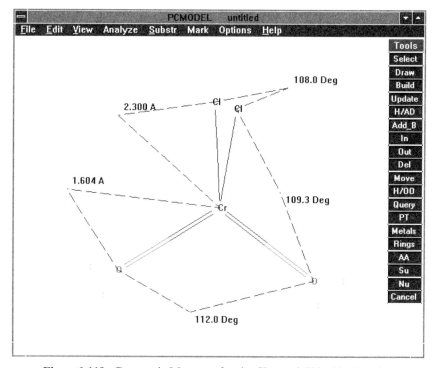

Figure 6-119. Geometric Measures for the Chromyl Chloride Exercise.

6.7. METAL ATOM BONDING AND COORDINATION

TABLE 6-17. Calculated and Empirical Parameters for CrO_2Cl_2

Cr (Å)	Cr—O Bond	Atom Type	Cr—O (Å)	Cr—Cl (Å)	O—Cr—O (deg)	Cl—Cr—Cl (deg)	Cl—Cr—O (deg)
1.25	2	@7 O	1.60	2.30	112.0	108.0	109.3
1.25	2	@46 O^+	1.70	2.30	110.2	109.1	109.4
1.25	3	@7 O	1.80	2.30	109.4	109.9	109.4
1.25	3	@46 O^+	1.70	2.30	111.1	109.0	109.2
1.19	2	@7 O	1.55	2.24	112.8	107.9	109.1
1.19	2	@46 O^+	1.64	2.24	110.3	109.0	109.5
1.19	3	@7 O	1.75	2.25	110.6	107.8	109.7
1.19	3	@46 O^+	1.64	2.24	110.4	109.0	109.3
1.30	2	@7 O	1.65	2.35	111.8	108.3	109.1
			1.57	2.12	105.1	113.3	109.6

The first three columns give the parameters used in the calculations: "Cr (Å)" is the value used for covalent radius; "Cr—O bond" is the multiplicity used for the metal–oxo bond; "Atom Type" is the specific oxygen atom type used, either normal sp^2 O (@7) or charged O^+ (@46). The last five columns give the calculated geometric values for bond lengths and bond angles from PCMODEL. The values in the last row come from a published electron diffraction study.[858] In this case, the error limits of the bond lengths are 0.02–0.03 Å, and those of the angle values are 3–4°.

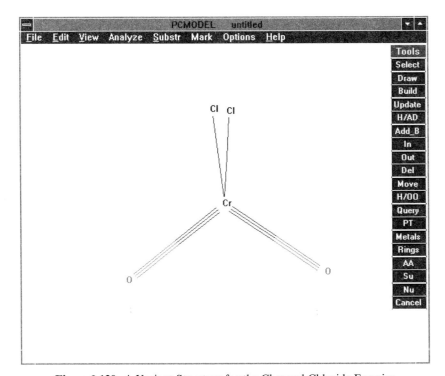

Figure 6-120. A Variant Structure for the Chromyl Chloride Exercise.

In order to test the effect of changing the covalent radius in PCMODEL from the standard 1.25 Å for chromium, the input file can be edited as shown in Fig. 6-121 (see Section 4.6). In line 5 of the input file, the covalent radius is the numerical data field immediately following the letter 'R' (for radius), and is shown in bold. The input file (a) is the one written for the structure in Fig. 6-119. In the modified input file (b), this value has been changed from 1.25 to 1.18 (Å). In line 5 the data field for the metal atom covalent radius is shown in bold, where the value is 1.25 (Å). A word-processing program was used to overwrite the metal atom covalent radius data field in line 5 to give a new value of 1.18 (Å). This input file was then subjected to minimization in PCMODEL.

Considering the values presented in Table 6-17, the best fit for the bond lengths and angles is obtained using the @7 oxygen atom and a value of 1.19 Å for the covalent radius of the metal. When a Cr—O bond multiplicity of two is used, the calculated Cr—O bond length is within the experimental error of the empirical value, but there is slightly over 5% discrepancy in the calculated value for the Cr—Cl bond. The O—Cr—O and Cl—Cr—Cl bond angles are modeled best with the resident value of 1.25 Å for the Cr covalent radius and a Cr—O bond multiplicity of three; both values are just outside the error limits in the empirical values. Good agreement of the Cl—Cr—O bond angle with the literature value was found in all of the models.

(a)
```
line
1    {PCM Chromyl chloride
2    NA 9
3    FL EINT4 UV1 PIPL1
4    AT 1 7    -1.295   -0.610   -0.083 B 2 2 6 1 7 1 S 0 C 0.105
5    AT 2 Cr   -0.020    0.362   -0.024 B 1 2 3 2 4 1 5 1 S 0 R 1.25
6    AT 3 7     1.357   -0.453    0.101 B 2 2 8 1 9 1 S 0 C 0.105
7    AT 4 12   -0.228    1.762    1.789 B 2 1 S 0
8    AT 5 12    0.036    1.657   -1.924 B 2 1 S 0
9    AT 6 20   -1.708   -0.537   -0.511 B 1 1 S 0 C -0.0525
10   AT 7 20   -1.386   -1.010    0.355 B 1 1 S 0 C -0.0525
11   AT 8 20    1.842   -0.266   -0.198 B 3 1 S 0 C -0.0525
12   AT 9 20    1.402   -0.905    0.493 B 3 1 S 0 C -0.0525
13   }
```

(b)
```
line
1    {PCM Chromyl chloride
2    NA 9
3    FL EINT4 UV1 PIPL1
4    AT 1 7    -1.269   -0.544   -0.067 B 2 2 6 1 7 1 S 0 C 0.105
5    AT 2 Cr   -0.009    0.344   -0.012 B 1 2 3 2 4 1 5 1 S 0 R 1.18
6    AT 3 7     1.298   -0.475    0.051 B 2 2 8 1 9 1 S 0 C 0.105
7    AT 4 12   -0.126    1.657    1.795 B 2 1 S 0
8    AT 5 12    0.041    1.663   -1.818 B 2 1 S 0
9    AT 6 20   -1.707   -0.418   -0.457 B 1 1 S 0 C -0.0525
10   AT 7 20   -1.343   -0.984    0.334 B 1 1 S 0 C -0.0525
11   AT 8 20    1.801   -0.253   -0.187 B 3 1 S 0 C -0.0525
12   AT 9 20    1.314   -0.989    0.361 B 3 1 S 0 C -0.0525
13   }
```

Figure 6-121. Changing the Metal Covalent Radius by Editing the Input File: (a) The Original Input File; (b) The Modified Input File.

The covalent and coordination modes of bonding, given respectively in Fig. 6-114 and 6-115, will now be examined in more detail. Atom type numbers are specified for the various metal–ligand combinations, and issues such as the number of lone pairs, valency about the ligand donor atoms, and charge are addressed as well. Metal atom types are referred to generically as @80.

6.7.3. Metal–Carbon Bonds

Metal–carbon covalent bonds can be used for the carbon atom types @1 (sp^3), @2 (alkene sp^2), @3 (carbonyl sp^2), @4 (sp), @22 (cyclopropyl), @29 (sp^2 carbon radical, C·, π-system only), @30 (sp^2 carbonium ion, C^+, π-system only), @40 (aromatic carbon, π-system only), and @48 (carbanion sp^3, C^-).

With @29, @30, and @48 atoms, the carbon keeps its charge or odd electron. Metal atoms may be coordinatively bonded to @2, @4, @29, @30, @40, and @48 carbons which are part of a conjugated π-system. In the metal complexes of such conjugated ligands, the **VESCF** calculations should be carried out on the π-system concurrently with the force field energy minimization process.

Alkylmetals in which the metal-bonded carbon is not part of a conjugated system use an @1—@80-specified bond. Covalent alkenylmetals use an @2—@80-specified bond, while a metal atom can be complexed to the @2 sp^2 carbons in a conjugated π-system.

Metal acyls or carbonyls bridging between metals will use a specified @3—@80 bond. The @4 sp carbon is used in a specified @4—@80 bond for metal acetylides, metal cyanides, or metal carbonyls, while this @4 carbon may also be a π-system donor ligand in a complex.

The @22 cyclopropyl carbon may be joined covalently to a metal with an explicit @22—@80 bond.

The coordination of a metal atom to a conjugated radical π-donor ligand is modeled using the @29 atom type, and in this case the doublet UHF option (odd electron) should be used in the π-calculation (see Section 6.5). A metal carbene/carbenoid can be modeled with the @30 sp^2 carbocation atom type using a specified @30—@80 bond, and a conjugated carbocation will use this same atom type to act as a π-donor ligand. The @30 carbon should not be used in an antiaromatic compound, such as cyclopentadienyl cation, because of the HOMO degeneracy.

The aromatic @40 can be used as a π-donor ligand for coordination to a metal atom. An advantage of using this atom type is that it is not necessary to carry out concurrent π-calculations, since the standard effects of aromaticity are built into the parametrization. This atom type will work best for simple aromatic ligands such as benzene and cyclopentadiene, but to model substitution-perturbed rings and heterocycles it is better to used an @2 carbon with π-calculations.

A covalent metal alkylidene can be modeled with a trivalent @48 sp^3 carbanion atom type using a specified @48—@80 bond. The lone pair on this carbon atom may be used in a coordinative bridging bond to another metal. The @48 carbanion can be used in an anionic conjugated π-donor ligand for a metal, but the lone pair must be removed and π-calculations must be done.

PCMODEL version 6.0 includes some new carbon atom types for organometallic modeling: the @61 metal carbene atom type, the @62 metal carbyne atom type, the @63 metal carbonyl atom type, and the @71 bridging carbonyl atom type.

6.7.4. Metal–Oxygen Bonds

Metal–oxygen bonds can be used for @6 (sp^3), @7 (carbonyl/oxo sp^2), @42 (oxy anion O^-), and @46 (trivalent oxonium O^+ sp^3) oxygen atom types.

In the @6 and @7, the oxygen is uncharged even though it may appear to be hypervalent, while the @42 and @46 oxygens are negatively and positively charged, respectively. A distinction is drawn in this discussion between an organic carbonyl, e.g. a ketone, ester, etc., and an inorganic carbonyl, as in $Ni(CO)_4$, as these are handled in different ways by the MMX force field.

The @6 oxygen atom type is used for divalent species such as metal hydroxides and alkoxides, in which a covalent @6—@80 bond is specified and the oxygen bears two lone pairs. The @6 is also used to model a tricovalent solvated species such as a metal–water, metal–alcohol, or metal–etherate system, in which a covalent @6—@80 bond is specified and oxygen bears only one lone pair (the bond to metal replacing the other). These latter species may also be modeled by invoking a metal–oxygen coordination bond in which both lone pairs on oxygen are allowed to remain.

The @7 sp^2-hybridized oxygen atom type is used for covalent metal–oxo species and for metal–(organic carbonyl) species. The metal–oxo oxygen should bear two lone pairs, and the organic carbonyl oxygen should bear only one, with the bond to metal taking the place of the other. A coordination bond between a metal and an organic carbonyl oxygen may be invoked, in which case the oxygen will bear two lone pairs.

The negatively charged monovalent @42 (O^-) with two lone pairs is used in a covalent bond with a coordinatively saturated metal to model a metal oxy anion. The positively charged trivalent @46 (O^+) with one lone pair is used in a covalent bond with a coordinatively unsaturated metal in modeling a metal oxoylidene.

6.7.5. Metal–Hydrogen Bonds

Only covalent metal–hydrogen bonds are parametrized in the MMX force field. The @5 is used for metal hydrides, and the transition state @45 (H*) is used for bridging hydrides of the type M—H^*—M.

PCMODEL version 6.0 includes the new @70 bridging hydrogen atom type for organometallic modeling.

6.7.6. Metal–Nitrogen Bonds

Metal–nitrogen bonds can be used with @8 (sp^3), @9 (amide/enamine sp^2), @10 (sp), @37 (imine sp^2) and @41 (tetravalent ammonium N^+ sp^3) nitrogen atom types. In all but the @41 atom type, the nitrogen is uncharged.

In covalent bonding to the @8 atom type, the lone pair is replaced by the metal in an @8–@80 bond. In coordination bonding to this type of nitrogen, the lone pair is allowed to remain. In covalent bonding to the @9 atom type, the metal is explicitly bonded to a trivalent nitrogen to give a metal amide or a metalloenamine; the metal is specified as coordinatively saturated, and the nitrogen atom bears no lone pair.

The @10 atom type is used for several types of bonds. Metal nitrides are modeled with an @10—@80 bond, and the lone pair remains on the nitrogen. A metal may be covalently bonded to an N_2 molecule or to a nitrile with an @10—@80 bond, in which the donor nitrogen is uncharged and the bond to metal replaces the lone pair. An N_2 or nitrile donor ligand may also be coordinatively bonded to a metal, in which case the lone pair remains.

The @37 atom type is used for metal bonding to an imine, either an @37—@80 covalent bond to a trivalent sp^2 imine nitrogen with no lone pair or a coordination bond to the same species with the lone pair included. This atom type is also used for a covalently bonded metalloimide or the bent metal nitrosyl, with a coordinatively saturated metal and a nitrogen lone pair. The positively charged @41 ammonium atom type is used with a coordinatively unsaturated metal to allow for nitrogen backbonding, for instance in a linear metal nitrosyl. This nitrogen bears no lone pair in this case.

PCMODEL version 6.0 includes the new @72 bridging nitrogen atom type for organometallic modeling.

6.7.7. Metal–Metal Bonds

Metal–metal bonds are drawn like any other, with an @80—@80 bond. The force field will alter the covalent radii parameters according to the specified bond order.

6.7.8. Metal–Sulfur Bonds

Metal–sulfur bonds can be used with @15 (sp^3), @16 (trivalent sulfonium S^+), and @38 (thiocarbonyl sp^2) sulfur atom types.

The @15 is used for divalent species such as metal hydrosulfides and alkylsulfides (or mercaptides), in which a covalent @15—@80 bond is specified and the sulfur bears two lone pairs. In analogy with the case of oxygen (*vide supra*), @15 is also used for tricovalent "solvated" species such as metal–(hydrogen sulfide), –mercaptan, and –(dialkyl sulfide) systems, in which a covalent @15–@80 bond is specified and the sulfur bears only one lone pair, with the bond to metal replacing the other. These latter species may also be modeled by invoking a metal–sulfur coordination bond, in which both lone pairs on sulfur are allowed to remain.

The positively charged trivalent @16 (S^+) with one lone pair is used in a covalent bond with a coordinatively unsaturated metal in modeling a metal thioylidene.

The @38 sp^2-hybridized sulfur atom type is used for double-bonded covalent metal–sulfide species. The sulfur in this bond should bear two lone pairs.

6.7.9. Metal–Phosphorus Bonds

Metal–phosphorus bonds can be used with the @23 (sp^3) phosphorus atom type, in either the covalent or the coordination format.

In covalent bonding to the @23 atom to make a tetravalent phosphorus species, the lone pair is replaced by an @23—@80 bond to the metal. In coordination bonding to phosphorus, the lone pair is allowed to remain. In order to model a covalent metal phosphide, the metal is explicitly bonded to a trivalent phosphorus; the metal is specified as coordinatively saturated, and the phosphorus atom bears a lone pair, in contrast to the case of metal amides.

6.7.10. Metal–Halogen Bonds

Metal halides are modeled as covalently bonded monovalent (i.e. terminal) or divalent (bridging) species. The bond format is @11—@80 for metal fluorides, @12—@80 for metal chlorides, @13—@80 for metal bromides, and @14—@80 for metal iodides, and either one or two bonds are explicitly drawn.

PCMODEL version 6.0 includes new bridging halogen atom types for organometallic modeling: the @73 bridging fluorine, the @74 bridging chlorine, the @75 bridging bromine, and the @76 bridging iodine.

6.7.11. MMX Algorithms for Metallic Compounds

Within the context of organometallic compounds in PCMODEL, the distinction between covalent and coordination bonds is made solely in order to determine which bonding algorithm will be used. The caveats against attributing physical reality to components of the force field model discussed in earlier chapters are as appropriate here as ever. It suffices to state that we have available two different modes to describe bonding to metals, and that either or both may be considered when refining the model for a particular system.

6.7. METAL ATOM BONDING AND COORDINATION   543

It should be reemphasized that the accuracy of the parameters and algorithms used to model organometallics have not been validated to the same extent as is the case for the main group nonmetallic elements, and must be considered provisional. Nevertheless, reasonable results are obtained in a wide variety of cases, regardless of the approximate nature of the MMX model for metal-containing molecule complexes. The integration of organometallic components into a mainly organic force field has been discussed by Comba and Hambley.[302]

The developers of PCMODEL have reported on the application of the MMX force field to transition metal complexes,[859] and the present discussion is adapted from that source. An in-depth discussion of organometallic bonding is beyond the scope of this work, and the reader is referred to a standard text in this area such as Cotton and Wilkinson.[860]

The energy contributions of metal covalent bonds are modeled with a subset of the parameters used in the standard bonds between nonmetals, where the metal is represented by the generic @80 atom type. The main parameter set contains twenty-eight entries for bond stretching, ten for angle bending, and one for out-of-plane bending. The generalized parameter set contains one entry for a generalized torsional parameter which includes the generic metal atom type, for the string @0—@6—@29—@80 (?—O—·CH—M). There are five entries for generalized "normal" angle bending parameters, one for angle bending in a four-membered ring, and two for angle bending in a three-membered ring.

The MMX covalent radius of the metal atom is used to derive the equilibrium bond lengths, and these may be modified with options available in the 'Metal Information' dialog box (see Fig. 6-116). A default value of 2.0 mdyn/Å is used for the stretching force constant.

No angle bending potential is evaluated in MMX when a metal is the central atom. When the metal is one of the terminal atoms, the bending force constants are kept low (2.0 × 10^{-5} mdyn/deg) to allow freedom of movement.

There is a generic out-of-plane bending potential for metal centers, with a force constant of 0.600 mdyn · Å/rad^2, which puts it on a par with sp^2 carbon and boron.

The MMX potential evaluates no stretch–bend cross terms, and with the single exception mentioned above no torsional terms are included. There are no heat parameters for metal bonds, and no VESCF parameters for π-calculations, so the absolute values for the heat of formation of organometallic molecules will be seriously in error.

The MMX metal atom covalent radius is also used in place of a van der Waals radius for nonbonded interactions. Metals are considered fairly soft, falling on the scale between bromine and iodine, and have a default non-bonding steric force constant ε of 0.400 kcal/mol. All of the normal van der Waals and electrostatic nonbonding interactions evaluated. In addition, unlike in the case of nonmetallic compounds, the 1,3-van der Waals interactions about a metal center (i.e. between X and Y of X—M—Y) are included in the

544 APPLICATIONS

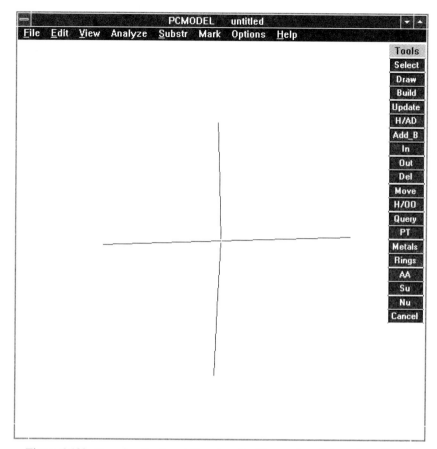

Figure 6-122. Drawing the Input Structure for Hexacarbonylchromium, Part 1.

potential energy. This use of the 1,3-nonbonding potential in place of the angle bending potential for modeling organometallic complexes has been made previously by other groups.[742–744,848]

Organometallic Modeling Exercise 2: Hexacarbonylchromium. In this exercise we will build a model of the simple octahedral complex hexacarbonylchromium. The geometry of the model will be checked against empirically derived values, and a slight adjustment in the one of the parameters will be made to improve the accuracy of the model.

With the **Draw** tool active in the PCMODEL window, the mouse cursor is used to draw from a central point in the work space to the right, and then clicked to place a horizontal bond. The **Draw** tool is clicked again, and starting from a point above the central atom, the mouse cursor is used to add an

6.7. METAL ATOM BONDING AND COORDINATION   545

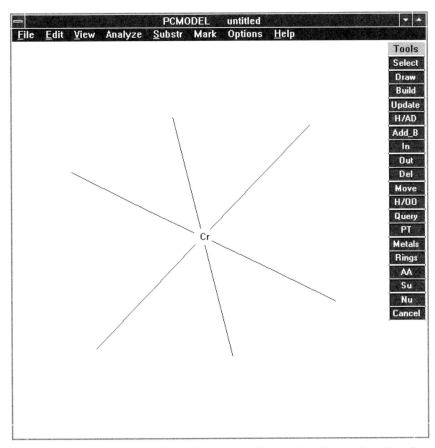

Figure 6-123. Drawing the Input Structure for Hexacarbonylchromium, Part 2.

atom above, click on the central atom again, and then click an equal distance below the central atom. The **Draw** tool is reset again by clicking on it, and a fourth atom is introduced to the left, and then connected to the central atom. Since all atoms are drawn in the same XY plane, we now have an input structure for square planar neopentane, as shown in Fig. 6-122.

Activate the **Metals** tool, pick 'Cr' from the 'Metals' menu, and place it on the central atom. Reactivate the **Draw** tool, and draw a bond from the metal a short way toward the upper right. Click on **Draw** again, and then draw a bond from the metal a short way toward the lower left. Activate the **In** tool, and click on the new upper right-hand carbon atom twice; then activate the **Out** tool, and click on the new lower left-hand carbon atom twice. Now, click on the **Analyze\Minimize** command to achieve a better-proportioned structure. The molecule may jump off screen from time to time, but clicking on the **Update** tool will return it to center screen. The resulting structure should resemble the display shown in Fig. 6-123.

Next reactivate the **Draw** tool, and from each of the carbon atoms draw a bond away from the metal center, and either click again on the first carbon and then on the new outermost one, or use the **Add_B** tool to increase the multiplicity of these carbon–carbon bonds to three. To finish the carbonyl ligands, activate the **PT** tool, pick 'O' from the 'Periodic Table' menu, and place it on each of the terminal atoms. Run the resulting input structure through several cycles of energy minimization, producing the final structure for hexacarbonylchromium shown in Fig. 6-124. The corresponding input file may be found on the input file library diskette in the CHAP6\SECT7 subdirectory as {CRCO6.PCM}, and is reproduced in Fig. 6-125.

Note that PCMODEL has replaced the carbonyl oxygens with the @46 O^+ atom type. Use the **Query** tool to sample some angles and bond lengths. As expected, all of the C—Cr—C angles will be either $\approx 90°$ or $180°$. The bond lengths from this calculation are slightly shorter than the corresponding values obtained from neutron diffraction: Fig. 6-124 shows a Cr—C bond length of

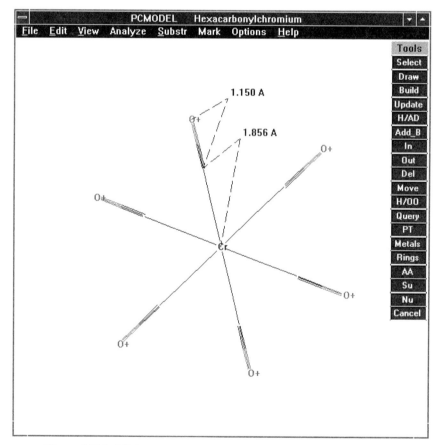

Figure 6-124. Minimum-Energy Structure for Hexacarbonylchromium.

6.7. METAL ATOM BONDING AND COORDINATION 547

(a)
```
 1  {PCM Hexacarbonylchromium
 2  NA 13
 3  FL EINT4 UV1 PIPL1
 4  AT  1 Cr    0.000    0.000    0.000 B 2 1 3 1 4 1 5 1 6 1 7 1 S 0 R 1.25
 5  AT  2  4    1.417   -0.537    1.071 B 1 1 10 3 S 0
 6  AT  3  4   -1.417    0.537   -1.071 B 1 1 13 3 S 0
 7  AT  4  4    0.346   -1.405   -1.162 B 1 1 11 3 S 0
 8  AT  5  4   -0.346    1.405    1.162 B 1 1  8 3 S 0
 9  AT  6  4    1.147    1.087   -0.973 B 1 1  9 3 S 0
10  AT  7  4   -1.147   -1.087    0.973 B 1 1 12 3 S 0
11  AT  8 46   -0.560    2.276    1.882 B 5 3 S 0
12  AT  9 46    1.858    1.760   -1.576 B 6 3 S 0
13  AT 10 46    2.295   -0.870    1.735 B 2 3 S 0
14  AT 11 46    0.560   -2.276   -1.882 B 4 3 S 0
15  AT 12 46   -1.858   -1.760    1.576 B 7 3 S 0
16  AT 13 46   -2.295    0.870   -1.735 B 3 3 S 0
17  }
```

(b)
```
 1  {PCM Hexacarbonylchromium_R=1.32
 2  NA 13
 3  FL EINT4 UV1 PIPL1
 4  AT  1 Cr    0.000    0.000    0.000 B 2 1 3 1 4 1 5 1 6 1 7 1 S 0 R 1.32
 5  AT  2  4    1.462   -0.554    1.106 B 1 1 10 3 S 0
 6  AT  3  4   -1.462    0.554   -1.106 B 1 1 13 3 S 0
 7  AT  4  4    0.356   -1.451   -1.198 B 1 1 11 3 S 0
 8  AT  5  4   -0.356    1.451    1.198 B 1 1  8 3 S 0
 9  AT  6  4    1.185    1.121   -1.005 B 1 1  9 3 S 0
10  AT  7  4   -1.185   -1.121    1.005 B 1 1 12 3 S 0
11  AT  8 46   -0.570    2.322    1.918 B 5 3 S 0
12  AT  9 46    1.896    1.793   -1.608 B 6 3 S 0
13  AT 10 46    2.340   -0.887    1.770 B 2 3 S 0
14  AT 11 46    0.570   -2.322   -1.918 B 4 3 S 0
15  AT 12 46   -1.896   -1.793    1.608 B 7 3 S 0
16  AT 13 46   -2.340    0.887   -1.770 B 3 3 S 0
17  }
```

Figure 6-125. Structure Files for Hexacarbonylchromium: (a) Original; (b) Modified.

1.856 Å and a C—O bond length of 1.150 Å, whereas average values of 1.914 and 1.140 Å are reported in the literature.[861] While the value for the organic ligand is not too bad, the metal–carbon bonds need some improvement.

The simplest way to modify this structure will be to use a different value for the covalent radius parameter. This may be effected by modifying the input file, as was done in the first exercise. The ASCII text for the input file {CRCO6.PCM} can be seen in part (a) of Fig. 6-125, and at the end of line 4 the default value of 'R 1.25' is given for the chromium atom. Use a text processor to create a new, modified file in which the covalent radius parameter is changed to 'R 1.32', as is seen in part (b) of Fig. 6-125, and write this to a new file. Such a file is available on the input file library diskette in the CHAP6\SECT7 subdirectory as {CRCO6B.PCM}. Read this new input file into PCMODEL, and run it through several cycles of energy minimization; then sample the geometry of the new complex. The final result is seen in Fig. 6-126, where the metal–carbon bonds are 1.915 Å, much closer to the empirically obtained measure.

The attractive potential energy algorithm for metal atom coordination is related to that used to evaluate the hydrogen bonding interaction (see Section

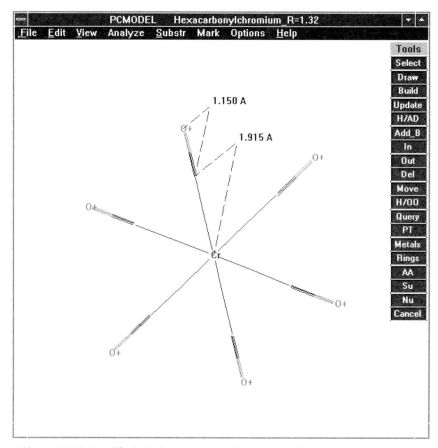

Figure 6-126. Modified Minimum-Energy Structure for Hexacarbonylchromium.

6.6). When metal atom coordination is flagged, an attractive potential proportional to $1/r^2$ is activated between the metal atom and all selected ligand atoms. This potential is a function of the distance, the covalent radius of the metal, a directionality factor, and a parameter for the **haptophilicity** of the donor atom (cf. the basicity parameter $\hat{k}_D$ discussed in Section 6.6). Qualified ligands are lone-pair-bearing atoms or unsaturated atoms, i.e. any atoms which could be flagged as part of a conjugated π-system (see Section 6.5). The user has no control over the parameters governing metal atom coordination, that is, they cannot be altered by supplying added constants.

In order to allow for directionality in metal atom coordination, the attractive potential between a metal and lone-pair-bearing donor atoms is weighted by the distance from the metal atom to the donor heavy atom bearing the lone pair divided by the distance from the metal atom to the lone pair. This is reminiscent of the analogous weighting factor used to weight directionality in hydrogen bonding (see Section 6.6). The attractive potential between metals and π-atoms is weighted by a function of the angle between the metal–donor-atom internuclear axis and the axis of the *p*-orbital. This later is calculated as the axis passing through the π-atom donor and normal to the plane containing the π-atom donor and the π-atoms attached to it. These weighting factors favor a collinear arrangement of metal, lone pair, and donor atom or of metal, *p*-orbital, and donor atom.

As was mentioned above, when ligand–metal coordination is set up, certain changes in the metal–ligand equilibrium bond length may be selected as appropriate for the complex under study. These changes are summarized in Table 6-18. Marking the metal center with **Metal_Coord** and selecting 'Square Planar' or '>18 electron' is how MMX implements the changes necessary to accurately model square planar, or lanthanide or actinide, complexes. Marking the metal center is also necessary for the **trans influence**[862–864] in square planar complexes to be implemented.

The default assumption implicit in the organometallic portion of the MMX force field is of an uncharged metal and a coordinatively saturated 18-electron system. Charge should be applied judiciously, and not only to mimic the actual electron count in the actual complex. A square planar geometry is attainable in MMX by placing lone pair dummy atoms in the two axial positions of a pseudo-octahedral geometry; otherwise, energy minimization will lead to a tetrahedral geometry. These two electron lone pairs on the metal atom have no correlation with measured or calculated electron density at those positions; their inclusion is simply a convenient technique to ensure a square planar geometry for these complexes. In addition to this method of using axial dummy atoms, other organometallic force fields enforce a square planar geometry by placing an out-of-plane bending potential on the metal, or by imposing stiff angle bending potentials on metal-centered atom triads, with 90° and 180° equilibrium angle values.[865] Selecting the 'Square Planar' in the

TABLE 6-18. Changes in Equilibrium M—X Bond Lengths Parametrized in MMX

Feature	Bond	Change (Å)
'Square Planar'	M—X	−0.2
'>18 electron'	M—X	+0.15
'Trans Influence'	M—X'	+0.2
'Trans Influence'	M—X''	−0.1

'Square Planar' and '>18 electron' features are set in the 'Metal Information' dialog box (see Fig. 6-116). The 'Trans Influence' feature will be operative whenever the metal center is marked for **Metal_Coord**. Here, X' is the affected ligand, either @12 chlorine or @13 bromine, and X'' is the directing ligand: @1, @2, @4, or @30 carbon, @5 hydrogen, or @25 phosphorus. When X'' is @4 carbon, the M—C bond is not changed.

'Metal Information' dialog box (see Fig. 6-116) does not affect the geometry, but reduces the equilibrium M—L bond length by 0.2 Å.

The **trans influence** is a structural phenomenon in organometallic complexes in which an activating ligand causes a weakening and an elongation of the bond from the metal to a *trans*-situated ligand. This influence is thought to arise from a competition between different ligands for interaction with a metal orbital. The more successful of the two ligands is termed the directing ligand, and the bond between the metal and the other affected ligand is consequently weakened and lengthened. This weaker bond may then undergo substitution with greater facility, a kinetic phenomenon known as the **trans effect**.[862–864] Some examples of ligands in decreasing order of directing ability are: H– > PR_3 > SCN– > I– ≈ –CR_3 ≈ C≡O ≈ –C≡N > Br– > Cl– > NR_3 > HO– (R may be hydrogen or alkyl).[863]

The MMX force field implements the trans influence in a limited way. When the affected ligand is a halogen, the metal–ligand bond lengths $r(M$—Cl) and $r(M$—Br) increase by 0.15 Å if the halogen is *trans* to a directly bonded carbon (@1, @2 or @4), hydrogen (@5), carbene (@30), or phosphorus (@25). The bond lengths for these *trans*-directing groups are decreased by 0.1 Å, except for @4 carbon, which is not affected. The trans influence on halogen by @25 phosphorus is recognized only if there is a covalent bond between it and the metal, not with a coordination bond.

In the input file, there will an indication of which option has been selected under 'Electron Count'. In the PCM format file (see Section 4.1), this flag for metal atom coordinative electron count comes in the atom description line for that metal, and will appear after the substructure membership flag and prior to the metal radius value. For the default 'Saturated_18_e' complex, no flag appears; for a 'Low Spin' complex, an 'M 1' appears; for a 'High Spin' complex, no flag appears; for a 'Square Planar' complex, an 'M 1 1' appears; and for a '>18 electron' complex, an 'M 2' appears.

A number of applications of the MMX force field, or variations of it, to the modeling of organometallic compounds[705,842–844,850,854,856,866–871] have appeared.

Organometallic Modeling Exercise 3: π-Allylpalladium dichloride. In this exercise we will model a simple π-allylpalladium complex in several ways, and compare our results with those from a literature diffraction structure. The π-allylpalladium dichloride anion forms part of a complex ion pair which was investigated by Hegedus *et al.*[872] In this exercise, we employ three different modeling strategies and compare the results from each.

With the **Draw** tool active in the PCMODEL window, the mouse cursor is used to draw from a central point in the work space to each of three points in an equilateral triangle, to the upper right, lower right, and left. The user must click on the **Draw** tool after each bond is drawn to avoid drawing bonds between all points. Next, activate the **Metals** tool, pick 'Pd' from the 'Metals' menu, and place it on the central atom. Then, activate the **PT** tool, pick 'Cl'

6.7. METAL ATOM BONDING AND COORDINATION 551

from the 'Periodic Table' menu, and place it on each of the two atoms to the right of center, producing the methyldichloropalladium structure shown in Fig. 6-127.

Activate the **View\Control_Panel** command, and use the 'Rotate X' scroll bar in the 'Dials' dialog box to rotate the structure so that the heavy atoms are all in a plane roughly orthogonal to the screen. Activate **Draw**, and, starting from the Pd atom, draw a bond straight up and then another straight down. Bring up the 'Periodic Table' menu again, and pick the lone pair ':' and place it on each of these two new atoms. Use the **H/AD** tool to add hydrogens to the methyl group; then ensure coplanarity between an atom string of H— C—Pd—: by using the **Select** tool to mark them in that order and then activating the **Edit\Rotate_Bond** command and using the scroll bar in the 'Rotate_bond' dialog box to bring the dihedral angle close to zero. The result should look as shown in Fig. 6-128.

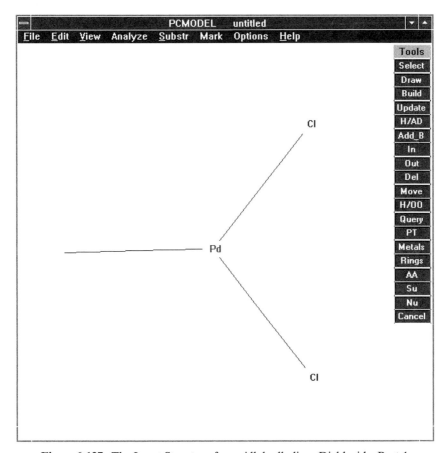

Figure 6-127. The Input Structure for π-Allylpalladium Dichloride, Part 1.

552 APPLICATIONS

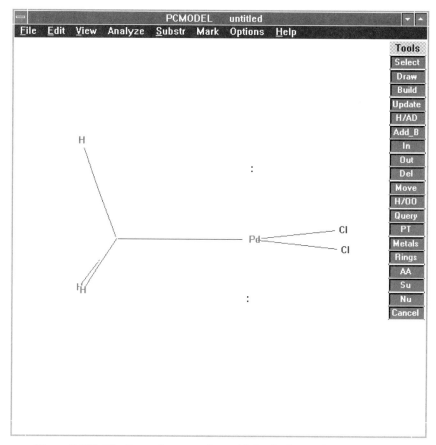

Figure 6-128. The Input Structure for π-Allylpalladium Dichloride, Part 2.

Run this input structure through several cycles of energy minimization. Then use the **Build** tool to convert the equatorially positioned hydrogens to methyl groups. Use the **Del** tool to remove from each of the new methyl groups that hydrogen which lies closest to the palladium center. Then, use the **Draw** tool to form bonds between each of these carbons and the Pd atom. Run this structure through several cycles of energy minimization, during the course of which one will see PCMODEL relabel the metal-bound carbons with the @22 cyclopropyl atom types (shown on screen as 'C'). Use the Query tool to sample some of the geometric properties of this model, which are collected with other relevant data in Table 6-19. The result should resemble the display shown in Fig. 6-129, and this input file may be found on the input file library diskette in the CHAP6\SECT7 subdirectory as the first of four structures in {ALPDCL2.PCM}.

An easy variation to try with this model is to flag the palladium as a metal coordination center with the 'Square Planar' treatment of bond lengths. With

6.7. METAL ATOM BONDING AND COORDINATION

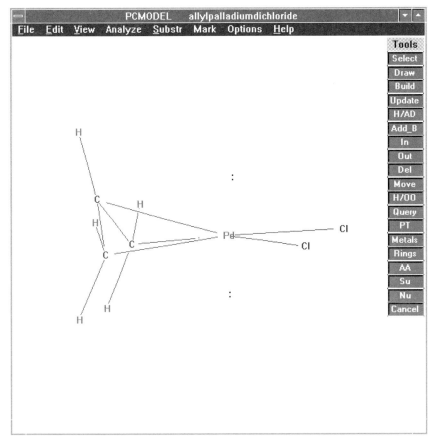

Figure 6-129. π-Allylpalladium Dichloride with Covalently Bonded Ligand, Structure 1.

the initial structure still on screen, use the **Select** tool to mark the palladium atom, then activate the **Mark\Metal_Coord** command and choose the 'Square Planar' option in the 'Metal Information' dialog box. Run this structure through several cycles of energy minimization, to produce a model which will resemble closely that shown in Fig. 6-129, and may be found on the input file library diskette in the CHAP6\SECT7 subdirectory as the second of four structures in {ALPDCL2.PCM}. Different geometric measures will be found upon examination of the structure with the **Query** tool, and these are collected with other data in Table 6-19.

Figure 6-130 shows the input files for these first two structures. They are essentially the same, save for the square planar metal coordination flag, 'M 1 1', in line 4 of the second structure.

For a contrast of the default (18-electron octahedral) model and the square planar model for organometallic bonding, consider the reproduction of the

```
1  {PCM allylpalladiumdichloride
2  NA 13
3  FL EINT4 UV1 PIPL1
4  AT  1 Pd   0.902    0.013    0.082 B 2 1 3 1 4 1 5 1 6 1 9 1 8 1 S 0 R 1.355
5  AT  2 12   2.320    0.016    2.041 B 1 1 S 0
6  AT  3 12   2.608   -0.037   -1.631 B 1 1 S 0
7  AT  4 22  -1.161    0.556   -0.091 B 1 1 7 1 8 1 9 1 S 0
8  AT  5 20   1.038    0.914    0.079 B 1 1 S 0
9  AT  6 20   0.987   -0.894    0.104 B 1 1 S 0
10 AT  7  5  -1.454    1.602   -0.136 B 4 1 S 0
11 AT  8 22  -0.720   -0.248   -1.306 B 1 1 4 1 10 1 11 1 S 0
12 AT  9 22  -0.917   -0.196    1.209 B 1 1 4 1 12 1 13 1 S 0
13 AT 10  5  -0.662    0.295   -2.246 B 8 1 S 0
14 AT 11  5  -1.136   -1.248   -1.405 B 8 1 S 0
15 AT 12  5  -1.005    0.386    2.124 B 9 1 S 0
16 AT 13  5  -1.347   -1.193    1.283 B 9 1 S 0
17 }
```
(a)

```
1  {PCM allylpalladiumdichloride
2  NA 13
3  FL EINT4 UV1 PIPL1
4  AT  1 Pd   0.990    0.018    0.077 B 2 1 3 1 4 1 5 1 6 1 9 1 8 1 S 0 M 1 1 R 1.355
5  AT  2 12   2.261    0.025    1.904 B 1 1 S 0
6  AT  3 12   2.529   -0.033   -1.530 B 1 1 S 0
7  AT  4 22  -1.080    0.556   -0.097 B 1 1 7 1 8 1 9 1 S 0
8  AT  5 20   1.021    0.928    0.066 B 1 1 S 0
9  AT  6 20   0.967   -0.892    0.091 B 1 1 S 0
10 AT  7  5  -1.377    1.602   -0.143 B 4 1 S 0
11 AT  8 22  -0.639   -0.249   -1.309 B 1 1 4 1 10 1 11 1 S 0
12 AT  9 22  -0.836   -0.195    1.203 B 1 1 4 1 12 1 13 1 S 0
13 AT 10  5  -0.589    0.294   -2.250 B 8 1 S 0
14 AT 11  5  -1.053   -1.250   -1.406 B 8 1 S 0
15 AT 12  5  -0.931    0.389    2.116 B 9 1 S 0
16 AT 13  5  -1.263   -1.192    1.277 B 9 1 S 0
17 }
```
(b)

Figure 6-130. Input Files for Allylpalladium Dichloride, Structures 1 and 2 in {ALPDCL2.PCM}.

'Bond and Non-Bonded Energies' portion from the 'Full' level energy minimization output from structure 1, found in Fig. 6-131. Note in particular the description of the bonded atoms, in lines 4–10. An equilibrium value of 2.395 Å is used for the Pd—Cl bond length, and the force constant for all bond stretch interactions is 2.000 mdyn/Å. Compare this with the corresponding output excerpt from the energy minimization of structure 2, found in Fig. 6-132. Here, the shorter value of 2.195 Å is used for the Pd—Cl bond length, and a larger force constant is used for all bond stretch interactions: 3.000 mdyn/Å.

Another item of note in Fig. 6-131 is the nonbonded interactions which are included (see Section 3.5). Lines 11–14, 16, 17, 22–24, etc., describe the van der Waals interactions between geminal atoms, i.e. the interactions which are not specifically evaluated in purely organic structures. The lines pertaining to these geminal interactions begin with a lowercase 'v'. Those lines (15, 18–21, 25, etc.) which pertain to van der Waals interactions between atoms which are vicinal

6.7. METAL ATOM BONDING AND COORDINATION

```
1      Bond and Non-Bonded Energies
2
3         at1        at2         r         r0        bk         eb
4    Pd(  1) - Cl(  2)       2.419      2.395     2.000      0.076
5    Pd(  1) - Cl(  3)       2.419      2.395     2.000      0.076
6    Pd(  1) - C  (  4)      2.140      2.145     2.000      0.003
7    Pd(  1) - :  (  5)      0.911      0.895     2.000      0.035
8    Pd(  1) - :  (  6)      0.911      0.895     2.000      0.036
9    Pd(  1) - C  (  8)      2.151      2.145     2.000      0.005
10   Pd(  1) - C  (  9)      2.151      2.145     2.000      0.005
11  V Cl(  2) - Cl(  3)      3.684      4.060     0.240     -0.142
12  V Cl(  2) - C  (  4)     4.116      3.930     0.103     -0.114
13  V Cl(  2) - :  (  5)     2.510      3.230     0.062      0.454
14  V Cl(  2) - :  (  6)     2.523      3.230     0.062      0.419
15  V Cl(  2) - H  (  7)     4.582      4.633     3.530      0.106    -0.047
16  V Cl(  2) - C  (  8)     4.530      3.930     0.103     -0.082
17  V Cl(  2) - C  (  9)     3.350      3.930     0.103      0.100
18  V Cl(  2) - H  ( 10)     5.165      5.230     3.530      0.106    -0.024
19  V Cl(  2) - H  ( 11)     4.994      5.044     3.530      0.106    -0.029
20  V Cl(  2) - H  ( 12)     3.333      3.346     3.530      0.106    -0.107
21  V Cl(  2) - H  ( 13)     3.880      3.938     3.530      0.106    -0.102
22  V Cl(  3) - C  (  4)     4.116      3.930     0.103     -0.114
23  V Cl(  3) - :  (  5)     2.510      3.230     0.062      0.454
24  V Cl(  3) - :  (  6)     2.523      3.230     0.062      0.418
25  V Cl(  3) - H  (  7)     4.581      4.632     3.530      0.106    -0.047
26  V Cl(  3) - C  (  8)     3.350      3.930     0.103      0.100
27  V Cl(  3) - C  (  9)     4.530      3.930     0.103     -0.082
28  V Cl(  3) - H  ( 10)     3.332      3.346     3.530      0.106    -0.107
29  V Cl(  3) - H  ( 11)     3.880      3.939     3.530      0.106    -0.102
30  V Cl(  3) - H  ( 12)     5.165      5.229     3.530      0.106    -0.024
31  V Cl(  3) - H  ( 13)     4.994      5.045     3.530      0.106    -0.029
32  V C (  4) - :  (  5)     2.234      3.100     0.027      0.516
33  V C (  4) - :  (  6)     2.599      3.100     0.027      0.044
34    C (  4) - H  (  7)     1.088      1.086     4.600      0.001
```

Figure 6-131. Excerpt from Output File from Energy Minimization of {ALPDCL2.PCM}, Structure 1.

or more distant from each other begin with an uppercase 'V'. Any lines describing electrostatic interactions would be prefaced with an uppercase 'Q', but there are none in this example.

Now we will try two models in which the allyl ligand is coordinated to the palladium rather than covalently bound. With either of the two earlier models for the allylpalladium dichloride structure on screen, delete all three of the carbon–palladium bonds. Bring up the 'Periodic Table' menu, select the @40 aromatic carbon atom type 'CA', and reassign the atom type of each of the three carbon atoms. Before setting up a metal coordination flag, we must be sure to clear any preexisting flags by activating the **Mark\Reset** command and selecting 'All'.

Next, use the **Select** tool to mark the metal atom and each of the three carbons. Use the **Mark\Metal_Coord** command to bring up the 'Metal Information' dialog box, accepting the default bond parameters. The use of the 'Square Planar' option in this model may render the structure unstable. Finally, activate the **Analyze\Minimize** command to carry out energy minimization of this structure. The structure may jump wholly or partly off screen,

								r	r0	bk	eb	
35	C	(	4)	-	C	(	8)	1.522	1.501	4.400	0.134	
36	C	(	4)	-	C	(	9)	1.522	1.501	4.400	0.135	
37	V	:	(5)	-	:	(	6)	1.809	2.400	0.016	0.180	
28	V	:	(5)	-	H	(	7)	2.547	2.594	2.700	0.027	-0.027
39	V	:	(5)	-	C	(	8)	2.519	3.100	0.027	0.091	
40	V	:	(5)	-	C	(	9)	2.519	3.100	0.027	0.091	
41	V	:	(5)	-	H	(	10)	2.893	2.942	2.700	0.027	-0.029
42	V	:	(5)	-	H	(	11)	3.324	3.404	2.700	0.027	-0.016
43	V	:	(5)	-	H	(	12)	2.892	2.942	2.700	0.027	-0.029
44	V	:	(5)	-	H	(	13)	3.324	3.404	2.700	0.027	-0.016
45	V	:	(6)	-	H	(	7)	3.419	3.500	2.700	0.027	-0.014
46	V	:	(6)	-	C	(	8)	2.308	3.100	0.027	0.349	
47	V	:	(6)	-	C	(	9)	2.308	3.100	0.027	0.348	
48	V	:	(6)	-	H	(	10)	3.035	3.110	2.700	0.027	-0.024
49	V	:	(6)	-	H	(	11)	2.586	2.630	2.700	0.027	-0.030
50	V	:	(6)	-	H	(	12)	3.035	3.110	2.700	0.027	-0.024
51	V	:	(6)	-	H	(	13)	2.587	2.630	2.700	0.027	-0.030
52	V	H	(7)	-	H	(	10)	2.513	2.605	3.000	0.047	0.080
53	V	H	(7)	-	H	(	11)	2.979	3.137	3.000	0.047	-0.055
54	V	H	(7)	-	H	(	12)	2.513	2.605	3.000	0.047	0.080
55	V	H	(7)	-	H	(	13)	2.979	3.137	3.000	0.047	-0.055
56	V	C	(8)	-	C	(	9)	2.524	3.800	0.044	2.012	
57		C	(8)	-	H	(	10)	1.088	1.086	4.600	0.001	
58		C	(8)	-	H	(	11)	1.087	1.086	4.600	0.001	
59	V	C	(8)	-	H	(	12)	3.414	3.500	3.400	0.045	-0.053
60	V	C	(8)	-	H	(	13)	2.786	2.827	3.400	0.045	0.132
61	V	C	(9)	-	H	(	10)	3.414	3.500	3.400	0.045	-0.053
62	V	C	(9)	-	H	(	11)	2.786	2.827	3.400	0.045	0.132
63		C	(9)	-	H	(	12)	1.088	1.086	4.600	0.001	
64		C	(9)	-	H	(	13)	1.087	1.086	4.600	0.001	
65	V	H	(10)	-	H	(	12)	4.226	4.384	3.000	0.047	-0.013
66	V	H	(10)	-	H	(	13)	3.757	3.891	3.000	0.047	-0.025
67	V	H	(11)	-	H	(	12)	3.757	3.891	3.000	0.047	-0.025
68	V	H	(11)	-	H	(	13)	2.683	2.698	3.000	0.047	-0.016

Figure 6-131. Continuation of Excerpt from Output File from Energy Minimization of {ALPDCL2.PCM}, Structure 1.

but may be retrieved by clicking on the **Update** tool. After several courses of energy minimization, the final structure should be as is depicted in Fig. 6-133; it may be found on the input file library diskette in the CHAP6\SECT7 subdirectory as the third of four structures in {ALPDCL2.PCM}. Use the **Query** tool to sample the geometric measures of the structure, which will be found in Table 6-19.

1	Bond and Non-Bonded Energies						
2							
3	at1		at2	r	r0	bk	eb
4	Pd(	1)	- Cl(2)	2.226	2.195	3.000	0.190
5	Pd(	1)	- Cl(3)	2.226	2.195	3.000	0.190
6	Pd(	1)	- C (4)	2.146	2.145	3.000	0.000
7	Pd(	1)	- : (5)	0.911	0.895	3.000	0.051
8	Pd(	1)	- : (6)	0.910	0.895	3.000	0.050
9	Pd(	1)	- C (8)	2.155	2.145	3.000	0.023
10	Pd(	1)	- C (9)	2.156	2.145	3.000	0.025

Figure 6-132. Excerpt from Output File from Energy Minimization of {ALPDCL2.PCM}, Structure 2.

6.7. METAL ATOM BONDING AND COORDINATION   557

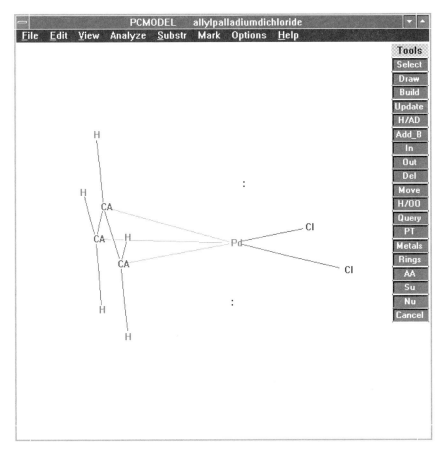

Figure 6-133. π-Allylpalladium Dichloride with Coordinated Ligand, Structure 3.

To prepare for the final variation, with structure 3 on screen, clear any preexisting flags by activating the **Mark\Reset** command and selecting 'All'. Then, use the appropriate atoms from the 'Periodic Table' menu to assign the @48 carbanion atom type (C−) to the nearest terminal carbon of the allyl fragment (this will be C^1 in Table 6-19), and default carbons to the other two. Use the **Add_B** tool to double the bond between these latter two carbons. Next, mark all three carbons with the **Select** tool, and use the **Mark\Piatoms** command to flag them as a conjugated system for Hückel calculations during the energy minimization phase. Then, use the **Select** tool to mark the metal atom and each of the three carbons. Use the **Mark\Metal_Coord** command to bring up the 'Metal Information' dialog box, accepting the default bond parameters. Finally, activate the **Analyze\Minimize** command to carry out energy minimization of this structure. After several courses of energy minimization, the final structure should be as is depicted in Fig. 6-134; it may be found on the input file library diskette in the CHAP6\SECT7 subdirectory as

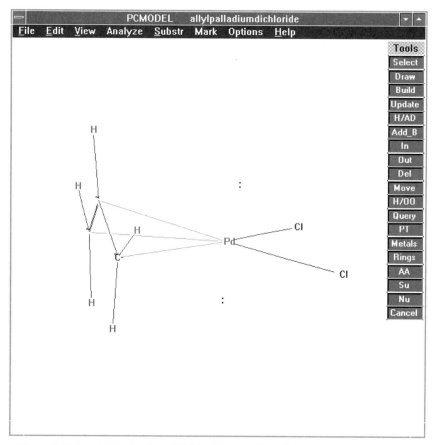

Figure 6-134. π-Allylpalladium Dichloride with Coordinated Ligand, Structure 4.

the fourth of four structures in {ALPDCL2.PCM}. Use the **Query** tool to sample the geometric measures of the structure, which will be found in Table 6-19.

The geometric values obtained from each of these models of π-allylpalladium dichloride, along with X-ray data from the literature, are collected in Table 6-19. As has been seen in the other examples where we compare calculated with empirical geometric measures, each of the models has strengths and weaknesses. Three of the four structures model the Pd–Cl bond with some accuracy, the exception being structure 2, the one which had been flagged for a square planar bond treatment! Interestingly, this latter model comes the closest to the X-ray value for the Cl—Pd—Cl angle. The Pd—C bonds and C—Pd—C angle are best emulated with the two covalent models, while the structure of the allyl ligand itself, as indicated by the C—C—C angle, is reproduced best in the two coordination models. Since PCMODEL doesn't recognize the equivalence of the two terminal allyl carbons in structure 4,

TABLE 6.19. Calculated and Empirical Parameters for π-Allylpalladium Dichloride

Model	Pd—Cl (Å)	Pd—C^1 (Å)	Pd—C^2 (Å)	Pd—C^3 (Å)	Cl—Pd—Cl (deg)	C^1—Pd—C^3 (deg)	C^1—C^2—C^3 (deg)
Covalent-1	2.419	2.151	2.140	2.151	92.0	71.8	112.0
Covalent-2	2.226	2.155	2.146	2.155	101.4	71.6	111.9
Coord-1	2.414	2.314	2.127	2.314	105.7	63.8	121.4
Coord-2	2.415	2.278	2.134	2.308	104.4	63.7	120.8
X-ray	2.395	2.15	2.09	2.15	97.9	71.2	121

The first column names the model used in the calculations: Covalent-1 has all bonds explicitly covalent and uses the default organometallic parameters; in Covalent-2 the palladium atom has been marked with the 'Square Planar' flag; in Coord-1 the allyl ligand is composed of @40 aromatic-type carbon atoms, with coordination bonds to the palladium; in Coord-2 the allyl ligand is composed of the atom-type string @2—@2—@48 with coordination bonds to the palladium. The values in the last row come from a published X-ray diffraction study.[872]

there is about a 1% difference in the bond lengths found for the Pd—(@2)C and Pd—(@48)C bond, 2.308 Å vs. 2.278 Å.

At this point, some major features of other organometallic empirical force field models will be discussed, particularly in the respects in which they differ significantly from the more standard organic formulations.

As was seen at the beginning of this section, the earliest formulation used simple nonbonded interaction potentials to calculate the structure of organometallic complexes.[43]

One early answer to the challenge of diverse valence geometries in inorganic compounds was to use the Points-On-a-Sphere (POS) model to simulate the effects of Valence Shell Electron Pair Repulsion (VSEPR) theory.[77] Briefly, this theory comes from the valence-bond formulation, and holds that valence geometry results from the drive to minimize the interactions between electron pairs in the valence shell. In the POS model, bonds to ligands are represented by mathematical points constrained to travel on the surface of a sphere, interacting with each other through repulsions proportional to a exponential power of the interpoint distance. The nature of these repulsions was allowed to include, but was not limited to, Coulombic interactions.

This concept was further elaborated by Lauher, who developed a surface force field model for transition metal carbonyl clusters.[848] A smooth, continuous surface was defined for the metal core as the region of space not penetrated by a probe sphere. The metal–ligand potential was of the MM2-type bond stretch type, a Hooke's law modified by the addition of a cubic term. MM2 potential forms were used for the carbonyl C—O bond stretch, M—C—O angle bend, and nonbonded interactions between ligands (steric and electrostatic).

A common formulation for modeling metal–ligand complexes is to employ some sort of harmonic bond stretch potential between the metal atom and the ligand donor atom(s). Some authors have referred to this as a **penalty function**, which exacts an energetic penalty related to the deviation of the M—L bond lengths from their ideal values.[114,873,874] The penalty function is added to the conformational energy and nonbonded interactions between the ligands, but the metal atom is considered as an otherwise noninteracting "floating point."

In this context, a generic penalty function will take the form

$$V_{\text{penalty}} = K_{\text{penalty}}(r_{M-L} - r^0_{M-L})^2$$

where K_{penalty} is a constant of large magnitude, r_{M-L} is the metal-atom–ligand-donor-atom distance, and r^0_{M-L} is the equilibrium value for this distance.

Vedani and coworkers have improved upon their earlier algorithm[522] and developed a sophisticated organometallic potential for their YETI force field (see Section 5.4.8).[524] It includes as variables the metal–ligand separations, the symmetry at the metal center, directionality of the metal–ligand bonds, ligand–metal charge transfer, and ligand field stabilization for transition metals. This potential bears some resemblance to the YETI hydrogen bond algorithm (see Section 6.6), was likewise parametrized from X-ray diffraction structural data, and is shown below:

$$V_{ML} = \left(\frac{C}{R_{M-L}^{12}} - \frac{D}{r_{M-L}^{10}}\right) + (E_{MC}^{\circ} + E_{LFS})$$

$$\times [\cos^2(\theta_{L-M-L'} - \theta^0) \cdot \frac{1}{n}\cos^n \theta_{M-L-lp}$$

where V_{ML} is the metal coordination potential; r_{M-L} is the internuclear distance between the metal atom and the ligand donor atom; C and D are constants which depend upon the equilibrium value for r_{M-L} and on the force constant for the particular metal–ligand combination; E_{MC}° is a weighting factor which depends upon the symmetry at the metal center and the directionality of the metal–ligand bonds; E_{LFS} is a weighting factor which depends on the ligand field stabilization for transition metals; $\theta_{L-M-L'}$ is the angle between the ligand donor atom L, the metal atom M, and a second ligand donor atom L'; θ^0 is the equilibrium angle for the specified L—M—L' set, obtained from X-ray structure data; n is a coefficient in the directional term analogous to that used in the hydrogen bonding potential, usually 3 or 4; and θ_{M-L-lp} is the angle between the metal atom M, the ligand donor atom L, and the closest lone pair lp. The latter portion of the potential exacts a penalty for deviation of the M—L radius from the nearest L—lp radius for that ligand donor atom.

In the early part of this decade, a description of the SHAPES empirical force field appeared.[309] In order to overcome the many shortfalls of retrofitting organic force fields to the task of modeling organometallics, Landis et al. made some basic changes in their formulation. They use spherical internal coordinates, to simplify the treatment of coordination geometry. A reference axis (equivalent to the Cartesian z-axis) passes through the metal center, and is defined as a line normal to the plane containing the ligands in trigonal planar, trigonal pyramidal, and square planar systems, and as the line passing through the metal center and the ligand defined by the user as axial in trigonal bipyramidal, square pyramidal, and octahedral systems.

Fourier angular potentials are used, which are functions of the angles θ, ϕ and the M—L distance R_{M-L}. The ϕ-plane is defined as the plane containing the metal center and orthogonal to the reference angle; the angle ϕ is the projection onto the ϕ-plane of the L—M—L' bond angle. This potential takes the form

$$V_{ang} = k_\theta[1 + \cos(n\theta_{M-L} - \theta_{M-L}^0)]$$
$$+ k_\# k_\phi[1 + \cos(n\phi_{L-M-L'} - \phi_{L-M-L'}^0)]$$

where k_θ is Fourier bond angle bend force constant; n is the periodicity of the cosine function; θ_{M-L} is the angle between the metal–ligand bond (M—L) and the reference axis, and θ^0_{M-L} is the equilibrium value for that bond; $k_\#$ is a weighting factor defined below; k_ϕ is equivalent to a Fourier out-of-plane bending force constant; $\phi_{L-M-L'}$ is the out-of-plane deformation angle for the ligand–metal–ligand fragment (L—M—L') with respect to the reference ϕ-plane, and $\phi^0_{L-M-L'}$ is the equilibrium value for that bond reference axis.

The weighting factor $k_\#$ is defined as

$$k_\# = \left(\frac{r_{M-L} r_{M-L'}}{R_{M-L} R_{M-L'}} \right)^2$$

The SHAPES formulation was introduced into the CHARMM force field for use in modeling inorganic and organometallic compounds.[309] Further work by this group has produced the VALBOND force field, which resembles the generic force fields (see Section 5.4.8) and applies the SHAPES-type angular potential to compounds of the Main Group elements of "normal" valency and hypervalency.[852,875]

A distinguishing feature of organometallic modeling is the more frequent use of **dummy atoms** or *pseudoatoms*.[705,827,838,840,846] These find utility especially in metal–ligand interactions where the ligand is viewed as bonding from two or more adjacent atoms, for example the dihapto olefin complex, the trihapto allyl complex (see Exercise 3), the hexahapto benzene complex (see Exercise 4), etc. A dummy atom can ensure the functional equivalence of the donor atoms within the ligand when it is appropriately situated at the **centroid** of those atoms. It is also helpful in avoiding the consequences of having a dihapto ligand seem to take up two coordination sites, which it will do if all metal–ligand bonds must be specified. The dummy atom often functions as a fixed point for bond stretch, angle bend, and torsional interactions, while the nonbonded interactions (steric and/or electrostatic) of the original atoms are left intact.

A published report discusses a modification of the MMX force field in which an @79 dummy atom is defined and situated at the centroid of π-bonded ligands such as ethylene, benzene, and cyclopentadienyl anion.[705] This @79 atom type has a covalent radius of 0 and a hardness parameter $\varepsilon = 0$. This allows the authors to carry out dihedral driver conformational searches with systematic rotation of the ligand with respect to the metal. In this work, it was found that axially oriented lone pairs were not effective in enforcing the square planar geometry in the trichloroplatinum olefin complex, but that axially oriented hydrogen atoms were.

The Allinger-type force fields MM2 and MM3 have been modified for use in modeling organometallics.[836,837,845,848,853,855,876–878]

In some work, the metal–ligand bonds involve a Hill-type nonbonded potential with a very strong attractive component, as was used to implement a hydrogen bonding potential. The appropriate van der Waals parameters were developed for individual metal–ligand atom pairs, angle bending interactions involving the metal were not evaluated, and the nonbonded interactions between the ligands determine the coordination geometry.[845]

In other cases, the metals were treated much like other atoms, and both bond stretch and angle bend were evaluated, though torsional interactions were assumed to be negligible.[876] The application to cobalt(III) amine complexes of an MM2 version, modified to encompass organometallic compounds, was used to predict whether unknown complexes could be synthesized.[837] A slightly modified MM2 was used to model the nickel tetrapyrrole complex factor F430.[877] MM2 calculations were used to model organometallic transition states in a study of the stereoselectivity of osmium-tetraoxide-mediated alkene dihydroxylation reactions.[855] A study of the influence of bis-ammonium compounds on the growth of MFI-type zeolites was carried out by combining the authors' zeolite-specific force field with MM3 to handle the organic ligands.[853] The MM3 force field was extended to include parameters for modeling crown ether complexes of the alkali metals and alkaline earths.[878]

In Section 6.4, we discussed briefly the use of a modified version of MMI to rationalize lanthanide-induced chemical shifts in ^{1}H-NMR; here, metal-centered angle bend interactions were not evaluated, but geminal nonbonded interactions were allowed to guide the geometry to an optimum conformation.[742–744] A later report described a metal-adapted MM2 variant (MM2MX) and its use in conformational analysis of some lanthanide complexes.[836]

A set of parameters has been developed for use in modeling platinum complexes of guanine derivatives within the AMBER force field.[427] Six new atom types were created to model these square planar species: the platinum atom, two guanine imidazole-type nitrogens bonded to platinum, two amine ligand atom types, and a hydrogen attached to the amino ligands. These were parametrized for bond stretch, angle bend, torsion strain, improper torsional strain, and nonbonded steric interactions, and a set of electrostatic charges was developed.

Another group developed AMBER-style parameters for modeling zinc complexes in metalloproteins.[421] In this case, only two new atom types were developed, the zinc atom and a oxygen for bound hydroxide and water. These were parametrized for bond stretch, angle bend, torsion strain, and nonbonded steric interactions, and a set of electrostatic charges was developed.

The AMBER force field has also been used for molecular dynamics simulations of a synthetic hexadentate iron(III) complex in an aqueous solvation environment,[879] and for the gas phase mechanism for the protonation of ferrocene, using parameters fitted from local spin density (LSD) and Hartree–Fock calculations.[841]

The MOMEC force field developed by Hambley is a more or less standard organic-type formulation which is heavily parametrized for organometallics.

An earlier incarnation had a four-component force field,[880] whereas the recent formulation consists of five components: harmonic bond stretch, angle bend, and out-of-plane bend potentials, a modified Buckingham van der Waals potential, and a cosine-based torsional potential with variable periodicity.[849] Electrostatics and a specific hydrogen bonding potential were not included in the reported version. No symmetry restrictions are placed on the local coordination sphere, and nonbonded interactions with metal atoms are neglected. There are resident parameters for Cu(II), Ni(II), Co(III), Fe(III), Cr(III), Zn(II), and Rh(III) complexes with amine, carboxylate, pyridine, and thioether ligands.

Organometallic structures which have been refined with force-field-driven geometry optimization may be analyzed within the **Angular Overlap Model** (AOM) to predict spectroscopic properties for these complexes.[20,851,881,882] This involves a simplified molecular orbital treatment in which the perturbations due to ligand interactions upon the metal d-orbitals are evaluated in terms of parameters which are related to σ-, π-, and δ-donor or -acceptor strength.

The **Jahn–Teller effect** has been modeled using dummy atoms, or by incorporating separate metal–ligand bond stretch parameters for axial and for equatorial ligands.[230,883,884] A relatively recent solution has been to determine the Jahn–Teller d-orbital splitting spectroscopically, and use this as a parameter to introduce a Jahn–Teller stabilization component into the force field.[883]

The generic force fields discussed in Section 5.4.8 have been used extensively to model organometallics. The UFF formulation uses all covalent-type interactions, but scales the bond order to implement certain aspects of metallic bonding.[512] Thus, increasing the bond order to 1.5 or 2.0 is used to take account of bond shortening due to backbonding. The bond order in planar or near planar metal amides is also increased to 1.5. Dative, or coordinative, bonding effects are mimicked with a decreased bond order of 0.5, and this technique is also used to model the trans influence, for instance in phosphine complexes. The angular distortion potentials are very close to those used in the SHAPES force field. The accuracy of the UFF molecular models of organometallics is not as high as with the purely organic molecules, but is still reasonable.

Several reports have appeared on organometallic applications of the DREIDING force field: modeling alkanethiol monolayers on a gold surface,[498] and aluminum triisopropoxide in ring-opening polymerization.[502]

Another recent arrival is the *Extensible Systematic Force Field* (ESFF), applications of which have been reported in binuclear copper(II) complexes[885] and in ferrocenylsilane oligomers.[847] This formulation has a six-component potential force field: a Morse potential for bond stretch; a choice of four angle bend potentials for normal, linear, perpendicular, and equatorial angles; a complex torsional potential function which combines some angle–torsion cross terms; a simple harmonic out-of-plane bending potential, a Lennard–Jones

6-9 nonbonding steric potential; and a Coulomb's law nonbonding electrostatic potential.[847]

Organometallic Modeling Exercise 4: Benzenechromiumtricarbonyl. For this exercise we will model the benzenechromiumtricarbonyl molecule in two ways, first by using π-calculations on the benzene ring, and then by using the parametrized aromatic carbon atoms. The structure will be prepared initially by drawing in the metal tricarbonyl moiety, then adding in benzene as a substructure and docking it manually onto the metal fragment. After energy minimization, the resulting geometric values are compared with physical measurements from X-ray, neutron, and electron diffraction studies. Several modeling studies on this molecule have been published.[886,887]

We will prepare the chromiumtricarbonyl moiety as a flat structure in the plane of the screen, using C≡O for the carbonyl groups. With the **Draw** tool active, the mouse cursor is used to draw from a central point in the work space upwards to a second point, then continue upwards to a third point. Next, this sequence is performed twice more at intervals of roughly 120° from the initial lines, so that the lines from the center will point to the vertices of an equilateral triangle. Then, the **Add_B** tool is activated, and the mouse cursor is used to click twice on each of the three legs, approximately three quarters of the way to the terminus. First a 'D' and then a 'T' will appear between the second and third atoms as the second and third multiple bonds are drawn in.

With the skeleton in place, we next assign the noncarbon atoms. Clicking on the **Metals** tool brings up the 'Metals' menu, from which the 'Cr' is selected and dropped on the central carbon. Next the 'Periodic Table' menu is brought up by clicking on the **PT** tool, and the 'O' is selected and dropped on all three terminal atoms. Rename the terminal atoms as oxygens (by clicking on oxygen in the atom menu and then clicking on each of the three atoms). Note from the updated structure that PCMODEL has assigned the @46 O^+ to these metal carbonyl oxygens. Clicking on the **H/AD** button will add a lone pair to each of the carbonyl oxygens. The **Analyze\Minimize** command is used several times give a geometry-optimized chromiumtricarbonyl moiety. The result should resemble Fig. 6-135. The user should write the structure to an input file; such a file may be found on the input file library diskette in the CHAP6\SECT7 subdirectory as {CRCO3.PCM}.

The chromiumtricarbonyl fragment is next rotated to a plane nearly orthogonal to the screen by using the 'X Rotate' Dial within **View\Control_Panel** on the menu bar. Clicking on the **Rings** command from the tool bar brings up the 'Rings' menu, from which the 'Ph' is selected from the template menu. This will shrink the $Cr(CO)_3$ structure and move it to the left, and will add a benzene ring of comparable size to the right-hand side of the work space, as a substructure. Note that the benzene ring is already flagged for π-atoms. The next step is to rotate this benzene ring until its plane too is nearly orthogonal to the screen. The **Select** tool is activated, and the mouse cursor is used to mark any atom in the benzene ring. Then, the 'X

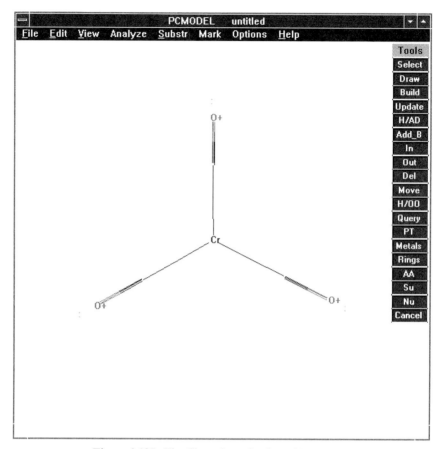

Figure 6-135. The Chromiumtricarbonyl Fragment.

Rotate' Dial within **View\Control_Panel** is used to rotate the benzene ring into roughly the same plane as the metal carbonyl fragment. The result should look like Fig. 6-136.

The judicious use of 'X Translate' and 'Y Translate' in the 'Dials' dialog box within **View\Control_Panel** will move two substructures into their approximately correct relative positions, with the benzene substructure just above the chromiumtricarbonyl. After an **Update** to resize the fragments, the **Select** tool is used to mark the chromium atom and each carbon atom of the benzene ring. Next, the **Mark\Metal_Coord** command from the menu bar is activated, which pops up the 'Metals Information' dialog box. The 'Metal' will already be identified as 'Cr', and the 'Charge' box will be empty. Under 'Electron Count', the user should select the default 'Saturated_18_e' option. In this case, chromium has six electrons, three carbonyl ligands contribute two electrons each, and the benzene ring contributes six electrons, giving us a total of eigh-

METAL ATOM BONDING AND COORDINATION 567

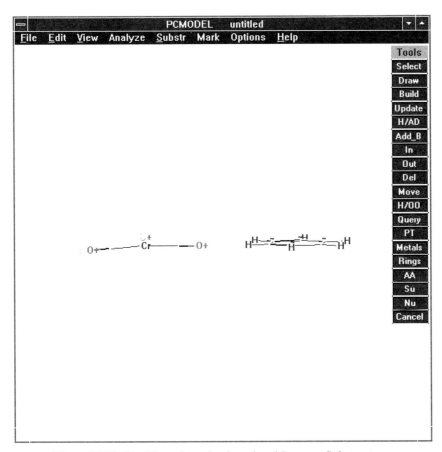

Figure 6-136. The Chromiumtricarbonyl and Benzene Substructures.

teen electrons. At this point, the screen should resemble Fig. 6-137. Clicking on the 'OK' button in the pop-up menu ends the setup dialog and displays the complex with coordination bonds as red lines between the chromium and the benzene ring carbons.

This complex is then run through energy minimization by activating the **Analyze\Minimize** command on the menu. This should be repeated several times until the final energy ceases to change significantly. The carbons of the benzene ring are already flagged for π-calculations, and in this case the default parameters for these calculations will suffice. These parameters could be modified by activating the **Options\MMX_PI_Calc** command on the menu bar, which pops up the 'MMX Pi Calc Options' dialog box (see Section 6.5). Figure 6-138 depicts how the minimum-energy structure should appear. The result should be written to an input file named according to the user's preference; a sample file for comparison may be found on the input file library diskette in the CHAP6\SECT7 subdirectory as {PHCRCOPI.PCM}. Use the **Query** tool

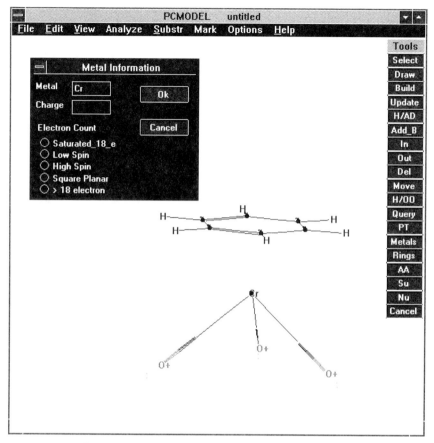

Figure 6-137. Setting Up the Metal Coordination Flags.

to sample some of the geometric features of this new structure, and compare them with the results given in Table 6-20.

With this structure still on screen, or after calling up the input file just written, we will amend the structure to use the aromatic-parametrized carbons to make up the benzene ring, in order to see what difference this would make to the outcome. Using this @40 'CA' carbon atom will decrease the time required for minimization, because concurrent π-calculations will no longer be carried out. If any **Query** reports remain on screen, these can be removed with **Update**. We first remove the flags for π-atoms by activating the **Mark\Reset** command, and in the 'Reset' dialog box click the box for 'Pi atoms'. Then, bring up the 'Periodic Table' menu by activating the **PT** tool, click on the 'CA' atom, and drop this on each of the benzene ring atom positions. Next, this new complex should be run through several cycles of energy minimization by activating the **Analyze\Minimize** command, until the final energy ceases to change significantly. The result should be written to an input file named

METAL ATOM BONDING AND COORDINATION   569

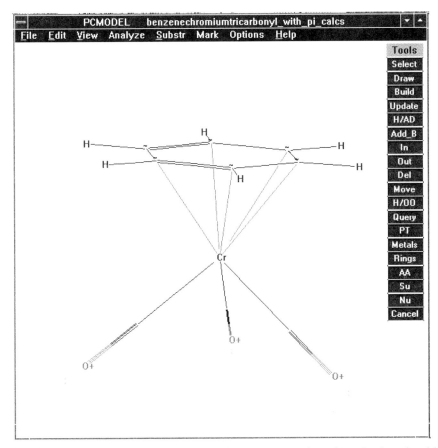

Figure 6-138. Minimum-Energy Structure for Benzenechromiumtricarbonyl with Benzene Composed of @2 Carbon Atoms Flagged for π-Calculations.

according to the user's preference; a sample file for comparison may be found on the input file library diskette in the CHAP6\SECT7 subdirectory as {PHCRCOAR.PCM}. The result should resemble Fig. 6-139. Use the **Query** tool to sample some of the geometric features of this new structure, for comparison with the results given in Table 6-20.

For the last part of this example, we will test the effect of changing the covalent radius of the metal. In a preferred text processor, open {PHCRCOPI.PCM} as an ASCII, text-only file; this is shown in Fig. 6-140. The covalent radius parameter is listed in the atom specification string for the chromium, toward the end of line 4, as 'R 1.25'.

Use a copy-and-paste operation to append two additional copies of the contents of this file to the original. In the second of the two, overwrite the covalent radius value with 1.19, and in the third, overwrite it with 1.30. Save the file, exit the text processor, and return to PCMODEL. Use the **Analyze\Batch**

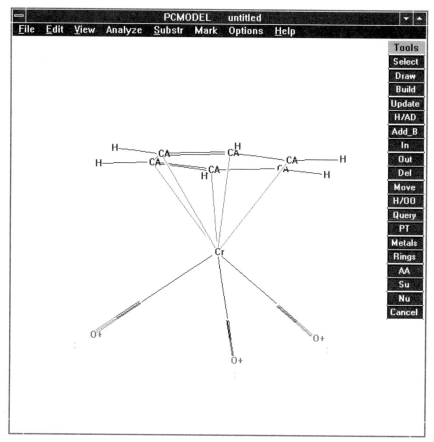

Figure 6-139. Minimum-Energy Structure for Benzenechromiumtricarbonyl with Benzene Composed of @40 Carbon Atoms.

command to run all three structure inputs through minimization. After each cycle of minimization, the new input file {PHCRCOPI.OUT} should be renamed to {PHCRCOPI.PCM}. Finally, use the **Query** tool to probe the geometric features of each structure. These results are given in Table 6-20.

We compare the geometry of these calculated structures with a structure calculated from density functional theory and with empirical values obtained from microwave spectra and from X-ray, neutron, and electron diffraction studies. The MMX force field calculations provide a good model in this class of complexes. As can be seen in the table, the agreement between the calculated and experimental values for the bond lengths in the two coordination models {PHCRCOPI} and {PHCRCOAR} is within ≈1%. For the bond angle OC—Cr—CO, the calculated values are almost 3% too large in the MMX models, with the least discrepancy in the case where a larger covalent radius was used for chromium.

```
1  {PCM Benzenechromiumtricarbonyl_with_pi_calcs
2  NA 22
3  FL EINT4 UV1 PIPL1
4  AT 1 Cr   -0.008   -0.284    0.011 B 2 1 4 1 6 1 15 9 16 9 14 9 11 9 13 9 12 9 S 0 R 1.25
5  AT 2 4    0.108   -1.396   -1.462 B 1 1 3 3 S 0
6  AT 3 46   0.181   -2.083   -2.381 B 2 3 8 1 S 0
7  AT 4 4    1.186   -1.363    0.921 B 1 1 5 3 S 0
8  AT 5 46   1.931   -2.031    1.488 B 4 3 9 1 S 0
9  AT 6 4   -1.416   -1.292    0.661 B 1 1 7 3 S 0
10 AT 7 46  -2.293   -1.915    1.067 B 6 3 10 1 S 0
11 AT 8 20   0.206   -2.322   -2.700 B 3 1 S 0
12 AT 9 20   2.190   -2.263    1.685 B 5 1 S 0
13 AT 10 20 -2.599   -2.131    1.207 B 7 1 S 0
14 AT 11 2  -0.131    1.392   -1.422 B 12 2 16 1 17 1 1 9 S  P  C -0.0381219
15 AT 12 2  -1.246    1.439   -0.582 B 11 2 13 1 18 1 1 9 S  P  C -0.0381769
16 AT 13 2  -1.073    1.481    0.804 B 12 1 14 2 19 1 1 9 S  P  C -0.0381131
17 AT 14 2   0.212    1.446    1.350 B 13 2 15 1 20 1 1 9 S  P  C -0.0381835
18 AT 15 2   1.326    1.399    0.508 B 14 1 16 2 21 1 1 9 S  P  C -0.0380957
19 AT 16 2   1.155    1.357   -0.877 B 11 1 15 2 22 1 1 9 S  P  C -0.0381917
20 AT 17 5  -0.267    1.377   -2.516 B 11 1 S       C 0.0381471
21 AT 18 5  -2.261    1.452   -1.012 B 12 1 S       C 0.0381471
22 AT 19 5  -1.952    1.536    1.468 B 13 1 S       C 0.0381471
23 AT 20 5   0.348    1.464    2.445 B 14 1 S       C 0.0381471
24 AT 21 5   2.342    1.391    0.938 B 15 1 S       C 0.0381471
25 AT 22 5   2.034    1.305   -1.540 B 16 1 S       C 0.0381471
26 }
```

Figure 6-140. The {PHCRCOPI.PCM} Structure Input File.

Since the MMX model works reasonably well for this molecule, one could envision its extension to other members of this series, or to related complexes such as those bearing cyclopentadienyl ligands. Especially in the latter case, involving charged ligands, a similar optimization of the model should take place, experimenting with charge on the metal, modes of ligand–metal bonding, etc. In the coordination examples here, the two modes of modeling the phenyl ring (π-calculations and the @40 carbons) give results which are very close to each other, indicating that, at least in the case of this very symmetrical aryl ligand, much calculation time can be saved by using the latter model.

6.8. DOCKING OF SUBSTRUCTURES

This section describes docking experiments, in which favorable relative geometries between two molecules are sought which maximize their attractive interactions. The area of docking experiments is briefly reviewed, and this is followed by an discussion of the PCMODEL implementation, available through the **Analyze\Dock** command, and its simulated annealing algorithm is explained.

Two exercises are provide in this section. The first exercise deals with docking two methanol molecules together, where hydrogen bonding drives the attractive interaction. The second exercise deals with docking the tricarbonylchromium fragment with a benzene ring, where metal atom coordination drives the attractive interaction.

TABLE 6-20. Geometric Parameters for Benzenechromiumtricarbonyl

r_{cov} (Å)	Ar@	Average Bond Length (Å)					OC—Cr—CO (deg)	Model	Ref.
		C—H	C—C	Cr—C$_{Ar}$	Cr—C$_{CO}$	C—O			
1.25	@2	1.103	1.397	2.204	1.849	1.150	90.0	MMX with π-calcs.	
1.25	@40	1.103	1.401	2.196	1.849	1.150	90.3	MMX with Ar atom	
1.19	@2	1.103	1.400	2.210	1.800	1.150	91.2	MMX with π-calcs.	
1.30	@2	1.103	1.380	2.215	1.894	1.150	89.6	MMX with π-calcs.	
		1.093	1.405	2.222	1.864	1.164	87.2	LDA/NL	887
		1.080	1.413	2.178	1.876	1.148	87.4	Microwave	888
		1.080	1.407	2.221	1.842	1.143	88.1	X-ray	889
		0.913	1.414	2.232	1.845	1.160	87.8	X-ray	890
		1.109	1.415	2.233	1.845	1.158	87.7	Neutron	890
		1.1011	1.417	2.208	1.863	1.1529	88.6	Electron	891

r_{cov} is the metal atom covalent radius; Ar@ gives the MMX atom type used for the benzene ring carbons; values given under OC—Cr—CO are average angle spans. The input parameters for the PCMODEL calculations are discussed in the text. Under "Model," the results for "MMX with π-calcs." refer to the input file {PHCRCOPI.PCM}; the results for "MMX with Ar atom" refer to he input file {PHCRCOAR.PCM}; the "LDA/NL" reference values come from a calculation using the local density approximation augmented with self-consistent nonlocal corrections[887]; the "Microwave" reference values come from microwave spectra,[888] the "X-ray" reference values come from X-ray diffraction,[889,890] the "Neutron" reference values come from neutron diffraction[890]; and the "Electron" reference values come from electron diffraction.[891]

These experiments work well, but docking experiments in the real world often deal with docking small-molecule substrates, ligands, or inhibitors into sites on synthetic catalysts, biomacromolecules such as enzymes, or other macromolecules. Such experiments would greatly exceed the atom limit on most desktop modeling programs, including PCMODEL, and must be carried out with other programs on higher level platforms.

The docking of one structure into or onto another is a way of finding the attractive forces which operate between the two structures in question, and seeking to maximize that attraction through manipulation of the relative orientation of the two structures. A simple definition of docking is the translation of two molecules toward each other, with concomitant reorientations of one or both partners in a manner which will maximize the attractive interactions. Docking can be done either manually, or by means of a systematic or stochastic computer algorithm. Rotations and translations of one molecule with respect to the other are carried out to discover a subset of relative geometries, and the conformations in this group are further analyzed to arrive at a prediction of optimized interaction. The phenomenon which is modeled here is referred to variously as docking or molecular recognition, and is currently generating keen interest with regard to the discovery of biologically active molecules, mechanisms of catalysis, and material science.[123,157,177,573,648]

Docking experiments were discussed in Section 6.2, and the operation of two sophisticated docking algorithms was discussed: the eponymous **Dock**, developed by I. D. Kuntz of UCSF,[667–669] and **Molecular Skins**, developed by B. Masek and coworkers at Zeneca.[670,671] Several other methods for this type of study have been reported.[497,664,666,892–897]

In PCMODEL, the **Analyze\Dock** command allows the user to dock a designated structure onto another.[665] For the purposes of this discussion, the pair of structures in a specified relative spatial arrangement is called an *ensemble*. The PCMODEL convention is that the first structure read in is the one which is manipulated about the second.

To prepare for a docking experiment, the first structure should be built and brought to minimum energy, or read in as a preminimized file. Then, a preminimized second structure is read in, and may be prepositioned with the **Substr\Move** command as the user wishes. The default for the **Substr\Move** command is to place the new (second) structure just to the right of the first. It is prudent to locate the second structure near the first, since the motion of the substructure is limited to the sum of the diameter of the main structure plus twice the diameter of the substructure plus 4.0 Å, even at the highest initial temperature. Larger separations may cause the two structures to lose each other. All desired interactions such as hydrogen bonding and metal atom coordination should be toggled on prior to initiating the **Dock** sequence.

The simulated annealing method is used to effect random translations and rotations of the first structure in a "box" around the second. Simulated annealing has been discussed earlier in Section 5.8.2, on conformation searching with the GMMX program. The PCMODEL implementation of this method is explained in the program documentation,[303,665] as the following series of steps:

1. The energy of the initial ensemble is evaluated.
2. The first structure suffers a translation and a rotation with respect to the second, where the magnitudes are determined from user-supplied parameters scaled by a random number.
3. The energy of the new ensemble is evaluated. If the new energy is *lower* than that of the previous one, the new ensemble is accepted, and the process moves to step 4. If the new energy is *higher* than that of the previous one, the Metropolis algorithm (see below) is used to determine whether the new ensemble should be accepted or the previous ensemble be retained.
4. A counter is incremented if the same ensemble is passing this step again. If a new ensemble is passing this step, the counter is reset to 1. When the counter reaches the value set in the 'Stop on No Change' option, the temperature of the Metropolis bath is decreased to

$$\frac{T^0}{1 + ni}$$

where T^0 is the initial temperature in kilocalories per mole (where a factor of R has been suppressed), n is the number of cooling cycles, and i is the fractional cooling decrement.

When the ensemble is ready for the docking experiment to begin, the **Analyze\Dock** command should be activated, which will bring up the 'Dock Setup' dialog box, as shown in Fig. 6-141. This dialog box provide six options for the user to set up for the docking experiment.

The option entitled 'XYZ Displacement,' with a default value of 1 Å, gives the base distance value for translation of the first structure about the second. The actual displacement is determined by the product of this value with a random number.

The option entitled 'Base Angle', with a default value of 15°, gives the base angle span value for rotations of the first structure. The actual rotation is determined by the product of this value with a random number.

The option entitled 'Starting Temperature', with a default value of 3 kcal/mol, sets the energy available to the molecule from the Metropolis bath, to help the ensemble in and out of local energy minima. The starting temperature is coded as T^0. When an ensemble of higher energy than the starting state is discovered, the Metropolis algorithm[42] uses a temperature-dependent, probabilistic function to determine whether that new ensemble should be discarded or retained.

The option entitled 'Cooling Decrement', with a default value of 0.1, sets the magnitude of the temperature step taken when enough cycles have passed

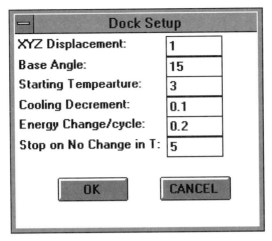

Figure 6-141. The 'Dock Setup' Dialog Box.

without reaching a new ensemble with a lower energy. The cooling decrement is coded as i.

The option entitled 'Energy Change/Cycle', with a default value of 0.2 kcal/mol, is the criterion used to determine whether a new ensemble has significantly (≥ 0.2 kcal/mol) lower energy than the reference structure for that cycle.

The option entitled 'Stop on No Change in T,' with a default value of 5, is the number of times the program may arrive at the same ensemble energy before undergoing a temperature decrement.

When the 'Dock Setup' dialog box options have been selected, a 'Write File for Batch, Dock and Dihedral Driver' dialog box (Fig. B-2 in Appendix B) pops up to allow the user to write a file to contain the final structure from the docking experiment.

Docking Exercise 1: Methanol Dimer. This experiment has the same setup as was used in Hydrogen Bonding Exercise 1 in Section 6.6. The procedure begins with the ensemble as depicted in Fig. 6-102. Ensure that the hydrogen bonding flag is turned on, and carry out an **Analyze\Single Point E** energy calculation.

Activate the **Analyze\Dock** command, and accept the default values by clicking on the 'OK' button. The first structure will begin to jump around the screen in an animated fashion, and the 'Output' window will document the progress of the docking experiment, as seen in Fig. 6-142.

Three columns of data are produced during the docking run: 'Trial' (or cycles), 'Temp' (or temperature, in kcal/mol), and 'Energy' (in kcal/mol). The

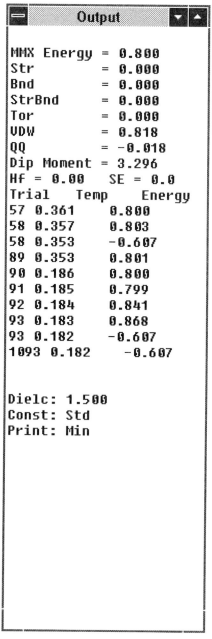

Figure 6-142. The 'Output' Window for the Methanol Dimer Docking Experiment.

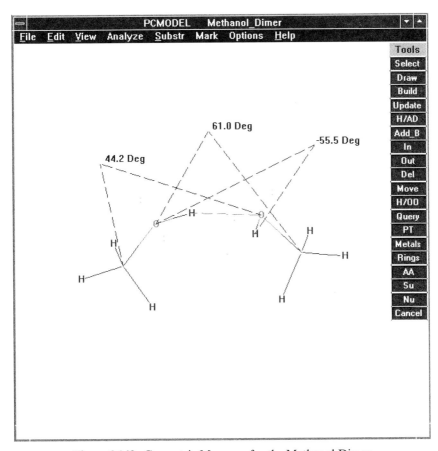

Figure 6-143. Geometric Measures for the Methanol Dimer.

report line for trials is not continuous; there is a break between trials 58 and 89, and between trials 93 and 1093. These indicate periods of nonproductive searching, that is, where no new lower energy ensembles are discovered.

The lowest energy ensemble will be saved in the file named during the setup procedure. If structure movements have placed some of the ensemble off screen, clicking on the **Update** tool will rescale it. The final structure should be subjected to several cycles of energy minimization, to ensure an optimized conformation, and then written to a file. Since this is a stochastic method, different results may be obtained in two successive runs, even if all the same parameters are used. The user may wish to try runs with variation of parameters to see their effect, such as changing the base distance and angle, starting temperature, etc. The final structures should all be run through several cycles of energy minimization and saved. The accumulated ensembles may be run through a **Batch** minimization afterwards, and the structures and energies compared.

TABLE 6-21. Comparison of Docked Methanol Dimer Ensembles

Energy (kcal/mol)	Dihedral angle (deg)		
	C—O—H—O	O—H—O—C	O—H—O—H
−4.128	81.0	−43.0	72.9
−4.124	−79.0	42.4	−73.7
−4.106	30.3	78.5	−165.9
−4.106	56.3	50.8	166.3
−4.103	106.5	−59.5	−175.0
−3.496	44.2	61.0	−55.5

Geometric values measured from methanol dimer ensembles discovered through docking experiments. The three dihedral angles are those depicted in Fig. 6-143.

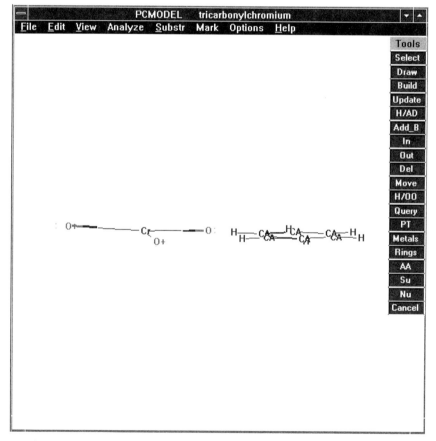

Figure 6-144. Setup for Docking Tricarbonylchromium with Benzene, Part 1.

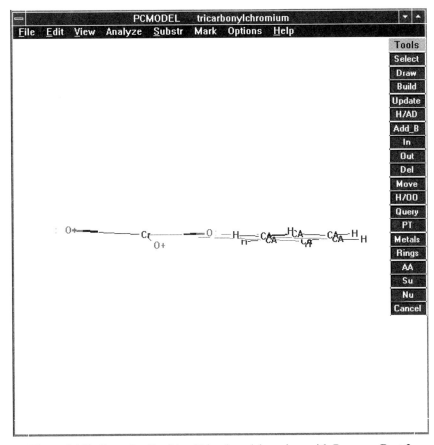

Figure 6-145. Setup for Docking Tricarbonylchromium with Benzene, Part 2.

The geometries of these methanol dimer ensembles may be analyzed by comparing three dihedral angles from the hydrogen bond donor molecule to the acceptor, corresponding to C—O—H···O, O—H···O—C, and O—H···O—H, as shown in Fig. 6-143. Some sample results are collected in Table 6-21.

Docking Exercise 2: Benzenechromiumtricarbonyl. This exercise is closely related to Organometallic Modeling Exercise 4. We will import the tricarbonylchromium fragment and a benzene molecule and let **Dock** form a complex between them.

To simplify the exercise, we will use a benzene ring composed of @40 atoms. To prepare this, use the **Rings** tool to bring up the 'Rings' menu and pick **Ph**. This benzene molecule is fitted for π-calculations, so we must use **Mark\Reset** to unflag the π-atoms, and then bring up the 'Periodic Table'

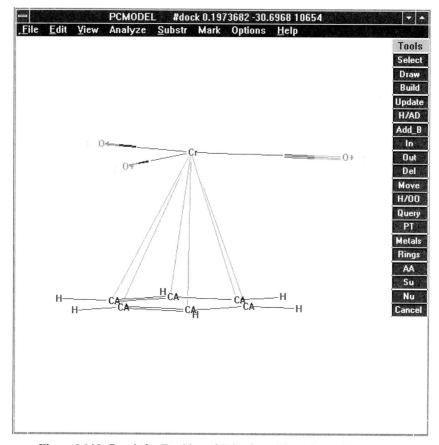

Figure 6-146. Result for Docking of Tricarbonylchromium with Benzene.

menu (**PT** tool) to get the 'CA' atom and replace each of the benzene ring carbons. Run the resulting structure through several energy minimizations, and use **View\Control_Panel** to rotate it to a nearly edge-on perspective. Save the result under a name like {BENZAR.PCM}, or obtain this file from the input file library diskette in the CHAP6\SECT8 subdirectory.

Read in the two files {CRCO3.PCM} and {BENZAR.PCM} to generate a screen display such as that shown in Fig. 6-144.

Use the **Select** tool to mark the chromium atom and each of the six benzene ring atoms. Set the flag for metal atom coordination with **Mark\Metal_Coord**, after which the screen should look like Fig. 6-145 (red dotted coordination bonds may appear faint).

An **Analyze\Single Point E** energy calculation should be carried out, and then the **Analyze\Dock** command should be activated. Accept the default values by clicking on the 'OK' button, and watch as the tricarbonylchromium

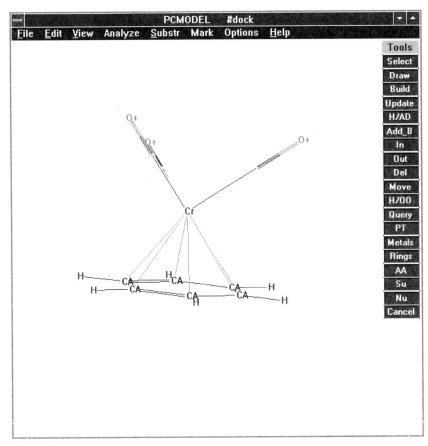

Figure 6-147. Minimum-Energy Structure for Docked Benzenetricarbonylchromium.

fragment is moved about in a search for the optimum complex. Strictly speaking, this fragment should be pyramidal instead of planar, but the experiment works well even with the planar carbonylchromium. This has the additional benefit of a plane of symmetry, which will increase the number of productive, low-energy encounters. After between 5000 and 10,000 cycles, the docking experiment will produce a good structure, such as that seen in Fig. 6-146.

This should be run through several cycles of energy minimization in order to arrive at a good final geometry for the complex, such as is shown in Fig. 6-147. The **Query** tool should be used to probe the measurements of this structure and compare these with the values given in Table 6-20 in Section 6.7. Other variations to try might include reading in the {BENZAR.PCM} file first, so that the benzene fragment will be docked onto the metal carbonyl. The use of other relative initial geometries and the variation of the **Dock** setup parameters are additional possibilities for further study.

6.9. REACTIVE INTERMEDIATES AND TRANSITION STATES

The modeling of reactive intermediates and transition states with the empirical force field technique has long been a controversial area. There are some conceptual similarities between these two areas, and they are treated together here because they both deal with atom types which are specially parametrized to simulate reactive and excited state species. It should be remembered that the force field representation of reactive intermediates, and especially of transition states, provides only a very crude model of the electronic features of these species. Only the conformational and nonbonded energy is evaluated. The method's applicability is dependent upon the assumption that in comparisons between isomeric reactive species and transition states, the bond electronics will be the same and so these contributions to the overall energy will cancel out. This assumption is not always valid and must always be kept in mind for studies of such species. Semiempirical or *ab initio* molecular orbital methods should be used when it is important to assess the bond electronics.

Some success has been obtained by using the hybrid technique **QM/MM** (quantum mechanics/molecular mechanics), where the *reaction zone* (those bonds in the molecules which are being formed and broken) is modeled with a molecular orbital method, and the remainder of the system is evaluated by force field methods.[633,898–902]

Force field models of reactive intermediates and transition states have been discussed in general terms elsewhere.[903,904] There are four in-depth reviews pertinent to this area: that of Houk, Tucker, and Dorigo on proximity effects on organic reactivity,[905] that of Eksterowicz and Houk on transition state modeling with force fields,[21] and those of Nakamura and Fukazawa[110] and of Lipkowitz and Peterson[22] on applications to synthetic design. In the latter review may be found an excellent discussion of the pitfalls inherent in studies of this type, such as the confusion between equilibrium constants and rate constants, and disregard for the implications of the **Curtin–Hammett principle**.

A comparison of several force field models for the Diels–Alder reaction has been reported.[906] Transition state modeling has been applied to asymmetric aldol reaction of enol borinates, with the goal of rational synthetic design.[907] The WIZARD program has been used to search over conformations of transition states to predict reaction outcomes.[908]

For general background on the mechanisms of organic reactions and transition states, the reader is referred to the standard textbook of March.[909] In particular there are discussions of aliphatic nucleophilic substitution,[910] aromatic electrophilic substitution,[911] aliphatic electrophilic substitution,[912] aromatic nucleophilic substitution,[913] free radical substitution,[914] addition to carbon–carbon multiple bonds (including the Diels–Alder reaction),[915] addition to carbon–heteroatom multiple bonds,[916] elimination reactions,[917] and electrocyclic and sigmatropic rearrangements (including the Cope and Claisen rearrangements).[918] March also has a good current treatment of reactive intermediates.[919]

TABLE 6-22. Transition State Bond Types Available in the MMX Force Field

Atom Types	Appearance	Application
@49—@49	C*—C*	Forming bonds in Diels–Alder reaction or in Cope rearrangement
@50—@50	C#—C#	Forming bonds in Diels–Alder reaction, or breaking bonds in a Cope rearrangement
@1—@45	C—H*	Breaking allylic C—H bond in ene reaction
@49—@45	C*—H*	Forming H—C bond in ene reaction or hydroboration
@50—@43	C#—B#	Forming B—C bond in hydroboration
@49—@51	C*—C$	Forming bond in nitrile oxide addition, dipole carbon to dipolarophile carbon
@50—@53	C#—O#	Breaking bond in Claisen rearrangement
@49—@52	C*—C%	Forming bond in S_N2, nucleophilic carbon to electrophilic carbon
@52—@54	C%—I%	Breaking bond in S_N2, electrophilic carbon to leaving group iodide
@53—@52	O#—C%	Forming bond in S_N2, nucleophilic oxygen to electrophilic carbon
@50—@55	C#—N#	Breaking bond in aza–Cope rearrangement

The classical work of Streitwieser on carbonium ions,[920] radicals,[921] carbanions,[922] transition states for aromatic substitution reactions,[923] and cyclization reactions[924] is still useful for reference.

Recent reports from the Schleyer group discuss modified geometry parameters for Allinger's MMP2 force field program, used to achieve good results in modeling localized and delocalized carbocations.[494,495]

6.9.1. Reactive Intermediates

PCMODEL has parametrized atoms types for the carbocation (@30, C^+), the carbon radical (@29, C ·) and the carbanion (@48, C^-). In addition, there is the ammonium nitrogen (@41, N^+), an oxonium oxygen (@46, O^+), and an oxyanion (@42, O^-). These may sometimes be part of a reactive intermediate, and other times belong to ground state structures such as common ammonium and alkoxide salts, and the less stable oxonium compounds such as Meerwein's salts (R_3O^+ BF_4^-). Metal alkoxides are best modeled with a covalent metal–oxygen bond.

Each of these atoms may be involved in conjugated systems for which π-calculations may be carried out. The two anionic species have a lone pair pseudoatom associate with them. The nonbonded parameters (van der Waals radius, hardness) are normally identical with those of the related neutral atom types. Out-of-plane bending interactions are evaluated for the carbon atom types, but not for these heteroatoms.

6.9.2. Transition State Atoms

The PCMODEL implementation of transition state atoms is discussed in detail in the program documentation and in the book chapter by the program developers.[904]

The MMX force field contains nine atoms parametrized as transition state atoms: B#, C*, C#, C$, C%, H*, I%, N#, O#. These may be accessed from the 'Periodic Table' menu, where each is indicated by the elemental symbol plus a character which codes for which kind of transition state atoms may be bonded together. When used in forming transition state models, the bonding pairs must follow the conventions dictated by the parameter set, or the user will need to import new parameters. The 11 parametrized transition state bonds are depicted in Table 6-22.

Prior to energy minimization, bonds containing one or two transition state atoms must be assigned a fractional bond order between 0.0 and 1.0. This parameter may be set by activating the **Mark\TS_Bond Orders** command, which brings up the dialog box, as shown in Fig. 6-148.

There are two columns in this dialog box: the first is entitled 'Bond Type' and normally will show a description of the transition state bonds contained in the active structure; the second is entitled 'Bond Order' and has data entry boxes to receive this parameter from the user. If these aren't assigned during construction of the input structure, the user will be prompted to make the

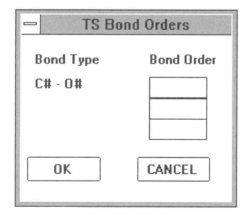

Figure 6-148. The 'TS Bond Orders' Dialog Box.

6.9. REACTIVE INTERMEDIATES AND TRANSITION STATES

assignment(s) either when writing an input file with **File\Save**, or when starting an **Analyze\Minimize** energy minimization.

MMX calculates bond lengths for transition state bonds from this 'Bond Order' parameter according to **Badger's rule**,[508–511] cast in Pauling bond order terms. Thus

$$r^{\ddagger}_{X-Y} = r_{X-Y} - 0.6 \ln \text{BO}$$

where $r^{\ddagger}_{X-Y}$ is the transition state bond length for a bond $X-Y$, r_{X-Y} is the normal equilibrium bond length, and BO is the bond order parameter supplied by the user. The constant 0.6 is twice as large as the Pauling constant for bond orders greater than 1. This value is used because it reproduces the structure of the Diels–Alder transition state calculated by Houk from the STO 3-21G basis set if the kinetic isotope effects determined at Indiana on the Diels–Alder reaction are used as input bond orders.

MMX scales the force constant for the transition state bond as the product of the ground state force constant with the user-supplied bond order.

Empirical evidence suggests that the bond order in such transition state bonds depends strongly upon substituent effects, and the user may optimize this parameter for a particular system by judicious comparisons to established cases. As a starting point, many reactions such as the Diels–Alder cycloaddition and the nitrile oxide cycloaddition have *early* transition states and can be modeled well using a value of 0.3 for the bond order. The bond order cannot be equal to 0 or 1, or the calculation will fail. To model very early or very late transition state stages, use values which approach the limits, such as 0.05 and 0.95.

There are a number of other limitations to keep in mind when preparing input for transition state modeling. The transition state atoms cannot be included in a π-atom array; for π-atoms in pericyclic transition states, the @29 C · should be used (see the Transition State Modeling Exercise, below). The @29—@29 bond in the transition state has a fixed length of 1.40 Å and a force constant $k_{\text{bnd}} = 7.0$ mdyn/Å. For transition state bonds including an @49 C* or an @50 C#, the angles between this transition state bond and the atoms attached to this carbon grow from 90° at a fractional bond order of ≈0.0 to 109° at a fractional bond order of ≈1.0. For a transition state bond including an @51 C$, the angles between the atoms attached to this carbon change from 180° at a fractional bond order of ≈0.0 to 120° at a fractional bond order of ≈1.0. For the hydroboration transition state, the parameters attempt to mimic the acute angle for the @43—@50—@49 B#—C#—C* with appropriate hydrogen angles and distances.

In an S_N^2 transition state complex involving an @52—@54 C%—I% bond in the electrophile, the angles between this bond and the atoms attached to the @54 C% begin at the tetrahedral value of 109° for an ≈1.0-bond-order C%—I% and ≈0.0 for the bond from C% to the O# or C* nucleophile. At the point where both bond orders are 0.5, the angles between the atoms

TABLE 6-23. Transition State Templates Available in {TRNZSTAT.SST} for the Diels–Alder Cyclization

Atom Labels	Atom Types	Diels–Alder Cycloaddition
C*, C*, C•, C#, C#, C•	@49, @49, @29, @50, @50, @29	All carbon
C*, C*, C•, N#, C#, C•	@49, @49, @29, @55, @50, @29	Aza-dienophile
C*, C*, C•, C#, N#, C•	@49, @49, @29, @50, @55, @29	Aza-diene
C*, C*, C•, O#, C#, C•	@49, @49, @29, @53, @50, @29	Oxa-dienophile
C*, C*, C•, C#, O#, C•	@49, @49, @29, @50, @53, @29	Oxa-diene

6.9. REACTIVE INTERMEDIATES AND TRANSITION STATES

TABLE 6-24. Transition State Templates Available in {TRNZSTAT.SST} for Electrocyclic and Sigmatropic Rearrangements

Atom Labels	Atom Types	Reaction
C*–C•–C# / C*–C•–C# (hexagon)	@29, @49, @50, @49, @50, @29	Cope rearrangement
C*–C•–O# / C*–C•–C# (hexagon)	@29, @49, @53, @49, @50, @29	Claisen rearrangement
H*–C*–C#–C#–C• (with ring)	@49, @45, @50, @1, @50, @29	Ene reaction
H*–C*–C•–C•–C• (with ring)	@49, @45, @29, @1, @29, @29	1,5 Hydrogen shift

attached to the C% are 120°, and these atoms form a plane, to which the (O#/C*)—C%—I% bond axis is orthogonal. As the bond order for the C%—I% bond approaches 0.0, and that for the C%—nucleophile bond approaches 1.0, the angles between the atoms attached to the C% return to 109°.

As was discussed in Section 3.2.6 on substructures, there is a input file of model transition states available, which the user may read in and modify with appropriate substituents. This file may be found on the input file library diskette in the CHAP6\SECT9 subdirectory as {TRNZSTAT.SST}. The transition state models available are depicted in Tables 6-23 to 6-26 for the different types of transition states.

Transition State Modeling Exercise: Diastereomeric Diels–Alder Transition States. For this exercise, we'll consider a Diels–Alder cyclization carried out

TABLE 6-25. Transition State Templates Available in {TRNZSTAT.SST} for Small-Ring Cycloadditions

Atom Labels	Atom Types	Reaction
C$–N=, O#, C*–C# (ring)	@37, @51, @53, @49—@50	1,3-Dipolar nitrile oxide cycloaddition
C*–N, O#, C*–C# (ring)	@8, @49, @53, @49—@50	1,3-Dipolar nitrone cycloaddition
B#–H*, C#–C*	@43—@45, @50—@49	Hydroboration

TABLE 6-26. Transition State Templates Available in {TRNZSTAT.SST} for Nucleophilic Substitutions

Atom Labels	Atom Types	Reaction
C* — C% – I%	@49 — @52 — @54	S_N2 displacement with carbon nucleophile
O# — C% – I%	@53 — @52 — @54	S_N2 displacement with oxygen nucleophile

6.9. REACTIVE INTERMEDIATES AND TRANSITION STATES 589

by Brown and Houk.[925] We will model the transition states leading to each diastereomer, compare the energies at different values for the bond order parameter, and then compare the modeling prediction with the experimental result.

The reaction to be modeled is depicted in Fig. 6-149. We will build structure **3** first, and then modify it to create structure **4**. Begin by reading in the transition state template for the Diels–Alder cyclization, which may be found on the input file library diskette in the CHAP6\SECT9 subdirectory as the first structure in the appended format file {TRNZSTAT.SST}. This will appear as is shown in Fig. 6-150.

Use the scroll bars in the 'Dials' dialog box to get a better perspective on the structure. Use the **Build** tool to convert the two hydrogens attached to the @29 carbon radical atoms to methyl groups. This will become a [2.2.1]bicycloheptyl moiety after a bit more tailoring. The downward-pointing hydrogens on the @49 (C*) and @50 (C#) atoms are likewise elaborated with the **Build** tool; these will become part of the maleic anhydride moiety. The result so far should look like the structure depicted in Fig. 6-151.

Manipulate the structure so that it is roughly in the plane of the screen, i.e. looking down at it from "above." Use the **H/AD** tool to remove the hydrogens; then fill in the maleic anhydride portion by activating **Draw** and clicking on one of the carbons, then outward from the ring and back again to make a double bond. Continuing with the same **Draw** operation, complete the five-membered ring of the maleic anhydride portion, and then out and back from the other carbon to form the second double bond. Load the cursor with O from the 'Periodic Table' menu, and reassign the appropriate atoms to complete the maleic anhydride portion, as in Fig. 6-152.

Now, complete the six-membered ring on the other side of the structure. Then, use the **H/AD** tool to replace hydrogens and add lone pairs to the oxygens. The next step is to do a preliminary energy minimization, but the transition state bond orders must be set first. Activate the **Mark\TS_Bond Orders** command, and enter the values 0.3, as shown in Fig. 6-153. Then proceed with the energy minimization.

Reposition the structure so that its plane is more orthogonal to the screen, to prepare for closing the methylene bridges. Use the **Build** tool to elaborate the hydrogen at either of the bridge carbons (C* or C#) to a methyl group; then remove hydrogens and draw the bond to make this bridge. At this point,

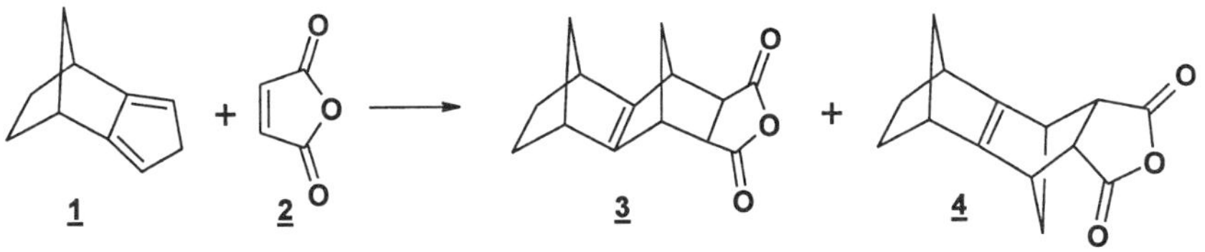

Figure 6-149. Diels–Alder Reaction for Transition State Modeling Exercise.

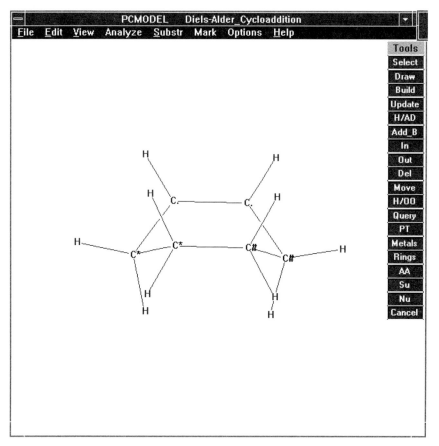

Figure 6-150. Transition State Modeling Exercise, Structure Preparation, Part 1: The Diels–Alder Transition State Template.

use **Draw** to add the remaining methylene bridge in the [2.2.1]bicycloheptyl moiety, to produce the structure shown in Fig. 6-154.

Use **H/AD** to add hydrogens, and carry out several cycles of energy minimization. Because of the very poor input geometry, some hydrogens may be placed inappropriately by **H/AD**, and the system may not minimize well. If there is an obvious problem, or if the energies remain high, interrupt the minimization with the ⟨Esc⟩ key, and remove and re-add the hydrogens. Continue this until the structure minimizes to an energy below ≈ 30 kcal/mol. It may also help to carry out a minimization cycle on the heavy atom skeleton alone. The final structure should resemble that shown in Fig. 6-155. The user should write the structure to an input file. If desired, a comparable structure is available from the input file library diskette in the CHAP6\SECT9 subdirectory as the first structure in the appended format file {DA_03.PCM}.

6.9. REACTIVE INTERMEDIATES AND TRANSITION STATES 591

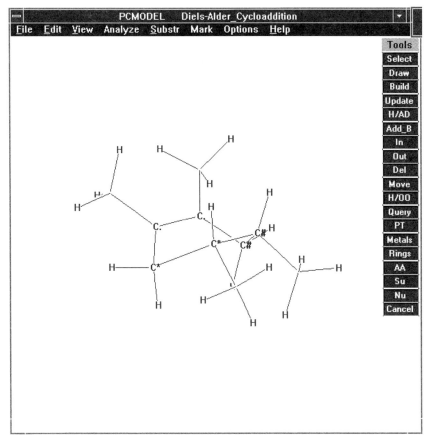

Figure 6-151. Transition State Modeling Exercise, Structure Preparation, Part 2.

592 APPLICATIONS

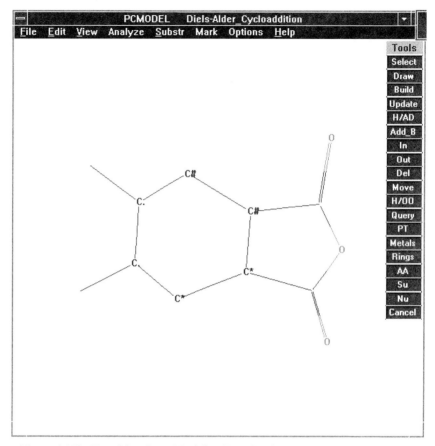

Figure 6-152. Transition State Modeling Exercise, Structure Preparation, Part 3.

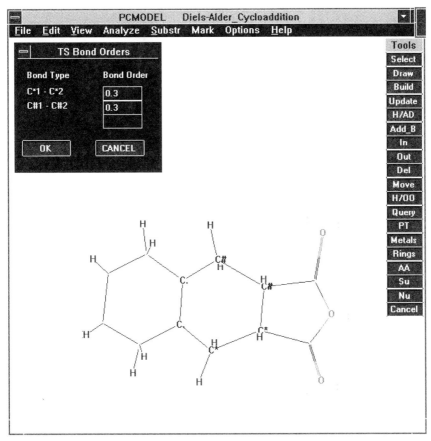

Figure 6-153. Transition State Modeling Exercise, Structure Preparation, Part 4: Preliminary Energy Minimization and Assigning Transition State Bond Orders.

594 APPLICATIONS

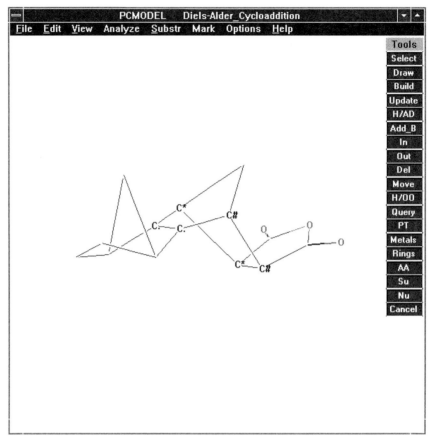

Figure 6-154. Transition State Modeling Exercise, Structure Preparation, Part 5.

6.9. REACTIVE INTERMEDIATES AND TRANSITION STATES 595

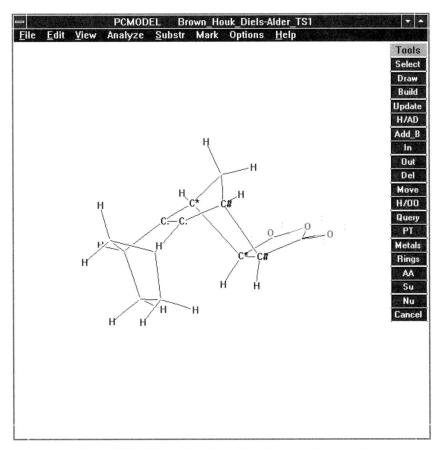

Figure 6-155. Diels–Alder Transition State for Structure **3**.

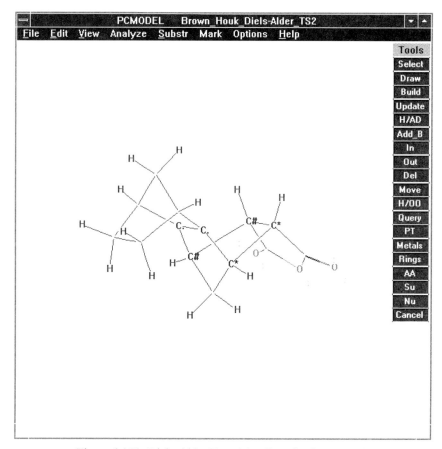

Figure 6-156. Diels–Alder Transition State for Structure **4**.

6.9. REACTIVE INTERMEDIATES AND TRANSITION STATES   597

```
1  {PCM Brown_Houk_Diels-Alder_TS1
2  NA 36
3  FL EINT4 UV1 PIPL1
4  AT 1  49    1.511   -0.451   -0.201 B 2 1 6 1 13 1 18 1 S 0
5  AT 2  49    0.277    1.157   -1.131 B 1 1 3 1 11 1 19 1 S 0
6  AT 3  29   -1.001    0.620   -1.113 B 2 1 4 1 7 1 S 0
7  AT 4  29   -1.489    0.608    0.197 B 3 1 5 1 8 1 S 0
8  AT 5  50   -0.533    1.134    1.053 B 4 1 6 1 11 1 20 1 S 0
9  AT 6  50    1.017   -0.470    1.113 B 1 1 5 1 12 1 21 1 S 0
10 AT 7  1   -1.882   -0.295   -1.917 B 3 1 9 1 10 1 22 1 S 0
11 AT 8  1   -2.679   -0.310    0.217 B 4 1 9 1 17 1 23 1 S 0
12 AT 9  1   -3.221    0.027   -1.199 B 7 1 8 1 24 1 25 1 S 0
13 AT 10 1   -1.668   -1.753   -1.458 B 7 1 17 1 26 1 27 1 S 0
14 AT 11 1    0.319    2.035    0.136 B 2 1 5 1 28 1 29 1 S 0
15 AT 12 3    2.051    0.083    1.941 B 6 1 14 1 15 2 S 0 C 0.4469196
16 AT 13 3    2.829    0.111   -0.125 B 1 1 14 1 16 2 S 0 H  C 0.4469196
17 AT 14 6    3.068    0.448    1.150 B 12 1 13 1 30 1 S 0 H  C 0.1158296
18 AT 15 7    2.034    0.200    3.143 B 12 2 31 1 32 1 S 0 H  C -0.2948344
19 AT 16 7    3.606    0.257   -1.037 B 13 2 33 1 34 1 S 0 C -0.2948344
20 AT 17 1   -2.214   -1.764   -0.007 B 8 1 10 1 35 1 36 1 S 0
21 AT 18 5    1.360   -1.315   -0.859 B 1 1 S 0 H
22 AT 19 5    0.761    1.459   -2.065 B 2 1 S 0 H
23 AT 20 5   -0.773    1.412    2.084 B 5 1 S 0
24 AT 21 5    0.478   -1.349    1.486 B 6 1 S 0
25 AT 22 5   -1.849   -0.118   -3.011 B 7 1 S 0
26 AT 23 5   -3.372   -0.144    1.068 B 8 1 S 0
27 AT 24 5   -4.056   -0.638   -1.516 B 9 1 S 0
28 AT 25 5   -3.545    1.088   -1.312 B 9 1 S 0
29 AT 26 5   -2.257   -2.444   -2.105 B 10 1 S 0
30 AT 27 5   -0.602   -2.061   -1.523 B 10 1 S 0
31 AT 28 5   -0.186    3.014   -0.041 B 11 1 S 0
32 AT 29 5    1.339    2.252    0.517 B 11 1 S 0
33 AT 30 20   3.573    0.708    1.343 B 14 1 S 0 C -0.21
34 AT 31 20   2.505    0.440    3.427 B 15 1 S 0 C -0.0525
35 AT 32 20   1.555    0.020    3.458 B 15 1 S 0 C -0.0525
36 AT 33 20   4.145    0.498   -0.936 B 16 1 S 0 C -0.0525
37 AT 34 20   3.452    0.088   -1.592 B 16 1 S 0 C -0.0525
38 AT 35 5   -3.084   -2.457    0.080 B 17 1 S 0
39 AT 36 5   -1.457   -2.083    0.741 B 17 1 S 0
40 FBND 0.300 0.300 0.000 0.000  0.000 0.000 0.000 0.000 0.000 0.000
41 }
```

Figure 6-157. Input File for the First Structure in {DA_TS01.PCM}.

Now, we will transmute structure **3** into an input for structure **4**. This will be most easily achieved by exchanging the one- and two-carbon bridges in the [2.2.1]bicycloheptyl moiety. Remove the hydrogens, and reposition the structure so that it is roughly in the plane of the screen, and then again so that the one- and two-carbon bridges are easily visible and separate from each other. Use the **Del** tool to remove a bond from the one-carbon bridge to the bridgehead, then draw from it to a new carbon and back to the bridge. Then use the **Del** tool to remove one of the carbons from the two-carbon bridge, and then draw a bond from it back to the bridge. Try a preliminary energy minimization on the heavy atom skeleton; then add hydrogens and complete the minimization. The same cautions about unnatural geometries mentioned earlier should be observed again at this stage. Keep applying cycles of energy minimization until the structure reaches a steady energy level below ≈ 30 kcal/mol. The user should also write this structure to an input file. If desired, a comparable structure is available from the input file library diskette in the CHAP6\SECT9 subdirectory as the second structure in the appended format file {DA_03.PCM}. This structure is shown in Fig. 6-156.

To get an accurate reading on the energies, the pair of structures may be run through energy minimization in **Batch** mode. This will also provide an energy summary file to be used in the analysis phase of this experiment.

As a further investigation, the user could evaluate this pair of diastereomeric transition states at other values for the bond order. The simplest way to do this would be to create copies of the input file and modify them in a text processor. The input file for the first structure is shown in Fig. 6-157.

The transition state bond order parameters are found in line 40, the 'FBND' statement:

```
FBND 0.300 0.300 0.000 0.000 0.000 0.000
     0.000 0.000 0.000 0.000 0.000
```

In a copy of this input file, the values 0.300 may be overwritten with values 0.100 and 0.500. These new input files may then be run through several cycles of energy minimization to arrive at final, optimized geometries. These pairs of structures are available from the input file library diskette in the CHAP6\SECT9 subdirectory as the appended format files {DA_01.PCM} and {DA_05.PCM}. The results are that at a bond order of 0.1, the energy difference is 0.562 kcal/mol; at a bond order of 0.3, the energy difference is 2.021 kcal/mol; and at a bond order of 0.5, the energy difference is 1.559 kcal/mol.

The literature reports the reaction of **1** and **2** at 25°C to give a 75:25 ratio for the products **3**:**4**.[925] Using the Boltzmann population analysis described in Appendix A, Section A.3, our results predict a ratio of 72:28 using the bond order value of 0.1, and a ratio of 97:3 using the bond order value of 0.3. If we are confident that steric energy differences at a bond order of 0.1 are significant and controlling, then this prediction comes fairly close. As the bond order 0.3 is more commonly used for predicting the outcome of Diels–Alder reactions, we would predict a much higher selectivity than was actually obtained.

APPENDIX A
General Information

A.1. PC FILENAMES

Every PC file will have a filename consisting of up to eight alphanumeric characters, sometimes with an extension of up to three alphanumerics. This is the name given to a particular file in order for the PC (or other computer) operating system and programs to recognize and operate upon it. This is different from any title or comments lines contained within the file.

In this book, filenames referred to in the text are capitalized and enclosed in braces, e.g. {FILENAME.EXT}. In the DOS and Windows 3.1 environment, a filename takes the generic form {FFFFFFFF.EEE}. The filename proper {FFFFFFFF} can be from one to eight characters from the recognized set; the **filename extension** {EEE} is separated from the filename by a period, and can be from zero to three characters from the recognized set. In cases where reference is made to a group of files which share a name or an extension, an asterisk (*) will be used as a wildcard to denote the common element. This is a standard DOS usage. For example, a group of data files could be referred to as {*.DAT}, and a family of files with the same name but different extensions could be called {FILENAME.*}. Some common function-based extensions are {*.BAK} = backup, {*.COM} = command file, {*.DAT} = data file, {*.DOC} = documentation or document, {*.EXE} = executable file, {*.INP} = input file, {*.OUT} = output file, {*.BAT} = batch file, {*.SYS} = system file, etc.

The recognized set of characters for DOS filenames includes all of the alphanumerics: ABCDEFGHIJKLMNOPQRSTUVWXYZ0123456789. In addition, the following symbols can be used: ~'!@#$&()_-{}'. There are two other symbols which DOS will accept in filenames, but which have specific use as wildcards which argues against their general use for filenames: ? is the single-character wildcard, and * is the wildcard for any number of characters.

It is best to give a file a filename which will suggest its contents, if possible. Then, if references to the filename are later lost, it may be rediscovered in a directory more easily.

A.2. UNIT ABBREVIATIONS, UNIT CONVERSIONS AND SAMPLE VALUES

TABLE A-1. Unit Abbreviations

Unit	Full Name (Quantity)
Å	Angstrom unit (distance)
C	Coulomb (charge)
°C	Degree Celsius (temperature)
cP	Centipoise (viscosity)
D	Debye (dipole moment)
dyn	Dyne (force)
e	Elementary charge
eu	Entropy unit
eV	Electron volt (work, heat)
fs	Femtosecond (time)
Hz	Hertz (frequency)
J	Joule (work, heat)
kcal	Kilocalorie (work, heat)
K	kelvin (temperature)
kJ	Kilojoule (work, heat)
mdyn	Millidyne (force)
mol	Mole (number)
N	Newton (force)
Pa	Pascal (pressure, stress)
ps	Picosecond (time)
s	Second (time)

The millidyne is a nonstandard unit of force. It comes originally from the spectroscopy field, but its use is strongly embedded in the molecular mechanics field.

TABLE A-2. Unit Conversions

0°C	=	273.15 K	1 fs	=	10^{-15} s
1 cP	=	0.01 dyn · s/cm^2	1 hartree	=	4.360×10^{-18} J
1 cP	=	0.01 g/cm · s	1 Hz	=	1 s^{-1}
1 cP	=	0.001 Pa · s	0K	=	$-273.15°C$
1 D	=	3.336×10^{-30} C · m	1 kcal/mol	=	4.187 kJ/mol
1 dyn	=	10^{-5} N	1 kcal/mol	=	0.0898 eV/mol
1 e	=	1.602×10^{-19} C	1 kJ/mol	=	0.239 kcal/mol
1 eu	=	J/mol · K	1 mdyn/Å	=	1×10^{-21} kJ/Å^2
1 eV	=	1.602×10^{-19} J	1 mdyn/Å	=	2.39×10^{-22} kcal/Å^2
1 eV/mol	=	2.66 kJ/mol	1 Pa	=	N · m^{-2}
1 eV/mol	=	11.14 kcal/mol	1 ps	=	10^{-12} s

TABLE A-3. Sample Values for Viscosity η

Medium	η (cP)			
	−25°C	0°C	25°C	100°C
Octane		0.700	0.508	0.243
Chloroform	0.988	0.706	0.537	
Toluene	1.165	0.778	0.560	0.270
Methanol	1.258	0.793	0.544	
Pyrrolidine	1.914	1.071	0.704	
Dimethylformamide		1.176	0.794	
Propanoic acid		1.499	1.030	0.449
Water		1.793	0.890	0.282
Ethanol	3.262	1.786	1.074	
Dimethylacetamide			1.956	0.661
1-Butanol	12.19	5.185	2.544	0.533
1-Octanol			7.288	0.991
Ethylene glycol			16.1	1.975

From reference 926.

TABLE A-4. Sample Dielectric Constant Values

Medium	D			
	20°C	25°C	70°C	100°C
Vacuum	1.000			
Octane	1.948		1.879	
Benzene	2.284	2.274		
Chloroform	4.806			
Ethyl acetate		6.02		
Acetic acid	6.15		6.62	
Dichloromethane	9.08			
1-Octanol	10.3			
1-Butanol	17.8	17.1		
Acetone		20.70		
Methanol	33.62	32.63		
Glycerol		42.50		
Water	80.37	78.54	63.73	55.51

From reference 927. The permittivity (dielectric constant) of a vacuum is defined as $D_0 = 8.854 \times 10^{-12}$ F · m^{-1}, but is arbitrarily set to a value of 1.00 for comparison with other media.[928] The values shown here are the unitless ratios of the permittivity of the liquid to that of a vacuum.

TABLE A-5. Boltzmann Population Analysis for the Two-Component System at Various Temperatures

δE (kcal/mol)	−75°C K	−75°C $N_i:N_j$	0°C K	0°C $N_i:N_j$	20°C K	20°C $N_i:N_j$	100°C K	100°C $N_i:N_j$
0.25	1.89	65:35	1.59	61:39	1.54	61:39	1.40	58:42
0.50	3.56	78:22	2.51	72:28	2.36	70:30	1.96	66:34
0.75	6.73	87:13	3.99	80:20	3.63	78:22	2.75	73:27
1.00	12.7	93:07	6.32	86:14	5.57	85:15	3.86	79:21
1.50	45.3	98:02	15.9	94:06	13.2	93:07	7.57	88:22
2.00	161	99:01	39.9	98:02	31.1	97:03	14.9	94:06
2.50	575	100:0	100	99:01	73.3	99:01	29.2	97:03
3.00	2047	100:0	252	100:0	173	100:0	57.3	98:02
3.50	7295	100:0	634	100:0	408	100:0	113	99:01
4.00	2.60×10^4	100:0	1.60×10^3	100:0	964	100:0	221	100:0
5.00	3.30×10^5	100:0	1.01×10^4	100:0	5.37×10^3	100:0	852	100:0
10.0	1.09×10^{11}	100:0	1.02×10^8	100:0	2.89×10^7	100:0	7.26×10^5	100:0

δE is the energy difference between the two components or two states. K is the equilibrium constant calculated from the expression given in the text, and $N_i:N_j$ is the ratio of the two components.

A.3. CONFORMER POPULATION ANALYSIS

When two or more products are formed under equilibrating conditions, or with two or more conformations which are interconvertible, the population distribution between these states depends upon their relative energies, and can be modeled with the Boltzmann distribution on a molar scale. Thus, with i, j, etc., representing the different states of this system, the fraction of the population in occupying the ith state is given by the equation

$$\frac{N_i}{N} = \frac{e^{\delta E_{i0}/RT}}{\sum e^{\delta E/RT}}$$

where N_i is the number of entities in the ith state, N is the total number of entities, δE_{i0} is difference in energy between the ith state and the lowest energy 0th state in joules per mole, R is the molar gas constant (8.3145 $J \cdot mol^{-1} \cdot K^{-1}$), and T is the temperature in Kelvin. The denominator represents the summation over all applicable states.

In the case of only two states, i and j, a simple equilibrium constant can be obtained as follows:

$$K = e^{\delta E_{ij}/RT}$$

where δE_{ij} is difference in energy between the ith and jth states.

For example, if the temperature is 20°C (=293.15 K) and the difference in energy between two states is 1 kcal/mol (=4187 J/mol), the exponent of our expression will be given by

$$\frac{\delta E_{ij}}{RT} = \frac{4187}{8.3145 \times 293.15} = 1.718$$

and

$$K = e^{\delta E_{ij}/RT} = e^{1.718} = 5.57$$

Some examples are given in Table A-5 for two-component product or conformation ratios at temperatures of interest to the organic chemist. The energy values have been converted to kilocalories per mole and the temperature values to degrees Celsius.

A.4. DERIVATIONS

A.4.1. Atomic Charges for Chloroethane Calculated from Bond Moments (Section 5.6)

The measured length of the C—Cl bond in chloroethane (Fig. 5-36) is 1.786 Å. The bond moment parameter for an @1—@12 [(sp^3)carbon–chlorine] bond is 1.940 D (from the MMX parameter file {MMXCONST.PAR} in the BND section]. Definitions for the Debye unit (D)[541] and the elementary electronic charge (e)[928] are

$$1 \text{ D} = 3.33564 \times 10^{-30} \text{ C} \cdot \text{m}$$

$$e = 1.6022 \times 10^{-19} \text{ C}$$

where C stands for coulomb. So

$$1 \text{ D} = 3.33564 \times 10^{-30} \text{ C} \cdot \text{m}$$

$$= (3.33564 \times 10^{-30} \text{ C} \cdot \text{m})\left(\frac{1}{1.6022} \times 10^{19} \, e \cdot \text{C}^{-1}\right)(10^{-10} \text{ Å} \cdot \text{m}^{-1})$$

$$= 0.20819 \, e \cdot \text{Å}$$

So, converting the bond dipole of 1.940 D to a value for charge, we have

$$(1.940 \text{ D})(0.20819 \, e \cdot \text{Å} \cdot \text{D}^{-1})\left(\frac{1}{1.786} \text{ Å}^{-1}\right) = 0.226e \approx 0.23e$$

This results in an assignment of −0.23 to the more electronegative chlorine atom, and +0.23 to the more electropositive carbon atom.

A.4.2. Distance between Ring Centroids from Distance Constraints (Section 6.1)

A.4.2.1. Derivation for the Face-to-Face Benzene Dimer From the values for the distance constraint and the other geometric parameters obtained from the minimum-energy structures of the benzene face-to-face dimer, we can triangulate to provide the perpendicular distance between the planes of the two rings. With the assumption that the two rings occupy parallel planes, this will be equal to the distance between the ring centroids. We define these values as shown in Figs. A-1 through A-4.

In Fig. A-1, we see a lateral view of the dimer. The distance between C1 and C4 of the benzene ring is designated by a, and is measured to be 2.804 Å. The distance between C1 and C3 of the benzene ring is designated by b, and is measured to be 2.435 Å. The distance between C1 and C2 of the benzene ring

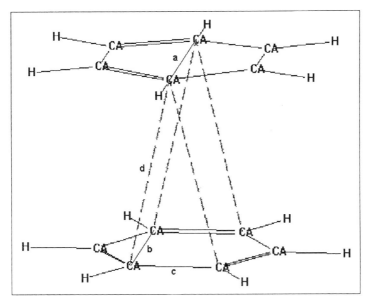

Figure A-1. Face-to-Face Benzene Dimer, Lateral View.

is designated by c, and is measured (and defined) to be 1.400 Å. In this analysis, the distance constraint is designated by d.

In Fig. A-2, we see a close-up of the lateral view of the dimer. Dropping a line from the upper ring carbon atom to the midpoint of the bond between the

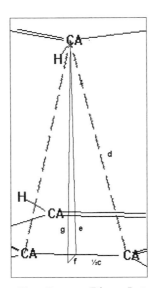

Figure A-2. Face-to-Face Benzene Dimer, Lateral View, Close-up.

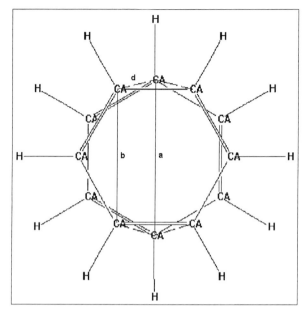

Figure A-3. Face-to-Face Benzene Dimer, Axial View from Beneath.

two ring carbons on the lower ring defines a distance designated by e. Dropping a line from the upper ring carbon atom perpendicular to the plane of the lower ring defines a distance designated by g. The termini of e and g are joined by the short segment f. The distance between the planes of the benzene rings, equivalent to the distance between centroids, will be given by g.

Figure A-3 gives an axial view of the dimer from beneath, and Fig. A-4 gives a close-up view of the dimer from this perspective. The line g is orthogonal to the plane of this view, and the line e is hidden by f; hence these are not visible in Fig. A-4.

There are two right triangles which will be used in this derivation: one is bounded by e and $\frac{1}{2}c$, with a hypotenuse of d, and the other is bounded by g

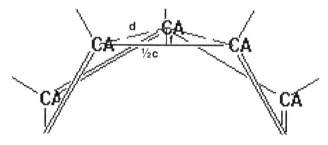

Figure A-4. Face-to-Face Benzene Dimer, Axial View from Beneath, Close-up.

and f, with a hypotenuse of e. The following relationships exist:

$$\tfrac{1}{2}c = \tfrac{1}{2}(1.400) = 0.700 \text{ Å}$$

$$f = \tfrac{1}{2}(a - b) = \tfrac{1}{2}(2.804 - 2.435) = 0.1845 \text{ Å}$$

$$d^2 = e^2 + (\tfrac{1}{2}c)^2 \quad \text{and} \quad e^2 = g^2 + f^2$$

Solving for g in terms of d, we get

$$g = (d^2 - 0.524)^{1/2} \text{ Å}$$

A.4.2.2. Derivation for the T-Shaped Perpendicular Benzene Dimer From the values for the distance constraint and the other geometric parameters obtained from the minimum-energy structures of the benzene T-shaped perpendicular dimer, we can triangulate to obtain the distance between the centroids of the two rings. We make the assumption that the long axis of one of the rings is perpendicular to the plane of the other. The necessary values are defined as shown in Fig. A-5.

The distance between C1 and C4 of the benzene ring is again designated by a is measured to be 2.804 Å. The length of the bond between the ring carbon and the hydrogen is designated by h, and is measured to be 1.104 Å. In the closest approaches, there is some stretching of the C—H bond for the hydrogen used as an anchor for the distance constraints. This variability in the value for h has been included in the data set. In this analysis, the distance constraint is again designated by d, although the anchors used are different than in the face-to-face dimer study.

Dropping a perpendicular line from the hydrogen anchor to the centroid of the bottom ring defines a distance i, and the distance between centroids is given by the sum $i + h + \tfrac{1}{2}a$. There is a right triangle which will be used in this derivation, which is bounded by i and $\tfrac{1}{2}a$, with a hypotenuse of d. The following relationships exist:

$$\tfrac{1}{2}a = \tfrac{1}{2}(2.804) = 1.402 \text{ Å}$$

$$h = 1.104 \text{ Å} \quad \text{(or value adjusted for stretching)}$$

$$d^2 = i^2 + (\tfrac{1}{2}a)^2$$

Solving for i in terms of d, we get

$$i = (d^2 - 1.966 \text{ Å}^2)^{0.5}$$

Thus, the distance between centroids is given by

$$i + h + \tfrac{1}{2}a = (d^2 - 1.966 \text{ Å}^2)^{0.5} + 2.506 \text{ Å}$$

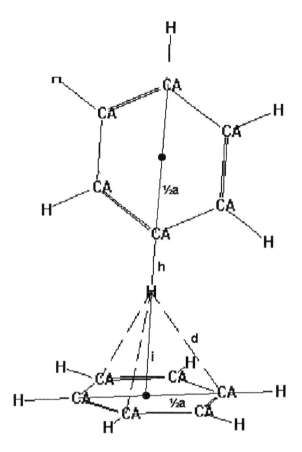

Figure A-5. Perpendicular Benzene Dimer, Side View.

APPENDIX B

Glossary of PCMODEL Commands

This appendix provides descriptions of all of the commands which appear in PCMODEL for Windows. For those commands which have been discussed at length in the main body of the book, a relatively brief description is provided here and the reader is referred to the earlier treatment for further information.

AA an abbreviation for *amino acid*, this command provides the user with a menu of amino acid templates for use in structure building. The **AA** command is found on the tool bar. Activating this command pops up a menu which contains buttons representing the available amino acid templates (see Fig. 3-14). Clicking on a button imports the selected amino acid into the work space. This template menu is discussed in Section 3.2.2.

About PCModel gives information about the current version of the program PCMODEL. This command is found under **Help** on the menu bar. Activating this command brings up the information box shown in Fig. B-1.

Add_B, i.e. *add bond*, increases the multiplicity of a bond by one, up to a total of three. This command appears on the tool bar. When this tool is activated, the mouse cursor may be clicked at the midpoint of the target bond. This bond is then overprinted with a symbol for the indicated change, a 'D' for a double and a 'T' for a triple bond. Clicking on the **Update** button redraws the structure with bond having the new multiplicity. Once the command is activated, any number of **Add_B** operations can be carried out sequentially until the screen is updated.

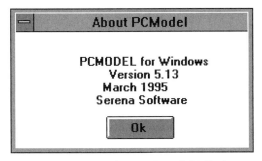

Figure B-1. The 'About PCModel' Window.

Added Constants sets a flag in the input file instructing PCMODEL to look for a file containing nonstandard constants to be read in for use in energy minimizations. This command is found under **Options** on the menu bar, and its operation is described in Section 4.7.

Analyze is a family of commands having to do with energy and property calculations. This command is found on the menu bar. Activating it pulls down a menu with nine options: **Minimize, Single Point E, Mopac, Dynam, Dock, Batch, Rot_E, Dihedral Driver, Surface Area,** and **Volume**. Figure 2-5 shows the **Analyze** command pulldown menu.

Ball_Stick, i.e. *ball and stick*, is a viewing option which renders the molecule as "balls" for the atoms and thickened "sticks" for the bonds. This command is found under **View** on the menu bar, and is discussed in Section 3.6. See also **Pluto**.

Batch carries out energy minimization in batch mode of two or more structure input data sets contained in a single, appended-format file. This command is found under **Analyze** on the menu bar. Activating this command brings up a dialog box entitled 'Write File for Batch, Dock and Dihedral Driver', as shown in Fig. B-2, for selecting the desired file. Note that only two file formats are readable for batch mode: 'PCModel' and 'MMX'. PCMODEL creates an appended format file during the **Dock** operation and during the **Dihedral Driver** operation. Otherwise, this file must be prepared by the user by consecutively preparing a number of different structures, or different conformations of the same structure, and then

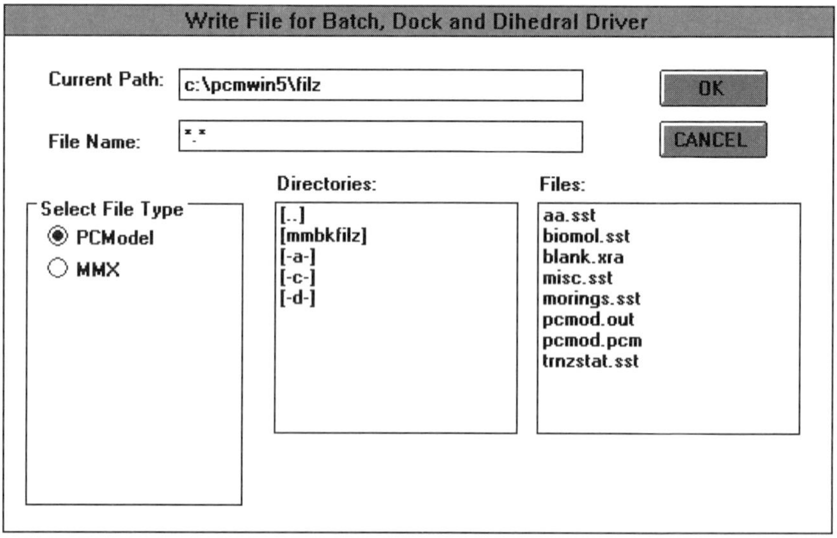

Figure B-2. The 'Write File Dialog Box for Batch, Dock and Dihedral Driver' Dialog Box.

writing them to the same filename with **File\Save**. When an existing filename is entered in the dialog box for the **File\Save** command, another dialog box entitled 'File Exists' pops up (shown in Fig. 3-26), asking the user to select between appending the new input data to the existing file and overwriting the latter. A maximum of 75 separate inputs may be processed in this way. In the actual input file, the complete data sets are simply concatenated, and there is no requirement that the contained files be related in any way. When the desired file has been selected, another dialog box entitled 'Batch' appears, as shown in Fig. B-3. The selected input file filename appears in the first line, and default filenames are suggested for the output file (i.e. the appended format file containing the energy-minimized versions of the structures in the input file) and the summary file (an ASCII file containing a table of the summary information from the batch calculation). This box also reports a default starting structure (structure 1) and the total number of structures. Any of the names or values in boxes may be changed by the user. For example, one may wish to give the output file or summary file different filenames, start with a structure other than structure 1, or batch-minimize fewer than the total number of structures. The summary file {*.SUM} is a table containing the file number ('Num'), the structure name (truncated after six characters), and values for the MMX energy ('Energy') and the heat of formation ('Hf'). The progress of the energy minimizations is contained in the {PCMOD.OUT} log file. Note that if the user opts to rename the output file to the same filename as the input file (i.e. same name and same extension), the input file will be overwritten and the batch process will terminate after the first input structure is processed.

Build allows the user to "grow" carbon atoms in place of hydrogens on a structure in the work space. This command is a drawing utility for rapidly

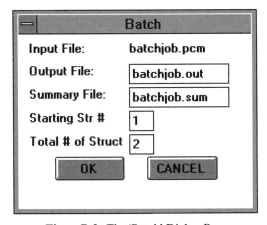

Figure B-3. The 'Batch' Dialog Box.

elaborating a carbon skeleton, and is found in the tool bar. Its operation is described in Section 3.4.

Cancel, found at the bottom of the tool bar, closes the tool bar menu window. The tool bar can be brought back by clicking on **Edit\Draw** on the menu bar.

Compare compares structures by calculating the best least squares fit between all or selected atoms within them. This command is found under **View** on the menu bar. Comparisons can be made between two or more molecules, or between conformations of the same molecule. The structures may be contained in separate files, or in the same file in appended format. The procedures for structural comparisons are discussed in detail in Section 3.6.3.

Connect joins two substructures by removing a selected hydrogen on each substructure and splicing together the two broken bonds. This command is found under **Substr** on the menu bar, and is discussed in Section 3.1.

Control_Panel allows the user to alter the position or perspective of the representation in the work space by changing the size, or by translating or rotating the structure(s). This command is found under **View** on the menu bar. If the representation in the work space is too crowded or confused to carry out operations with the **Query** tool, it may be altered with **Control_Panel**. The 'Scale' function may be used to expand or shrink the molecule onscreen, or the 'Translate' and/or 'Rotate' functions may be used to change their perspective, and allow easier viewing and access to the portion of the molecule under scrutiny. The operation of this command is described in more detail in Section 3.6.1.

Copy_BW_To_Clipboard, i.e. *copy black-and-white image to the clipboard*: This command is found under **Edit** on the menu bar, and copies a rendering of the work space in black and white to the Windows clipboard. The operation of this command is described in Section 3.6.

Copy_CO_To_Clipboard, i.e. *copy color image to the clipboard*: This command is found under **Edit** on the menu bar, and copies the work space in color to the Windows clipboard. The operation of this command is described in Section 3.6.

CPK_Surface draws a surface of solid, overlapping spheres to represent the atoms of the structure in the work area. The **Query** command may be used while a CPK surface is active. This command is found under **View** on the menu bar, and is discussed in Sections 3.6 and 6.2.

Create defines all or a portion of a structure in the work space as a substructure. This command is found under **Substr** on the menu bar, and is discussed in Section 3.2.6.

Del, i.e. *delete*, removes undesired atoms or bonds during a drawing operation. This command is found on the tool bar, and is discussed in Section 3.1.

GLOSSARY OF PCMODEL COMMANDS   613

Dielc,   i.e. *dielectric*, allows the user to change the dielectric constant used in the force field calculation to a value other than the default of 1.5. This command is found under **Options** in the menu bar, and is discussed in Section 5.6.

Dihed Map   displays the results of a **Dihedral Driver** operation in graphical format. This command is found **View** in the menu bar, and is discussed in Section 6.3.

Dihedral Driver   sets up the input file to do a dihedral driver calculation on one or two dihedral angles of the molecule in the work space. This command is found under **Analyze** on the menu bar, and is discussed in Section 6.3.2. See also **Dihed Map** and **Rot_E**.

Dock   sets up a docking experiment which seeks to maximize the attractive interactions between two molecules by optimizing their relative geometry by means of a simulated annealing algorithm. This command is found under **Analyze** on the menu bar, and is discussed in Section 6.8.

Don't Minimize   exempts a designated substructure from the geometry optimization process, although its energy is still calculated on a static basis. This command is found under **Substr** on the menu bar, and its use is described in Sections 3.2.6 and 6.1.

Dot_Surface   draws a diaphanous surface around the structure, or a defined substructure, in the work area. This command is found under **View** on the menu bar, and is discussed in Sections 3.6 and 6.2.

DP_DP,   i.e. *dipole–dipole*, allows the user to toggle between the models employed by the MMX force field to calculate the electrostatic contribution to the energy the charge–charge (default) and the bond-dipole–bond-dipole interaction. This command is found under **Options** in the menu bar, and is discussed in Section 5.6.

Dynam,   i.e. *molecular dynamics*, runs a molecular dynamics simulation on the molecule(s) in the work space. This command is found under **Analyze** in the menu bar, and is discussed in Section 5.9.

Edit   is a family of commands having to do mostly with structure modification. This command is found on the menu bar. Activating this command pulls down a menu with eleven options: **Draw, Erase, Structure Name, Rotate_Bond, Epimer, Enantiomer, Copy_BW_To_Clipboard, Copy_CO_To_Clipboard, Orient_XY_Plane, Orient_XZ_Plane,** and **Orient_YZ_Plane**. Figure 2-3 shows the **Edit** command pulldown menu.

Enantiomer   generates the mirror image of the structure(s) in the work space by reflection about a user-selected Cartesian plane. This command is found under **Edit** in the menu bar, and is discussed in Section 3.1.

Epimer   carries out an epimerization of two substituents bonded to a common stereocenter. This command is found under **Edit** in the menu bar, and is discussed in Section 3.1.

Erase   names two commands for erasing the structure(s) in the work space.

One instance of this command is **Edit\Erase**, for global erasure, and this is discussed in Section 3.1. The other is **Substr\Erase**, for selective erasures of substructures, and this command is discussed in Section 3.2.3.

Exit   terminates program operation. This command is found under **File** in the menu bar. Activating this command will bring up a prompt asking whether the user really wants to exit from PCMODEL (see Fig. B-4), a precaution to avoid the loss of work if this button is clicked on by accident. Upon exiting PCMODEL, the current structure, updated as of the last time either **Minimize** or **Single Point E** was run, will be saved under the filename of {PCMOD.BAK}. If no energy calculations have been carried out, and if the contents of the work space have not been written to a file with the **Save** procedure, the structure will be lost.

File   is a family of commands having to do mostly with file management. This command is found on the menu bar. Activating it pulls down a menu with six options: **Open, Save, Save Graphic, Print, Printer Setup, Exit**. Figure 2-2 shows the **File** command pulldown menu.

Fix_Atoms   allows the user to fix the coordinates of one or more marked atoms in a structure, in any one or all three of the Cartesian axes. This command is found under the **Mark** command on the menu bar, and is discussed in Section 6.1.

Fix_Distance   allows the user to fix the relative position in a maximum of 10 pairs of atoms in a structure. This command is found under the **Mark** command on the menu bar, and is discussed in Section 6.1.

Fix_Torsions   allows the user to fix the value for one or two torsional angles in a structure. This command is found under the **Mark** command on the menu bar, and is discussed in Section 6.1.

H/AD,   i.e. *hydrogens add or delete*, toggles between adding hydrogens and lone pairs to a structure which has none, and deleting hydrogens and lone pairs from a structure which already has them. This command is found on the tool bar, and is discussed in Sections 3.1 and 3.4.4.

H-Bonds,   i.e. *hydrogen bonds*, turns on the flag for hydrogen bonding in the

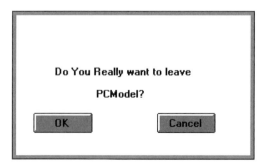

Figure B-4. The Exit Program Confirmation Prompt.

current structure. It is found under the **Mark** command on the menu bar, and is discussed in Section 6.6.

Help: a family of one command. The **Help** command pulldown menu is shown in Fig. 2-9. This command is found on the menu bar, and provides access to the **About PCMODEL** information window, shown in Fig. B-1.

Hide is a toggle to render visible or invisible the graphic representation of a selected substructure in the work space. This command is found under **Substr** on the menu bar, and its operation is described in Section 3.2.6.

H/OO, i.e. *hydrogens on*/off, toggles on and off the depictions of the skeletal (but not functional group) hydrogens on structures in the work space. This command is found on the tool bar, and its operation is discussed in Section 3.6.

In moves an atom into the plane of the screen by ≈ 0.275-Å increments. This command is found on the tool bar, and is discussed in Section 3.1. See **Out**.

Labels allows the user to employ different formats for displaying the atoms or substructures present in the work area. It is found under **View** in the menu bar, and its operations and the options available are discussed in Section 3.6.

Mark is a family of commands having to do with setting flags and restraints in molecules. This command is found on the menu bar. Activating it pulls down a menu with eight options: **H-Bonds, Piatoms, Metal_Coord, TS_Bond Orders, Fix_Atoms, Fix_Distance, Fix_Torsions**, and **Reset**. Figure 2-7 shows the **Mark** command pulldown menu.

Metal_Coord, i.e. *metal coordination*, sets up a flag in the input file for using the algorithm for the organometallic coordination bonding potential between a lone pair or *p*-orbital donor ligand and a metal atom. This command is found under **Mark** in the menu bar, and its operation is discussed in Section 6.7.

Metals provides the user with a menu of metal atoms for use in structure building, and is found on the tool bar. The 'Metals' menu is depicted in Fig. 2-12 and is discussed in Section 2.4.3.2 (see also Section 6.7).

Minimize initiates an energy minimization on the structure(s) in the work space. The **Minimize** command is found under **Analyze** on the menu bar, and its operation is discussed in Section 5.7.2. See also **Single Point E**.

MMX_PI_Calc, i.e. *MMX π-calculations*, sets up the parameters to be used in the VESCF π-calculations on appropriate conjugated molecules during energy minimization runs. This command is found under **Options** on the menu bar, and its operation is discussed in Section 6.5. See also **Piatoms**.

Mono/Stereo toggles the display between monoscopic and stereoscopic graphic representations of molecules in the work space. This command is found under **View** in the menu bar, and its operation is discussed in Section 3.6.

Mopac initiates a semiempirical quantum mechanical calculation on the structure in the work space (see **MOPAC** in Appendix D). This command is found under **Analyze** in the menu bar, and its use requires the presence of the separate MOPAC program {MOPAC.EXE} in the PCMODEL directory. When this command is activated, PCMODEL carries out a write file sequence to generate a MOPAC input file for the structure in the work space, with the default name {SCRATCH.MOP}. Then, if the desktop platform has sufficient volatile RAM available, it opens a DOS window, passes this filename to MOPAC in a command line, and executes the calculation. When finished, PCMODEL will close the DOS window and ask whether the user wishes to read back the MOPAC archive file along with the atomic charges generated from the calculation.

Move comprises two commands: one (**Substr\Move**) enables the movement of a substructure to another position in the work space; the other (**Move** on the tool bar) allows the user to move the atoms in a structure in the work space within the XY plane of the screen. **Substr\Move** is discussed in Section 3.2.3, and the **Move** tool is discussed in Section 3.1.

Nu, an abbreviation for *nucleosides*, provides the user with a menu of nucleoside templates for use in structure building. The **Nu** command is found on the tool bar, and the use of this template menu is discussed in Section 3.2.4.

Open initiates a read file sequence, for reading in an molecular structure input file. This command appears under the **File** command in the menu bar. Its operation is covered in Section 3.2.5, and input file formats are discussed in greater detail in Chapter 4. See also **Save** and **Read**.

Options is a family of commands having to do with program options. This command is found on the menu bar. Activating it pulls down a menu with seven options: **Printout, Dielc, DP_DP, MMX_PI_Calc, Standard Constants, Added Constants, Stereo**, and **Pluto**. Figure 2-8 shows the **Options** command pulldown menu.

Orient_XY_Plane, i.e. *orient in the XY plane*, allows the user to reorient a structure by locating three selected atoms in the XY plane (plane of the screen). This command is found under **Edit** in the menu bar, and its operation is discussed in Section 3.6.1.

Orient_XZ_Plane, i.e. *orient in the XZ plane*, allows the user to reorient a structure by locating three selected atoms in the XZ plane (plane horizontal and orthogonal to the screen). This command is found under **Edit** in the menu bar, and its operation is discussed in Section 3.6.1.

Orient_YZ_Plane, i.e. *orient in the YZ plane*, allows the user to reorient a structure by locating three selected atoms in the YZ plane (plane vertical and orthogonal to the screen). It is found under **Edit** in the menu bar, and its operation is discussed in Section 3.6.1.

Ortep is one of the options for graphical presentation of structures found under the **View** command on the menu bar (see **ORTEP** in Appendix D). Its operation is discussed in Section 3.6.

Out   moves an atom out of the plane of the screen by ≈ 0.275-Å increments. This command is found on the tool bar, and is discussed in Section 3.1. See **In**.

Piatoms   flags either all or a subset of the appropriate atoms for inclusion in the π-atom array for the VESCF π-calculations during energy minimization runs. This command is found under **Mark** in the menu bar, and the application of this feature is described in detail in Section 6.5. See also **MMX_PI_Calc**.

Pluto   is used to set the parameters and options for the **PLUTO**-style 'Ball and Stick' structure display format available with the **View\Ball_Stick** command (see **PLUTO** in Appendix D). It is found under the **Options** command on the menu bar, and is discussed in Section 3.6. See also **Ball_Stick**.

Print   initiates a screen dump of selected contents of the monitor screen to the default printing device. This command is found under **File** on the menu bar, and its operation is discussed in Section 3.6. See also **Printer Setup**.

Printer Setup   allows the user to change the designated printer from the default, and to modify other printing parameters. This command is found under **File** on the menu bar. Activation of this command pops up a standard Windows™ Printer Setup dialog box, as shown in Fig. B-5. The title bar gives the name of the default printer, and other options include the 'Resolution', 'Paper Size', 'Paper Source', and 'Orientation'.

Printout   allows the user to reset the flag for the level of printout from an energy minimization run. This command is present under **Options** in the menu bar, and its operation is discussed in Section 3.5.

PT,   i.e. *periodic table*, provides the user with a menu of Main Group nonmetal MMX atom types for use in structure building. The **PT** command is found on the tool bar, and its use is discussed in Section 2.4.3.1.

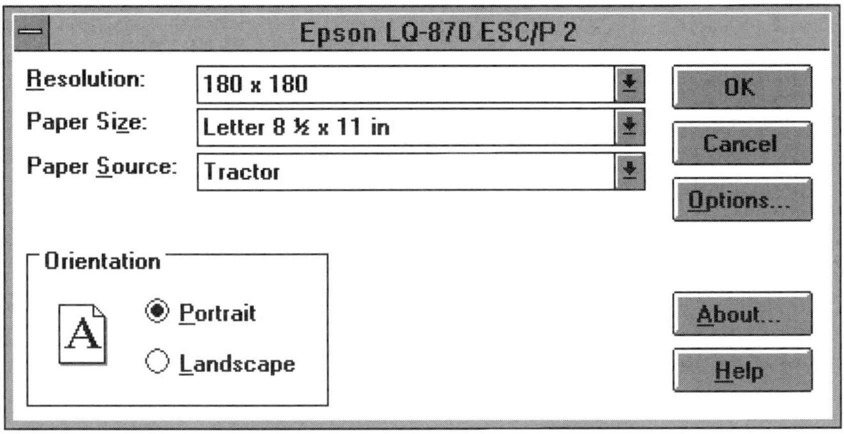

Figure B-5. The Printer Setup Dialog Box.

Query allows user to extract quantitative information regarding the position and charge of an atom, or the relative geometry between the atoms of one or more molecules displayed on the screen. This command appears on the tool bar, and its use is discussed in Section 3.6.2.

Read reads a substructure into the work space. It is found under **Substr** in the menu bar, and its operation is discussed in Section 3.2.6. See also **Open**.

Reset allows options and flags in the input file to be reset to default values or to zero. This command is found under **Mark** on the menu bar. Activating this command brings up the dialog box shown in Fig. B-6. Any combination of the items can be selected by clicking in the appropriate check box with the mouse cursor. The end of the selection process is signalled by clicking on the **OK** button, after which all of the selected resets will be executed. 'All' resets all of the named options. 'Pi Atoms' removes all of the π-atom flags introduced with **Mark\Piatoms** (see Section 6.5). 'Coordinated Atoms' removes the flags for metal atom coordination introduced with **Mark\Metal_Coord** (see Section 6.7). 'Dihedral Driver' removes all flags and instructions for the dihedral driver calculation which were set up with **Analyze\Dihedral Driver** (see Section 6.3.2). 'Hydrogen Bonds' removes the flags for hydrogen bonding which were set up with **Mark\H-Bonds** (see Section 6.6). 'Fixed Distances' removes spatial constraints imposed on atoms with **Mark\Fix_Distance** (see Section 6.1). 'Fixed Torsions' removes constraints imposed upon dihedral angles with **Mark\Fix_Torsions** (see Section 6.1). 'Substructure Membership' will remove the identification of molecules or fragments as substructures which remain after using **Substr\Create** or adding a structure from one of the the template menus (see

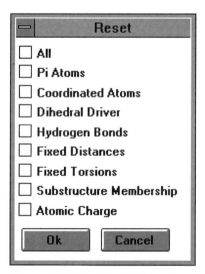

Figure B-6. The 'Reset' Dialog Box.

Section 3.2.6). 'Atomic Charge' removes any charges that were placed on metal atoms introduced as part of the setup for **Metal_Coord** (see Section 6.7).

Ribbon draws a four-line ribbon representing the backbone of amino acid residues in a polypeptide, or the nucleic acid residues in a polynucleotide. It is found under **View** on the menu bar, and its operation is discussed in Section 3.6.

Rings provides the user with a menu of mono- and polycyclic ring system templates for use in structure building. The **Rings** command is found on the tool bar, and its use is discussed in Section 3.2.1.

Rotate_Bond carries out a measured rotation on a chosen bond. This command is found under **Edit** on the menu bar, and its use is discussed in Section 3.1.

Rot_E, i.e. *rotational energy*, carries out a rigid rotation of a specified torsional angle over a specified range, with no intermediate energy minimization. This command is found under **Analyze** on the menu bar, and its operation is discussed in Section 6.3.1. See also **Dihedral Driver**.

Save initiates a write file operation on the current contents of the work space, one or more molecules. This command is found under **File** in the menu bar. Its use is discussed in Section 3.2.5, and the file formats are discussed in greater detail in Chapter 4. See also **Open**.

Save Graphic saves the image contained in the PCMODEL window to a graphics format file. This command is found under **File** in the menu bar, and its operation and the available graphics formats are discussed in Section 3.6.

Single Point E, i.e. *single-point energy*, calculates the static energy of the structure in the work space without carrying out any energy minimization. This command is found under **Analyze** on the menu bar, and its operation is discussed in Section 5.7.1. See also **Minimize**.

Standard Constants reinstates the standard MMX force field parameters, by removing the flag for added constants in the input file. This command is found under **Options** in the menu bar, and its operation is discussed in Section 4.7.

Stereo is a toggle for the perspective used in creating stereoscopic representations, allowing the user to choose between 'Rotate Out' (the default) and 'Rotate In' format for the stereo pair. This command is found under **Options** in the menu bar, and is discussed in Section 3.6.

Stick Figure is a viewing option which renders the bonds in a molecule as lines drawn between atoms, with all noncarbon atoms indicated by their elemental symbols. This option is PCMODEL's default graphical representation of a molecule. The command is found under **View** on the menu bar, and is discussed in Section 3.6.

Structure Name allows the user to enter or modify a descriptive structure name or title for the structure in the work space. This command is found under **Edit** in the menu bar. Activating it brings up a dialog box entitled 'Edit Structure Name', as shown in Fig. B-7. The new name may be entered, or the existing name edited, in the text box labeled 'Enter Structure Name:'. In the absence of a defined structure name, the default name will be 'untitled'. There need not be any structure present in the work space to assign a name. Up to 60 characters can be entered for the name, with no restrictions on format beyond the space limitation, except that empty spaces may lead to name truncation after some operations. The name will be displayed in the title bar of the PCMODEL window, and will also be written on the first line of any MMX or PCM format input file that is generated in the **Save** option. The *structure name* should not be confused with the *filename*, which is the name given to the input file consistent with the user's software environment (see Appendix A.1).

Su an abbreviation for *sugars*, brings up a menu of sugar templates for use in structure building. The **Su** command is found on the tool bar, and the use of this template menu is discussed in Section 3.2.3.

Substr is a family of commands having to do with substructures. This command is found on the menu bar, and substructure operations are discussed in Section 3.2.6. Activating this command pulls down a menu (see Fig. 2-6) with seven options: **Read, Create, Move, Connect, Erase, Hide**, and **Don't Minimize**.

Surface Area computes the van der Waals molecular surface area of the molecule in the work space. This command is found under **Analyze** on the menu bar, and its operation is discussed in Section 6.2. See also **Volume**. Activating this command brings up a dialog window entitled 'Dot Surface Calculation' (see Fig. 6-42) which allows the user to change the resolution at which the surface will be cast.

Tools refers to the floating tool bar menu in the PCMODEL work space, shown in Fig. 2-10. The **Tools** menu contains buttons for thirteen commands: **Select, Draw, Build, Update, H/AD, Add_B, In, Out, Del, Move, H/OO, Query**, and **Cancel**; and for six pop-up menus: **PT, Metals, Rings, AA, Su,** and **Nu**.

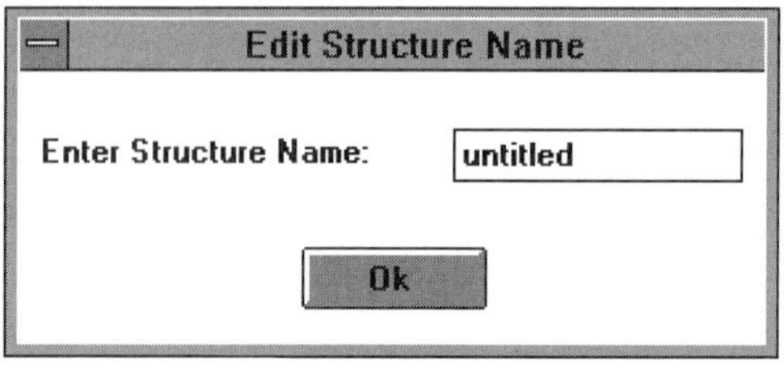

Figure B-7. The 'Edit Structure Name' Dialog Box.

TS_Bond Orders, i.e. *transition state bond orders*, allows the user to specify the bond order parameter for transition state bonds in a structure. This command is found under **Mark** in the menu bar, and its use is discussed in Section 6.9.

Tubular Bonds is one of the options for graphical presentation of structures found under the **View** command on the menu bar, and its operation is discussed in Section 3.6.

Update cleans up the work space and redraws structure(s) in the work space, putting into effect any additions, deletions, atom retypings or movements, or any other changes that have been carried out. This command is found on the tool bar.

View is a family of commands governing graphic representations. This command is found on the menu bar, and the **View** command pulldown menu is shown in Fig. 2-4. Activating this command pulls down a menu with twelve options: **Control_Panel, Labels, Mono/Stereo, Stick Figure, Ball_Stick, Ortep, Tubular Bonds, CPK_Surface, Dot_Surface, Ribbon, Compare**, and **Dihed Map**.

Volume computes the volume of the molecule in the work space. This command is found under **Analyze** on the menu bar, and its use is discussed in Section 6.2. See also **Surface Area**.

APPENDIX C
Atom Types in PCMODEL

TABLE C-1. Atom Types in PCMODEL

Type	Symbol	Description	Color	r_{vdw} (Å)	ε (kcal/mol)
@1	C	sp^3-Hybridized carbon	Cyan	1.900	0.044
@2$^{\pi,oop}$	C	sp^2-Hybridized carbon, C=C	Cyan	1.940	0.044
@3$^{\pi,oop}$	C	sp^2-Hybridized carbon, C=O	Cyan	1.940	0.044
@4$^\pi$	C	sp-Hybridized carbon	Cyan	1.940	0.044
@5	H	Hydrogen	Gray	1.500	0.047
@6$^{\pi,\cdot\cdot,oop}$	O	sp^3-Hybridized oxygen	Red	1.740	0.050
@7$^{\pi,\cdot\cdot}$	O	sp^2-Hybridized oxygen	Red	1.740	0.050
@8$^{\cdot\cdot}$	N	sp^3-Hybridized nitrogen	Dark blue	1.820	0.055
@9$^{\pi,oop}$	N	Amide/enamine $sp^{2.5}$-nitrogen	Dark blue	1.820	0.055
@10$^{\pi,\cdot\cdot}$	N	sp-Hybridized nitrogen	Dark blue	1.820	0.055
@11	F	Fluorine	Light blue	1.650	0.078
@12	Cl	Chlorine	Light green	2.030	0.240
@13	Br	Bromine	Purple	2.180	0.320
@14	I	Iodine	Gray	2.320	0.424
@15$^{\cdot\cdot}$	S	Divalent sulfide sulfur	Light yellow	2.110	0.202
@16	S+	Trivalent sulfonium sulfur	Light yellow	2.110	0.202
@17	S	Trivalent sulfoxide sulfur	Light yellow	2.110	0.202
@18	S	Tetravalent sulfone sulfur	Light yellow	2.110	0.202
@19	Si	Silane	Gray	2.250	0.140
@20	:	Lone pair	Dark blue	1.200	0.016
@21	H	Hydroxyl hydrogen	Gray	1.100	0.036
@22	C	Cyclopropane carbon	Cyan	1.900	0.044
@23	H	Amino hydrogen	Gray	1.125	0.034
@24	H	Carboxylic acid hydrogen	Gray	0.800	0.015
@25$^{\cdot\cdot}$	P	Trivalent phosphorus	Gray	2.050	0.157
@26^{oop}	B	Trigonal boron	Green	1.980	0.034
@27	B−	Tetrahedral boron	Green	1.980	0.034
@28	H	Enolic/phenolic hydrogen	Gray	0.900	0.015

TABLE C-1—Continued

Type	Symbol	Description	Color	r_{vdw} (Å)	ε (kcal/mol)
@29$^{\pi,oop}$	C ·	Carbon radical	Cyan	1.940	0.044
@30$^{\pi,oop}$	C+	Carbonium ion	Cyan	1.940	0.044
@31	Ge	Germanium	Gray	2.400	0.200
@32	Sn	Tin	Gray	2.550	0.270
@33	Pb	Lead	Gray	2.700	0.340
@34¨	Se	Selenium	Red	2.250	0.276
@35	Te	Tellurium	Gray	2.400	0.240
@36	D	Deuterium	Gray	1.497	0.047
@37$^{\pi,\cdots,oop}$	N	sp^2-Hybridized nitrogen	Dark blue	1.820	0.055
@38¨	S	Thione sulfur	Light yellow	2.110	0.202
@39	Se	Selenoxide selenium	Red	2.250	0.276
@40^{oop}	C	Aromatic carbon	Cyan	1.900	0.044
@41$^\pi$	N+	Ammonium nitrogen	Dark blue	1.820	0.055
@42$^{\pi,\cdots}$	O−	Oxygen anion	Red	2.200	0.050
@43‡	B#	Transition state boron	Green	1.980	0.034
@44	Al	Tricoordinate aluminum	Gray	1.900	0.050
@45‡	H*	Transition state hydrogen	Gray	1.500	0.047
@46$^{\pi,\cdots}$	O+	Oxonium ion	Red	1.740	0.050
@47	P	Pentavalent phosphorus	Gray	2.050	0.157
@48$^{\pi,\cdots,oop}$	C−	Carbanion	Cyan	1.940	0.044
@49‡	C*	Transition state carbon	Cyan	1.940	0.044
@50‡	C#	Transition state carbon	Cyan	1.940	0.044
@51‡	C$	Transition state trigonal carbon	Cyan	1.900	0.044
@52‡	C%	Transition state pentavalent carbon	Cyan	1.900	0.044
@53$^{\cdots,\ddagger}$	O#	Transition state oxygen	Red	1.740	0.050
@54‡	I%	Transition state iodine	Gray	2.320	0.424
@55$^{\cdots,\ddagger}$	N#	Transition state nitrogen	Dark blue	1.820	0.055
@56	C	Cyclobutane carbon	Cyan	1.900	0.044
@57		Metal	Gray	0.000	0.400
@58	Al	Tetracoordinate aluminum	Gray	1.900	0.030
@60	aq	Water pseudoatom		1.530	0.500
@61	C	Metal carbene	Cyan	1.940	0.044
@62	C	Metal carbyne	Cyan	1.940	0.044
@63	C	Metal carbonyl	Cyan	1.940	0.044
@70	H	Bridging hydrogen	Gray	1.500	0.047
@71	C	Bridging carbon	Cyan	1.900	0.044
@72	N	Bridging nitrogen	Dark blue	1.820	0.055

TABLE C-1—Continued

Type	Symbol	Description	Color	r_{vdw} (Å)	ε (kcal/mol)
@73	F	Bridging fluorine	Light blue	1.650	0.078
@74	Cl	Bridging chlorine	Light green	2.030	0.240
@75	Br	Bridging bromine	Purple	2.180	0.320
@76	I	Bridging iodine	Gray	2.320	0.420
@80^{oop}		Generalized metal	Gray	0.000	0.400

The information contained in this appendix is taken from the {MMXCONST.PAR} file, which contains the constants which PCMODEL uses in the MMX force field equations to compute the potential energy of a molecule or an ensemble of molecules (see Section 5.1). Updates contained in the {MMXCONST.PAR} which accompanies PCMODEL v. 6 for Windows (dated 5/16/96) are included.

Under "Type," the MMX atom type number is given preceeded by an @-symbol. This is the convention adopted for this book to distinguish atom type numbers from other numbers associated with the atoms. Under "Symbol" is given the atom label which will appear in molecules on screen, except for carbon, which is generally unlabeled. Under "Description" is given a brief description of the atom type (see Section 5.3). Under "Color" is given the color code assigned to that atom type (the color palette may vary from one hardware configuration to another). Under "r_{vdw}" is given the van der Waals radius. Under "ε" is given the hardness parameter for the van der Waals potential equation. This is a measure of the stiffness of the van der Waals repulsive component, with smaller values indicating harder atoms (see Section 5.1.6).

π-Atoms: The following 14 atoms can be designated as π-atoms for MMX calculations and thus be included in VESCF π-calculations; these atom types are flagged in this appendix with a $^\pi$ (see Section 6.5): @2, @3, @4, @6, @7, @9, @10, @29, @30, @37, @41, @42, @46, and @48. The MMX force field will accurately model most conjugated functional groups without running π-calculations, for example allyl anion, allyl cation, allyl radical, carboxylate anion, carboxylic acid, carboxylic amide, carboxylic ester, enamine, enol ether, nitro, and α,β-unsaturated carbonyl. For any conjugated functional group not listed here, or any of the above which are involved with further conjugation, a π-calculation should be carried out to model them accurately. An @6 oxygen atom used in a π-calculation will bear only one lone pair, because the other lone pair is included in the π-calculation.

Lone pairs: The following 15 atoms will bear one or two electron lone pair dummy atoms (@20), and are flagged in this Appendix with a ¨ (see Section 6.5): @6 (2), @7 (2), @8 (1), @9 (1), @10 (1), @15 (2), @25 (1), @34 (2), @37 (1), @38 (1), @42 (2), @46 (1), @48 (1), @53 (2), and @55 (1). When an @6 oxygen is attached to a double bond (carboxylic acid, phenol, enol ether, etc.), one lone pair is removed and specific parameters account for resonance effects, so it will bear only one lone pair. For the same reason, the @9 nitrogen bears no lone pair. The inclusion of lone electron pairs is necessary for accurate modeling of the first-row elements oxygen and nitrogen, and for the carbanion carbon. In addition, lone pairs are included on these atoms, and on the second-row elements sulfur and phosphorus, and on the third-row element selenium, to provide for accepting a hydrogen or deuterium bond, or to serve as a donor atom in metal atom coordination. The donor-atom-lone-pair "bond" dipoles are factored into the calculation of the attractive potential for hydrogen bonding and metal atom coordination.

Out-of-plane bending potential: For the following 11 atoms types, an out-of-plane angle bending potential component is added to the overall energy (see Section 5.1.3); these atom types are flagged in this Appendix with an ^{oop}: @2, @3, @6 (when included in a conjugated π-system), @9, @26, @29, @30, @37, @40, @48 (when included in a conjugated π-system), and metals @80, @81, etc.

For atom type @33 (lead), only the van der Waals parameters are resident—all others must be supplied by the user.

For atom type @35 (tellurium), only a limited parameter set is available (see Section 5.3). All

others must be supplied by the user.
The parameters available for @36 (deuterium) limit its usage to bonds with carbon ((@1-@4, @49, and @50).
Earlier versions of PCMODEL would assign the atom types @44, @56, and @57 to metal atoms. The @44 atom type is now (as of v. 6) used for tricoordinate aluminum, @56 is now used for cyclobutane carbon, and @57 is no longer explicitly used for a metal in current versions. The @80 is used for metals; as different metallic atoms are added to a structure or ensemble, the atom type number is incremented by 1, so that the second metal introduced is @81, the third is @82, etc. The treatment of metal complexes and organometallics is discussed in Section 6.7.
There are ten atom types which are wildcards and may be defined by the user: @59, @64, @65, @66, @67, @68, @69, @77, @78, and @79. Parameters must be supplied in order to do calculations on structures which incorporate these atoms.
The @60 is a pseudoatom or dummy atom which is a monatomic spherical approximation to a water solvent molecule and bears the atom label 'aq'. It may be employed by editing of the input file (see Section 4.6), but it is not currently implemented in the menus for building molecules within PCMODEL. The @60 atom should not be bonded to anything but itself. Its parameters are: $r_{vdW} = 1.53$ Å; $\varepsilon = 0.5$; @60—@60 equilibrium bond length 2.8 Å; charge 0.20; @60—@60—@60 equilibrium bond angle 80°.
There are nine transition state atom types indicated by a ‡, which are used for modeling transition state: @43, @45, and @49-@55. The use of transition state atoms is discussed in Section 6.9.

APPENDIX D
Glossary of Computer and Molecular Modeling Terms

> In combating definitions it is always one of the chief elementary principles to take by oneself a happy shot at a definition of the object before one, or to adopt some correctly expressed definition. For one is bound, with the model (as it were) before one's eyes, to discern both any shortcoming in any features that the definition ought to have, and also any superfluous addition, so that one is better supplied with lines of attack.
> —Aristotle, *Topics*, Book VII, Part 1 (ca. 350 B.C.)

2D an abbreviation for two-dimensional.

3D an abbreviation for three-dimensional.

3DSEARCH a program for searching in databases of atomic coordinates for specified 3D substructures, written by Venkataraghavan *et al.*[930] Searches are divided into two parts: a fast prescreen using an inverted key system, and a slower atom-by-atom geometric search.

δ in NMR spectroscopy, the symbol for the **chemical shift**.

ϕ a stereochemical descriptor in biomolecule conformational analysis. In peptides, it indicates the torsional angle between the nitrogen and the C_α of an amino acid unit (see Fig. 3-15).[240] In saccharides, it indicates the torsional angle between the anomeric carbon of the glycosyl donor and the glycosidic oxygen (see Fig. 3-20).[241,931]

ψ a stereochemical descriptor in biomolecule conformational analysis. In peptides, it indicates the torsional angle between the C_α and the carbonyl carbon of an amino acid unit (see Fig. 3-15).[240] In saccharides, it indicates the torsional angle between the glycosidic oxygen and the attached carbon of the glycosyl acceptor (see Fig. 3-20).[241,931]

ω a stereochemical descriptor in biomolecule conformational analysis. In peptides, it indicates the torsional angle of the peptide linkage, between the nitrogen of one amino acid unit and the carbonyl carbon of the next (see Fig. 3-15).[240]

***ab initio* method** See **molecular modeling**.

AccuModel a molecular modeling package distributed by MicroSimulations, and currently available for a variety of desktop platforms. AccuModel has a

2D → 3D conversion algorithm based on a fragment library, and the conversion can be carried out directly on imported 2D MDL ISIS {*.SKC} files. It employs the MM3 force field for energy minimizations with dynamic monitoring of computational process, and enables the analysis of geometric features (distances, angles, and torsions) in the resulting structures. The molecules may be visualized as wire-frame, ball-and-stick, or CPK models, and AccuModel enables side-by-side and red–green stereo.

Alchemy® a desktop molecular modeling system developed by Tripos, Inc. The current version is Alchemy® 2000, which offers several force field and semiempirical calculation methods and provides structure building, comparison, energy optimization, and display capabilities.

algorithm a set of rules for determining a result; a step-by-step procedure for solving a problem or accomplishing some end in a finite number of steps; a detailed sequence of actions to perform to accomplish some task. The word comes from the name of the ninth century Arab mathematician Al-Khuwarizmi. An algorithm frequently involves repetition of an operation, and technically must reach a result after a finite number of steps. This would seem to rule out **brute force** search methods for certain problems, yet a brute force search is also a valid algorithm.[932]

An *evolutionary algorithm* incorporates aspects of natural selection or survival of the fittest. It maintains a population of *individuals* (usually randomly generated at first), which evolve according to genetic operators (rules of selection, recombination, mutation, and survival). A shared *environment* determines the fitness or performance of each individual in the population. The fittest individuals are more likely to be selected for reproduction (retention or duplication), while recombination and mutation also modify those individuals, yielding potentially superior ones. Evolutionary algorithms are useful for optimization when other techniques such as gradient descent or direct, analytical discovery are not possible, such as when the fitness landscape is rugged, possessing many locally optimal solutions.[932]

The *genetic algorithm*, inspired by the mechanisms of genetics, has been applied to global optimization, especially combinatorial optimization problems. Genetic algorithms randomly build individuals (sets of solutions) from some encoded forms known as a *chromosomes* or *genome*. The genetic algorithm requires the specification of three operations on objects, each typically probabilistic, called *strings*, which may be real-valued vectors: (1) reproduction: combining strings in the population to create a new string, or *offspring*; (2) mutation: spontaneous alteration of characters in a string; and (3) crossover: combining strings to exchange values, creating new strings in their place. Thus, the genomes of individuals are combined or mutated to breed new individuals, influenced by an ongoing process of competition (natural selection) within populations.

Genetic algorithms are useful for multidimensional optimization problems in which the chromosome can encode the values for the different variables being optimized.[121,932,933]

all atom a **force field** representation of molecules in which all atoms, including the hydrogens attached to carbons, are treated explicitly. Contrasts with the **united atom** or **extended atom** representation, in which a combination of a carbon with its attached hydrogens is treated as the building block for a molecule.

AMBER acronym for *assisted model building with energy refinement*, AMBER is the name of a suite of programs for molecular modeling and molecular dynamics, and also refers to the force field used with these programs. AMBER was initially developed by P. K. Weiner and P. A. Kollman, with subsequent improvements (see Section 5.4.3). These force fields were designed primarily for proteins and nucleic acids.[419]

AMBER* is an adaptation of the Kollman AMBER force field, created by the Still group at Columbia University, for use in the modeling program MacroModel.[285,286]

AMBER/OPLS is an amalgamation of certain components of the Kollman AMBER force field (bond stretch, angle bend, torsional strain) with the nonbonding potential from OPLS (Coulomb electrostatics and Lennard–Jones sterics) to provide a force field formulation for modeling biomacromolecules.[435]

Angular Overlap Model (also **AOM**) a calculational method based on a simplified MO treatment of the metal–ligand interaction, and used for the analysis and prediction of electronic spectra in organometallic complexes. This definition is taken from Comba and Hambley.[882]

The AOM is based on the molecular symmetry of the complex, and involves parameters e_σ, e_π, and e_δ which have chemical significance in terms of σ-, π-, and δ-donor or -acceptor strength. These parameters are a function of the metal, the ligand donor atom, and the metal–ligand bond strength, and may be derived from the corresponding ligand field parameters D_q, D_s, and D_t. The model is based upon idea that the metal d-orbitals are perturbed by weak covalent interactions with ligand orbitals which have suitable symmetry, where the magnitude of the destabilization is proportional to the two-atom overlap integral. The energy change for a particular d-orbital is a function of the e-parameters and of the angular geometry.

annealing See **simulated annealing**.

anomer An example of an **epimer**, in this case a configurational isomer at the acetal (ketal) carbon of a sugar in its cyclic form. The word comes from the Greek $\alpha\nu\omega$ (upper) and $\mu\varepsilon\rho\sigma$ (part), and refers to C1 of the sugar, which in most cases correspond to the acetal carbon. In the **Fischer projection** of sugars, C1 is drawn at the top of the structure.

An *anomeric carbon* is the (hemi)acetal carbon of a sugar adjacent to the ring oxygen. The two configurational isomers at this position are referred to as anomers.

The *anomeric effect* is the preference for the axial conformer exhibited by six-membered ring heterocycles substituted with an electronegative group at the carbon vicinal to the heteroatom.[934] This enhanced stability for the axial anomer over the equatorial in a sugar or sugarlike molecule is counter to the prediction from classical conformational analysis of a greater stability for the equatorial anomer, and is rationalized on stereoelectronic grounds.

The two major conceptual approaches for explaining the anomeric effect are (1) the stabilizing effect of interactions between the lone pair electrons on the endocyclic oxygen (or other heteroatom) and the σ^* antibonding orbital of the bond connecting an electronegative axial substituent to the anomeric carbon, and (2) a destabilizing dipole–dipole repulsion between the two electronegative atoms and their associated electron pairs when that substituent is equatorial.[935,936]

anti a stereochemical term for the relationship between two **vicinal** substituents in a conformation characterized by a dihedral angle of approximately 180°. It is generally used to refer to the two largest or highest priority substituents attached to the two central atoms, and is usually an energy minimum in *conformation space* (See **conformation**). Anti is a special case of **staggered**. See **antiperiplanar**.

anticlinal a stereochemical term proposed by W. Klyne and V. Prelog[929] to describe the relative **conformation** of two **vicinal** substituents about a *dihedral angle* (See **torsion**). The two substituents A and B in the system A—X—Y—B bear an anticlinal relationship to each other if the dihedral angle between them is $+120° \pm 30°$ or $-120° \pm 30°$ (see Fig. D-1). See also **antiperiplanar, synclinal, synperiplanar**.

antiperiplanar a stereochemical term proposed by W. Klyne and V. Prelog[929] to describe the relative **conformation** of two **vicinal** substituents about a *dihedral angle* (See **torsion**). The two substituents A and B in the system A—X—Y—B bear an antiperiplanar relationship to each other if the dihedral angle between them is $+180° \pm 30°$ (see Fig. D-1). See also **anticlinal, synclinal, synperiplanar**.

AOM See **angular overlap model**.

append in the context of data on electronic media, to add new material to the existing data in a file or data field, as opposed to writing over the existing data.

aromatic a class of molecules which have the following characteristics: a planar, cyclic, conjugated π-electron system, where the number of π-electrons is equal to $4n + 2$ where $n = 1, 2, 3, \ldots$ (i.e. obey the *Hückel rule*—see **Hückel calculation**); unusual stability compared to similar but nonaromatic molecules; the tendency to undergo substitution rather than addition reactions with electrophiles; a diamagnetic ring current demonstrable through NMR or other physical methods. Benzene is a simple

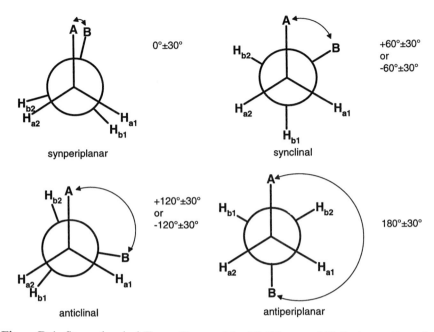

Figure D-1. Stereochemical Terms Proposed by W. Klyne and V. Prelog to Describe the Relative Conformation of Two **Vicinal** Substituents A and B about the Torsional Angle in the System $A-X-Y-B$.[929]

example of an aromatic molecule.[937]

Aromaticity in the molecular context is the quality or state of being aromatic. Every proposed definition of aromaticity has encountered difficulties, in addition to which there is a conflict between using the term to refer to purely structural features and using it to refer to chemical reactivity.[938]

ASCII acronym for *American Standard Code for Information Interchange*. *ASCII characters* are 30 control characters plus the 96 standard keyboard characters: alphabetic, numeric, and symbolic (punctuation and symbols). An *ASCII file* is one which contains only these 126 characters, and no higher level (127–255) characters. An ASCII file can be read by or transmitted through any software medium.

asymmetric center a structural feature which makes a molecule non-superimposable upon its mirror image. This structural feature is most commonly an atom with a valency of at least four, each of the valences of which binds a different substituent. For example, a tetravalent carbon bonded to four different groups is asymmetric. Some other structural features which constitute an asymmetric center include: a cumulated diene or allene, each of the termini of which bears two different groups; an otherwise symmetrical molecule with steric constraints which prevent conformational averaging (such as hindered biphenyls); and a molecule which can adopt one of two helical conformations which are not easily interconvertable (such as a helicene, or a peptide α-helix).

asymmetry See **chirality**.

BABEL a file format interconversion program developed by A. Shah and coworkers, and updated by P. Walters and M. Stahl.[650,939] The name derives from the babel fish in D. Adams's *Hitchhiker's Guide to the Galaxy*. BABEL converts the input format to a *universal molecule structure* (UMS), and then converts this to the desired output format. For more information on the formats handled by BABEL, see Section 4.5.

BABEL is also the name of an electronic glossary of computer-oriented abbreviations and acronyms.[940]

Badger's rule an empirically derived relationship between the bond stretching force constant and the bond length.[508,509] The original Badger equation had the form

$$k_{bnd} = 1.86(r_0 - d_{ij})^{-3}$$

where k_{bnd} is the bond stretching constant in dynes per centimeter, r_0 is the equilibrium bond length in angstroms, and d_{ij} is a constant for any pair of atoms in the ith and jth rows of the periodic table. Later workers[510,511] modified the equation to enlarge the scope of application. In rearranged form, it becomes

$$r_0 = d_{ij} + (a_{ij} - d_{ij})[(-1)^n k_{bnd/n}]^{-1/(n+1)} \quad \text{for} \quad n = 2, 3$$

where now d_{ij} can be considered a distance of closest approach, and in addition to the variables in the first equation, a_{ij} can be considered a standard bond length and $k_{bnd/n}$ is the force constant for the quadratic ($n = 2$) or cubic ($n = 3$) term of the stretching potential equation.

In the MMX force field, Badger's rule is used to generate a force constant for a bond between two transition state atoms given a particular partial bond order. The rule can also be used to approximate a force constant for pairs of atoms for which no parameters are resident.

basis set in the **force field** modeling context, a representative group of molecules chosen to provide **parameters** to be used in developing a force field. The basis set usually contains a variety of simple molecules for which physical properties are well known or can be easily studied.

In the context of **molecular orbital** calculations, the basis set is the collection of orbital functions which are combined and scaled in order to model molecules.

batch relating to a software device which enables a series of related or unrelated commands and operations to be carried out consecutively without intercession by the user. A *batch job* is a job or task being run in this way;

batch mode describes the state in the which the computer or PC is executing a batch job; a *batch file* or *batch program* is a program set up to execute a batch job.

BATCH is a PCMODEL command (see Appendix B).

beautify See **clean**.

bent bond a bond in which the path of maximum electron density does not coincide with a straight line between the two bonded atoms.[244]

betaine an overall neutral molecule which contains two opposite charges on nonadjacent atoms.

binary the standard medium of computer language, consisting of numbers in a base-two counting system. This is the simplest way of representing data electronically: to detect or recognize an on or off state, either the presence or absence of a charge or pulse.

The smallest unit in this language is the *bit*, and eight bits taken together constitute a *byte*. In the earlier versions of the **DOS** operating system, the eight-bit byte formed a *word* in the programming language, but in most newer generation PCs there are now 16-bit and 32-bit processors and higher.

Files in binary format cannot be read visually as easily as **ASCII** files can be, unless one is familiar with the binary language.

biophore a neologism meaning the structural and chemical features of a molecule which convey its biological activity, where this activity is defined broadly. The concept of biophore is similar to that of **pharmacophore**, but is not limited to medicinal or pharmacological activity. See **pharmacophore**.

bit a contraction of *binary digit*: something which can have the value 0 or 1. The bit is the smallest unit of information for a computer. See **binary**.

bitmap See **graphics**.

blocking in the molecular graphics context, visualizing *slices* or *blocks* of a large structure, in order to eliminate a confusing excess of detail. Similar to a *cutaway view* in macroscopic representations. Forming a molecular *cross section*. Synonymous with **Z-clipping**.

bond length a geometric measure of the distance between two atoms which are assumed to be bonded. There are many conventions for determining bond length experimentally or by calculation, which may give slightly different values: $r_0, r_\alpha, r_\alpha^0, r_a, r_e, r_g, r_s, r_v, r_z$.

bond multiplicity synonymous with **bond order**.

bond order the multiplicity of a bond, related to its strength. For ground state nonconjugated organic systems, the bond order may be single, double, or triple. Some excited state species and bond units within a conjugated system may be described as having fractional bond order. See **Badger's rule**.

Born–Oppenheimer approximation the assumption that the motions of electrons and those of nuclei may be considered separately. This approximation, which greatly simplifies the calculation of molecular structure, is named for the two physicists who formulated it, M. Born and R. Oppenheimer.[941]

In the context of molecular structure calculations, this means that the nuclei may be considered in an average field comprised of the electronic effects in aggregate. A complete theoretical treatment of the Born–Oppenheimer approximation may be found in electronic book form on the World Wide Web.[942]

browser See **web browser**.

brute force method See **grid search**.

button in the graphical user interface, may be a *mouse button*, a *screen button*, or a *radio button*.

A *mouse button* is a physical button on the mouse used to address or activate a point on the screen where the mouse **cursor** is positioned.

A *screen button* is a defined image or shape on the screen (e.g. a circle, square or rectangle) which may contain a word or short phrase, or be located adjacent to a word or phrase, which describes its function. The screen button is activated when it is clicked on with the **mouse**, and this has the effect of selecting an option or initiating some command or operation.

A *radio button* is one type of screen button, and appears as an empty circle with an associated label or name denoting one of several choices or options for program operation. The desired option is selected by clicking inside the circle with the mouse cursor, which occupies the circle with a black dot. Often, one of the several choices is preselected as a default. Radio buttons are used for selection between several mutually exclusive options in a program window or **dialog box**.

byte a quantity in the **binary** language consisting of eight **bits**. In earlier versions of the **DOS** operating system, one byte was the amount of memory needed to store one character, also called a *word*. More recently, the processors in desktop platforms use 16-, and increasingly 32-bit words. **RAM** is measured in **bytes** or multiples thereof, for example, 1 kilobyte (kB) = 1024 bytes; 1 megabyte (MB) = 1024 kB = 1,048,576 bytes. See **binary**.

CACAO an acronym for *Computer Aided Composition of Atomic Orbitals*, this program is part of a suite of electronic structure programs for the calculation and analysis of molecular orbitals, which includes an extended Hückel calculation program (SIMCON), a molecular orbital analysis program (MOAN) and a display program for atomic orbitals (CACAO). They were developed by D. M. Proserpio of the Università di Milano and C. Mealli of the CNR Firenze, Italy.[214]

CADD an acronym for *computer-aided drug design*. The use of computer molecular modeling to build, visualize, analyze and store models of complex molecular systems which help to interpret structure–activity relationships, followed by the use of those models to predict one or more novel molecular structures expected to exhibit the desired biological activity.[88] See **CAMD**.

CADPAC an acronym for *Cambridge Analytic Derivatives Package*: a quantum chemistry program designed for calculating molecular structures, vibrational frequencies and molecular properties. The program utilizes s, p, d, and f Gaussian basis sets, and there is an extensive basis set library. Density functional theory has been included in the latest edition of the program, and it is possible to calculate molecular energies and structures using local and nonlocal functionals for both closed-shell and open-shell systems. It was written by R. D. Amos, I. L. Alberts, J. S. Andrews, S. M. Colwell, N. C. Handy, D. Jayatilaka, P. J. Knowles, R. Kobayashi, N. Koga, K. E. Laidig, P. E. Maslen, C. W. Murray, J. E. Rice, J. Sanz, E. D. Simandiras, A. J. Stone, and M-D Su, and is distributed by R. D. Amos of the University of Cambridge.

calipers in the context of this book, a computer graphics term which describes the dotted indicator lines which appear in the graphic display to designate atoms upon which some operation is being carried out.

Cambridge Crystallographic Data Centre (CCDC) the organization based in Cambridge (U.K.) which manages the **Cambridge Structural Database** of structural coordinates derived from diffraction methods (X-ray and neutron) for molecules and molecular ensembles.

Cambridge Structural Database (CSD) a database containing crystal structure information for over 150,000 organic and organometallic compounds for crystal structures analyzed using X-ray or neutron diffraction techniques. For each crystallographic entry in the CSD there are three distinct types of information stored, categorized by their dimensionality:

The 1D information incorporates all of the bibliographic material for the entry and summarizes the structural and experimental information for the crystal structure. The text and numerical information includes the authors' names and the full journal reference, as well as the crystallographic cell dimensions and space group.

The 2D information is a conventional chemical diagram of the molecule. This is encoded as a chemical **connection table** comprising atom properties (element symbol, number of connected nonhydrogen atoms, number of connected hydrogen atoms, and net charge) and bond properties (seven different bond types can be specified).

The 3D information includes the atomic coordinates, the space group symmetry, the covalent radii, and the crystallographic connectivity established by using those radii, from which a 3D representation of the molecule can be generated. Each 3D representation is matched with the 2D chemical structure.

Cambridge Structural Database System (CSDS) currently a suite of three software programs developed at the **Cambridge Crystallographic Data Centre** for use in searching the **Cambridge Structural Database** and displaying and analyzing the found structures. The three program modules are **QUEST3D**, **PLUTO**, and **VISTA**. The functionality of an earlier module,

GSTAT, is now available within a combination of QUEST3D and VISTA.

CAMD   an abbreviation for *computer-aided molecular design*, or for *computer-assisted molecular design*. The use of computer molecular modeling to build, visualize, analyze, and store models of molecular systems which help to interpret structure–property, structure–reactivity, or structure–activity relationships, followed by the use of those models to predict one or more novel molecular structures expected to exhibit the desired property, reactivity, or activity. See Section 1.5.

CAMEO   an acronym for computer-assisted mechanistic evaluation of organic reactions, is a program to model reactivity and predict the products of organic reactions, developed by Jorgensen *et al.* at Yale University.[265] Chemical transformations are processed by applying mechanistic rules to 3D structural representations of the reactants. These 3D structures are generated from 2D input with the STRFIT program of Saunders.

Catalyst   a program by Molecular Simulations, Inc., which analyzes numerous active analogs and their conformations and formulates a hypothesis about the **pharmacophore**, or the 3D structure and properties of the corresponding active site. It uses an appended-style structure input file format to store the generated conformations, which use the filename extension TPL.

CAVE   an acronym for *Cave Automatic Virtual Environment*: a virtual reality theater for molecular modeling developed through a group collaboration.[97] It is part of the Virtual Biomolecular Environment (see **VIBE**).

CAVEAT   an acronym for *computer assisted vector evaluation and target design*, a computer program for generating novel structures which conform to a given **pharmacophore** model.[943,944] The innovative feature of this program is that it analyzes molecular structure in terms of the relationships between bond vectors of important functional groups or skeletal fragments. Many other programs with the same goal use point-to-point distance geometry for structural analysis.

A given, user-selected set of bonds in a target molecule are treated as vectors, and the relationship between all the pairs is calculated and this information is coded into a *vector index* in which the molecule is characterized by bond relationships instead of atomic coordinates and connectivities. This index may be used as a fingerprint to search over an appropriately coded database for similar molecules that have in common a particular relationship among bonds. A CAVEAT search may be used in any sort of similarity comparison experiment, such as structure homology of proteins.

CAVEAT was developed by G. Lauri, G. Shea, S. Waterman, S. Telfer, and P. A. Bartlett of the University of California, Berkeley. Licenses for CAVEAT and related programs are available from the UCB Office of Technology Licensing and from Molecular Simulations, Inc.

CCDC   an abbreviation for **Cambridge Crystallographic Data Centre**.

centroid   the hypothetical center of a cyclic molecule, or of some other grouping of atoms, calculated as the average position of the atoms making up the ring(s) in a cyclic molecule. The centroid is a convenient geometric marker for the position of the ring, for determining relative distances to other parts of the molecule. An example is the distance between a side chain nitrogen atom and the centroid of the benzene ring in a series of phenylalkylamine analogs.

CFF an abbreviation for *consistent force field*, an empirical force field originally developed by S. Lifson, A. Warshel and coworkers, of which there are a number of extant versions today (see Section 5.4.2).

CGI an abbreviation for *common gateway interface*, a computing term that describes a set of programming techniques which facilitate text entry by a user into a computer form or page within the context of a server–client system, especially the **World Wide Web**.

CGM an abbreviation for **Computer Graphics Metafile**.

CHARMM an acronym for *chemistry at Harvard Macromolecular Mechanics*, a molecular modeling program and attendant force field developed by M. Karplus and coworkers, predominantly for carrying out structure calculations on biomacromolecules (see Section 5.4.4).[439] More recently, CHARMM has come to indicate the program and force field available from the academic source, and CHARMm refers to the commercial product available through MSI.

CHEAT an acronym from *carbohydrate hydroxyl groups represented by extended atoms*, a specialized set of parameters for the **CHARMm** force field to aid in modeling carbohydrates. CHEAT employs a distance-dependent dielectric, and has no cutoff for nonbonded interactions. CHEAT was developed by P. D. J Grootenhuis, C. A. G. Haasnoot, and M. L. C. E. Kouwijzer, and there are CHEAT93[447] and CHEAT95[448,449] versions; the latter is distributed through MSI.

Chem123 a file format interconversion program distributed by Micro-Simulations, and currently available for Windows 95, and Windows NT. Chem123 interconverts **SMILES** strings and 2D MDL MOL files. It can also carry out 2D → 3D conversions of MOL files based upon fragment library, and refines the 3D coordinates with an MM3 force field calculation. This is a batch conversion option which accepts MDL SDFiles as input. Chem123 can simultaneously display the SMILES string with the 2D and 3D representations.

Chem3D a program developed by CambridgeSoft Corp. for molecular modeling and graphic display on desktop and workstation platforms. Chem3D can visualize structures of > 5000 atoms in space filling (CPK), cylindrical bonds, ball-and-stick, and wire frame representations. It performs real-time rotation and animation, and calculates MM2 energy minimization, partial charge distributions, and molecular dynamics. The term also refers to the native structure file format for this modeling program.

ChemDraw   a program for preparing graphic images of chemical structures and reaction schemes for the Macintosh platform. These graphic image files are binary in nature, but ChemDraw can also prepare ASCII plain text connection tables which contain the atom and bonding information from which structure files for the molecule can be assembled.

Chemeleon   is a file format conversion program produced by ExoGraphics. It functions in the Windows environment as a **TSR**, and will carry out file format conversion "on the fly" when structure files are copied to the clipboard. Chemeleon will automatically detect the input format by recognizing the program within which the file originates, and it will likewise recognize the proper file format for the target application. See Section 4.5.

chemical shift   the term used in NMR spectroscopy to describe how the same magnetic nucleus (e.g. ^{1}H or ^{13}C) in different chemical environments will resonate at different radio frequencies in the same applied magnetic field. It is measured in parts per million (ppm) of the applied magnetic field, and is thus unitless. Expressed relative to some standard, it is often symbolized by δ.

Differences in the local chemical environment of the nucleus will cause magnetic anisotropy; the applied magnetic field is perturbed, enhanced, or attenuated enough to shift apart the absorptions of the nuclei in different molecules or in different parts of the same molecule. The chemical shift is the reason that NMR spectroscopy is such a powerful tool for structure elucidation.

child window   a secondary window spawned by, and partially dependent upon, the main or *parent* window. The child window differs from a **dialog box** or a *menu window* in several ways. A child window may appear partially or wholly outside the parent window, unless the parent window occupies the full screen. A child window will govern or monitor a subprocess (child program) of the main or parent program. Iconizing (or minimizing) the parent window will not necessarily affect the child window unless it is wholly bounded by the parent window. However, if the parent window is closed or canceled, the child window will likewise close. (This is a working interpretation, which differs somewhat from the official Microsoft definition.)

One common example is the 'Print' child window which controls printing jobs in Windows applications. An example within PCMODEL is the 'Output' child window which appears to the right of the PCMODEL main program window during an energy calculation or an energy minimization.

Chime™   a World Wide Web browser plug-in for visualizing molecular structures, developed by MDL, currently in the **DLL** format. The name is a contraction of *chemical MIME*.

Chime contains the **RASMol** 3D rendering code, and thus is intended for enabling the visualization and manipulation of protein, nucleic acid, and small-molecule structures which are embedded in HTML documents or tables available on the World Wide Web. It can also be used to visualize structures stored on the local platform by pointing the browser at these files. It supports multipart XYZ format files which are visualized as animated images.

Chime's file format input/output capabilities are discussed in Section 4.5.

chirality a property of structural asymmetry in a molecule which can exist as two nonsuperimposable structures which are mirror images of each other. The property of nonidentity of a molecule with its mirror image.[240]

CHUCKLES a program developed by M. A. Siani *et al.* which can convert between the monomer representations in macromolecules containing peptides and peptoids, and the corresponding **SMILES** code strings, to aid in sequence searching.[945]

CICADA acronym for *channels in conformational space analyzed by driver approach*, a program for performing conformational analysis on oligosaccharides developed by J. Koca.[946,947]

CIP an abbreviation for Cahn, Ingold, and Prelog, who developed sequence or priority rules for the unambiguous determination of **configuration**.[948,949] This system was later amended and expanded by the IUPAC Commission on the Nomenclature of Organic Chemistry.[240] In the CIP convention, the attached atoms, or the bonded atoms of the attached groups, are compared on the basis of their atomic number—the higher atomic number has higher priority. For example, this convention is used to assign the *E* or *Z* configuration to a substituted alkene, the *trans* or *cis* relative configuration to a substituted cyclic compound, or the *R* or *S* configuration to a chiral center.

cis the descriptor applied to stereoisomers in which the substituents (atoms or groups) lie on the same side of a reference plane identifiable as common among the stereoisomers.[240] This descriptor is best reserved for application to the relative stereochemistry of two substituents on a cyclic compound, and its proper application requires that the ring be drawn in its most extended form, i.e. with no reentrant angles. See *trans*.

Class I (or Class 1) force field See **force field**.

Class II (or Class 2) force field See **force field**.

Class III (or Class 3) force field See **force field**.

clean in the molecular graphics context, to make alterations in the starting geometry of the depiction (either 2D or 3D) of a molecular structure so that it conforms with standard rules of molecular configuration, geometry, and valence.

Performing this operation will reduce the amount of time necessary to carry out an energy minimization since it fixes the improper bond lengths, angles, and torsional angles. This is generally done by comparison with a reference library or database of values for these quantities for a specific collection of atoms. Cleaning is generally not available in PC-based modeling programs because of the difficulties of handling such a large database on the smaller platform.

In the Tripos software, the **CONCORD®** utility can be used to clean; in the CAChe® modeling environment cleaning is done with the **beautify** command, according to a proprietary algorithm. **Clean** is also a command in the MDL program ISISDraw®, which can take a poorly drawn 2D structure and improve its appearance by similarly applying a set of rules from proper bond angles and lengths.

clipboard a computer term describing the temporary storage area for a selection of text and/or graphics imported from a program window or file using the **copy** or **cut** procedures. This imported text and/or graphics selection may be transplanted to a new location within the program window or file, or into a new program window or file, using the **paste** procedure.

CLogP an acronym for *calculated* log*P*, a program developed by C. Hantsch of Pomona College to calculate a surrogate for the logarithm of the *n*-octanol–water partition coefficient (a measure of a molecule's **lipophilic** character), based on a library of empirically determined contributions for various structural fragments.

The program will parse a structure input file (usually as a **SMILES** code string), analyze it in terms of the structural fragments in the library, and sum the contributions to generate a CLogP value.

COBRA a program to search the *conformation space* of a molecule and identify a group of low-energy *conformations*, developed by Leach of the University of Califfornia, San Francisco.[264] *COBRA* will **parse** a 2D representation of the molecule and select appropriate conformational units from a library. The energy of the 3D structure is evaluated using a variant of the **COSMIC** force field, after which *COBRA* methodically joins the conformational units in various ways. The generated *conformations* are organized in a tree structure for analysis. *COBRA* is available commercially through Oxford Molecular.

computer-aided molecular desgn   See **CAMD**.

computer-assisted molecular design   See **CAMD**.

computer graphics   See **graphics**.

Computer Graphics Metafile (CGM) a standard file format for storage and communication of graphical information, widely used on personal computers and accepted by desktop publishing and technical illustration systems.[932]

CONCERTS an acronym for *creation of novel compounds by evaluation of residues at target sites*, a program for *de novo* molecular design developed by D. A. Pearlman and M. A. Murcko.[173]

CONCORD® a program which generates a reasonable three-dimensional structure for a molecule starting from a one- or two-dimensional representation. The program was written by Dr. Robert Pearlman and coworkers at the University of Texas at Austin, and is distributed by Tripos.

The program, as implemented by Tripos, accepts SMILES code or MDL MOL files (or equivalent) as input. The structure input is parsed and analyzed, and two algorithms are used to assemble the 3D structure: an expert system for the acyclic parts and a simplified force field treatment for the cyclic parts. The acyclic bond lengths, bond angles, and torsion angles are assigned approximate values by comparison with a reference library or database of values, which is indexed on specified groups of atoms. For the cyclic portions of the structure (if any), a combined angle–torsional-angle potential function is employed used to carry out a rapid minimization.

The standard CONCORD is parametrized for up to tetravalent compounds containing H, C, N, O, F, Si, P, S, Cl, and I atoms, including charged species. It will accept SMILES stereochemical descriptors, and can be run in batch mode to convert compound databases. CONCORD is a standard utility in Tripos modeling packages such as SYBYL and **Alchemy® 2000**.

Concordize is a neologism meaning to generate a 3D structure from a 2D representation by making comparison with a reference library of standard molecular geometric values—more specifically, to employ the CONCORD program to carry out this conversion. An example of this type of database conversion has been published.[950]

configuration a molecular structure term, meaning the arrangement of atoms or groups of atoms about a stereochemical center in a molecule. Usually one of two possible arrangements: e.g. *R* or *S* for a three-dimensional asymmetric center such as a chiral carbon or an allene, *E* or *Z* for a two-dimensional asymmetric center such as an alkene. Assignment of such configuration markers is made on the basis of a convention such as the **CIP** sequence rules. In this sense, configuration is closely related to the **stereochemistry** of a center. Bond breaking is necessary to interchange configurations.
IUPAC;

The configuration of a molecule of defined constitution is the arrangement of its atoms in space without regard to arrangements that differ only as after rotation about one or more single bonds.[240]

Configuration can also refer to the arrangement of electrons in the allowable atomic or molecular orbitals (electron configuration).

conformation an arrangement of the parts of a molecule in three-dimensional space, arrived at by rotating single bonds, which may be accompanied by the stretching or compressing of bonds or of two-bond (dihedral) angles. Bond breaking is not necessary to interchange conformations. Conformation may be described by key words or symbols such as **gauche** or **anti**, or in terms of coordinates, either internal or Cartesian.

The first use of the term conformation in a stereochemical sense is attributed to the British chemist W. Haworth,[951] who employed it in 1929 as a descriptor for the shapes of sugar molecules. Some other definitions are:

... the conformations of a molecule of defined configuration are the various arrangements of its atoms in space that differ only as after rotation about single bonds.[240]

... the non-identical arrangements of the atoms in a molecule obtained by rotation about one or more single bonds.[952]

The word conformation is used to denote differing strainless arrangements in space of a set of bonded atoms. In accordance with the tenets of classical stereochemistry, these arrangements represent only one molecular species.[39]

Conformation space is an $n + 1$-dimensional representation or model of the allowed conformations of a particular molecule, where n is the number of degrees of rotational freedom in the molecule, and the additional dimension is the potential energy associated with each of the conformations. The length of time needed to analyze conformation space rigorously, i.e. by a **grid search** (**brute force** method), increases exponentially with n. A number of alternative, more rapid searching methods have been developed to analyze conformation space without searching it exhaustively. These methods rely upon **stochastic** sampling techniques to provide a model of the entire space from analysis of a representative portion of it.

Conformational analysis is an analysis of the physical and chemical properties of a compound in terms of the conformation (or conformations) of the pertinent ground states, transition states, and (in the case of spectra) excited states.[952]

A *conformer* is a conformational isomer: a molecule in a conformation into which its atoms return spontaneously after small displacements.[240] "Conformer" is synonymous with **rotamer**. The first use of the term conformer as shorthand for conformational isomer is attributed to the American chemist P. Hay, who employed it in his PhD dissertation at Ohio State University in 1952.[953]

conjugate gradient method   a mathematical method first applied to molecular mechanics strain energy minimization by Engler, Andose, and Schleyer.[7] It is related to the **steepest descent method**, but is designed to point the system more rapidly toward and energy minimum. It is discussed in the context of other minimization engines in Section 5.7.3.

conjugation   a molecular structure term meaning the effect in which the π-orbital of a double bond is linked to, or forms a continuous system with, an un-, singly, or doubly occupied p- or p-hybrid orbital on an adjacent atom. Such a system is said to be *conjugated*, and the π-electrons in the system are said to be **delocalized**, or spread throughout that system.

This linkage confers an energy stabilization on the molecule, and the components of the conjugated system will react as a whole rather than as discrete units. The structures of conjugated molecules are characterized by alternating multiple and single bonds and/or a multiple bond separated by a single bond from a lone pair donor atom, carbocation, carbon radical, or carbanion.

connection table   a portion of a molecular structure input file, a logical description of the structure and bonding in the molecule. In the strictest sense, it is a table describing the atoms involved (i.e. atomic number, hybridization, etc.), and which are connected or bonded to which other atoms. As such, it is one of the four main parts of a molecular structure input file (see Chapter 4). In this sense, the connection table gives the **constitution** of the molecule.

MDL Information Systems, Inc., use the term "Ctab" or "connection table" more broadly to refer to the main portion of their MOL format structure input file, minus the header. In this sense,

a connection table (Ctab) contains information describing the structural relationships and properties of a collection of atoms. The atoms may be wholly or partially connected by bonds. Such collections may, for example, describe molecules, molecular fragments, substructures, substituent groups, polymers, alloys, formulations, mixtures, and unconnected atoms. The connection table is fundamental to all of MDL's file formats.[274]

Thus, the Ctab incorporates three of the four main parts of a molecular structure input file described in Chapter 4: Cartesian coordinates, connection table, and property values and flags.

Connolly surface   See **molecular surface**.
Consistent Force Field   See **CFF**.
constitution   an aspect of molecular structure, given by naming all of the atoms involved and stipulating which are connected or bonded to which other atoms and what is the bond multiplicity in each case. In this sense, it is roughly equivalent to the **connection table** in molecular modeling input files.
ConSystant   a file format conversion program produced by ExoGraphics. It can operate in either a DOS or a Windows environment. ConSystant may be used to convert specific files from one file format to another, or it may be used to carry out a file format conversion on a structure file residing on the Windows clipboard. The input and output formats are selected from a menu, and are discussed in Section 4.5.
coordination bond   in the narrowest interpretation, a bond between two partners (atoms or molecules) in which one partner donates two electrons, and the other accepts them into an unoccupied orbital. This is very close to the definition of a Lewis-acid–Lewis-base complex.

While it is not universally accepted as correct to do so, many (especially organic) chemists draw a distinction between a **covalent bond**, such as is found in the nonionic compounds of the Main Group elements, and a *coordination bond*, such as is found between transition or other metals and their ligands, including solvent molecules. Within the context of organometallic compounds in PCMODEL, this distinction is made solely in order to determine which bonding algorithm will be used. See Section 6.7.

A metal center is said to be *coordinatively saturated* when all of the available coordination sites (a function of the geometry and electronic structure of the metal) are occupied by a **ligand**. A metal center is said to be *coordinatively unsaturated* if there is at least one coordination site which is not occupied by a ligand. Coordinatively unsaturated complexes are far more labile, since the open coordination serves as a site of attack by other ligands or reactive species.

copy a computer term with several meanings. In the **graphical user interface** context, it describes the capture of a selection of text and/or graphics from a program window or file, and its placement on the **clipboard**. From there, it is available to **paste** into other locations. With copy, the text or graphics selection remains active in the original location, in contrast to **cut**.

Copying can also mean the creation of a duplicate of an electronic file.

CORINA an acronym for *coordinates*, a program developed by the Computational Chemistry group of at Universität Erlangen to convert chemical structures from a 2D format into a 3D format (see Section 3.4.4). The CORINA algorithm utilizes ring patterns and a pseudo-force-field optimization for this conversion,[140,263] and is applicable to a wide variety of structural types, including polycyclic, macrocyclic, and strained cage compounds; radicals and reactive intermediates; and metal complexes or organometallic systems. Atoms with up to six neighbors are allowed.

COSMIC an acronym for *computation and structure manipulation in chemistry*, a molecular modeling and graphical display software package developed by Vinter *et al.*[271]

Coulomb potential a measure of the energy of the nonbonding electrostatic interaction between two charges. In molecular modeling where the electrostatic potential employs a charge–charge model, the following electrostatic term is included in the energy calculation:

$$V_{\text{elec}/qq} = \frac{(q_I q_J)}{D r_{IJ}}$$

where q_I, q_J are the charges on atoms I and J, D is the **dielectric constant** of the medium, and r_{IJ} is the distance between the two charges (see Fig. 5-21). This is the basis of the QQ term in MMX. See also **Jeans's formula**.

coupling constant a numerical constant obtained from a nuclear magnetic resonance spectroscopy experiment, measured in hertz (Hz). The coupling constant measures the magnitude of the spin–spin splitting of an NMR signal due to the coupling of the spin of the nucleus being observed with the spin of a nearby nucleus, of either the same element (homonuclear coupling) or another magnetically active element (heteronuclear coupling). The coupling of the spins takes place only through bonds, i.e. is mediated by coupling with the electronic spins, and is attenuated rapidly with more intervening bonds. Also known as *J-value*.

covalent bond a molecular structure term, meaning a bond between two atoms in which each donates one electron to be shared in a two-electron bond between them. In terms of simplified molecular orbital theory, a covalent molecular bonding orbital is formed by combining one atomic orbital each from the two involved atoms. A covalent bond will generally be stronger than an **ionic bond** or a **coordination bond**, and is most often not highly polarized. Some bonds are of intermediate type, and are said to have partial covalent character.

CPK models™ abbreviation for *Corey–Pauling–Koltun* models, a set of space-filling molecular models which accurately depict van der Waals radii, bond lengths, and bond angles. They are best for visualizing nonbonded interactions and surfaces. The CPK models have become the industry standard of this type.

CRYSTAL a program for calculating the wave function and properties of bulk, crystalline, or periodic materials using a Hartree–Fock linear-combination-of-atomic-orbitals (HF-LCAO) approximation. The program was developed jointly by N. M. Harrison of the Daresbury Laboratory (U.K.) and R. Dovesi of the University of Torino (Italy).

CSD an abbreviation for **Cambridge Structural Database**.

CSDS an abbreviation for **Cambridge Structural Database System**.

Ctab an acronym used by MDL for **connection table**.

cubic stretch catastrophe the effect in force field modeling wherein a bond stretching potential which consists of a Taylor expansion out to the cubic term will provide a good approximation of the Morse curve in the vicinity of the equilibrium bond length, but will invert at greater lengths and cause the constituent atoms to fly apart. This effect has been known since the earliest days of force field modeling, but the phrase seems to have been coined recently by Halgren.[321]

This catastrophe is avoided in older force fields by withholding the cubic term until the bond is in the vicinity of its equilibrium length, and in newer force fields by adding quartic or higher terms, or by employing a different potential form (see Chapter 5).

Curtin–Hammett principle the conclusion by the American chemists D. Y. Curtin and L. P. Hammett regarding the course of thermodynamically controlled reactions where there are two or more different conformations of a reactant or intermediate, or two or more different reaction intermediates formed from the reactant, each of which leads to a different product. The ratio of products obtained does not depend upon the ratio of the corresponding conformations or intermediates which react to form those products, but will depend only upon the relative energies of the two (or more) transition states leading to each product.

This is most simply illustrated with the hypothetical reaction of a molecule which exists as two interconverting conformers, A and B, which are separated by a small free energy difference, and have a small activation energy for the interconversion. The conformers A and B will each react to form distinct final products, C and D respectively. Each of these conversions has an energy of activation which is large compared to that of the conformational interconversion, that is $\Delta H^{\ddagger}_{A \to C} \gg \Delta H^{\ddagger}_{A \leftrightarrow B}$ and $\Delta H^{\ddagger}_{B \to D} \gg \Delta H^{\ddagger}_{A \leftrightarrow B}$, so that the ratio of products will be determined only by the difference between the two activation energies for the conversion to product. The product ratio is thus independent of the relative populations of the intermediate conformers, for which the rate of interconversion will be far more rapid than the two rate-determining steps for product formation.

The principle may also be stated as follows:

... if two or more isomeric forms of a compound which are in rapid equilibrium (such as two different conformations of a compound) undergo a reaction in which each isomeric form gives rise to its own characteristic product, the ratio of the products so formed is independent of the relative energy levels of the various starting forms and depends only on the relative energy of the transition states by which the products are formed—provided that the activation energy for product formation is large compared to the activation energy for the interconversion of the two isomeric starting materials.[952]

... although one conformation of a molecule is more stable than other possible conformations, this does not mean that the molecule is compelled to react as if it were in this conformation or that it is rigidly fixed in any way.[39]

cut a computer term describing the capture of a selection of text and/or graphics from a program window or file, and its placement on the **clipboard**. From there, it is available to **paste** into other locations. With a cut, the text or graphics selection is no longer active in the original location, in contrast to **copy**.

default a computer programming term meaning the answer to a query which is assumed by an interactive computer program if none other is supplied by the user. For example: 'Enter # of years experience with molecular modeling (Default = 1):_.' A carriage return here with no other keyboard entry would enter the value of 1 year of experience, which is the default answer to this query.

deformation density the difference between the electron density in a compound or substructure, determined empirically or by calculation, and that represented by a set of spherically symmetrical atomic **electron clouds** which are centered at the corresponding nuclear positions.[244] The deformation density indicates how the electron density shifts as a result of forming bonds.

delocalization   a term from molecular structure referring to the randomization of the positions of electrons in a compound which exhibits **conjugation**. The electrons are no longer localized between any two bonded atoms, but occupy the whole of the conjugated system. The enhanced stability due to increased entropy is what drives delocalization.

DGEOM   an implementation of **distance geometry** developed by Blaney, Crippen, Dearing, and Dixon, and available as QCPE #590.[593] DGEOM is a distance geometry program for molecular model building, receptor modeling, conformational analysis, and solution structure determination from 2D NMR data using algorithms developed by Crippen, Havel, and Kuntz. Distance geometry is a general method for converting a set of $N^2 - N$ distance bounds into a set of $3N$ Cartesian coordinates for configurations consistent with these bounds.

Molecular structures can be described by the set of all pairs of interatomic distances set up as an N by N matrix, along with connectivity information. The upper triangular matrix contains the upper bounds (maximum distance) and the lower triangular matrix contains the lower bounds (minimum distance) allowed between each pair of atoms. Generally, the upper bound is determined as the length of a fully extended chain connecting the two atoms, while the lower bound is set to the sum of the van der Waals radii, or to the hydrogen bonding distance for polar interactions. See **distance geometry**.

diagonal force field   See **force field**.

dialog box   a small window which appears on the screen in a computer application, allowing the user to adjust some operating parameters or to input some data or values which will be stored or used by the current program.

A dialog box may contain one or more of the following features: command **buttons** labeled 'OK', 'Cancel', or 'Help'; text input boxes, giving a rectangular box which will accept text entry from the keyboard; list boxes, giving a list of selectable items, often with scroll bars to access more choices than will fit in the list box space (examples being printer selection, or directory and file selection); **radio buttons** for selecting program options or parameters; and check boxes for selecting one or more options or parameters simultaneously.

diastereomer or **diastereoisomer**   one of a pair of stereoisomers whose members are not mirror image isomers and are not superimposable upon each other. The most general usage is for molecules which contain at least two **asymmetric centers**. Diastereomers may have different chemical and physical properties.

dielectric constant   a macroscopic property of a substance which measures its ability to increase the capacitance of a capacitor relative to a vacuum. The characteristic dielectric constant ε is unitless (see Table A-4 in Appendix A for representative values). The dielectric constant appears in the denominator of the **Coulomb potential** expression. In the discussions of force field potentials in this book, D is used to symbolize the dielectric constant to avoid confusion with the hardness parameter ε.

The electrostatic potential in the MMX force field uses a default dielectric constant of 1.5 to mimic a hydrocarbon environment rather than a vacuum. The dielectric constant of a medium is a function of the permanent dipole moments of the constituent molecules, their **polarizability**, and the temperature.

dihedral angle See *torsional angle* under **torsion**.

dihedral driver calculation a computational operation in conformational analysis in which an energy minimization of a molecule is calculated iteratively during rotation through one or more of its torsional angles at specified intervals. This operation, also known as driving the angle, was first published by Wiberg and Boyd.[82] The resulting data can be graphed (energy vs. angle) to demonstrate a *torsional barrier* (see **torsion**), with the appropriate caveats (see Section 6.3).

dipole a vector defined by two opposite charges. In molecular context, dipoles may be permanent or temporary. The magnitude of a dipole is measured by its *dipole moment*, μ, in debyes (for the Dutch physicist P. J. W. Debye), and is given by the following equation:

$$\mu = er \times 10^{18}$$

where e is the electric charge in electrostatic units (esu), and r is the distance between the charges in centimeters.

A permanent dipole results from a separation of charge, which is a function of the relative electronegativity of the constituent atoms and how they are bonded together. For molecules of more than two atoms, the net dipole moment is found from the vector sum of the constituent bond dipoles.

For temporary dipoles, see the discussion of **van der Waals forces**.

In molecular modeling, the interactions between bond dipoles may be used to model the **electrostatic** interactions within molecules by **Jeans's Formula** (the DD option in MMX).

Directed Tweak a three-dimensional structure searching technique designed by T. Hurst to facilitate the discovery of matches between a 3D structure query and accessible conformations of flexible molecules contained in a structure database.[620]

In Directed Tweak, the rotatable bonds of a conformationally flexible structure are adjusted at search time with a torsional space minimizer, to produce a conformation which most closely matches the 3D structure query. The tolerances of the query, and an assessment of conformational energy based upon the sum of van der Waals interactions, are used to score search hits.

dispersion forces See **van der Waals forces**.

Distance Geometry a method which uses the N by N internuclear distance matrix for the set of N atoms comprising a molecule to generate sets of Cartesian coordinates which satisfy the distance constraints. The constraints may be either specific or generic. The upper right triangle of the matrix is used for the upper bounds, and the lower left triangle of the matrix is used for the lower bounds.

Specific constraints may be derived from experimental data such as NMR NOE studies, or from the requirements of a hypothetical biophore. For the generic constraints, the upper bounds will be equal to the length of a fully extended chain connecting the two atoms, and the lower bound will be set to the sum of the van der Waals radii, or the hydrogen bonding distance for polar interactions.

A set of intramolecular distances is then chosen at random, and eigenvalue techniques are used to find the best three-dimensional fit to those distances. Energy minimization of the molecule, either alone or interacting with a second molecule, is then employed to rank the found conformation.[264,619,954,955] See **DGEOM**.

DLL a computer term, an abbreviation for *dynamic link library*. It is used as a filename extension, so that files of this type follow the format {*.DLL}. Such files contain one or more functions which may be invoked, or *called*, by an operating program or module. When called, the function(s) will load dynamically in order to support a particular program operation, for example a graphics, text, or numerical function. Using DLL files gives programmers more flexibility and can simplify the logistics of the application.

A DLL file is an executable file that provides code, data, or routines to applications running in the Microsoft Windows® operating system, allowing them to perform particular tasks, such as opening a new window, linking files, etc. Some DLL files carry an EXE filename extension.[956] For example, {PCWIN.EXE} requires access to the DLL-type files {GDI.EXE}, {USER.EXE}, and {KRNL386.EXE} in order to run.

dock See **docking**.

Dock a suite of programs used to produce a negative image or cast of the binding pockets on a biomacromolecule, developed by I. D. Kuntz of UCSF.[667-669] This cast consists of a set of overlapping spheres which complement the molecular surface of the binding site. This image is matched to positive images of putative ligands having a known 3D structure. Molecules which fit the receptor are scored and prioritized, and stored in **PDB** format. The program has been applied to protein–small-molecule docking, protein–protein docking, and assaying large groups of small-molecule ligands against a protein binding site (or **pharmacophore**) in order to propose new actives for pharmaceutical development.

Dock starts with the crystal coordinates of target receptor, and then generates molecular surface for the active site using the *Connolly surface* algorithm (see **molecular surface**). The next step is to use the shapes of cavities in the receptor surface to define spheroids which are generated to fill the active site. The centers of these spheres become potential locations for ligand atoms, and the next step is to match these sphere centers to the ligand atoms, to determine possible orientations for the ligand. Typically, tens of thousands of orientations are generated for each ligand molecule. Each oriented molecule is then scored for fit by one of three schemes: (1) shape scoring, which uses a loose approximation to the Lennard–Jones potential; (2) electrostatic scoring, which uses the program DELPHI to calculate electrostatic potential; or (3) force field scoring, which uses the AMBER potential (see Section 5.4.3). The top-scoring orientation for each molecule is then saved and compared with the scores of other molecules, and the final result is ordered by score.

docking translating two molecules toward each other, with concomitant reorientations of one or both partners in a manner which will maximize the attractive interactions.[123,157,158,177,286,573,645,648,649,660–666] Docking can be done either manually, by means of a computer algorithm, or both.

The generic terms *dock* and *docking* are distinct from, and have a more general meaning than, the computer program named **Dock**, which employs a specific algorithm for docking two species.

DOS an acronym for *disk operating system*. DOS is the basic operating system used by IBM and related "clone" desktop computers. DOS was created by MicroSoft, and further information may be obtained from the manufacturer,[957] or from any of various monographs on the subject. MS-DOS is the specific version of DOS for IBM products, and PC-DOS is the virtually identical version for PC "clones".

DOS extender a tool to turn MS-DOS or PC-DOS into a true 32-bit operating system. DOS-Extender can run in protected mode and eliminate the 640-kB memory barrier of **DOS**, allowing the application to access directly up to 16 MB of memory. A DOS-extended application looks just like any other DOS application, and retains access to all MS-DOS functions.

The Phar Lap DOS-extender[958] is used to run the Serena Software GMMX program (see Section 5.8.2).

DREIDING a generic force field developed by Goddard *et al.*[496] (see Section 5.4.8).

Dreiding stereomodels™ a set of skeletal molecular models which accurately depict bond lengths and bond angles. They are best for visualizing bond and torsional angles and interatomic distances. The Dreiding models have become the industry standard of this type.

DRV a computer term, a filename extension for driver files. For example, {PCWIN.EXE} requires access to {SOUND.DRV}, and {KEYBOARD.DRV} in order to run.

dummy atom in molecular modeling, a mathematical construct which doesn't represent any actual atoms, which is added to a force field to provide a reference point for accurate modeling in some systems which are not modeled well by the unenhanced force field. A dummy atom may serve as a *fixed point* in the potential functions for bond stretch/compression, angle bend, and torsional bend, or for nonbonding interactions including solvation. Also referred to as a *virtual atom* or *pseudoatom*.

An example is the use of a **lone pair** on oxygen or nitrogen in the MMX force field to model anisotropic **van der Waals** interactions. Since more "atoms" require more computation, a minimum number of dummy atoms should be employed consistent with their usefulness in improving the molecular model.

Several literature reports discuss the use of dummy atoms, especially in the context of organometallic modeling.[705,827,838,840,846,865]

E a stereochemical descriptor for that isomer of an unsymmetrically substituted alkene in which the two highest priority substituents are on opposite sides of a plane which contains the axes of the π-orbitals. The **CIP** sequence rules are used to rank the priority of the substituents. Earlier convention called this the *trans* isomer.[240]

The term is an abbreviation of the German *Entgegen*, which means "opposite".

EAS from the initial of the names of the program developers, is a designation for a hydrocarbon force field reported by Engler, Andose and Schleyer,[7] which built upon the earlier work of the Schleyer group on the STRAIN program.[73,489] The *EAS* force field is discussed in Section 5.4.8.

ECEPP an acronym for *empirical conformational energy program for peptides*, a molecular mechanics modeling program and the associated force field for biomacromolecules developed by H. A. Scheraga et al.[68,307,308,482,483] The source is available from QCPE as program 286. **UNICEPP** is a united atom version of the ECEPP force field. See Section 5.4.8.

eclipsed a stereochemical term for the relationship between two **vicinal** substituents in a conformation characterized by a dihedral angle of approximately 0°. Another description is that the substituents on two adjacent saturated carbon atoms overlap when the molecule is viewed along the bond in question as a Newman projection. This will usually be an energy maximum in **conformation** space, unless there is a compensatory attractive interaction between the two substituents.

EFF an abbreviation for *empirical force field*, a hydrocarbon formulation developed by Dillen at the University of Pretoria.[514–516] See Section 5.4.8.

electron cloud a metaphor to describe the volume surrounding an atom or group of atoms within which the electrons are most probably located. The electron cloud corresponds to the atomic or molecular orbitals, and is not rigid but somewhat polarizable under the influence of other charges, atoms, ions, or molecules. This metaphor derives from the wave nature of electrons.

electronegativity the ability of some atoms to attract more strongly their constituent or shared electrons toward their nucleus. The electrons of highly *electronegative* atoms have very small **polarizability**. The more electronegative atoms will have one or more lone or nonbonding electron pairs, and will more readily accept a **hydrogen bond** or a metal atom **coordination bond**. Electronegativity increases toward the right and toward the top of the periodic table of the elements.

electrophile a chemical species (atom, molecule, or ion) which is attracted to a region of high electron density on another chemical species. A second definition is a chemical species able to accept an electron pair from another chemical species. This latter partner of the electrophile is termed a **nucleophile**. The name comes from combining "electron" with Greek φιλος (loving), and was coined by the British chemist C. K. Ingold.[973]

electrostatic having to do with the interactions between permanent stationary charges. Electrostatic effects in force field modeling are considered part of the nonbonded interactions.

In PCMODEL there are two models available to handle electrostatic effects: the charge–charge model (**Coulomb potential**) and the dipole–dipole model (**Jeans's formula**). The former is the default for PCMODEL and for most empirical force fields, while the latter is employed in the Allinger force fields (see Section 5.4.1).

enactive a term applied to a representation or model of a molecular structure which embodies the physical or chemical behavior of the molecule.[127] For example, physical characteristics such as bond lengths, bond angles, dipole moment, electron distributions could be obtained from an *enactive* model, which contrasts with an **iconic** model.

enantiomer a stereochemical term meaning one of a pair of **stereoisomers** which are mirror images of each other but are not superimposable upon each other. An enantiomeric molecule will contain at least one **asymmetric center**, which differentiates one enantiomer from the other, even though they have an identical chemical formula and the same set of bonds. Enantiomers will have the same chemical and physical properties as each other, except when those properties involve an interaction with other an asymmetric molecules or plane-polarized light.

The term comes from Greek εναντιος meaning opposite.

Encapsulated PostScript (EPS) an extension of the **Postscript** graphics file format developed by Adobe Systems, and used for such files that are to be incorporated into other documents. An EPS file includes pragmas (special PostScript comments) giving information such as the bounding box, page number, and fonts used. EPS files may include a low-resolution version of the PostScript image. On the Macintosh this is in PICT format, while on the IBM PC it is in TIFF or Microsoft Windows metafile format (WMF).[932]

endothermic a term used for a reaction in which energy or heat is absorbed.

engine a computer programming term, meaning a portion of a program which is supplied with certain parameters as input and then proceeds to carry out a multistage task, making necessary adjustments and decisions with no further input from the user. This software entity is often a part of a larger program. Examples are a search engine, which will search through one or more databases with a set of constraints supplied by the user, or the minimization engine in PCMODEL which takes a molecule in an initial geometry as input, and begins the energy minimization process of optimizing potential energy by varying atomic positions.

enthalpy the thermodynamic state function which comprises the sum of the bond energies, resonance energy, strain energy, and solvation energy in a chemical species (molecule or ion) or ensemble, and takes the symbol H. In the organic chemical context, this quantity is generally measured in kilocalories per mole or kilojoules per mole.

Enthalpy is related to the **free energy** (G) and the **entropy** (S) by the equation:

$$G = H + TS$$

where T is the temperature in kelvin.

A synonym for enthalpy in some contexts is "heat," as in "heat of reaction," "heat of formation," etc.

entropy the thermodynamic state function which measures the disorder, or randomness, or degrees of freedom in a chemical species (molecule or ion) or ensemble, and takes the symbol S. In the organic chemical context, this quantity is generally measured in cal/mol · K or J/mol · K, this last being equal to an *entropy unit* (eu).

Entropy is related to the **free energy** G and the **enthalpy** H by the equation

$$G = H + TS$$

where T is the temperature in kelvin.

epimer one of a pair of stereoisomers which differ from each other in the configuration at a single chiral center. A partial synonym is **anomer**.

Epimerization is a process in which the configuration is changed at only one chiral center of a molecule.

EPS an abbreviation for **Encapsulated Postscript**; also, the filename extension for files in that graphics format.

evolutionary algorithm See **algorithm**.

exothermic a term used for a reaction in which energy or heat is produced.

extended atom a **force field** representation of molecules in the **CHARMM** force field, in which a carbon with its attached hydrogens is treated as a single unit, rather than treating the attached hydrogens as discrete atoms. For example, groups of atoms such as a CH_3 or an aromatic CH are treated as single atoms with regard to calculating their nonbonding interactions. This simplifies and shortens the calculations necessary for energy minimization and dynamics experiments.

Extended atom is synonymous with **united atom**, and contrasts with the **all atom** representation in which hydrogens are treated explicitly.

extension See **filename**.

FANTOM an acronym for *fast Newton–Raphson torsion angle minimizer*, a modeling program for Unix platforms which carries out calculations of low-energy conformations of linear and cyclic polypeptides and proteins. FANTOM enables energy minimizations and/or **Monte Carlo** simulations of an empirical energy function, and accepts distance and dihedral angle constraints from NMR. Protein–solvent interaction is included, with a fast routine for the calculation of accessible surface areas of individual atoms and their gradients. It employs the ECEPP/2 amino acid library.

FANTOM was developed by W. Braun, B. von Freyberg, and T. Schaumann of the University of Texas Medical Branch, and C. Mumenthaler of the ETH.[959]

field a computer term meaning a location in a record or file, or on a monitor screen, where a particular type of data or information can be found or displayed, or may be entered by the user. Also, a term for such a unit of data or information. Fixed-format data files are constructed to follow a format in which the exact location of the data defines what type of data it is. When programs read a file, they use this format to interpret and select the particular data they seek from the available fields.

In the example of an MMX input file, columns 1–60 of the first line constitute the ID field, for the name of the molecule or comments. Columns 1–60 of the second line constitute the LOGARY field, where T's and F's indicate whether those atom numbers correspond to π-atoms. (See Chapter 4.)

Similarly, a spreadsheet on a screen may have a number of labeled *prompt* spaces or small *windows*; these are data fields which are to be filled in by the user, and the data entered is classified by which field it is entered into.

filename or **file name** the name given to a particular file in order for the computer operating system and programs to recognize and operate upon it. This is different from any title or comment lines contained within the file. In DOS, a filename takes the following generic form: {ffffffff.eee}. The filename proper {ffffffff} can be from 1 to 8 characters from the recognized set; the *filename extension* {eee} is separated from the filename by a period, and can be from 0 to 3 characters from the recognized set. Filenames are discussed in greater detail in Section A.1.

first row refers to the row of the periodic table which begins with lithium and ends with neon, and contains many of the elements common to organic chemistry: boron, carbon, nitrogen, oxygen and fluorine.

Fischer projection a convention for drawing carbon chains to portray the relative three-dimensional stereochemistry of the carbon atoms on a two-dimensional drawing. By this convention, the longest carbon chain is oriented vertically and the vertically oriented bonds from an atom will project into the plane of the page from that atom, while the horizontally oriented bonds from an atom will project out of the plane of the page from that atom.[960]
IUPAC:

In a Fischer projection the atoms or groups attached to a tetrahedral center are projected on to the plane of the paper from such an orientation that atoms or groups appearing above or below the central atom lie behind the plane of the paper and those appearing to the left or the right of the central atom lie in front of the plane of the paper, and that the principal chain appears vertical with the lowest numbered chain member at the top.[240]

flag a computer programming term, meaning a small **field** (typically one byte) which acts as a status indicator, i.e. which is used to indicate if a certain condition, or which of several conditions, is in effect or not. Similar to **toggle**.

In the molecular mechanics input file, flags are used to trigger a particular feature or option to be operative during the minimization process. For example, hydrogen bonding or metal atom coordination can be flagged so that these attractive forces will be recognized during the energy calculations. These flags are set by activating certain commands via the graphic user interface during input file generation, often with options selected from a pop-up menu. In PCMODEL, for example, the flag for hydrogen bonding is set by activating the **Mark\H-Bonds** command.

flap angle a type of improper torsional angle. For a group of four atoms in a molecule such as I, J, K, and L, when the pairs I—J and K—L are joined by chemical bonds but the pair J—K is not, the span between the bond vectors I—J and L—K sighting along an axis between atoms J and K is the flap angle. See *improper torsional angle* and *torsional angle* under **improper torsion** and **torsion**.

force constant in the molecular modeling context, a coefficient in the mathematical expression for the potential which acts to constrain the relative displacement of the constituent atoms of a molecule. Molecular modeling force constants must be applicable (transferable) to any molecule within the scope of the **force field** (see Chapter 5).

The force constants in molecular mechanics, or empirical force fields, are derived by consideration of observables from a large basis set of molecules. These observables may be empirical data such as from absorption spectroscopy or diffraction crystallography, or may be derived from the potential surface calculated by *ab initio* molecular orbital methods.

Force constants often require refinement before they fit well into the force field; they are the hypothetical coefficients which are necessary to enable the force field equations to produce reasonable structures.[35]

force field a mathematical model used to describe the potential energy of molecules and related species in terms of the attractive and repulsive interactions between the constituent atoms or molecular fragments. A force field is embodied in a set of potential energy equations, used to relate the relative motions and energies of small fragments (two to four atoms) of the molecule to simple models from classical physics, such as the hard sphere, the ball and spring, and the rigid rotor, and their associated sets of numerical parameters.[10,88] The parameters (constants, coefficients, and variables) in the equations are derived from a representative group of molecules known as the **basis set**. In a process known as optimizing the force field, these parameters are adjusted to give the best overall fit for the basis set,. The assumption is then made that the resulting force field will apply to any conceivable molecule related to the basis set (see Chapter 5).

A *central force field* is a model employed by spectroscopists in which the off-diagonal terms in the force field matrix are assumed to have a negligible contribution, and the component potentials consist of vibrations which are a function only of internuclear distances for pairs of atoms.[34]

A *Class I* (or *Class 1*) *force field* is one which contains only harmonic potentials and utilizes explicit diagonal terms from the force constant matrix.[90,92,93]

A *Class II* (or *Class 2*) *force field* is one which contains anharmonic potentials (through the addition of cubic and higher terms) and utilizes explicit off-diagonal terms from the force constant matrix.[90,92,93] More specifically, a Class II force field may be applied with equal success to isolated small molecules, to condensed phases, and to macromolecular systems; it fits highly strained molecules including small rings with the same parameters; it employs a Morse or quartic bond stretching potential and a quartic angle bending potential; it has well-characterized one-, two-, and threefold torsional terms; it contains cross terms; and it employs an exponential or a ninth power nonbonded repulsion.[92]

A *Class III* (or *Class 3*) *force field* is one which can model the influence of chemical effects, electronegativity, and hyperconjugation on molecular structure and properties.[93]

A *diagonal force field* is one which employs simple harmonic potentials to model bond and angle perturbations, and does not include any cross terms such as stretch–bend or torsion–stretch. The name comes from the fact that in the force field matrix, only the matrix diagonal terms are selected for inclusion in the molecular potential energy equations. In common usage, *diagonal force field* is closely synonymous with *valence force field*.

A *generic force field* is one in which the number of atom types is minimized and the treatment of different bonding configurations is standardized in order to simplify the calculations and broaden the applicability across many classes of molecules. In this environment, a lower accuracy for any particular subset of molecules is accepted in exchange for broader applicability.

format a computer term with several meanings:

In memory organization, formatting a physical storage medium (hard disk or diskette) involves dividing it up into addressable portions so that data can be written and retrieved readily. This is accomplished by running the appropriate utility program.

In an electronic data file, the data is arranged according to a format so that data is written to the file will be found in the location designated for that type of data.

In a programming language, variables are assigned a format, which the values supplied for that variable must adhere to in order to be recognized and operated upon.

free energy the thermodynamic state function which comprises the sum of the **enthalpy** (the bond energies, resonance energy, strain energy, and solvation energy) and the temperature-adjusted **entropy** (the disorder, or randomness, or degrees of freedom) in a chemical species (molecule or ion) or ensemble, and takes the symbol G. In the organic chemical context, this quantity is generally measured in units of kilocalories per mole or kilojoules per mole.

The change in free energy for a particular chemical process is a function of the equilibrium constant for that reaction and the temperature. For the reaction

$$A + B \to C + D$$

the equilibrium constant is given by

$$K_{eq} = \frac{[C][D]}{[A][B]}$$

and

$$\Delta G = RT \ln K_{eq}$$

where K_{eq} is the equilibrium constant for the indicated reaction, $\ln K_{eq}$ is its natural logarithm, ΔG is the change in free energy, R is the molar gas constant, and T is the temperature.

The free energy is related to the **enthalpy** (H) and the **entropy** (S) by the equation

$$G = H + TS$$

where T is the temperature in kelvin. The thermodynamic changes over the course of a chemical reaction take the form

$$\Delta G = \Delta H + T \, \Delta S$$

front end programming added to a preexisting software package, program, or algorithm which provides a simple or simpler interface to that application. A front end might be a graphical interface to a numerically based program; it might present simpler menus or lists of options than exist in the original program; it might present more informative prompts and relieve the need to master a command vocabulary; it might also provide help to inexperienced users.

frontier molecular orbital theory the theory, developed primarily by Fukui, that the course of chemical reactions between two molecules will be dictated by the characteristics of certain molecular orbitals of each partner, in bimolecular reactions.[813,814] In this analysis, "... chemical reactions should take place at the position and in the direction of maximum overlapping of the **HOMO** and **LUMO** of the reacting species ..."[813] This theory provides an easy heuristic which generates the same predictions as would arise from the more complete molecular orbital analysis of a correlation diagram.

Only those orbitals corresponding to bonds being formed or broken are considered. The reaction is seen as taking place through charge transfer from the occupied molecular orbital of one component which is highest in energy (HOMO) to the unoccupied molecular orbital of the other component which is lowest in energy (LUMO). The descriptive term "frontier" comes from the orbitals of interest being at the frontier, or interface, between the two reacting species.

There are several sources of graphic representations of molecular orbitals, such as the monograph by Jorgensen and Salem.[815]

GAMESS an acronym for *general atomic and molecular electronic structure system*, a general *ab initio* quantum chemistry package which is maintained by the members of the M. S. Gordon research group at Iowa State University.

gauche a stereochemical term for the relationship between two **vicinal** substituents in a conformation characterized by a dihedral angle of approximately 60°. It is generally used to refer to the two largest or most important substituents attached to the two central atoms, and is generally a local energy minimum in **conformation space**. Gauche is a special case of **staggered**.

The term comes from French, literally meaning "left" or "left-hand" and figuratively meaning "clumsy, awkward."

The *gauche effect* is the observation in the area of conformational analysis that in a molecule possessing two electronegative groups in a vicinal relationship, the gauche rotamer will be more stable that the *anti* rotamer, despite the increase in nonbonded repulsions. A more specific example of the gauche effect is the **anomeric** effect.

Early explanations held this effect to be a result of the balance of attractive and repulsive forces between the substituents.[961] A more recent explanation for this effect is that the C—C bond is subject to bending by each of the electronegative substituents, and this bending causes more of a net weakening of the C—C bond when the two electronegative substituents are **anti** than when they are gauche.[244,292]

Gaussian an *ab initio* quantum chemical molecular orbital calculation package which is available in many forms and versions for various hardware platforms. It is the most widely used program of its kind. It utilizes a Gaussian-type orbital (GTO) basis set, and interfaces to many other software applications are available. A good description of Gaussian is available in Clark's monograph on computational chemistry.[962]

GB/SA an acronym for generalized Born equation/surface area, *GB/SA* is a continuum treatment of solvation effects for use ion empirical force field modeling, developed by Still and coworders.[568] The solvation free energy is calculated as the sum of a solvent-solvent cavity term and a solute-solvent van der Waals term, which are calculated from the molecule's surface area, and a solute-solvent electrostatic polarization term, which is calculated from the Born solvation equation and a Coulomb electrostatic component. The *GB/SA* treatment is discussed in Section 5.6.

GEGOP an acronym for *geometry of glycoproteins*, a program which carries out a conformational analysis of a glycoprotein, within a specified force field, by simultaneous variance of the torsional angles.[963,964]

geometric isomer one of a group of two or more stereoisomers which cannot be interconverted because of hindered bond rotation. For example, this type of isomerism produces the *E* or *Z* (*trans* or *cis*) isomers about a double bond.

GERM an acronym for *genetically evolved receptor models*, a program for generating a 3D description of a hypothetical receptor derived from structure–activity relationships, developed by Walters and Hinds.[162] The models are evolved by application of a genetic **algorithm**, and may be used to predict bioactivity for structures not used in the formulation of the receptor hypothesis.

GIF an acronym for *graphics interchange format*, a standard graphics format defined in 1987 by CompuServe Inc. for digitized images, converted for transmission on the Internet with the Unisys, Inc., LZW compression algorithm. GIF is currently an open (i.e. not proprietary) format.

global minimum that **conformation** of a molecule which has the lowest relative potential energy. In some cases of conformational analysis, a local minimum may be confused with the global minimum. There is no method to determine whether a discovered conformational energy minimum is the global minimum except a rigorous, exhaustive search of **conformation space**. A number of searching techniques can be used to arrive at the global minimum with varying degrees of confidence, and the level of confidence decreases with increased complexity of the molecule (see Section 5.8).

GLYCAM the name of an additional parameter set of the **AMBER** force field, developed by Woods *et al.* at Oxford University,[429] to enable the treatment of oligosaccharides and glycoproteins. The *GLYCAM* parameter set is discussed in Section 5.4.3, and the parameters may be downloaded from the AMBER Web site.[419]

glycosidic linkage a term from carbohydrate chemistry, also referred to as the glycosidic bond. When two sugar residues are joined by a bridging oxygen between the anomeric carbon of one residue and a nonanomeric carbon of the other, the C—O—C unit is called a glycosidic linkage.

glycosyl acceptor a term from synthetic carbohydrate chemistry: in the case of a reaction in which two sugar residues are being joined through the formation of a glycosidic bond, the glycosyl acceptor is that sugar residue in which a carbon other than the anomeric center forms part of the linkage.

glycosyl donor a term from synthetic carbohydrate chemistry: in the case of a reaction in which two sugar residues are being joined through the formation of a glycosidic bond, the glycosyl donor is that sugar residue in which the anomeric carbon forms part of the linkage.

GOPT an acronym for *graphics optimization tool*, a graphics front end for the Lyngby version of the **CFF** modeling program, developed by Rasmussen and coworkers.[392,395]

graphics specifically *computer graphics*, images produced on a monitor on a computer equipped with the proper hardware and software. The hardware requirements are a monitor and a video display card, and the software must incorporate a driver which allows the program to address the graphics hardware.

Earlier generation graphics were of the *vector graphics* type, where the image was produced as a series of straight or curved lines, polygons, etc., and groups of such objects. The lined features are mapped from beginning point to end point within the xy coordinate system corresponding to the screen (whence the name), and the images are continuously refreshed. An advantage of vector graphics is that any element of the picture (the features mentioned above) may be changed at any time, since each element is stored as an independent object.

The modern standard is the *raster graphics* type, in which an image is composed of an array of pixels arranged in rows and columns. The term comes from the Latin *rastrum*, meaning rake, from the motion of the cathode ray scanning the screen elements. In this case, a continuous image is produced from a pattern of small, closely spaced square or rectangular elements (monochrome) or triads of red–blue–green elements (color). These elements are called **pixels**, and larger numbers of rows and columns of pixels leads to higher resolution graphics. Important features besides the resolution (usually quantified by the column and row dimensions such as 640 by 480, 800 by 600, 1024 by 768, etc.), are the refresh rate and, in color graphics, the number of colors (e.g. 8, 16, 256). The area of the screen is scanned from side to side in lines from top to bottom and certain of the elements are activated, which forms the graphical display.

In *bitmap graphics*, a type of raster graphics, each pixel in the display is mapped literally bit for bit from data in the image file. Strictly speaking, bit-for-bit file-to-image translation is limited to black-and-white graphics, since a color image requires more that one bit for each pixel; but the term bitmap is generally used for color images as well. Such a display can be updated extremely rapidly, since changing a pixel involves only a single processor write to memory, whereas with a terminal connected via a serial line the speed of the serial line limits the speed at which the display can be changed.[932]

Most modern personal computers and workstations have bitmapped graphics, allowing the efficient use of graphical user interfaces, interactive graphics, and a choice of on-screen fonts. Because of the large amount of information required in such a specification, bitmapped graphics files tend to be large in size. Some more expensive systems still delegate graphics operations to dedicated hardware such as graphic accelerators.

Some graphics displays have elements of both vector and raster format.

Graphics Optimization Tool   See **GOPT**.

graphic user interface   a type of computer operating environment which is characterized by the use of pictorial or visual screen objects to represent the input and output of a program, rather than only text on a command line. It is also known by the abbreviation GUI. A discussion of the GUI may be found in a recent review article.[88]

A program with a GUI generally runs under a windowing system such as the Apple Macintosh™ operating system or Microsoft Windows™. In this environment, the user controls the program by using a pointing device (such as a mouse, trackball, keyboard-driven cursor, etc.) to indicate objects (such as icons, buttons, dialog boxes, etc.) in the screen windows by directing the pointing device cursor at them and pressing a button to select it. Apple Computers™ popularized the GUI with their Macintosh™ operating system, but the concept originated in the early 1970s at Xerox's PARC laboratory.[932]

grid search   a search conducted systematically over all available coordinates or values, an antonym for **stochastic** and **Monte Carlo**.

The *grid search method* involves the systematic selection or generation of input data or coefficient values, or systematic sampling techniques to obtain solutions to problems which are exact or nearly so. In a conformational analysis context, the grid search method may be used to find the minimum energy conformation as a function of all rotatable bonds. For n rotatable bonds, and after selecting a desired precision, this results in an *n*-dimensional grid, each point of which will be sampled. The grid search method is also referred to as the **brute force** method, because it painstakingly examines each possible value before arriving at an answer.

GSTAT a multifunctional geometry program with the **Cambridge Structural Database** Software (**CSDS**). It will perform geometric calculation for complete CSD entries, location of 3D substructure fragments within the crystallographic connectivity record of the CSD, geometric calculations for substructure fragments, and statistical and numerical analysis of fragment geometry. Most of the functionality of GSTAT is now available within a combination of the other CSDS components **QUEST3D** and **VISTA**.

GUI an abbreviation for **Graphic User Interface**.

Hammond postulate the observation (by the American chemist G. S. Hammond) that the structure of a transition state can be approximated best by the structure of that ground state species which is nearest in energy. More specifically, the transition state for an exothermic reaction will look more like the reactant(s), and the transition state for an endothermic reaction will look more like the product(s).

Applying the Hammond postulate aids the modeling of transition states by perturbation theory by showing whether the reactant or the product structure is the best starting point for applying perturbation, based on a knowledge of the energetics of the reaction.

haptophilic a term used for a molecule, or a functional group within a molecule, which may act as a donor in forming a **coordination bond** with a metal atom partner. A haptophilic group usually consists or one or more atoms which bear lone electron pairs, or two or more atoms which have a π-orbital system. The word comes from Greek $\alpha\pi\tau\omega$ (to fasten) and $\phi\iota\lambda o\varsigma$ (loving), and was coined by H. W. Thompson.[986]

Haptophilicity is the quality of being haptophilic. *Monohapto, dihapto,* and *trihapto* refer to ligands which will interact with the metal center via one, two, or three atoms, respectively.

hard acid an acid with small acceptor atoms and high positive charge which does not contain unshared lone pairs of electrons in its valence shells. A hard acid has low polarizability and high electronegativity, and will react preferentially with a **hard base**. In hard acids, the acceptor atom does not have easily excited outer electrons.[965] See **HSAB principle**.

hard base a base with donor atoms of high electronegativity and low polarizability. A hard base is difficult to oxidize, holds its valence electrons tightly, and will react preferentially with a **hard acid**. In hard bases, the donor atom is associated with empty orbitals which are of high energy and hence inaccessible.[965] See **HSAB principle**.

hardness in the molecular modeling context, the hardness is equivalent to a force constant for the van der Waals nonbonded interaction potential. This parameter is symbolized by ε in the nonbonding potential of many force fields, and its use is described in Section 5.1.3.1. The hardness characterizes how easily the corresponding atom can be deformed as it approaches other atoms. The hardness can be understood in mathematical terms as the steepness of the potential curve for the van der Waals repulsion. See **van der Waals**.

For hardness in the physical organic context, see **HSAB principle**.

Hard-Sphere *exo*-Anomeric method   See **HSEA method**.

Hartree–Fock   an approach to molecular orbital (MO) calculations in which the energy of a set of molecular orbitals can be derived from the basis set functions used to define each orbital, and a set of adjustable coefficients which are used to minimize the total energy of the system.[121,781] Thus, in Hartree's original method the wave function for a number of electrons N was found as the product of N one-electron wave functions. Fock modified the method to use Slater determinants (or matrices) to combine the one-electron wave functions instead of using their products. This method took account of most of the effects of electron spin. Electron–electron repulsions are addressed by considering the effect of the averaged Coulombic repulsions of the other electrons on each electron in turn.

Essentially, the *Hartree–Fock method* employs antisymmetric Slater determinants for combining the one-electron wave functions, and through an iterative process the coefficients of the orbitals are modified from cycle to cycle until the electronic energy achieves a constant minimum value and the orbitals no longer change, i.e., a **self-consistent field (SCF)** has been reached.

The *restricted Hartree–Fock* (RHF) calculation is used for a closed-shell system. It carries the restriction that there must be an even number of electrons, and that they must all be spin-paired. The *restriction* is that every orbital is either doubly occupied or unoccupied. The RHF calculation is appropriate for neutral systems, anions, or cations.

The *unrestricted Hartree–Fock* (UHF) calculation is for an open-shell system, where at least one orbital will be singly occupied. It calculates two sets of molecular orbitals, one for each type of spin, named *alpha* and *beta*. It thus considers each doubly occupied MO to consist of two spin-paired singly occupied MOs. The energies of these latter are not necessarily the same, as would be required in the RHF formulation. The UHF calculation allows for the perturbation effect of an unpaired spin upon the formally paired spins, and generates more realistic spin densities.

The UHF calculation is appropriate for radicals or diradicals. A radical (doublet) will have one more alpha-type electron than beta-type, and a diradical (triplet) will have an excess of two alpha-type electrons. The UHF calculation causes the electronic energy of the system to be lower than does the RHF calculation for the same system, so that the results from the two types of calculation cannot be compared with each other.

heavy atom   a nonhydrogen atom. Any atom from the periodic table except hydrogen and its isotopes (e.g. deuterium).

Heisenberg uncertainty   the principle or relation that the position and the velocity of a subatomic particle cannot both be measured exactly, at the same time, because the act of measuring one alters the other. The statement of this principle was made in 1927 by the German physicist W. Heisenberg.

Hessian   the matrix of second partial derivatives of a function (assumed to be twice differentiable). This is often denoted $H_F(x)$, where F is the function and x is a point in its domain.[933] The Hessian determinant is named for the German mathematician L. O. Hesse, who introduced it in 1842.

In molecular modeling, for a molecule of N atoms, the N by N Hessian matrix will contain the force constants which govern the interaction of each atom with each other atom.

heteroaromatic   a molecule which is **aromatic**, and in which one or more of the ring members is a **heteroatom**.

heteroatom   an atom other than carbon or hydrogen, or their isotopes.

heterocyclic   being a cyclic molecule in which one or more of the ring members is a **heteroatom**.

heuristic   a logic term meaning a "rule of thumb," a simplification or an educated guess that reduces or limits the search for solutions in domains that are difficult and poorly understood. Unlike an **algorithm**, a heuristic does not guarantee a solution.[932]

HEV   an abbreviation for *hydrogen-extension vector*, a parameter used in the YETI force field to represent the linearity of the hydrogen bond.[168,525] The HEV will originate at the hydrogen bond donor, and its end point marks the ideal position for a hydrogen bond acceptor atom relative to that hydrogen bond donor fragment. It is one of the descriptors used in building pseudoreceptors in the **YAK** program. See also **YAK** and **LPV**.

HMO   an abbreviation for Hückel molecular orbital; see **Hückel calculation**.

HOMO   an acronym for highest occupied molecular orbital.

Hondo   an *ab initio* HFMO (Hückel frontier molecular orbital) calculation package developed by M. Dupuis, J. D. Watts, H. O. Villar, and G. J. B. Hurst of IBM Corp. The program uses Gaussian-type basis functions up to d type, and can handle up to 50 atoms and 120 shells for a total of 440 unique Gaussian exponential functions. The code is available as QCPE #544.[966] A brief account of some of its calculational capabilities includes:

Single-configuration self-consistent-field wave functions (closed-shell RHF, spin-unrestricted UHF, restricted open-shell ROHF); generalized valence bond and general multiconfiguration self-consistent-field; and configuration interaction, with full use of molecular symmetry.

Calculation of the electron correlation correction to the energy of closed-shell RHF wave functions; use of the effective core potential approximation;

optimization of molecular geometries using the gradient of the energy with respect to nuclear coordinates for all SCF wavefunctions; calculation of the force constant matrix in the Cartesian or internal coordinates space and of the vibrational spectrum (including infrared and Raman intensities) for all SCF wave functions; calculation of the dipole moment and polarizability with respect to the nuclear coordinates; determination of transition state structures by taking advantage of the energy gradients for all SCF wave functions.

Hooke's law a relationship from classical physics which describes harmonic oscillation, stretching, bending etc. It has the form

$$V = K \cdot \tfrac{1}{2} k_r (r - r_0)^2$$

where V is the potential energy, K is a constant for unit conversion, k_r is the force constant for the oscillator, r_0 is the equilibrium value of the physical quantity (distance, angle, etc.), and $r - r_0$ gives the displacement from equilibrium. This function, or a slight modification of it including a cubic term, is used in the diagonal **force field** to model the energy of bond stretching, angle bending, and torsional strain.

HPGL or **HP-GL** an abbreviation for Hewlett–Packard Graphics Language, a vector graphics language used by Hewlett–Packard plotters.[932]

HPV an abbreviation for *hydrophobicity vector*, a parameter used in the YETI force field to represent a hydrophobic region.[168] The HPV will originate at an apolar hydrogen atom, and its end point marks the approximate position of another hydrophobic moiety relative to the apolar hydrogen atom. It is one of the descriptors used in building pseudoreceptors in the YAK program. See also **YAK**.

HSAB principle the principle of hard and soft acids and bases: a **hard acid** will react preferentially with a **hard base** in a largely ionic fashion, and a **soft acid** will react preferentially with a **soft base** in a largely covalent fashion.

This principle has been placed on a firm theoretical basis, where the hardness of a chemical species is given as half the derivative of its chemical potential with respect to the number of electrons. The chemical potential is the negative of the Pauling–Mulliken electronegativity, which is the average of the ionization potential and the electron affinity.[965] Softness is indicated by a low value of hardness.

HSEA force field an abbreviation for *hard-sphere exo-anomeric*, an approach to calculating conformational preferences about glycosidic linkages, developed by R. Lemieux *et al.*[84,85]

The method combines hard-sphere molecular mechanics calculations with a correction factor for the *exo*-anomeric effect, which is a function of the ϕ-angle. These correction terms are derived from the results of *ab initio* calculations on the model system dimethoxymethane.

Hückel calculation a semiempirical molecular orbital calculation of the π-orbital energies in a conjugated molecule.[967] These calculations bear the name of Erich Hückel, the German physicist who developed them. Hückel calculations assume that the electronics of the π-system can be treated independently of that of the σ-framework for the atoms constituting the conjugated system, and further that each electron is affected only by its parent and immediately neighboring nuclei.

This method calculates the energy of each molecular orbital (E) as follows:

$$E = \alpha + m_j \beta$$

and

$$m_j = 2 \cos \frac{j\pi}{n+1}$$

for $j = 1, 2, 3, \ldots, n$, where α is the Coulomb integral related to the binding of an electron in its atomic p-orbital and β is the resonance integral related to the energy of an electron in the field of two or more nuclei. From this calculation we also get the coefficient c_{rj} for the contribution of the p-orbital of the rth atom (out of n total constituent atoms) to the jth molecular orbital (out of n total molecular orbitals) as follows:

$$c_{rj} = \left(\frac{2}{n+1}\right)^{1/2} \sin \frac{rj\pi}{n+1}$$

The calculations involve minimizing the total π-orbital energy of the molecule by variation of the atomic p-orbital contribution coefficients c_{rj} until the total energy reaches a consistent lower limit. This *Hückel method* is discussed in Section 6.5.

The *Hückel rule* is an empirical method used to determine whether a cyclic conjugated π-system should be **aromatic**. The rule states that a cyclic conjugated π-system will be aromatic only if the number of π-electrons is equal to $4n + 2$ where $n = 1, 2, 3, \ldots$. The rule is particularly useful in determining aromaticity in some heterocycles, or where orbitals bearing charge are included in the ring (such as cyclopentadienyl anion), or in systems where valence-bond depictions break down and cannot be used to determine conjugation (such as some benzpyrenes).[937]

hybridization the process of combining a certain number of atomic orbitals of different type but similar energies to form a set of the same number of equivalent composite or *hybrid* orbitals. Such hybrid orbitals exist only when the parent atom is part of a molecule or ion. An example in the case of carbon is the combination of the 2s orbital with the three 2p orbitals to produce four sp^3 hybrid orbitals for bonding to other chemical species.

hydrogen bond an attractive interaction between a functional group A—H (where A is generally an electronegative atom) and an atom or group of atoms B in the same or a different molecule. The hydrogen bond is a special case of the electrostatic interaction between two fixed dipoles. A hydrogen atom bonded to a small, highly electronegative atom A forms one of the dipoles, and the other is defined by a lone or nonbonding electron pair associated with another such electronegative atom B. In most cases, the atoms A and B are nitrogen, oxygen, fluorine, or sulfur. Hydrogen bonds generally range in strength from 3 to 8 kcal/mol.[791]

The implementation of the hydrogen bonding potential in molecular mechanics[804–807] is discussed in Section 6.6.

hydrophilic a descriptive term connoting an affinity for water, usually in the context of solvation. The word comes from the Greek υδωρ (water) and φιλος (love). "Hydrophilic" may be used to describe the favorable interactions of polar substances with water or with each other, generally driven by the enthalpic contribution of reciprocal **hydrogen bond** and **electrostatic** effects.

hydrophobic a descriptive term meaning a lack of affinity for water, usually in the context of solvation. The word comes from the Greek υδωρ (water) and φοβος (fear). "Hydrophobic" may be used to describe the interactions of nonpolar substances with water, as in hydrophobic interactions or hydrophobic effects. The hydrophobic effect of poor solubility of nonpolar solutes in water is ascribed largely to the entropic contribution, having to do with the disruption of the bulk water medium by the formation of a structured solvation shell around the solute.[565]

hyperconjugation a molecular structure term, which is an extension of the **conjugation** concept to σ-bonds. In this case, polarization effects would deform the electron density in a single bond, in effect giving a fractional **bond order** less than one, while increasing the bond order between a neighboring pair of atoms, or at a neighboring cation or radical.[968] This effect arises naturally from a simple molecular orbital analysis. Hyperconjugation is invoked today mainly to explain stabilization of reactive intermediates.

The Allinger MM3 and MM4 force fields contain components designed to model the effect of hyperconjugation.[332,381]

hypertext refers to words, phrases or text strings in a document which serve as links to different documents, graphic images, or other sources of information in the **World Wide Web** environment. *Hypertext* documents or pages are viewed with a **web browser** running on a *GUI*, and the mouse cursor or other graphical pointing device is used to trigger the switch to the new document or page. *Hypertext* documents are prepared in a format called *hypertext markup language* (HTML), which enables switching between **internet** addresses or **URLs**.

hypervalent a term used to describe a molecule, or an atom within such a molecule, in which the number of bonds to one or more of the atoms exceeds that allowed by the traditional Lewis–Langmuir theory, which is based on sharing valence electron pairs.[969] Hypervalent atoms in the periodic table's Main Group will bear net charge.

For example, a quaternary ammonium salt is hypervalent on nitrogen, because it has four bonds to other groups instead of the normal three.

ICM an abbreviation for *internal coordinate modeling*. It refers to a modeling methodology developed by Abagyan and coworkers for treating biological macromolecules, in which the structure is described by the internal coordinates of bond lengths, bond angles, and torsional angles, rather than by a Cartesian coordinate representation.[894] It is useful for calculating the structure and properties of proteins, as well as for docking experiments.

iconic a term applied to a simple representation or picture of a molecule, which does not embody any physical or chemical behavior of the molecule.[127] This contrasts with an **enactive** model or representation, which does embody such information.

improper torsion (sometimes referred to as a virtual torsional angle) most generally a dihedral angle moiety defined by a group of four atoms in a molecule, say *I, J, K*, and *L*, when those atoms are not all joined in a linear fashion by chemical bonds (see **flap angle** and **torsional angle**). In the molecular modeling context, the improper torsion is a measure of the molecular vibration in which a trigonal atom, such as sp^2 carbon or boron, or a **united atom** tetrahedral atom, assumes a nonplanar or pyramidal geometry (see Section 5.1.1.5).

For the four atoms *I, J, K*, and *L*, the *improper torsional angle* has been defined in two different ways. In Fig. 5-7, the angles ω_1, ω_2, and ω_3 show the definitions used in Allinger's MM4 force field, while the angles $\omega_{1'}$, $\omega_{2'}$, and $\omega_{3'}$ show the definitions used in Kollman's AMBER force field. Force fields incorporate an improper torsional potential either to reinforce planarity at a trigonal atom, or to reinforce a tetrahedral geometry (especially asymmetry) when the central atom is a trisubstituted **united atom** group, depending upon the parameters employed.

An *improper torsional potential* in force field molecular modeling is a potential energy term applied to a group of four atoms as defined above. See *torsional potential* under **torsion**.

This distortion can also be quantified as the **out-of-plane angle**, the **pyramid height**, or the **Wilson angle**.

inductive effect an electrostatic effect transmitted through the bonds of an organic molecule, and manifested in the formation of a **dipole**. This effect arises from the permanent polarization due to the electronegativity or electropositivity of substituents concerned.

Internet an interconnected system of computers worldwide that allows ready access to information, files, and communications between any two (or more) points in the network. Also called the Information Superhighway. See **World Wide Web**.

ionic bond an electrostatic attraction between two oppositely charged or strongly polarized atoms, fragments, or molecules. The strength of the interaction may be calculated as the **Coulomb potential**.

Jahn–Teller effect also known as **Jahn–Teller distortion** a structural distortion which occurs in nonlinear molecules with an electronically degenerate state. This distortion lowers the symmetry, which removes the degeneracy and results in a lowering of the energy of the system.

The Jahn–Teller distortion is evident in a number of organometallic complexes, and attempts have been used to model this effect by force field methods.[883,884]

Jeans's formula an equation for determining the electrostatic interaction potential between two dipoles. In the context of bond dipoles on the molecular scale (see Fig. 5-22), this relationship takes the form

$$V_{elec/dd} = \frac{\mu_a \mu_b}{Dr_{ab}^3} (\cos \chi - 3 \cos \alpha_a \cos \alpha_b)$$

where μ_a and μ_b are the magnitudes of the two dipole vectors in debyes; D is the dielectric constant of the medium, with a default value of 1.5 D; r_{ab} is the distance between the centers of the two dipole vectors in angstroms; χ is the angle between the two dipole vectors; and α_a and α_b are the angles between the axes of the vectors **a** and **b** (respectively) and the intervector radius r_{ab}. Strictly speaking, Jeans's formula is only applicable when the distance between the dipoles is large compared with their effective lengths.[534]

The bond-dipole–bond-dipole interaction potential is used to model the electrostatic effects in Allinger's force fields (see Section 5.4.1). See also **Coulomb potential**.

***J*-value** a coupling constant in nuclear magnetic resonance spectroscopy, measured in hertz. See **coupling constant**.

Kekulé an optical chemical structure recognition program, which converts scanned chemical structures into structure file formats with connection tables. Kekulé produces chemical structures through its drawing utility, or by scanning printed structures and converting them to connection table format and editable structure diagrams. It contains import and export filters for reading and writing connection tables in other formats.

The program is named after Friedrich August Kekulé von Stradonitz, the nineteenth century French chemist who contributed to the structural theory for organic compounds.

kinetic control a term applied to reactions or the products formed from reactions, meaning that the product(s) with the highest rate of formation will predominate. See **thermodynamic control**.

Lewis-acid–Lewis-base a formalism for acid–base reactions developed by G. N. Lewis. The *Lewis acid* is capable of using an unoccupied orbital to accept a pair of electrons from the *Lewis base* to form a **coordination bond**.

ligand in the context of organometallic chemistry, a molecule or molecular fragment, either charged or neutral, which will form a bond with a metal; the bond may have ionic, covalent, or coordination characteristics. The ionic contribution to bonding of a charged ligand depends upon electrostatic attraction. The covalent contribution depends on the extent of suitable overlap between the valence orbitals on the metal and on the ligand. The coordination contribution can be considered as a **Lewis-acid–Lewis-base** interaction between a vacant orbital on the metal and a nonbonding electron pair on the organic ligand, sometimes with a back-bonding component involving the interaction of occupied metal orbitals into low-lying unoccupied orbitals of the ligand.[970]

PCMODEL models metal–ligand interactions as follows: the ionic contribution (if any) is treated with the **Coulomb potential**; the covalent contribution is governed by the parametrized valence bonding potentials; and the coordination contribution is modeled by a Coulombic attraction weighted by a dipole–dipole contribution (see Section 6.7).

lipophilic a descriptive term connoting an affinity for a nonpolar medium, usually in the context of solvation or more specifically in transport through biological membranes. The word comes from the Greek λιπος (fat) and φιλος (love).

Generally used to describe the propensity of nonpolar substances to associate with other nonpolar substances in preference to polar ones. The classical physical measurement used to quantify this behavior is the *n*-octanol–water partition coefficient. This physical quantity (partition coefficient) is symbolized by the letter P, and since the measurements vary over such a large range, comparisons are usually made on a logarithmic scale; hence log P is used.

An appreciation of the importance of this quantity to computer-aided molecular design has led to the development of empirically based algorithms to calculate it as a combination of contributions from identifiable structural fragments. The **CLogP** program is one such method.

London forces (named for F. London) See **van der Waals** forces.

lone pair nonbonding electron pair associated with a **heteroatom** or a carbanion. In the first row of the periodic table, nitrogen bears one lone pair, oxygen two, and fluorine three. The elements in a family (column in the periodic table) will have the same number of lone pairs in their outer shell.

In the MMX force field,[343] molecules containing nitrogen and oxygen will not minimize correctly unless the lone pairs are explicitly considered, so provision is made for adding lone pairs to these atoms during the file preparation. Second-row elements (P, S, Cl) and lower do not require explicit inclusion of lone pairs to minimize correctly. In molecules containing N and O which are conjugated, one lone pair is considered to be involved in the conjugation, and so is not shown, since its presence is already allowed for by the appropriate parameters.

LPV   an abbreviation for *lone pair-extension vector*, a parameter used in the **YETI** force field to represent the directionality of the hydrogen bond.[168,525] The LPV will originate at the hydrogen-bond acceptor, and its end point marks the ideal position for a hydrogen-bond donor atom relative to that hydrogen-bond acceptor fragment. It is one of the descriptors used in building pseudoreceptors in the **YAK** program. See also **YAK** and **HEV**.

LUMO   an acronym for lowest unoccupied molecular orbital.

MacMolecule   an interactive 3D molecular visualization program for the Macintosh platform, featuring a video animation generation program and a molecular image database. It is designed for creating 3D images of molecules for use in teaching molecular structure to students of biology, chemistry, and allied fields. It was written by E. Myers, C. Blanco, R. B. Hallick, and J. Jahnke, of the University of Arizona.

MacroModel   a workstation and VAX platform molecular modeling program for biomacromolecules as well as small molecules, developed by the research group of W. C. Still at Columbia University.[285,286] It has a graphic interface to prepare molecular structures, and can perform molecular mechanics minimization and molecular dynamics simulations to evaluate the energies and geometries of molecules in vacuo or in solution. Macromodel can use the MM2*, MM3*, AMBER*, AMBER94, OPLSA*, or MMFF94 force fields, and the analytical **GB/SA** continuum treatment for solvation.

Main Group   that group of elements in the periodic table comprising the groups III A, IV A, V A, VI A, VII A, and VIII A. The collection of the boron family, the carbon family, the nitrogen family (or pnicogens), the oxygen family (or chalcogens), the halogens, and the inert or noble gases.

Markownikoff rule   an empirical relationship dealing with the regioselectivity of addition of a hydrogen halide across an unsymmetrically substituted alkene (see **regioselectivity, electrophile**). In its more general form, the rule predicts the formation of that product in which the more electronegative portion of an unsymmetrical electrophile bonds to the less substituted side of the carbon–carbon double bond (see **electronegative**).

The term derives from the name of the nineteenth century Russian chemist V. V. Markownikoff.

MEDLA   an acronym for *molecular electron-density Lego assembler*, a technique for constructing reasonably accurate electron densities for larger molecules from a library of precalculated fragments. The building blocks are "fuzzy" electron distributions without boundaries, calculated for fragments and functional groups by *ab initio* molecular orbital methods. These units are attached on the basis of the molecular atom skeleton, whereby the electron clouds interpenetrate to form a continuous, cumulative electron distribution. Proteins and DNA have been modeled successfully in this way.[677,949] Part of the name derives from the popular children's construction toy, Lego®.

MEP an abbreviation for **molecular electrostatic potential.**

Metropolis algorithm a mathematical treatment for calculating the properties of an ensemble of interacting entities as a function of the potential between them and of the temperature, based on a **stochastic** or **Monte Carlo** sampling of those entities.[42,573,606-616] The term derives from the name of the American physicist N. Metropolis.

In molecular modeling, this algorithm can be applied in the **simulated annealing** approach to geometry optimization of a molecule based on the **force field** which governs interactions between its constituent parts.

microwave spectroscopy the study of the interaction of molecules with electromagnetic radiation in the wavelength range of 10^{-2}–1 m. This energy envelope corresponds to the rotational frequencies of organic molecules, and is useful for structure characterization.

MicroWorld a molecular display program for the Macintosh platform, for three-dimensional model renderings (Dreiding, ball-and-Stick, space-filling) from atomic space coordinates, e.g. from X-ray data, databases, or energy optimization programs.

MIME an acronym for *multipurpose Internet mail extension.* A feature of electronic files transmitted over the Internet, the MIME is a filename extension which describes the format of a file's contents and signals the local client platform, or browser, to launch any necessary software or helper programs to aid in interpreting or displaying the contents. The MIME may currently be formulated as belonging to any of seven primary types: text, application, image, audio, video, message, or multipart. Each primary type may comprise many secondary types.[3]

MM2 a molecular mechanics force field developed by the Allinger group, first in published in 1977 and updated often since (see Section 5.4.1). Arguably the most popular, most widely used, and most often imitated force field in existence.

MM2* an adaptation of the Allinger MM2 force field created by the Still group at Columbia University, for use in the modeling program MacroModel.[285,286]

MM2X an adaptation of the Allinger MM2 force field created at Merck,[456] a precursor to the **MMFF** formulation (see Section 5.8.6).

MM3 a molecular mechanics force field developed by the Allinger group, and published in 1989 (see Section 5.4.1).

MM4 a molecular mechanics force field developed by the Allinger group, and published in 1996 (see Section 5.4.1).

MMADS an abbreviation for *molecular modeling and display system*.

MMFF an abbreviation for *Merck Molecular Force Field*, a force field developed by T. A. Halgren[321] with a focus more on **molecular dynamics** than on energy minimization. The current version is MMFF94. There is a companion version, named MMFF94s, which is intended for use solely in energy minimization.

MMFF94 an abbreviation for *Merck Molecular Force Field 1994*. This force field is described in a series of publications by the Merck group headed by T. A. Halgren,[314,321,453–455] and is discussed in Section 5.8.6.

MMI a molecular mechanics force field developed by the Allinger group, and first published in 1974 (see Section 5.4.1).

MMX a molecular mechanics force field developed by K. Gilbert, J. Gajewski, M. Midland, and K. Steliou, used in the PCMODEL program by Serena Software (see Section 5.2).

MO an abbreviation for **molecular orbital**.

MOL a filename extension used for structure input files. This name is used to describe several unrelated formats, including files for the Tripos modeling program Alchemy-III and the 2D or 3D structure files for MDL molecule database applications such as the ISISBase/ISISDraw suite.

Mol2Mol a molecular structure input file conversion program which runs in the Microsoft Windows environment. Mol2Mol enables standard molecule structure editing as well as standard manipulations of the structure such as translation, rotation, and stereo modes. It also is possible to calculate basic geometrical data such as atom–atom distances, bond angles, planes, pyramidality, proton–proton and proton–methyl centroid distances, etc. This and other file conversion programs are discussed in Section 4.5.

Mol2Mol is commercially available through Cherwell Scientific Publishing.

molecular dynamics a family of techniques which model the dynamic behavior of molecules by applying the Newtonian laws of motion to individual atoms and tracking the effect over time in order to simulate the gross movement of the molecule. There are many methods of carrying out molecular dynamics.[113,581]

In a molecular dynamics experiment, a static molecular structure derived from force field calculations is perturbed in a series of small steps (on the order of 10^{-15} s, or 1 fs) by the assignment of a force vector to each atom of the molecule. The acceleration of each atom will be a function of this force vector and of the mass of the atom, and the velocity of each atom is calculated from its position and acceleration by some algorithm. The next dynamic step takes the new resulting static structure through a second, analogous perturbation. The static structure from each step is stored, and when presented in sequence constitute a frame-by-frame "movie" of simulated molecular movement.

Molecular dynamics simulations have a number of applications: to correlate molecular structure with experimental data, such as diffraction crystallography or NMR experiments; as a conformational search method; to study molecules in solution (see Section 5.6); to simulate intermolecular interactions such as docking and molecular recognition; and to simulate macroscopic processes and properties such as melting, diffusion, and the mechanical properties of macromolecules, among many others.

molecular electrostatic potential (MEP) a physical property of a molecule or ensemble, which may be determined experimentally by X-ray diffraction, or may be calculated from the wave functions or from some other function which reflects the electron charge distribution.[222] It is most often represented as a two-dimensional contour map of isopotential lines superimposed upon the structure of the molecule under study. MEPs may be used to study such phenomena as protonation processes, molecular recognition, and excited electronic states and their influence on chemical reactivity.

The MEP function can be calculated rigorously for a molecule from the corresponding nuclear charges, the nuclear position vectors, and the electronic charge density function. It may be approximated as a multipolar expansion using the fractional atomic charges obtained from *ab initio* or semiempirical MO calculations. Molecular surfaces or contours derived from the MEP are used to represent molecular shapes or to model non-bonded interactions between the polar regions of molecules.[645]

molecular mechanics method See **molecular modeling**.

molecular modeling the use of graphical, mathematical, or physical representations of molecules to help understand and predict their structure and properties. Computer-aided molecular modeling is generally divided into three areas depending upon the level of theory employed: the *ab initio molecular orbital method*, the *semiempirical molecular orbital method*, and the *molecular mechanics method*. Each is described briefly below, and there are several in depth discussions of these three methodologies available.[1,88,119] Some general definitions follow:

... A general, all-comprising term including perception, manipulation, physicochemical parametrization, and visual reproduction of molecular structures ...[113]

... the use of theoretical models (methods) for the interpretation and/or prediction of molecular properties.[119]

... the science (or art) of representing molecular structures numerically and simulating their behavior with the equations of quantum and classical physics.[121]

High-precision molecular modeling has been defined as a method which, on average, reproduces heats of formation to within 1 kcal/mol, bond lengths to within 0.002 Å, valency angles to within 0.5°, and torsional angles to within 1°.[112]

In the *ab initio molecular orbital method* (from the phrase in Latin meaning "from first principles"), the molecular structure is calculated from mathematical descriptions of the component orbitals, or basis sets, which are combined and scaled. These equations cannot be solved exactly, and so the energy is calculated for a trial combination of basis set orbitals, and then a second combination with varied coefficients is calculated, and this process is continued iteratively to reach the lowest energy, which will approximate the "true" ground state energy. This is the most exact method of computer molecular modeling, and also the most expensive in computer CPU time. The complexities of this method limit it to smaller molecules.

The *semiempirical molecular orbital method* also involves solving equations for molecular orbitals, but in this case approximations are used. The first is to consider only the valence electrons (those involved in bonding). Another is to avoid the full calculation of the overlap integrals involving different atomic orbitals; they are replaced by simpler expressions, or by standard values obtained from experimental data or from other calculations. The usefulness of this method, then, will depend upon the quality of the empirical parametrization. Improvements in these methods generally involve additional correctional functions, or more efficient energy optimization procedures.

In the *molecular mechanics method*, or empirical force field calculation, the electrons are not involved in the calculation *per se*, but their properties are included in the parametrization for their parent atoms. The **force field** is a collection of potential functions from Newtonian physics, used to model bonded interactions (bond stretch, angle bend, and torsional strain, with optional cross terms) and nonbonded interactions (van der Waals repulsion/attraction, electrostatic repulsion/attraction, with optional specifically parametrized interactions such as hydrogen bonding or metal atom complexation). A simplified molecular orbital method may be used to take account of the special properties of conjugated π-electron systems. The potential energy functions are selected and harmonized with each other, to be able to reproduce experimental data of model compounds as closely as possible. The strength and limitation of the method are reflected in the character of the molecules in the basis set used to derive those parameters.[121,395]

molecular orbital (MO) that portion of the space around a molecule in which electrons of a specified energy may be found to a certain degree of probability. Alternatively, an electronic orbital derived from the combination of atomic orbitals of the constituent atoms of a molecule.

Molecular orbitals are calculated by a *molecular orbital method*. There are several sources of graphic representations of molecular orbitals, such as the monograph by Jorgensen and Salem.[815]

GLOSSARY OF COMPUTER TERMS 675

molecular recognition a term which describes the operation of elements of complementarity in the nonbonding interaction between two or more molecules, such as shape or surface complementarity, lipophilic attraction, electrostatic complementarity, dipolar attraction, and the complementarity of specific interactions such as between hydrogen bond donors and acceptors or between metals and ligands. The more of such interactions and complementarity are present, the higher the degree and the more specific is the molecular recognition.

Molecular skins an analytical method for the comparison of molecular shapes by optimizing the intersection of the respective *molecular skins*, which are molecular surfaces with thickness, i.e. a pair of surfaces (inner and outer) which are characteristic of the molecule (see Section 6.2).[670]

molecular statics that branch of molecular modeling which deals with determining a minimum-energy, resting geometry of a molecule. This molecular geometry may be a local or the global energy minimum.[112]

molecular surface the limit or boundary of the space occupied by a molecule (see Section 6.2). Several types of surface have been defined, and the character of the surface depends strongly upon the algorithm used to generate it and upon the choice of parameters (usually van der Waals radii) which serve as input to that algorithm. A quantitative conceptualization of the surface of a molecule was first introduced by Lee and Richards in the context of proteins.[80] Both outer and inner (i.e. cavity) surfaces were calculated from the van der Waals radii of the exposed atoms.

One model for the surface of a molecule is the *solvent-accessible surface*, defined as a smooth network of convex and reentrant surfaces traced by a probe sphere as it rolls over the molecule. In the Lee–Richards formulation,[80] the surface is defined by the *center* of the probe sphere; in the later Richards formulation[644] and the Connolly formulation,[651,652] the surface is defined by the inward-facing part of the probe sphere. In attempts to reduce confusion, this latter type of surface (defined by the inward-facing part of the probe sphere) has been referred to as the *molecular surface*, the *Connolly surface*, or the *solvent-exclusion surface*.

The *van der Waals surface* is another model for the surface of a molecule. It comprises the sum of the nonintersecting parts of the atomic spheroids determined by the van der Waals radii of the constituent atoms.[80] In mathematical terms, it corresponds to the outer limits (to a high level of probability) of the sum of all molecular orbitals. A van der Waals surface can also be thought of as a solvent-accessible surface where the probe sphere has a radius of zero. It is composed of two parts, the *contact surface* and the *reentrant surface*. The *contact surface* is the part of the van der Waals surface that is accessible to the probe sphere, i.e. that makes a single contact with it. The *reentrant surface* is the inward-facing part of the surface of the probe sphere when it is in contact with more than one atom.[80]

See also **potential energy surface**.

Molecular Surface Package also known as MSP, a suite of two programs developed by M. L. Connolly (**MSRoll** and **MSDraw**) which calculate and display molecular surfaces.[689] It is related to the earlier program MS (QCPE 429).

MOLGEN a molecular structure manipulation package developed by A. Kerber and R. Laue of the Universität Bayreuth.[225,971,972] MOLGEN functionality includes the generation of connectivity isomers, a graphical molecule editor and 2D display, a display for 3D placements using energy optimization, and the generation of all configurational isomers.

Monte Carlo random, or involving a random variable; probabilistic, involving chance or probability. Synonym for **stochastic**; in the molecular modeling conformational search context, an antonym for **grid search**.

A Monte Carlo method or model involves the random selection or generation of input data or coefficient values, or the use of random sampling techniques, to obtain approximate solutions to problems which cannot easily be solved exactly or quickly.[42,616] The term derives from the name of the main city of the European principality of Monaco, which is famous for games of chance. See also **Metropolis algorithm**.

MOPAC an abbreviation for *molecular orbital package*. MOPAC is a generalpurpose semiempirical molecular orbital package for the study of chemical structures and reactions. The semiempirical Hamiltonians MNDO, MINDO/3, AM1, and PM3 are used in the electronic part of the calculation to obtain molecular orbitals and the heat of formation. From these results the vibrational spectra, thermodynamic quantities, isotopic substitution effects, and force constants for molecules, radicals, ions, and polymers can be calculated. There is a transition state location routine, and two transition state optimizing routines are available for studying chemical reactions.[284]

The format of MOPAC input files is discussed in Section 4.5.

Morse potential an empirical anharmonic potential function useful in force field modeling of the bond stretch vibration, with an exponential dependence upon distance.[318–329] It was devised by the American physicist P. M. Morse,[317] and is discussed in Section 5.1.1.1.

mouse a device which translates motion on a two-dimensional real-world surface into motion on the two-dimensional electronic surface of the monitor. It is connected by a cable to a board in the system unit.

The mouse contains a tracking device (with a moving ball or directional rollers), the motion of which is cued electronically to address specific points within the display area of the monitor, or **pixels**. The user regards which part of the screen is being addressed by following the movement of the cursor.

MS-DOS see **DOS**.

MSDraw an abbreviation for *molecular surface draw*, a program written by

M. L. Connolly, one of a suite of two programs in the **Molecular Surface Package**.[689] MSDraw renders and plots molecular surfaces, computes interface-bisecting surfaces for protein complexes, and measures protein surface shape by solid angles.

MSRoll an abbreviation for *molecular surface roll*, a program written by M. L. Connolly, one of a suite of two programs in the **Molecular Surface Package**.[689] MSRoll reads a PDB file and rolls a water-sized probe sphere over the molecule. The program writes a piecemeal quartic molecular surfaces, a dot surface, a polyhedral surface, and a solvent-excluded density.

MUB an abbreviation for *modified Urey–Bradley*, a type of force field developed by L. S. Bartell and coworkers.[78]

Newman projection a stereochemical convention in drawn chemical structures in which two atoms bonded together are represented by a dot inside a circle. The substituents attached to each of the atoms may be indicated by lines radiating from either the central dot or the circumference of the circle, respectively. This convention is particularly clear in presenting **eclipsed**, **gauche**, and **anti** stereochemical relationships. The term derives from the name of the American chemist M. Newman.
IUPAC:

To prepare a Newman projection, a molecules is viewed along the bond between two atoms; a circle is used to represent these atoms, with lines from outside the circle toward its center to represent bonds to other atoms; the lines that represent bonds to the nearer and further atom end at, respectively, the center and the circumference of the circle. When two such bonds would be coincident in the projection, they are drawn at a small angle to each other.[240]

Newton–Raphson algorithm a mathematical minimization engine first applied to molecular mechanics strain energy minimization by Boyd,[58] this algorithm applies both the first and second derivative of the change in energy in order to guide the system to equilibrium. A modified version of this method, the *block-diagonal Newton–Raphson method*, was primarily developed for molecular mechanics strain energy minimization by Allinger and coworkers.[572,576] These methods are discussed in the context of other minimization engines in Section 5.7.3.

NMR spectroscopy NMR is an abbreviation for *nuclear magnetic resonance*. This type of spectroscopy involves the interaction of the nuclei of certain atoms within a sample molecule with a magnetic field and radio-frequency electromagnetic radiation. Magnetically active nuclei are those which have a nonzero spin. Such nuclei may occupy one of two spin states in the presence of an external magnetic field, and absorption of the correct quantum of radio-frequency electromagnetic radiation allows the nuclei to jump from the lower state to the higher. The exact frequency necessary for the transition depends on the local environment of the nucleus. The relaxation of the nuclei back to their original state can be detected, and the frequency spectrum obtained provides information about the structure of the sample molecule. Examples of magnetically active nuclei are ^{1}H, ^{13}C, and ^{19}F.

NOE an abbreviation for *nuclear Overhauser enhancement*. The NOE effect is employed in a technique in **NMR spectroscopy** in which energy absorbed by one magnetically active nucleus is transferred through space to, and causes a spin state transition in, another nucleus. The nuclei can be of the same element (homonuclear, e.g. both hydrogen) or different (heteronuclear, e.g. hydrogen and carbon). Since the energy transfer occurs through space, its efficiency drops off rapidly with increasing distance. The magnitude of the homonuclear NOE can be used to provide information about the distance between the two atoms involved. In the case of conformationally flexible molecules, this effect will predict the time-averaged distance between the two atoms.

nucleophile a chemical species (atom, molecule, or ion) which is attracted to a region of low electron density on another chemical species. A second definition is a chemical species able to donate an electron pair to another chemical species. This latter partner of the nucleophile is termed an **electrophile**. The name comes from combining "nucleus" with Greek $\varphi\iota\lambda o\varsigma$ (loving), and was coined by C. K. Ingold.[973]

OPLS an abbreviation for *Optimized Potentials for Liquid Simulations*. OPLS is a set of potential functions and parameters for solvent molecules developed by the Jorgensen group to model different liquid phases,[462] in order to predict thermodynamic and physical properties of the liquid medium (see Section 5.4.7). The water model **TIP4P** is considered part of the OPLS suite.

OPLS-AA is a new, all atom formulation which incorporates the bond stretch and angle bend potentials from **AMBER**, torsional potentials which have been obtained from *ab initio* calculations, and the OPLS nonbonding potential terms (i.e. Coulomb electrostatics and Lennard–Jones 6–12 sterics).[476]

OPTIMOL a host molecular modeling program developed at Merck, which can employ a user-selected force field to model ligand–macromolecule interactions.[456]

option button See *radio button* under **button**.

ORTEP an acronym for the *Oak Ridge Thermal Ellipsoid Plot* program. ORTEP is a Fortran molecular graphics program used to produce a type of ball-and-stick presentation of a molecule from its diffraction crystal structure data. It can produce either a monoscopic graphic or a stereoscopic pair of representations. The senior author is C. K. Johnson of the Oak Ridge National Laboratory. ORTEP-I was distributed in 1965, ORTEP-II in 1976, and ORTEP-III in 1996. Information, examples, and a manual for ORTEP-III may be accessed at a Web page.[64]

The "balls" in this case may be spheroids or ellipsoids which encompass the area swept out by the constituent atoms in the course of their averaged thermal motions. The ellipsoid parameters are derived from anisotropic temperature factors on the atomic sites. These atom representations are generally drawn with a one-eighth section (octant) removed and the concavity shaded as an aid to three-dimensional visualization of each ellipsoid.

For graphical purposes in ORTEP, an ellipsoid is considered to be composed of ellipses and straight lines. The ellipses are of two types—principal ellipses and boundary ellipses. Relative to the viewpoint, a principal ellipse is further subdivided into a front half and a back, or hidden, half. There are three principal ellipses per ellipsoid, corresponding to the three principal planes. The boundary ellipse is the edge of the ellipsoid as seen from the viewpoint. The front and back halves of the principal ellipses meet at the boundary ellipse. The straight-line segments of the ORTEP ellipsoid are the forward principal axes, reverse principal axes, and octant-shading lines. Various combinations of these components are seen in Fig. D-2, along with the ORTEP instruction number and parameter values used to produce each.[64]

The structure rendering codes in ORTEP have been implemented in many modeling applications, including PCMODEL (see Section 3.6).

out-of-plane angle or **out-of-plane bend** or **out-of-plane distortion**   a measure of the molecular vibration in which a trigonal atom, such as sp^2 carbon or boron, assumes a nonplanar or pyramidal geometry. It is defined as the angle between a bond from the central trigonal atom to one substituent, and the plane determined by the three substituents (see Section 5.1.1.4 and Fig. 5-6 for a geometric definition).[329]

Many force fields incorporate a restoring potential to reinforce planarity in these systems. This distortion can also be quantified as an **improper torsion**, the **pyramid height**, or the **Wilson angle**.

parameters   in the molecular modeling context, the constants and coefficients fed into the force field equations to determine the energy of a system. Parameters for empirical force fields may be calculated from experimental data, but they often require some optimization or modification to be applicable to a wide variety of systems. See Section 5.5.

In general mathematical terms, a parameter is a constant in a mathematical program, not subject to choice in the decision problem, but that could be varied outside the control of the decisions.[933]

parse   to analyze a string of data, resolving it into component parts and testing its conformity to a "grammar" or logic, in order to assign a meaning to these components.

A *parser* is a programming device or **engine** designed to parse data files.

paste   an opertion which returns text or graphics from the clipboard to a new location within the same program window or file, or into a different program or file. Generally, the mouse cursor is used to designate the posi-

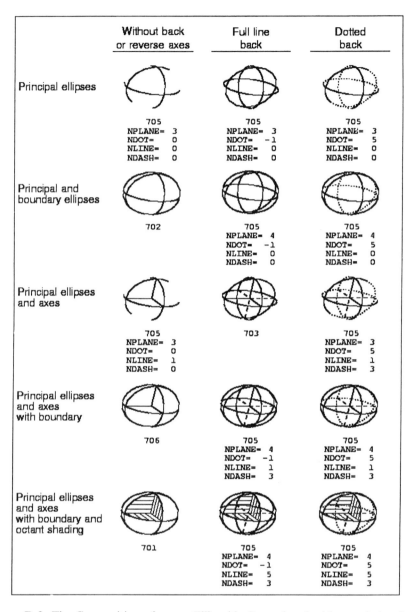

Figure D-2. The Composition of ORTEP Ellipsoids. Reproduced with permission from Michael N. Burnett and C. K. Johnson, ORTEP-III: Oak Ridge Thermal Ellipsoid Plot Program For Crystal Structure Illustrations, Oak Ridge National Laboratory Report ORNL-6895, July 1996.

tion on the screen where a paste takes place.

PDB the **file extension** and **MIME** type for structure files in the format promulgated by the **Protein Data Bank** of the Brookhaven National Laboratory. This format is discussed in detail in Section 4.5. The vast majority of structure files available on the **World Wide Web** are found in this format. The **RasMol** program is a convenient web browser plug-in to visualize structures in this format.

PEF an abbreviation for potential energy function.

penalty function in the molecular modeling context, a potential function which exacts an energetic penalty related to the deviation of a quantity from its ideal value. It is closely related to the Hooke's law harmonic potential functions often applied to bond stretch and angle bend interactions. A generic penalty function V_{penalty} takes the form

$$V_{\text{penalty}} = K_{\text{penalty}}(r - r_0)^2$$

where K_{penalty} is a constant; r is the quantity of interest, such as an internuclear distance; and r^0 is the equilibrium value for this quantity. Penalty functions have been used in organometallic force field models (see Section 6.7).

In mathematics, the penalty function is traditionally composed of a function and one positive constant, which augments an objective so that the original mathematical program is replaced by solving a parametric family of the form $\max\{f(x)\text{-}uP(x)\}$. The function P is called a penalty function if it satisfies $P(x) > 0$ for x not feasible and $P(x) = 0$ if x is feasible.[933]

PEOE an abbreviation for *Partial Equalization of Orbital Electronegativity*, an algorithm to calculate atomic charges developed by Gasteiger and Marsili.[452]

In this method, an equation expresses the electronegativity of a bonding orbital on an atom as a quadratic polynomial function of the total charge on that atom, with coefficients which depend upon the valence state of the atom. These derived orbital electronegativities are then used to generate partial charges on the two bonded atoms. The partial charges on an atom are then summed to provide a new total charge. A new iteration then begins in which the orbital electronegativities are recalculated from the new charges, etc., and this continues until the charge distribution becomes self-consistent (see Section 5.6).

PES an abbreviation for *potential energy surface*.

pharmacophore a term used in **computer-aided molecular design**, meaning a collection of 3D structural features which together confer a particular pharmacologic or medicinal effect. The implication is that a molecule whose properties match the features of the pharmacophore should exhibit the associated effect.

Nearly all pharmacophores will be putative, that is, inferred from known structural features of the biochemical site of action, or hypothesized from some data reduction analysis carried out on a large number of known active molecules.

Pharmacophores can be coded in a variety of ways: as a set of interaction or affinity sites (**electronegative, hydrophobic**, etc.), separated from each other by a set of geometric constraints with associated tolerances, or as a set of complementary **molecular surfaces**, or even as a specific molecular functional group or substructure.

A pharmacophore which is derived from a data reduction analysis carried out on a large number of known active molecules, looking for binding interactions which are common to the more active analogs, without recourse to any structural data for the actual binding site, may be termed a *virtual pharmacophore* or a *pseudoreceptor*.

PIMM is a force formulation which combines a π-SCF-LCAO-MO treatment of conjugated systems with a molecular mechanics treatment of the σ-framework of the molecule, developed by the Lindner group at the Technische Hochschule Darmstadt in Germany.[517,518] The *PIMM* force field is discussed in Section 5.4.8.

pixel a neologism derived from the phrase *picture element*. The smallest visual element of the display on a monitor. All text and graphical images are made of pixels, and the number of them limits the resolution of the monitor. Typical resolutions run from 640 by 200 to 1024 by 768.

A pixel is the smallest resolvable rectangular area of an image, either on a screen or stored in memory. Each pixel in a monochrome image has its own brightness, from 0 for black to the maximum value for white (e.g. 255 for an eight-bit pixel). In a color image, along with brightness each pixel has its own color, usually represented as a triad of red, green, and blue intensities with a numerical value (0–255) for each.[932]

In Fig. D-3 is shown (above) a screen capture of the structure of alanine from an 800 by 600 monitor, and (below) an approximately 6 × expansion of the N–H region which shows individual pixels filled with differing gray tones.

plug-in a computer term meaning a program or utility which can mesh its function within the operation of a larger main or host program. A plug-in can be an improvement of some feature from the main program, such as an in-line update, or it can add functionality originally absent in the main program. In the Windows™ environment, a plug-in can function from within the main program's window, or through a child window. The plug-in program may be in the EXE (executable), **DLL**, or other format.

A plug-in can also be a file containing data used to alter, enhance, or extend the operation of a parent application program. The plug-in is usually specific to a particular operating system, and aids in the display or interpretation of a particular file format.[932] In the context of the **World**

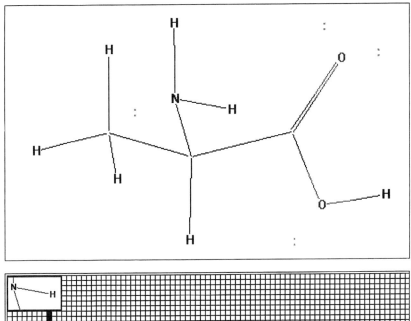
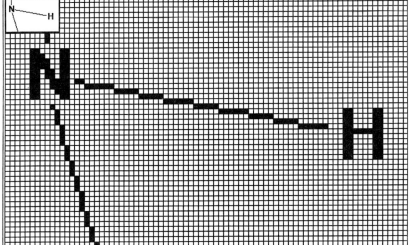

Figure D-3. ≈5.5 × Expansion of Alanine Structure Graphic to Show Pixels.

Wide Web, plug-ins help a browser to interpret certain **MIME** types and subtypes.

PLUTO part of the **Cambridge Structural Database System** suite of programs: a visualization tool for displaying the 3D structures found in the **Cambridge Structural Database**. It incorporates many features not available within the QUEST program, including the ability to explore the nonbonded networks present in crystal structures down user-defined atomic pathways.

PLUTO generates mono- or stereoscopic representations of molecules or crystal structures in three basic styles: stick or line structures, ball-and-stick structures, or CPK-like space-filling structures. The user has complete control over the perspective and the sizes of the bonds and atoms, has the option to exclude any atoms or elements, and can print the results to a **Postscript** file.

The structure rendering codes in PLUTO have been implemented in other applications, such as PCMODEL (see Section 3.6).

PNG an abbreviation for **Portable Network Graphics.**

Points-on-a-Sphere (POS) is a variant of **Valence Shell Electron Pair Repulsion** theory, in which pairwise interactions energies on the surface of a sphere are taken to simulate *valence shell electron pair repulsions*.[77] This approach has been applied with some success to predicting the structure of organometallics and compounds of the Main Group such as pentavalent phosphorus.

POS See **Points-on-a-Sphere.**

polarizability in the context of atoms, fragments, or molecules, the property of being able to distort an otherwise symmetrical outer electron distribution or electron cloud, and to assume dipolar character in response to a local electric field, such as is produced by another nearby dipole.

poling a technique developed by a research group at MSI for ensuring uniform coverage in a conformational search by associating a **penalty function**, or poling function, with each found conformation. The poling function modifies the potential surface by raising the energy drastically in the vicinity of a known conformation in order to avoid discovering another close by on the potential surface. Only those new conformations outside the *penalized conformational space* surrounding known conformations will have sufficiently low energy to be kept.[622–624]

Portable Network Graphics (PNG) a portable, well-compressed, well-specified **raster** image graphics standard for lossless bitmapped image files, still under active development by the World Wide Web Consortium.[974]

PNG was developed as a replacement for **GIF**, because of concerns over the legal status of free use of the latter. Currently, the specification contains a number of useful new features not available in GIF.

Postscript a graphics file format, or *page description language* (PDL), which can communicate a document description from a composition system to a printing system independent of the device used. Its primary application is to describe the appearance of text, graphical shapes, and sampled images on printed or displayed pages. It is an unusually powerful printer language in that it is a full programming language, rather than the series of low-level escape sequences which constitute most printing applications.

The development of Postscript began with J. Gaffney at Evans & Sutherland in 1976, continued through M. Newell at Xerox PARC, and implemented in its current form by J. Warnock *et al.* after he cofounded Adobe Systems, Inc. in 1982. It was first used broadly by the Apple LaserWriter, but is now widely implemented on many laser printers and on-screen graphics systems.[932]

potential energy function (PEF) a mathematical equation which relates a component of the overall potential energy of a molecule to a function describing a particular molecular vibration in a bonded moiety, such as bond stretch in a two-atom unit, angle bending in a three-atom unit, or attraction/repulsion in a nonbonded interaction such as steric or **van der Waals** interactions or **electrostatic** interactions (see Chapter 5).

The form of a potential energy function will contain one or more variables for experimental or derived physical quantities, parametrized constants such as **force constants**, coefficients for scaling the functional components, and possibly additional constants for unit conversion.

potential energy surface (PES) a mathematical surface associated with a molecule. For a molecule of N atoms, the potential energy surface will have $3N + 1$ dimensions, corresponding to three spatial coordinates for each atom and one additional dimension for the energy.[305] Simplified surfaces may be generated for a particular molecule if some of the spatial coordinates are correlated, or if only a subset of the atoms or motions are being related to the energy.

PowerFit a program for the structural comparison of different molecules or of different conformations of the same molecule, distributed by MicroSimulations, Inc., and currently available for Windows 95, Windows NT, and PowerMac platforms. It uses a fitting potential based on the SEAL (steric and electrostatic alignment) method. It offers manual fitting with rotation, translation, and torsional changes, or automated rigid body fitting, conformationally flexible fitting, and fitting with distance constraints. It enables the calculation of fitting energy as a quantitative measurement of the fit.

probe sphere a theoretical entity in the computer modeling of molecules, primarily used to construct **molecular surfaces**. A probe sphere is often, but not always, an idealized solvent molecule. For example, water may be represented as a sphere of radius 1.4 Å. The probe sphere can be driven computationally over the exposed portions of a molecule, or over an interior cavity, and from some aspect of its path a surface is assembled by summation. See **molecular surfaces**.

Property3D a program for the calculation of physical properties for a molecule with the 3D structure (AccuModel or MDL MOL file) as input, distributed by MicroSimulations, Inc., and currently available for Windows 95 and Windows NT platforms. Property3D can calculate molecular volume, molecular surface area, molecular size and principle axes in three dimensions, molecular dipole moment, molecular radius of rotation, molecular shape index, molecular branch index, and the molecular flexibility index. It can create a "flower plot" for visualizing molecular properties, and it can process a large number of structures in batch mode.

Protein Data Bank (PDB) an archival computer database of three-dimensional structures of biological macromolecules.[280] The database contains atomic coordinates, bibliographic citations, and primary sequence and secondary structure information, as well as crystallographic structure factors and 2D NMR experimental data. Information is available on 5638 proteins, 447 nucleic acids, and 12 carbohydrates as of July 16, 1997.[137,281]

pseudoatom   See **dummy atom**.

pseudoreceptor   See **pharmacophore**.

pyramid height   a measure of the molecular vibration in which a trigonal atom, such as sp^2 carbon or boron, assumes a nonplanar or pyramidal geometry. It is defined as the distance at which a nonplanar trigonal atom is displaced from the plane determined by its three attached atoms[327] (see Section 5.1.3 and Fig. 5-6).

Many force fields incorporate a restoring potential to reinforce planarity in these systems. This distortion can also be quantified as an **improper torsion**, the **out-of-plane angle**, or the **Wilson angle**.

QCPE   is an abbreviation for the *Quantum Chemistry Program Exchange*, a clearinghouse, based at Indiana University, for freely available, noncommercial, well-known computational chemistry software programs widely used in the field of molecular chemistry and general-purpose programs in mathematics, physics, and chemistry.

Established in the early 1950s, QCPE has served as a collection and distribution center for software for use in understanding the structure and energetics of molecules. Initially established for quantum chemistry programs, the center now collects and distributes a variety of software of use to all chemists. Current holdings are more than 850 systems, programs, and routines.[966]

The archives are searchable on the **World Wide Web**, and links to these sites are provided in the HTML document " Molecular Modeling Resources on the World Wide Web" accompanying this book.

QMFF   an abbreviation for *Quantum Mechanical Force Field*, an updated version of the CFF molecular modeling formulation developed by Hagler *et al.* at Biosym Technologies.[90,92] The potential energy surface from quantum mechanical calculations was used in the parametrization, and extensive use is made of off-diagonal or cross-term components in the force field equations which characterize this *Class II force field*. See **force field** and Section 5.4.2.4.

QM/MM   an abbreviation for *quantum mechanics/molecular mechanics*, A hybrid molecular modeling technique, especially for transition state molecules, where the *reaction zone* (those bonds in the molecules which are being formed and broken) is modeled with a molecular orbital method, and the remainder of the system is evaluated by force field methods.[633,898-902]

Quanta a suite of programs for 2D and 3D modeling, simulation, and analysis of macromolecules and small organics, including structural and similarity analysis tools, cluster analysis, and flexible fitting. Quanta has access to all standard file formats, and interfaces to the **CHARMm** force field and other MSI modeling programs. Distributed by Molecular Simulations, Inc.

Quantum Mechanics/Molecular Mechanics See **QM/MM**.

QUEST3D a module of the **Cambridge Structural Database System** suite: a program to generate 3D queries to search over structures or substructures in the **Cambridge Structural Database**. It will display text and 2D and 3D graphical data on the found structures, and creates the subfiles of database search hits required as input for the **VISTA** and **PLUTO** modules.

R a stereochemical descriptor for one of the configurational isomers of a molecule containing an asymmetric center. To assign the configuration, the molecule is held in a standard orientation, which for tetrahedral carbon involves placing the lowest priority substituent on the far side of the asymmetric center from the viewer. The R-isomer is the one for which a clockwise path leads from the highest to second-lowest priority substituents. The other isomer is S. The **CIP** sequence rules are used to rank the priority of the substituents.[240]

In the case of asymmetric carbons, earlier conventions called such **stereopairs** d/l, D/L, or $+/-$.

The term comes from Latin *rectus*, which means "right."

r_0 a measure of bond length obtained from microwave spectroscopy, defined as the distance between nuclear positions derived from rotational constants A_0, B_0, and C_0 for the ground vibrational state.[109]

r_α a measure of bond length obtained from electron diffraction, defined as the distance between mean nuclear positions at thermal equilibrium for a specified temperature, and derived by applying a vibrational correction to r_g.[109,243]

r_α^0 a measure of bond length, defined as the distance between average nuclear positions at thermal equilibrium in the vibrational ground state, extrapolated to 0 K. The quantity r_α^0 is equal to r_z, and is the one used for comparisons with X-ray diffraction data.[109,243]

r_a a measure of bond length, obtained directly from the electron diffraction radial distribution function. It is an argument in the scattering intensity derived from scattering vectors. It comes from the thermal average values of internuclear distances evaluated with a weight factor of r^{-1} in the averaging.[109,243]

r_e a measure of bond length obtained from microwave spectroscopy, defined as the distance between nuclear positions derived from the rotational constants A_e, B_e, and C_e for the hypothetical vibrationless state. The values of

A_e, B_e, and C_e are derived from rotational constants for at least two vibrational states, one of which is always the ground state.[109,243]

r_g a measure of bond length obtained from electron diffraction, defined as the thermal-average value of internuclear distances. It is equivalent to the average of r_a over all of the molecular vibrations.[109,243]

r_s a measure of bond length obtained from microwave spectroscopy, defined from the substitution structure for a set of isotopomers obtained from Kraitchman's formulae.[109,243]

r_v a measure of bond length obtained from microwave spectroscopy, defined as the distance between average nuclear positions of the vibrational state v.[109,243]

r_z a measure of bond length obtained from microwave spectroscopy, defined as the distance between average nuclear positions (i.e. r_v) at thermal equilibrium in the vibrational ground state. The quantity r_z is equivalent to r_α^0.[109,243]

radical   a chemical species having one or more free valences, or singly occupied orbitals. A radical may be charged or uncharged.

radio button   See **button**.

RASMol   a public domain molecular visualization program developed by R. A. Sayle of Glaxo Wellcome Res. & Dev. The name is considered to be derived from a contraction of *raster molecule*, or from the initials of the developer plus "mol."

RASMol is intended for the visualization of proteins, nucleic acids and small molecules, and is aimed at display, teaching, and generation of publication quality images. The program runs on Microsoft Windows, Apple Macintosh, Unix, and VMS systems. The program reads in a molecule coordinate file and interactively displays the molecule on the screen in a variety of color schemes and molecule representations, such as depth-cued wireframes, Dreiding sticks, space-filling (CPK) spheres, ball and stick, solid and strand biomolecular ribbons, atom labels, and dot surfaces. It supports molecular animation through the use of script files.

RASMol may be used as a plug-in to a Web browser such as Netscape, in order to visualize molecules available from that medium. RASMol now comes bundled with MDL's ISISDraw/ISISBase suite, for use in visualizing structures in this medium. It can also operate as a standalone program.

The RasMol manual is available in electronic book form.[975] For a description of the file format capabilities of RASMol, see Section 4.5.

RASSE   an acronym for *Rational Space Searching*, a program for *de novo* molecular design developed by L. Lai and coworkers.[172] It is based on atom growing with combinatorial search of atom types and conformations.

Raster3D   a program for preparing high-quality images from molecular graphics representations. It can prepare illustrations of space-filling models, ball-and-stick models, and ribbon-and-cylinder representations.[976]

Raster3D was developed by E. A. Merritt and M. E. P. Murphy of the University of Washington, and is currently available gratis.

raster graphics See **graphics**.

real time immediate occurrence of any computer operation, especially changes in a graphic image, as the user executes the corresponding command. For example, a real-time rotation of a structure will occur at the same time that the rotation command is executed.

regioselectivity a term coined by the American chemist A. Hassner to denote a relative preference for one product among several possible products of a chemical reaction in terms of orientation or direction.[993] The term is derived from the Latin word regio meaning 'direction'. For example, regioselectivity may be observed in the addition of an unsymmetrical **electrophile** to produce an unsymmetrical alkene and alkyne, or in an elimination reaction to produce an alkene. A reaction which produces any product distribution other than that dictated by statistics is described as **regioselective**. A degree of regioselectivity is ascribed to a reaction which is not completely **regiospecific**. See **regiospecificity**.

regiospecificity a term coined by the American chemist A. Hassner to denote a preference for a single product among several possible products of a chemical reaction in terms of orientation or direction.[993] The term is derived from the Latin word *regio* meaning 'direction'. For example, regiospecificity may be observed in the addition of an unsymmetrical **electrophile** to produce an unsymmetrical alkene or alkyne, or in an elimination reaction to produce an alkene. When a single product results from a reaction in which a distribution of products would be dictated by statistics is described as **regiospecific**. If the product preference in not complete, the reaction may be described as **regioselective**. See **regioselectivity**.

resonance a concept used to describe the bonding in molecules which have a *conjugated* π-electron system (see **conjugation**) and for which a single-valence-bond format structure is insufficient. In this view, the **delocalization** of the π-electrons leads to chemical and physical behavior which can only be predicted when the molecule is thought of as a hybrid or composite of two or more canonical structures, or resonance structures.

restricted Hartree–Fock See **Hartee–Fock**.

RHF an abbreviation for *restricted Hartree–Fock*; see **Hartree–Fock**.

RINGMAKER a program for carrying out conformation search on cyclic structures developed by the research group of W. C. Still. Using this technique, a single ring bond is broken, and a systematic search is carried out over the rotatable bonds of the resulting acyclic structure at a specified angle increment. Only those conformations which bring the two atoms of the broken bond within a specified distance are kept and subjected to energy minimization.[477,478]

ring strain a combination of torsional strain and angle strain (and transannular **van der Waals repulsion** in larger rings) which raises the potential energy of certain cyclic structures. For example, small rings (three- and four-membered) are more strained than related, slightly larger rings, and carbocyclic rings having an odd number of members are more strained than homologous rings having an even number of members. Transannular repulsions between hydrogens and other substituents become important in rings having more than six members.

ROSDAL   an acronym for *representation of structure description arranged linearly*. ROSDAL is a chemical line notation format (one-dimensional structure code) developed for organic molecules by S. Welford, J. Barnard, and M. F. Lynch for the Beilstein Institute. B. Roth extended ROSDAL to inorganic notation for the Gmelin database. Atom charge, isotope, and coordinate values are interpreted if present.[259]

rotamer   a rotational isomer. Rotamers can be separated when the magnitude of the *torsional barrier* (see **torsion**) is sufficient to hinder free rotation. A synonym is *conformer* (see **conformation**).

rotational barrier   see *torsional barrier* under **torsion**.

S   a stereochemical descriptor for one of the configurational isomers of a molecule containing an asymmetric center. To assign the configuration, the molecule is held in a standard orientation, which for tetrahedral carbon involves placing the lowest priority substituent on the far side of the asymmetric center from the viewer. The *S*-isomer the one for which a counterclockwise path leads from the highest to second-lowest priority substituents. The other isomer is *R*. The **CIP** sequence rules are used to rank the priority of the substituents.[240]

In the case of asymmetric carbons, earlier conventions called such stereopairs d/l, D/L or +/−.

The term comes from Latin *sinister* which means "left."

saturated   a term applied to organic molecules, which indicates no open valences and no multiple bonds. Saturated molecules are relatively unreactive toward addition and some substitution reactions. See also **unsaturated** and **coordination**.

SCF   an abbreviation for **self-consistent field**.

SCULPT   a molecular modeling program for the Apple Macintosh and SGI hardware environments, which allows the user to grab and part of a protein model and move it to a specific place. Initially, these movements are made without regard to changes caused by steric clashes and other physical interactions; in a separate step the structure is submitted to a batch mechanics program for repair of the induced structural defects. The program supports electrostatic and van der Waals interactions, and can also model amino and nucleic acids, cofactors, and arbitrary small molecules.

SCULPT enables the interactive manipulation of atoms and secondary structure, interactive docking of a ligand into a flexible receptor, on-the-fly specification of movable and frozen side chains and backbones, and *de novo* protein design. It was developed by M. Surles, J. Richardson, and D. Richardson of Duke University and F. Brooks of UNC—Chapel Hill, and is available through Interactive Simulations, Inc.[977]

second row refers to the row of the periodic table which begins with sodium and ends with argon, and contains some of the elements common to organic chemistry: silicon, phosphorus, sulfur and chlorine.

self-consistent field (**SCF**) an approach to **molecular orbital** calculations through successive approximations or estimates. Through an iterative process, the coefficients of the orbitals are modified from cycle to cycle until the electronic energy reaches a constant minimum value and the orbitals no longer change, resulting in a self-consistent field.

In the context of PCMODEL π-calculations, the *variable-electronegativity self consistent field* (VESCF) method is a Hückel MO treatment in which the π-atom parameters are not based solely on atom type, but include a contribution from the electronegativity of the substituents attached to the atom and the π-electron density on the atom.[751,774] This method gives good structures and accurate electron distributions, but the calculated heats of formation aren't consistent with those for unconjugated molecules.

semiempirical method See **molecular modeling**.

sequence number a number found in an MMX parameter file line, which occurs after the parameter type code word (see Section 4.7). It is merely for bookkeeping purposes, and is not used in any calculation. Generally, sequence numbers will increase monotonically in the parameter file lines for a particular parameter type.

sequence rules a system of rules which permit the determination of priority of substituent atoms or groups in a molecule capable of existing as stereoisomers. When the priorities of the substituents are known, then the proper stereochemical descriptor may be assigned.[240]

SHAKE a molecular dynamics algorithm developed by Ryckaert, Ciccotti, and Berendsen[594,595] in which the bond lengths and angle spans of the subject molecule are frozen so that the dynamic trajectory may be calculated without interference by the most rapid internal vibrations. Newer applications of SHAKE have frozen hydrogen bonds.[628]

simulated annealing a stochastic computational method for searching for minimum-energy conformations of a molecule. Essentially, this method places a molecule in a very hot *bath* in which much conformational flexibility is possible. The program then samples the allowable conformations while gradually cooling the bath until a single conformation is "frozen out."

In this way, a molecule slides into a potential energy well, much as it does in a standard energy minimization protocol. The difference is that built-in probabilistic factors provide a way out of the well. Over several iterations, this method generally provides a satisfactory conformational search. See also **Metropolis algorithm**.

Simulated annealing is an algorithm for solving hard combinatorial optimization problems, based on the metaphor of physical annealing: reach a minimum-energy state upon cooling a substance, but not too quickly, in order to avoid reaching an undesirable final state. As a heuristic search, it allows a nonimproving move to a neighboring state with a probability that decreases over time. The rate of this decrease is determined by the cooling schedule, often a parameter used in an exponential decay, in keeping with the thermodynamic metaphor. With some assumptions about the cooling schedule, this will probably converge to a global optimum.[933]

SMILES   an acronym for *simplified molecular input line entry system*. It is an ASCII text-only line notation which codes for the connectivity of chemical structures.[255–258]

SMILES code and two-dimensional structures may be interconverted readily, following the rules for coding. Atoms are indicated by their element symbols with an uppercase (or an initial uppercase) letter, and hydrogens are generally not specified. Aromatic atoms are indicated in lowercase letters. Single bonds are generally not specified; a double bond is indicated by = and a triple bond by #. Branches are specified by enclosure within parentheses. Cyclic structures are depicted by breaking open the ring to a linear string, and denoting the two formerly bonded atoms by appending a number to their atom designators, using the same number for both formerly bonded atoms.

SmilogP   a program for the rapid evaluation of **lipophilicity** (log P) from the **SMILES** code string of a particular molecule, created by T. Convard *et al.*[978] SmilogP creates an extended connectivity matrix from the SMILES string, and then matches these elements with lipophilicity contribution codes which are summed to provide the estimate of log P.

soft acid   an acid with large acceptor atoms having low positive charge and containing unshared lone pairs of electrons (*p*- or *d*-orbitals) in their valence shells. A soft acid has high polarizability and low electronegativity, and will react preferentially with a **soft base**. In soft acids, the acceptor atom has several easily excited outer electrons.[965] See **HSAB principle**.

soft base   a base with donor atoms of low electronegativity and high polarizability. A soft base is easy to oxidize, holds its valence electrons loosely, and will react preferentially with a **soft acid**. In soft bases, the donor atom is easily oxidized and is associated with empty, low-lying orbitals.[965] See **HSAB principle**.

solvent-accessible surface   See **molecular surfaces**.

SPROUT   a computer program developed by A. P. Johnson (University of Leeds) to carry out *de novo* molecular design. It accepts user-supplied approximations and constraints, and uses these conditions to guide the assembly of structural fragments or templates with the goal of producing a novel molecule which satisfies the criteria.[156]

staggered a stereochemical term for the relationship between two **vicinal** substituents in a conformation characterized by a dihedral angle of approximately 60°. It is most often used to describe hydrogen substituents, or in cases where the substituents are the same. See also **gauche**.

start token See **token**.

steepest descent algorithm a mathematical method first applied to molecular mechanics strain energy minimization by K. Wiberg.[57] This method drives a system to an energy minimum by examining how small changes in the positions of the constituent atoms affect the energy, and implementing those changes which reduce the energy. It is discussed in the context of other minimization engines in Section 5.7.3.

stereo a prefix indicating an involvement or importance of the three-dimensional aspect, specifically at the molecular level. The term derives from the Greek στερεος (solid).

stereochemistry the study of the three-dimensional aspects of chemistry. One may also speak of the stereochemistry of a molecule, or the stereochemistry at a particular center or atom; this usage is synonymous with **configuration**.

stereoelectronic concerning the three-dimensional relationship of atomic or molecular orbitals; having to do with the overlap or vectorial alignment of the axes of such orbitals, especially as a rationale for the operation of a selective or specific reaction pathway or mechanism, when other effects such as innate reactivity or nonbonded interactions are insufficient to explain empirical results.

stereoisomer one of a group of two or more molecules which consist of the same number and types of atoms with the same connectivity but different configurations.
IUPAC:

Isomers are termed stereoisomers when they differ only in the arrangement of their atoms in space.[240]

A stereoisomer may be a **geometric isomer**, a **diastereomer**, or an **enantiomer**.

stereolithography a technology which converts the 3D image of an object resident in a computer-aided design (CAD) program to a solid object. This is most commonly implemented with stepwise photopolymerization. In each step, the CAD program directs a laser beam to cure a 2D pattern on the surface of a monomer solution (the XY plane). This process is used iteratively by appropriate steps along the Z-axis, by a sequence of adding more monomer solution, curing, etc., until the fabrication of the object is complete.[102]

stereopair a pair of stereoisomers.

stereoselective a term applied to a reaction in which there are at least two possible stereoisomeric products, and where predominantly a single stereoisomeric product is obtained.

stereospecific a term applied to a reaction in which there are at least two possible stereoisomeric products, and where only a single stereoisomeric product is obtained.

steric hindrance a term from conformational analysis which refers to nonbonding through-space interactions which alter, slow down, or prevent the approach of a reagent or reactant to a substrate in a chemical reaction.

stochastic random, or involving a random variable; probabilistic, involving chance or probability. Synonym for **Monte Carlo**, antonym for **grid search** in the molecular modeling context.

A *stochastic method* or *model* involves the random selection or generation of input data or coefficient values, or random sampling techniques, to obtain approximate solutions to problems which cannot easily be solved exactly or quickly. See also **Metropolis algorithm**.

stop token See **token**.

substructure a separate fragment or a portion of an intact structure. In PCMODEL, substructure specifically means a designated collection of atoms, which may or may not be connected and which are defined as a separate group, differentiated from the remaining molecule(s) in the work space. See **Substr** in Appendix B, and Section 3.2.6.

surface See **molecular surfaces**.

synclinal a stereochemical term proposed by W. Klyne and V. Prelog[929] to describe the relative **conformation** of two **vicinal** substituents about a **dihedral angle**. The two substituents A and B in the system $A—X—Y—B$ bear an synclinal relationship to each other if the dihedral angle between them is $+60° \pm 30°$ or $-60° \pm 30°$ (see Fig. D-1). See also **anticlinal, antiperiplanar, synperiplanar**.

synperiplanar a stereochemical term proposed by W. Klyne and V. Prelog[929] to describe the relative **conformation** of two **vicinal** substituents about a **dihedral angle**. The two substituents A and B in the system $A—X—Y—B$ bear an synperiplanar relationship to each other if the dihedral angle between them is $0° \pm 30°$ (see Fig. D-1). See also **anticlinal, antiperiplanar, synclinal**.

tautomerism a form of structural isomerism where the two or more structures are interconvertible by means of migrating a proton. An example is the interconversion between the keto and enol forms of a carbonyl compound.

thermodynamic control a term applied to reactions or the products formed from reactions, meaning that the product(s) with the largest decrease in enthalpy and/or the largest increase in entropy will predominate. See **kinetic control**.

TINKER a molecular modeling software package for Unix platforms, for carrying out molecular mechanics and dynamics calculations on polypeptides and other molecules. TINKER has a native parameter set, and can also use the AMBER/OPLS, CHARMM22, MM2/MM3, and AMBER-95 parameter sets; support for the newer ENCAD, MMFF-94, and MM4 parameter sets is under development. TINKER also implements algorithms such as Elber's reaction path methods, modified versions of Scheraga's and Straub's global optimization via potential smoothing, multipole expansion treatment of electrostatics and polarizability, Eisenberg–McLachlan ASP and Macromodel **GB/SA** solvation, a novel fast distance geometry method, a new truncated Newton (TNCG) local optimizer, a new energy cutoff smoothing scheme, calculation of surface areas and volumes with derivatives, a simple free energy perturbation facility, normal mode analysis, and minimization in torsional space.

TINKER was developed by the research group of J. W. Ponder of Washington University.[702]

TIP3P an acronym for *transferable intermolecular potential, 3-point*, a force field formulation for liquid water developed by Jorgensen,[461] the three points being the two hydrogens and the oxygen. In this system, the H—O bond length and the H—O—H bond angle are fixed, and a combined nonbonding potential (Coulomb + Lennard–Jones 6–12) describes the intermolecular interactions during molecular dynamics simulation of liquid water structure (see Section 5.4.7).

TIP4P an acronym for *transferable intermolecular potential, 4-point*, a force field formulation for liquid water developed by Jorgensen,[461-463] the four points being the two hydrogens, the oxygen, and a virtual point between these three atoms where the oxygens' negative charge is localized. In this system, the H—O bond length and the H—O—H bond angle are fixed, and a combined nonbonding potential (Coulomb + Lennard–Jones 6–12) describes the intermolecular interactions during molecular dynamics simulation of liquid water structure (see Section 5.4.7).

TIPS an acronym for *transferable intermolecular potential functions*, force field formulations for solvent molecules, developed by Jorgensen.[458-460] In this system, bond lengths and angles are fixed and a combined nonbonding potential (Coulomb + Lennard–Jones 6–12) describes the intermolecular interactions during molecular dynamics simulation of bulk liquid structure (see Section 5.4.7). An intramolecular torsional potential is also added for appropriate solvent molecules such as ethanol.

toggle to switch back and forth between two or more options or parameter values. Sometimes used in the more restricted sense of a toggle between an on and an off state for some function. Similar to **flag**.

In PCMODEL, for example, the user may toggle the display of hydrogens on or off with the **H/OO** tool, or may toggle between six different coloring or labeling schemes in the **View/Labels** pop-up menu entitled 'Labels'.

token in the programming context, a symbol at the beginning or end of a data file which signals the **parser** to start, pause, or stop reading data.

In the context of PCMODEL, the *start token* is the left-hand brace '{', which appears at the beginning of a structure input file and signals the parser to begin reading parameters for the structure file. The start token must appear in the first line of input. The *stop token* is the right-hand brace '}', which at the end of a structure input file signals the parser to stop reading parameters for the structure file. The stop token must appear alone in the last line of input.

torsion a term taken from vibrational spectroscopy and used in molecular modeling. For a group of four atoms in a molecule such as I—J—K—L, when the pairs I—J, J—K, and K—L are all joined by chemical bonds, a rotation about an axis between atoms J and K of the units I—J and K—L with respect to each other is referred to as a torsion.

A *torsional angle* in a molecule with four atoms I—J—K-L with the pairs I—J, J—K, and K—L all joined by chemical bonds is defined as the angle between the bond vectors I—J and K—L along the axis between atoms J and K (see Figure 5-5). A synonym is *dihedral angle*.

A *torsional barrier* is a potential energy barrier to free rotation about a torsional angle, normally caused by torsional strain between bonding electron pairs and **van der Waals** repulsions between atoms or groups of atoms on either side of the dihedral. A synonym is *rotational barrier*.

A *torsional potential* in force field molecular modeling is a potential energy term applied to rotation about the central atoms J—K as defined above, which has the effect of restricting the rotation of the units I—J and L—K with respect to each other.

Torsional strain comes from one or more repulsive interactions which act as a barrier to free rotation through the torsional angle defined by the four atoms I—J—K—L. The origin of torsional strain is a subject of debate, and certainly depends upon the nature of the substituents. It may arise from repulsive interactions between the I—J and K—L bonding electrons, or from repulsions between the groups (atoms, molecular fragments, lone electron pairs, or hydrogen-bonded protons or complexed metals) bonded to I and those bonded to J.

TPL the filename extension used for an appended-style structure input file format used by the Catalyst program (Molecular Simulations, Inc.) to store the generated conformations of a molecule.

trans the descriptor applied to stereoisomers in which the substituents (atoms or groups) lie on opposite sides of a reference plane identifiable as common among the stereoisomers.[240] This descriptor is best reserved for application to the relative stereochemistry of two substituents on a cyclic compound, and its proper application requires that the ring be drawn in its most extended form, i.e. with no reentrant angles. See *cis*.

trans effect a kinetic phenomenon in the organometallic ligand substitution reactions of square planar and octahedral complexes in which a rate enhancement is found for substitution of a ligand at the position *trans* to certain activating ligands. This rate acceleration is ascribed to the **trans influence** of the activating ligands.[862–864]

Some examples of activating ligands for the trans effect, in decreasing order of directing ability, are: $C\equiv O \approx C\equiv N \approx H_2C=CH_2 > PR_3 \approx H- > CH_3 > C_6H_5 \approx NO_2- \approx I \approx SCN- > Cl^-$, $Br^- >$ pyridine, H_2O, HO^-, NH_3.[863]

trans influence a structural anomaly in organometallic complexes in which an activating ligand causes a weakening and an elongation of the bond from the metal to a *trans*-situated ligand. This influence is thought to arise from a competition between different ligands for interacting with a metal orbital. The more successful of the two ligands is termed the directing ligand, and the bond between the metal and the other ligand is consequently weakened and lengthened. This weaker bond may then undergo substitution with greater facility, which is known as the **trans effect**.[862–864]

Some examples of ligands in decreasing order of directing ability are: $H- > PR_3 > SCN- > I^- \approx -CH_3 \approx C\equiv O \approx -C\equiv N > Br^- > Cl^- > NH_3 > HO^-$.[863]

TSR an abbreviation for *terminate stay resident*. A TSR is a device driver or a program which may loaded from the {AUTOEXEC.BAT} or {CONFIG.SYS} files during system startup, or from the keyboard soon after boot. TSRs may not appear on the screen until a single hot key, or combination of hot keys, is entered. TSRs take up a certain amount of RAM, and may interfere with or prevent the operation of other programs, particularly memory-intensive programs such as PCMODEL. When there is expanded RAM available, it may be possible to load TSRs into high memory. It takes some experimenting to determine which programs and/or drivers should be loaded into high memory, and in what order, to create an efficient and workable configuration.

UBCFF an abbreviation for *Urey–Bradley Consistent Force Field*. See **CFF**.

UFF an abbreviation for *Universal Force Field*, a generic force field developed by A. K. Rappé et al. (see Section 5.4.8).[505–507]

UHF an abbreviation for *unrestricted Hartree–Fock*. See **Hartree–Fock**.

UNICEPP an acronym for *united atom conformational energy program for peptides*, an empirical force field specifically developed by the research group of H. A. Scheraga at Cornell University in 1977 for carrying out conformational analysis on polypeptides.[69,70] It is a **united atom** implementation of the **ECEPP** force field, and is available from QCPE as program 361 (see Section 5.4.8).

Uniform resource locator See **URL**.

united atom a **force field** representation of molecules in the **AMBER** and **UNICEPP** force fields, in which a carbon with its attached hydrogens is treated as a single unit, rather than treating the attached hydrogens as discreet atoms. This simplifies and shortens the calculations necessary for energy minimization and dynamics experiments. For example, groups of atoms such as a CH_3 or an aromatic CH are treated as single "atoms" with regard to calculating their nonbonding interactions. This simplifies and shortens the calculations necessary for minimization and dynamics experiments. **Extended atom** is synonymous with united atom, and contrasts with the **all atom** representation in which hydrogens are treated explicitly.

Some force fields use the united atom approximation during a part of the energy minimization process (such as **MM2** and **MMX**). These latter force fields normally need to treat some subset of the interactions more rigorously for accurate modeling, for example hydrogen bonding.

unrestricted Hartree–Fock   See **Hartree–Fock**.

unsaturated   a term applied to organic molecules, which indicates the presence of open valences, or of multiple bonds. Unsaturated molecules are more labile toward addition or substitution reactions. See **saturated** and **coordination**.

Urey–Bradley force field   a type of molecular force field formulation, developed from early work by Urey and Bradley,[19,33,34] which added a repulsive term for nonbonded interactions between geminal substituents (the *Urey–Bradley term*) to the simple **valence force field**, thus providing an improved model which includes the nonbonded interactions between geminal substituents. This term allows for the fact that in geminal systems the bond stretch is coupled to the angle bending, but uses a direct geminal nonbonding potential rather than a cross term (see Fig. 5-20 and Section 5.1.3.2).

URL   an abbreviation for *uniform resource locator*, a descriptor used to connect with a remote server or computer resource on the **Internet**. A URL consists of resource type descriptor and an electronic address for the resource, often with directories, subdirectories, and filenames to pinpoint an exact resource. For example, in the address http://www.nih.gov/molecular_modeling/gateway.html, "http" is the resource type (hypertext transport protocol), "www.nih.gov" is the general electronic address of the U.S. government National Institutes of Health, and "molecular_modeling/gateway.html" means that the resource is in the "gateway" document of the "molecular_modeling" directory, and "html" means that the document one will see is rendered in Hypertext Markup Language.

valence force field   a simple force field model which calculates the potential energy of a molecule as a set of functions of the distances between atoms, bond angles, and torsional angles. It entails the assumption that molecular vibrations are not coupled to each other, so that it ignores the cross terms in the force component matrix. The parameters for this type of force field do not generalize well from one molecule or group of molecules to the next, because of the neglect of cross terms.

Valence Shell Electron Pair Repulsion (VSEPR)   is the theory that molecular structure may be predicted on the basis of the electrostatic repulsion between pairs of valence electrons, and that the most stable structure is the one is which these electron pairs are the most distant from each other. The theory also encompasses non-bonding electron pairs, which are held further from the nucleus than the bonding pairs.

For example, a four-coordinate structure such as methane assumes a tetrahedral geometry because this arrangement maximizes the separation of the electron pairs which make up the C—H bonds. Ammonia, with three bonding pairs and one nonbonding pair, and water, with two bonding pairs and two nonbonding pairs, both assume distorted tetrahedral geometries which maximize the separation of the bonding and nonbonding electron pairs.

van der Waals (attraction, energy, forces, radius, repulsion, surface) descriptive terms arising from the solid sphere model of the atom, and used in connection with weak forces acting between atoms. These phenomena are named after the Dutch physicist J. D. van der Waals.

A weak *van der Waals attraction* force is felt between two nonbonded atoms when the distance between their nuclei falls to values slightly greater than the sum of the van der Waals radii. This force arises from temporary, mutually induced complementary dipoles in the respective **electron clouds** of the two atoms. This is also referred to as **London force** or dispersion force.

A discussion of the potential forms used to calculate the *van der Waals energy* in the MMX force field is given in Section 5.1.3.1.

The term *van der Waals forces* is used generally to describe the relatively weak forces acting between atoms or molecules including hydrogen bonding, dipole–dipole attractions, and dispersion forces. In a stricter reading "van der Waals force" refers specifically to the dispersion forces.

The *van der Waals radius* is the largest distance from the nucleus where the electrons are likely to be found, with a particular probability cutoff. These values are derived from physical methods such as X-ray diffraction crystallography.

A strong *van der Waals repulsion* is felt between two nonbonded atoms when the internuclear distance falls to less than the sum of their van der Waals radii. This force originates in a mutual repulsion of the **electron clouds** of the corresponding atoms as the corresponding populated electronic orbitals attempt to occupy the same space.

For *van der Waals surface*, see **molecular surfaces**.

VCFF an abbreviation for *Valence Consistent Force Field*; see **CFF**.

VDM an abbreviation for *Visual Molecular Dynamics*, a molecular graphics program designed for the display and analysis of molecular assemblies, in particular biopolymers such as proteins and nucleic acids. It was written by W. Humphrey, A. Dalke, and K. Schulten.[979]

vector graphics See **graphics**.

VESCF an abbreviation for *Variable Electronegativity Self-Consistent Field*. See **self-consistent field**.

VIBE an acronym for *Virtual Biomolecular Environment*, a system for virtual reality molecular modeling developed through a group collaboration.[97] VIBE consists of three parts: a massively parallel computing environment to simulate the physical and chemical properties of a molecular system (using the **CHARMM** force field); the Cave Automatic Virtual Environment (CAVE) modeling theater for immersive display and interaction with the molecular system; and a high-speed network interface to exchange data between the simulation and the CAVE.

vicinal a term for the configurational relationship between two substituents bonded to adjacent atoms of a molecule. It comes from Latin *vicinalis* meaning "neighboring."

virtual atom See **dummy atom**.

virtual pharmacophore See **pharmacophore**.

virtual torsion See **improper torsion**.

virtual torsional angle See *improper torsional angle* under **improper torsion**.

VISTA the visual statistics module of the **Cambridge Structural Database Software** suite. It is an interactive program for the analysis and display of statistics following the results of a search of the **Cambridge Structural Database**. VISTA provides for exploring the dataset, checking on the actual data entry in normal 1D, 2D, or 3D displays, excluding entries as required, and relating geometrical results to their chemical and structural context. There are PostScript editing features for output of histograms and scatter diagrams for printing.

VRML an abbreviation for *virtual reality modeling language*, a format for encoding the description of a molecular model for transfer via the World Wide Web. It strives to be a platform- and viewer-independent format for the exchange of molecular information, synthesizing data from various sources via hyperlinks to present the most complete electronic molecular model possible.[96]

VSEPR See **Valence Shell Electron Pair Repulsion**.

web browser a software application which provides a **graphic user interface** to the **World Wide Web** system. The browser is also referred to as the *client* in the client–server relationship between the end user platform (client) and the host platform (server) which holds information and provides connectivity to other servers. For desktop computers, the most popular web browsers are Netscape (available commercially from Netscape Communications Corp.) and Internet Explorer (available commercially from Microsoft Inc.).

Westheimer method a term used by early workers to mean the use of the force field or molecular mechanics model to describe the structure of molecules. Named for F. H. Westheimer following his pioneering paper in the field.[37] See Section 1.1.

WHAT IF a protein structure analysis program developed by G. Vriend of the Universität Heidelberg.[980,981] It can be used for mutant prediction, structure verification, and protein molecular graphics.

Wilson angle a measure of the molecular vibration in which a trigonal atom, such as sp^2 carbon or boron, assumes a nonplanar or pyramidal geometry. The Wilson angle is defined as the angle between the bond from the central trigonal atom to one of its substituents, and the plane defined by the bonds to the other two substituents (see Section 5.1.3 and Fig. 5-6). The Wilson angle is named for the American physical chemist E. B. Wilson who coauthored the monograph in which this molecular vibration was described.[327,328] Many force fields incorporate a restoring potential to reinforce planarity in these systems. This distortion can also be quantified as an **improper torsion**, the **out-of-plane angle**, or the **pyramid height**.

WIZARD a program which generates three-dimensional molecular structures from "flat" or two-dimensional input files.[270] WIZARD **parses** structure input in **SMILES** or other common file formats, and intelligent algorithms recognize chain and ring patterns from a library base. Three-dimensional fragments from the library are assembled, and then a limited force field optimization is employed to fine-tune the resulting three-dimensional structures. WIZARD may be used in conformational search experiments.[602,603]

World Wide Web an **Internet** client–server hypermedia (text, image, animation, and audio) distributed information retrieval system. More generally, it is an interconnected system of computers worldwide that allows ready access to information and files and communication between any two (or more) points in the network. The World Wide Web originated from the CERN high-energy physics laboratories in Geneva, Switzerland, and was publicly introduced to scientific audiences in 1991. Its use was largely limited to the government and educational spheres until 1993, when **web browser** programs became publicly available, and has grown astronomically ever since.

Hypermedia links, embedded in Web pages with HTML or related codings, refer to other documents or files by their **URL**s, which are accessed from local or remote resources by protocols such as HTTP (for hypertext documents), FTP (for file transfer in native format), or gopher (for file transfer in ASCII text format).

While also used to refer to the Internet, or Information Superhighway, as a whole, the term "World Wide Web" is most often applied to those resources of the Internet which are available with a graphics and text interface, and are driven by hypertext connections.

WWW see **World Wide Web**.

XMOL a program for graphic display and manipulation of 3D molecular models in a variety of formats on an X-terminal, developed at the Minnesota Supercomputer Center, Inc. Geometric quantities such as atom-to-atom distances, bond angles, and torsion angles can be calculated, and animations of multistep datafiles are possible. Xmol can create encapsulated PostScript graphics format output files. It can import the following file types : Alchemy MOL, Gaussian input, Gaussian output, MOLSIM, MOPAC files, Brookhaven Protein Databank PDB, and Xmol XYZ.

X-ray diffraction the technique of elucidating structural information by interpreting the interference patterns produced when electromagnetic radiation from the soft X-ray portion of the spectrum is diffracted when passing through a substance. Among several diffraction techniques, the one of interest here is the determination of molecular structure by diffraction with a single crystal of the pure substance, with X-ray wavelengths on the order of 0.1–1 nm (1–10 Å), which is in the same range as common molecular bond lengths. See Section 3.3.1.

YAeHMOP an acronym for *yet another extended Hückel molecular orbital program*, a group of programs developed by the computational chemistry group of R. Hoffmann at Cornell for performing extended Hückel molecular orbital calculations and analyzing and visualizing the results. It consists of two main programs, BIND and VIEWKEL. BIND is the program which performs the actual extended Hückel calculations. It can be used to perform calculations on both isolated molecules and extended systems of one, two, or three dimensions. VIEWKEL is an X-Windows-based, interactive program for displaying and printing the results obtained using BIND.

YAK a program for deriving a pseudoreceptor, or virtual **pharmacophore**, from commonalities in the structural data for a series of compounds active at the putative active site.[168] The structural cues used are hydrogen extension vectors (**HEV**s), lone pair vectors (**LPV**s), and hydrophobicity vectors (**HPV**s). YAK is a module within the **YETI** force field modeling program.

YETI a molecular mechanics program and attendant force field, developed by A. Vedani for use in modeling the interactions between small molecules and macromolecules.[521–525] YETI employs the AMBER force field for bond stretch and angle bend interactions, but incorporates independent potential functions for torsional, electrostatic, van der Waals, hydrogen bonding, and metal complexation interactions.

Z a stereochemical descriptor for that isomer of an unsymmetrically substituted alkene in which the two highest priority substituents are on the same side of a plane which contains the axes of the π-orbitals. The **CIP** sequence rules are used to rank the priority of the substituents. Earlier convention called this the *cis*-isomer.[240]

The term comes from German *zusammen* which means "together".

Z-clipping visualizing *slices* or *blocks* of a large structure, in order to eliminate a confusing excess of detail. Producing a molecular "cross-section". The term comes from clipping away sections along the Cartesian Z-axis, the one orthogonal to the normal (i.e. *XY*) viewing plane of the monitor. Synonymous with **blocking**.

Z-matrix a structure input file format commonly used for molecular orbital calculation packages such as MOPAC. The Z-matrix sets up an internal coordinate system, consisting of one line for each atom. Each line starts with the atom label, followed by a series of three numbers, which give values for bond length, bond angle and dihedral angle (see Section 4.5).

A number of sources were consulted during the preparation of some of these definitions: *Free On-Line Dictionary of Computing*,[932] *BABEL: A Glossary of Computer Oriented Abbreviations and Actronyms*,[940] *Organic Chemistry—The Name Game*,[982] *Common Definitions and Terms in Organic Chemistry*,[983] *Network Science—Scientific Software Listing*,[984] *Hyperglossary of Terminology*,[985] *IUPAC Glossary of Terms Used in Computational Drug Design*,[987] *IUPAC Glossary of Terms Used in Physical Organic Chemistry*,[988] *Acronyms Used in Theoretical Chemistry*,[989] *Basic Terminology of Stereochemistry*,[990] *Glossary of Basic Terms in Polymer Science*,[991] and *Glossary of Terms used in Bioinorganic Chemistry*.[992]

REFERENCES

1. Clark, T. *A Handbook of Computational Chemistry.* J. Wiley & Sons, New York (1985).
2. Rzepa, H. S.; Whitaker, B. J.; Winter, M. J. *J. Chem. Soc. Chem. Commun.* **1994** 1907.
3. Casher, O.; Chandramohan, G. K.; Hargreaves, M. J.; Leach, C.; Murray-Rust, P.; Rzepa, H. S.; Sayle, R.; Whitaker, B. J. *J. Chem. Soc. Perkin Trans. 2* **1995** 7.
4. Krieger, J. H. *Chem. Eng. News* (Nov. 13) **1995** 35.
5. *The Internet. A Guide for Chemists* (ed. Bachrach, S. M.). Am. Chem. Soc., Washington (1996).
6. Brickmann, J.; Vollhardt, H. *Trends Biotechnol.* **1996** *14* 167.
7. Engler, E. M.; Andose, J. D.; Schleyer, P. v. R. *J. Am. Chem. Soc.* **1973** *95* 8005.
8. Allinger, N. L. *Advances in Physical Organic Chemistry* **1976** *13* 1–82.
9. Niketic, S. R.; Rasmussen, K. *The Consistent Force Field: A Documentation.* Springer-Verlag, New York (1977), pp. 1–9.
10. Burkert, U.; Allinger, N. L. *Molecular Mechanics (ACS Monograph 177).* American Chemical Society, Washington (1982).
11. Brubaker, G. R.; Johnson, D. W. *Coord. Chem. Rev.* **1984** *53* 1.
12. Morawetz, H. *Polymers, The Origins and Growth of a Science.* J. Wiley & Sons, New York (1985).
13. Rasmussen, K., in *Potential Energy Functions in Conformational Analysis (Lecture Notes in Chemistry 35)* (ed. Berthier, G.; Dewar, M. J. S.; Fischer, H.; Fukui, K.; Hall, G. G.; Hinze, J.; Jaffé, H. H.; Jortner, J.; Kutzelnigg, W.; Ruedenberg, K.). Springer-Verlag, New York (1985), pp. 17–37.
14. Saito, Y. *J. Mol. Struct.* **1985** *126* 461.
15. Lifson, S. *Gazz. Chim. Ital.* **1986** *116* 687.
16. Saunders, M.; Jarret, R. M. *J. Comput. Chem.* **1986** *7* 578.
17. Hancock, R. D., in *Prog. Inorg. Chem. Vol. 37* (ed. Lippard, S. J.). Wiley-Interscience, New York (1989), pp. 187–291.
18. Langridge, R.; Klein, T. E., in *Comprehensive Medicinal Chemistry, Vol. 4. Quantitative Drug Design* (ed. Ramsden, C. A.). Pergamon Press, New York (1990), pp. 413–429.
19. Bowen, J. P.; Allinger, N. L., in *Reviews of Computational Chemistry, Vol. 2* (ed. Lipkowitz, K. B.; Boyd, D. B.). VCH, New York (1991), pp. 81–97.
20. Comba, P. *Coord. Chem. Rev.* **1993** *123* 1.
21. Eksterowicz, J. E.; Houk, K. N. *Chem. Rev.* **1993** *93* 2439.
22. Lipkowitz, K. B.; Peterson, M. A. *Chem. Rev.* **1993** *93* 2463.

23. Comba, P.; Hambley, T. W. *Molecular Modeling of Inorganic Compounds*. VCH, New York (1995), pp. 3–5.
24. Zimmer, M. *Chem. Rev.* **1995** *95* 2629.
25. Connolly, M. L. "Molecular Surfaces: A Review," *Network Sci.* [http://www.awod.com/netsci/Science/Compchem/feature14.html]. (Apr. 1996).
26. Rappé, A. K.; Casewit, C. J. *Molecular Mechanics Across Chemistry*. University Science Books, Fort Collins, CO (1996).
27. Morawetz, H. *Polymers, The Origins and Growth of a Science*. J. Wiley & Sons, New York (1985), pp. 7–15.
28. Solomons, T. W. G. *Organic Chemistry, 3rd Edition*. J. Wiley & Sons, New York (1984).
29. Bader, A. *Chem. Brit.* (Sept.) **1996** 41.
30. Wiswesser, W. J. *Aldrichim. Acta* **1989** *22* 17 [*Chem. Abstr.* **1989** *111* 38474].
31. Nickon, A.; Silversmith, E. F. *Organic Chemistry—The Name Game*. Pergamon Press, New York (1987), pp. 289–290.
32. Walton, A. *Molecular and Crystal Structure Models*. Ellis Horwood, Chichester, U.K. (1978).
33. Urey, H. C.; Bradley, C. A., Jr. *Phys. Rev.* **1931** *38* 1969.
34. Burkert, U.; Allinger, N. L. *Molecular Mechanics* (*ACS Monograph 177*). American Chemical Society, Washington, 1982), pp. 17–58.
35. Hill, T. L. *J. Chem. Phys.* **1946** *14* 465.
36. Dostrovsky, I.; Hughes, E. D.; Ingold, C. K. *J. Chem. Soc.* **1946** 173.
37. Westheimer, F. H.; Mayer, J. E. *J. Chem. Phys.* **1946** *14* 733.
38. Westheimer, F. H. *J. Chem. Phys.* **1947** *15* 252.
39. Barton, D. H. R. *Experientia* **1950** *6* 316.
40. Pitzer, K. S. *Disc. Faraday Soc.* **1951** *10* 66.
41. Watson, J. D.; Crick, F. H. C. *Nature* (*London*) **1953** *171* 737.
42. Metropolis, N.; Rosenbluth, A. W.; Rosenbluth, M. N.; Teller, A. H.; Teller, E. *J. Chem. Phys.* **1953** *21* 1087.
43. Corey, E. J.; Bailar, J. C., Jr. *J. Am. Chem. Soc.* **1959** *81* 2620.
44. Corey, E. J.; Sneen, R. A. *J. Am. Chem. Soc.* **1955** *77* 2505.
45. Mason, E. A.; Kreevoy, M. M. *J. Am. Chem. Soc.* **1955** *77* 5808.
46. Kitaigorodsky, A. I. *Tetrahedron* **1960** *9* 183.
47. Kitaigorodsky, A. I. *Tetrahedron* **1961** *14* 230.
48. Kitaigorodsky, A. I. *Chem. Soc. Rev.* **1978** *7* 133.
49. Hendrickson, J. B. *J. Am. Chem. Soc.* **1961** *83* 4537.
50. Hendrickson, J. B. *J. Am. Chem. Soc.* **1962** *84* 3355.
51. Hendrickson, J. B. *J. Am. Chem. Soc.* **1964** *86* 4854.
52. Hendrickson, J. B. *J. Org. Chem.* **1964** *29* 991.
53. Hendrickson, J. B. *J. Am. Chem. Soc.* **1967** *89* 7036.
54. Hendrickson, J. B. *J. Am. Chem. Soc.* **1967** *89* 7043.
55. Hendrickson, J. B. *J. Am. Chem. Soc.* **1967** *89* 7047.
56. Schachtschneider, J. H.; Snyder, R. G. *Spectrochim. Acta* **1963** *19* 117.

57. Wiberg, K. B. *J. Am. Chem. Soc.* **1965** *87* 1070.
58. Boyd, R. H. *J. Chem. Phys.* **1968** *49* 2574.
59. Snow, M. R.; Buckingham, D. A.; Marzili, P. A.; Sargeson, A. M. *J. Chem. Soc. Chem. Commun.* **1969** 801.
60. Buckingham, D. A.; Maxwell, I. E.; Sargeson, A. M.; Snow, M. R. *J. Am. Chem. Soc.* **1970** *92* 3617.
61. Snow, M. R. *J. Am. Chem. Soc.* **1970** *92* 3610.
62. Allinger, N. L. *J. Org. Chem.* **1962** *27* 443.
63. Johnson, C. K. *Oak Ridge National Laboratory Rept. ORNL-3794* **1965** [*Chem. Abstr.* **1966** *64* 18545f].
64. Burnett, M. N.; Johnson, C. K. *ORTEP-III World Wide Web Site* [http://www.ornl.gov/ortep/ortep.html] (revised Apr. 19, 1996).
65. Hafner, K.; Lyon, M. *The Sciences* **1996** (Sept./Oct.) 32.
66. Scott, R. A.; Scheraga, H. A. *J. Chem. Phys.* **1966** *44* 3054.
67. Scheraga, H. A. *Adv. Phys. Org. Chem.* **1968** *6* 103.
68. Momany, F. A.; McGuire, R. F.; Burgess, A. W.; Scheraga, H. A. *J. Phys. Chem.* **1975** *79* 2361.
69. Browman, M. J.; Burgess, A. W.; Dunfield, L. G.; Rumsey, S. M.; Endres, G. F.; Scheraga, H. A. *Quantum Chemistry Program Exchange Program* #361 Documentation (1977).
70. Dunfield, L. G.; Burgess, A. W.; Scheraga, H. A. *J. Phys. Chem.* **1978** *82* 2609.
71. Lifson, S.; Warshel, A. *J. Chem. Phys.* **1968** *49* 5116.
72. Wertz, D. H.; Allinger, N. L. *Tetrahedron* **1974** *30* 1579.
73. Gleicher, G. J.; Schleyer, P. v. R. *J. Am. Chem. Soc.* **1967** *89* 582.
74. Chang, S.-J.; McNally, D.; Shary-Tehrany, S.; Hickey, M. J., Sr.; Boyd, R. H. *J. Am. Chem. Soc.* **1970** *92* 3109.
75. Burgi, H. B.; Bartell, L. S. *J. Am. Chem. Soc.* **1972** *94* 5236.
76. Bartell, L. S. Burgi, H. B. *J. Am. Chem. Soc.* **1972** *94* 5239.
77. Bartell, L. S.; Plato, V. *J. Am. Chem. Soc.* **1973** *95* 3097.
78. Fitzwater, S.; Bartell, L. S. *J. Am. Chem. Soc.* **1976** *98* 5107.
79. Bartell, L. S. *J. Am. Chem. Soc.* **1977** *99* 3279.
80. Lee, B.; Richards, F. M. *J. Mol. Biol.* **1971** *55* 379.
81. Shrake, A.; Rupley, J. A. *J. Mol. Biol.* **1973** *79* 351.
82. Wiberg, K. B.; Boyd, R. H. *J. Am. Chem. Soc.* **1972** *94* 8426.
83. Lemieux, R. U.; Koto, S. *Tetrahedron* **1974** *30* 1933.
84. Lemieux, R. U.; Bock, K.; Delbaere, L. T. J.; Koto, S.; Rao, V. S. *Can. J. Chem.* **1980** *58* 631.
85. Thøgersen, H.; Lemieux, R. U.; Bock, K.; Meyer, B. *Can. J. Chem.* **1982** *60* 44.
86. Suter, U. W.; Flory, P. J. *Macromolecules* **1975** *8* 765.
87. Connolly, M. L. *J. Appl. Cryst.* **1983** *16* 548.
88. Cohen, N. C.; Blaney, J. M.; Humblet, C.; Gund, P.; Barry, D. C. *J. Med. Chem.* **1990** *33* 883.
89. Saunders, M. *J. Am. Chem. Soc.* **1987** *109* 3150.

90. Maple, J. R.; Hwang, M.-J.; Stockfisch, T. P., Dinur, U.; Waldman, M.; Ewig, C. S.; Hagler, A. T. *J. Comput. Chem.* **1994** *15* 162.
91. Hagler, A. T.; Ewig, C. S. *Comput. Phys. Commun.* **1994** *84* 131.
92. Hwang, M. J.; Stockfisch, T. P.; Hagler, A. T. *J. Am. Chem. Soc.* **1994** *116* 2515.
93. Allinger, N. L.; Chen, K. S.; Lii, J.-H. *J. Comput. Chem.* **1996** *17* 642.
94. Maple, J. R.; Hwang, M.-J.; Stockfisch, T. P.; Hagler, A. T. *Israel J. Chem.* **1994** *34* 195.
95. Hartshorn, M. J.; Herzyk, P.; Hubbard, R. E. *Trends Biotechnol.* **1995** *13* 84.
96. Vollhardt, H.; Henn, C.; Moeckel, G.; Teschner, M.; Brickmann, J. *J. Mol. Graph.* **1995** *13* 368.
97. Cruz-Neira, C.; Langley, R.; Bash, P. A. *Computers Chem.* **1996** *20* 469.
98. March, J. *Advanced Organic Chemistry. Reactions, Mechanisms and Structure.* 4th Edition. J. Wiley & Sons, New York (1992), pp. 138–149.
99. Gordon, A. J.; Ford, R. A. *The Chemist's Companion.* J. Wiley & Sons, New York (1972), pp. 499–506.
100. Hernandez, S. A.; Rodriguez, N. M.; Quinzani, O. V. *J. Chem. Educ.* **1996** *73* 748.
101. Skawinski, W. J.; Busanic, T. J.; Ofsievich, A. D.; Venanzi, T. J.; Luzhkov, V. B.; Venanzi, C. A. *J. Mol. Graphics* **1995** *13* 126.
102. Neckers, D. C. *CHEMTECH* **1990** 615.
103. Renard, A.; Dormal, T.; Czajkowski, V. *Eur. Pat. Appl. EP 691194* (Jan. 10, 1996) [*Chem. Abstr.* **1996** *124* 144856k].
104. Altona, C.; Faber, D. H. *Top. Curr. Chem.* **1974** *45* 1.
105. Niketic, S. R.; Rasmussen, K. *The Consistent Force Field: A Documentation.* Springer-Verlag, New York (1977), pp. 78–95.
106. Boyd, D. B.; Lipkowitz, K. B. *J. Chem. Educ.* **1982** *59* 269.
107. Cox, P. J. *J. Chem. Educ.* **1982** *59* 275.
108. Meyer, A. Y., in *The Chemistry of Functional Groups, Supplement D. The Chemistry of Halides, Pseudo-Halides and Azides, Part 1* (ed. Patai, S.; Rappoport, Z.). J. Wiley & Sons, New York (1983), pp. 1–47.
109. Rasmussen, K., in *Potential Energy Functions in Conformational Analysis* (*Lecture Notes in Chemistry*, 35) (ed. Berthier, G.; Dewar, M. J. S., Fischer, H.; Fukui, K.; Hall, G. G.; Hinze, J.; Jaffé, H. H.; Jortner, J.; Kutzelnigg, W.; Ruedenberg, K.). Springer-Verlag, New York (1985), pp. 5–16.
110. Nakamura, E.; Fukuzawa, Y. *Yuki Gosei Kagaku* **1987** *45* 1044 [*Chem. Abstr.* **1988** *109* 539936b].
111. Lipkowitz, K. B., in *Conformational Analysis of Cyclohexenes, Cyclohexadienes, and Related Hydroaromatic Compounds* (ed. Rabideau, P. W.). VCH, New York (1989), pp. 301–319.
112. White, D. N. J.; Ruddock, J. N.; Edgington, P. R., in *Computer-Aided Molecular Design* (ed. Richards, W. G.). VCH, New York (1989), pp. 23–41.
113. Marsili, M. *Computer Chemistry.* CRC Press, Boca Raton, FL (1990), pp. 35–93.
114. Pozigun, D. V.; Kuz'min, V. E.; Kamalov, G. L. *Russ. Chem. Rev.* **1990** *59* 1867.
115. Seibel, G. L.; Kollman, P. A., in *Comprehensive Medicinal Chemistry*, Vol. 4, *Quantitative Drug Design* (ed. Ramsden, C. A.). Pergamon Press, New York (1990),

pp. 125–138.
116. Hay, B. J. *Coord. Chem. Rev.* **1993** *126* 177.
117. Goodman, J. M., in *Molecular Modelling and Drug Design; Topics in Molecular and Structural Biology* (ed. Vinter, J. G.; Gardner, M.). CRC Press, Boca Raton, FL (1994), pp. 53–88.
118. Tollenaere, J. P., in *Adv. Drug Des. Dev.* (ed. Kourounakis, P. N.; Rekka, E.). Horwood, London (1994), pp. 109–121.
119. Wylie, W. A., in *Molecular Modelling and Drug Design; Topics in Molecular and Structural Biology* (ed. Vinter, J. G.; Gardner, M.). CRC Press, Boca Raton, FL (1994), pp. 1–52.
120. Kollman, P. A. *Acc. Chem. Res.* **1996** *29* 461.
121. Richon, A. B. *An Introduction to Molecular Modeling* [http://www.awod.com/netsci/Science/Compchem/feature01.html] [*Chem. Abstr.* **1996** *124* 331664].
122. Kollman, P. A. *Acc. Chem. Res.* **1977** *10* 365.
123. Andrews, P. A.; Tintelnot, M., in *Comprehensive Medicinal Chemistry, Vol. 4, Quantitative Drug Design* (ed. Ramsden, C. A.). Pergamon Press, New York (1990), pp. 321–347.
124. Gund, P.; Barry, D. C.; Blaney, J. M.; Cohen, N. C. *J. Med. Chem.* **1988** *31* 2230.
125. Osawa, E.; Musso, H. *Angew. Chem. Int. Ed. Engl.* **1983** *22* 1.
126. Frühbeis, H.; Klein, R.; Wallmeier, H. *Angew. Chem. Int. Ed. Engl.* **1987** *26* 403.
127. Suckling, C. J. in *Comprehensive Medicinal Chemistry, Vol. 4, Quantitative Drug Design* (ed. Ramsden, C. A.). Pergamon Press, New York (1990), pp. 83–104.
128. Jaime, C. *Comput. Aided Innovation New Mater. 2, Proc. Int. Conf. Exhib. Comput. Appl. Mater. Mol. Sci. Eng.*, 2nd, Pt. 1 (1992; publ. 1993). p. 823.
129. Kollman, P. *Acc. Chem. Res.* **1985** *18* 105.
130. Carey, F. A.; Sundberg, R. J. *Advanced Organic Chemistry, 3rd Edition, Part A: Structure and Mechanisms*. Plenum Press, New York (1990).
131. Comba, P.; Hambley, T. W. *Molecular Modeling of Inorganic Compounds*. VCH, New York (1995).
132. Boyd, D. B., in *Reviews of Computational Chemistry, Vol. 2.* (ed. Lipkowitz, K. B.; Boyd, D. B.). VCH, New York (1991), pp. 461–479.
133. Varveri, F. S. *J. Chem. Educ.* **1993** *70* 204.
134. Harkanyi, K.; Carande, R. *J. Chem. Educ.* **1995** *72* 812.
135. Heller, S. R. *J. Chem. Inf. Comput. Sci.* **1996** *36* 205.
136. *Cambridge Structural Database* [http://csdvx2.ccdc.cam.ac.uk/].
137. *Protein Data Bank* [http://www.pdb.bnl.gov/], (updated July 16, 1997); contact Nancy Oeder Manning.
138. Ihlenfeldt, W.-D. *The WWW Chemical Structures Database* [http://schiele.organik.uni-erlangen.de/services/webmol.html] (Sept. 1996).
139. Ihlenfeldt, W.-D *GIFs and PNGs of 2D-Plots of Chemical Structures* [http://schiele.organik.uni-erlangen.de/services/gif.html].
140. *CORINA 3D Coordinates* [http://schiele.organik.uni-erlangen.de/services/3d.html].

141. Ambos, M. M.; Richon, A. B. *1995 Analysis of the Computational Chemistry Marketplace* [http://www.awod.com/netsci/Science/Special/feature09.html] (1996).
142. Wimmer, E. *Analusis* **1995** *23* M12 [*Chem. Abstr.* **1996** *124* 116263w].
143. Gund, P.; Andose, J. D.; Rhodes, J. B.; Smith, G. M. *Science* **1980** *208* 1425.
144. Hassall, C. H. *Chem. Britain* (Jan.) **1985** 39.
145. McCammon, J. A. *Science* **1987** *238* 486.
146. Sheridan, R. P.; Venkataraghavan, R. *Acc. Chem. Res.* **1987** *20* 322.
147. Topliss, J. G. *J. Med. Chem.* **1988** *31* 2229.
148. Hancock, R. D.; Martell, A. E. *Chem. Rev.* **1989** *89* 1875.
149. Van Drie, J. H.; Weininger, D.; Martin, Y. C. *J. Comput.-Aided Mol. Design* **1989** *3* 225.
150. Moon, J. B.; Howe, W. J. *Proteins Struct. Funct. Genet.* **1991** *11* 314.
151. Nilakantan, R.; Bauman, N.; Venkataraghavan, R. *J. Chem. Inf. Comput. Sci.* **1991** *31* 527.
152. Kuntz, I. D. *Science* **1987** *257* 1078.
153. Lewis, R. A.; Roe, D. C.; Huang, C.; Ferrin, T. E.; Langridge, R.; Kuntz, I. D. *J. Mol. Graph.* **1992** *10* 66.
154. Bugg, C. E.; Carson, W. M.; Montgomery, J. A. *Sci. Amer.* (Dec.) **1993** 92.
155. Dean, P. M. *BioEssays* **1994** *16* 683 [*Chem. Abstr.* **1995** *121* 291820k].
156. Gillet, V. J.; Newell, W.; Mata, P.; Myatt, G.; Sike, S.; Zsoldos, Z.; Johnson, A. P. *J. Chem. Inf. Comput. Sci.* **1994** *34* 207.
157. Huml, K.; Remko, M. *Chem. Listy* **1994** *88* 711 [*Chem. Abstr.* **1995** *122* 23077g].
158. Kuntz, I. D.; Meng, E. C.; Shoichet, B. K. *Acc. Chem. Res.* **1994** *27* 117.
159. Leach, A. R.; Lewis, R. A. *J. Comput. Chem.* **1994** *15* 233.
160. Nikiforovich, G. V. *Int. J. Peptide Protein Res.* **1994** *44* 513.
161. Tapia, O.; Paulino, M.; Stamato, F. M. L. G. *Mol. Eng.* **1994** *4* 377. [*Chem. Abstr.* **1995** *122* 45481m].
162. Walters, D. E.; Hinds, R. M. *J. Med. Chem.* **1994** *37* 2527.
163. Carotti, A.; Altomare, C. *Chim. Ind.* (*Milan*) **1995** *77* 13 [*Chem. Abstr.* **1995** *122* 229955a].
164. Gálvez, J.; García-Domenech, R.; de Julián-Ortiz, J. V.; Soler, R. *J. Chem. Inf. Comput. Sci.* **1995** *35* 272.
165. Kolinski, A.; Galazka, W.; Skolnick, J. *J. Chem. Phys.* **1995** *103* 10286.
166. Li, S.; Jiao, K. F.; Zheng, Z. B.; Lu, Z. Z.; Xie, Y. D. *Jisuanji Yu Yingyong Huaxue* **1995** *12* 38 [*Chem. Abstr.* **1995** *122* 305762c].
167. Venkatasubramanian, V.; Chan, K.; Caruthers, J. M. *J. Chem. Inf. Comput. Sci.* **1995** *35* 188.
168. Vedani, A.; Zbinden, P.; Snyder, J. P.; Greenidge, P. A. *J. Am. Chem. Soc.* **1995** *117* 4987.
169. Bohacek, R. S.; McMartin, C.; Guida, W. C. *Med. Res. Rev.* **1996** *16* 3 [*Chem. Abstr.* **1996** *124* 192976w].
170. Clark, D. E.; Firth, M. A.; Murray, C. W. *J. Chem. Inf. Comput. Sci.* **1996** *36* 137.
171. Liao, D.-W.; Huang, Z.-N.; Lin, Y.-Z.; Wan, H.-L.; Zhang, H.-B.; Tsai, K.-R. *J.*

Chem. Inf. Comput. Sci. **1996** *36* 1178.
172. Luo, Z.; Wang, R.; Lai, L. *J. Chem. Inf. Comput. Sci.* **1996** *36* 1187.
173. Pearlman, D. A.; Murcko, M. A. *J. Med. Chem.* **1996** *39* 1651.
174. Potenzone, R., Jr.; Doherty, D. C., in *Computer Applications in the Polymer Laboratory, ACS Symposium Series 313*. American Chemical Society, Washington (1986), pp. 31–38.
175. Gund, P.; Halgren, T. A.; Smith, G. M., in *Annual Reports in Medicinal Chemistry, Vol. 22* (ed. Bailey, D. M.). Academic Press, New York (1987), pp. 269–279.
176. Dearing, A., in *Pesticide Chemistry, Proc. Intl. Congr. Pest. Chem. 7th*, (1990), p. 61.
177. Lewis, R. A.; Meng, E. C., in *Molecular Modelling and Drug Design: Topics in Molecular and Structural Biology* (ed. Vinter, J. G.; Gardner, M.). CRC Press, Boca Raton, FL (1994), pp. 170–210.
178. Gelin, B. R., in *Computer-Aided Molecular Design (ACS Symposium Series 589), Applications in Agrochemicals, Materials, and Pharmaceuticals* (ed. Reynolds, C. H.; Holloway, M. K.; Cox, H. K.). American Chemical Society, Washington (1995), pp. 1–11.
179. Venkatasubramanian, V.; Chan, K.; Caruthers, J. M., in *Computer-Aided Molecular Design. Applications in Agrochemicals, Materials, and Pharmaceuticals. ACS Symposium Series 589* (ed. Reynolds, C. H.; Holloway, M. K.; Cox, H. K.), Am. Chem. Soc., Washington (1995), pp. 396–414.
180. Marshall, G. R., in *Computer-Aided Molecular Design* (ed. Richards, W. G.). VCH, New York (1989), pp. 91–104.
181. *Computer-Aided Molecular-Design. Applications in Agrochemicals, Materials, and Pharmaceuticals. ACS Symposium Series 589* (ed. Reynolds, C. H.; Holloway, M. K.; Cox, H. K., Am. Chem. Soc., Washington (1995).
182. Dugas, H. *J. Chem. Educ.* **1992** *69* 533.
183. Hornack, F. M. *J. Chem. Educ.* **1988** *65* 24.
184. Lipkowitz, K. B. *J. Chem. Educ.* **1989** *66* 275.
185. Farrell, J. J.; Haddon, H. H. *J. Chem. Educ.* **1989** *66* 839.
186. Bailey, R. A. *J. Chem. Educ.* **1989** *66* 836.
187. Gilliom, R. D. *J. Chem. Educ.* **1989** *66* 47.
188. Simpson, J. M. *J. Chem. Educ.* **1989** *66* 406.
189. Lillie, T. S.; Yeager, K. *J. Chem. Educ.* **1989**, *66* 675.
190. Jarret, R. M.; Sin, N. *J. Chem. Educ.* **1990** *67* 153.
191. Biali, S. E. *J. Chem. Educ.* **1990** *67* 1038.
192. Box, V. G. S. *J. Chem. Educ.* **1991** *68* 662.
193. Blankespoor, R. L.; Piers, K. *J. Chem. Educ.* **1991** *68* 693.
194. Canales, C.; Egan, L.; Zimmer, M. *J. Chem. Educ.* **1992** *69* 21.
195. Anderson, E. L.; Li, D.; Owen, N. L. *J. Chem. Educ.* **1992** *69* 846.
196. Box, V. G. S. *J. Chem. Educ.* **1993** *70* 236.
197. Buss, D. C.; Fountain, K. R. *J. Chem. Educ.* **1993** *70* 295.
198. Fitzgerald, J. P. *J. Chem. Educ.* **1993** *70* 988.
199. Bellomo, K. A.; Bush, R. C.; Sands, R. D. *J. Chem. Educ.* **1993** *70* A122.

200. Bahnick, D. A. *J. Chem. Educ.* **1994** *71* 171.
201. Hanks, T. W. *J. Chem. Educ.* **1994** *71* 62.
202. Pietro, W. J. *J. Chem. Educ.* **1994** *71* 416.
203. Simpson, J. *J. Chem. Educ.* **1994** *71* 607.
204. Fountain, K. R.; McGuire, B. R. *J. Chem. Educ.* **1994** *71* 938.
205. Mencarelli, P. *J. Chem. Educ.* **1995** *72* 511.
206. Lee, M.; Garbiras, B.; Preti, C. *J. Chem. Educ.* **1995** *72* 378.
207. Lipkowitz, K. B.; Pearl, G. M.; Robertson, D. H.; Schultz, F. A. *J. Chem. Educ.* **1996** *73* 105.
208. Sauers, R. R. *J. Chem. Educ.* **1996** *73* 114.
209. Lee, M. *J. Chem. Educ.* **1996** *73* 184.
210. Wolfson, A. J.; Hall, M. L.; Branham, T. R. *J. Chem. Educ.* **1996** *73* 1026.
211. Lipkowitz, K. *J. Chem. Educ.* **1984** *61* 1051.
212. Colwell, S. M.; Handy, N. C. *J. Chem. Educ.* **1988** *65* 21.
213. Sannigrahi, A. B.; Kar, T. *J. Chem. Educ.* **1988** *65* 674.
214. Mealli, C.; Proserpio, D. M. *J. Chem. Educ.* **1990** *67* 399.
215. Tanner, J. J. *J. Chem. Educ.* **1990** *67* 917.
216. Rosenfeld, S. *J. Chem. Educ.* **1991** *68* 488.
217. Sauers, R. R. *J. Chem. Educ.* **1991** *68* 816.
218. Aduldecha, S.; Akhter, P.; Field, P.; Nagle, P.; O'Sullivan, E.; O'Connor, K.; Hathaway, B. J. *J. Chem. Educ.* **1991** *68* 576.
219. Bays, J. P. *J. Chem. Educ.* **1992** *69* 209.
220. Harvey, S. C.; Tan, R. K.-Z. *Biophys. J.* **1992** *63* 1683.
221. Casanova, J. *J. Chem. Educ.* **1993** *70* 904.
222. Campanario, J. M.; Bronchalo, E.; Hildago, M. A. *J. Chem. Educ.* **1994** *71* 761.
223. Hanks, T. W.; Hallford, R.; Wright, G. *J. Chem. Educ.* **1995** *72* 329.
224. Hyde, R. R.; Shaw, P. N.; Jackson, D. E.; Woods, K. *J. Chem. Educ.* **1995** *72* 699.
225. Benecke, C.; Grund, R.; Kerber, A.; Laue, R.; Wieland, T. *J. Chem. Educ.* **1995** *72* 403.
226. Rhodes, G. *J. Chem. Educ.* **1995** *72* A178.
227. Taylor, E. R.; Sonnier, R.; Doré, B. *J. Chem. Educ.* **1995** *72* 236.
228. Barnea, N.; Dori, Y. J. *J. Chem. Inf. Comput. Sci.* **1996** *36* 629.
229. Delaware, D. L.; Fountain, K. R. *J. Chem. Educ.* **1996** *73* 116.
230. Comba, P.; Zimmer, M. *J. Chem. Educ.* **1996**, *73* 108.
231. Sadek, M.; Munro, S. *J. Comput.-Aided Mol. Des.* **1988** *2* 81.
232. Boyd, D. B., in *Reviews, of Computational Chemistry, Vol. 1* (ed Lipkowitz, K. B.; Boyd, D. B.). VCH, New York (1990), pp. 383–392.
233. Boyd, D. B., in *Reviews of Computational Chemistry, Vol. 2* (ed. Lipkowitz, K. B.; Boyd, D. B.). VCH, New York (1991), pp. 461–479.
234. Boyd, D. B., in *Reviews of Computational Chemistry, Vol. 3* (ed. Lipkowitz, K. B.; Boyd, D. B.). VCH, New York (1992), pp. 223–247.
235. Endres, M. *Today's Chemist at Work* (Oct.) **1992** 29.

236. Boyd, D. B., in *Reviews of Computational Chemistry*, Vol. 4 (ed. Lipkowitz, K. B.; Boyd, D. B.). VCH, New York, (1993), pp. 229–257.
237. Boyd, D. B., in *Reviews of Computational Chemistry*, Vol. 5 (ed. Lipkowitz, K. B.; Boyd, D. B.). VCH, New York (1994), pp. 381–428.
238. Boyd, D. B., in *Reviews of Computational Chemistry*, Vol. 6 (ed. Lipkowitz, K. B.; Boyd, D. B.). VCH, New York (1995), pp. 383–437.
239. Boyd, D. B., in *Reviews of Computational Chemistry*, Vol. 7 (ed. Lipkowitz, K. B.; Boyd, D. B.). VCH, New York (1995), pp. 303–380.
240. IUPAC–IUB Commission on Biochemical Nomenclature. *Biochem.* **1970** *9* 3471.
241. Karlson, P.; Dixon, H. B. F.; Liébecq, C.; Loening, K. L.; Moss, G. P.; Reedjik, J.; Velick, S. F.; Vliegenthart, J. F. G. *Eur. J. Biochem.* **1983** *131* 5.
242. Eliel, E. L.; Allinger, N. L.; Angyal, S. J.; Morrison, G. A. *Conformational Analysis.* Wiley, New York (1965), pp. 129–188.
243. Burkett, U.; Allinger, N. L. *Molecular Mechanics* (*ACS Monograph 177*). American Chemical Society, Washington (1982), pp. 6–10.
244. Wiberg, K. B. *Acc. Chem. Res.* **1996** *29* 229.
245. Allen, F. H.; Bellard, S.; Brice, M. D.; Cartwright, B. A.; Doubleday, A.; Higgs, H.; Hummelink, T.; Hummerlink-Peters, B. G.; Kennard, O.; Motherwell, W. D. S.; Rodgers, J. R.; Watson, D. G. *Acta Cryst. B* **1979** *B35* 2331.
246. Allen, F. H.; Davies, J. E.; Galloy, J. J.; Johnson, O.; Kennard, O.; MacRae, C. F.; Mitchell, G. F.; Smith, J. M.; Watson, D. G. *J. Chem. Inf. Comput. Sci.* **1991** *31* 187.
247. *McGraw-Hill Encyclopedia of Science and Technology, 7th Edition.* McGraw-Hill, New York (1992), Vol. 19, pp. 551ff.
248. Omel'chenko, Yu. A.; Kondrashev, Yu, D. *Sov. Phys.—Cryst.* (*Engl. Transl. of Kristallografiya*) **1971** *16* 88.
249. *McGraw-Hill Encyclopedia of Science and Technology, 7th Edition.* McGraw-Hill, New York (1992), Vol. 11, pp. 693ff.
250. Allen, P. W.; Sutton, L. E. *Acta Cryst.* **1950** *3* 46.
251. *McGraw-Hill Encyclopedia of Science and Technology, 7th Edition.* McGraw-Hill, New York (1992), Vol. 6, pp. 181ff.
252. Dommen, J.; Brupbacher, T.; Grassi, G.; Bauder, A. *J. Am. Chem. Soc.* **1990** *112* 953.
253. Ibison, P.; Jacquot, M.; Kam, F.; Neville, A. G.; Simpson, R. W.; Tonnelier, C.; Venczel, T.; Johnson, A. P. *J. Chem. Inf. Comput. Sci.* **1993** *33* 338.
254. Simon, A.; Johnson, A. P.; Pret, J.-C.; Vizkeleti, F. *CLiDE Chemical Literature Data Extraction* [http://www.chem.leeds.ac.uk:80/ICAMS/CLiDE.html] (1996).
255. Weininger, D. *J. Chem. Inf. Comput. Sci.* **1988** *28* 31.
256. Weininger, D.; Weininger, A.; Weininger, J. L. *J. Chem. Inf. Comput. Sci.* **1989** *29* 97.
257. Weininger, D.; Weininger, J. L., in *Comprehensive Medicinal Chemistry*, Vol. 4, *Quantitative Drug Design* (ed. Ramsden, C. A.). Pergamon Press, New York (1990), pp. 59–82.
258. *SMILES Home Page* [http://www.daylight.com/dayhtml/smiles/index.html]. Maintained by Daylight Chemical Information Systems, Inc.

259. Rohbeck, H. G. in *Software Dev. Chem. 5, Proc. Workshop "Comput. Chem.," 5th* (1991) p. 49 [*Chem. Abstr.* **1992** *116* 105223s].
260. Greene, J.; Kahn, S.; Savoj, H.; Sprague, P.; Teig, S. *J. Chem. Inf. Comput. Sci.* **1994** *34* 1297.
261. Randic, M.; Razinger, M. *J. Chem. Inf. Comput. Sci.* **1995** *35* 140.
262. Schuur, J. H.; Selzer, P.; Gasteiger, J. *J. Chem. Inf. Comput. Sci.* **1996** *36* 334.
263. Sadowski, J.; Gasteiger, J. *Chem. Rev.* **1993** *93* 2567.
264. Leach, A. R. *Pestic. Sci.* **1991** *33* 87.
265. Gothe, S. A.; Helson, H. E.; Houdaverdis, I.; Lagerstedt, I.; Sinclair, S.; Jorgensen, W. L. *J. Org. Chem.* **1993** *58* 5081.
266. Sadowski, J.; Gasteiger, J.; Klebe, G. *J. Chem. Inf. Comput. Sci.* **1996** *34* 1000.
267. Blundell, T.; Carney, D.; Gardner, S.; Hayes, F.; Howlin, B.; Hubbard, T.; Overington, J.; Singh, D. A.; Sibanda, B. L.; Sutcliffe, M. *Eur. J. Biochem.* **1988** *172* 513.
268. Miller, K. J. *Biopolym.* **1979** *18* 959.
269. Taylor, E. R.; Miller, K. J. *Biopolym.* **1984** *23* 2853.
270. Dolata, D. P.; Leach, A. R.; Prout, K. *J. Comput.-Aided Mol. Des.* **1987** *1* 73.
271. Vinter, J. G.; Davis, A.; Saunders, M. R. *J. Comput.-Aided Mol. Des.* **1987** *1* 31.
272. Allinger, N. L.; Yuh, Y. H. *Quantum Chemistry Program Exchange* (*Indiana University, Bloomington IN*) *Program* #395 Documentation (1980).
273. Dalby, A.; Nourse, J. G.; Hounshell. W. D.; Gushurst, A. K. I.; Grier, D. L.; Leland, B. A.; Laufer, J. *J. Chem. Inf. Comput. Sci.* **1992** *32* 244.
274. *MDL CTfile Formats.* MDL Information Systems San Leandro, CA (1993).
275. MDL Information Systems, Inc. *CTfile Formats* [http://www.mdli.com/prod/ctfile.pdf] (Aug. 1996).
276. Koetzle, T. F.; Lehmann, M. S.; Hamilton, W. C. *Acta Cryst.* **1973** *B29* 231.
277. Allinger, N. L.; Yuh, Y. H.; Lii, J.-H. *J. Am. Chem. Soc.* **1989** *111* 8551.
278. Lii, J.-H.; Allinger, N. L. *J. Am. Chem. Soc.* **1989** *111* 8566.
279. Lii, J.-H.; Allinger, N. L. *J. Am. Chem. Soc.* **1989** *111* 8576.
280. Bernstein, F. C.; Koetzle, T. F.; Williams, G. J. B.; Meyer, E. F., Jr.; Brice, M. D.; Rodgers, J. R.; Kennard, O.; Shimanouchi, T.; Tasumi, M. *J. Mol. Biol.* **1977** *112* 535.
281. Martz, E. *The Protein Data Bank* (*PDB*) *Format for Atomic Coordinate Files* [http://klaatu.oit.umass.edu/microbio/rasmol/pdb.htm] (1996).
282. Stewart, J. J. P. *J. Comput. Chem.* **1989** *10* 209.
283. Stewart, J. J. P. *J. Comput. Chem.* **1989** *10* 221.
284. Stewart, J. J. P. *MOPAC Manual* (*6th Edition*) [http://tutor.oc.chemie.th-darmstadt.de/Mopac/mopac.html] (1990).
285. Mohamadi, F.; Richards, N. G. J.; Guida, W. C.; Liskamp, R.; Lipton, M.; Caufield, C.; Chang, G.; Hendrickson, T.; Still, W. C. *J. Comput. Chem.* **1990** *11* 440.
286. Still, W. C. *MacroModel Home Page* [http://www.columbia.edu/cu/chemistry/mmod/mmod.html].
287. Friesen, D.; Hedberg, K. *J. Am. Chem. Soc.* **1980** *102* 3987.

288. Radom, L.; Lathan, W. A.; Hehre, W. J.; Pople, J. A. *J. Am. Chem. Soc.* **1973** *95* 693.
289. Kveseth, K. *Acta Chem. Scand.* **1978** *A32* 51.
290. Wiberg, K. B.; Murcko, M. A. *J. Phys. Chem.* **1987** *91* 3616.
291. Miyajima, T.; Kurita, Y.; Hirano, T. *J. Phys. Chem.* **1987** *91* 3954.
292. Wiberg, K. B.; Murcko, M. A.; Laidig, K. E.; MacDougall, P. J. *J. Phys. Chem.* **1990** *94* 6956.
293. Durig, J. R.; Liu, J.; Little, T. S.; Kalasinsky, V. F. *J. Phys. Chem.* **1992** *96* 8224.
294. Topol, I. A.; Burt, S. K. *Chem. Phys. Lett.* **1993** *204* 611.
295. Papasavva, S.; Illinger, K. H.; Kenny, J. E. *J. Phys. Chem.* **1996** *100* 10100.
296. Gundertofte, K.; Liljefors, T.; Norrby P.-O.; Pettersson, I. *J. Comput. Chem.* **1996** *17* 429.
297. Liang, G.; Fox, P. C.; Bowen, J. P. *J. Comput. Chem.* **1996** *17* 940.
298. Harris, W. C.; Holtzclaw, J. R.; Kalasinsky, V. F. *J. Chem. Phys.* **1977** *67* 3330.
299. Eliel, E. L.; Allinger, N. L.; Angyal, S. J.; Morrison, G. A. *Conformational Analysis.* J. Wiley & Sons, New York (1965), pp. 433–486.
300. Bixon, M.; Lifson, S. *Tetrahedron* **1967** *23* 769.
301. Clark, T. *A Handbook of Computational Chemistry.* J. Wiley & Sons, New York (1985), pp. 12–25.
302. Comba, P.; Hambley, T. W. *Molecular Modeling of Inorganic Compounds.* VCH, New York (1995), pp. 14–28.
303. Gajewski, J. J.; Gilbert, K. E. *PCMODEL for Windows, Molecular Modeling Software for IBM and Compatible PCs under the Windows Operating System.* Serena Software, Bloomington, IN (1993).
304. Gajewski, J. J.; Gilbert, K. E.; McKelvey, J., in *Advances in Molecular Modeling, Vol. 2* (ed. Liotta, D). JAI Press, Greenwich, CT (1990), pp. 65–92.
305. Rasmussen, K., in *Potential Energy Functions in Conformational Analysis* (*Lecture Notes in Chemistry, 35*) (ed. Berthier, G.; Dewar, M. J. S.; Fischer, H.; Fukui, K.; Hall, G. G.; Hinze, J.; Jaffé, H. H.; Jortner, J.; Kutzelnigg, W.; Ruedenberg, K.). Springer-Verlag, New York (1985).
306. Clark, M.; Cramer, R. D., III; Van Opdenbosch, N. *J. Comput. Chem.* **1989** *10* 982.
307. Roterman, I. K.; Gibson, K. D.; Scheraga, H. A. *J. Biomol. Struct. Dynam.* **1989** *7* 391.
308. Roterman, I. K.; Lambert, M. H.; Gibson, K. D.; Scheraga, H. A. *J. Biomol. Struct. Dynam.* **1989** *7* 421.
309. Allured, V. S.; Kelly, C. M.; Landis, C. R. *J. Am. Chem. Soc.* **1991** *113* 1.
310. Saunders, M. *J. Comput Chem.* **1991** *12* 645.
311. Gundertofte, K.; Palm, J.; Pettersson, I.; Stamvik, A. *J. Comput. Chem.* **1991** *12* 200.
312. Ferguson, D. M.; Gould, I. R.; Glauser, W. A.; Schroeder, S.; Kollman, P. A. *J. Comput. Chem.* **1992** *13* 525.
313. Cornell, W. D.; Ha, M. P.; Sun, Y.; Kollman, P. A. *J. Comput. Chem.* **1996** *17* 1541.

314. Halgren, T. A. *J. Comput. Chem.* **1996** *17* 520.
315. Burkert, U.; Allinger, N. L. *Molecular Mechanics (ACS Monograph 177)*. American Chemical Society, Washington (1982), pp. 22–27.
316. Comba, P.; Hambley, T. W. *Molecular Modeling of Inorganic Compounds*. VCH, New York (1995), pp. 16–18.
317. Morse, P. M. *Phys. Rev.* **1929** *34* 57.
318. Steele, D.; Lippincott, E. R.; Vanderslice, J. T. *Rev. Mod. Phys.* **1962** *34* 239.
319. Atkins, P. *Molecular Quantum Mechanics, 2nd Edition*. Oxford University Press, New York (1983), pp. 298–301.
320. Daudel, R.; Leroy, G.; Peeters, D.; Sana, M. *Quantum Chemistry*. Wiley-Interscience, New York (1983), pp. 148–153.
321. Halgren, T. A. *J. Comput. Chem.* **1996** *17* 490.
322. Dinur, U.; Hagler, A. T. *J. Comput. Chem.* **1994** *15* 919.
323. Comba, P.; Hambley, T. W. *Molecular Modeling of Inorganic Compounds*. VCH New York (1995), pp. 18–22.
324. Burkert, U.; Allinger, N. L. *Molecular Mechanics (ACS Monograph 177)*. American Chemical Society, Washington (1982), pp. 32–36.
325. Comba, P.; Hambley, T. W. *Molecular Modeling of Inorganic Compounds*. VCH, New York (1995). pp. 22–23.
326. Dunitz, J. D., in *Perspectives in Structural Chemistry, Vol. 2* (ed. Dunitz, J. D.; Ibers, J. A.), Wiley, New York (1968), pp. 57–60.
327. Maple, J. R.; Dinur, U.; Hagler, A. T. *Proc. Natl. Acad. Sci. USA* **1988** *85* 5350.
328. Wilson, E. B.; Decius, J. C.; Cross, P. C. *Molecular Vibrations*, Dover, New York (1980), p. 59.
329. Comba, P.; Hambley, T. W. *Molecular Modeling of Inorganic Compounds*. VCH, New York (1995), p. 28.
330. Miller, K. *J. Comput. Chem* **1990** *11* 336.
331. Comba, P.; Hambley, T. W. *Molecular Modeling of Inorganic Compounds*. VCH, New York (1995), p. 24.
332. Nevins, N.; Chen, K. S.; Allinger, N. L. *J. Comput Chem.* **1996** *17* 669.
333. Nevins, N.; Allinger, N. L. *J. Comput. Chem.* **1996** *17* 730.
334. Burkert, U.; Allinger, N. L. *Molecular Mechanics (ACS Monograph 177)*, American Chemical Society, Washington (1982), pp. 27–32.
335. Comba, P.; Hambley, T. W. *Molecular Modeling of Inorganic Compounds*. VCH, New York (1995), pp. 24–25.
336. Allinger, N. L.; Chung, D. Y. *J. Am. Chem. Soc.* **1976** *98* 6798.
337. Eisenstein, O.; Anh, N. T.; Jean, Y.; Devaquet, A.; Cantacuzène, J.; Salem, L. *Tetrahedron* **1974** *30* 1717.
338. Hill, J.-R. *J. Comput. Chem.* **1997** *18* 211.
339. Califano, S. *Pure Appl. Chem.* **1969** *18* 353.
340. Comba, P.; Hambley, T. W. *Molecular Modeling of Inorganic Compounds*. VCH, New York (1995), p. 26.
341. Burkert, U.; Allinger, N. L. *Molecular Mechanics (ACS Monograph 177)*, American Chemical Society, Washington (1982), pp. 156–157.

342. Allinger, N. L.; Flanagan, H. L. *J. Comput. Chem.* **1983** *4* 399.
343. Gajewski, J. J.; Gilbert, K. E.; McKelvey, J., in *Advances in Molecular Modeling*, Vol. 2 (ed. Liotta, D.). JAI Press, Greenwich, CT (1990), pp. 76–77.
344. Emsley, J. *The Elements.* Clarendon Press, Oxford (1989).
345. *The CRC Handbook of Chemistry and Physics, 73rd Edition* (ed. in chief Lide, D. R.). CRC Press, Ann Arbor, MI (1992), p. 9-1ff.
346. Allinger, N. L. *J. Am. Chem. Soc.* **1977** *99* 8127.
347. Kontoyianni, M.; Bowen, J. P. *J. Comput. Chem.* **1992** *13* 657.
348. Allinger, N. L.; Yan, L.; Chen, K. *J. Comput. Chem.* **1994** *15* 1321.
349. Allinger, N. L.; Hickey, M. J.; Kao, J. *J. Am. Chem. Soc.* **1976** *98* 2741.
350. Došen-Mićović, L.; Jeremić, D.; Allinger, N. L. *J. Am. Chem. soc.* **1983** *105* 1716.
351. Došen-Mićović, L.; Jeremić, D.; Allinger, N. L. *J. Am. Chem. Soc.* **1983** *105* 1723.
352. Došen-Mićović, Li, S.; Allinger, N. L. *J. Phys. Org. Chem.* **1991** *4* 467.
353. Stewart, E. L.; Bowen, J. P. *J. Comput. Chem.* **1992** *13* 1125.
354. Allinger, N. L.; Kuang, J.; Thomas, H. D. *J. Mol. Struct. (THEOCHEM)* **1990** *209* 125.
355. Tribble, M. T.; Allinger, N. L. *Tetrahedron* **1972** *28* 2147.
356. Frierson, M. R.; Imam, M. R.; Zalkow, V. B.; Allinger, N. L. *J. Org. Chem.* **1988** *53* 5248.
357. Frierson, M. R.; Allinger, N. L. *J. Phys. Org. Chem.* **1989** *2* 573.
358. Allinger, N. L.; Hindman, D.; Hönig, H. *J. Am. Chem. Soc.* **1977** *99* 3282.
359. Allinger, N. L.; Lii, J.-H. *J. Comput. Chem.* **1987** *8* 1146.
360. Allinger, N. L.; Geise, H. J.; Pyckhout, W.; Paquette, L. A.; Gallucci, J. C. *J. Am. Chem. Soc.* **1989** *111* 1106.
361. Allinger, N. L.; Li, F.; Yan, L. *J. Comput. Chem.* **1990** *11* 848.
362. Allinger, N. L.; Li, F.; Yan, L.; Tai, J. C. *J. Comput. Chem.* **1990** *11* 868.
363. Allinger, N. L.; Rahman, M.; Lii, J.-H. *J. Am. Chem. Soc.* **1990** *112* 8293.
364. Schmitz, L. R.; Allinger, N. L. *J. Am. Chem. Soc.* **1990** *112* 8307.
365. Allinger, N. L.; Chen, K.; Raham, M.; Pathiaseril, A. *J. Am. Chem. Soc.* **1991** *113* 4505.
366. Allinger, N. L.; Zhu, Z.-q. S.; Chen, K. *J. Am. Chem. Soc.* **1992** *114* 6120.
367. Aped, P.; Allinger, N. L. *J. Am. Chem. Soc.* **1992** *114* 1.
368. Fan, Y.; Allinger, N. L. *J. Comput. Chem.* **1993** *14* 655.
369. Zhou, X.; Allinger, N. L. *J. Phys. Org. Chem.* **1994** *7* 420.
370. Goldstein, E.; Ma, B.; Lii, J.-H.; Allinger, N. L. *J. Phys. Org. Chem.* **1996** *9* 191.
371. Fan, Y.; Allinger, N. L. *J. Comput. Chem.* **1994** *15* 1446.
372. Fernández, B.; Ríos, M. A. *J. Comput. Chem.* **1994** *15* 455.
373. Liu, R.; Zhou, X.; Allinger, N. L. *J. Phys. Org. Chem.* **1994** *7* 551.
374. Lii, J.-H.; Allinger, N. L. *J. Phys. Org. Chem.* **1994** *7* 591.
375. Chen, K.; Allinger, N. L. *J. Phys. Org. Chem.* **1991** *4* 659.
376. Liang, G.; Bowen, J. P.; Bentley, J. A. *J. Comput. Chem.* **1994** *15* 866.
377. Stewart, E. L.; Mazurek, U.; Bowen, J. P. *J. Phys. Org. Chem.* **1996** *9* 66.

378. Allinger, N. L.; Quinn, M.; Rahman, M.; Chen, K. *J. Phys. Org. Chem.* **1991** *4* 647.
379. Allinger, N. L.; Fan, Y.; Varnali, T. *J. Phys. Org. Chem.* **1996** *9* 159.
380. Nevins, N.; Lii, J.-H.; Allinger, N. L. *J. Comput. Chem.* **1996** *17* 695.
381. Allinger, N. L.; Chen, K. S.; Katzenellenbogen, J. A.; Wilson, S. R.; Anstead, G. M. *J. Comput Chem.* **1996** *17* 747.
382. Warshel, A.; Lifson, S. *Chem. Phys. Lett.* **1969** *4* 255.
383. Warshel, A.; Levitt, M.; Lifson, S. *J. Mol. Spectr.* **1970** *33* 84.
384. Warshel, A.; Lifson, S. *J. Chem. Phys.* **1970** *53* 582.
385. Karplus, S.; Lifson, S. *Biopolym.* **1971** *10* 1973.
386. Ermer, O.; Lifson, S. *J. Am. Chem. Soc.* **1973** *95* 4121.
387. Ermer, O. *Tetrahedron* **1975** *31* 1849.
388. Niketić, S. R.; Rasmussen, K. *The Consistent Force Field: A Documentation.* Springer-Verlag, New York (1977).
389. Hagler, A. T.; Stern, P. S.; Lifson, S.; Ariel, S. *J. Am. Chem. Soc.* **1979** *101* 813.
390. Lifson, S.; Hagler, A. T.; Dauber, P. *J. Am. Chem. Soc.* **1979** *101* 5111.
391. Huige, C. J. M.; Hezemans, A. M. F.; Rasmussen, K. *J. Comput. Chem.* **1987** *8* 204.
392. Rasmussen, K.; Engelsen, S. B.; Fabricius, J.; Rasmussen, B., in *Recent Experimental and Computational Advances in Molecular Spectroscopy. NATO ASI Series C, Vol. 406* (ed. Fausto, R.). Kluwer, Boston (1993), p. 381.
393. Engelsen, S. B.; Fabricius, J.; Rasmussen, K. *Acta Chem. Scand.* **1994** *48* 548.
394. Engelsen, S. B.; Fabricius, J.; Rasmussen, K. *Acta Chem. Scand.* **1994** *48* 553.
395. Engelsen, S. B.; Fabricius, J.; Rasmussen, K. *Computers Chem.* **1994** *18* 397.
396. Jónsdóttir, S. Ó.; Rasmussen, K. *New J. Chem.* **1995** *19* 1113.
397. Fabricius, J.; Engelsen, S. B.; Rasmussen, K. *New J. Chem.* **1995** *19* 1123.
398. Hagler, A. T.; Huler, E.; Lifson, S. *J. Am. Chem. Soc.* **1974** *96* 5319.
399. Hagler, A. T.; Lifson, S. *J. Am. Chem. Soc.* **1974** *96* 5327.
400. Warshel, A.; Karplus, M. *J. Am. Chem. Soc.* **1972** *94* 5612.
401. Hagler, A. T.; Lifson, S.; Dauber, P. *J. Am. Chem. Soc.* **1979** *101* 5122.
402. Hagler, A. T.; Dauber, P.; Lifson, S. *J. Am. Chem. Soc.* **1979** *101* 5131.
403. Dauber, P.; Hagler, A. T. *Acc. Chem. Res.* **1980** *13* 105.
404. Niketić, S. R.; Rasmussen, K. *The Consistent Force Field: A Documentation.* Springer-Verlag, New York (1977), p. 49.
405. Niketić, S. R.; Rasmussen, K. *The Consistent Force Field: A Documentation.* Springer-Verlag, New York (1977), pp. 180–197.
406. Rasmussen, K., in *Potential Energy Functions in Conformational Analysis* (*Lecture Notes in Chemistry, 35*) (ed. Berthier, G.; Dewar, M. J. S.; Fischer, H.; Fukui, K.; Hall, G. G.; Hinze, J.; Jaffé, H. H.; Jortner, J.; Kutzelnigg, W.; Ruedenberg, K.). Springer-Verlag, New York (1985), pp. 95–117.
407. Rasmussen, K. in *Potential Energy Functions in Conformational Analysis* (*Lecture Notes in Chemistry, 35*) (ed. Berthier, G.; Dewar, M. J. S.; Fischer, H.; Fukui, K.; Hall, G. G.; Hinze, J.; Jaffé, H. H.; Jortner, J.; Kutzelnigg, W.; Ruedenberg, K.).

408. Rasmussen, K., in *Potential Energy Functions in Conformational Analysis* (*Lecture Notes in Chemistry*, 35) (ed. Berthier, G.; Dewar, M. J. S.; Fischer, H.; Fukui, K.; Hall, G. G.; Hinze, J.; Jaffé, H. H.; Jortner, J.; Kutzelnigg, W.; Ruedenberg, K.). Springer-Verlag, New York (1985), pp. 119–130.

Springer-Verlag, New York (1985), pp. 39–48.

409. Rasmussen, K., in *Potential Energy Functions in Conformational Analysis* (*Lecture Notes in Chemistry*, 35) (ed. Berthier, G.; Dewar, M. J. S.; Fischer, H.; Fukui, K.; Hall, G. G.; Hinze, J.; Jaffé, H. H.; Jortner, J.; Kutzelnigg, W.; Ruedenberg, K.). Springer-Verlag, New York (1985), pp. 49–61.

410. Rasmussen, K., in *Potential Energy Functions in Conformational Analysis* (*Lecture Notes in Chemistry*, 35) (ed. Berthier, G.; Dewar, M. J. S.; Fischer, H.; Fukui, K.; Hall, G. G.; Hinze, J.; Jaffé, H. H.; Jortner, J.; Kutzelnigg, W.; Ruedenberg, K.). Springer-Verlag, New York (1985), pp. 63–65.

411. Sun, H.; Mumby, S. J.; Maple, J. R.; Hagler, A. T. *J. Am. Chem. Soc.* **1994** *116* 2978.

412. Meier, R. J.; Maple, J. R.; Hwang, M.-J.; Hagler, A. T. *J. Phys. Chem.* **1995** *99* 5445.

413. Molecular Simulations, Inc. *CFF95* [http://www.msi.com/corp/pdf/cff95.pdf] (1996).

414. Weiner, P. K.; Kollman, P. A. *J. Comput. Chem.* **1981** *2* 287.

415. Weiner, S. J.; Kollman, P. A.; Case, D. A.; Singh, U. C.; Ghio, C.; Alagona, G.; Profeta, S., Jr.; Weiner, P. *J. Am. Chem. Soc.* **1984** *106* 765.

416. Weiner, S. J.; Kollman, P. A.; Nguyen, D. T.; Case, D. A. *J. Comput. Chem.* **1986** *7* 230.

417. Ferguson, D. M.; Kollman, P. A. *J. Comput. Chem.* **1991** *12* 620.

418. Cornell, W. D.; Cieplak, P.; Bayly, C. I.; Kollman, P. A. *J. Am. Chem. Soc.* **1993** *115* 9620.

419. Cheatham, T.; Ross, B. *Amber Home Page* [http://www.amber.ucsf.edu/amber/amber.html] (1995).

420. Cornell, W. D.; Cieplak, P.; Bayly, C. I.; Gould, I. R.; Merz, K. M., Jr.; Ferguson, D. M.; Spellmeyer, D. C.; Fox, T.; Caldwell, J. W.; Kollman, P. A. *J. Am. Chem. Soc.* **1995** *117* 5179 [correction, *J. Am. Chem. Soc.* **1996** *118* 2309].

421. Hoops, S. C.; Anderson, K. W.; Merz, K. M., Jr. *J. Am. Chem. Soc.* **1991** *113* 8262.

422. Mitchell, M. J.; McCammon, J. A. *Computers Chem.* **1991** *15* 79.

423. McDonald, D. Q.; Still, W. C. *Tetrahedron Lett.* **1992** *33* 7743.

424. Cannon, J. F. *J. Comput. Chem.* **1993** *14* 995.

425. Edvardsen, Ø. *Computers Chem.* **1994** *18* 433.

426. Glennon, T. M.; Zheng, Y.-J.; Le Grande, S. M.; Shutzberg, B. A.; Merz, K. M., Jr. *J. Comput. Chem.* **1994** *15* 1019.

427. Yao, S.; Plastaras, J. P.; Marzilli, L. G. *Inorg. Chem.* **1994** *33* 6061.

428. Huige, C. J. M.; Altona, C. *J. Comput. Chem.* **1995** *16* 56.

429. Woods, R. J.; Dwek, R. A.; Edge, C. J.; Fraser-Reid, B. *J. Phys. Chem.* **1995** *99* 3832.

430. Senderowitz, H.; Parish, C.; Still, W. C. *J. Am. Chem. Soc.* **1996** *118* 2078

[correction, *J. Am. Chem. Soc.* **1996** *118* 8985].
431. Warshel, A.; Levitt, M. *QCFF/PI: Quantum Mechanical Extension of the Consistent Force-Field Method (QCPE* #247) [http://ccinfo.ims.ac.jp:80/cgi-bin/qcpe] (1982).
432. Cox, S. R.; Williams, D. E. *J. Comput. Chem.* **1981** *2* 304.
433. Singh, U. C.; Kollman, P. A. *J. Comput. Chem.* **1984** *5* 129.
434. Bayly, C. I.; Cieplak, P.; Cornell, W. D.; Kollman, P. A. *J. Phys. Chem.* **1993** *97* 10269.
435. Jorgensen, W. L.; Tirado-Rives, J. *J. Am. Chem. Soc.* **1988** *110* 1657.
436. Cheatham, T.; Ross, B. *Amber Home Page* [http://munin.ucsf.edu/amber/ff94_glycam.html] (1995).
437. Edvardsen, Ø. *Comput. Chem.* **1996** *20* 483 [http://atf1.fagmed.uit.no/farma/ampar.html].
438. Gelin, B. R.; Karplus, M. *Biochem.* **1979** *18* 1256.
439. Brooks, B. R.; Bruccoleri, R. E.; Olafson, B. D.; States, D. J.; Swaminathan, S.; Karplus, M. *J. Comput. Chem.* **1983** *4* 187.
440. Nilsson, L.; Karplus, M. *J. Comput. Chem.* **1986** *7* 591.
441. Smith, J. C.; Karplus, M. *J. Am. Chem. Soc.* **1992** *114* 801.
442. Schmidt, J.; Brüschweiler, R.; Ernst, R. R.; Dunbrack, R. L., Jr.; Joseph, D.; Karplus, M. *J. Am. Chem. Soc.* **1993** *115* 8747.
443. MacKerell, A. D., Jr.; Wiórkiewicz-Kuczero, J.; Karplus, M. *J. Am. Chem. Soc.* **1995** *117* 11946.
444. Blondel, A.; Karplus, M. *J. Comput. Chem.* **1996** *17* 1132.
445. Nagle, R. *Chemistry at HARvard Macromolecular Mechanics* [http://yuri.harvard.edu/charmm/charmm.html] (1996).
446. Momany, F. A.; Rone, R. *J. Comput. Chem.* **1992** *13* 888.
447. Grootenhuis, P. D. J.; Haasnoot, C. A. G. *Mol. Simul.* **1993** *10* 75.
448. Kouwijzer, M. L. C. E.; Grootenhuis, P. D. J. *J. Phys. Chem.* **1995** *99* 13426.
449. Kouwijzer, M. L. C. E.; Grootenhuis, P. D. J. *The CHEAT95 Force Field: A Carbohydrate Force Field in Which the Carbohydrate Hydroxyls are Represented by Extended Atoms* [http://www.msi.com/support/life/quanta/cheat95.html] (1996).
450. Pavelites, J. J.; Gao, J.; Bash, P. A.; Mackerell, A. D., Jr. *J. Comput. Chem.* **1997** *18* 221.
451. White, D. N. J.; Bovill, M. J. *J. Chem. Soc. Perkin Trans. 2* **1977** 1610.
452. Gasteiger, J.; Marsili, M. *Tetrahedron* **1980** *36* 3219.
453. Halgren, T. A. *J. Comput. Chem.* **1996** *17* 553.
454. Halgren, T. A.; Nachbar, R. B. *J. Comput. Chem.* **1996** *17* 587.
455. Halgren, T. A. *J. Comput. Chem.* **1996** *17* 616.
456. Holloway, M. K.; Wai, J. M.; Halgren, T. A.; Fitzgerald, P. M. D.; Vacca, J. P.; Dorsey, B. D.; Levin, R. B.; Thompson, W. J.; Chen, L. J.; deSolms, S. J.; Gaffin, N.; Ghosh, A. K.; Giuliani, E. A.; Graham, S. L.; Guare, J. P.; Hungate, R. W.; Lyle, T. A.; Sanders, W. M.; Tucker, T. J.; Wiggins, M.; Wiscount, C. M.; Woltersdorf, O. W.; Young, S. D.; Darke, P. L.; Zugay, J. A. *J. Med. Chem.* **1995** *38*

305.
457. *Journal of Computational Chemistry Home Page* [http://journals.wiley.com/0192-8651/] (1996).
458. Jorgensen, W. J. *J. Am. Chem. Soc.* **1981** *103* 335.
459. Jorgensen, W. J. *J. Am. Chem. Soc.* **1981** *103* 341.
460. Jorgensen, W. J. *J. Am. Chem. Soc.* **1981** *103* 345.
461. Jorgensen, W. L.; Chandrasekhar, J.; Madura, J. D.; Impey, R. W.; Klein, M. L. *J. Chem. Phys.* **1983** *79* 926.
462. Jorgensen, W. L.; Madura, J. D.; Swenson, C. J. *J. Am. Chem. Soc.* **1984** *106* 6638.
463. Jorgensen, W. L.; Madura, J. D. *Mol. Phys.* **1985** *56* 1381.
464. Jorgensen, W. L.; Swenson, C. J. *J. Am. Chem. Soc.* **1985** *107* 569.
465. Jorgensen, W. L. *J. Phys. Chem.* **1986** *90* 1276.
466. Jorgensen, W. L. *J. Phys. Chem.* **1986** *90* 6379.
467. Jorgensen, W. L.; Briggs, J. M.; Contreras, M. L. *J. Phys. Chem.* **1990** *94* 1683.
468. Jorgensen, W. L.; Severance, D. L. *J. Am. Chem. Soc.* **1990** *112* 4768.
469. Jorgensen, W. L.; Pranata, J. *J. Am. Chem. Soc.* **1990** *112* 2008.
470. Pranata, J.; Wierschke, S. G.; Jorgensen, W. L. *J. Am. Chem. Soc.* **1991** *113* 2810.
471. Briggs, J. M.; Nguyen, T. B.; Jorgensen, W. L. *J. Phys. Chem.* **1991** *95* 3315.
472. Tirado-Rives, J.; Jorgensen, W. L. *J. Am. Chem. Soc.* **1990** *112* 2773.
473. Orozco, M.; Tirado-Rives, J.; Jorgensen, W. L. *Biochemistry* **1993** *32* 12864.
474. Jorgensen, W. L. *ChemTracts—Org. Chem.* **1991** 91.
475. Jorgensen, W. L. *Science* **1981** *254* 954.
476. Jorgensen, W. L.; Maxwell, D. S.; Tirado-Rives, J. *J. Am. Chem. Soc.* **1996** *118* 11225.
477. Still, W. C.; Galynker, I. *Tetrahedron* **1981** *37* 3981.
478. Still, W. C.; MacPherson, L. J.; Harada, T.; Callahan, J. F.; Rheingold, A. L. *Tetrahedron* **1981** *40* 2275.
479. Browman, M. J.; Carruthers, L. M.; Kashuba, K. L.; Momany, F. A.; Pottle, M. S.; Rosen, S. P.; Rumsey, S. M.; Endres, G. F.; Scheraga, H. A. *ECEPP: Empirical Conformational Energy Programs for Peptides (QCPE #286)* [http://ccinfo.ims.ac.jp:80/cgi-bin/qcpe] (1975).
480. Browman, M. J.; Carruthers, L. M.; Kashuba, K. L.; Momany, F. A.; Pottle, M. S.; Rosen, S. P.; Rumsey, S. M.; Endres, G. F.; Scheraga, H. A. *ECEPP/2: Empirical Conformational Energy Programs for Peptides (QCPE #454)* [http://ccinfo.ims.ac.jp:80/cgi-bin/qcpe] (1983).
481. Zimmerman, S. S.; Pottle, M. S.; Némethy, G.; Scheraga, H. A. *Macromol.* **1977** *10* 1.
482. Némethy, G.; Pottle, M. S.; Scheraga, H. A. *J. Phys. Chem.* **1983** *87* 1883.
483. Sippl, M. J.; Némethy, G.; Scheraga, H. A. *J. Phys. Chem.* **1984** *88* 6231.
484. Pincus, M. R.; Scheraga, H. A. *Acc. Chem. Res.* **1985** *18* 372.
485. Chuman, H. Momany, F. A.; Schäfer, L. *Int. J. Peptide Protein Res.* **1984** *24* 233.
486. Chuman, H. Momany, F. A. *Int. J. Peptide Protein Res.* **1984** *24* 249.

487. Némethy, G.; Gibson, K. D.; Yoon, C. N.; Paterlini, G.; Zagari, A.; Rumsey, S.; Scheraga, H. A. *J. Phys. Chem.* **1992** *96* 6472.
488. Duben, A. J.; Hricovini, M.; Tvaroška, I. *Carbohydr. Res.* **1993** *247* 71.
489. Bingham, R. C.; Schleyer, P. v. R. *J. Am. Chem. Soc.* **1971** *93* 3189.
490. Andose, J. D.; Engler, E. M.; Collins, J. B.; Hummel, J. P.; Mislow, K.; Schleyer, P. v. R. *BIGSTRN: Empirical Force Field Calculations* (*QCPE* #348) [http://ccinfo.ims.ac.jp:80/cgi-bin/qcpe] (1978).
491. Iverson, D. J.; Mislow, K. *BIGSTRN-2.: Empirical Force-Field Calculation* (*QCPE* #410) [http://ccinfo.ims.ac.jp:80/cgi-bin/qcpe] (1981).
492. Barborak, J. C.; Khoury, D.; Maier, W. F.; Schleyer, P. v. R.; Smith, E. C.; Smith, W. F., Jr.; Wyrick, C. *J. Org. Chem.* **1979** *44* 4761.
493. Maier, W. F.; Schleyer, P. v. R. *J. Am. Chem. Soc.* **1981** *103* 1891 (correction, *ibid.* **1987** 7037).
494. Reindl, B.; Clark, T.; Schleyer, P. v. R. *J. Comput. Chem.* **1996** *17* 1406.
495. Reindl, B.; Clark, T.; Schleyer, P. v. R. *J. Comput. Chem.* **1997** *18* 28.
496. Mayo, S. L.; Olafson, B. D.; Goddard, W. A., III. *J. Phys. Chem.* **1990** *94* 8897.
497. Gabriel, J. L.; Mitchell, W. M. *Proc. Natl. Acad. Sci. USA* **1993** *90* 4186.
498. Gerdy, J. J.; Goddard, W. A., III. *J. Am. Chem. Soc.* **1996** *118* 3233.
499. Fan, C. F.; Çagin, T.; Chen, Z. M.; Smith, K. A. *Macromolecules* **1994** *27* 2383.
500. Jones, M.-C. G.; Martin, D. C. *Macromolecules* **1995** *28* 6161.
501. Kitano, Y.; Usami, I.; Obata, Y.; Okuyama, K.; Jinda, T. *Polymer* **1995** *36* 1123.
502. Parenté, V.; Brédas, J.-L.; Dubois, P.; Ropson, N.; Jérôme, R. *Macromol. Theory Simul.* **1996** *5* 525.
503. Kazmaier, P. M.; McKerrow, A. J.; Buncel, E. *J. Imaging Sci. Tech.* **1992** *36* 373.
504. Wang, Q.; Stidham, H. D.; Papadimitrakopolos, F. *Spectrochim. Acta* **1994** *50A* 421.
505. Rappé, A. K.; Casewit, C. J.; Colwell, K. S.; Goddard, W. A., III; Skiff, W. M. *J. Am. Chem. Soc.* **1992** *114* 10024.
506. Casewit, C. J.; Colwell, K. S.; Rappé, A. K. *J. Am. Chem. Soc.* **1992** *114* 10035.
507. Casewit, C. J.; Colwell, K. S.; Rappé, A. K. *J. Am. Chem. Soc.* **1992** *114* 10046.
508. Badger, R. M. *J. Chem. Phys.* **1934** *1* 128.
509. Badger, R. M. *J. Chem. Phys.* **1935** *3* 710.
510. Herschbach, D. R.; Laurie, V. W. *J. Chem. Phys.* **1961** *35* 458.
511. Halgren, T. A. *J. Am. Chem. Soc.* **1990** *112* 4710.
512. Rappé, A. K.; Colwell, K. S.; Casewit, C. J. *Inorg. Chem.* **1993** *32* 3438.
513. Grigoras, S.; Qian, C.; Crowder, C.; Harkness, B.; Mita, I. *Macromolecules* **1995** *28* 7370.
514. Dillen, J. L. M. *J. Comput. Chem.* **1990** *11* 1125.
515. Dillen, J. L. M. *J. Comput. Chem.* **1995** *16* 595.
516. Dillen, J. L. M. *J. Comput. Chem.* **1995** *16* 610.
517. Lindner, H. J. *Tetrahedron* **1974** *30* 1127.
518. Smith, A. E.; Lindner, H. J. *J. Comput.-Aided Mol. Des.* **1991** *5* 235.
519. Kroeker, M. *Entwicklung und Erprobung des PIMM91-Kraftfeldes zur Model-*

lierung von Molekülstrukturen (Development and Validation of the PIMM91 Force Field for Modeling Molecular Structures). [http://tutor.oc.chemie.th-darmstadt.de/TZ/AKLindner/MK/node1_e.html] (1994).
520. Lehmann, J.; Kleinpeter, E. Monatsh. Chem. **1995** *126* 1051.
521. Vedani, A.; Dunitz, J. D. *J. Am. Chem. Soc.* **1985** *107* 7653.
522. Vedani, A.; Dobler, M.; Dunitz, J. D. *J. Comput. Chem.* **1986** *7* 701.
523. Vedani, A. *J. Comput. Chem.* **1988** *9* 269.
524. Vedani, A.; Huhta, D. W. *J. Am. Chem. Soc.* **1990** *112* 4759.
525. Vedani, A.; Huhta, D. W. *J. Am. Chem. Soc.* **1991** *113* 5860.
526. Kovach, I. M. *J. Mol. Struct. (THEOCHEM)* **1988** *170* 159.
527. Kovach, I. M.; Huhta, D.; Baptist, S. *J. Mol. Struct. (THEOCHEM)* **1991** *226* 99.
528. Burkert, U.; Allinger, N. L. *Molecular Mechanics (ACS Monograph No. 177)*. American Chemical Society. Washington (1982), pp. 36–52.
529. Jaime, C.; Osawa, E. Tetrahedron **1983** *39* 2769.
530. Rasmussen, K., in *Potential Energy Functions in Conformational Analysis (Lecture Notes in Chemistry, 35)* (ed. Berthier, G.; Dewar, M. J. S.; Fischer, H.; Fukui, K.; Hall, G. G.; Hinze, J.; Jaffé, H. H.; Jortner, J.; Kutzelnigg, W.; Ruedenberg, K.). Springer-Verlag, New York (1985), pp. 67–80.
531. Comba, P.; Hambley, T. W. *Molecular Modeling of Inorganic Compounds*. VCH, New York (1995), pp. 28–31.
532. Osawa, E.; Lipkowitz, K. B., in *Reviews of Computational Chemistry, Vol. 6.* (ed. Lipkowitz, K. B.; Boyd, D. B.). VCH, New York (1995), pp. 355–381.
533. Cheatham, T.; Ross, B. *Parameter Development* [http://munin.ucsf.edu/amber/newparams.html] (1995).
534. Collins, L. J.; Kirk, D. N. *Tetrahedron Lett.* **1970** *11* 1547.
535. Burkert, U.; Allinger, N. L. *Molecular Mechanics (ACS Monograph No. 177)*. American Chemical Society, Washington (1982), pp. 195–202.
536. Allinger, N. L.; Schäfer, L.; Siam, K.; Klimkowski, V. J.; Van Alsenoy, C. *J. Comput. Chem.* **1985** *6* 331.
537. Dykstra, C. E. *Chem. Rev.* **1993** *93* 2339.
538. Finn, P. W., in *Molecular Modelling and Drug Design; Topics in Molecular and Structural Biology* (ed. Vinter, J. G.; Gardner, M.), CRC Press, Boca Raton, FL (1994), pp. 266–304.
539. Honig, B.; Nicholls, A. *Science* **1995** *268* 1144.
540. Wiberg, K. B. *J. Chem. Educ.* **1996** *73* 1089.
541. *The CRC Handbook of Chemistry and Physics, 73rd Edition* (ed. in chief). Lide, D. R. CRC Press, Ann Arbor, MI (1992), p. 9-42ff.
542. Abraham, R. J.; Grant, G. H.; Haworth, I. S.; Smith, P. E. *J. Comput.-Aid. Mol. Des.* **1991** *5* 21.
543. Del Re, G. *J. Chem. Soc.* **1958** 4031.
544. Abraham, R. J.; Hudson, B. *J. Comput. Chem.* **1985** *6* 173.
545. Rasmussen, K., in *Potential Energy Functions in Conformational Analysis (Lecture Notes in Chemistry, 35)* (ed. Berthier, G.; Dewar, M. J. S.; Fischer, H.; Fukui, K.; Hall, G. G.; Hinze, J.; Jaffé, H. H.; Jortner, J.; Kutzelnigg, W.; Ruedenberg, K.).

Springer-Verlag, New York (1985), pp. 81–94.
546. Abraham, R. J.; Smith, P. E. *J. Comput. Chem.* **1987** *9* 288.
547. Allen, L. C. *J. Am. Chem. Soc.* **1989** *111* 9115.
548. Craw, J. S.; Hinchliffe, A.; Perez, J. J. "Electric and Magnetic Properties," in *Self-Consistent Field. Theory and Applications* (*Studies in Physical and Theoretical Chemistry 70*) (ed. Carbó, R.; Klobukowski, M.). Elsevier, New York (1990), pp. 866–910.
549. Rappé, A. K.; Goddard, W. A., III. *J. Phys. Chem.* **1991** *95* 3358.
550. Dinur, U.; Hagler, A. T. *J. Comput. Chem.* **1995** *16* 154.
551. St. Amant, A.; Cornell, W. D.; Kollman, P.; Halgren, T. A. *J. Comput. Chem.* **1995** *16* 1483.
552. Bakowies, D.; Thiel, W. *J. Comput. Chem.* **1996** *17* 87.
553. Politzer, P.; Whittenberg, S. L.; Wärnhelm, T. *J. Phys. Chem.* **1982** *86* 2609.
554. Ohta, T.; Kuroda, H. *Bull. Chem. Soc. Japan* **1976** *49* 2939.
555. Mark, J. E.; Sutton, C. *J. Am. Chem. Soc.* **1972** *94* 1083.
556. Meyer, A. Y. *J. Chem. Soc. Perkin Trans. 2* **1982** 1199.
557. Smith, R. P.; Ree, T.; Magee, J. L.; Eyring, H. *J. Am. Chem. Soc.* **1951** *73* 2263.
558. Smith, R. P.; Mortensen, E. M. *J. Am. Chem. Soc.* **1956** *78* 3932.
559. Allinger, N. L.; Wuesthoff, M. T. *Tetrahedron* **1977** *33* 3.
560. Došen-Mićović, L.; Allinger, N. L. *Tetrahedron* **1978** *34* 3385.
561. Gasteiger, J.; Marsili, M. *Org. Magn. Res.* **1981** *15* 353.
562. Meyer, A. Y.; Forrest, F. R. F. *J. Comput. Chem.* **1985** *6* 1.
563. Williams, D. E. *J. Comput. Chem.* **1994** *15* 719.
564. Wong, M. W.; Frisch, M. J.; Wiberg, K. B. *J. Am. Chem. Soc.* **1991** *113* 4776.
565. Blokzijl, W.; Engberts, J. B. F. N. *Angew. Chem. Int. Ed. Engl.* **1993** *32* 1545.
566. Tomasi, J.; Persico, M. *Chem. Rev.* **1994** *94* 2027.
567. Cramer, C. J.; Truhlar, D. G., in *Reviews of Computational Chemistry, Vol 6* (ed. Lipkowitz, K. B.; Boyd, D. B.). VCH, New York, 1995, pp. 1–72.
568. Still, W. C.; Tempczyk, A.; Hawley, R. C.; Hendrickson, T. *J. Am. Chem. Soc.* **1990** *112* 6127.
569. Williams, D. J.; Hall, K. B. *J. Phys. Chem.* **1996** *100* 8224.
570. Sefler, A. M.; Lauri, G.; Bartlett, P. A. *Int. J. Pept. Prot. Res.* **1996** *48* 129.
571. Nagy, P. I.; Bitar, J. E.; Smith, D. A. *J. Comput. Chem.* **1994** *15* 1228.
572. Burkert, U.; Allinger, N. L. *Molecular Mechanics* (*ACS Monograph 177*). American Chemical Society, Washington (1982), pp. 64–72.
573. Morley, S. D., in *Molecular Modelling and Drug Design: Topics in Molecular and Structural Biology* (ed. Vinter, J. G.; Gardner, M.). CRC Press, Boca Raton, FL (1994), pp. 53–88.
574. *Molecular Modelling and Drug Design: Topics in Molecular and Structural Biology.* (ed. Vinter, J. G.; Gardner, M.). CRC Press, Boca Raton, FL (1994).
575. Comba, P.; Hambley, T. W. *Molecular Modeling of Inorganic Compounds.* VCH, New York (1995), pp. 42–47.
576. Allinger, N. L.; Tribble, M. T.; Miller, M. A.; Wertz, D. H. *J. Am. Chem. Soc.* **1971**

93 1637.
577. Comba, P.; Hambley, T. W. *Molecular Modeling of Inorganic Compounds.* VCH, New York (1995), pp. 48–52.
578. Anet, F. A. L. *Top. Curr. Chem.* **1974** *45* 169.
579. Howard, A. E.; Kollman, P. A. *J. Med. Chem.* **1988** *31* 1669.
580. Lipkowitz, K. B., in *Conformational Analysis of Cyclohexenes, Cyclohexadienes, and Related Hydroaromatic Compounds* (ed. Rabideau, P. W.). VCH, New York (1989), pp. 213–265.
581. van Gunsteren, W. F.; Berendsen, H. J. C. *Angew. Chem. Int. Ed. Engl.* **1990** *29* 992.
582. Leach, A. R., in *Reviews of Computational Chemistry, Vol. 2* (ed. Lipkowitz, K. B.; Boyd, D. B.). VCH, New York, 1991, pp. 1–55.
583. Scheraga, H. A. *Biophys. Chem.* **1996** *59* 329.
584. Mastryukov, V. S.; Popik, M. V.; Dorofeeva, O. V.; Golubinskii, A. V.; Belikova, N. A.; Allinger, N. L. *J. Am. Chem. Soc.* **1981** *103* 1333.
585. Porter, T. L.; Maxka, J.; Abes, J. *J. Chem. Educ.* **1995** *72* A236.
586. Clark, D. E.; Jones, G.; Willet, P.; Kenny, P. W.; Glen, R. C. *J. Chem. Inf. Comput. Sci.* **1994** *34* 197.
587. Moock, T. E.; Henry, D. R.; Ozkabak, A. G.; Alamgir, M. *J. Chem. Inf. Comput. Sci.* **1994** *34* 184.
588. Saunders, M.; Houk, K. N.; Wu, Y.-D.; Still, W. C.; Lipton, M.; Chang, G.; Guida, W. C. *J. Am. Chem. Soc.* **1990** *112* 1419.
589. Ferguson, D. M.; Raber, D. J. *J. Am. Chem. Soc.* **1989** *111* 4371.
590. Saunders, M.; Jiménez-Vásquez, H. A. *J. Comput. Chem.* **1993** *14* 330.
591. Lipton, M.; Still, W. C. *J. Comput. Chem.* **1988** *9* 343.
592. Chang, G.; Guida, W. C.; Still, W. C. *J. Am. Chem. Soc.* **1989** *111* 4379.
593. Blaney, J. M.; Crippen, G. M.; Dearing, A.; Dixon, J. S. *DGEOM: Distance Geometry Program (QCPE #590)* [http://ccinfo.ims.ac.jp:80/cgi-bin/qcpe] (1990).
594. Ryckaert, J.-P.; Ciccotti, G.; Berendsen, H. J. C. *J. Comput. Phys.* **1977** *23* 327.
595. Ryckaert, J. P. *Molec. Phys.* **1985** *55* 549.
596. Treasurywala, A. M.; Jaeger, E. P.; Peterson, M. L. *J. Comput. Chem.* **1996** *17* 1171.
597. Meza, J. C.; Judson, R. S.; Faulkner, T. R.; Treasurywala, A. M. *J. Comput. Chem.* **1996** *17* 1142.
598. Ivanov, P. M.; Osawa, E. *J. Comput. Chem.* **1984** *5* 307.
599. Bruccoleri, R. E.; Karplus, M. *Biopolym.* **1987** *26* 137.
600. Luke, B. T. *J. Comput. Chem.* **1994** *15* 1176.
601. Goodman, J. M.; Still, W. C. *J. Comput. Chem.* **1991** *12* 1110.
602. Dolata, D. P.; Carter, R. E. *J. Chem. Inf. Comput. Sci.* **1987** *27* 36.
603. Leach, A. R.; Prout, K.; Dolata, D. P. *J. Comput. Chem.* **1990** *11* 680.
604. Yan, B.; Lai, C.-M.; Lin, S.-F.; Li, Z.-M. *Gaodeng Xuexiao Huaxue Xuebao* **1992** *13* 1555 [*Chem. Abstr.* **1993** *118* 228100v].
605. Xie, Y.; Schafer, H. F., III; Liang, G.; Bowen, J. P. *J. Am. Chem. Soc.* **1994** *116*

606. Kirkpatrick, S.; Gelatt, C. D. Jr.; Vecchi, M. P. *Science* **1983** *220* 671.
607. Moskowitz, J. W.; Schmidt, K. E.; Wilson, S. R.; Cui, W. *Intl. J. Quantum Chem: Quantum Chem. Symp.* **1988** *22* 611.
608. Wilson, S. R.; Cui, W.; Moskowitz, J. W.; Schmidt, K. E. *Tetrahedron Lett.* **1988** *29* 4373.
609. Wilson, S. R.; Cui, W. *Biopolym.* **1990** *29* 225.
610. Wilson, S. R.; Guarnieri, F. *Tetrahedron Lett.* **1991** *32* 3601.
611. Guarnieri, F.; Cui, W.; Wilson, S. R. *J. Chem. Soc. Chem. Commun.* **1991** 1542.
612. Wilson, S. R.; Cui, W.; Moskowitz, J. W.; Schmidt, K. E. *J. Comput. Chem.* **1991** *12* 342.
613. Peters, T.; Meyer, B.; Stuike-Prill, R.; Somorjai, R.; Brisson, J.-R. *Carbohydr. Res.* **1993** *238* 49.
614. Guarnieri, F.; Wilson, S. R. *J. Comput. Chem.* **1995** *16* 648.
615. Kerr, I. D. *Simulated Annealing via Restrained Molecular Dynamics* [http://indigo1.biop.ox.ac.uk/ian/sa_md.d/sim_anneal.html] (1996).
616. Owicki, J. C., in *Computer Modeling of Matter, ACS Symposium Series 78* (ed. Lykos, P.) American Chemical Society, Washington (1978), pp. 159–171.
617. Sun, Y.; Kollman, P. A. *J. Comput. Chem.* **1992** *13* 33.
618. Mestres, J.; Scuseria, G. E. *J. Comput. Chem.* **1995** *16* 729.
619. Weiner, P. K.; Profeta, S., Jr.; Wipff, G.; Havel, T.; Kuntz, I. D.; Langridge, R.; Kollman, P. A. *Tetrahedron* **1983** *39* 1113.
620. Hurst, T. *J. Chem. Inf. Comput. Sci.* **1994** *34* 190.
621. Uiterwijk, J. W. H. M.; Harkema, S.; van de Waal, B. W.; Göbel, F.; Nibbeling, H. T. M. *J. Chem. Soc. Perkin Trans. 2* **1983** 1843.
622. Smellie, A.; Teig, S. T.; Towbin, P. *J. Comput. Chem.* **1995** *16* 171.
623. Smellie, A.; Kahn, S. D.; Teig, S. T. *J. Chem. Inf. Comput. Sci.* **1995** *35* 285.
624. Smellie, A.; Kahn, S. D.; Teig, S. T. *J. Chem. Inf. Comput. Sci.* **1995** *35* 295.
625. E!ber, R.; Roitberg, A.; Simmerling, C.; Goldstein, R.; Li, H.; Verkhivker, G.; Keasar, C.; Zhang, J.; Ulitsky, A. *Comput. Phys. Commun.* **1995** *91* 159.
626. Li, F.; Cui, W.; Allinger, N. L. *J. Comput. Chem.* **1994** *15* 769.
627. Makara, G. M.; Keserû, G. M.; Kajtár-Peredy, M.; Anderson, W. K. *J. Med. Chem.* **1996** *39* 1236.
628. Morita, H.; Yoshida, N.; Takeya, K.; Itokawa, H.; Shirota, O. *Tetrahedron* **1996** *52* 2795.
629. Engelsen, S. B.; Brady, J. W.; Sherbon, J. W. *J. Agric. Food Chem.* **1994** *42* 2099.
630. Jorgensen, W. L. *ChemTracts—Org. Chem.* **1995** *8* 374.
631. Sok, R. M.; Berendsen, H. J. C.; van Gunsteren, W. F. *J. Chem. Phys.* **1992** *96* 4699.
632. Hagler, A. T.; Osguthorpe, D. J.; Dauber-Osguthorpe, P.; Hempel, J. C. *Science* **1985** *227* 1309.
633. Field, M. J.; Bash, P. A.; Karplus, M. *J. Comput. Chem.* **1990** *11* 700.
634. Verlet, L. *Phys. Rev.* **1967** *159* 98.

635. Stork, G.; Still, W. C.; Singh, J. *Tetrahedron Lett.* **1979** *20* 5077.
636. Dewan, J. *Acta Cryst.* **1979** *B35* 3111.
637. Ungaro, R.; Pochini, A.; Andreeti, G. D.; Sangermano, V. *J. Chem. Soc. Perkin Trans. 2* **1984** 1979.
638. Hunter, C. A.; Sanders, J. K. M. *J. Am. Chem. Soc.* **1990** *112* 5525.
639. Karlström, G.; Linse, P.; Wallqvist, A.; Jönsson, B. *J. Am. Chem. Soc.* **1983** *105* 3777.
640. Chipot, C.; Jaffe, R.; Maigret, B.; Pearlman, D. A.; Kollman, P. A. *J. Am. Chem. Soc.* **1996** *118* 11217.
641. Hobza, P.; Selzle, H. L.; Schlag, E. W. *Chem. Rev.* **1994** *94* 1767.
642. Janda, K. C.; Hemminger, J. C.; Winn, J. S.; Novick, S. E.; Harris, S. J.; Klemperer, W. *J. Chem. Phys.* **1975** *63* 1419.
643. Steed, J. M.; Dixon, T. A.; Klemperer, W. *J. Chem. Phys.* **1979** *70* 4940.
644. Richards, F. M. *Ann. Rev. Biophys. Bioeng.* **1977** *6* 151.
645. Mezey, P. G., in *Reviews of Computational Chemistry, Vol. 1* (ed. Lipkowitz, K. B.; Boyd, D. B.). VCH, New York, 1990, pp. 265–294.
646. Good, A. C., in *Molecular Similarity in Drug Design* (ed. Dean, P. M.). Blackie, Glasgow, U.K. (1995), pp. 24–56 [*Chem. Abstr.* **1996** *124* 249385j].
647. Jurs, P. C.; Dixon, S. L.; Egolf, L. M. *Methods Princ. Med. Chem.* **1995** *2* 15.
648. Masek, B. B., in *Molecular Similarity in Drug Design* (ed. Dean, P. M.). Blackie, Glasgow, U.K. (1995), pp. 163–186 [*Chem. Abstr.* **1996** *124* 249388n].
649. Mezey, P. G., in *Molecular Similarity in Drug Design* (ed. Dean, P. M.). Blackie, Glasgow, U.K. (1995), pp. 241–268 [*Chem. Abstr.* **1996** *124* 249389p].
650. Warde, S. *Molecular Modeling and Simulation of Surfaces* [http://www.awod.com/netsci/Science/Compchem/feature16.html] (1996).
651. Connolly, M. L. *Science* **1983** *221* 709.
652. Connolly, M. L. *J. Am. Chem. Soc.* **1985** *107* 1118.
653. Greer, J.; Bush, B. L. *Proc. Natl. Acad. Sci. USA* **1978** *75* 303.
654. Pascual-Ahuir, J. L.; Silla, E. *J. Comput. Chem.* **1990** *11* 1047.
655. Le Grand, S. M.; Merz, K. M. *J. Comput. Chem.* **1993** *14* 349.
656. Gavezzotti, A. *J. Am. Chem. Soc.* **1985** *107* 962.
657. Hall, L. H.; Kier, L. B. *Eur. J. Med. Chem.* **1981** *16* 399.
658. Marsili, M.; Floersheim, P.; Dreiding, A. S. *Comput. Chem.* **1983** *7* 175.
659. Grant, J. A.; Gallardo, M. A.; Pickup, B. T. *J. Comput. Chem.* **1996** *17* 1653.
660. Mecozzi, S.; West, A. P., Jr.; Dougherty, D. A. *Proc. Natl. Acad. Sci. USA* **1996** *93* 10566.
661. Meyer, A. Y.; Richards, W. G. *J. Comput.-Aided Mol. Des.* **1991** *5* 427.
662. Zachmann, C.-D.; Heiden, W.; Schlenkrich, M.; Brickmann, J. *J. Comput. Chem.* **1992** *13* 76.
663. Masuya, M.; Doi, J. *J. Mol. Graph.* **1995** *13* 331.
664. Ivanov, A. S.; Skortsov, V. S.; Lyulkin, Y. A.; Rumyantsev, A. B., in *Cytochrome P450 Int. Conf. 8th* (ed. Lechner, M. C.). Libbey, Montrouge, France, (1993; publ. 1994), p. 493.

665. Gajewski, J. J.; Gilbert, K. E.; McKelvey, J., in *Advances in Molecular Modeling*, *Vol. 2* (ed. Liotta, D.), JAI Press, Greenwich, CT (1990), pp. 80–81.
666. Testa, B.; Carrupt, P.-A.; Gaillard, P.; Billois, F.; Weber, P. *Pharm. Res.* **1996** *13* 335.
667. Kuntz, I. D.; Blaney, J. M.; Oatley, S. J.; Langridge, R.; Ferrin, T. E. *J. Mol. Biol.* **1982** *161* 269.
668. Leach, A. R.; Kuntz, I. D. *J. Comput. Chem.* **1992** *13* 730.
669. Meng, E. C.; Shoichet, B. K.; Kuntz, I. D. *J. Comput. Chem.* **1992** *13* 505.
670. Masek, B. B.; Merchant, A.; Matthew, J. B. *Prot. Struct. Funct. Genet.* **1993** *17* 193.
671. Masek, B. B.; Merchant, A.; Matthew, J. B. *J. Med. Chem.* **1993** *36* 1230.
672. Richards, F. M. *J. Mol. Biol.* **1974** *82* 1.
673. Goodford, P. J. *J. Med. Chem.* **1985** *28* 849.
674. Meyer, A. Y. *J. Comput. Chem.* **1986** *7* 144.
675. Bohacek, R. S.; Guida, W. C. *J. Mol. Graph.* **1989** *7* 113.
676. Heiden, W.; Goetze, T.; Brickmann, J. *J. Comput. Chem.* **1993** *14* 246.
677. Walker, P. D.; Mezey, P. G. *J. Am. Chem. Soc.* **1993** *115* 12423.
678. Rozas, I.; Du, Q.; Artcca, G. A. *J. Mol. Graph.* **1995** *13* 98.
679. Wong, M. W.; Wiberg, K. B.; Frisch, M. J. *J. Comput. Chem.* **1995** *16* 385.
680. Bliznyuk, A. A.; Gready, J. E. *J. Comput. Chem.* **1996** *17* 962.
681. Bliznyuk, A. A.; Gready, J. E. *J. Comput. Chem.* **1996** *17* 970.
682. Immel, S.; Lichtenthaler, F. W. *Liebigs Ann.* **1996** 39.
683. Lichtenthaler, F. W.; Immel, S. *Liebigs Ann.* **1996** 27.
684. Rozas, I.; Martín, M. *J. Chem. Inf. Comput. Sci.* **1996** *36* 872.
685. Richmond, T. J. *J. Mol. Biol.* **1984** *178* 63.
686. Perrot, G.; Cheng, B.; Gibson, K. D.; Vila, J.; Palmer, K. A.; Nayeem, A.; Maigret, B.; Scheraga, H. A. *J. Comput. Chem.* **1992** *13* 1.
687. You, T.; Bashford, D. *J. Comput. Chem.* **1995** *16* 743.
688. Taverner, B. C. *J. Comput. Chem.* **1996** *17* 1612.
689. Connolly, M. L., *Molecular Surface Package* [http://www.biohedron.com/msp.html].
690. Petitjean, M. *J. Comput. Chem.* **1994** *15* 507.
691. Schulman, J. M.; Venanzi, T.; Disch, R. L. *J. Am. Chem. Soc.* **1975** *97* 5335.
692. Ermer, O. *Angew. Chem. Int. Ed. Engl.* **1977** *16* 411.
693. Schulman, J. M.; Disch, R. L. *J. Am. Chem. Soc.* **1978** *100* 5677.
694. Disch, R. L.; Schulman, J. M. *J. Am. Chem. Soc.* **1981** *103* 3297.
695. Ternansky, R. J.; Balogh, D. W.; Paquette, L. A. *J. Am. Chem. Soc.* **1982** *104* 4503.
696. Paquette, L. A; Ternansky, R. J.; Balogh, D. W.; Kentgen, G. *J. Am. Chem. Soc.* **1983** *105* 5446.
697. Eaton, P. *Tetrahedron* **1979** *35* 2189.
698. Gavezzotti, A. *J. Am. Chem. Soc.* **1983** *105* 5220.
699. Gallucci, J. C.; Doecke, C. W.; Paquette, L. A. *J. Am. Chem. Soc.* **1986** *108* 1343.

700. Estrada, E. *J. Chem. Inf. Comput. Sci.* **1995** *35* 31.
701. Laskowski, R. A. *J. Mol. Graph.* **1995** *13* 323.
702. Ponder, J. W., *TINKER Home Page* [http://dasher.wustl.edu/tinker] (updated Sept. 16, 1996).
703. *The CRC Handbook of Chemistry and Physics, 73rd Edition* (ed. in chief). Lide, D. R. CRC Press, Ann Arbor, MI (1992), pp. 3-12–3-523.
704. Boyd, R. H.; Breitling, S. M. *Macromolecules* **1972** *5* 1.
705. Polowin, J.; Poe, R.; Baird, M. C. *Can. J. Chem.* **1995** *73* 1078.
706. Osawa, E.; Aigami, K.; Inamoto, Y. *J. Chem. Soc. Perkin Trans. 2* **1979** 172.
707. Osawa, E.; Shirahama, H.; Matsumoto, T. *J. Am. Chem. Soc.* **1979** *101* 4824.
708. Osawa, E. *J. Comput. Chem.* **1982** *3* 400.
709. Burkert, U.; Allinger, N. L. *J. Comput. Chem.* **1982** *3* 40.
710. Silverstein, R. M.; Bassler, G. C.; Morrill, T. C. *Spectrometric Identification of Organic Compounds, 4th Edition.* J. Wiley & Sons, New York (1981).
711. Günther, H. *NMR Spectroscopy, an Introduction.* J. Wiley & Sons, New York (1973).
712. Pretsch, E.; Clerc, T.; Seibl, J.; Simon, W., in *Tables of Spectral Data for Structure Determination of Organic Compounds* (ed. Boschke, F. L.; Fresenius, W.; Huber, J. F. K.; Pungor, E.; Rechnitz, G. A.; Simon, W.; West, T. S.). Springer-Verlag, New York (1983).
713. Derome, A. E. *Modern NMR Techniques for Chemistry Research.* Pergamon Press, New York (1987).
714. Marchand, A. P. *Stereochemical Applications of NMR Studies in Rigid Bicyclic Systems.* Verlag Chemie Intl., Dearfield Beach, FL (1982).
715. Levy, G. C.; Lichter, R. L.; Nelson, G. L. *Carbon-13 Nuclear Magnetic Resonance Spectroscopy, 2nd Edition.* Wiley-Interscience, New York (1980).
716. Whitesell, J. K.; Minton, M. A. *Stereochemical Analysis of Compounds by C-13 NMR Spectroscopy.* Chapman & Hall, New York (1987).
717. Breitmaier, E.; Voelter, W. *Carbon-13 NMR Spectroscopy.* VCH, New York (1987).
718. Haasnoot, C. A. G.; De Leeuw, F. A. A. M.; Altona, C. *Tetrahedron* **1980** *36* 2783.
719. Garbisch, E. A., Jr. *J. Am. Chem. Soc.* **1964** *86* 5561.
720. Cristol, S. J.; Russell, T. W.; Mohrig, J. R.; Plorde, D. E. *J. Org. Chem.* **1966** *31* 581.
721. Tori, K.; Takano, Y.; Kitahonoki, K. *Chem. Ber.* **1964** *97* 2798.
722. Karplus, M. *J. Chem. Phys.* **1959** *30* 11.
723. Marchand, A. P. *Stereochemical Applications of NMR Studies in Rigid Bicyclic Systems.* Verlag Chemie Intl., Dearfield Beach, FL (1982).
724. Tribble, M. T.; Traynham, J. G., in *Advances in Linear Free-Energy Relationships* (ed. Chapman, N. B.; Shorter, J.). Plenum Press, New York, 1972, pp. 143–201.
725. Pierson, G. O.; Runquist, O. A. *J. Org. Chem.* **1968** *33* 2572.
726. Eliel, E. L. *Stereochemistry of Carbon Compounds*, McGraw-Hill, New York (1962), pp. 124–179.
727. Eliel, E. L.; Allinger, N. L.; Angyal, S. J.; Morrison, G. A. *Conformational*

Analysis. J. Wiley & Sons, New York (1965), pp. 5–35.
728. Schlecht, M. F., unpublished results.
729. Hehre, W. J.; Taft, R. W.; Topsom, R. D., in *Progress in Physical Organic Chemistry, Vol. 12* (ed. Taft, R. W.). Wiley-Interscience, New York (1976), pp. 159–187.
730. Mahanti, M. K. *Indian J. Chem.* **1977** *15B* 168.
731. Tsujimoto, T.; Kobayashi, C.; Nomura, T.; Iifuru, M.; Sasaki, Y. *Chem. Pharm. Bull.* **1979** *27* 2105.
732. Levy, G. C.; Lichter, R. L.; Nelson, G. L. *Carbon-13 Nuclear Magnetic Resonance Spectroscopy, 2nd Edition.* Wiley-Interscience, New York (1980), pp. 29–33.
733. Li, S.; Allinger, N. L. *Tetrahedron* **1988** *44* 1339.
734. Whitesell, J. K.; LaCour, T.; Lovell, R. L.; Pojman, J.; Ryan, P.; Yamada-Nosaka, A. *J. Am. Chem. Soc.* **1988** *110* 991.
735. Bloor, J. E.; Breen, D. L. *J. Am. Chem. Soc.* **1967** *89* 6835.
736. Adam, W.; Grimison, A.; Rodríguez, G. *Tetrahedron* **1967** *23* 2513.
737. Pugmire, R. J.; Grant, D. M. *J. Am. Chem. Soc.* **1968** *90* 697.
738. Pugmire, R. J.; Grant, D. M.; Robins, M. J.; Robins, R. K. *J. Am. Chem. Soc.* **1969** *91* 6381.
739. Levy, G. C.; Lichter, R. L.; Nelson, G. L. *Carbon-13 Nuclear Magnetic Resonance Spectroscopy, 2nd Edition.* Wiley-Interscience, New York (1980), pp. 102–135.
740. Boulton, A. J.; McKillop, A. "Structure of Six-Membered Rings," in *Comprehensive Heterocyclic Chemistry, Vol 2.* (ed. Boulton, A. J.; McKillop, A.). Pergamon Press, New York (1984), p. 1ff.
741. Ohsawa, A.; Arai, H.; Ohnishi, H.; Igeta, H. *J. Chem. Soc. Chem. Commun.* **1981** 1174.
742. Raber, D. J.; Janks, C. M.; Johnston, M. D., Jr.; Raber, N. K. *J. Am. Chem. Soc.* **1980** *102* 6591.
743. Raber, D. J.; Janks, C. M.; Johnston, M. D., Jr.; Raber, N. K. *Tetrahedron Lett.* **1980** *21* 677.
744. Raber, D. J.; Janks, C. M.; Johnston, M. D., Jr.; Schwalke, M. A.; Shapiro, B. L.; Behelfer, G. L. *J. Org. Chem.* **1980** *46* 2528.
745. Gajewski, J. J.; Gilbert, K. E.; McKelvey, J., in *Advances in Molecular Modeling, Vol. 2* (ed. Liotta, D.), JAI Press, Greenwich, CT (1990), pp. 70–76.
746. Streitwieser, A., Jr. *Molecular Orbital Theory for Organic Chemists.* J. Wiley & Sons, New York (1961).
747. Heilbronner, E.; Straub, P. A. *Hückel Molecular Orbitals, HMO.* Springer-Verlag, New York (1966).
748. Zahradník, R.; Pancír, J. *HMO Energy Characteristics.* Plenum, New York (1970).
749. Burkert, U.; Allinger, N. L. *Molecular Mechanics (ACS Monograph No. 177).* American Chemical Society, Washington (1982), pp. 52–55.
750. Burkert, U.; Allinger, N. L. *Molecular Mechanics (ACS Monograph No. 177).* American Chemical Society, Washington (1982), pp. 144–156.
751. Allinger, N. L.; Sprague, J. T. *J. Am. Chem. Soc.* **1973** *95* 3893.
752. Favini, G.; Simonetta, M.; Sottocornola, M.; Todeschini, R. *J. Chem. Phys.* **1981** *74* 3953.

753. Warshel, A.; Lappicirella, A. *J. Am. Chem. Soc.* **1981** *103* 4664.
754. Herndon, W. C., Connor, D. A.; Lin, P. *Pure Appl. Chem.* **1990** *62* 435.
755. Froimowitz, M. *J. Comput. Chem.* **1991** *12* 1129.
756. Roth, W. R.; Adamczak, O.; Breuckmann, R.; Lennartz, H.-W.; Boese, R. *Chem. Ber.* **1991** *124* 2499.
757. Janke, R. H.; Haufe, G.; Würthwein, E.-U.; Borkent, J. H. *J. Am. Chem. Soc.* **1996** *118* 6031.
758. McGaughey, G. B.; Stewart, E. L.; Bowen, J. P. *J. Comput. Chem.* **1996** *17* 1395.
759. Ma, B.; Lii, J.-H.; Schaefer, H. F., III; Allinger, N. L. *J. Phys. Chem.* **1996** *100* 8763.
760. Allinger, N. L. *MMI*/MMPI: Calculations by the Method of Molecular Mechanics (Quantum Chemistry Program Exchange Program #318 Documentation, QCPE, Indiana Uinversity, Bloomington IN.
761. Allinger, N. L.; Youngdale, G. A. *J. Org. Chem.* **1960** *25* 1509.
762. Allinger, N. L.; Miller, M. A. *J. Am. Chem. Soc.* **1964** *86* 2811.
763. Allinger, N. L.; Tai, J. C. *J. Am. Chem. Soc.* **1965** *87* 2081.
764. Allinger, N. L.; Miller, M. A.; Chow, L. W.; Ford, R. A.; Graham, J. C. *J. Am. Chem. Soc.* **1965** *87* 3430.
765. Allinger, N. L.; Tai, J. C.; Stuart, T. W. *Theoret. Chim. Acta* **1967** *8* 101.
766. Allinger, N. L.; Stuart, T. W.; Tai, J. C. *J. Am. Chem. Soc.* **1968** *90* 2809.
767. Tai, J. C.; Allinger, N. L. *Theoret. Chim. Acta* **1968** *12* 261.
768. Tai, J. C.; Allinger, N. L. *Theoret. Chim. Acta* **1969** *15* 133.
769. Liljefors, T.; Allinger, N. L. *J. Am. Chem. Soc.* **1976** *98* 2745.
770. Kao, J.; Allinger, N. L. *J. Am. Chem. Soc.* **1977** *99* 975.
771. Allinger, N. L.; Tai, J. C. *J. Am. Chem. Soc.* **1977** *99* 4256.
772. Tai, J. C.; Allinger, N. L. *J. Am. Chem. Soc.* **1988** *110* 2050.
773. Lipkowitz, K. B.; Naylor, A. M.; Melchior, W. B. *Tetrahedron Lett.* **1984** *25* 2297.
774. Brown, R. D.; Heffernan, M. L. *Aust. J. Chem.* **1959** *12* 319.
775. Tai, J. C.; Allinger, N. L. *VESCF: Variable Electronegativity Self-Consistent Field* (*QCPE* #443) [http://ccinfo.ims.ac.jp:80/cgi-bin/qcpe] (1986).
776. Kao, J. *J. Am. Chem. Soc.* **1987** *109* 3817.
777. Lo, D. H.; Whitehead, M. A. *Can. J. Chem.* **1968** *46* 2027.
778. Pople, J. A. *Trans. Faraday Soc.* **1953** 1375.
779. Brickstock, A.; Pople, J. A. *Trans. Faraday Soc.*, **1954** 901.
780. Pariser, R. ; Parr. R. G. *J. Chem. Phys.* **1953** *21* 767.
781. Clark, T. *A Handbook of Computational Chemistry.* J. Wiley & Sons, New York (1985), pp. 93–99.
782. Anderson, D. R. *Hückel Molecular Programs* [http://www.softshell.com/FREE/HUCKEL/1-HUCKEL.html] (1996).
783. Linker, G.-J. *Explanation of the (HMO) Theory* [http://home.pi.net/ ~ linker/theory.htm] (1996).
784. Salter-Duke, B. *The Hückel Method* [http://lacebark.ntu.edu.au/demo3/huckel.html] (1996).

785. Cross, G. *Hückel Calculator* [http://www.chem.swin.edu.au/misc/huckel/index.html] (1995).
786. Linker, G.-J. *Description of the (Hückel calculator) Program* [http://home.pi.net/~linker/desc.htm] (1996).
787. Craven, B. M.; McMullan, R. K.; Bell, J. D.; Freeman, H. C. *Acta Cryst.* **1977** *B33* 2585.
788. Katritzky, A. R.; Lagowski, J. M. "Structure of Five-Membered Rings with Two or More Heteroatoms," in *Comprehensive Heterocyclic Chemistry, Vol 5.* (ed. Potts, K. T.). Pergamon Press, New York (1984), p. 1ff.
789. Berthou, J.; Elguero, J.; Rérat, C. *Acta Cryst.* **1970** *B26* 1880.
790. Larsen, F. K.; Lehmann, M. S.; Søtofte, L.; Rasmussen, S. E. *Acta Chem. Scand.* **1970** *24* 1880.
791. Huggins, M. L. *Angew. Chem. Int. Ed. Engl.* **1971** *10* 147.
792. Latimer, W. M.; Rodebush, W. H. *J. Am. Chem. Soc.* **1920** *42* 1419.
793. Suzuki, M; Shimanouchi, T. *J. Mol. Spectrosc.* **1968** *28* 394.
794. Kollman, P. A.; Allen, L. C. *J. Chem. Phys.* **1969** *51* 3286.
795. Murthy. A. S. N.; Rao, C. N. R. *J. Mol. Struct.* **1970** *6* 253.
796. Kollman, P. A.; Allen, L. C. *Chem. Rev.* **1972**, *72*, 283.
797. Sherry, A. D.; Purcell, K. F. *J. Am. Chem. Soc.* **1972** *94* 1853.
798. Allen, L. C. *J. Am. Chem. Soc.* **1975** *97* 6921.
799. Alagona, G.; Ghio, C.; Kollman, P. *J. Am. Chem. Soc.* **1983** *105* 5226.
800. Taylor, R.; Kennard, O. *Acc. Chem. Res.* **1984** *17* 320.
801. Legon, A. C.; Millen, D. J. *Acc. Chem. Res.* **1987** *20* 39.
802. Hasenein, A. A.; Hinchliffe, A. "Ab Initio Studies of Hydrogen Bonding," in *Self-Consistent Field. Theory and Applications (Studies in Physical and Theoretical Chemistry 70)* (ed. Carbó, R.; Klobukowski, M.). Elsevier, New York (1990), pp. 670–705.
803. Gordon, M. S.; Jensen, J. H. *Acc. Chem. Res.* **1996** *29* 536.
804. Niketić, S. R.; Rasmussen, K. *The Consistent Force Field: A Documentation.* Springer-Verlag, New York (1977), p. 85.
805. Burkert, U.; Allinger, N. L. *Molecular Mechanics (ACS Monograph No. 177).* American Chemical Society, Washington (1982), pp. 222–224.
806. Comba, P.; Hambley, T. W. *Molecular Modeling of Inorganic Compounds.* VCH, New York (1995), p. 27.
807. Gajewski, J. J.; Gilbert, K. E.; McKelvey, J., in *Advances in Molecular Modeling, Vol. 2* (ed. Liotta, D.). JAI Press, Greenwich, CT (1990), pp. 77–80.
808. Teixeira-Dias, J. J. C.; Fausto, R. *NATO ASI Ser., Ser. C 1993 406 (Recent Experimental and Computational Advances in Molecular Spectroscopy)*, p. 131.
809. da Silva, J. B. P.; Moura, G. L. C.; Ramos, M. N.; Teixeira-Dias, J. J. C.; Fausto, R. *J. Mol. Struct.* **1994** *317* 157 [*Chem. Abstr.* **1994** *120* 133348n].
810. Janda, K. C.; Steed, J. M.; Novick, S. E.; Klemperer, W. *J. Chem. Phys.* **1977** *67* 5162.
811. Odutola, J. A.; Viswanathan, R.; Dyke, T. R. *J. Am. Chem. Soc.* **1979** *101* 4787.
812. Legon, A. C.; Millen, D. J. *Can. J. Chem.* **1989** *67* 1683.

813. Fukui, K. *Acc. Chem. Res.* **1971** *4* 57.
814. Fleming, I. *Frontier Orbitals and Organic Chemical Reactions.* J. Wiley & Sons, New York (1976).
815. Jorgensen, W. L.; Salem, L. *The Organic Chemist's Book of Orbitals.* Academic Press, New York (1973).
816. Jorgensen, W. L.; Salem, L. *The Organic Chemist's Book of Orbitals.* Academic Press, New York (1973), pp. 105–106.
817. Taylor, R. *J. Mol. Struct.* **1981** *71* 311.
818. Boobbyer, D. N. A.; Goodford, P. J.; McWhinnie, P. M.; Wade, R. C. *J. Med. Chem.* **1989** *32* 1083.
819. Jeffrey, G. A. *J. Mol. Struct.* **1990** *237* 75.
820. Dauchez, M.; Derreumaux, P.; Lagante, P.; Vergoten, G., in *Modelling of Molecular Structures and Properties, Studies in Physical and Theoretical Chemistry, Vol. 7* (ed. Rivail, J.-L.). Elsevier, Amsterdam (1990), pp. 45–62.
821. Yokoyama, I.; Miwa, Y.; Machida, K. *J. Am. Chem. Soc.* **1991** *113* 6458.
822. Masella, M.; Lefour, J.-M.; Flament, J. P. *Bull. Soc. Chim. France* **1995** *132* 224.
823. Masella, M.; Lefour, J.-M.; Flament, J. P. *Bull. Soc. Chim. France* **1996** *133* 405.
824. Suzuki, K.; Green, P. G.; Bumgarner, R. E.; Dasgupta, S.; Goddard, W. A., III; Blake, G. A. *Science* **1992** *257* 942.
825. Desiraju, G. R. *Acc. Chem. Res.* **1996** *29* 441.
826. Hancock, R. D. *Acc. Chem. Res.* **1990** *23* 253.
827. Bosnich, B. *Chem. Soc. Rev.* **1994** *23* 387.
828. Niketić, S. R.; Rasmussen, K. *The Consistent Force Field: A Documentation.* Springer-Verlag, New York (1977), pp. 68–69.
829. Laier, T.; Larsen, E. *Acta Chem. Scand. A* **1979** *A33* 257.
830. Rasmussen, K.; Woldbye, F. *Coord. Chem. Rev.* **1980** *20* 219.
831. Raos, N.; Niketić, S. R.; Simeon, V. *J. Inorg. Biochem.* **1982** *16* 1.
832. Wipff, G.; Weiner, P.; Kollman, P. *J. Am. Chem. Soc.* **1982** *104* 3249.
833. Thöm, V. J.; Fox, C. C.; Boeyens, J. C. A.; Hancock, R. D. *J. Am. Chem. Soc.* **1984** *106* 5947.
834. Hancock, R. D. *Pure Appl. Chem.* **1986** *58* 1445.
835. Morley, J. O. *Monatsh. Chem.* **1988** *119* 1263.
836. Ferguson, D. M.; Raber, D. J. *J. Comput. Chem.* **1990** *11* 1061.
837. Yoshikawa, Y. *J. Comput. Chem.* **1990** *11* 326.
838. Norrby, P.-O.; Åkermark, B.; Hæffner, F.; Hansson, S.; Blomberg, M. *J. Am. Chem. Soc.* **1993** *115* 4859.
839. Ranghino, G.; Antonini, G.; Fantucci, P. *Israel J. Chem.* **1994** *34* 239.
840. Doman, T. N.; Landis, C. R.; Bosnich, B. *J. Am. Chem. Soc.* **1992** *114* 7264.
841. Jungwirth, P.; Stussi, D.; Weber, J. *Chem. Phys. Lett.* **1992** *190* 29.
842. Mackie, S. C.; Baird, M. C. *Organomet.* **1992** *11* 3712.
843. Polowin, J.; Mackie, S. C.; Baird, M. C. *Organomet.* **1992** *11* 3724.
844. Perjéssy, A.; Ertl, P.; Prónayová, N.; Gautheron, B.; Broussier, R. *J. Organomet. Chem.* **1994** *466* 133.

REFERENCES 733

845. Timofeeva, T. V.; Lii, J.-H.; Allinger, N. L. *J. Am. Chem. Soc.* **1995** *117* 7452.
846. White, D. P.; Brown, T. L. *Inorg. Chem.* **1995** *34* 2718.
847. Barlow, S.; Rohl, A. L. Shi, S.; Freeman, C. M.; O'Hare, D. *J. Am. Chem. Soc.* **1996** *118* 7578.
848. Lauher, J. W. *J. Am. Chem. Soc.* **1986** *108* 1521.
849. Bernhardt, P. V.; Comba, P. *Inorg. Chem.* **1992** *31* 2638.
850. Bencze, L.; Szilagyi, R. *J. Organomet. Chem.* **1994** *475* 183.
851. Comba, P.; Hambley, T. W.; Ströhle, M. *Helv. Chim. Acta* **1995** *78* 2042.
852. Cleveland, T.; Landis, C. R. *J. Am. Chem. Soc.* **1996** *118* 6020.
853. Burchart, E. d. V.; Jansen, J. C.; van de Graaf, B.; van Bekkum, H. *Zeolites* **1993** *13* 216.
854. Poch, M.; Vallenti, E.; Moyano, A.; Pericàs, M. A.; Castro, J.; DeNicola, A.; Greene, A. E. *Tetrahedron Lett.* **1990** *31* 7505.
855. Wu, Y.-D.; Wang, Y.; Houk, K. N. *J. Org. Chem.* **1992** *57* 1362.
856. Castro, J.; Moyano, A.; Pericàs, M. A.; Riera, A. *Tetrahedron* **1995** *51* 6541.
857. Winter, M. *WebElements 2.0* [http://www.shef.ac.uk./ ~ chem/web-elements/] or [http://www.cchem.berkeley.edu/Table/index.html] (1997).
858. Palmer, K. J. *J. Am. Chem. Soc.* **1938** *60* 2360.
859. Gajewski, J. J.; Gilbert, K. E.; McKelvey, J., In *Advances in Molecular Modeling, Vol. 2* (ed. Liotta, D.). JAI Press, Greenwich, CT (1990), pp. 85–86.
860. Cotton, F. A.; Wilkinson, G. *Advanced Inorganic Chemistry, 5th Edition.* Wiley-Interscience, New York (1988).
861. Jost, A.; Rees, B.; Yelon, W. B. *Acta Cryst.* **1975** *B31* 2649.
862. Heslop, R. B.; Jones, K. *Inorganic Chemistry, a Guide to Advanced Study.* Elsevier, New York (1976), pp. 587–589.
863. Cotton, F. A.; Wilkinson, G. *Advanced Inorganic Chemistry, 5th Edition.* Wiley-Interscience, New York (1988), pp. 1299–1300.
864. Comba, P.; Hambley, T. W. *Molecular Modeling of Inorganic Compounds.* VCH, New York (1995), pp. 117–118.
865. Comba, P.; Hambley, T. W. *Molecular Modeling of Inorganic Compounds.* VCH, New York (1995), p. 35.
866. Anderson, W. P.; Behm, P., Jr.; Glennon, T. M.; Graves, S. H.; Zerner, M. C. *Preprints Am. Chem. Soc. Div. Petrol. Chem.* **1992** *37(2)* 580.
867. Dominguez-Vera, J.-M.; Colacio, E.; Ruiz, J.; Moreno, J. M.; Sundberg, M. R. *Acta Chem. Scand.* **1992** *46* 1055.
868. Farkas, E.; Kozma, E.; Gunda, T. E. *Polyhedron* **1992** *11* 3069.
869. Tjaden, E. B.; Schwiebert, K. E.; Stryker, J. M. *J. Am. Chem. Soc.* **1992** *114* 1100.
870. Sundberg, M. R.; Sillanpää, R. *Acta Chem. Scand.* **1993**, *47* 1173.
871. Gao, Y.-D.; Lipkowitz, K. B.; Schultz, F. A. *J. Am. Chem. Soc.* **1995** *117* 11932.
872. Hegedus, L. S.; Åkermark, B.; Olsen, D. J.; Anderson, O. P.; Zetterberg, K. *J. Am. Chem. Soc.* **1982** *104* 697.
873. Dashevskii, V. G.; Baranov, A. P.; Medved', T. Ya.; Kabachnik, M. I. *Proc. Acad. Sci. USSR, Phys. Chem. (Engl. Transl.)* **1983** *270* 335 [*Chem. Abstr.* **1983** *99* 81551z].

874. Baranov, A. P.; Dashevskii, V. G.; Medved', T. Ya.; Kabachnik, M. I. *Proc. Acad. Sci. USSR, Phys. Chem. (Engl. Transl.)* **1985** *281* 283 [*Chem. Abstr.* **1985** *103* 110252m].
875. Root, D. M.; Landis, C. R.; Cleveland, T. *J. Am. Chem. Soc.* **1993** *115* 4201.
876. Drew, M. G. B.; Yates, P. C. *Pure Appl. Chem.* **1989** *61* 835.
877. Zimmer, M.; Crabtree, R. H. *J. Am. Chem. Soc.* **1990** *112* 1062.
878. Hay, B. P.; Rustad, J. R. *J. Am. Chem. Soc.* **1994** *116* 6316.
879. Xiao, G.; van der Helm, D.; Hider, R. C.; Rai, B. L. *J. Phys. Chem.* **1996** *100* 2345.
880. Hambley, T. W., Hawkins, C. J.; Palmer, J. A.; Snow, M. R. *Aust. J. Chem.* **1981** *34* 45.
881. Bernhardt, P. V.; Comba, P. *Inorg. Chem.* **1993** *32* 2798.
882. Comba, P.; Hambley, T. W. *Molecular Modeling of Inorganic Compounds.* VCH, New York (1995). p. 96 and footnote 3.
883. Comba, P.; Zimmer, M. *Inorg. Chem.* **1994** *33* 5368.
884. Comba, P.; Hambley, T. W. *Molecular Modeling of Inorganic Compounds.* VCH, New York (1995), pp. 118–122.
885. Brandès, S.; Gros, C.; Denat, F.; Pullumbi, P.; Guilard, R. *Bull. Soc. Chim. France* **1996** *133* 65.
886. Savary, F.; Weber, J.; Calzaferri, G. *J. Phys. Chem.* **1993** *97* 3722.
887. Bérces, A.; Ziegler, T. *J. Phys. Chem.* **1994** *98* 13233.
888. Sickafoose, S. M.; Breckenridge, S. M.; Kukolich, S. G. *Inorg. Chem.* **1994** *33* 5176.
889. Bailey, M. F.; Dahl, L. F. *Inorg. Chem.* **1965** *4* 1314.
890. Rees, B.; Coppens, P. *J. Organomet. Chem.* **1972** *42* C102.
891. Chiu, N.-S.; Schäfer, L.; Seip, R. *J. Organomet. Chem.* **1975** *101* 331.
892. Wang, H. *J. Comput. Chem.* **1991** *12* 746.
893. Katchalski-Katzir, E.; Shariv, I.; Eisenstein, M.; Friesem, A. A.; Aflalo, C.; Vakser, I. A. *Proc. Natl. Acad. Sci. USA* **1992** *89* 2195.
894. Abagyan, R.; Totrov, M.; Kuznetsov, D. *J. Comput. Chem.* **1994** *15* 488.
895. Luty, B. A.; Wasserman, Z. R.; Stouten, P. F. W.; Hodge, C. N.; Zacharias, M.; McCammon, J. A. *J. Comput. Chem.* **1995** *16* 454.
896. McMartin, C.; Bohacek, R. S. *J. Comput.-Aided Mol. Des.* **1997** *11* 333.
897. Sobolev, V.; Wade, R. C.; Vriend, G.; Edelman, M. *Prot. Struct. Funct. Genet.* **1996** *25* 120.
898. Gao, J.; Xia, X. *Science* **1992** *258* 631.
899. Liu, H.; Shi, Y. *J. Comput. Chem.* **1994** *15* 1311.
900. Lyne, P. D.; Mulholland, A. J.; Richards, W. G. *J. Am. Chem. Soc.* **1995** *117* 11345.
901. Barnes, J. A.; Williams, I. H. *Biochem. Soc. Trans.* **1996** *24* 263.
902. Gao, J. *Acc. Chem. Res.* **1996** *29* 298.
903. Burkert, U.; Allinger, N. L. *Molecular Mechanics (ACS Monograph No. 177)*, American Chemical Society, Washington (1982), pp. 285–305.
904. Gajewski, J. J.; Gilbert, K. E.; McKelvey, J., in *Advances in Molecular Modeling,*

Vol. 2 (ed. Liotta, D.). JAI Press, Greenwich, CT (1990), pp. 81–85.
905. Houk, K. N.; Tucker, J. A.; Dorigo, A. E. *Acc. Chem. Res.* **1990** *23* 107.
906. Lenz, T. G.; Vaughan, J. D. *J. Comput. Chem.* **1990** *11* 351.
907. Bernardi, A.; Gennari, C.; Goodman, J. M.; Paterson, I. *Tetrahedron: Asym.* **1995** *6* 2613.
908. Dolata, D. P.; Spina, D. R.; Stahl, M. T. *J. Chem. Inf. Comput. Sci.* **1996** *36* 228.
909. March, J. *Advanced Organic Chemistry. Reactions, Mechanisms and Structure. 4th Edition.* J. Wiley & Sons, New York (1992), pp. 205–230.
910. March, J. *Advanced Organic Chemistry. Reactions, Mechanisms and Structure. 4th Edition.* J. Wiley & Sons, New York (1992), pp. 293–369.
911. March, J. *Advanced Organic Chemistry. Reactions, Mechanisms and Structure. 4th Edition.* J. Wiley & Sons, New York (1992), pp. 501–521.
912. March, J. *Advanced Organic Chemistry. Reactions, Mechanisms and Structure. 4th Edition.* J. Wiley & Sons, New York (1992), pp. 569–580.
913. March, J. *Advanced Organic Chemistry. Reactions, Mechanisms and Structure. 4th Edition.* J. Wiley & Sons, New York (1992), pp. 641–653.
914. March, J. *Advanced Organic Chemistry. Reactions, Mechanisms and Structure. 4th Edition.* J. Wiley & Sons, New York (1992), pp. 677–689.
915. March, J. *Advanced Organic Chemistry. Reactions, Mechanisms and Structure. 4th Edition.* J. Wiley & Sons, New York (1992), pp. 734–757, 839–852.
916. March, J. *Advanced Organic Chemistry. Reactions, Mechanisms and Structure. 4th Edition.* J. Wiley & Sons, New York (1992), pp. 879–881.
917. March, J. *Advanced Organic Chemistry. Reactions, Mechanisms and Structure. 4th Edition.* J. Wiley & Sons, New York (1992), pp. 982–1010.
918. March, J. *Advanced Organic Chemistry. Reactions, Mechanisms and Structure. 4th Edition.* J. Wiley & Sons, New York (1992), pp. 1051–1067, 1110–1133.
919. March, J. *Advanced Organic Chemistry. Reactions, Mechanisms and Structure. 4th Edition.* J. Wiley & Sons, New York (1992), pp. 165–204.
920. Streitwieser, A., Jr. *Molecular Orbital Theory for Organic Chemists.* J. Wiley & Sons, New York (1961), pp. 357–391.
921. Streitwieser, A., Jr. *Molecular Orbital Theory for Organic Chemists.* J. Wiley & Sons, New York (1961), pp. 392–412.
922. Streitwieser, A., Jr. *Molecular Orbital Theory for Organic Chemists.* J. Wiley & Sons, New York (1961), pp. 413–448.
923. Streitwieser, A., Jr. *Molecular Orbital Theory for Organic Chemists.* J. Wiley & Sons, New York (1961), pp. 307–356.
924. Streitwieser, A., Jr. *Molecular Orbital Theory for Organic Chemists.* J. Wiley & Sons, New York (1961), pp. 432–448.
925. Brown, F. K.; Houk, K. N. *J. Am. Chem. Soc.* **1985** *107* 1971.
926. *The CRC Handbook of Chemistry and Physics, 73rd Edition* (ed. in chief Lide, D. R.). CRC Press, Ann Arbor MI (1992), pp. 6-166ff.
927. *The CRC Handbook of Chemistry and Physics, 73rd Edition* (ed. in chief Lide, D. R.). CRC Press, Ann Arbor MI (1992), pp. 9-51ff.
928. *The CRC Handbook of Chemistry and Physics, 73rd Edition* (ed. in chief Lide, D.

R.). CRC Press, Ann Arbor MI (1992), pp. 1-1ff.
929. Klyne, W.; Prelog, V. *Experientia* **1960** *16* 521.
930. Sheridan, R. P.; Nilakantan, R.; Rusinko, A., III; Bauman, N.; Haraki, K. S.; Venkataraghavan, R. *J. Chem. Inf. Comput. Sci.* **1989** *29* 255.
931. Bader, G.; Danzandarjaa, T.; Hiller, K.; Reznicek, G.; Jurenitsch, J.; Golly, M.; Schröder, H.; Schubert-Zsilavecz, M.; Haslinger, E. *Helv. Chim. Acta* **1994** *77* 1861.
932. Tóth, G. J. *Free On-Line Dictionary of Computing* 1996 [http://wfn-shop.Princeton.EDU/foldoc/] (1996).
933. Greenberg, H. J. *Mathematical Programming Glossary* [http://www-math.cudenver.edu/ ~ hgreenbe/glossary/glossary.html] (1996).
934. Perrin, C. L.; Armstrong, K. B.; Fabian, M. A. *J. Am. Chem. Soc.* **1994** *116* 715.
935. Franck, R. W., in *Conformational Behavior of Six-Membered Rings. Analysis, Dynamics, and Stereoelectronic Effects* (ed. Juaristi, E.). VCH, New York (1995), p. 181.
936. Salzner, U. *J. Org. Chem.* **1995** *60* 986.
937. Streitwieser, A., Jr. *Molecular Orbital Theory for Organic Chemists*. J. Wiley & Sons, New York (1961), pp. 256-304.
938. Lloyd, D. *J. Chem. Inf. Comput. Sci.* **1996** *36* 442.
939. Shah, A. V.; Walters, W. P.; Shah, R.; Dolata, D. P., in *ASTM Spec. Tech. Publ.* (*STP 1214*) (ed. Lysakowski, R.; Gragg, C. E.). Am. Soc. Test. Mat., Philadelphia (1994), p. 45 [*Chem. Abstr.* **1995** *123* 197882b].
940. Kind, I. *BABEL: A Glossary of Computer Oriented Abbreviations and Acronyms, Version 97B* [http://www.access.digex.net/ ~ ikind/babel.html] (1997).
941. Born, M.; Oppenheimer, R. *Ann. Phys.* **1927** *84* 457.
942. Sherrill, C. D. *A Brief Review of Elementary Quantum Chemistry. The Born–Oppenheimer Approximation* [http://zopyros.ccqc.uga.edu/Docs/Knowledge/ Fundamental _Theory/quantrev/ node28.html # SECTION00072000000000000000] (1996).
943. Lauri, G.; Bartlett, P. A. *J. Comput.-Aided Mol. Des.* **1994** *8* 51.
944. Lauri, G.; Shea, G.; Waterman, S.; Telfer, S.; Bartlett, P. A. *CAVEAT* [http://www.cchem.berkeley.edu:80/ ~ pabgrp/Data/caveat.html] (1996).
945. Siani, M. A.; Weininger, D.; Blaney, J. M. *J. Chem. Inf. Comput. Sci.* **1994** *34* 588.
946. Koca, J.; Pérez, S.; Imberty, A. *J. Comput. Chem.* **1995** *16* 296.
947. Robijn, G. W.; Imberty, A.; van den Berg, D. J. C.; Ledeboer, A. M.; Kamerling, J. P.; Vliegenthart, J. F. G.; Pérez, S. *Carbohydr. Res.* **1996** *288* 57.
948. Cahn, R. S.; Ingold, C.; Prelog, V. *Angew. Chem. Int. Ed. Engl.* **1996** *5* 385.
949. March, J. *Advanced Organic Chemistry. Reactions, Mechanisms and Structure. 4th Edition* J. Wiley & Sons, New York (1992), pp. 109–111.
950. Rusinko, A., III; Sheridan, R. P.; Nilakantan, R.; Haraki, K. S.; Bauman, N.; Venkataraghavan, R. *J. Chem. Inf. Comput. Sci.* **1989** *29* 251.
951. Eliel, E. L., in *Conformational Behavior of Six-Membered Rings. Analysis, Dynamics, and Stereoelectronic Effects* (ed. Juaristi, E.). VCH, New York (1995), p. 3.

952. Eliel, E. L.; Allinger, N. L.; Angyal, S. J.; Morrison, G. A. *Conformational Analysis.* J. Wiley & Sons, New York (1965).
953. Nickon, A.; Silversmith, E. F. *Organic Chemistry—The Name Game,* Pergamon Press, New York (1987), p. 32.
954. Crippen, G. M., in *Computer-Aided Molecular Design* (ed. Richards, W. G.). VCH, New York (1989), pp. 61–65.
955. Ghose, A. K.; Crippen, G. M., in *Comprehensive Medicinal Chemistry, Vol. 4* (ed. Hansch, C.; Sammes, P. G.; Taylor, J. B.). Pergamon Press, New York (1990), pp. 715–733.
956. Microsoft, Inc. *Definition and Explanation of a .DLL file.* Article ID: Q87934 [http://www.microsoft.com/kb/softlib/Office/q_word.htm] (revision date: Nov. 21, 1994).
957. MicroSoft, Inc. *MS-DOS Support Home Page* [http://www.microsoft.com/Support/Products/Windows95/MS-DOS/] (1997).
958. MicroWay Pty Ltd. *The Phar Lap DOS Extender* [http://www.microway.com.au/catalog/pharlap/pharlap.html]. (1996).
959. von Freyberg, B.; Mumenthaler; C.; Schaumann, T.; Braun, W., *FANTOM Home Page* [http://www.mol.biol.ethz.ch/wuthrich/people/braun/fm_home.html] (updated Apr. 26, 1995).
960. March, J. *Advanced Organic Chemistry. Reactions, Mechanisms and Structure. 4th Edition.* J. Wiley & Sons, New York (1992), pp. 106–107.
961. Wolfe, S. *Acc. Chem. Res.* **1972** *5* 102.
962. Clark, T. A. *Handbook of Computational Chemistry.* J. Wiley & Sons, New York (1985), pp. 242–314.
963. Stuike-Prill, R.; Meyer, B. *Eur. J. Biochem.* **1990** *194* 903.
964. Jansson, R.-E.; Kjellberg, A.; Rundlöf, T.; Widmalm, G. *J. Chem. Soc. Perkin Trans. 2* **1996** 33.
965. Parr, R. G.; Pearson, R. G. *J. Am. Chem. Soc.* **1983** *105* 7512.
966. *Search Engine for the QCPE Archives* [http://ccinfo.ims.ac.jp:80/cgi-bin/qcpe].
967. Streitwieser, A., Jr. *Molecular Orbital Theory for Organic Chemists.* J. Wiley & Sons, New York (1961), pp. 33–63.
968. Laube, T. *Acc. Chem. Res.* **1995** *28* 399.
969. Musher, J. I. *Angew. Chem. Int. Ed. Engl.* **1969** *8* 54.
970. Cotton, F. A.; Wilkinson, G. *Advanced Inorganic Chemistry, 5th Edition.* Wiley-Interscience, New York (1988), pp. 35–83.
971. Baricic, P.; Mackov, M. *J. Mol. Graph.* **1995** *13* 184.
972. Kerber, A.; Laue, R. *MOLGEN—Automatic Structure Elucidation in Chemistry* [http://btm2xd.mat.uni-bayreuth.de/molgen] (1996).
973. Ingold, C. K. *Structure and Mechanism in Organic Chemistry.* G. Bell & Sons, London (1953), pp. 200–201.
974. Roelofs, G. *Portable Network Graphics Home Page* [http://www.wco.com/ ~ png/] (July 5, 1997).
975. Sayle, R. A. *The RasMol Manual, v. 2.6* [http://klaatu.oit.umass.edu/microbio/rasmol/distrib/rasman.htm] (July 25, 1997).

976. Merritt, E. A.; Murphy, M. E. P. *Acta Cryst.* **1994** *D50* 869.
977. Surles, M. *Real-Time Mechanics in Molecular Modeling* [http://www.awod.com/netsci/Science/Compchem/feature05.html] (1996).
978. Convard, T.; Dubost, J.-P.; le Solleu, H.; Kummer, E. *Quant. Struct.-Act. Relat.* **1994** *13* 34.
979. Humphrey, W.; Dalke, A.; Schulten, K. *J. Mol. Graph.* **1996** *14* 33.
980. Vriend, G. *J. Mol. Graph.* **1990** *8* 52.
981. Vriend, G. *WHAT IF Home Page* [http://swift.embl-heidelberg.de/whatif/] (July 30, 1997).
982. Nickon, A.; Silversmith, E. F. *Organic Chemistry—The Name Game.* Pergamon Press, New York (1987).
983. Gosper, J. J.; Sammes, P. *Common Definitions and Terms in Organic Chemistry* [http://http2.brunel.ac.uk:8080/depts/chem/definit/definit.htm] (Feb. 1995).
984. *Network Science—Molecular Modeling Programs Software Listing* [http://www.awod.com/netsci/Resources/Software/Modeling/top.html] (1996).
985. Tollenaere, J. P.; Moret, E. E. *Hyperglossary of Terminology* [http://cmcind.far.ruu.nl/webcmc/glossary.html] (1996).
986. Thompson, H. W.; Naipawer, R. E. *J. Am. Chem. Soc.* **1973** *95* 6379.
987. Van de Waterbeemd, H.; Carter, R. E.; Grassy, G.; Kubinyi, H.; Martin, Y. C.; Tute, M. S.; Willett, P. *Pure Appl. Chem.* **1997** *69* 1137.
988. Müller, P. *Pure Appl. Chem.* **1994** *66* 1077 [http://www.chem.qmw.ac.uk/iupac/gtpoc/].
989. Brown, R. D.; Boggs, J. E.; Hilderbrandt, R.; Lim, K.; Mills, I. M.; Nikitin, E.; Palmer, M. H. *Pure Appl. Chem.* **1996** *68* 387.
990. Moss, G. P. *Pure Appl. Chem.* **1996** *68* 2193 [http://alpha.qmw.ac.uk/ ~ ugca000/iupac/stereo/].
991. Jenkins, A. D.; Kratochvíl, P.; Stepto, R. F. T.; Suter, U. W. *Pure Appl. Chem.* **1996** *68* 2287.
992. De Bolster, M. W. G. *Pure Appl. Chem.* **1997** *69* 1251.
993. Hassner, A. *J. Org. Chem.* **1968** *33* 2684.

INDEX

α-anomer (saccharide), 54
α-helix conformation (peptide), 52–53, 113, 398
β-anomer (saccharide), 54
β-sheet conformation (peptide), 52–53, 113, 398
β-turn conformation (peptide), 52–53, 113, 398
γ-shielding, in NMR, 480
ε (hardness parameter), 94, 191, 217, 218, 232, 241, 243, 247, 254, 257, 267, 272, 279, 286, 288, 298, 302, 303, 305, 513, 521, 525, 543, 562, 583, 622–625, 647, 662, 664
π-allylpalladium dichloride organometallic exercise, 550–560
π-atom, 15, 98, 131, 137, 138, 223, 489, 491, 493, 549, 565, 568, 585, 617, 622–625, 653
π-atom array, 489, 491, 617
π-atom flag, 175, 489, 502, 503, 504, 565, 568, 579, 617, 618
π-calculation, 2, 13, 15, 124, 132, 133, 137, 176, 246, 267, 313, 363, 480–486, 488–508, 513, 539, 540, 543, 565, 583, 615, 617, 665, 691
  atom types in MMX, 238–240
  modified Pople-Pariser-Parr, 488
  Pariser-Parr-Pople, 267, 313, 493, 495
  RHF, 132, 138, 490, 496
  SCF-LCAO-MO, 313
  UHF, 132, 138, 490, 496
π-calculation exercise
  modeling 1,3-butadiene, 491–492, 497–501
  modeling benzene, 490–491, 497–501
  modeling imidazole, 502–508
  modeling pyrazole, 502–508
π-donor ligand, 539, 540
π-electron density, 465, 481–486, 489, 513, 691
π-electrons, 529, 539, 629, 641, 689
π-system planarity flag, 137
ϕ-angle (peptide), 7, 52, 53, 288, 293, 626
ϕ-angle (saccharide), 55, 626, 665
χ-angle (nucleoside), 288
ψ-angle (peptide), 7, 52, 53, 288, 293, 626

ψ-angle (saccharide), 55, 626
ω-angle (peptide), 52, 53, 293, 626
@ notation for atom type, 238, 624
1-chlorobutan-3-one, NMR coupling constant analysis, 476–480
1D structure, 80, 85, 634, 639, 690, 692
1,2-dichloroethane
  dihedral driver exercise, 461–464
  electrostatic exercise, 325–329
  rigid rotor exercise, 458–459
  structure comparison exercise, 122–125
  torsion potential, 227–230
1,2-difluoroethane
  added constants exercise, 195–199
  electrostatic interactions, 236–237
1,2,3-triazine, NMR chemical shift prediction exercise, 481–486
1,2,4-triazine, NMR chemical shift prediction exercise, 481–486
1,3-butadiene, π-calculation exercise, 491–492
1,3,5-triazine, NMR chemical shift prediction exercise, 481–486
2D structure, 18, 80, 142, 346, 634, 635, 639, 672
2D to 3D conversion, 37, 627, 636, 639–640, 643, 701
2-methoxy-4-methyltetrahydropyran, NMR coupling constants, 470–476
2-methyl-1,3-dioxolane, conformation search exercise, 350–376
2-methylpyridine, NMR chemical shift prediction exercise, 481–486
2-(2'-phenolylmethyl)phenol
  Don't Minimize exercise, 403–406
  fixed torsion exercise, 398–401
  geometric constraint exercise, 398–401, 403–406
2,6-naphthyridine, NMR chemical shift prediction exercise, 481–486
3-cyanopyridine, NMR chemical shift prediction exercise, 481–486
3D structure, 18, 19, 36, 40, 50, 82–89, 142, 346, 349, 442, 634, 635, 639, 647, 672, 683

739

3D structure databases, 17, 18, 50, 346, 349, 634
3D structure generation, 6, 37, 40, 85–89, 535, 639
3D-QSAR, 85, 658
4-bromopyridine, NMR chemical shift prediction exercise, 481–486
4-methyl-1,2,3-triazine, NMR chemical shift prediction exercise, 481–486
ab initio calculation, 8, 60, 204, 262, 271, 276, 299, 301, 302, 306, 314, 316, 321, 322, 330, 346, 407, 408, 454, 512, 525, 582, 657, 658, 663, 665, 670, 673, 674, 678
Acetic acid solvation model, 306
Acidic hydrogen, 108, 114, 238, 267, 272, 296, 304, 330, 433, 509, 525
Actinides, 240, 530, 549
Add__B command, 33, 37, 38, 626
Added constants, 31, 134, 186, 548, 610
Added constants exercise, 1,2-difluoroethane, 195–199
Added constants file, 186–194
Added Constants File dialog box, 188
Added constants flag, 136, 188, 339
AIMB, 87
AL2 file format (Tripos Alchemy 2000), 128
Alchemy 2000 file format, 128, 159–160
Alchemy program, 61, 152, 157, 201, 296, 640
Alchemy-III, 201
Alchemy-III file format, 157–159
  bond count, 159
  charge count, 159
Alcohols solvation model, 304, 305
Algorithm, 627–628
  2D to 3D conversion, 37, 627, 636, 639–640, 643, 701
  Altona, 465, 469–470
  block-diagonal Newton-Raphson, 339, 344, 677
  charge equilibration , 323–324
  charge equilibration-potential derived charge, 323–324
  conjugate gradient, 283, 341–342, 641
  Connolly, 322–323, 437, 442, 443–444, 447, 649, 676–677
  DOCK, 348, 441, 442, 648–649
  energy minimization, 339, 341–344, 385
  evolutionary, 627
  Garbisch, 465, 470
  genetic, 349, 627–628, 658
  Karplus, 468–469, 470
  leapfrog, 380
  Lee-Richards, 7, 425, 431–434, 435–437, 675
  Metropolis, 5, 348, 385, 574, 671

molecular free surface, 438
Molecular Skins, 441–444, 675
molecular surface, 7, 322, 425, 428–429, 431–437, 438, 675, 676–677
molecular volume, 428, 429, 437–439, 445–447
Newton-Raphson, 6, 342–344, 677
Petitjean, 454–455
random kick, 8, 348, 350, 354, 355, 366
simulated annealing, 5, 28, 348, 533, 571, 573–574, 613, 671, 691–692
steepest descent, 6, 341, 343, 641, 693
Verlet, 380
Alkali metals, 240
Alkaline earths, 240
All atom model, 283, 289, 304, 443, 628
Allinger, N. L., 4, 7, 15, 132, 200, 205, 207, 219, 222, 232, 242, 262, 299, 309, 312, 319, 322, 344, 469, 488, 493, 521, 562, 651, 666, 667, 668, 671, 672, 677
Altona, C., 469
Altona algorithm, 465, 469–470
AM1 calculation, 346
AMBER force field, 282–288, 291, 306, 307, 308, 309, 314, 316, 322, 442, 524, 563, 628, 649, 659, 667, 678, 695, 697
  angle bend potential, 284, 287
  bond stretch potential, 284, 287
  electrostatic potential, 285–286, 288, 524
  hydrogen bonding potential, 286–287, 524
  improper torsion potential, 285
  torsion potential, 284–285, 287–288
  van der Waals potential, 285–286, 288
AMBER* force field, 307, 628, 670
AMBER/OPLS force field, 306, 628, 695
AMBER94 force field, 307, 670
Amino acids, 283, 306, 307, 347
Amino Acids menu (PCMODEL), 39, 51–54, 609
Analytical calculation of molecular surface, 445
Analytical calculation of molecular volume, 445
Andose, J. D., 7, 308, 342, 641
ANG parameter, in MMX, 188, 189, 192–193
ANG3 parameter, in MMX, 188, 189, 192–193
ANG4 parameter, in MMX, 188, 189, 192–193
Angle
  ϕ (peptide), 7, 52, 53, 288, 293, 626
  ϕ (saccharide), 55, 626, 665
  χ (nucleoside), 288
  ψ (peptide), 7, 52, 53, 288, 293, 626
  ψ (saccharide), 55, 626
  ω (peptide), 52, 53, 293, 626
  bond, 9, 13, 79, 119, 156

INDEX   **741**

flap, 119, 222, 248, 396, 654, 667
out-of-plane, 208, 209
torsional, 4, 9, 13, 41, 42, 43, 119, 156, 696
Angle bend
  organometallic, 543, 549, 561, 564
  out-of-plane, 95, 194, 208-210, 216, 253, 266-267, 268-269, 276, 298, 301
Angle-bend-angle-bend potential, 213, 253, 257, 270, 279-280
Angle bend force constant, organometallic, 543
Angle bend potential, 4, 5, 6, 12, 22, 90, 94-95, 192-193, 223, 225, 244, 248, 251-252, 255-256, 263-264, 266, 268, 273, 275, 277-278, 284, 287, 289-290, 292, 297, 300, 309, 310, 311, 564
Angle-bend-torsion-angle-bend potential, 215, 245, 260-261, 265-266, 270, 280
Angular Overlap Model, organometallic model, 564, 628
Anharmonic potential, 8, 281
Anisotropy, neighboring group, 480
Anomer, 7, 54, 275, 628-629, 652
  α (saccharide), 54
  β (saccharide), 54
Anomeric carbon, 55, 56, 57, 275, 276, 626, 628-629, 659
Anomeric effect, 7, 629, 658, 664-665
anti conformation, 122, 194, 195-199, 229, 258, 260, 326-329, 458, 461, 463, 476-480, 629, 657, 658
Anticlinal, 629, 630
Antiperiplanar, 629, 630
AOM organometallic model, 564, 628
Appended format input file, 129, 183, 610
Aristotle, 626
Aromatic bonds, 160
Aromatic compounds, 261, 480-488, 665
Aromaticity, 539, 629-630, 665
Artificial intelligence, 347
ASCII characters, 149, 174, 630, 632, 637, 692
Asymmetric center, 630-631, 646, 651
ATEQ mode, GMMX, 357-358
Atm field (PDB input file), 168
Atom
  π, 15, 98, 131, 137, 138, 223, 565, 568, 585, 622-625
  color scheme, 4
  deleting, 37, 38
  drawing, 38
  haptophilic, 237
  marking, 37, 41
  moving, 40-41
  transition state, 238-240

Atom block (MDL MOL input file), 145
Atom file numbers, 90, 99, 101, 129, 133, 134, 140, 142, 143-147, 151, 153, 458, 461
Atomic charges
  chloroethane, 321, 604
  fluoroethane, 322-323
  methanol, 323-324
Atomic solvation parameters, 330
Atom labels, 98, 142, 146, 150, 151, 153, 160
Atoms
  display options (in PCMODEL), 98-101
  restricted, 136
Atom type
  @ notation, 238, 624
  wild card, 266, 297, 543, 625
Atom-to-atom mapping in PCMODEL, 121
Atom types, 22, 36, 90, 100, 102, 132, 159, 160, 166, 222, 238-241, 246, 250, 254, 256, 272, 283, 287, 289, 291, 296, 307, 309, 310-311, 313, 314, 423, 425, 432, 433, 434, 488, 489, 509, 513, 515, 526, 527, 529, 531, 536, 549-550, 582, 583-588, 622-625
  for coordination organometallic bonding, 531, 539-542, 549-550
  for covalent organometallic bonding, 530, 539-542
  for hydrogen bonding, 509, 513, 515, 622-625
  for metals, 527, 529, 531, 622-625
  for MMX π-calculation, 238-240, 622-625
  for reactive intermediates, 583-584, 622-625
  for transition states, 540, 583-588, 622-625
  in MMX force field, 238-241, 622-625
Attached atoms, 139, 141, 166
Available Chemicals Directory, 17
aza-Cope rearrangement, 239, 583

BABEL (file format interconversion), 128, 169-171, 631
Backbonding, 541
Badger's rule, 311, 585, 631
Bailar, J. C., Jr., 5
Ball and stick model, 4, 9, 18, 104-106, 425, 610, 617, 627, 636, 671, 678, 688
Bartell, L. S., 7, 309, 677
Bartlett, P. A., 635
Barton, Sir D. H. R., 5
Batch dialog box, 611
Batch minimization, 450, 491, 493, 505, 569-570, 577, 592, 610, 631-632
Bath Temperature in PCMODEL molecular dynamics, 379
BDHF parameter, in MMX, 188, 189, 194

Beautify, 639
Bent bond, 632
Benzene
 π-calculation exercise, 490-491
 molecular surface calculations, 437-439
 structure comparison exercise, 123-127
Benzenechromiumtricarbonyl
 organometallic exercise, 565-572
 docking exercise, 579-581
Benzene dimer, 234-235, 250, 306, 389-396, 406-421, 604-608
 fixed atoms exercise, 389-396, 604-608
 geometric constraints exercise, 389-396, 409-421, 604-608
 leading corner perpendicular, 409
 orientations, 406-408
 staggered stacked, 409
Betaine, 632
Bicyclo[2.2.2]octadienone, NMR coupling constants in, 466-468
BIGSTRN program, 308
Biophore, see also Pharmacophore, 83, 632, 658
Bischoff, C. A., 4
Bitmap graphics format, 80, 660
Block-diagonal Newton-Raphson algorithm, 339, 344, 677
Blocking, see also z-clipping), 632
BND parameter, in MMX, 187, 190-191
Bohacek, R. S., 454
Bohlmann effect, 259, 260
Boltzmann population analysis, 14, 363, 367, 371, 465, 475, 476, 479-480, 593, 602-603
Bond
 bent, 75
 coordination, 527, 531-533, 567
 covalent, 527, 529-531
 deleting, 37, 38
 drawing, 38
 rotating, 41-43
 stereo, 147
Bond angle, 9, 13, 79, 119, 156, 527, 644
Bond block (MDL MOL input file), 147
Bond-by-bond iterative conformation search, 347
Bond compression, 203
Bond count (Alchemy-III input file), 159
Bond dissociation energy, 204, 273, 311
Bonded interactions, 94
Bonding, metal atom, 5, 6, 12, 13
Bond length, 6, 9, 13, 79, 119, 495, 631, 632, 644, 687-688
 electronegativity correction, 251
Bond moment, 190-191

Bond multiplicity, 37, 80-82, 129, 133, 141, 147, 152, 157, 159, 161, 165, 166-167, 269, 527, 536, 609, 632, 666
Bond order, 495, 631, 632
 fractional, 584, 585, 589, 593, 631, 632
Bond stretch, 4, 5, 12, 22, 90, 94, 190, 543
 organometallic, 543, 554, 560, 563, 564
Bond-stretch-angle-bend potential, 12, 90, 95, 210, 212, 219, 230-232, 244-245, 248-249, 252, 257, 270-271, 279, 300-301
Bond-stretch-bond-stretch potential, 213-214, 259, 269-270, 279
Bond stretch force constant
 organometallic, 543, 554
 transition state, 585
Bond stretch potential, 202-205, 222-224, 243-244, 247, 255, 263, 266, 268, 272-273, 275, 277, 284, 287, 289, 292, 296-297, 300, 309, 495, 631
Bond topology, 147
Bond type flag, 147, 634
Bond types
 for organometallics, 529-532, 539-542
 for transition states, 583-588
Bond vector, 635
Born, M., 633
Born equation, 330-331
Born-Oppenheimer approximation, 11, 488, 632-633
Boyd, R. H., 6, 7, 343, 456, 647, 677
Bradley, C. A., Jr., 4
Brookhaven National Laboratory, 168
Brown, A. C., 4
Browser, web, 8, 17, 18, 172
Brute force conformation search, 347, 661
Buckingham potential, 232. See also Modified
 Buckingham potential
Buckminsterfullerene, molecular surface calculations, 439-441
Build tool, in PCMODEL, 87-88, 611-612
Burkert, U., 4, 200, 207
Butlerov, A. M., 3

C3D file format (CambridgeSoft Chem-3D), 128
CACAO program, 633
CADD, 633
CADPAC program, 634
Calculation
 π, 2, 13, 15, 124, 132, 133, 137, 246, 267, 313, 363, 480-486, 488-508, 539, 540, 565, 583, 615, 617, 665

ab initio, 8, 204, 262, 271, 276, 299, 301, 302, 306, 314, 316, 321, 322, 330, 346, 407, 408, 454, 512, 525, 582, 657, 658, 663, 665, 670, 673; 674, 678
AM1, 346
energy, 90, 334-345
heat of formation, 222, 254, 339, 495, 543
Hückel, 481, 488-508
MO, 153, 197, 321, 481, 488, 631, 662
PCILO, 308
semiempirical MO, 152, 481, 582, 615, 627, 674, 676
VESCF, 137, 222, 465, 480-486, 488-508, 539, 615, 617, 691
Calipers, 118, 119, 634
Calix[4]arene, 80, 81-82, 398-401, 403-406
MDL MOL file, 81-82
SMILES structure file, 80
Cambridge Crystallographic Structural Database, 17, 18, 76, 87, 148, 149, 310, 314, 524, 634, 661
CAMD, 14, 635, 639, 681, 683-684, 687, 688, 690-691, 692
CAMEO, 85, 635
CAMSEQ force field, 283
Carbocations, 309, 583
Carbohydrate parameters, 288, 295
Carbohydrates, 7, 9, 275, 295, 525, 658, 664-665
Carbonyl, 540
Cartesian coordinates, 141, 145, 150, 159, 160, 165, 168, 169, 218, 319, 339, 341, 349, 350, 386, 387, 640, 646, 702
Cartesian stochastic conformation search, 346
Catalyst program, 635
Catalysts, 310
Cauchy generating function, 348
CAVE, 9, 635, 700
Cave Automatic Virtual Environment, 9, 635, 700
CAVEAT program, 635
Cenco-Petersen model, 9
Centroid, 408, 411, 419, 562, 604-608, 635-636
Cerius program, 310
CFF force field, 262-282, 636, 659
CFF93 force field, 276, 281-282, 316
CFF95 force field, 282
CGI, 636
Charge, 8, 90, 119, 120, 129, 131, 139, 157, 160, 165, 265, 267, 271-272, 274, 276, 278, 286, 294, 298, 303, 304, 305, 319-329, 422, 423, 427, 527, 532-533, 604, 636, 681
on metal, 527, 530, 532-533

Charge count (Alchemy-III input file), 159
Charge equilibration algorithm, 323-324
Charge equilibration-potential, derived charge algorithm, 323-324
CHARGE2 electrostatic model, 323-324
Charge-charge electrostatic model, 138, 139, 220, 235-237, 246, 265, 267, 274, 278, 285, 303, 317, 319, 324-329, 513, 643, 651
Charge flag, 133, 146
CHARMm, 289, 299, 636, 687
CHARMM force field, 9, 16, 210, 219, 283, 286, 288-296, 307, 308, 309, 314, 522, 524, 562, 636, 652, 687, 695, 700
angle bend potential, 289-290, 292
bond stretch potential, 289, 292
electrostatic potential, 290-291, 294, 324-325
hydrogen bonding potential, 291, 294-295, 522-523, 524
improper torsion potential, 293
torsion potential, 290, 292-293
Urey-Bradley potential, 295
van der Waals potential, 290-291, 293-294
CHEAT parameters, 295, 636
Chem123 program, 636
Chem-3D file format, 165-166, 636
Chem-3D program, 61, 152, 165-166, 201, 636
ChemDraw, 82, 637
CHEMELEON, 128, 169, 637
ChemFinder, 18
Chemical Literature Data Extraction, 80
Chemical shift, 481-486, 637
Chem-X Builder, 87
Chime, 18, 128, 169, 172-174, 637
Chiral flag, 144
Chloroethane
atomic charges, 321, 604
dipole moment exercise, 319-320
Chromyl chloride
organometallic exercise, 533-538
CHUCKLES program, 638
CICADA program, 638
CIP system, 638, 640, 687, 690
Claisen rearrangement, 239, 583, 587
Clean, 638-639
CLiDE, 80
CLogP, 639, 669
COBRA, 86, 87, 639
Color scheme for atoms, 4
Color scheme for molecular models, 4
Comba, P., 342, 543, 628
CoMFA, 296
Compare dialog box, 120-121, 124, 127, 612
Compare window, 121-122, 125, 127, 612

COMP mode, GMMX, 359–360
Computer-aided drug design, 633
Computer-aided molecular design, 14, 635, 639, 681, 683–684, 687, 688, 690–691, 692
Concatenate
  amino acids, 52, 380
  sugars, 54
CONCERTS program, 639
CONCORD, 36, 86, 87, 639–640
Conformation, 4, 6, 11, 629, 640–641
  α-helix, 52–53, 113, 398
  β-sheet, 52–53, 398
  β-turn, 52–53, 398
  anti, 122, 194, 195–199, 229, 258, 260, 326–329, 458, 461, 463, 476–480, 640
  gauche, 7, 122, 194, 195–199, 229, 242, 258, 260, 326–329, 458, 463, 476–480, 640
  minimum energy, 11, 345–376, 465
Conformational analysis, 5, 6, 8, 9, 18, 242, 289, 295, 307, 465, 480, 563, 629, 638, 639, 641, 646, 658, 664–665, 678
Conformational enantiomers, 355
Conformational energy, 201
Conformation search, 7, 8, 14, 15, 307, 345–376, 456, 639, 689
  bond-by-bond iterative method, 347
  brute force method, 347, 627, 641, 661
  Cartesian stochastic method, 346
  constrained molecular dynamics method, 346, 348–349, 409
  cycloheptadecane, 346
  Directed Tweak method, 349
  Distance Geometry method, 346, 349
  duplicates check, 357–360
  genetic algorithm method, 349
  grid search, 641, 660–661
  molecular dynamics method, 346, 348–349, 378, 673
  Monte Carlo method, 346, 347–348, 407–408
  poling method, 350, 684
  random kick method, 346, 354, 366
  simulated annealing method, 348, 691–692
  stochastic method, 347–348, 641
  SUMM method, 347
  systematic method, 347
  Systematic Unbounded Multiple, Minimum method, 347
  torsional Monte Carlo method, 346, 347–348
  torsional tree method, 346, 347
Conformation search exercise
  2-methyl-1,3-dioxolane, 350–376
  pentane, 350–376
Conformation space, 347, 348, 629, 639, 641, 650, 657, 659, 684

Conjugated system, 13, 201, 216, 251, 255, 262, 267, 299, 300, 301, 313, 488–508, 548, 583, 629, 641, 646, 665, 689
Conjugate gradient algorithm, 283, 341–342, 641
Conjugation, 488, 624, 629, 641–642, 646, 666.
  See also Delocalization
  effect on torsion potential, 207
Conjugation flag, 129
Connected atom list, 138, 140
Connection table, 142, 159, 160, 161, 165, 634, 637, 642
Connectivity (in file format), 129, 152, 527
Connect tool, 612
Connolly, M. L., 8, 9, 322, 437, 442, 676, 677
Connolly algorithm, 437, 442, 443–444, 447, 649, 675
Constrained molecular dynamics, conformation search, 346, 348–349
CONSYSTANT program, 128, 169, 642
Continuum solvation model, 307, 330–331
Conversion, 2D to 3D, 37
Coordinate information (Alchemy-III input file), 159, 160, 165
Coordinates
  Cartesian, 141, 145, 150, 159, 160, 165, 168, 169, 218, 319, 339, 341, 349, 350, 386, 387, 640, 646
  internal, 640, 667
  spherical, 561
Coordination bonding, 527, 531–533, 573, 642–643, 651, 661, 669, 675
Coordinative bridging bond, 540
Cope rearrangement, 583, 587
Copy graphics to Windows Clipboard, 97–98, 612
Corey, E. J., 5
CORINA, 18, 83, 86, 87, 643
COSMIC force field, 87, 639, 643
Coulomb integral, 495, 500, 665
Coulomb potential, 220, 235–236, 265, 267, 274, 276, 278, 285, 290, 298, 304, 309, 311, 314, 325, 330, 509, 513, 522, 565, 628, 643, 647, 651, 669, 678
  buffered, 303
Counts line (MDL MOL input file), 144
Couper, A. S., 3
Coupling constants, 465, 468–480, 643–644
Covalent bonding, 527, 529–531, 643, 644
Covalent radius, 129, 178, 241, 529–530, 536, 538, 543, 634
Covalent radius flag, 538, 547
CPK model, 10, 106–107, 423, 636, 644, 684, 688
CPK Models dialog box, 111, 423–424
CPK surface, 111, 423–427

INDEX   **745**

histidine, 426–427
CPK__Surface command, 423–427, 612
Crick, F. H. C., 5, 509
Cross term, 12, 90, 95, 210–216, 242, 245, 250, 252, 268, 271, 277
  angle-bend–angle-bend, 213
  angle-bend–torsion–angle-bend, 215
  bond-stretch–bond-stretch, 213–214
  improper-torsion–torsion–improper-torsion, 215–216, 271
  out-of-plane–out-of-plane, 271
  stretch–bend, 12, 90, 95, 252
  torsion–angle-bend, 214, 245, 250
  torsion–bond-stretch, 213, 214, 252
  torsion–improper-torsion, 214–215
  torsion–torsion, 214, 215
CRYSTAL program, 644
Ctab, 142, 642
Cubic correction term, 204, 222–223, 243–244, 247, 255, 273, 277
Cubic stretch catastrophe, 204, 251, 644
Curtin-Hammett principle, 582, 644–645
CVFF force field, 268–272, 279
  angle-bend–angle-bend potential, 270
  angle bend potential, 268
  angle-bend–torsion–angle-bend potential, 270
  bond-stretch–angle-bend potential, 270–271
  bond-stretch–bond-stretch potential, 269–270
  bond stretch potential, 268
  hydrogen bonding potential, 271–272
  out-of-plane angle bend potential, 268–269
  out-of-plane out-of-plane potential, 271
  torsion potential, 269
  van der Waals potential, 271–272
Cycloheptadecane, conformation search, 346
Cyclohexane, molecular volume calculation, 452–456
Cyclophanes, 6

Darling model, 9
Data address, 134
Databases
  3D structure, 17, 18, 50
Data fields, 134
Data prefix, 129
DD electrostatic model, 139
Debye, P. J. W., 647
Deformation density, 75, 217, 234, 645–646.
  See also Electron density
Deleting atom in PCMODEL, 37, 38
Delocalization, 495, 583, 641, 646, 689. *See also* Conjugation
DELPHI program, 442

del Re electrostatic model, 321–322, 323
de novo molecular design, 639, 688, 690–691, 692
Density functional theory, 570, 572, 673
Depth-cuing, 427
Deuterium, 217, 238, 245, 257, 509, 623, 624, 625
DGEOM program, 346, 646, 648
Diagonal force field, 201, 283
Dialog box, 646
  Added Constants File, 188
  Batch, 611
  Compare, 120–121, 124, 127, 612
  CPK Models, 111, 423–424
  Dials, 67, 70, 116–117
  Dielectric Constant, 325
  Dihedral Driver Setup, 461
  Dock Setup, 574–575
  Dot Surface Display, 427
  Dynamics Setup, 375, 379
  Electrostatic of Dipole D, 139, 326
  Enantiomer, 38
  File Exists, 63, 611
  Fix Atoms, 386–387
  Fix Torsions, 397–398
  Labels, 98
  Metal Information, 532–533, 549–550
  MMX Pi Calc Options, 138, 490, 496
  Mopac Data, 153, 155
  Pi Atom Selection, 489–490
  Pluto Options, 105–106
  Print, 97
  Printer Setup, 617
  Print Level, 90
  Read Dihedral Driver File, 464
  Ribbon Information, 112, 115
  Rot__E, 457–458
  Rotate__Bond, 41, 48
  Save File, 60
  Save Graphic, 97
  Stereo Options, 103
  Structure List, 66, 71
  Substructure Name, 66
  TS Bond Orders, 584–585, 589
  Write File for Batch, Dock and Dihedral Driver, 379, 411, 418, 610
Dials dialog box, 67, 70, 116–117
Diamagnetic shielding, in NMR, 480
Diborane, 310
Dielectric, distance dependent, 294, 303, 317
Dielectric constant, 138, 220, 236, 265, 274–275, 285, 290, 298, 303, 317, 325–329, 331, 601, 613, 643, 646–647
Dielectric Constant dialog box, 325
DIELE field (MMX input file), 138

Diels-Alder reaction, 582, 585, 586, 588-598
  transition state exercise, 588-598
Diffraction
  electron, 78-79, 196-197, 262, 316, 536, 537, 565, 570, 572
  neutron, 10, 77-78, 148-152, 506, 546-547, 565, 570, 572, 634
  X-ray, 19, 75-77, 148-152, 316, 506, 558, 561, 565, 570, 572, 634, 699, 702
Dihed Map command, 463, 464, 613
Dihedral driver, 7, 42, 43, 140, 396, 455-464, 613, 647
  metal complex, 456-457, 562
  minimized step, 459-464
  polymer, 456
  rigid rotation, 457-459
Dihedral driver exercise, 1,2-dichloroethane, 461-464
Dihedral Driver Setup dialog box, 461
Dillen, J. L. M., 312, 650
Dipole, 647
Dipole-charge interactions, 257
Dipole-dipole electrostatic model, 138, 139, 220-222, 237, 246, 248, 254, 257, 319, 324-329, 521, 613, 651, 666, 668
Dipole-dipole interactions, effect on torsion potential, 207
Dipole moment, 13, 90, 236, 319-320, 322, 324-329, 604, 647, 668
Dirac, P. A. M., 10
Diradical, 496
Directed Tweak conformation search, 349, 647
Directionality
  of hydrogen bonds, 315, 513-515, 520-524, 663, 670
  of metal coordination bonding, 548, 560-561
DISCOVER, 201, 282
Display options for atoms, 98-101
Display tools, 96, 127
Distance constraint flag, 392, 406, 409
Distance dependent dielectric, 294, 303, 317
Distance dependent dieletric model, 220
Distance Geometry, 349, 646, 648
Distance Geometry conformation search, 346, 349, 646
Distributed multipole analysis electrostatic model, 520
Di-t-butyl ether, energy minimization exercise, 340-341
Di-t-butyl ether, single point energy exercise, 336-339
DLL, 648
DMA electrostatic model, 520
DMC program, 348-349
DNA, 5, 509

DOCK algorithm, 348, 441, 442, 573, 648-649
Dock command (PCMODEL), 533, 613
Docking, 13, 66, 136, 138, 307, 378, 422, 441-444, 565, 571-581, 649, 667, 673, 690
Docking exercise
  benzenechromiumtricarbonyl, 579-581
  methanol dimer, 575-579
DOCK program, 348, 441, 442, 648-649
Dock Setup dialog box, 574-575
Dodecahedrane, molecular volume calculation, 447-452
Donor ligand, 526, 529, 532, 548, 549
Don't Minimize command, 401-406, 613
Don't Minimize exercise, 2-(2'-phenolylmethyl)phenol, 403-406
DOS extender, 353, 649
Dostrovsky, I., 5
Dot surface, 7, 112, 427, 428-429, 432, 434, 435, 436, 441, 613, 688
Dot Surface command, 427-428, 434, 435, 613
Dot Surface Display dialog box, 427
Drawing atom in PCMODEL, 38
Drawing exercise
  1,1-difluoroethene, 43-45
  adamantane, 45-49
  lactose, 55-58
  octapeptide, 52-53
Drawing tools in PCMODEL, 24, 33-34, 37-49
DREIDING force field, 201, 208, 300, 309-310, 311, 322, 524, 564, 649
  angle bend potential, 310
  out-of-plane angle bend potential, 310
Dreiding model, 9, 98, 649, 671
DRV, 650
Dummy atom, 250, 296, 549, 562, 564, 650. See also Pseudoatom; Virtual atom
Dunitz, J. D., 208, 314
Duppa, B. F., 4
DuPuis, M., 60, 166
Dynamic link library, 648
Dynamics, molecular, 5, 138, 287, 288, 291, 295, 299, 304, 307, 329, 330, 348-349, 376-384, 516-520, 563, 613, 636, 672-673
Dynamics Setup dialog box, 375, 379

EAS force field, 7, 308-309, 650
ECEPP force field, 7, 307-308, 650, 653
Editing an input file, 174-185, 410-411, 415-418, 449-450, 536-538, 547, 569-570, 592-593
EFF force field, 312-313, 650
EHT treatment, 481
Eigenvalues, 497, 499-500
Electron cloud, 216, 217, 218, 430, 645, 650-651, 671, 684, 699

Electron density, 75, 216, 321, 480, 507, 549, 632, 646, 670. *See also* Deformation density
Electron density-bond order matrix, 482, 497, 499-500, 507-508
Electron diffraction, 78-79, 196-197, 262, 316, 536, 537, 565, 570, 572
Electronegativity, 8, 322, 469, 470, 480, 493, 508, 629, 651, 655, 657, 664, 666, 681, 682, 687, 688, 692
variable, 481
Electronegativity correction to bond length, 251, 311
Electrophiles, 507-508, 583, 629, 651, 678, 689
Electropositivity, 480
Electrostatic interactions, 12, 13, 22, 90, 93, 94, 131, 138, 139, 207-208, 227, 228, 230, 243, 245, 246, 288, 304, 316-329, 480, 508, 520, 523, 530, 555, 560, 666, 675, 685
effect on torsion potential, 207-208, 227, 228, 230
Electrostatic model, 138, 139, 220-222, 235-237, 246, 265, 267, 274, 278, 285, 286, 287, 288, 294, 303, 313, 317, 319, 321-329, 520, 643, 651, 668, 695
charge-charge, 138, 139, 220, 235-237, 246, 265, 267, 274, 278, 285, 294, 303, 317, 319, 324-329, 643, 651
CHARGE2, 323-324
del Re method, 321-322, 323
dipole-dipole, 138, 139, 220-222, 613, 651, 668
distributed multipole analysis, 520
DMA, 520
ESP, 286, 287, 322-323, 324
functional group dependent, 294
Gasteiger-Marsili, 322
LCD, 520
localized charge distribution, 520
modified Smith-Eyring, 322
Mulliken population analysis, 321, 323
multipole expansion, 294, 695
PEOE, 313, 322, 681
QEq, 323-324
QEq-PD, 323-324
RESP, 287, 288, 322-323, 324
shifted dielectric, 294
Smith-Eyring, 322
Electrostatic of Dipole D dialog box, 139, 326
Electrostatic potential, 220-222, 235-237, 242, 254, 257, 265, 267, 274, 276, 278-279, 285-286, 288, 290-291, 294, 298-299, 303, 308, 309, 427, 509, 513, 651
Electrostatic scoring, in Dock program, 442, 649
Electrostatics exercise
1,2-dichloroethane, 325-329

dipole moment of chloroethane, 319-320
electrostatic surface of formamide, 332-334
Electrostatic surface, 296, 325, 331-334
formamide, 332-334
Elimination reaction, 5
Enactive molecular model, 13, 79, 651
Enantiomer, 38-39, 613, 651
Enantiomer command, 38-39, 613
Enantiomer dialog box, 38
Enantiomers, conformational, 355
Encapsulated postscript graphics format, 651
Ene reaction, 583, 587
Energy, strain, 91
Energy calculations, 90, 334-345
Energy minimization, 136, 339-345, 615, 627, 665
Energy minimization exercise, di-t-butyl ether, 340-341
Energy minimum
global, 11, 14, 15, 344-376
local, 15
Engler, E. M., 7, 308, 341, 641
Ensemble, 573, 577
Enthalpy, 315, 652
Entropy, 91, 652, 666
Epimer, 39, 472, 628, 652
Epimer command, 39, 472, 613
Epimerization, 39, 472
EPS graphics format, 651, 652
Equilibration Time in PCMODEL, molecular dynamics, 379
Erase command, 39, 613-614
ESFF force field, 564-565
ESP electrostatic model, 286, 287, 322-323, 324
Ethanol solvation model, 304
Ethers solvation model, 304
EVHF parameter, in MMX, 188, 189, 194
Explicit solvation model, 329-330
Extended atom model, 289, 628, 652-653. *See also* United atom
Extended Hückel Theory, 481, 702
Extensible Systematic Force Field, 564-565

Fachinformationszentrum, 17
FANTOM program, 653
Fieser model, 9
File Exists dialog box, 63, 611
File format
AL2 (Tripos Alchemy 2000), 128, 159-160
C3D (Chem-3D), 128, 636
connection tables, 129, 138
coordinates, 128
header/footer information, 128
interconversion, 169-174, 631, 636, 737, 642, 643, 668, 672
MAC (MacroModel), 128
MMX (PCMODEL), 128, 133-142, 610, 653

File format (*Continued*)
  MOL (MDL), 128, 142–148, 636, 640, 672
  MOL (Tripos Alchemy-III), 157–159, 672, 702
  MOL (Tripos Alchemy), 128, 157–159, 672
  MOL/SYB (Tripos SYBYL), 128, 672
  MOPAC, 152–157, 702
  PCM (PCMODEL), 128, 129–133, 610
  PDB (Brookhaven), 128, 648, 677, 681, 702
  property values and flags, 129
  SMILES, 80, 86, 636, 639, 640
Filenames, 21, 599, 653
First-row elements, 217, 231, 249, 653–654
Fischer projection, 628, 654
Fix Atoms dialog box, 386–387
Fix_Atoms command, 386–387, 614
Fix_Distance command, 387–396, 406–421, 614
Fix_Distance command, force constant, 388
Fixed atoms exercise
  benzene dimer, 389–396
  2-(2′-phenolylmethyl)phenol, 398–401
Fix Torsions dialog box, 397–398
Fix_Torsions command, 396–401, 614
Flag, 131, 132, 134, 139, 144, 174, 617, 618, 654, 695
  π atom, 175, 489, 502, 503, 504, 565, 568, 579, 617, 618
  π-system planarity, 137
  added constants, 136, 188, 610
  charge, 133, 619
  conjugation, 129
  covalent radius, 538, 547
  distance constraint, 392, 406, 409, 618
  hydrogen bonding, 129, 133, 137, 142, 175, 181, 509, 510, 512, 518, 575, 618
  metal atom electronic state, 133
  metal coordination, 129, 137–138, 549–550, 553–554, 580, 618
  reset, 618
  substructure membership, 129, 175, 618
  transition state bond order, 592
Flap angle, 119, 222, 248, 396, 654, 667
Flory, P. J., 8
Fluoroethane, atomic charges, 322–323
F-matrix, 497, 499–500
FMO theory, 520, 657
Force constant, 654–655
  angle bend, 205, 223, 244, 248, 251–252, 255–256, 263–264, 266, 275, 278, 284, 287, 289–290, 292, 297, 300, 543
  angle-bend–angle-bend potential, 253, 270, 279–280
  angle-bend–torsion–angle-bend potential, 245, 260–261, 265–266, 270, 280
  bond-stretch–angle-bend potential, 212, 230–231, 244, 248–249, 252, 257, 270–271, 279, 300–301
  bond-stretch–bond-stretch potential, 259, 269–270, 279
  bond stretch potential, 202, 204, 222–223, 243, 244, 247, 251, 255, 263, 266, 268, 275, 277, 284, 287, 289, 292, 296–297, 300, 495, 543, 585, 631
  cubic angle bend, 273, 277–278, 300
  cubic bond stretch, 204, 222–223, 243–244, 251, 255, 273, 277, 300
  electrostatic potential, 304
  hydrogen bonding potential, 266, 267, 286–287, 291, 294–295, 302–303, 513
  improper torsion potential, 210, 257–258, 285, 293
  improper-torsion–torsion–improper-torsion potential, 262
  organometallic angle bend, 543
  organometallic bond stretch, 543, 554
  organometallic out-of-plane angle bend, 543
  out-of-plane angle bend potential, 208, 223–225, 253, 266–267, 268–269, 276, 298, 301
  out-of-plane–out-of-plane angle bend potential, 271
  quartic angle bend, 251, 277–278
  quartic bond stretch, 251, 255, 277, 300
  quintic angle bend, 251, 255
  sextic angle bend, 223, 248, 251, 255
  stretch-bend potential, 212, 230–231, 244, 248–249, 252, 257, 270–271
  torsion–angle-bend potential, 259–260, 281
  torsion bond stretch potential, 252, 258–259, 281
  torsion–improper-torsion potential, 261–262
  torsion potential, 206, 225–228, 256, 264, 266, 269, 275–276, 278, 284–285, 287–288, 290, 292–293, 297, 301, 495
  torsion–torsion potential, 261
  Urey-Bradley potential, 219, 264–265, 266
  van der Waals potential, 217, 218, 232, 243, 246–247, 253, 256–257, 264, 267, 271–272, 276, 278–279, 285–286, 288, 290–291, 292–293, 297–298, 301–303, 304, 522
Force constant calculation, 288, 299
Force constant for Fix_Distance command, 388
Force field, 201–222, 655–656, 674
  AMBER, 16, 201, 207, 282–288, 291, 306, 307, 308, 309, 314, 316, 322, 442, 524, 563, 628, 649, 659, 667, 678, 695, 697
  AMBER*, 201, 307, 628, 670

INDEX

AMBER/OPLS, 306, 628, 695
AMBER94, 307, 670
CAMSEQ, 283
central, 655
CFF, 7, 201, 219, 262-282, 521, 636, 659
CFF93, 281-282
CFF95, 282
CHARMM, 9, 16, 210, 219, 283, 286, 288-296, 307, 308, 309, 314, 522, 524, 636, 652, 687, 695, 700
class I, 655
class II, 8, 201, 281, 655, 686
class III, 8, 655
comparisons, 201
components, 201-222
COSMIC, 87, 639, 643
CVFF, 201, 268-272, 279
diagonal, 201, 283, 655
DREIDING, 201, 208, 300, 309-310, 311, 322, 524, 564, 649
EAS, 7, 308-309, 650
ECEPP, 7, 307-308, 650, 653
EFF, 312-313, 650
empirical, 1, 5, 6, 407, 408, 582
ESFF, 564-565
Extensible Systematic Force Field, 564-565
generic, 309-312, 655-656
HSEA, 7-8, 308
hydrocarbon, 6, 242, 254, 263, 268, 308, 312
Lyngby CFF, 272-276, 659
MM2, 7, 8, 16, 136, 139, 194, 208, 217, 222, 232, 234, 242, 245-249, 251, 252, 253, 256, 257, 258, 299, 305, 308, 312, 313, 408, 493, 521, 562, 563, 671, 695, 697
MM2*, 201, 307, 670, 671
MM2CARB, 308
MM2MX, 563
MM2X, 299, 671
MM3, 15, 16, 100, 152, 194, 217, 246, 250-254, 255, 256, 257, 258, 287, 299, 307, 312, 313, 512, 521, 562, 563, 627, 636, 666, 695
MM3*, 201, 307, 670
MM4, 15, 207, 212, 217, 254-262, 265, 271, 280, 312, 313, 521, 666, 667, 695
MMFF, 307, 672
MMFF93, 201
MMFF94, 194, 299-303, 316, 378, 525, 670, 673, 695
MMI, 7, 242-245, 246, 247, 248, 252, 263, 283, 308, 309, 469, 488, 521, 563, 672
MMP2, 246, 309, 583
MMX, 11, 14, 200, 201, 205, 217, 219, 222-238, 246, 409, 512, 669, 672, 697, 699
MOMEC, 563-564
MUB, 7, 309, 677
nonbonding terms, 216-222
OPLS, 287, 288, 303-306, 524, 525, 628, 678
OPLS-AA , 306
OPLSA*, 306, 670
organometallic, 526-571
PIMM, 313-314, 488, 682
QCFF-PI , 283
QM/MM, 379, 686
QMFF, 276-281, 282, 686
SHAPES, 561-562, 564
Tripos, 201, 208, 296-299, 525
UBCFF, 263-268, 272, 522
UFF, 201, 300, 310-312, 378, 564, 697
UNICEPP, 7, 283, 307-308, 650, 697
VALBOND, 562
valence, 6, 201, 268, 655, 698
valence cross terms, 210-216
valence terms, 202-210
YETI, 314-315, 524, 560-561, 663, 664, 670, 702
Force field scoring, in Dock program, 442, 649
Formamide
  electrostatic surface exercise, 332-334
  solvation model, 305
Format, input file, 22, 59-61, 80-85, 128-199
Fortran format, 134, 143, 168
Fractional bond order, 584
Frankland, Sir E., 4
Free energy, 656-657. *See also* Enthalpy; Entropy
  solvation, 330-331
Free format, 148, 160
Front end, 657
Frontier molecular orbitals, 520, 657
Fukui, K., 520, 657
Fuller, R. B., xiii
Full printout output file, 92-95
Functional group, 3, 50, 495, 515, 624, 670, 682
Functional group dependent electrostatic model, 294

Gajewski, J. J., 222, 672
GAMESS program, 657
Garbisch algorithm, 465, 470
Gasteiger, J., 299, 313, 322, 681
Gasteiger-Marsili electrostatic model, 322, 681
gauche conformation, 7, 122, 194, 195-199, 229, 242, 258, 260, 326-329, 458, 463, 476-480, 657
gauche effect, 194, 657-658
Gauss-Bonnet theorem, 445-447

Gaussian, 152, 167, 658, 663, 702
Gavezzotti, A., 454
Gay-Lussac, J. L., 10
GB/SA solvation model, 307, 330–331, 658, 670, 695
GEGOP program, 658
Gelin, B., 289, 290, 291
Geminal van der Waals interactions, 543, 554–555
Generalized Born equation, 330–331
Generalized parameters, 335, 543
Generation, 3D structure, 6, 37, 40, 85–89
Generic force field, 309–312
Generic parameters, 529, 543
Genetic algorithm, 349, 627–628, 658
Genetic algorithm conformation search, 349
Geometric constraints, 13, 385–421, 614, 682
Geometric constraints exercise
  2-(2'-phenolylmethyl)phenol, 398–401, 403–406
  benzene dimer, 389–396, 409–421, 604–608
  giberrellin intermediate, 386–389
Geometry optimization, 6
GERM program, 658
Giberrellin intermediate
  geometric constraints exercise, 386–389
  XRA format file, 386–389
Gilbert, K. E., 222, 672
Global minimum, 344–376, 658–659
GLYCAM parameters, 288, 659
Glycosyl acceptor, 55, 659
Glycosyl donor, 55, 659
GMMX program, 99, 347, 350–376, 573
  ATEQ mode, 357–358
  COMP mode, 359–360
  NSEQ mode, 358–359
  NSRF mode, 359–360
  NSRO mode, 359
GMMXINP program, 351–367
Goddard, W. A., 309, 323, 524, 649
GOPT program, for Lyngby CFF force field, 275, 659, 660
Graphics
  copying to Windows Clipboard, 97–98
  printing, 96–97
Graphics capture utilities, 96–98
Graphics format
  bitmap, 80, 660
  encapsulated postscript, 651
  EPS, 651, 652
  HPGL, 96, 97, 664
  Postscript, 96, 97, 651, 684
  raster, 80, 659–660
  vector, 80, 659
  Windows Metafile, 96–98

Graphics Optimization Tool for Lyngby CFF force field, 275, 659, 660
Graphics output, 96–127
Grid search conformation search, 641, 660–661
GSTAT program, 635, 661
Guida, W. C., 454

H/AD tool, 88–89, 152, 614
Hagler, A. T., 686
Halgren, T. A., 299, 644, 672
Hambley, T. W., 342, 543, 628
Hamiltonian, 154
Hammond postulate, 661
Hantsch, C., 639
Haptophilic atom, 237, 661
Haptophilicity, 237, 548, 661
Hard acid, 661, 664. See also HSAB principle
Hard base, 661, 664. See also HSAB principle
Hard sphere, 6, 7
  exo-anomeric force field, 7–8, 308, 664–665
Hardness, 202, 216, 217, 661–662
Hardness parameter ε, 94, 191, 217, 218, 232, 241, 243, 247, 254, 257, 267, 272, 279, 286, 288, 298, 302, 303, 305, 513, 521, 525, 543, 562, 583, 622–625, 647, 662, 664
Harmonic oscillators, 200
Harmonic potential, 219, 283, 527
Hartree-Fock, 276, 281, 662
Hartree-Fock *ab initio* potential energy surface, 276, 281
Haworth, W., 9, 641
Haworth projection, 9
Hazlitt, W., 12
Header, in PDB input file, 168
Header (MOPAC input file), 153
Header block (MDL MOL input file), 143
Header information
  in Alchemy-III input file, 159
  in MacroModel input file, 162, 165
Header section, in Alchemy 2000 input file, 160
Heat of formation, 14, 15, 19, 91, 140, 194, 222, 254, 339, 363, 495, 543, 652
Heat Transfer Time in PCMODEL, molecular dynamics, 379
Heavy halogen substituent effect, 480
Heisenberg uncertainty, 429, 663
Hendrickson, J. B., 6
Hessian matrix, 342, 663
Heteroaromatic compounds, 480–488, 496, 502, 539, 663
Heteroatoms, 481, 488, 500, 502, 629, 663, 669
Heteroconjugated system, 488, 500

INDEX   **751**

Heterocyclic compounds, 629, 663
HEV, 315, 663, 702
Hexacarbonylchromium organometallic exercise, 544-548
HFORM field (MMX input file), 140
HGS-Maruzen model, 9
Hill, T. L., 5, 315-316, 562
Histidine
 CPK surface exercise, 426-427
 surface area calculation, 429
 van der Waals dot surface, 427-428
 volume calculation, 429
HMO theory, 493, 663
Ho, H., 3
Hofmann, A. W., 4
Hondo, 60, 166-167, 663-664
Hooke's law potential, 4, 203-206, 664, 681. *See also* Modified Hooke's law
Houk, K. N., 346, 350, 582, 585, 589
HPGL graphics format, 96, 97, 664
HPV, 664, 702
HSAB principle, 661, 664, 692
HSEA force field, 7-8, 308, 664-665
Hückel calculation, 481, 488-508, 629, 665
Hückel Molecular Orbital theory, 493, 665, 691
Hückel rule, 629, 665
Huggins, M. L., 508
Hughes, E. D., 5
Hurst, T., 349, 647
Hybridization, 100, 309, 310, 311, 480, 527, 665-666
Hydride
 bridging, 540
 metal, 540
Hydroboration, 583, 587
Hydrocarbon force field, 6, 242, 254, 263, 268, 308, 312
Hydrogen atom
 adding, 40
 deleting, 40
Hydrogen atom types, 176, 256, 509, 513, 515, 622-625
Hydrogen bond
 acceptor, 509, 520, 521, 523, 524
 acceptor antecedent, 523, 524
Hydrogen bond acceptor, MMX parameter, 513-515
Hydrogen bond directionality, 315, 513-515, 520-524, 663, 670
Hydrogen bond donor, 406, 509, 512, 520, 521, 522, 523, 524
 MMX parameter, 513-515
Hydrogen bonding, 12, 13, 22, 217, 250, 289, 304, 399, 401, 508-526, 529, 573, 575,
   614-615, 651, 663, 666, 670, 675, 691
Hydrogen bonding exercise
 dynamics on methanol dimer, 516-520
 methanol dimer, 510-512
Hydrogen bonding flag, 129, 133, 137, 142, 175, 181, 509, 510, 512, 518
Hydrogen bonding interactions, 142, 237
Hydrogen bonding potential, 266, 267, 271-272, 286-287, 291, 294-295, 298, 302-303, 308, 309, 314-315, 509, 520-526, 547, 666
 angle dependence factor, 513-515, 521, 523
 attractive component, 286
 repulsive component, 286
 weak carbon acids, 525-526
Hydrogen extension vector, 315, 663, 702
Hydrophilicity, 422, 666
Hydrophobicity, 8, 664, 666, 682
Hydrophobicity vector, 664
Hyperconjugation, 8, 207, 228, 245, 259, 480, 655, 666
 effect on torsion potential, 207, 228, 245, 259, 666
Hypervalency, 562, 667
Hypervalent atom, 529, 667

ICM, 667
Iconic molecular model, 13, 667
ICONN field (MMX input file), 140
ICOV field (MMX input file), 137
ID field (MMX input file), 134
IDOCK field (MMX input file), 136
IDYN field (MMX input file), 138
IHBD field (MMX input file), 137
IHDON parameter (MMX input file), 192
IHTYP parameter (MMX input file), 192
IHUCK field (MMX input file), 138
Imidazole, π-calculation exercise, 502-508
Improper torsion angle, 210, 667, 679, 686, 701
Improper torsion potential, 210, 257-258, 285, 293, 295, 524
Improper-torsion-torsion-improper-torsion potential, 215-216, 262, 271
Inductive effects, 480, 667
Information window
 Surface Area Results, 429-430, 433-434, 438
 Volume Results, 429, 431
Infrared spectroscopy, 316
Ingold, C. K., 5, 638, 651, 678
INIT field (MMX input file), 136
Inorganic molecules, 526-571
Inorganic potential, 526-571

In-plane angle bend, 4, 5, 6, 12, 22, 90, 94–95, 192–193
Input file
 appended format, 129, 183
 editing, 174 185, 410 411, 415 418, 449 450, 536–538, 547, 569–570, 592–593
Input file modification exercise
 hydroxyethyl pyridine, 180–185
 magnesium ethanediolate, 177–180
Interactions
 bonded, 94
 electrostatic, 12, 13, 22, 90, 93, 94, 131
 gauche, 7
 nonbonded, 5, 6, 22
 van der Waals, 12, 22, 90, 93, 94, 140, 191
Intermediates, 13, 582–584
Internal coordinates, 640, 667
Internet, 1, 7, 8, 16, 20, 87, 168, 667, 668, 671, 698, 701
IPRINT field (MMX input file), 136
ISIS, 61, 627
ISISBase, 142, 672, 688
ISISDraw, 142, 639, 672, 688
Isostere, 124
ITER field (MMX input file), 137
ITYPE field (MMX input file), 141
IUV field (MMX input file), 138

Jahn-Teller effect, 527, 564, 668
JATTCH field (MMX input file), 141
Jeans's potential, 220–222, 248, 257, 319, 325, 651, 668
Jorgensen, W. J., 303, 304, 305, 306, 635, 657, 678, 695
JPRINT field (MMX input file), 137
JSTART field (MMX input file), 138

Karplus, M., 288, 289, 290, 291, 292, 379, 636
Karplus algorithm, 468–469, 470
KATTCH field (MMX input file), 141
Kekulé, F. A., 3, 4, 668
Kekulé program, 80, 668
Ketene, molecular volume calculation, 452–456
Keywords, in MOPAC input file, 153, 154
Kitaigorodsky, A. I., 6, 7
Klyne, W., 629, 694
Kollman, P., 282, 283, 524, 628, 667
Kreevoy, M. M., 5
Kroeker, M., 313
Kuntz, I. D., 441, 573, 648

L-4-hydroxyproline
 ALCHEMY 2000 format file, 159–160
 ALCHEMY-III format file, 157–159

MDL MOL format file, 142–148
MMX format file, 133–142
MOPAC format file, 152–157
PCM format file, 129–133
XRA format file, 148–152
LABEL field (MMX input file), 139
Labels, atom, 98
Labels dialog box, 98
Lagging, torsional, 464
Lanthanides, 240, 486–488, 530, 549
Lanthanide shift effect in NMR spectra, 486–488, 563
LCD electrostatic model, 520
Leach, A., 639
Leapfrog algorithm, 380
LeBel, J. A., 4
Lee, B., 7, 425
Lee-Richards algorithm, 425, 431–434, 435–437
Lemieux, R. U., 7, 664–665
Lennard-Jones potential, 264, 267, 274, 275, 276, 278, 285, 290, 293, 297, 304, 305, 308, 309, 311, 313, 314, 330, 442, 522, 524, 564–565, 628, 678
Lifson S., 7, 262, 636
Ligand field stabilization, 560–561
LIMIT field (MMX input file), 137
Lindner, H. J., 313, 682
Lipkowitz, K. B., 582
Lipophilicity, 422, 639, 669, 675, 692
Literature
 electronic, 16
 print, 16
LMO formalism, 520
Localized charge distribution electrostatic model, 520
Localized molecular orbitals formalism, 520
LOGARY field (MMX input file), 137, 653
London dispersion force, 431, 699
Lone pair
 adding, 40
 deleting, 40
Lone pair extension vector, 315
Lone pairs, 40, 153, 158, 168, 217, 239, 240, 241, 246, 250, 283, 286, 287, 296, 304, 324, 325, 508, 509, 513, 515, 520, 521, 524, 529, 532, 533, 539–542, 548, 549, 561, 562, 583, 615, 622–625, 629, 661, 669–670
Loschmidt, J. J., 3, 4
LPV, 315, 670, 702
Lyngby CFF force field, 272–276, 659
 angle bend potential, 273, 275
 bond stretch potential, 272–273, 275
 electrostatic potential, 274, 276
 out-of-plane angle bend potential, 276

INDEX **753**

PEF91L version, 275, 276
torsion potential, 273-274, 275-276
Urey-Bradley potential, 274
van der Waals potential, 274, 276

MACCS, 142
MAC file format (MacroModel), 128
MACMIMIC, 201
MACMODEL program, 670
MacroModel file format, 161-165
MACROMODEL program, 8, 152, 201, 299, 306-307, 670, 671
Macromolecules, 8, 242, 422, 673
Main Group, 238, 526, 529, 562, 643, 667, 670, 684
MANIAC computer, 5
Mapping, atom-to-atom (PCMODEL), 121
Marking atom in PCMODEL, 37, 41
Marsili, M., 96, 299, 313, 322, 681
Masek, B. B., 442, 573
Mason, E. A., 5
MATOM field (MMX input file), 141
Mayer, J. E., 5
McKelvey, J., 222
MDL MOL file format, 142-148, 636, 640
  bond block, 147
  L-4-hydroxyproline, 42-148
MEDLA formulation, 670-671
Menu
  Amino Acids, 39, 51-54, 609
  Metals, 35, 529, 535, 615
  Nucleosides, 59, 616
  Periodic Table, 34-35, 584, 617
  Rings, 51, 619
  Sugars, 39, 53-58, 620
  Tools, 33, 620
Metal alkoxides, 583
Metal-amine complexes, 526
Metal atom electronic state flag, 133, 141
Metal atom types, 527, 529, 531
Metal-carbon bonds, 540
Metal complex, 5, 6, 217, 237-238, 274, 275, 291, 312, 314-315, 486-488, 526-571, 668, 696
  dihedral driver, 456-457
  Metal__Coord command, 529-531, 532-533, 566, 615
Metal coordination bonding, 527, 531-533, 573, 615, 642-643, 651, 661, 669, 675
  directionality, 548, 560-561
Metal coordination compounds, 526-571
Metal coordination flag, 129, 137-138, 141, 549-550, 553-554
Metal coordinative bridging bond, 540
Metal-halogen bonds, 542

Metal hydride, 540
Metal-hydrogen bonds, 540
Metal Information dialog box, 532-533, 549-550, 566-567
Metal-metal bonds, 541
Metal-nitrogen bonds, 541
Metal-oxygen bonds, 540
Metal-π-complexes, 526
Metal-phosphorus bonds, 542
Metals menu (PCMODEL), 35, 529, 535, 615
Metal-sulfur bonds, 541
Methanol
  atomic charges, 323-324
  molecular volume calculation, 452-456
  solvation model, 304
Methanol dimer
  docking exercise, 575-579
  hydrogen bonding and dynamics, 516-520
  hydrogen bonding exercise, 510-512
Methyl acetate solvation model, 306
Metropolis, N., 5, 671
Metropolis algorithm, 5, 348, 574, 671
Microwave spectroscopy, 79, 316, 506, 671, 687, 688
MICROWORLD program, 671
MIME, 671
Minimal printout output file, 90-91
Minimization
  energy, 136, 385, 615
  nonsimultaneous local energy, 344
  valence force, 344
Minimization algorithms, 339, 341-344, 385
Minimization engine, 335, 339, 641, 652, 677, 693
Minimum energy structures, 11, 345-376, 465
Mirror image, 38
Mislow, K., 308
MM2 force field, 7, 8, 16, 136, 139, 194, 208, 217, 222, 232, 234, 242, 245-249, 251, 252, 253, 256, 257, 258, 299, 305, 308, 312, 313, 493, 521, 562, 563, 636, 671, 695, 697
  angle bend potential, 248
  bond-stretch-angle-bend potential, 248-249
  bond stretch potential, 247, 560
  electrostatic potential, 248
  out-of-plane angle bend potential, 249
  torsion potential, 249
  van der Waals potential, 246-247
MM2* force field, 201, 307, 670
MM2CARB force field, 308
MM2MX force field, 563
MM2X force field, 299, 671

754 INDEX

MM3 force field, 15, 16, 100, 152, 194, 217, 246, 250-254, 255, 256, 257, 258, 287, 299, 307, 312, 313, 521, 562, 563, 627, 636, 666, 671, 695
  angle-bend-angle-bend potential, 253
  angle bend potential, 251-252
  bond-stretch-angle-bend potential, 252
  bond stretch potential, 251
  electrostatic potential, 254
  hydrogen bonding potential, 521
  out-of-plane angle bend potential, 253
  torsion-bond-stretch potential, 252
  torsion potential, 252
  van der Waals potential, 253-254
MM3* force field, 201, 307, 670
MM4 force field, 15, 207, 212, 217, 254-262, 265, 271, 280, 312, 313, 316, 521, 666, 667, 671, 695
  angle-bend-angle-bend potential, 257
  angle bend potential, 255-256
  angle-bend-torsion-angle-bend potential, 260-261
  bond-stretch-angle-bend potential, 257
  bond-stretch-bond-stretch potential, 259
  bond stretch potential, 255
  electrostatic potential, 257
  improper torsion potential, 257-258
  improper-torsion-torsion-improper-torsion potential, 262
  torsion-angle-bend potential, 259-260
  torsion-bond-stretch potential, 258-259
  torsion-improper-torsion potential, 261-262
  torsion potential, 256
  torsion-torsion potential, 261
  van der Waals potential, 256-257
MMFF force field, 307
MMFF93 force field, 201
MMFF94 force field, 194, 299-303, 316, 378, 525, 670, 695
  angle bend potential, 300
  bond-stretch-angle bend potential, 300-301
  bond stretch potential, 300
  electrostatic potential, 303
  hydrogen bonding potential, 302-303, 525
  out-of-plane angle bend potential, 301
  torsion potential, 301
  van der Waals potential, 301-303
MMI force field, 7, 242-245, 246, 247, 248, 252, 263, 283, 308, 309, 469, 488, 563, 672
  angle bend potential, 244
  angle-bend-torsion-angle-bend potential, 245
  bond-stretch-angle-bend potential, 244-245
  bond stretch potential, 243-244
  torsion potential, 245
  van der Waals potential, 243
MMP1 program, 222, 493
MMP2 program, 246, 309, 583
MMX atom types, 222
  boron family, 238
  carbon family, 238-239
  chalcogen family, 239-240
  halogen family, 240
  hydrogen, 238
  lone pair, 241
  metals, 240-241
  pnicogen family, 239
  transition state, 238-240
  water molecule pseudoatom, 238, 239
  wild card, 241
MMX BakMod, 371
MMX file format (PCMODEL), 128, 133-142, 610, 653
MMX force field, 11, 14, 200, 201, 205, 217, 219, 222-238, 246, 512, 669, 672, 697, 699
  angle bend potential, 223, 225
  atom types, 238-241
  bond-stretch-angle-bend potential, 230-232
  bond stretch potential, 222-223, 224
  electrostatic potential, 235-237, 513, 647
  hydrogen bonding potential, 237, 238, 325, 513-516
  organometallic potential, 237-238, 325, 542-544, 547-550
  out-of-plane angle bend potential, 223-225, 226
  torsion potential, 225-230
  van der Waals potential, 232-235, 513, 699
MMX format file, L-4-hydroxyproline, 133-142
MMX parameter file, 187-194
MMX parameters
  ANG, 188, 189, 192-193
  ANG3, 188, 189, 192-193
  ANG4, 188, 189, 192-193
  BDHF, 188, 189, 194
  BND, 187, 190-191
MMX Pi Calc Options dialog box, 138, 490, 496, 567
MMX program, 134
MO calculation, 153, 197, 321, 488, 631, 662
Model, *see also* Electrostatic model; Molecular model; Solvation model
  all atom, 283, 289, 304, 443, 628
  extended atom, 289, 628, 652-653
  united atom, 283, 304, 305, 307, 329, 431, 443, 628, 667, 697-698
Model program, 8, 306

INDEX **755**

Modified Buckingham potential, 267, 274, 308, 309, 312–313, 314
Modified Hooke's law potential, 203–205, 222–223, 225, 244, 248, 251, 312
Modified Pople-Pariser-Parr π-calculation method, 488
Modifying an input file, 174–185, 410–411, 415–418, 449–450, 536–538, 547, 567–570, 592–593
MOL file format (MDL), 128, 142–148, 636, 640, 672
MOL file format (Tripos Alchemy), 128, 157–159, 672
MOL file format (Tripos SYBYL), 128, 672
MOL2MOL, 169, 170–172, 672
Molecular dynamics, 5, 138, 287, 288, 291, 295, 299, 304, 307, 329, 330, 348–349, 376–384, 516–520, 563, 613, 636, 672–673, 691
   Bath Temperature (PCMODEL), 379
   Equilibration Time (PCMODEL), 379
   Heat Transfer Time (PCMODEL), 379
   Sample Temperature (PCMODEL), 379
   Time Step (PCMODEL), 379
   Viscosity (PCMODEL), 379
Molecular dynamics conformation search, 346, 348–349, 378
Molecular dynamics exercise, octaglycine, 380–384
Molecular electrostatic potential surface, 441, 673
Molecular free surface algorithm, 438
Molecular graph, 87, 349
Molecular mechanics, 201–222, 655–656
Molecular model
   ball and stick, 4, 9, 18, 104–106, 425, 610, 617, 627, 636, 671, 678, 688
   Cenco-Petersen, 9
   color scheme, 4
   CPK, 10, 106–107, 423, 636, 684
   Darling, 9
   Dreiding, 9, 98, 671
   enactive, 13, 79, 651
   Fieser, 9
   HGS-Maruzen, 9
   iconic, 13, 667
   physical, 4, 5, 9, 10
   space-filling, 10, 18, 671
Molecular modeling, 673–674
Molecular orbital, 674
Molecular recognition, 10, 13, 306, 378, 422, 573, 673, 675
Molecular similarity, 441, 442, 443
Molecular Skins, 441, 573, 675
Molecular surface, 7, 111–112, 307, 331, 421–455. *See also* Surface

Molecular surface algorithm, 7, 322, 425, 428–429, 431–437
Molecular surface area, 429, 431–434, 695
Molecular surface calculations
   benzene, 437–439
   buckminsterfullerene, 439–441
Molecular surface exercise
   CPK surface of histidine, 426–427
   surface area calculation for histidine, 429
   surface calculations for benzene, 437–439
   surface calculations for buckminsterfullerene, 439–441
   van der Waals dot surface of histidine, 427–428
Molecular Surface program, 8, 437, 442, 676
Molecular trajectory, 348, 378
Molecular volume, 421–455, 695
Molecular volume algorithm, 428, 429
Molecular volume calculation
   ab initio Monte Carlo, 454
   cyclohexane, 452–456
   dodecahedrane, 447–452
   histidine, 429
   ketene, 452–456
   methanol, 452–456
   molecular connectivity, 454
MOLGEN program, 676
MOMEC force field, 563–564
Monoscopic representation, 101–104, 615, 684
Monte Carlo calculation of molecular surface, 445, 447
Monte Carlo conformation search, 346, 347–348, 407–408
Monte Carlo method, 5, 6, 330, 653, 671, 676
MOPAC, 61, 152–157, 615–616, 676, 702
Mopac Data dialog box (PCMODEL), 153, 155
MOPAC file format, 152–157, 702
Morse potential, 4, 203–204, 267, 273, 275, 282, 311, 522, 564, 676
Moving atom, 40–41
MSDRAW program, 676
MSROLL program, 676–677
MUB force field, 7, 309, 677
Mulliken population analysis, 321, 323
Multiplicity, of bond, 37, 80–82, 129, 133, 141, 147, 152, 157, 159, 161, 165, 166–167, 269, 527, 536, 565, 609, 632
Multipole expansion electrostatic model, 294, 695
MVDW field (MMX input file), 140

Nakamura, E., 582
NATTCH field (MMX input file), 139
NBUT field (MMX input file), 139
NCALC field (MMX input file), 140

NCON field (MMX input file), 138
NCONST field (MMX input file), 136
NDC field (MMX input file), 139
NDRIVE field (MMX input file), 140
Neighboring group anisotropy, 480
Neutron diffraction, 10, 77–78, 148–152, 506, 546–547, 565, 570, 579, 634
Newman projection, 9, 229, 650, 677
Newton-Raphson algorithm, 6, 342–344, 677
Newton's force law, 378
N field (MMX input file), 134
Nitrile oxide cycloaddition, 583, 585, 588
Nitrone cycloaddition, 588
N-methylacetamide solvation model, 305
NMR
  chemical shift, 481–486, 637
  lanthanide shift effect, 486–488, 563
NMR chemical shift prediction, 465, 488–508, 637
  exercise, 481–486
NMR correlations, 2, 8, 13, 19, 85, 119, 307, 314, 315, 316, 346, 378, 464–488, 643–644, 673, 677–678
NMR coupling constants
  analysis for 1-chlorobutan-3-one, 476–480
  2-methoxy-4-methyltetrahydropyran, 470–476
  bicyclo[2.2.2]octadienone, 466–468
NMR exercise, predicting heteroaromatic chemical shifts, 481–486
NMR NOE measurements, 315, 316, 349, 465, 648, 678
N,N-dimethylacetamide solvation model, 305
Nonbonding potential, 5, 6, 7, 278–279, 304, 330. *See also* Electrostatic; Urey-Bradley; van der Waals
Nonsimultaneous local energy minimization, 344
NPLANE field (MMX input file), 137
NROT field (MMX input file), 139
NRSTR field (MMX input file), 136
NSEQ mode, GMMX, 358–359
NSETAT field (MMX input file), 138
NSRF mode, GMMX, 359–360
NSRO mode, GMMX, 359
NSYMM field (MMX input file), 139
Nucleophiles, 507, 583, 651, 678
Nucleosides menu (PCMODEL), 59, 616
Nucleotides, 283
Numbers, atom file, 90, 99, 101, 129, 133, 134
Numerical calculation of molecular surface, 445
Num field (PDB input file), 168
NX field (MMX input file), 139

Occ field (PDB input file), 168
Octaglycine, molecular dynamics exercise, 380–384
Octapeptide drawing exercise, 52–53
Off-diagonal term, 8, 281
Oligosaccharides, 7
One Angle Plot window, 457–458, 460, 463–464
OOP parameter, in MMX, 188, 189, 194
OPLSA* force field, 306, 670
OPLS-AA force field, 306, 678
OPLS force field, 287, 288, 303–306, 524, 525, 628, 678
  nonbonding potential, 304, 305
  torsion potential, 304, 305
Oppenheimer, J. R., 633
Optimization flags, in MOPAC input file, 154
OPTIMOL program, 299, 678
Organometallic attractive potential, 532
Organometallic bond length
  in greater than 18 electron complex, 549–550
  in square planar complex, 549–550
  with *trans* influence, 549–550
Organometallic exercise
  π-allylpalladium dichloride, 550–560
  benzenechromiumtricarbonyl, 565–572
  chromyl chloride, 533–538
  hexacarbonylchromium, 544–548
Organometallic force field, 526–571
Organometallic model
  Angular Overlap Model, 564, 628
  AOM, 564, 628
  Points-on-a-Sphere, 560
  POS, 560
  surface force field, 560
Organometallic molecules, 5, 6, 217, 237–238, 238–241, 242, 272, 526–571, 668, 696
Organometallic potential, 237–238, 272, 309, 314–315, 526–571, 681
  YETI force field, 560–561
Organometallic reactions, 526
ORTEP, 6, 106–108, 616, 678–680
Out-of-plane angle, 208, 209, 226, 667, 679, 686, 701
Out-of-plane angle bend, 95, 194, 208–210, 216, 249, 261, 262, 584, 622–625, 679
  organometallic, 543, 549, 562, 564
Out-of-plane angle bend force constant, organometallic, 543
Out-of-plane angle bending potential, 208–210, 223–225, 226, 249, 253, 258,

266–267, 268–269, 276, 298, 301, 309, 310, 312, 543, 564
Out-of-plane–out-of-plane potential, 271
Output file, 89–95, 136, 176, 482, 553–556, 617
  full printout, 92–95
  minimal printout, 90–91
Output window, 335–336, 461

Parameter, polarizability, 291, 293, 302
Parameter code word, 187–194
Parameters, 6, 7, 11, 14, 22, 75, 129, 144, 185–186, 200, 242, 275, 287, 299, 309, 310, 315–316, 631, 655, 679. *See also* Force constant; Potential
  atomic solvation, 330
  carbohydrate, 288, 295
  CHEAT, 295
  generic, 529
  GLYCAM_93, 288
  nucleoside, 286
  peptides, 288
  unit cell, 150–151, 634
Parameter sequence number, 188
Pariser-Parr-Pople MO treatment, 267, 313
Parity, stereo, 146
Parse, parser, 61, 129, 130, 133, 160, 161, 176, 639, 679, 696, 701
Partition function, 140
Pasteur, L., 4
Paterno, E., 4
Pattern search method, 343
PCILO calculation, 308
PCM file format (PCMODEL), 128, 129–133, 610
PCM format file
  L-4-hydroxyproline, 129–133
  modifying, 174–185, 410–411, 415–418, 449–450
PDB, *see* Protein Data Bank
PDB file format (Brookhaven), 128, 152, 167–169, 648, 677, 681
PDB input file
  Atm field, 168
  header, 168
  Num field, 168
  Occ field, 168
  Res field, 168
  Rtyp field, 168
  Temp field, 168
Pearlman, R., 640
PEF91L version of Lyngby CFF force field, 275, 276
Penalty function, 560, 681, 684

Pentane, conformation search exercise, 350–376
PEOE electrostatic model, 313, 322, 681
Peptides, 307–308, 525
Peptide solvation model, 305
Periodicity, torsional, 206, 207, 269, 275, 284, 290, 292, 297, 308, 312, 564
Periodic Table menu (PCMODEL), 34–35, 584, 617
Petitjean, M., 454–455
Petitjean algorithm, 454–455
Pharmacophore, 315, 635, 648, 658, 681–682. *See also* Biophore
  virtual, 315, 658
Pi Atom Selection dialog box, 489–490
PIHF parameter, in MMX, 188, 189, 194
PIMM force field, 313–314, 488, 682
Pitzer, K. S., 5
Pixel, 659, 676, 682, 683
Plug-in, 172, 637, 682–683
PLUTO, 104–106, 617, 634, 683–684, 687
Pluto Options dialog box, 105–106, 617
PM3, in MOPAC input file, 154
Points-on-a-Sphere organometallic model, 560, 684
Polarizability, 647, 651, 684, 692, 694
Polarizability parameter, 291, 293, 302
Polarization, 12
Polar molecular surface area, 429, 433
Poling conformation search, 350, 684
Polygraf program, 310
Polymer, dihedral driver, 456
Polymers, 310, 312, 456
Polypeptides, 307–308
Pople, J. A., 61, 167
Pople-Pariser-Parr π-calculation method, modified, 488
p-orbital, 532, 615
POS organometallic model, 560, 684
Postscript graphics format, 96, 97, 651, 684–685
Potential
  angle bend, 205–206, 223, 225, 244, 248, 255–256, 263–264, 266, 268, 273, 275, 277–278, 284, 287, 289–290, 292, 297, 300, 308, 310, 311, 564
  angle-bend–angle-bend, 213, 253, 257, 270, 279–280
  angle-bend–torsion–angle-bend, 260–261, 265–266, 270, 280
  anharmonic, 8, 281
  bond stretch, 202–205, 222–224, 243–244, 247, 251, 255, 263, 266, 268, 272–273,

Potential (*Continued*)
275, 277, 284, 287, 289, 292, 296–297, 300, 308, 309
bond-stretch–angle-bend, 230–232, 244–245, 248–249, 252, 257, 270–271, 279, 300–301
bond-stretch–bond-stretch, 259, 269–270, 279
Buckingham, 232, 267. *See also* Modified Buckingham potential
Coulomb, 220, 235–236, 265, 267, 274, 276, 278, 285, 290, 298, 303, 304, 309, 311, 314, 317, 325, 330, 522, 565, 628, 643, 647, 651, 669, 678
electrostatic, 220–222, 235–237, 248, 257, 265, 267, 274, 276, 278–279, 285–286, 288, 290–291, 294, 298–299, 303, 304, 308, 309, 651
hard sphere nonbonding, 6
harmonic, 219, 283, 527
Hooke's Law, 4, 203–206, 664, 681
hydrogen bonding, 266, 267, 271–272, 286–287, 291, 294–295, 298, 302–303, 308, 309, 314–315, 509, 520–526, 547
improper torsion, 210, 257–258, 285, 293
improper-torsion–torsion–improper-torsion, 262, 271
inorganic, 526–571
Jeans's, 220–222, 248, 257, 319, 325, 651
Lennard-Jones, 5, 264, 267, 274, 275, 276, 278, 285, 290, 293, 297, 304, 305, 308, 309, 311, 313, 314, 330, 522, 564–565, 628, 678
modified Buckingham, 267, 274, 308, 309, 312–313, 314
modified Hooke's Law, 203–205, 222–223, 225, 244, 248, 251, 312, 560
Morse, 4, 203–204, 267, 273, 275, 282, 311, 522, 564, 676
nonbonding, 5, 6, 7, 278–279, 304, 330, 562
organometallic, 237–238, 272, 299, 309, 314–315, 526–257, 615
out-of-plane angle bend, 223–225, 226, 249, 253, 258, 266–267, 268–269, 276, 298, 301, 310, 312, 564
out-of-plane–out-of-plane, 271
stretch-bend, 12, 90, 95, 210, 212, 219, 230–232, 244–245, 248–249, 252, 257, 270–271, 279, 300–301
torsion, 5, 6, 7, 95, 206–208, 225–230, 245, 249, 252, 256, 264, 266, 269, 273–274, 275–276, 278, 284–285, 287–288, 290, 292–293, 297, 301, 304, 305, 308, 309, 311–312, 564, 696
torsion–angle-bend, 214, 245, 250, 259–260, 281, 564
torsion–bond-stretch, 252, 258–259, 280–281
torsion–improper-torsion, 261–262
torsion–torsion, 261
Urey-Bradley, 4, 219, 264, 266, 274, 295, 698
van der Waals, 216–218, 232–235, 246–247, 253–254, 257, 264, 267, 274, 276, 278–279, 285–286, 288, 290–291, 293–294, 297–298, 301–303, 304, 308, 309
Potential energy surface, 201
Hartree-Fock *ab initio*, 276, 281
Potential function, 7, 12
POWERFIT program, 685
Prefix, data, 129
Prelog, V., 629, 638, 694
Print dialog box (PCMODEL), 97
Printer Setup dialog box, 617
Printing graphics, 96–97
Print Level dialog box (PCMODEL), 90
Probe atom, 447, 675
PROBE program, 299
Probe sphere, 435, 448, 675, 685
Projection
Fischer, 628, 654
Newman, 9, 229, 650, 677
sawhorse, 9, 229
PROPERTY 3D program, 685–686
Protein Data Bank, 18, 76, 152, 681, 686
Pseudoatom, 153, 158, 165, 217, 246, 250, 283, 286, 287, 296, 304, 324, 325, 521, 549, 562, 583, 623, 650. *See also* Dummy atom; Virtual atom
Pseudoreceptor, 315, 635, 648, 658, 681–682
Pyramidalization, 215, 262
Pyramid height, 208, 209, 298, 667, 679, 686, 701
Pyrazine, NMR chemical shift prediction exercise, 481–486
Pyrazole, π-calculation exercise, 502–508
Pyridine, NMR chemical shift prediction exercise, 481–486
Pyrimidine, NMR chemical shift prediction exercise, 481–486

QCFF-PI force field, 283
QCPE, 307, 308, 309, 686
QEq electrostatic model, 323–324
QEq-PD electrostatic model, 323–324
QMFF force field, 276–281, 282, 686
angle-bend–angle-bend potential, 279–280
angle bend potential, 277–278
angle-bend–torsion–angle-bend potential, 280
bond-stretch–angle-bend potential, 279
bond-stretch–bond-stretch potential, 279
bond stretch potential, 277

INDEX **759**

electrostatic potential, 278–279
torsion–angle-bend potential, 281
torsion–bond-stretch potential, 280–281
torsion potential, 278
van der Waals potential, 278–279
QM/MM force field, 379, 582, 686–687
QQ electrostatic model, 139
QUANTA program, 295, 687
Queries, 119, 618
Query display exercise, lactose, 119
QUEST program, 683
QUEST3D program, 634, 687
Quinoline, NMR chemical shift prediction exercise, 481–486

Radical, 496, 688
Radius
 covalent, 129, 178, 241, 543, 634
 van der Waals, 7, 129, 217, 218, 232, 240, 423, 431, 513, 522, 525, 543, 622–625, 644, 675
Raman spectroscopy, 316
Random kick conformation search, 8, 346, 348, 354, 366
Random number generator, 353
Rappé, A. K., 310, 311, 323, 697
RASMOL program, 18, 128, 167, 169, 172–173, 637, 681, 688
RASSE program, 688
RASTER 3D program, 688–689
Raster graphics format, 80, 659–660
Rational formula, 3
REACCS, 142
Reacting center, 147
Reaction coordinate, 16
Reaction zone, 582, 686
Reactive intermediates, 582–584
 atoms types, 583–584
Read Dihedral Driver File dialog box, 464
Reduced hydrogen correction, 75, 94, 218, 233–235, 236–237, 243, 247, 254, 257, 303
Reentrant surface, 435, 675
Regioselectivity, 670, 689
Regiospecificity, 670, 689
Reset Atomic Charge in PCMODEL, 325
Res field (PDB input file), 168
Resonance effects, 480, 689
Resonance integral, 495, 500, 665
RESP electrostatic model, 287, 288, 322–323, 324
Restrained Electrostatic Potential, 287, 288
Restricted atoms, 136
RHF, 662, 664
π-calculation, 132, 138, 490, 496
Ribbon Information dialog box, 112, 115
Ribbon representation, 112–116, 619, 688

Richards, F. M., 7, 425
Rigid rotor, 457–459
Rigid rotor exercise, 1,2-dichloroethane, 458–459
Ring, medium-sized, 6
RINGMAKER program, 306, 347, 689
Rings menu (PCMODEL), 51, 619
ROSDAL, 80, 690
Rot_E command, 457–459, 619
Rot_E dialog box, 457–458
Rotate_Bond command, 41, 48, 619
Rotate_Bond dialog box, 41, 48
Rtyp descriptor (PDB input file), 168
Rupley, J. A., 7

Saccharides, *see* Carbohydrates
Saddle point, 345, 441
Sample Temperature in PCMODEL molecular dynamics, 379
Saturated molecular surface area, 429, 433
Saunders, M., 8, 346, 350
Save File dialog box, 60
Save Graphic dialog box, 97
Sawhorse projection, 9, 229
Scaling factors, in deriving CFF93, 281
SCF iteration, 137
SCF-LCAO-MO treatment, 313
Schachtshneider, J. H., 6
Scheraga, H. A., 7, 307, 347, 650, 697
Schleyer, P. v. R., 7, 308, 309, 342, 583, 641
SCULPT program, 690–691
Second row elements, 231, 249, 669, 691
Semiempirical MO calculations, 69, 152, 582, 615, 627, 674, 676
Sextic angle bend force constant, 223, 248
SHAKE program, 346, 691
Shakespeare, W., 385
Shape scoring, in Dock program, 442, 649
SHAPES force field, 561–562, 564
Shifted dielectric electrostatic model, 294
Shrake, A., 7
Silhouette, van der Waals, 434
Similarity, molecular, 441, 442, 443
Simulated annealing, 5, 348, 533, 571, 573–574, 613, 671, 691–692
Simulated annealing conformation search, 348
Single point energy, 336–339, 619
Single point energy exercise, di-t-butyl ether, 336–339
Slater determinants, 496, 662
SLPS, in MMX, 191
SLPT, in MMX, 191
Small rings, 139
SMILES, 80, 86, 636, 639, 640, 692, 701

SMILOGP program, 692
Smith-Eyring electrostatic model, 322
S_N2 reaction, 239, 240, 583, 585–587, 588
Snow, M. R., 6
Snyder, R. G., 6
Soft acid, 664, 692
Soft base, 664, 692
Solute-solvent electrostatic polarization term, 330–331, 658
Solute-solvent van der Waals term, 330–331, 658
Solvation, 15, 288, 295, 299, 303, 304, 316–317, 329–334, 422, 666
Solvation free energy, 330–331
Solvation model
  acetic acid, 306
  alcohols, 304, 305, 525
  continuum, 307, 330–331
  ethanol, 304
  ethers, 304
  explicit, 329–330
  formamide, 305
  GB/SA, 307, 330–331, 695
  methyl acetate, 306
  N,N-dimethylacetamide, 305
  N-methylacetamide, 305
  peptides, 305
  thioethers, 305
  thiols, 305
  TIP3P, 287, 288, 305, 524, 695
  TIP4P, 305, 678, 695
  TIPS2, 305
  water, 287, 288, 304, 305, 525
Solvent accessible surface, 435, 437, 442–443, 675
Solvent accessible surface area, 330
Solvent-excluded surface, 437, 675
Solvent-solvent cavity term, 330–331, 658
Space-filling model, 10, 18, 671
Spectroscopy
  infrared, 316
  microwave, 79, 316, 506
  Raman, 316
Spherical coordinates, 561
Spin-spin coupling constants, 465, 468–480
SPROUT program, 692–693
Square planar complex, 530, 549–550
  organometallic bond length, 549–550
Stacking interaction, 396, 406–408
Start token, 160, 175, 696
Steepest descent algorithm, 6, 341, 641, 693
Stereoatom, 146
Stereo bond, 147
Stereolithography, 10, 684
Stereo Options dialog box, 103
Stereo parity flag, 146

Stereoscopic representation, 101–104, 615, 619
Stereoselectivity, 694
Stereospecificity, 694
Stewart, J. J. P., 61
Still, W. C., 8, 61, 162, 307, 346, 347, 350, 658, 670, 671
Stochastic calculation of molecular surface, 445
Stochastic conformation search, 347–348, 409
Stochastic ensemble search, 573, 577, 671
Stop token, 147, 160, 161, 169, 696
Strained olefins, 309
Strain energy, 91
STRAIN program, 308
Stretch-bend potential, 12, 90, 95, 210, 212, 219, 230–232, 244–245, 248–249, 252, 257, 270–271
STRFIT program, 8, 635
Structure
  1D, 80
  2D, 18, 80, 142
  3D, 18, 19, 36, 82–89, 142
Structure comparison exercise
  1,2-dichloroethane, 122–125
  1-phenylbicyclo[2.2.2]octane, 73–75
  benzene and thiophene, 123–127
Structure file input exercise, X-ray format, imidazole, 76–77
Structure files, 22
Structure List dialog box (PCMODEL), 66, 71
Structure queries, 117–119
Structures, written, 3, 82
Structure templates, 49–63
Substitution, 5
Substructure, Don't Minimize, 401–406
Substructure manipulation exercise, 1-phenylbicyclo[2.2.2]octane, 69–75
Substructure membership flag, 129, 175, 512
Substructure Name dialog box (PCMODEL), 66
Substructures, 2, 22, 37, 40, 63–75, 109–113, 133, 138, 423, 510-512, 532–533, 571–581, 587–588, 614, 615, 620, 694
  docking, 571–581
Sugars menu (PCMODEL), 39, 53–58, 620
SUMM conformation search, 347
Surface, *see also* Molecular surface
  concave section, 445–446
  concavity, 441
  convexity, 441
  convex section, 445–446
  CPK, 111, 423–427, 612
  dot, 7, 112, 427, 428–429, 432, 434, 435, 436, 441, 613, 688

electrostatic, 296, 325, 331-334, 427, 673
molecular, 7, 111-112, 307, 421-455
molecular electrostatic potential, 441
reentrant, 435, 675
saddle point, 441
solvent accessible, 435, 437, 442-443, 675
solvent-excluded, 437, 675
toroidal section, 445-446
van der Waals, 218, 425, 427, 431-434, 437, 441, 442, 675
Surface area, solvent accessible, 330, 442-443
Surface area calculation for histidine, 429
Surface Area command, 428-429, 620
Surface Area Results information window, 429-430, 433-434, 438
Surface force field organometallic model, 560
Suter, U. W., 8
SYB file format (Tripos SYBYL), 128
SYBYL file format, 160-161
SYBYL program, 61, 152, 296, 640
Synclinal, 694
Synperiplanar, 694
Systematic conformation search, 347
Systematic Unbounded Multiple Minimum conformation search, 347

Temp field (PDB input file), 168
Thioethers solvation model, 305
Thiols solvation model, 305
Thiophene, structure comparison exercise, 123-127
Time Step in PCMODEL molecular dynamics, 379
TINKER program, 695
TIP3P solvation model, 287, 288, 305, 695
TIP4P solvation model, 305, 678, 695
TIPS, 304, 329-330, 512, 695
TIPS2 solvation model, 305
TMAX field (MMX input file), 136
Toggle, 134, 402, 613, 615, 654, 695
Token, 130, 172, 696
  start, 160, 175, 696
  stop, 147, 160, 161, 169, 696
Tool
  Build, 87-88
  H/AD, 88-89
Tools menu (PCMODEL), 33, 620
TOR parameter, in MMX, 187, 188-190
Torsion
  constrained, 396-401
  organometallic, 543, 563, 564
Torsional angle, 4, 9, 13, 41, 42, 43, 119, 156, 225-230, 667, 696
Torsional barrier, 4, 13, 170, 201, 206, 250, 254, 456, 457, 647, 690

Torsional interaction force constant, 206, 225-228, 245, 249
Torsional interactions, 188, 190, 225-230
Torsional lagging, 464
Torsional Monte Carlo conformation search, 346, 347-348
Torsional periodicity, 206, 207, 269, 275, 284, 290, 292, 297, 308, 312, 564
Torsional strain, 6, 12, 22, 90, 95, 188, 190, 696
Torsional tree conformation search, 346, 347
Torsion-angle-bend potential, 214, 250, 259-260, 281, 564
Torsion-bond-stretch potential, 213, 214, 252, 258-259, 280-281
Torsion-improper-torsion potential, 214-215, 261-262
Torsion potential, 5, 6, 7, 95, 205-208, 225-230, 245, 249, 252, 256, 264, 266, 269, 273-274, 275-276, 278, 284-285, 287-288, 290, 292-293, 295, 297, 301, 308, 309, 311-312, 495, 564, 696
  effect of conjugation, 207
  effect of electrostatic interactions, 207-208, 227, 228, 230
  effect of hyperconjugation, 207, 228, 245
  effect of van der Waals interactions, 207-208, 227, 228-229, 230
  four-fold term, 256
  six-fold term, 256, 273
Torsion-torsion potential, 214, 215, 261
Transannular interactions, 309, 690
trans effect, 550, 696-697
trans influence, 527, 549-550, 697
  effect on organometallic bond length, 549-550
Transition metal complexes, 543
Transition metals, 222, 240, 643
Transition state bond order flag, 592
Transition state bond stretch force constant, 585
Transition state exercise, Diels-Alder reaction, 588-598
Transition states, 10, 13, 14, 15, 222, 582-583, 584-598, 661
  atoms types, 583-588
  bond types, 583-588
  limitations, 585-587
Transition state templates, 586-588
Tripos force field, 201, 208, 296-299, 525
  angle bend potential, 297
  bond stretch potential, 296-297
  electrostatic potential, 298-299
  out-of-plane angle bend potential, 298
  torsion potential, 297
  van der Waals potential, 297-298

TS Bond Orders dialog box, 584–585, 589
TS_Bond Orders command, 584–585, 589, 621
Tubular Bonds command, 621

UBCFF force field, 263–268, 272, 522
  angle bend potential, 263–264, 266
  angle-bend–torsion–angle-bend potential, 265–266
  bond stretch potential, 263, 266
  electrostatic potential, 265, 267
  hydrogen bonding potential, 266, 267, 522
  out-of-plane angle bend potential, 266–267
  torsion potential, 264, 266
  Urey-Bradley potential, 264–265, 266
  van der Waals potential, 264, 267
UFF force field, 300, 310–312, 378–379, 697
  angle bend potential, 311
  out-of-plane angle bend potential, 312
  torsion potential, 311–312
UHF, 662, 664
  π-calculation, 132, 138, 490, 496
UNICEPP force field, 7, 283, 307–308, 650, 697
Unit cell parameters, 150–151, 634
United atom model, 283, 304, 305, 307, 329, 431, 443, 628, 667, 697–698. *See also* Extended atom
Unsaturated molecular surface area, 429, 433
Urey, H. C., 4
Urey-Bradley force constant, 219
Urey-Bradley potential, 219, 264–265, 266, 274, 295, 698
Utilities, graphics capture, 96–98

VALBOND force field, 562
Valence cross terms, 210–216, 268
Valence force field, 201, 698
Valence force minimization, 344
Valence shell electron pair repulsion theory, 560, 684, 698–699
van der Waals attractive force, 217, 431, 699
van der Waals dot surface of histidine, 427–428
van der Waals force constant, 217, 218, 232, 243
van der Waals interactions, 207–208, 217, 227, 228–229, 230, 232–234, 431, 448, 480, 521, 525, 647, 685
  between two fluorine atoms, 232–234
  effect on torsion potential, 207–208, 227, 228–229, 230
  geminal, 543, 554–555

van der Waals potential, 216–218, 232–235, 243, 246–247, 253–254, 256–257, 264, 267, 271–272, 274, 276, 278–279, 285–286, 288, 290–291, 293–294, 297–298, 301–303, 308, 309, 513, 522
  attractive component, 232–233, 243, 246–247, 267, 271
  buffered 14-7, 301–302
  force constant derivation, 278–279, 285–286, 298
  repulsive component, 232–233, 243, 246–247
van der Waals radius, 7, 129, 217, 218, 232, 240, 247, 267, 271, 513, 522, 525, 543, 622–625, 644, 675
van der Waals repulsive force, 448, 662, 690, 699
van der Waals silhouette, 434
van der Waals surface, 218, 425, 427, 431–434, 437, 441, 442, 675, 699
van't Hoff, J. H., 4
Variable electronegativity, 481
VCFF, *see* CVFF
VDM program, 699
VDW parameter, in MMX, 188, 189, 191–192
Vector, *see* Graphics format
Vector analysis method, 5
Vedani, A., 314
Verlet algorithm, 380
VESCF calculation, 137, 222, 465, 480–486, 488–508, 539, 615, 617, 691
VIBE, 9, 700
Vibrational enthalpy, 262
Vibrational spectra, 250, 254, 259, 262, 265, 315, 644
Virtual atom, 650. *See also* Dummy atom; Pseudoatom
Virtual Biomolecular Environment, 9, 700
Virtual pharmacophore, 315, 658
Virtual position, 218
Virtual reality, 8, 9, 700
Viscosity, 601
Viscosity in PCMODEL molecular dynamics, 379
VISTA program, 634, 687, 700
Volume, molecular, 421–455
Volume calculation for histidine, 429
Volume command, 29, 429, 621
Volume Results information window, 429, 431
VSEPR theory, 560, 684, 698–699

Warshel, A., 7, 637

INDEX **763**

Water accessible surface, 427, 435–437
Water solvation model, 287, 288, 304, 305
Watson, J. D., 5, 509
Web browser, 8, 17, 18, 172, 637, 666, 671, 700, 701
WebElements, 528
Westheimer, F. H., 5, 313, 456, 700
Westheimer method, 5, 700
WHATIF program, 700–701
Wiberg, K. B., 6, 7, 342, 456, 647, 693
Wild card atom type, 266, 297, 543
Wilson angle, 208, 209, 310, 312, 667, 679, 686, 701
Window
   Compare, 121–122, 125, 127
   One Angle Plot, 457–458, 460, 463–464
   Output, 335–336, 461
Windows Metafile graphics format, 96–98
WIZARD program, 86, 347, 582, 701
World Wide Web, xiii, 1, 3, 8, 11, 16–18, 152, 167, 168, 169, 174, 283, 288, 289, 296, 299, 307, 313, 316, 377, 528, 633, 636, 637, 638, 666, 667, 671, 678, 681, 682–683, 684, 686, 698, 700, 701

Write File for Batch, Dock and Dihedral Driver dialog box, 379, 411, 418, 610
WWW Chemical Structures Database, 18

XMOL program, 701
XRA file format, 148–152
XRA format file
   giberrellin intermediate, 386–389
   imidazole, 76–77
   L-4-hydroxyproline, 148–152
X-ray diffraction, 19, 75–77, 148–152, 316, 506, 558, 561, 565, 570, 572, 634, 699, 702

YAEHMOP program, 702
YAK pharmacophore search, 315, 663, 664, 670, 702
YETI force field, 314–315, 524, 560–561, 663, 664, 670, 702
   hydrogen bonding potential, 524–525
   organometallic potential, 560–561

Z-clipping, 702. *See also* Blocking
Zeolites, 526, 563
Z-matrix, 152–154, 702–703

CUSTOMER NOTE: IF THIS BOOK IS ACCOMPANIED BY SOFTWARE, PLEASE READ THE FOLLOWING BEFORE OPENING THE PACKAGE.

This software contains files to help you utilize the models described in the accompanying book. By opening the package, you are agreeing to be bound by the following agreement:

This software product is protected by copyright and all rights are reserved by the author and John Wiley & Sons, Inc. You are licensed to use this software on a single computer. Copying the software to another medium or format for use on a single computer does not violate the U. S. Copyright Law. Copying the software for any other purpose is a violation of the U. S. Copyright Law.

This software product is sold as is without warranty of any kind, either express or implied, including but not limited to the implied warranty of merchantability and fitness for a particular purpose. Neither Wiley nor its dealers or distributors assumes any liability of any alleged or actual damages arising from the use of or the inability to use this software. (Some states do not allow the exclusion of implied warranties, so the exclusion may not apply to you.)

WILEY